The Dopamine Receptors

The Receptors

The Dopamine Receptors,
edited by ***Kim A. Neve and Rachael L. Neve,*** 1997
The Metabotropic Glutamate Receptors,
edited by ***P. Jeffrey Conn and Jitendra Patel,*** 1994
The Tachykinin Receptors,
edited by ***Stephen H. Buck,*** 1994
The Beta-Adrenergic Receptors,
edited by ***John P. Perkins,*** 1991
Adenosine and Adenosine Receptors,
edited by ***Michael Williams,*** 1990
The Muscarinic Receptors,
edited by ***Joan Heller Brown,*** 1989
The Serotonin Receptors,
edited by ***Elaine Sanders-Bush,*** 1988
The Alpha-2 Adrenergic Receptors,
edited by ***Lee Limbird,*** 1988
The Opiate Receptors,
edited by ***Gavril W. Pasternak,*** 1988
The Alpha-1 Adrenergic Receptors,
edited by ***Robert R. Ruffolo, Jr.,*** 1987
The GABA Receptors,
edited by ***S. J. Enna,*** 1983

The Dopamine Receptors

Edited by

Kim A. Neve

Oregon Health Sciences University and VA Medical Center, Portland, OR

Rachael L. Neve

McLean Hospital, Harvard Medical School, Belmont, MA

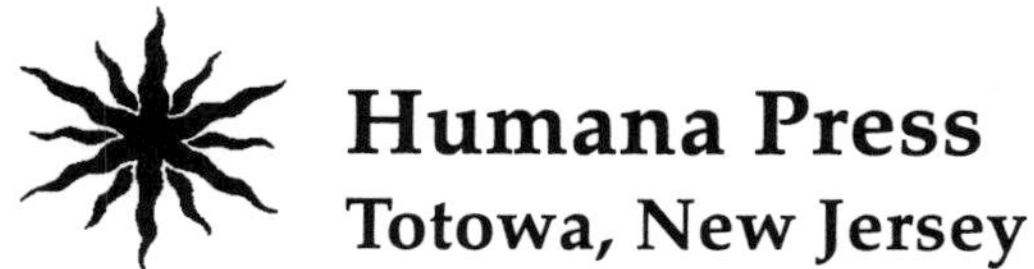

999 Riverview Drive, Suite 208
Totowa, New Jersey 07512

This publication is printed on acid-free paper. ∞
ANSI Z39.48-1984 (American National Standards Institute) Permanence of Paper for Printed Library Materials.

For additional copies, pricing for bulk purchases, and/or information about other Humana titles, contact Humana at the above address or at any of the following numbers: Tel.: 201-256-1699; Fax: 201-256-8341; E-mail: humana@interramp.com

Printed in the United States of America. 10 9 8 7 6 5 4 3 2 1

Library of Congress Cataloging in Publication Data

The Dopamine receptors / edited by Kim A. Neve, Rachael L. Neve.
p. cm. -- (The Receptors)
Includes index.
ISBN 0-89603-433-X (alk. paper)
1. Dopamine--Receptors. I. Neve, Kim A. II. Neve, Rachael L. III. Series.
[DNLM: 1. Receptors, Dopamine--physiology. WL 102.8D6915 1996]
QP563.D66D662 1997
612'.8'042--dc20
DNLM/DLC
for Library of Congress 96-31720
CIP

Preface

Seven years after the cloning of the rat dopamine D_2 receptor, and four years after the cloning of the last mammalian dopamine receptor identified to date, this seems to be an excellent time to put together the present *The Dopamine Receptors* volume of this series, *The Receptors.* There has been time for considerable characterization of the novel receptor subtypes, and new, exciting lines of research from the molecular to the behavioral levels are taking shape.

We asked the contributors to *The Dopamine Receptors* to follow the superb example set by the previous volumes in this series by writing comprehensive, historical reviews that will comprise an essential resource for nonspecialists and newcomers to the dopamine receptor field, while at the same time providing up-to-date summaries of the most active areas of research.

It is difficult these days to write about receptors without addressing the issue of receptor nomenclature. For dopamine receptors, valid arguments can be made for a system in which the subtypes are classified as belonging to the D1 or D2 classes, with letters assigned in the order of cloning (D_{1A}, D_{1B}, D_{2A}, D_{2B}, D_{2C}). We decided, however, that common usage counts for something, and chose to use D_2, D_3, and D_4 for the D2-like receptors because these names are nearly unanimously used in the literature. To be in agreement with this system for the D2-like receptors, we chose to use D_1 and D_5 for the D1-like receptors, in lieu of D_{1A} and D_{1B}, or the awkward compromise, D_1/D_{1A} and D_5/D_{1B}. The alternatively spliced D_2 receptors are referred to as D_{2L} and D_{2S}. We passed over the terms $D2_{444}$ and $D2_{415}$, which are unsatisfactory because the number of amino acids in the receptors varies among species, and D_{2A} and D_{2B}, which would set the precedent of naming receptor subtypes in order of their overall abundance rather than in the order of cloning.

In this book, the molecular subtypes are named D_1, D_2, D_3, D_4, and D_5, the nomenclature used in the Sixth Edition of the *TIPS Receptor & Ion Channel Nomenclature Supplement.* Nonmammalian homologs of the mammalian subtypes are given the same names, whereas nonmammalian subtypes for which mammalian homologs have not yet been found are referred to by the names used in the primary literature. When it is not clear which genetic subtype is being characterized, the receptors are referred to as D1 or D1-like and D2 or D2-like, i.e., without subscripted numerals.

Some of the contributors to this book have said that the dopamine receptor field is too large to be comprehensively reviewed in a single volume, and we have probably omitted some important topics. We hope, however, that the great majority of our readers will feel that *The Dopamine Receptors* covers most of the field.

Kim A. Neve
Rachael L. Neve

Contents

Contributors

MONIQUE R. ADAMS • *Department of Pharmacology, University of Washington, Seattle, WA*

PAUL R. ALBERT • *Neuroscience Research Institute, University of Ottawa, Ontario, Canada*

MARJORIE A. ARIANO • *Department of Neuroscience, The Chicago Medical School, University of Health Sciences, Chicago, IL*

ROSS J. BALDESSARINI • *Department of Psychiatry, Harvard Medical School, Boston, MA and Bipolar and Psychotic Disorders Program, McLean Hospital, Belmont, MA*

CLARE BERGSON • *Department of Pharmacology, Pennsylvania State College of Medicine, Hershey, PA*

MARC G. CARON • *Department of Cell Biology, Howard Hughes Medical Institute, Duke University Medical Center, Durham, NC*

IAN CREESE • *Center for Molecular and Behavioral Neuroscience, Rutgers University, Newark, NJ*

DANIEL M. DORSA • *Department of Psychiatry and Behavioral Sciences, University of Washington, Seattle, WA*

JOHN D. ELSWORTH • *Departments of Pharmacology and Psychiatry, Yale University, New Haven, CT*

MOHAMMAD H. GHAHREMANI • *Department of Pharmacology and Therapeutics, McGill University, Montreal, Quebec, Canada*

PATRICIA S. GOLDMAN-RAKIC • *Department of Neurobiology, Yale University School of Medicine, New Haven, CT*

JOHAN GRENHOFF • *Department of Physiology and Pharmacology, Karolinska Institute, Stockholm, Sweden*

RITA M. HUFF • *Department of Cell Biology, The Upjohn Company, Kalamazoo, MI*

STEVEN W. JOHNSON • *Departments of Pharmacology and Neurology, Oregon Health Sciences University, Portland, OR*

GERALD J. LAHOSTE • *Department of Psychobiology, University of California, Irvine, CA*

JEAN PHILIPPE LOEFFLER • *Physiology Laboratory, Strasbourg, France*

RICHARD B. MAILMAN • *Division of Medicinal Chemistry and Natural Products, Departments of Pharmacology and Psychiatry, Neuroscience Center, University of North Carolina School of Medicine, Chapel Hill, NC*

JOHN F. MARSHALL • *Department of Psychobiology, University of California, Irvine, CA*

STEPHEN J. MORRIS • *Department of Pharmacology and Therapeutics, McGill University, Montreal, Quebec, Canada*

LADISLAV MRZLJAK • *Section of Neurobiology, Yale University School of Medicine, New Haven, CT*

KIM A. NEVE • *Department of Behavioral Neuroscience, Oregon Health Sciences University and Veterans Affairs Medical Center, Portland, OR*

RACHAEL L. NEVE • *Department of Genetics, Harvard Medical School and McLean Hospital, Belmont, MA*

DAVID E. NICHOLS • *Department of Medicinal Chemistry and Pharmacognosy, School of Pharmacy and Pharmacal Sciences, Purdue University, West Lafayette, IN*

ABDEL-MOUTTALIB OUAGAZZAL • *Center for Molecular and Behavioral Neuroscience, Rutgers University, Newark, NJ*

JAMES L. ROBERTS • *Dr. Arthur M. Fishberg Research Center for Neurobiology, Mount Sinai School of Medicine, New York, NY*

SUSAN W. ROBINSON • *Department of Cell Biology, Howard Hughes Medical Institute, Duke University Medical Center, Durham, NC*

ROBERT H. ROTH • *Departments of Pharmacology and Psychiatry, Yale University School of Medicine, New Haven, CT*

DAVID N. RUSKIN • *Department of Psychobiology, University of California, Irvine, CA*

STUART C. SEALFON • *Dr. Arthur M. Fishberg Research Center for Neurobiology, Mount Sinai School of Medicine, New York, NY*

DAVID R. SIBLEY • *Molecular Neuropharmacology Section, Experimental Therapeutics Branch, National Institute of Neurological Disorders and Stroke, National Institutes of Health, Bethesda, MD*

PHILIP STRANGE • *The Biological Laboratory, University of Kent at Canterbury, UK*

BAO-CUN SUN • *Center for Molecular and Behavioral Neuroscience, Rutgers University, Newark, NJ*

Alexander Tropsha • *Division of Medicinal Chemistry and Natural Products, University of North Carolina School of Pharmacy, Chapel Hill, NC*
Raymond P. Ward • *Department of Psychiatry and Behavioral Sciences, University of Washington, Seattle, WA*
Graham V. Williams • *Section of Neurobiology, Yale University School of Medicine, New Haven, CT*
Ming Zhang • *Center for Molecular and Behavioral Neuroscience, Rutgers University, Newark, NJ*

Part I

Characterization of Dopamine Receptors

CHAPTER 1

Biochemical Characterization of Dopamine Receptors

Philip Strange

1. Introduction

Following the recognition of dopamine as a neurotransmitter in the brain, independent of its role as a precursor of noradrenaline, there has been great interest in the sites of action of dopamine, namely the receptors. This interest was fueled by studies on certain brain diseases such as Parkinson's disease and schizophrenia, which showed that dopamine had a role either in the pathogenesis or drug treatment of the disease and that substances acting at the receptors might act as therapeutic agents *(1)*. Dopamine receptors were, therefore, studied intensively using a number of techniques including electrophysiology, animal behavioral experiments, and biochemical studies. In 1976 Cools and van Rossum reviewed a large body of data on the actions of dopamine and concluded that there might be more than one receptor for dopamine in the brain *(2)*. Their conclusions were based on anatomical, electrophysiological, and pharmacological data, and they proposed two receptors—one of which was insensitive to the butyrophenone, haloperidol.

Biochemical studies on dopamine receptors began in the 1970s with the description of the stimulation of cyclic AMP (cAMP) production in brain tissue by dopamine by several groups *(3–6)*. The dopamine response could be prevented by many antipsychotics and it was suggested that this was an important normal site of the action of dopamine. On the whole, there was a good correlation between the ability of drugs to treat psychosis and their activity to inhibit the dopamine stimulation of cAMP production, but there were notable exceptions including the butyrophenones (e.g., haloperidol) and substituted benzamides (e.g., sulpiride).

The Dopamine Receptors Eds.: K. A. Neve and R. L. Neve
Humana Press Inc., Totowa, NJ

In the mid-to-late 1970s considerable effort was invested in developing radioligand-binding assays for dopamine receptors particularly by the groups of Snyder *(7)* and Seeman *(8,9)*. This became possible owing to the availability of tritium-labeled drugs with a specific activity high enough to detect these low-abundance receptors. For the dopamine receptors [^{3}H]dopamine, the natural agonist, and [^{3}H]haloperidol, the butyrophenone antipsychotic, were initially tested. These studies showed that [^{3}H]haloperidol labels a class of binding sites in the brain which have a high affinity for the majority of antipsychotic drugs including the butyrophenones. The affinities of drugs for these sites are correlated with the doses of drugs that are used to treat the psychosis of schizophrenia, and it was concluded that these sites represent dopamine receptors *(10,11)*.

Kebabian and Calne *(12)* provided a rationalization for these data in their 1979 review, in which they extended an earlier suggestion from Spano et al. *(13)* that there are two subtypes of receptors for dopamine, D1 and D2.* The properties of the two receptors are summarized in Table 1. The D1 receptor was defined as the receptor linked to the stimulation of adenylate cyclase, and with a lower affinity for the butyrophenones and substituted benzamides, whereas the D2 receptor was the receptor with a high affinity for the butyrophenones (and labeled with high affinity by radioligands such as [^{3}H]haloperidol) and substituted benzamides, but which either did not affect adenylate cyclase or inhibited the enzyme. These studies and ideas then provided a basis for the definition of the properties of the two receptor subtypes in biochemical and pharmacological studies.

In parallel with the development of these ideas, dopamine receptors in peripheral tissues were being characterized, mostly using functional tests *(15,16)*. The data obtained were broadly consistent with the categorization of the peripheral dopamine receptors into two subgroups. These were termed DA1 and DA2 receptors, but seem to be mostly equivalent to the central D1 and D2 receptors. DA1 receptors are activated by SKF 38393 and blocked by bulbocapnine, whereas DA2 receptors are stimulated by dopamine and *N,N* dipropyl dopamine and blocked by substituted benzamides and haloperidol. DA1 receptors are generally postjunctional in smooth muscle, whereas DA2 receptors are found prejunctionally on neurons. There are some discrepancies in the detailed pharmacological properties of the DA1/DA2 and D1/D2 groups and it is not clear whether this reflects real differences or the problems of defining pharmacological properties in isolated peripheral tissues.

*As detailed in the Preface, the terms D1 or D1-like and D2 or D2-like are used to refer to the subfamilies of DA receptors, or used when the genetic subtype (D_1 or D_5; D_2, D_3, or D_4) is uncertain.

Table 1
Dopamine Receptor Subtypes Defined from Physiological, Pharmacological, and Biochemical Studies[a]

	D1	D2
Pharmacological characteristics		
Selective antagonists	SCH 23390	(–)-Sulpiride YM 09151-2
Selective agonists	SKF 38393	Quinpirole N-0437
Specific radioligands	[^{3}H]SCH 23390[b]	[^{3}H]YM 09151-2 [^{3}H]Spiperone[c]
Physiological functions	Aspects of motor function (brain), cardiovascular function	Aspects of motor function and behavior (brain), control of prolactin and αMSH secretion from pituitary, cardiovascular function
Biochemical responses	Adenylate cyclase ↑ phospholipase C ↑	Adenylate cyclase ↓ K^+ channel ↑ Ca^{2+} channel ↓
Localization	Caudate nucleus, putamen, nucleus accumbens, olfactory tubercle, cerebral cortex (brain), cardiovascular system	Caudate nucleus, putamen, nucleus accumbens, olfactory tubercle, cerebral cortex (brain) anterior and neurointermediate lobes of pituitary gland, cardiovascular system
Size of protein (Daltons)		
Affinity labeling	72,000	85,000–150,000 (major species of 94,000, 118,000, 140,000)
Purification	—	92,000–95,000 (brain), 120,000 (pituitary)

[a]With the advent of molecular biological studies (Chapter 2), these subtypes should be termed D_1-like and D_2-like receptors. The localization data are from functional and ligand-binding studies on dispersed tissues and tissue slices. The data on the size of the receptor proteins are from affinity-labeling studies of tissues and from the purification of receptors by affinity chromatography. For more details, *see* ref. *14*.

[b][^{3}H]SCH 23390 can also bind to $5HT_{1C}$ and $5HT_2$ receptors.

[c][^{3}H]Spiperone can also bind to $5HT_{1A}$, $5HT_2$ receptors, and alpha-1 adrenergic receptors.

2. Ligand Binding and Second Messenger Assays for D1 and D2 Receptors

The definition of D1 and D2 dopamine receptors led to the search for more selective ligands for the two receptors. For the D2 receptor, the synthesis of [^{3}H]spiperone *(17)* provided an advance in that the radioligand has a high affinity for the D2 subtype and suitable properties for use in radioligand-binding assays. This radioligand does, however, bind to the serotonin $5HT_2$ receptor and the alpha-1 adrenergic receptor with a moderate affinity *(18–20),* but these components of binding can be eliminated using suitable concentrations of selective blocking agents (e.g., mianserin and prazosin, ref. *14*). [^{3}H]Spiperone is still a popular ligand, but newer ligands have been developed based on the substituted benzamide structure, with these ligands offering a better selectivity for D2 receptors. Examples are [^{3}H]YM 09151-2 (nemonapride), [^{3}H]raclopride, and [^{125}I]iodosulpride *(21–23)*. For the D1 receptor, initial ligand-binding studies were performed using the mixed D1/D2 antagonist [^{3}H]flupentixol *(24)* with the binding to D2 receptors blocked with a D2-selective antagonist such as domperidone *(25)*. The provision of selective ligands for D1 receptors depended on the synthesis of the benzazepine SCH 23390 *(26,27)* which provides a useful ligand as [^{3}H]SCH 23390 *(28–30)*. Although this ligand does bind to serotonin $5HT_2$ and $5HT_{1C}$ receptors with a moderate affinity *(31,32),* this does not seem to interfere with its use in tissues such as the brain *(33)*. The related [^{125}I]SCH 23982 has also been described for labeling the D1 receptors *(34)*.

Using these ligands, the basic pharmacological properties of the two receptor subtypes were defined (Table 1). When the binding of ligands to membranes containing the relevant receptor subtype is examined in detail, antagonist binding for both subtypes can be modeled generally in terms of interaction at a single class of binding sites. In contrast, the binding of agonists appears to be more complex, and can be modeled as interaction at two classes of sites with higher and lower affinities for the agonists *(35,36)*. This complexity is reduced or eliminated by the addition of a guanine nucleotide such as GTP, which yields a single class of agonist binding sites whose affinity corresponds to the lower affinity class of sites seen in the absence of added guanine nucleotides. This behavior is typical of a receptor that signals via coupling to a guanine nucleotide regulatory protein (G protein) and is thought to reflect the reversible formation of an agonist/receptor/G-protein complex (ARG, ternary complex). The agonist has a higher affinity for the RG form of the receptor than for the free receptor, whereas antagonists generally have equal affinities for the two receptor forms. This interaction has been extensively modeled by Wreggett and De Léan *(37)* and the group of Lefkowitz et al. *(38)*, and is considered in more detail in Section 5. (*see also* Chapter 5).

For a short period in the 1980s there were indications that additional receptor subtypes could be identified using ligand-binding assays *(9,39)*. These were termed D_3 and D_4 receptors (although there was some confusion over nomenclature, *see* refs. *9,39*) and showed higher affinities for agonists. It was subsequently recognized that these "subtypes" correspond to the agonist high-affinity states of the D1 and D2 receptors defined previously. Nevertheless there were further tantalizing indications of receptor heterogeneity. These suggestions of further heterogeneity came from studies of [^{3}H]domperidone binding in the striatum and pituitary gland *(40)*. Whereas in the striatum, agonist inhibition of [^{3}H]domperidone binding conforms to a two-site model and is sensitive to GTP as expected for a typical D2 receptor, in the pituitary, agonist binding was insensitive to GTP. In addition, in the striatum, the substituted benzamide antagonists also showed biphasic inhibition curves which were not seen in the pituitary. The binding of the substituted benzamides in the striatum was not, however, sensitive to GTP. This heterogeneity of substituted benzamide binding has not been reported by other workers in the field *(41)*. Others have, however, reported that agonist binding to D1 and D2 receptors shows variable sensitivity to GTP in different tissues *(42,43)*. There is no simple explanation for these observations, although the differences in GTP sensitivity could represent differences in posttranslational modification such as phosphorylation. The application of molecular biology techniques has redefined the field (*see* Chapter 2), but these observations may also have been indications of the heterogeneity that has subsequently been uncovered using the molecular biological approach.

Extensive studies were also performed to characterize the second messenger systems linked to the D1 and D2 receptors (*see* Chapter 6). For the D1 receptor, stimulation of adenylate cyclase has been demonstrated in a variety of tissues (*see* Table 1 and ref. *44*), and this response presumably involves the G protein G_s. In the brain there is some indication that the D1 receptor will stimulate phospholipase C, leading to the activation of the inositol phospholipid pathway *(45)*. For the D2 receptor, inhibition of adenylate cyclase has been demonstrated in both the brain and anterior pituitary, but this receptor alters the activity of both K^+ and Ca^{2+} channels, stimulating the former and inhibiting the latter *(46)*. Presumably the G_i/G_o isoforms of G proteins are involved in these effects.

This relatively simple picture of dopamine receptor subtypes was swept aside in the late 1980s by the application of molecular biological techniques to the receptors, demonstrating that there are at least five receptor subtypes as defined by the identification of cloned genes (for reviews *see* refs. *47–49*). These could be grouped, however, into subfamilies (D_1, D_5 and D_2, D_3, D_4) with properties resembling the original D1 and D2 receptors. D_1 and D_5 can be termed the D1-like receptors, and D_2, D_3, and D_4 can be termed the D2-like receptors (*see* Chapter 2). Sections 3.–5. refer to D1-like and D2-like receptors for studies that were performed before the diversity of receptor subtypes was recognized or when the subtype is not clearly defined, and D_1–D_5 in reference to subtypes defined from cloned genes.

3. Biochemical Characterization of the Dopamine Receptor Proteins

Once the two dopamine receptor subtypes were defined using pharmacological and biochemical techniques, a substantial amount of work was performed to determine the nature of the proteins and their properties. Much of this work was aimed at the purification of the proteins so their amino acid sequences could be determined and the amino acid sequence data used to design oligonucleotide probes to identify the corresponding cDNA, as had been performed successfully for the beta-2 adrenergic receptor *(50)*. This aim was superseded by the isolation of the genes for the dopamine receptors using homology cloning *(51)* but, nevertheless, purification techniques will be required to produce pure receptor derived from recombinant systems for biophysical and structural analysis.

3.1. Solubilization of Dopamine Receptors

The purification of a membrane protein, such as a dopamine receptor, requires the solubilization of the receptor in an active form, generally necessitating the use of a nondenaturing detergent. Work on this topic began in the late 1970s using the detergent digitonin, which was shown to solubilize D2-like dopamine receptors from canine striatum *(52,53)*. Many studies were subsequently published on the solubilization of D2-like dopamine receptors using different detergents and different sources of receptor *(54),* and the results are quite variable in terms of the yields of solubilized receptor and its properties. It became clear that it is important to characterize the soluble preparations very carefully, using physical biochemical techniques to ensure true solubilization and using extensive ligand-binding assays with rigorous definitions of specific radioligand binding, to avoid artefactual results *(55)*. Once these criteria had been satisfied, soluble preparations of D2-like dopamine receptors became available that provided a reasonable yield of soluble receptor (~40%). The preparations obtained using digitonin, cholate, and 3-[(3-cholamidopropyl)-dimethylammonio]-1-propane-sulphonate (CHAPS) have been exploited for purification studies *(56–58)*. In these preparations, the binding of ligands is generally of high affinity and with the pharmacological profile expected for D2-like receptors as compared with results obtained on membrane-bound receptors. In some cases, however, there are reductions in the affinities of ligands for binding to the soluble preparations when compared to the membrane-bound receptors *(59)*. This presumably reflects detergent-induced alterations in the precise environment of the receptor. It is interesting that in studies on the membrane-bound D2 receptor expressed in some recombinant systems (e.g., insect cells and yeast), reductions in binding affinities are also seen, which may reflect the different environment of the receptor protein in the membrane of the recombinant cell host. In the soluble preparation obtained from brain using cholate, attempts were made to optimize the stability of the preparation on the assumption that during the lengthy purification procedure a better yield of pure receptor would be obtained with a more stable preparation. This involves a 1.3-fold dilution of the preparation and the extensive use of protease inhibitors *(58)*.

The solubilization of D1-like receptors has been achieved using the detergents digitonin *(60–62)* and cholate *(63,64)*. The pharmacological properties of the receptor are preserved in the soluble preparations, although there are some reductions in ligand affinities relative to the membrane-bound receptor. The yields of soluble receptor are similar to yields obtained for the D2 receptor (~40%).

3.2. Purification of Dopamine Receptors

The purification of a receptor for a neurotransmitter such as dopamine poses a formidable challenge in terms of protein biochemistry. The receptors are present at relatively low concentrations even in tissue such as the striatum, which is the richest source of dopamine receptors in the brain. Elsewhere, I have calculated that the caudate nuclei from 10 bovine brains contain about 200 μg of D2-like receptor protein, and it would require a purification of about 50,000-fold to obtain pure receptor from this tissue *(54,55)*. Despite this, extensive work was carried out on both the D1-like and D2-like receptors using various types of affinity chromatography to purify the receptors. These methods exploit the glycoprotein nature of the receptors by using lectin-affinity chromatography and the specificity of ligand–receptor interaction with immobilized ligands in ligand-affinity chromatography.

For D1-like receptors, there is one report of the partial (200- to 250-fold) purification of the receptor on affinity columns based on a derivative of SCH 23390 (SCH 39111); this has not been pursued further *(60)*. Sidhu *(65)* took advantage of the presence of sulfhydryl groups in the binding site of the D1 receptor *(66)* to achieve ~8000-fold purification using a novel procedure in which membranes are first treated with the sulfhydryl-alkylating reagent *N*-ethylmaleimide (NEM), in the presence of a D1-receptor agonist to protect the ligand binding site. The protected, intact sulfhydryl groups in the solubilized receptor are then exploited to selectively immobilize the D1 receptor on a mercury-agarose matrix, followed by elution from the matrix using β-mercaptoethanol.

In the case of D2-like receptors, adsorption of soluble receptor to wheat germ agglutinin-sepharose columns and elution with *N*-acetyl glucosamine was reported by several groups *(54,58,67)*. The use of *Datura stramonium* agglutinin lectin has also been successful *(57)*. This shows that D2-like receptors are glycoproteins, and provides a method for the partial purification (to 30-fold) of the soluble receptors with high yield. Studies on the elution of the receptor from lectin affinity columns showed that the oligosaccharide chains carried by D2-like receptors are heterogeneous and contain sialic acid and *N*-acetyl glucosamine residues *(68)*.

Columns for ligand-affinity chromatography have been synthesized based on the high-affinity antagonists haloperidol or spiperone *(58,63,69–72)* or the agonist N-0434 *(67)* covalently linked to sepharose through spacer arms of varying length. Using these columns, partial purification of the D2-like dopamine receptor (to 2000-fold) was achieved. The combination of ligand and lectin affinity chromatography steps then led to the full purification of D2-like receptors, and in 1988 three reports appeared describing purification by up to 25,000-fold. Two of these reported the purification of the receptor from brain, and the pure receptor ran as a protein on SDS-PAGE of M_r 92–95k, with clear evidence in one report for the existence of a doublet of bands centered on 95 kDa *(56,58)*. The third study, using sequential ligand-, lectin-, and hydroxylapatite-chromatography, reported purification of a protein of M_r 120k from anterior pituitary *(57)*. A comparative study of the purification of D2-like receptors from different brain regions and the lobes of the pituitary gland confirmed that the M_r 95k species can be purified from several regions of bovine brain, whereas from the anterior pituitary gland, species of M_r 142–145k are purified *(73)*. These different species probably reflect differential glycosylation of a common core protein (*see* Section 3.3.).

The interaction of D2-like receptors with G proteins was studied by purifying the complex of receptor and G protein using the ligand affinity columns described, and in reconstitution studies of receptors and G proteins *(74–77)*. D2-like receptors interact with G_i and G_o forms of G proteins, with some preference for the α_{i2} form. Studies with G protein α subunit-specific antibodies showed that D2-like receptors interact with the G_o and G_{i3} forms of G proteins to modulate the activities of Ca^{2+} and K^+ channels, respectively *(78)*.

In 1988, the first report of the cloning of the D_2 receptor appeared *(51)*, leading to a temporary diminution of interest in protein chemical work on these receptors. Nevertheless, the methods described will be invaluable in the purification of large amounts of receptors for biophysical and structural analyses, such as the application of spectroscopic techniques with the ultimate goal of crystallization and determination of the three-dimensional structure of the receptor.

To obtain the large amounts of receptor needed for purification, it seems advantageous to use a highly expressing recombinant system such as baculovirus-infected insect cells, or microorganisms such as bacteria or yeast. For other receptors, these systems have been applied with some success to the purification of recombinant protein *(79)*. Mammalian cell expression systems tend not to generate the high levels of recombinant protein required. In the case of the dopamine receptors, the D_2 and D_3 receptors have been expressed at high levels in insect cells using the baculovirus system *(80–83),* but purification of the recombinant receptors has not yet been reported. In our own studies, we have applied the purification technique derived successfully for brain (*see* Section 3.2.), but the yield of the procedure was surprisingly low (Sanderson and Strange, unpublished results). The D_2 receptor has also been expressed in the yeasts *Saccharomyces cerevisiae* and *Schizosaccharomyces pombe,* with higher levels of expression being achieved in *S. pombe (84–86)*. In one case the expression levels reported in *S. pombe* were very high *(86),* although the use of an identical construct in the author's lab gave much lower levels of expression *(84)*. The reasons for these discrepancies are unclear at present.

One clear feature of these recombinant systems is that compared to the recombinant mammalian systems such as Chinese hamster ovary (CHO) cells in which the rat D_2 receptor expressed shows very similar properties to that in the native rat brain *(87),* both the yeast and insect cell systems have significant differences in the detailed pharmacological properties of the expressed receptor *(82–85)*. The overall pharmacological properties of the recombinant D_2 receptor are similar to those of the native receptor but there are significant reductions in the affinities of the receptor for certain drugs. There are many reasons why this might be, but one possibility is that the membranes of the insect cells and of the yeast are different in composition from those of rat brain and that this affects the ability of the receptor to bind drugs. There is some precedent for effects of this nature, attributed to differences in the amounts of sterols in the membrane *(88)* which alter the properties of the expressed protein *(89–91)*.

3.3. Affinity Labels for Dopamine Receptors

Substances that bind with high affinity and specificity to a receptor and subsequently form a covalent linkage to the receptor (affinity labels) can be of great utility in studying its properties, particularly if the substance can also be radiolabeled to a high specific activity. Several substances have been synthesized with this aim in mind and the properties of the earlier ligands tested are summarized in ref. *54*. In many cases, these substances did not have optimal properties for extensive use.

More recently, substances have been synthesized that have more useful properties. For D1-like receptors the ligand [^{125}I]IMAB, (±)-7-[^{125}I]iodo-8-hydroxy-3-methyl-1-(4-azidophenyl)-2,3,4,5-tetrahydro-1*H*-3-benzazepine, was synthesized and shown to label a protein of M_r 74k in brain and 62k in the parathyroid gland; this labeling can be prevented by typical D1-like receptor antagonists *(92,93)*. These species of different molecular size were examined by peptide mapping, demonstrating that the two species are very similar, and probably are glycosylation variants of the same core protein. In a complementary study, the ligand ^{125}I-SCH 38548 was crosslinked to a protein of M_r 72k in brain, and again this labeling showed the specificity of a D1-like receptor *(94)*. The species labeled by [^{125}I]IMAB were studied in more detail using glycosidases and lectin affinity columns. Glycosylation of D1-like receptors is heterogeneous and includes some sialic acid and terminal mannose residues *(95)*. Complete deglycosylation yields a species of M_r 46k from different tissues which presumably represents the core protein and correspond reasonably with the size of the receptor from cloning (*see* Chapter 2).

For D2-like receptors, the ligand ^{125}I-NAPS, ^{125}I-*N*-azidophenethylspiperone *(96),* specifically labels species of M_r 120–140k and 94k in several tissues *(96–99)*. In some studies this affinity label was used to study the nature of the glycosylation of the D2-like receptors. Removal of sialic acids with neuraminidase yields a species of M_r ~50–54k, and this is further reduced to M_r ~40–44k by complete deglycosylation with glycopeptidase F *(97–99)*. Thus the receptor is substantially sialylated, and it seems that the D2-like receptors are unusually highly glycosylated. The species of M_r 140k and 94k are probably glycosylation variants as they both yield the same core protein on deglycosylation. In cell lines expressing a recombinant D_2 receptor, the receptor exists as a glycosylated protein of M_r 75k *(100)*. Interestingly, removal of sialic acid does not affect the binding of [^{3}H]spiperone to D2-like receptors *(98,99),* nor does it prevent GTP-sensitive high-affinity binding of agonists *(101)*. The species of M_r 44k probably represents the core protein, in reasonable agreement with values obtained from cloning *(51)*. This core protein also retains the pharmacological profile of the glycosylated D_2 receptor *(97)*. There is some correspondence between these results and those reported for purified receptors (*see* Section 3.2.). ^{125}I-azido-NAPS has also been shown to label the $5HT_{1A}$ serotonin receptor expressed in recombinant COS cells (African green monkey kidney cells) *(102)*. Despite this, it has proven a useful affinity label for the D2-like receptors as indicated.

3.4. Fluorescent Ligands for Dopamine Receptors.

In enzymology, and to some extent in the study of receptors, the use of fluorescent-labeled ligands has proven to be of great utility in studying the detailed properties of the receptors and in particular in probing conformational changes undergone by these proteins. For example, for the nicotinic acetylcholine receptor, a fluorescent-labeled ligand (NBD-5-acyl choline) was used to probe conformational transitions in this receptor *(103)*. For the G protein-coupled receptors, the neurokinin-1 receptor has been examined with a fluorescent-labeled neurokinin peptide, and this has exposed certain conformational transitions undergone by the receptor when receptor–G protein interaction is manipulated *(104–105)*.

For the dopamine receptors, fluorescent-labeled ligands with high affinity and selectivity for the D1-like and D2-like receptors have been synthesized and characterized *(106,107)*. A range of fluorescent-labeled agonists and antagonists is available from these studies with high affinities (< 10 n*M*) for the respective receptors. Examples of compounds that have been evaluated in ligand binding studies are NBD-SKF 83566 (D1-selective) and NBD-NAPS (D2-selective); some fluorescent ligands are available commercially. These have not been used to study the biochemical properties of the receptors, but they have been used to label receptors in tissue slices as a means of exploring receptor localization (*see* Chapter 3).

3.5. Antibodies to Dopamine Receptors

An additional motivation for purification of dopamine receptors was the desire to use pure or enriched receptor as an antigen for the production of receptor-specific antibodies. This proved to be difficult until determination of the sequence of the receptors created the possibility of generating antipeptide antibodies or antibodies generated using fusion protein that would recognize the receptor from which the peptide sequence was derived. Several groups have developed antibodies that selectively recognize the D_1 receptor *(108–110)*, the D_3 receptor *(80,108,111)*, and the D_2 receptor *(110,112–114)*, including some antibodies that distinguish D_{2L} from D_{2S} *(115–117)*, and antiserum that recognizes in vitro-translated D_{2S} but not D_{2L} *(111)*. To date, the primary use for these dopamine receptor antibodies has been for immunohistochemical localization of dopamine receptors. Many of these antibodies are also capable of precipitating solubilized receptors, and have some capacity to label dopamine receptors by Western blot analysis. In some studies, antibodies have been used to identify regions of the D2 receptor that are involved in coupling to G proteins. Three groups of researchers have determined that antibodies directed against peptides from the third cytoplasmic loop *(112,114,115)*, but not the second cytoplasmic loop *(112)*, prevent coupling to G proteins, as assessed by high-affinity binding of agonists. Interestingly, antibodies directed against peptides from approximately the middle of the third cytoplasmic loop *(114,115)* or from immediately after the alternatively spliced insert *(112)* interfere with coupling, whereas antiserum against a peptide comprising most of the alternatively spliced insert has no effect on the coupling of D_{2L} to G proteins *(80)*.

4. Chemical Modification Studies on Dopamine Receptors

Limited studies have been performed using the technique of chemical modification with group-specific reagents to infer the participation of certain amino acid residues in the function of the receptors. The receptor preparation is treated with a reagent that selectively modifies certain amino acid side chains. If the binding of ligands to the receptor is affected and the effect can be prevented by previous occupancy of the receptor with a selective ligand, this is evidence that the particular amino acid side chain is present at the ligand binding site. The amino acid side chain may then also be involved in ligand binding. It is important to recognize that some of the reagents used have a limited specificity and could give misleading results.

For the D1-like receptors, these studies have been performed in brain tissue using reagents specific for carboxyl residues (*N,N'* dicyclohexyl carbodiimide), sulfhydryl residues (5,5'-dithiobis[2-nitrobenzoic acid] or DTNB, NEM), tyrosine residues (*N*-acetyl imidazole, tetranitromethane, 4-nitrobenzene sulphonyl fluoride), amino groups (ethyl acetimidate), and histidine residues (diethyl pyrocarbonate). Modification of carboxyl residues attenuates the binding of ligands to D1-like receptors *(33),* consistent with the participation of an aspartic acid residue in the binding of the cationic ligands (*see* Chapter 2). Further circumstantial evidence for the participation of carboxyl residues in the binding of ligands comes from a study of the pH-dependence of the binding of [^{3}H]SCH 23390 to D1-like receptors in bovine brain *(33),* where ligand binding depends on the ionization of a group of pK_a 6.9, which could correspond to the ionization of a carboxyl residue; although the pK_a value is high for a carboxyl residue, this may reflect the microenvironment of the receptor. Modification of sulfhydryl residues with DTNB or NEM attenuates the binding of antagonists, and this effect can be prevented by occupancy of the receptors by ligands, indicating that the modified sulfhydryl is at the ligand binding site *(33,66,118,119)*. Experiments on the modification of tyrosine residues gave conflicting results, and it is unclear whether this reflects the selectivity or accessibility of the reagents used *(33,120)*. Modification of histidine and amino groups does not provide any evidence for their participation in the binding of ligands *(33)*. From these experiments some evidence was obtained for the presence of carboxyl and sulfhydryl residues in the ligand-binding site, and these may participate in the binding of ligands to

D1-like receptors. Verification of the roles of these residues awaits the results of mutagenesis experiments.

For the D2-like receptors, studies have been performed in brain, anterior pituitary, and cells expressing the cloned D_2 receptor. The carboxyl-specific reagent *N,N'*dicyclohexyl carbodiimide inhibits the binding of both [^{3}H]spiperone and the substituted benzamide [^{3}H]nemonapride (YM-09151-2) to the D2-like receptors, implying the participation of carboxyl groups in the ligand binding process *(121)*. As will be discussed in Chapter 2, this is consistent with the interaction of cationic ligands with an aspartic acid residue at the binding site. For tyrosyl residues, studies with the reagent *N*-acetyl imidazole indicate no effect of the modification, whereas studies with the reagents tetranitromethane and 4-nitrobenzene sulphonyl fluoride show effects of the modification *(120)*. The reasons for these differences are unclear, but there are several tyrosine residues potentially at the ligand binding site. As in the case of the D1-like receptors, this could reflect differences in the specificity or accessibility of the reagents used.

Several studies have been performed on D2-like receptors with reagents that modify sulfhydryl residues. In studies on brain and pituitary tissue, treatment with NEM inhibits the binding of [^{3}H]spiperone and [^{3}H]sulpiride by decreasing the density of binding sites. Inhibition depends on the concentration of Na^+ in the assay, with high Na^+ inhibiting the effect of NEM *(122–125)*. Similar effects of NEM have been reported on the binding of [^{125}I]epidepride to recombinant D_2 receptors expressed in L-cells *(126)*. It seems likely that the effects of Na^+ reflect a conformational change in the receptor that makes a sulfhydryl residue less accessible. These effects of Na^+ are discussed in more detail in Section 5. The more selective, polar sulfhydryl reagents DTNB and methanethiosulfonate (MTS) have also been used with conflicting results. DTNB has no effect on the binding of [^{3}H]spiperone to D2-like receptors in bovine brain or to recombinant D_2 receptors expressed in CHO cells *(127,128)*. Nevertheless, DTNB does inhibit the binding of [^{3}H]spiperone to D2-like receptors in anterior pituitary. On the other hand, MTS irreversibly inhibits radioligand binding to the recombinant D_2 receptor *(129)*. The reasons for this difference are unclear. Modification of histidine residues using diethyl pyrocarbonate inhibits [^{3}H]spiperone binding to D2-like receptors in bovine brain, but the effect cannot be prevented by previous occupancy of the receptors by an antagonist, and is unlikely to reflect the modification of a histidine residue at the binding site. Hence, similar to D1-like receptors, there is evidence that D2-like receptors have carboxyl and sulfhydryl residues at the ligand binding site that may participate in ligand binding.

These data on the chemical modification of D2-like receptors have been complemented by studies on the pH-dependence of the binding of ligands to these receptors. Initial studies on the pH-dependence of antagonist binding showed that the binding of classical antagonists is much less sensitive than the binding of drugs of the substituted benzamide class to changes in pH *(121,126,130)*. The binding of classical antagonists such as spiperone and haloperidol depends on the ionization of one ionizing group on the receptor of pK_a ~6, whereas the binding of the substituted benzamides depends on the ionization of two groups whose pK_a values are in the range 6–7 *(128)*. Thus it is clear that the two classes of antagonist bind in different ways to the receptors. The nature of the ionizing groups cannot be determined from the pH-dependence data alone, but studies on these and other receptors suggest that the single residue affecting the binding of the classical antagonists is Asp-114, the aspartic acid residue that is important in forming an electrostatic interaction with the cationic drugs. This residue is also likely to be important for the binding of the substituted benzamides, but for these drugs the ionization of a second aspartic acid residue (Asp-80) may be important in influencing binding.

5. Regulation of Dopamine Receptors

In the process of the characterization of the ligand binding properties of the dopamine receptors it has become apparent that these receptors are subject to acute regulation by several influences such as guanine nucleotides, ions, and small organic molecules. The dopamine receptors are also subject to longer-term regulation including changes in the number of receptors on cells (*see* Chapter 13).

As outlined in Section 2., agonist binding to both the D1-like and D2-like receptors is sensitive to the addition of guanine nucleotides. Typically, this is seen as a reduction in agonist affinity in the presence of GTP or poorly hydrolyzable analogs such as GppNHp (*see* Chapter 5). More detailed analysis of agonist binding data shows that these effects can generally be described as a conversion of higher and lower agonist affinity states into a single affinity state in the presence of GTP. This is thought to reflect a GTP-induced destabilization of a ternary complex (ARG) containing the agonist (A), the receptor (R), and the G protein (G). This has been extensively modeled *(37)* and more recent modifications of this model have been suggested including an allosteric transition in the AR complex *(38)*. In terms of the observations on agonist binding, the basic tenet of the models is that agonists have a higher affinity for RG than for R, and this seems to apply for both the D1-like and the D2-like receptors. Antagonists generally do not show large differences in their affinities for the two states of receptor, although there are indications that some antagonists may have a higher affinity for R than for RG *(37)*. This may be a genuine reflection of differences in affinities of antagonists for the free and G protein-coupled forms of dopamine receptors as has been described for other G protein-coupled receptors *(131)*.

The ergot alkaloids are a class of compounds that display unusual properties with respect to these models. The ergots are functional agonists at D2-like receptors but do not show the differential affinities for the receptors in the absence and presence of GTP *(132)*. This suggests that these compounds bind to the receptors in a different manner from other agonists, having similar affinities for the R and RG states of the receptor. Similar observations have been made for the $5HT_{1A}$ receptor for the ergot agonist lisuride *(133)*.

Ligand-binding to D2-like receptors is also regulated by Na^+ and Mg^{2+} ions. These cations generally have only small effects on the binding of antagonists, with Na^+ increasing and Mg^{2+} reducing the affinity of antagonists *(134)*, although not all workers have observed these effects *(124)*. One class of dopamine antagonist, the substituted benzamides, is much more sensitive to the effects of Na^+, with removal of Na^+ reducing affinities for these compounds by up to 30-fold *(135)*. Agonist binding to D2-like receptors is regulated in a reciprocal manner, with Na^+ ions reducing agonist affinity and Mg^{2+} ions increasing agonist affinity *(134)*.

Regulation of ligand binding by Na^+ ions is also seen for the alpha-2 adrenergic receptor. Further, it was shown by mutagenesis that the site of this regulation is an aspartic acid residue in the second putative transmembrane region *(136)*. A similar mutational study has been performed for the D_2 receptor showing that the corresponding aspartic acid residue, Asp-80, is the site of the Na^+ regulation *(137)*. Regulation of agonist and antagonist (substituted benzamide) binding by Na^+ ions is abolished by the mutation so that Asp-80 is probably a common site for the regulation, by Na^+ ions, of both agonist and antagonist binding. As indicated in Section 4., this aspartic acid residue is also probably one of the sites of regulation of the binding of the substituted benzamides by H^+ ions. It seems likely, therefore that the D_2 receptor may exist in different conformations regulated by Na^+ and H^+ ions. In these different conformations, modification by NEM occurs at different rates, as outlined. Whereas the effects of Na^+ and H^+ are on the receptor itself, the effects of Mg^{2+} ions are likely to reflect the stabilization of the ARG state of the receptor by an effect on the G protein *(138)*. The effects of Na^+ on the D_2 receptor seem to be independent of the effects of guanine nucleotides *(139,140)*.

A further form of regulation of the D_2 receptor has been described and is again based on observations on the alpha-2 adrenergic receptor which has been reported to be regulated by the diuretic amiloride and analogs thereof *(141)*. These observations were extended to the D_2 receptor and it was shown that amiloride accelerates the dissociation of the radioligands [^{3}H]spiperone and [^{125}I]epidepride from this receptor so that it is acting in an allosteric manner *(126)*. Much greater accelerations of the rate of dissociation of [^{3}H]spiperone are seen with analogs of amiloride such as methyl-isopropyl-amiloride *(142)*. The site of action of amiloride and its analogs is unknown at present, but the allosteric regulation of G protein-coupled receptors may be a fruitful area for the design of novel drugs *(143)*.

It seems that the D_2 receptor can be regulated by several different influences (ions, guanine nucleotides, amiloride and its analogs) and that these regulatory effects relate to the ability of the receptor to exist in multiple conformational states. At present it does not seem that these conformational states accessed by these different influences are related to one another but more experimentation is required.

References

1. Strange, P. G. (1992) *Brain Biochemistry and Brain Disorders,* Oxford University Press, Oxford, UK.
2. Cools, A. R. and van Rossum, J. M. (1976) Excitation-mediating and inhibition-mediating dopamine receptors: a new concept towards a better understanding of electrophysiological, biochemical, pharmacological, functional and clinical data. *Psychopharmacologia* **45,** 243–254.
3. Brown, J. H. and Makman, M. H. (1972) Stimulation by dopamine of adenylate cyclase in retinal homogenates and of adenosine-3':5'-cyclic monophosphate formation in intact retina. *Proc. Natl. Acad. Sci. USA* **69,** 539–543.
4. Greengard, P. (1976) Possible role for cyclic nucleotides and phosphorylated membrane proteins in postsynaptic actions of neurotransmitters. *Nature* **260,** 101–108.
5. Iversen, L. L. (1975) Dopamine receptors in the brain. *Science* **188,** 1084–1089.
6. Kebabian, J. W., Petzold, G. L., and Greengard, P. (1972) Dopamine-sensitive adenylate cyclase in caudate nucleus of rat brain, and its similarity to the "dopamine receptor." *Proc. Natl. Acad. Sci. USA* **69,** 2145–2149.
7. Burt, D. R., Creese, I., and Snyder, S. H. (1976) Properties of [^{3}H]haloperidol and [^{3}H]dopamine binding associated with dopamine receptors in calf brain membranes. *Mol. Pharmacol.* **12,** 800–812.
8. Seeman, P., Chau-Wong, M., Tedesco, J., and Wong, K. (1975) Brain receptors for antipsychotic drugs and dopamine: direct binding assays. *Proc. Natl. Acad. Sci. USA* **72,** 4376–4380.
9. Seeman, P. (1980) Brain dopamine receptors. *Pharmacol. Rev.* **32,** 229–313.
10. Creese, I., Burt, D. R., and Snyder, S. H. (1976) Dopamine receptor binding predicts clinical and pharmacological potencies of antischizophrenic drugs. *Science* **192,** 481–483.
11. Seeman, P., Lee, T., Chau-Wong, M., and Wong, K. (1976) Antipsychotic drug doses and neuroleptic/dopamine receptors. *Nature* **261,** 717–719.
12. Kebabian, J. W. and Calne, D. B. (1979) Multiple receptors for dopamine. *Nature* **277,** 93–96.

13. Spano, P. F., Govoni, S., and Trabucchi, M. (1978) Studies on the pharmacological properties of dopamine receptors in various areas of the central nervous system. *Adv. Biochem. Psychopharm.* **19,** 155–165.
14. Strange, P. G. (1987) Dopamine receptors in the brain and periphery: "state of the art." *Neurochem. Int.* **10,** 27–33.
15. Cavero, I., Massingham, R., and Lefevre-Borg, F. (1982) Peripheral dopamine receptors, potential targets for a new class of antihypertensive agents. Part I: subclassification and functional description. *Life Sci.* **31,** 939–948.
16. Cavero, I., Massingham, R., and Lefevre-Borg, F. (1982) Peripheral dopamine receptors, potential targets for a new class of antihypertensive agents. Part II: sites and mechanisms of action of dopamine receptor agonists. *Life Sci.* **31,** 1059–1069.
17. Leysen, J. E., Gommeren, W., and Laduron, P. M. (1977) Spiperone: a ligand of choice for neuroleptic receptors. 1. Kinetics and characteristics of in vitro binding. *Biochem. Pharmacol.* **27,** 307–316.
18. Leysen, J. E., Niemegeers, C. J. E., Tollenaere, J. P., and Laduron, P. M. (1978) Serotonergic component of neuroleptic receptors. *Nature* **272,** 168–171.
19. Morgan, D. G., Marcusson, J. O., and Finch, C. E. (1984) Contamination of serotonin-2 binding sites by an α-1 adrenergic component in assays with [^{3}H]spiperone. *Life Sci.* **34,** 2507–2514.
20. Withy, R. M., Mayer, R. J., and Strange, P. G. (1981) Use of [^{3}H]spiperone for labelling dopaminergic and serotonergic receptors in bovine caudate nucleus. *Neurochem. Int.* **10,** 27–33.
21. Högberg, T., Rämsby, S., Ögren, S.-P., and Norinder, U. (1987) New selective dopamine D-2 antagonists as antipsychotic agents. Pharmacological, chemical, structural and theoretical considerations. *Acta Pharm. Suec.* **24,** 289–328.
22. Martres, M.-P., Sales, N., Bouthenet, M.-L., and Schwartz, J.-C. (1985) Localisation and pharmacological characterisation of D-2 dopamine receptors in rat cerebral neocortex and cerebellum using [^{125}I]iodosulpride. *Eur. J. Pharmacol.* **118,** 211–219.
23. Niznik, H. B., Grigoriadis, D. E., Pri-Bar, I., Buchman, O., and Seeman, P. (1985) Dopamine D_2 receptors selectively labeled by a benzamide neuroleptic: [^{3}H]-YM-0915-2. *Naunyn-Schmiedeberg's Arch. Pharmacol.* **329,** 333–343.
24. Hyttel, J. (1978) Effects of neuroleptics on ^{3}H-haloperidol and ^{3}H-cis(Z)-flupenthixol binding and on adenylate cyclase activity in vitro. *Life Sci.* **23,** 551–556.
25. Leff, S. E., Hamblin, M. W., and Creese, I. (1984) Interactions of dopamine agonists with brain D_1 receptors labeled by ^{3}H-antagonists. *Mol. Pharmacol.* **27,** 171–183.
26. Hyttel, J. (1983) SCH 23390—The first selective dopamine D-1 antagonist. *Eur. J. Pharmacol.* **91,** 153,154.
27. Iorio, L. C. (1981) SCH 23390, a benzazepine with atypical effects on dopaminergic systems. Pharmacologist **23,** 136.
28. Andersen, P. H., Grønvald, F. C., and Jansen, J. A. (1985) A comparison between dopamine-stimulated adenylate cyclase and ^{3}H-SCH 23390 binding in rat striatum. *Life Sci.* **37,** 1971–1983.
29. Billard, W., Ruperto, V., Crosby, G., Iorio, L. C., and Barnett, A. (1984) Characterization of the binding of ^{3}H-SCH 23390, a selective D-1 receptor antagonist ligand, in rat striatum. *Life Sci.* **35,** 1885–1893.
30. Wyrick, S. D. and Mailman, R. B. (1985) Tritium labelled (±)-7-chloro-8-hydroxy-3-methyl-1-phenyl-2,3,4,5-tetrahydro-1$\underline{H}$-3-benzazepine (SCH23390). *J. Labelled Compounds Radiopharmaceut.* **22,** 189–195.

31. Bischoff, S., Heinrich, M., Sonntag, J. M., and Krauss, J. (1986) The D-1 dopamine receptor antagonist SCH 23390 also interacts potently with brain serotonin (5-HT_2) receptors. *Eur. J. Pharmacol.* **129,** 367–370.
32. Nicklaus, K. J., McGonigle, P., and Molinoff, P. B. (1988) [^{3}H]SCH 23390 labels both dopamine-1 and 5-hydroxytryptamine$_{1C}$ receptors in the choroid plexus. *J. Pharmacol. Exper. Ther.* **247,** 343–348.
33. Hollis, C. M. and Strange, P. G. (1992) Studies on the structure of the ligand-binding site of the brain D_1 dopamine receptor. *Biochem. Pharmacol.* **44,** 325–334.
34. Sidhu, A., van Oene, J. C., Dandridge, P., Kaiser, C., and Kebabian, J. W. (1986) [^{125}I]SCH 23893: the ligand of choice for identifying the D-1 dopamine receptor. *Eur. J. Pharmacol.* **128,** 213–220.
35. Hess, E. J. and Creese, I. (1987) Biochemical characterization of dopamine receptors, in *Dopamine Receptors* (Creese, I. and Fraser, C., eds.), Liss, New York, pp. 1–27.
36. Zahniser, N. R. and Molinoff, P. B. (1978) Effect of guanine nucleotides on striatal dopamine receptors. *Nature* **275,** 453–455.
37. Wreggett, K. A. and De Léan, A. (1984) The ternary complex model: its properties and application to ligand interactions with the D_2-dopamine receptor of the anterior pituitary gland. *Mol. Pharmacol.* **26,** 214–227.
38. Lefkowitz, R. J., Cotecchia, S., Samama, P., and Costa, T. (1993) Constitutive activity of receptors coupled to guanine nucleotide regulatory proteins. *Trends Pharmacol. Sci.* **14,** 303–307.
39. Sokoloff, P., Martres, M. P., and Schwartz, J. C. (1980) Three classes of dopamine receptor (D-2, D-3, D-4) identified by binding studies with ^{3}H-apomorphine and ^{3}H-domperidone. *Naunyn-Schmiedeberg's Arch. Pharmacol.* **315,** 89–103.
40. Sokoloff, P., Martres, M.-P., Delandre, M., Redouane, K., and Schwartz, J.-C. (1984) ^{3}H-domperidone binding sites differ in rat striatum and pituitary. *Naunyn-Schmiedeberg's Arch. Pharmacol.* **327,** 221–227.
41. Leonard, M. N., Halliday, C. A., Marriott, A. S., and Strange, P. G. (1988) D_2 dopamine receptors in rat striatum are homogeneous as revealed by ligand-binding studies. *Biochem. Pharmacol.* **37,** 4335–4339.
42. De Keyser, J., De Backer, J.-P., Convents, A., Ebinger, G., and Vauquelin, G. (1985) D_2 dopamine receptors in calf globus pallidus: agonist high- and low-affinity sites not regulated by guanine nucleotide. *J. Neurochem.* **45,** 977–979.
43. Leonard, M. N., Macey, C. A., and Strange, P. G. (1987) Heterogeneity of D_2 dopamine receptors in different brain regions. *Biochem. J.* **248,** 595–602.
44. Kebabian, J. W., Agui, T., van Oene, J. C., Shigematsu, K., and Saavedra, J. M. (1986) The D_1 dopamine receptor: new perspectives. *Trends Pharmacol. Sci.* **7,** 96–99.
45. Andersen, P. H., Gingrich, J. A., Bates, M. D., Dearry, A., Falardeau, P., Senogles, S. E., and Caron, M. G. (1990) Dopamine receptor subtypes: beyond the D1/D2 classification. *Trends Pharmacol. Sci.* **11,** 231–236.
46. Vallar, L. and Meldolesi, J. (1989) Mechanisms of signal transduction at the dopamine D_2 receptor. *Trends Pharmacol. Sci.* **10,** 74–77.
47. Civelli, O., Bunzow, J. R., and Grandy, D. K. (1993) Molecular diversity of the dopamine receptors. *Annu. Rev. Pharmacol. Toxicol.* **32,** 281–307.
48. O'Dowd, B. F. (1993) Structures of dopamine receptors. *J. Neurochem.* **60,** 804–816.
49. Sibley, D. R. and Monsma, F. J. (1992) Molecular biology of dopamine receptors. *Trends. Pharmacol. Sci.* **13,** 61–69.

50. Dixon, R. A. F., Kobilka, B. K., Strader, D. J., Benovic, J. L., Dohlman, H. G., Frielle, T., et al. (1986) Cloning of the gene and cDNA for mammalian β-adrenergic receptor and homology with rhodopsin. *Nature (Lond.)* **321,** 75–79.
51. Bunzow, J. R., Van Tol, H. H. M., Grandy, D. K., Albert, P., Salon, J., Christie, M., Machida, C. A., Neve, K. A., and Civelli, O. (1988) Cloning and expression of a rat D_2 dopamine receptor cDNA. *Nature* **336,** 783–787.
52. Gorissen, H. and Laduron, P. (1979) Solubilisation of high-affinity dopamine receptors. *Nature* **279,** 72–74.
53. Laduron, P. M. and Ilien, B. (1982) Solubilization of brain muscarinic, dopaminergic and serotonergic receptors: a critical analysis. *Biochem. Pharmacol.* **31,** 2145–2151.
54. Strange, P. G. (1987) Isolation and molecular characterisation of dopamine receptors, in *Dopamine Receptors* (Creese, I. and Fraser, C., eds.), Liss, New York, pp. 29–43.
55. Strange, P. G. (1983) Isolation and characterisation of dopamine D2 receptors. *Trends Pharmacol. Sci.* **4,** 188–190.
56. Elazar, Z., Kanety, H., David, C., and Fuchs, S. (1988) Purification of the D-2 dopamine receptor from bovine striatum. *Biochem. Biophys. Res. Commun.* **156,** 602–609.
57. Senogles, S. E., Amlaiky, N., Falardeau, P., and Caron, M. G. (1988) Purification and characterization of the D_2-dopamine receptor from bovine anterior pituitary. *J. Biol. Chem.* **263,** 18,996–19,002.
58. Williamson, R. A., Worrall, S., Chazot, P.L., and Strange, P. G. (1988) Purification of brain D_2 dopamine receptor. *EMBO J.* **7,** 4129–4133.
59. Hall, J. M., Frankham, P. A., and Strange, P. G. (1983) Use of cholate/sodium chloride for solubilisation of brain D_2 dopamine receptors. *J. Neurochem.* **41,** 1526–1532.
60. Gingrich, J. A., Amlaiky, N., Senogles, S. E., Chang, W. K., McQuade, R. D., Berger, J. G., and Caron, M. G. (1988) Affinity chromatography of the D_1 dopamine receptor from rat corpus striatum. *Biochemistry* **27,** 3907–3912.
61. Niznik, H. B., Grigoriadis, D. E., Otsuka, N. Y., Dumbrille-Ross, A., and Seeman, P. (1986) The dopamine D_1 receptor: partial purification of a digitonin-solubilized receptor-guanine nucleotide binding complex. *Biochem. Pharmacol.* **35,** 2974–2977.
62. Zhang, X. and Segawa, T. (1989) Investigation of rat striatal dopamine D-1 receptors solubilized by digitonin with a precipitation method. *Eur. J. Pharmacol.* **166,** 401–410.
63. Ramwani, J. and Mishra, R. K. (1986) Purification of bovine striatal dopamine D-2 receptor by affinity chromatography. *J. Biol. Chem.* **261,** 8894–8898.
64. Sidhu, A. and Fishman, P. H. (1986) Solubilization of the D-1 dopamine receptor from rat striatum. *Biochem. Biophys. Res. Commun.* **137,** 943–949.
65. Sidhu, A. (1990) A novel affinity purification of D-1 dopamine receptors from rat striatum. *J. Biol. Chem.* **265,** 10,065–10,072.
66. Sidhu, A., Kassis, S., Kebabian, J., and Fishman, P. H. (1986) Sulfhydryl group(s) in the ligand binding site of the D-1 dopamine receptor: specific protection by agonist and antagonist. *Biochemistry* **25,** 6695–6701.
67. Bosker, F. J., Van Bussel, F. J., Thielen, A. P. G. M., Soei, Y. L., Sieswerda, G. T., Dijk, J., Tepper, P. G., Horn, A. S., and Möller, W. (1989) Affinity chromatography with the immobilized agonist N-0434 yields an active and highly purified preparation of the dopamine D-2 receptor from bovine striatum. *Eur. J. Pharmacol.* **163,** 319–326.

68. Leonard, M. N., Williamson, R. A., and Strange, P. G. (1988) The glycosylation properties of D_2 dopamine receptors from striatal and limbic areas of bovine brain. *Biochem. J.* **255,** 877–883.
69. Antonian, L., Antonian, E., Murphy, R. B., and Schuster, D. I. (1986) Studies on the use of a novel affinity matrix, sepharose amine-succinyl-amine haloperidol hemisuccinate, ASA-HHS, for purification of canine dopamine (D_2) receptor. *Life Sci.* **38,** 1847–1858.
70. Clagett-Dame, M., Schoenleber, R., Chung, C., and McKelvy, J. F. (1989) Preparation of an affinity chromatography matrix for the selective purification of the dopamine D2 receptor from bovine striatal membranes. *Biochim. Biophys. Acta* **986,** 271–280.
71. Senogles, S. E., Amlaiky, N., Johnson, A. L., and Caron, M. G. (1986) Affinity chromatography of the anterior pituitary D_2-dopamine receptor. *Biochemistry* **25,** 749–753.
72. Soskic, V. and Petrovic, J. (1986) Purification of dopamine D_2 receptors of the bovine caudate nucleus by spiperone-sepharose 4B affinity beads. *Iugoslav. Physiol. Pharmacol. Acta* **22,** 329–338.
73. Chazot, P. L. and Strange, P. G. (1992) Molecular characterization of D2 dopamine-like receptors from brain and from the pituitary gland. *Neurochem. Int.* **21,** 159–169.
74. Elazar, Z., Siegel, G., and Fuchs, S. (1989) Association of two pertussis toxin-sensitive G-proteins with the D2-dopamine receptor from bovine striatum. *EMBO J.* **8,** 2353–2357.
75. Ohara, K., Haga, K., Berstein, G., Haga, T., and Ichiyama, A. (1988) The interaction between D-2 dopamine receptors and GTP-binding proteins. *Mol. Pharmacol.* **33,** 290–296.
76. Senogles, S. E., Benovic, J. L., Amlaiky, N., Unson, C., Milligan, G., Vinitsky, R., Spiegel, A. M., and Caron, M. G. (1987) The D_2-dopamine receptor of anterior pituitary is functionally associated with a pertussis toxin-sensitive guanine nucleotide binding protein. *J. Biol. Chem.* **262,** 4860–4867.
77. Senogles, S. E., Spiegel, A. M., Padrell, E., Iyengar, R., and Caron, M. G. (1990) Specificity of receptor-G protein interactions: discrimination of G_i subtypes by the D_2 dopamine receptor in a reconstituted system. *J. Biol. Chem.* **265,** 4507–4514.
78. Lledo, P. M., Homburger, V., Bockaert, J., and Vincent, J.-D. (1992) Differential G protein-mediated coupling of D_2 dopamine receptors to K^+ and Ca^{2+} currents in rat anterior pituitary cells. *Neuron* **8,** 455–463.
79. Parker, E. M., Kameyama, K., Higashijima, T., and Ross, E. M. (1991) Reconstitution of active G protein-coupled receptors purified from baculovirus-infected insect cells. *J. Biol. Chem.* **266,** 519–527.
80. Boundy, V. A., Luedtke, R. R., Gallitano, A. L., Smith, J. E., Filtz, T. M., Kallen, R. G., and Molinoff, P. B. (1993) Expression and characterization of the rat D_3 dopamine receptor: pharmacologic properties and development of antibodies. *J. Pharmacol. Exp. Ther.* **264,** 1002–1011.
81. Javitch, J. A., Kaback, J., Li, X., and Karlin, A. (1994) Expression and characterization of human dopamine D_2 receptor in baculovirus-infected insect cells. *J. Recept. Res.* **14,** 99–117.
82. Ng, G. Y. K., O'Dowd, B. F., Caron, M., Dennis, M., Brann, M. R., and George, S. R. (1994) Phosphorylation and palmitoylation of the human $D2_L$ dopamine receptor in Sf9 cells. *J. Neurochem.* **63,** 1589–1595.

83. Woodcock, C., Graber, S. G., Strange, P. G., and Rooney, B. C. (1994) Characterisation of the D2long and D3 dopamine receptors expressed in insect cells. *Biochem. Soc. Trans.* **22,** 935.
84. Presland, J. and Strange, P. G. (1994) Expression of D2 dopamine receptors in the yeast *S. pombe. Br. Pharmacol. Soc. Proc.* London Meeting, 50.
85. Sander, P., Grünewald, S., Maul, G., Reiländer, H., and Michel, H. (1994) Constitutive expression of the human D2S-dopamine receptor in the unicellular yeast *Saccharomyces cerevisiae. Biochim. Biophys. Acta* **1193,** 255–262.
86. Sander, P., Grünewald, S., Reiländer, H., and Michel, H. (1994) Expression of the human D_{2S} dopamine receptor in the yeasts *Saccharomyces cerevisiae* and *Schizosaccharomyces pombe*: a comparative study. *FEBS Lett.* **344,** 41–46.
87. Castro, S. W. and Strange, P. G. (1993) Differences in the ligand binding properties of the short and long versions of the D_2 dopamine receptor. *J. Neurochem.* **60,** 372–375.
88. Ziaser, E., Sperka-Gottlieb, C. D. M., Frasch, E., Kohlwein, S. D., Paltauf, F., and Daum, G. (1991) Phospholipid synthesis and lipid composition of subcellular membranes in the unicellular eukaryote *S. Cerevisiae. J. Bact.* **173,** 2026–2034.
89. Hildebrandt, V., Fendles, K., Heberle, J., Hoffman, A., Bamberg, E., and Buldt, G. (1993) Bacteriorhodopsin expressed in *S. pombe* pumps protons through the plasma membrane. *Proc. Natl. Acad. Sci. USA* **90,** 3578–3582.
90. Kiritovsky, J. and Schramm, M. (1983) Delipidation of β-adrenergic receptor preparation and reconstitution by specific lipids. *J. Biol. Chem.* **258,** 6841–6849.
91. Mitchell, D. C., Straume, M., Miller, J. L., and Litman, B. J. (1990) Modulation of metarhodopsin formation by cholesterol-induced ordering of bilayer lipids. *Biochemistry* **29,** 9143–9149.
92. Niznik, H. B., Jarvie, K. R., Bzowej, N. H., Seeman, P., Garlick, R. K., Miller, J. J., Baindur, N., and Neumeyer, J. L. (1988) Photoaffinity labeling of dopamine D1 receptors. *Biochemistry* **27,** 7594–7599.
93. Niznik, H. B., Jarvie, K. R., and Brown, E. M. (1989) Dopamine D1 receptors of the calf parathyroid gland: identification of the ligand binding subunit with lower apparent molecular weight but similar primary structure to neuronal D1 receptors. *Biochemistry* **28,** 6925–6930.
94. Amlaiky, N., Berger, J. G., Chang, W., McQuade, R. J., and Caron, M. G. (1987) Identification of the binding subunit of the D_1-dopamine receptor by photoaffinity crosslinking. *Mol. Pharmacol.* **31,** 129–134.
95. Jarvie, K. R., Booth, G., Brown, E. M., and Niznik, H. B. (1989) Glycoprotein nature of dopamine D1 receptors in the brain and parathyroid gland. *Mol. Pharmacol.* **36,** 566–574.
96. Amlaiky, N. and Caron, M. G. (1986) Identification of the D_2-dopamine receptor binding subunit in several mammalian tissues and species by photoaffinity labeling. *J. Neurochem.* **47,** 196–204.
97. Clagett-Dame, M. and McKelvy, J. F. (1989) N-linked oligosaccharides are responsible for rat striatal dopamine D2 receptor heterogeneity. *Arch. Biochem. Biophys.* **274,** 145–154.
98. Grigoriadis, D. E., Niznik, H. B., Jarvie, K. R., and Seeman, P. (1988) Glycoprotein nature of D_2 dopamine receptors. *FEBS Lett.* **227,** 220–224.
99. Jarvie, K. R., Niznik, H. B., and Seeman, P. (1988) Dopamine D_2 receptor binding subunits of $M_r \cong 140,000$ and 94,000 in brain: deglycosylation yields a common unit of $M_r \cong 44,000$. *Mol. Pharmacol.* **34,** 91–97.

100. David, C., Fishburn, C. S., Monsma, F. J., Jr., Sibley, D. R., and Fuchs, S. (1993) Synthesis and processing of D_2 dopamine receptors. *Biochemistry* **32,** 8179–8183.
101. Jarvie, K. R., Niznik, H. B., Bzowej, N. H., and Seeman, P. (1988) Dopamine D_2 receptors retain agonist high-affinity form and guanine nucleotide sensitivity after removal of sialic acid. *J. Biochem.* **104,** 791–794.
102. Raymond, J. R., Fargin, A., Lohse, M., Regan, J. R., Senogles, S. E., Lefkowitz, R. J., and Caron, M. G. (1989) Identification of the ligand binding subunit of the human 5-hydroxytryptamine 1A receptor with N-(p-azido-m[^{125}I]iodophenethyl)spiperone, a high affinity radioiodinated photoaffinity probe. *Mol. Pharmacol.* **36,** 15–21.
103. Covarrubias, M., Prinz, H., Meyers, H., and Maelicke, A. (1986) Equilibrium binding of cholinergic ligands to the membrane bound acetylcholine receptor. *J. Biol. Chem.* **261,** 14,955–14,961.
104. Strader, C. D., Fong, T. M., Tota, M. R., Underwood, D., and Dixon, R. A. F. (1994) Structure and function of G protein-coupled receptors. *Annu. Rev. Biochem.* **63,** 101–132.
105. Tota, M. R., Daniel, S., Sirotina, A., Mazina, K. E., Fong, T. M., Longmore, J., and Strader, C. D. (1994) Characterization of a fluorescent substance P analogue. *Biochemistry* **33,** 13,079–13,086.
106. Madras, B. K., Canfield, D. R., Pfaelzer, C., Vittimberga, F. J., DiFiglia, M., Aronin, N., Bakthavachalam, V., Baindur, N., and Neumeyer, J. L. (1990) Fluorescent and biotin probes for dopamine receptors: D_1 and D_2 receptor affinity and selectivity. *Mol. Pharmacol.* **37,** 833–839.
107. Monsma, F. J., Barton, A. C., Kang, H. C., Brassard, D. L., Haugland, R. P., and Sibley, D. R. (1989) Molecular characterization of novel fluorescent ligands with high affinity for D_1 and D_2 dopamine receptors. *J. Neurochem.* **52,** 1641–1644.
108. Ariano, M. A. and Sibley, D. R. (1994) Dopamine receptor distribution in the rat CNS: elucidation using anti-peptide antisera directed against D_{1A} and D_3 subtypes. *Brain Res.* **649,** 95–110.
109. Huang, Q., Zhou, D., Chase, K., Gusella, J. F., Aronin, N., and DiFiglia, M. (1992) Immunohistochemical localization of the D_1 dopamine receptor in rat brain reveals its axonal transport, presynaptic and postsynaptic localization, and prevalence in the basal ganglia, limbic system, and thalamic reticular nucleus. *Proc. Natl. Acad. Sci. USA* **89,** 11,988–11,992.
110. Levey, A. I., Hersch, S. M., Rye, D. B., Sunahara, R. K., Niznik, H. B., Kitt, C. A., Price, D. L., Maggio, R., Brann, M. R., and Ciliax, B. J. (1993) Localization of D_1 and D_2 dopamine receptors in brain with subtype-specific antibodies. *Proc. Natl. Acad. Sci. USA* **90,** 8861–8865.
111. Fishburn, C. S., David, C., Carmon, S., Wein, C., and Fuchs, S. (1994) In vitro translation of D2 dopamine receptors and their chimaeras: analysis by subtype-specific antibodies. *Biochem. Biophys. Res. Commun.* **205,** 1460–1466.
112. Chazot, P. L., Doherty, A. J., and Strange, P. G. (1993) Antisera specific for D_2 dopamine receptors. *Biochem. J.* **289,** 789–794.
113. Farooqui, S. M. and Prasad, C. (1992) An antibody to dopamine D2 receptor inhibits dopamine antagonist and agonist binding to dopamine D2 receptor cDNA transfected mouse fibroblast cells. *Life Sci.* **51,** 1509–1516.
114. Plug, M. J., Dijk, J., Maassen, J. A., and Möller, W. (1992) An anti-peptide antibody that recognizes the dopamine D2 receptor from bovine striatum. *Eur. J. Biochem.* **206,** 123–130.

115. Boundy, V. A., Luedtke, R. R., Artymyshyn, R. P., Filtz, T. M., and Molinoff, P. B. (1993) Development of polyclonal anti-D_2 dopamine receptor antibodies using sequence-specific peptides. *Mol. Pharmacol.* **43,** 666–676.
116. David, C., Ewert, M., Seeburg, P. H., and Fuchs, S. (1991) Antipeptide antibodies differentiate between long and short isoforms of the D_2 dopamine receptor. *Biochem. Biophys. Res. Commun.* **179,** 824–829.
117. McVittie, L. D., Ariano, M. A., and Sibley, D. R. (1991) Characterization of antipeptide antibodies for the localization of D_2 dopamine receptors in rat striatum. *Proc. Natl. Acad. Sci. USA* **88,** 1441–1445.
118. Braestrup, C. and Andersen, P. H. (1987) Effects of heavy metal cations and other sulfhydryl reagents on brain dopamine D1 receptors: evidence for involvement of a thiol group in the conformation of the active site. *J. Neurochem.* **48,** 1667–1672.
119. Dewar, K. M. and Reader, T. A. (1989) Specific [^{3}H]SCH 23390 binding to dopamine D_1 receptors in cerebral cortex and neostriatum: role of disulfide and sulfhydryl groups. *J. Neurochem.* **52,** 472–482.
120. Srivastava, L. K. and Mishra, R. K. (1990) Chemical modification reveals involvement of tyrosine in ligand binding to dopamine D_1 and D_2 receptors. *Biochem. Int.* **21,** 705–714.
121. Williamson, R. A. and Strange, P. G. (1990) Evidence for the importance of a carboxyl group in the binding of ligands to the D2 dopamine receptor. *J. Neurochem.* **55,** 1357–1365.
122. Freedman, S. B., Poat, J. A., and Woodruff, G. N. (1982) Influence of sodium and sulphydryl groups on [^{3}H]sulpiride binding sites in rat striatal membranes. *J. Neurochem.* **38,** 1459–1465.
123. Reader, T. A., Molina-Holgado, E., Lima, L., Boulianne, S., and Dewar, K. M. (1992) Specific [^{3}H]raclopride binding to neostriatal dopamine D2 receptors: role of disulfide and sulfhydryl groups. *Neurochem. Res.* **17,** 749–759.
124. Sibley, D. R. and Creese, I. (1983) Regulation of ligand binding to pituitary D-2 dopaminergic receptors: effects of divalent cations and functional group modification. *J. Biol. Chem.* **258,** 4957–4965.
125. Suen, E. T., Stefanini, E., and Clement-Cormier, Y. C. (1980) Evidence for essential thiol groups and disulfide bonds in agonist and antagonist binding to the dopamine receptor. *Biochem. Biophys. Res. Commun.* **96,** 953–960.
126. Neve, K. A. (1991) Regulation of dopamine D2 receptors by sodium and pH. *Mol. Pharmacol.* **39,** 570–578.
127. Chazot, P. L. and Strange, P. G. (1992) Importance of thiol groups in ligand binding to D_2 dopamine receptors from brain and anterior pituitary gland. *Biochem. J.* **281,** 377–380.
128. D'Souza, U. and Strange, P. G. (1995) pH dependence of ligand binding to D2 dopamine receptors. *Biochemistry* **34,** 13,635–13,641.
129. Javitch, J. A., Li, X., Kaback, J., and Karlin, A. (1994) A cysteine residue in the third membrane-spanning segment of the human D_2 dopamine receptor is exposed in the binding-site crevice. *Proc. Natl. Acad. Sci. USA* **91,** 10,355–10,359.
130. Presland, J. P. and Strange, P. G. (1991) pH dependence of sulpiride binding to the D2 dopamine receptor. *Biochem. Pharmacol.* **41,** R9–R12.
131. Sundaram, H., Newman-Tancredi, A., and Strange, P. G. (1993) Characterization of recombinant human serotonin $5HT_{1A}$ receptors expressed in Chinese hamster ovary cells. *Biochem. Pharmacol.* **45,** 1003–1009.

132. Sibley, D. R. and Creese, I. (1983) Interactions of ergot alkaloids with anterior pituitary D-2 dopamine receptors. *Mol. Pharmacol.* **23,** 585–593.
133. Sundaram, H., Turner, J., and Strange, P. G. (1995) [^{3}H]Lisuride binding to 5HT1A serotonin receptors. *J. Neurochem.* **65,** 1909–1916.
134. Watanabe, M., George, S. R., and Seeman, P. (1985) Regulation of anterior pituitary D_2 dopamine receptors by magnesium and sodium ions. *J. Neurochem.* **45,** 1842–1849.
135. Theodorou, A. E., Hall, M. D., Jenner, P., and Marsden, C. D. (1980) Cation regulation differentiates specific binding of [^{3}H]sulpiride and [^{3}H]spiperone to rat striatal preparations. *J. Pharm. Pharmacol.* **32,** 441–444.
136. Horstman, D. A., Brandon, S., Wilson, A. L., Guyer, C. A., Cragoe, E. J., and Limbird, L. E. (1990) An aspartate conserved among G-protein receptors confers allosteric regulation of α_2-adrenergic receptors by sodium. *J. Biol. Chem.* **265,** 21,590–21,595.
137. Neve, K. A., Cox, B. A., Henningsen, R. A., Spanoyannis, A., and Neve, R. L. (1991) Pivotal role for aspartate-80 in the regulation of D2 receptor affinity for drugs and inhibition of adenylyl cyclase. *Mol. Pharmacol.* **39,** 733–739.
138. Birnbaumer, L. (1990) G proteins in signal transduction. *Annu. Rev. Pharmacol. Toxicol.* **30,** 675–705.
139. Neve, K. A., Henningsen, R. A., Bunzow, J. R., and Civelli, O. (1989) Functional characterization of a rat dopamine D-2 receptor cDNA expressed in a mammalian cell line. *Mol. Pharmacol.* **36,** 446–451.
140. Urwyler, S. (1989) Mono- and divalent cations modulate the affinities of brain D1 and D2 receptors for dopamine by a mechanism independent of receptor coupling to guanyl nucleotide binding proteins. *Naunyn-Schmiedeberg's Arch. Pharmacol.* **339,** 374–382.
141. Nunnari, J. M., Repaske, M. G., Brandon, S., Cragoe, E. J., and Limbird, L. E. (1987) Regulation of porcine brain α2-adrenergic receptors by Na^+, H^+, and inhibitors of Na^+/H^+ exchange. *J. Biol. Chem.* **262,** 12,387–12,392.
142. Hoare, S. and Strange, P. G. (1994) Allosteric regulation of D2 dopamine receptors by amiloride analogues. *Br. Pharmacol. Soc. Proc.* Manchester Meeting, P17.
143. Birdsall, N. J. M., Cohen, F., Lazareno, S., and Matsui, H. (1995) Allosteric regulation of G-protein linked receptors. *Biochem. Soc. Trans.* **23,** 108–111.

CHAPTER 2

Molecular Biology of Dopamine Receptors

Kim A. Neve and Rachael L. Neve

1. Molecular Cloning of the Dopamine Receptors

The molecular biological characterization of the dopamine receptors began with the cloning of the gene and cDNA for the hamster beta-2 adrenergic receptor in 1986 *(1)*, followed closely by the isolation of cDNAs encoding the turkey erythrocyte beta-adrenergic receptor *(2)* and the porcine m1 muscarinic receptor *(3)*. The realization that G protein-coupled receptors form a family of proteins related by primary structure and predicted secondary structural features, a family big enough to include the light-activated receptor rhodopsin, opened up the possibility of exploiting this homology for the molecular cloning of additional members of the gene family. Indeed, all the dopamine receptors have been cloned because of their homology to other dopamine or monoamine receptors.

*1.1. Dopamine D_2 Receptors**

The first dopamine receptor cDNA was cloned using the hamster beta-2 adrenergic receptor gene to screen a rat genomic library under conditions of reduced stringency. This procedure identified a 0.8 kb *Eco*RI-*Pst*I fragment that had a region with significant homology in one reading frame to the amino acid sequence of transmembrane (TM) domains 6 and 7 of the hamster beta-2 adrenergic receptor *(4)*. The genomic fragment was used to probe a rat brain cDNA library, yielding a 2,455 bp cDNA encoding a 415 amino-acid

*As detailed in the Preface, the terms D1 or D1-like and D2 or D2-like are used to refer to the subfamilies of DA receptors, or used when the genetic subtype (D_1 or D_5; D_2, D_3, or D_4) is uncertain.

The Dopamine Receptors Eds.: K. A. Neve and R. L. Neve
Humana Press Inc., Totowa, NJ

protein (GenBank accession no. M36831). The predicted amino acid sequence of this protein was consistent with the presence of seven TM domains, and was more closely related to alpha-2 adrenergic receptors than to any other receptor whose sequence was known at the time. The likelihood that the cDNA encoded a dopamine D2 receptor was suggested both by predicted structural features that included the long third cytoplasmic loop and short C-terminus characteristic of G_i-coupled receptors, and by the presence of its cognate mRNA in brain regions known to express D2 receptors at high densities. Thus, the abundance of mRNA is highest in the intermediate lobe of the pituitary and the anterior basal ganglia, also high in the septum, posterior basal ganglia, and the anterior lobe of the pituitary, and lower or undetectable in other brain regions *(4)*.

The identity of the cDNA was confirmed by expression in Ltk^- mouse fibroblast and GH_4C_1 rat pituitary cell lines *(4–6)*. The pharmacological profile of the recombinant receptors is identical to that of rat neostriatal D2 receptors, with the rank order of potency: spiperone > (+)-butaclamol > haloperidol > sulpiride >> SCH 23390 = ketanserin > (–)-butaclamol *(4)*. The recombinant receptor has affinity values in the low micromolar range for dopamine and other D2 receptor agonists such as quinpirole and (+)3-PPP, and nanomolar affinity for bromocriptine (Table 1). Furthermore, in the absence of GTP, the recombinant receptor displays high- and low-affinity states for dopamine. Addition of GTP or inactivation of pertussis toxin-sensitive G proteins reduces or eliminates the high-affinity component of binding. D2 receptors have classically been defined as inhibiting adenylate cyclase activity, and transfection of the cloned cDNA into either Ltk^- or GH_4C_1 cells confers on the cells dopamine-inhibited adenylate cyclase activity, and also, for GH_4C_1 cells, dopamine inhibition of prolactin secretion. Both of these responses are also inhibited by treatment with pertussis toxin, indicating that the receptors are activating the G proteins G_i/G_o.

Soon after the publication of the sequence of the rat D_2 receptor cDNA, numerous reports described the existence of a second, longer form of the D_2 receptor, generated by alternative splicing of the same gene that encodes the short form of the receptor *(7–15)*. The receptor corresponding to the cDNA that was first cloned by Bunzow et al. *(4)* is designated D_{2S}, for the short form of D_2 receptors. The longer form, with 29 additional amino acids inserted in the third cytoplasmic loop between amino acids 241 and 242 of D_{2S}, is designated D_{2L}. D_{2L} is the predominant form in vivo, although there is some variability among brain regions in the relative proportions of the two forms *(16,17)*. Initial studies indicated that the pharmacological profile of the two forms is virtually identical *(8–10,18)*, but a few recent studies have reported subtle differences between the affinities of D_{2S} and D_{2L} for some drugs, with D_{2S} having two- to threefold higher affinity for some substituted benzamide

Table 1
Affinity of Mammalian Recombinant Dopamine Receptors for Drugs[a]

Drug	Receptor				
	D_1	D_5	D_2	D_3	D_4
Antagonists					
(+)-Butaclamol	0.2–5	0.3–32	0.3–4	0.2–2	40–87
Chlorpromazine	16–73	33–133	1–8	1.2–6.1	8–37
Clozapine	18–150	35–400	17–158	74–620	9–32
Domperidone			0.3–1.3	3–9.5	
Eticlopride	18,000	19,000	0.03–0.04	0.02	2.00
Flupentixol, *cis*	0.7–21	0.7–24	0.1–5	0.1–1.1	
Fluphenazine	21–28	14			46
Haloperidol	10–110	27–151	0.3–3	1–10	1–5
Ketanserin	90–420	2500	6000		90–148
Pimozide			0.4–10	0.4–11	43
Raclopride	>72,000		1–10	2	240–2000
Remoxipride			50–300	970–2300	3690
Risperidone			1–6	7–70	
SCH 23390	0.1–0.6	0.1–0.5	267–2000	314–780	3560
Spiperone	100–540	500–4500	0.02–0.14	0.03–0.61	0.08–0.4
Sulpiride, S	36,000	77,000	5–80	8–570	52–570
Thioridazine	100	300	1–16	2–8	12–14
YM-09151-2			0.02–0.09	0.04–0.06	0.04–0.6
Agonists					
6,7-ADTN					
High					2
Low	4600	909	55	2	4
Apomorphine					
High			0.4–2	0.2–70	
Low	210–680	140–360	40–250	12–75	
Bromocriptine	670	450	2–13	7	340
Cl-APB	20–83		510		
Dopamine					
High	10–32	10–24	4–50	0.8–4	3–28
Low	1000–18,000	228–3900	650–17,000	22–99	400–19,000
DPAT, 7-OH					
High			1.4–3.6	0.3–0.5	6
Low	5300		31–66	2–4	651
Fenoldopam					
High	2	0.6	3		321
Low	17–28	11–27	1000		
N-0437					
High			1.8	0.5	122
Low	2170	990			

(continued)

Table 1 *(continued)*

Drug	Receptor				
	D_1	D_5	D_2	D_3	D_4
NPA					
High		92			6
Low	300–1540	755–1050	0.1–20	0.2	
Quinpirole					
High			6	0.4–40	2–46
Low	≥14,000	>20000	2500–9000	16–25	71–800
SKF 38393					
High	1–6	0.5	200		
Low	61–300	43–470	9500	5000	1095

[a]The data shown represent the range of K_I or K_D values, in nanomolar, for the indicated drug, as reported in studies cited in this chapter. Only data from studies in which mammalian recombinant receptors were expressed in mammalian cells are included. For the agonists, there is often considerable variability arising from buffer conditions (e.g., the presence or absence of sodium, pH) and from the lack of information provided regarding the presence or absence of two affinity states. The high-affinity values (High) are from studies in which binding curves were analyzed for the presence of high- and low-affinity classes of binding sites. The low-affinity values (Low) are from studies in which high- and low-affinity binding was determined, studies carried out with GTP or GTP analogues in the assay, and studies that were carried out using receptors expressed in cell lines in which two affinity states were not detected.

derivatives *(19,20)*. The observation of Leysen et al. *(18),* that D_{2S} is much more sensitive than D_{2L} to GTPγS-induced inhibition of 2'-iodospiperone binding, suggests that the binding pockets of D_{2S} and D_{2L} may differ. The inhibition of binding is hypothesized to be caused by a reduction of disulphide bonds in the receptor, and can be partially prevented by the presence of dopamine.

Because the alternatively spliced exon that differentiates D_{2S} and D_{2L} lies in the putative third cytoplasmic loop of the receptors, the speculation was that differences in coupling to G proteins and signaling pathways might be more pronounced and more significant than the small differences in drug affinity that are observed. Indeed, several groups have reported that, in the absence of GTP, a higher proportion of D_{2S} receptors than D_{2L} receptors have high affinity for dopamine, indicating either a greater efficiency of coupling for D_{2S} receptors or differential selection of G proteins *(9,21–23)*. Either mechanism is consistent with, and could explain, the finding that D_{2S} inhibits adenylate cyclase more efficiently than does D_{2L} *(8,24,25)*. These are quantitative rather than qualitative differences, however, and most signaling pathways that are modulated by D_2 receptors appear to be activated similarly by D_{2S} and D_{2L} receptors (*see* Chapter 6). In some measures of receptor regulation, too, differences between D_{2S} and D_{2L} receptors are equivocal or quantitative. In vivo, mRNA for both forms is elevated after 6-hydroxydopamine-induced denervation of dopamine receptors *(16)*. With regard to the up-regulation of D_2 receptors that occurs after treatment of D_2 receptor-expressing cell lines with agonists (*see* Chapter 13), we have described differences between D_{2S} and D_{2L} in the extent of up-regulation *(26)*. On the other hand, Zhang et al. *(27)* reported that only D_{2L} receptors are up-regulated by agonists, whereas the density of D_{2S} receptors is slightly decreased.

One particularly exciting area of research is beginning to accumulate evidence of qualitative differences in the selection of G proteins by the two forms of D_2 receptors. Two studies used either selective antisense inhibition of G-protein expression *(28)* or receptor/G protein cotransfection *(23)* to demonstrate coupling of D_{2L}, but not D_{2S} receptors, to $G\alpha_{i2}$ for inhibition of adenylate cyclase. A separate study used pertussis-toxin treatment to obtain a more complete ablation of $G\alpha_i$ proteins than was observed after antisense inhibition, then assessed the selective interactions of mutant pertussis toxin-resistant $G\alpha_i$ subunits with the two forms of D_2 receptors. The results from the latter study support the existence of differences between the two receptor forms, but indicate that D_{2S} receptors selectively couple to $G\alpha_{i2}$, whereas D_{2L} receptors selectively couple to $G\alpha_{i3}$ to inhibit adenylate cyclase *(29)*. Finally, Chiodo and colleagues reported that D_{2S} receptor inhibition of a voltage-dependent potassium current is not sensitive to pertussis toxin *(30)* and may be mediated via $G\alpha_s$ *(31)*, whereas D_{2L} inhibition of the potassium current is inhibited by pertussis toxin or treatment with an antibody to $G\alpha_o$ (*see* Chapter 5, for additional discussion of the coupling of D_2 receptors to G proteins).

Following the cloning of the rat (*4,9,12*; GenBank accession no. X17458), human (*8,10,15,32*; GenBank accession no. X51645), and bovine (*7*; GenBank accession no. X51657) D_2 receptor cDNAs, the murine (*33,34*; GenBank accession no. X55674) and *Xenopus* (*35*; GenBank accession no. X59500) cDNAs were cloned and sequenced. Interestingly, there are two highly homologous *Xenopus* D_2 receptor genes, probably as a result of duplication of the entire *Xenopus* genome. Only a partial cDNA sequence is available for the *Xenopus* D_{2B} receptor (*36*; GenBank accession no. X72902). Both *Xenopus* D_2 receptors correspond to mammalian D_{2L}, and there is no evidence for alternative splicing of either gene to produce a shorter receptor.

1.2. Dopamine D_1 Receptors

The first report of a partial cDNA sequence for the rat dopamine D_1 receptor was published in 1990 *(37)*, although the cDNA was not identified as a dopamine receptor until the subsequent cloning of full-length cDNAs and genes by several groups *(38–41)*. Dearry et al. *(38)* screened a human cDNA library under low-stringency conditions with an oligonucleotide derived from TM 2 of the rat dopamine D_2 receptor. The other three groups first amplified partial cDNAs by the polymerase chain reaction (PCR), using either two sets of degenerate primers corresponding to sequence in TM 3 and 6 of other catecholamine receptors *(39,41),* or one degenerate primer coding for a particularly highly conserved region of TM 3 together with a vector primer *(37)*. The partial cDNAs were then used to isolate full-length clones from cDNA *(39)* or genomic *(40,41)* libraries.

The identity of the D_1 receptor cDNAs/genes was confirmed on the basis of several criteria. The distribution of mRNA that hybridizes with the probes resembles the distribution of D1 receptor radioligand binding. Cognate mRNA is most abundant in the neostriatum and other nuclei of the basal ganglia, and less abundant in other brain regions such as the frontal cortex. Transfection of any of the cloned cDNAs or genes into mammalian cells results in the expression of a dopamine D_1 receptor with high affinity (<1 n*M*) for the D1 antagonist [^{3}H]SCH 23390, and n*M* affinity for other dopamine D1 receptor antagonists such as (+)-butaclamol, *cis*-flupentixol, and *cis*-piflutixol. The recombinant receptors have lower affinity for nondopaminergic or D2-selective antagonists. Furthermore, the recombinant receptors have higher affinity for D1-selective agonists such as fenoldepam and SKF 38393 than for D2-selective or nondopaminergic agonists. D1 receptors have classically been differentiated from D2 receptors not only on the basis of drug potency, but also by the ability of D1 receptors to mediate stimulation of adenylate cyclase activity *(42,43)*. The recombinant D_1 receptor confers this property on HEK293 cells *(41),* COS-7 cells *(39),* and Ltk$^-$ cells *(38)*.

The rat (*39,41*; GenBank accession no. M35077), human (*38,40,41*; GenBank accession no. X55760), opossum (*45*; GenBank accession no. S67258), rhesus macaque *(46),* and porcine *(47)* D_1 receptors have been cloned. Nonmammalian D_1 or D1-like receptors that have been cloned include the goldfish gene (*48*; GenBank accession no. L08602), a D_1 receptor cDNA from the fish *Tilapia mossambica* (GenBank accession no. X81969), a D1-like cDNA from *Drosophila* (*49*; GenBank accession no. X77234), the *Xenopus* D_1 gene (*50*; GenBank accession no. U07863), and the chicken *(Gallus domesticus)* D_1 receptor gene (*51*; GenBank accession no. L36877). In addition, two novel D1-like subtypes, designated D_{1C} (GenBank accession no. U07865) and D_{1D} (GenBank accession no. L36879), have been cloned from *Xenopus* and *Gallus,* respectively *(50,51)*. The pharmacological profiles of the goldfish, frog, and chicken D_1 receptors are similar to that of the mammalian receptors, and the nonmammalian receptors also mediate stimulation of adenylate cyclase. The novel D_{1C} (frog) and D_{1D} (chicken) subtypes, on the other hand, display agonist-binding profiles that more closely resemble the mammalian and nonmammalian D_5 receptors because of their higher affinity for dopamine and some agonists (*see* Section 1.5.). The *Drosophila* D1-like receptor may have a pharmacological profile quite different from the other D_1 receptors. The *Drosophila* receptor mediates stimulation of adenylate cyclase by dopamine and the D1 receptor agonist SKF 38393, but appears to have low affinity for the prototypical D1 receptor ligand SCH 23390 *(49)*.

1.3. Dopamine D_3 Receptors

In September of 1990, at approximately the same time that reports of the cloning of D_1 receptors were published, Sokoloff et al. *(52)* published the first report of the cloning of a cDNA encoding a novel dopamine receptor subtype, termed D_3. An iterative cloning protocol was used in which oligonucleotides derived from the sequence of the D_2 receptor were used to screen a rat cDNA library, yielding a partial cDNA that hybridized weakly with a D_2 receptor cDNA. The new partial cDNA was used to probe a rat genomic library, yielding a clone that had a potential translational start codon followed by an open reading frame encoding a protein fragment with strong homology to TM 1 and 2 of the D_2 receptor. PCR primers were designed, one (primer 1) using the sequence upstream of the potential start codon and the other a degenerate primer corresponding to TM 7 of the D_2 receptors, and used to amplify products from a rat brain cDNA preparation. A second rat genomic library was probed with the PCR products, resulting in the cloning of a fragment that included the 3'-end of the coding region. A third oligonucleotide primer was designed that was complementary to sequence immediately downstream from the stop codon. This primer was used together with primer 1 to amplify a full-length D_3 receptor cDNA from mRNA isolated from the rat olfactory tubercle (GenBank accession no. X53944).

The rat D_3 cDNA was determined to encode a subtype of the dopamine D2 receptors, based on its sequence and pharmacological profile. The sequence of the D_3 receptor is more similar to that of the D_2 receptor (52% overall amino acid identity, increasing to >70% within the TM regions) than to any other receptor. The pharmacological profile of the receptor is also extremely D2-like, with high affinity for every D2 receptor ligand that has been tested (Table 1). There are, however, differences between the two receptors in pharmacology and distribution. D_3 receptors are present in greatest abundance in the ventral forebrain, in the islands of Calleja, and in the nucleus accumbens, and are also present at lower levels in the hypothalamus, hippocampus, and substantia nigra, pars compacta *(53)*. The distribution of D_3 receptors and D_3 receptor message is more restricted than that of D_2 receptors. Two regions with the highest density of D_2 receptors, the pituitary and neostriatum, have a conspicuously low abundance of D_3

receptor mRNA. In addition to the rat D_3 receptor, cDNAs encoding human *(54,55)* and mouse (*56*; GenBank accession no. X67274) D_3 receptors have been cloned.

Sokoloff et al. *(52)* found that, with a few exceptions, the rat D_3 receptor has lower affinity than does the D_2 receptor for D2 receptor antagonists. The magnitude of the difference in potency varies from less than 2-fold, to 10- to 30-fold for some classical D2 antagonists such as domperidone, haloperidol, and spiperone. Our data comparing the affinity of rat recombinant or endogenous D_2 and D_3 receptors for domperidone, haloperidol, spiperone, and (+)-butaclamol (Cox and Neve, unpublished observations) are in qualitative agreement with the initial characterization of D_3 receptors. Several quantitative autoradiographic studies also suggest that some D2-receptor antagonists have markedly lower affinity for the D_3 receptor *(57–59)*. Other studies, however, have found that differences in the affinity of the two receptor subtypes for classical D2 antagonists were either less pronounced than originally reported or less consistently in the direction of D_2 receptor selectivity *(20,55,60–63)*. In contrast to our work and that of Sokoloff et al. *(52)* using the rat D_3 receptor, the latter studies characterized the human recombinant D_3 receptor, which could be the basis for the discrepant findings. For example, Freedman et al. *(60)* reported that haloperidol, sulpiride, and chlorpromazine are all nonselective or selective for *D3* receptors, and in the extensive list of compounds tested by Kula et al. *(61)*, (+)-butaclamol, (–)-3-PPP, and chlorprothixene are reported to be more potent at D_3 than at D_2 receptors. Interestingly, the dopamine transporter ligand GBR12909 *(64)* is approximately as potent at the human D_3 receptor (K_i = 4 n*M*) as at the dopamine transporter, and 100-fold more potent at the D_3 receptor than at the D_2 receptor *(61)*.

There is more agreement in the literature that some agonists, including dopamine, 7-OH-DPAT, dihydrexidine, and quinpirole, have higher affinity for the D_3 receptor than for the D_2 receptor *(21,52,55,60–62,65)*. This seems to be owing, at least in part, to high-affinity binding of these agonists to the D_3 receptor that resembles formation of an agonist/receptor/G protein ternary complex, except that it is resistant to the addition of GTP or GTP analogs. This led to the development and use of several presumptive agonists as radioligands for the D_3 receptor *(57,60,66–68)*. Recent reports have emphasized the need for careful control of assay conditions to ensure that only D_3 receptors, and not D_2 receptors in an agonist high-affinity conformation *(69,70)* or σ receptors *(71)*, are labeled by an agonist radioligand.

The D_3 receptor inhibits adenylate cyclase and many other pathways modulated by the D_2 receptor only weakly or not at all *(52,60,62,63)*. The protein does, however, appear to be an active receptor, because several cellular responses to stimulation of the D_3 receptor have been demonstrated, including inhibition of dopamine release *(63)*, activation of Na^+/H^+ exchange *(72,73)*, stimulation of mitogenesis *(72,74)*, activation of c-*fos* and neurotensin expression *(74,75)*, inhibition of calcium currents *(76)*, and neurite branching and extension *(77)*. In most cases, the D_3 receptor appears to modulate cellular responses less efficiently than does the D_2 receptor (*see* Chapter 6).

An unresolved and potentially important issue concerns the apparently anomalous coupling of the D_3 receptor to G proteins (*see* Chapter 5) and signaling pathways. Does high-affinity binding of agonists to the D_3 receptor reflect formation of the ternary complex, and if so, why is it reversed poorly or not at all by GTP? Furthermore, if the D_3 receptor is "locked" in a ternary complex with a G protein, why does the D_3 receptor interact less efficiently than the D_2 receptor with signaling pathways? The answers to these questions could be informative about the mechanisms of activation of G proteins by receptors.

1.4. Dopamine D_4 Receptors

In early 1991, the first description of the cloning of DNA encoding a novel dopamine receptor subtype termed D_4 was published *(78)*. Several species of RNA were identified in the neuroblastoma cell line SK-N-MC that hybridized weakly to a D_2 receptor partial cDNA, suggesting the presence of receptors related to the D_2 receptor. A cDNA library was prepared from SK-N-MC mRNA and screened with a full-length D_2 receptor cDNA. One of the clones isolated had considerable homology to TM 5–7 of the D_2 receptor. The novel clone was used to screen a human genomic library, and a genomic clone was isolated that contained the entire coding region of the novel receptor, together with intronic sequence. Initial efforts to clone the full-length cDNA were unsuccessful, and expression of the full-length gene in COS-7 cells resulted in very low levels of specific [^{3}H]spiperone binding. Therefore, a gene/cDNA hybrid was constructed and expressed. The D_4 receptor (GenBank accession no. X58497) was identified as a dopaminergic D2-like receptor based on its predicted amino acid sequence, which is 41% homologous to the D_2 receptor and 39% homologous to the D_3 receptor *(78)*. Amino acid sequence identity is greatest within the TM regions, being particularly high in TM 2, 3, and 7, and lowest in TM 1 and 4. Furthermore, the pharmacological profile of the D_4 receptor is similar, but not identical, to that of the D_2 receptor, with high affinity for [^{3}H]spiperone (70 p*M*), eticlopride, and clozapine (9 n*M*). Several D2-selective antagonists, including (+)-butaclamol, fluphenazine, and the substituted benzamide raclopride *(78)*, have markedly lower affinity for the D_4 receptor. The greatest abundance of D_4 receptor mRNA in humans is in the retina *(79)*. In primate brain, message is highest in the frontal cortex, mesencephalon, amygdala, thalamus, cerebellum, and medulla *(78,79)*. D_4 receptor message is high in the neostriatum and midbrain of the rat brain, with higher levels observed in the rat heart *(80)*. In brain, even where highest, D_4 receptor message is estimated to be one to two orders of magnitude less abundant than D_2 receptor message. The rat D_4 receptor gene was cloned independently of the human receptor, using degenerate oligonucleotides based on the D_2/D_3 receptor TM 6 and 7 to screen a genomic library *(80),* and also cloned by Asghari et al. (*81*; GenBank accession no. U03551). The mouse D_4 receptor gene (GenBank accession no. U19880) has also been cloned *(82)*.

Because of the difficulty of cloning a full-length human D_4 receptor cDNA and the low receptor expression after transfection with the human genomic clone, several approaches have been used to obtain cDNAs. Some groups constructed synthetic D_4 receptor cDNAs, designed to have a reduced content of G + C without changing the amino acid sequence of the protein *(83–85)*. Van Tol et al. *(86)* isolated a full-length cDNA from a cDNA library synthesized using mRNA from COS-7 cells that had been transfected with a D_4 receptor gene. Additional cDNAs encoding a number of allelic variants (*see* following paragraphs) were created by substituting fragments from partial cDNAs *(86)* or cloned PCR products *(81)*. More recently, Matsumoto et al. *(79)* cloned a "native" human D_4 receptor cDNA by the PCR, using *Pfu* DNA polymerase and formamide to deal with the high G + C content of the cDNA.

Several characteristics of the D_4 receptor have combined to focus considerable interest on this novel D2-like receptor. The original report described a receptor with an affinity for clozapine anywhere from 6–15 times higher than that of the D_2 receptor *(78)*. This is noteworthy because of the unique efficacy of clozapine in a subpopulation of patients with treatment-resistant schizophrenia *(87),* and because of the lower incidence of extrapyramidal side effects observed during treatment with clozapine, compared to the effect of classical antipsychotic drugs such as haloperidol *(88)*. It is possible to make too much of the D_4 receptor selectivity of clozapine, since numerous receptors have higher affinity for clozapine than the D_2 receptor, including several 5HT receptor subtypes *(89–92)* and alpha-1 adrenergic receptors *(89)*. Furthermore, other studies have not consistently confirmed the higher affinity of the D_4 receptor for clozapine, or have found the affinity of the recombinant D_2 receptor for clozapine to be as high as reported for the D_4 receptor *(20,62,80,93)*.

More interesting, particularly from a genetic point of view, is the unusual structure of the D_4 receptor gene (*see* Section 2.1.3.). Many allelic variants of the human D_4 receptor gene have been identified that differ in the number and order of copies of a direct, imperfect 48-bp repeat *(79,81,86,94),* giving rise to proteins that vary in amino acid sequence. Because the polymorphism is in the third cytoplasmic loop, there is the exciting possibility that the D_4 receptor variants may differ in signal transduction properties. At this time, however, observed differences in affinity for ligands and efficiency of coupling to adenylate cyclase have been minor or nonexistent *(81,86,95)*. On the other hand, there is considerable interest in the recent observation that the density of D_4 receptors is selectively and markedly increased in the caudate-putamen from schizophrenics *(96–98)*.

The endogenous dopamine D_4 receptor inhibits adenylate cyclase activity in the mouse retina *(99)* and may mediate the light-evoked inhibition of melatonin biosynthesis in the chick retina *(100,101)*, presumably via inhibition of adenylate cyclase. Recombinant D_4 receptors display GTP-sensitive binding of agonists *(63,78,84)*, activate G proteins *(102)*, inhibit adenylate cyclase *(63,83,84)*, and also regulate other pathways that affect arachidonate release, Na^+/H^+ exchange *(83)*, and neurite branching and extension *(77)*.

1.5. Dopamine D_5 Receptors

At approximately the same time as the initial cloning of the dopamine D_4 receptor, the first description of the cloning of the gene for a novel D1-like receptor was published (*103*; GenBank accession no. X58454), followed by three additional reports *(104–106)* of the receptor named D_5 or D_{1B}. The human gene was isolated by probing human genomic libraries with D_1 receptor gene fragments *(103,104)* or a $5HT_{1A}$ receptor gene fragment *(106)*, whereas the rat gene was cloned by screening a genomic library with a PCR fragment that was obtained by amplification of genomic DNA with degenerate primers corresponding to TM 5 and 6 of the human D_1 receptor *(105)*.

The D_5 receptor gene product was identified as a D1-like receptor based on its sequence and pharmacological profile. The sequence of the D_5 receptor is more closely related to that of the D_1 receptor than to other dopamine or catecholamine receptors, and the predicted secondary structure matches that of the D_1 receptor and other G_s-coupled receptors, with a short (~50 amino acids) predicted third cytoplasmic loop, and a long (~120 amino acids) intracellular C-terminus *(103)*. The D_5 receptor has a pharmacological profile similar to that of the D_1 receptors, with a high affinity for [^{3}H]SCH 23390, and the following rank order of drug potency: SCH 23390 > *cis*-flupentixol ≅ (+)-butaclamol > haloperidol ≅ clozapine > spiperone > sulpiride. There is little difference between the affinity of D_1 and D_5 receptors for most antagonists (Table 1). On the other hand, substantial differences are observed in the affinity of the D1-like receptors for some agonists. In particular, the D_5 receptor has markedly higher affinity than does the D_1 receptor for dopamine and the conformationally restricted dopamine analog 6,7-ADTN *(103,105–107)*, as well as moderately higher affinity than does the D_1 receptor for several other agonists *(103)*. Like the D_1 receptor, the D_5 receptor exhibits GTP-sensitive high-affinity binding of agonists and stimulates adenylate cyclase activity *(103–108)*.

The distribution of D_5 receptor mRNA differs markedly from that of D_1 receptor mRNA. Whereas D_1 receptor mRNA is most abundant in the neostriatum, D_5 receptor mRNA is present in the striatum, but at about 1/10 the abundance of D_1 receptor message *(103)*. D_5 receptor message is more abundant in other rat brain regions, including the hippocampus, hypothalamus, and midbrain *(103,105,109)*. Compared to the rat, in the primate brain D_5 message is lower in the hippocampus, but higher in the frontal cortex *(103,106,109)*. It is noteworthy that some brain areas with relatively abundant D_5 receptor mRNA have the lowest amounts of D_1/D_5 radioligand binding in the brain *(110)*.

One interesting characteristic of the D_5 receptor, which may account for its higher affinity for some agonists, is that it behaves like a constitutively active receptor *(111)*. Thus, basal adenylate cyclase activity in cells expressing the D_5 receptor is higher than in cells expressing the D_1 receptor, suggesting that the unoccupied D_5 receptor has a greater capacity to activate G_s and stimulate adenylate cyclase. Many G protein-coupled receptors, if not all, have some constitutive activity, so that increasing the receptor density will increase "basal" activity of the appropriate signaling pathway. Indeed, adenylate cyclase activity in the absence of agonist increases with increasing receptor density for both D_1 and D_5 receptors, but the slope of the increase is substantially greater for D_5 than for D_1 receptors *(111)*. Furthermore, two antagonists, (+)-butaclamol and *cis*-flupentixol, possess negative intrinsic activity, as indicated by their ability to inhibit agonist-independent stimulation of adenylate cyclase by either D_1 or D_5 receptors. As observed for a constitutively active mutant of the beta-2 adrenergic receptor *(112)*, the D_5 receptor may be constitutively desensitized, since maximal stimulation of adenylate cyclase is always less for the D_5 receptor than for the D_1 receptor, and pretreatment with dopamine decreases its EC_{50} for stimulation of adenylate cyclase via the D_1 receptor, but not the D_5 receptor *(108,111)*.

In addition to the rat (GenBank accession no. M69118) and human (GenBank accession no. X58454) D_5 receptor genes, the frog (*50*; GenBank accession no. U07864) and chicken (*51*; GenBank accession no. L36878) D_5 receptor genes have been cloned. In contrast to the similar antagonist-binding profiles of mammalian and chicken D_1 and D_5 receptors, the frog D_5 receptor has substantially lower affinity than the frog D_1 receptor for several antagonists, including (+)-butaclamol, *cis*-flupentixol, lisuride, and chlorpromazine. Like the mammalian receptors, the frog and chicken D_5 receptors have relatively high affinity for dopamine and 6,7-ADTN, and both of the nonmammalian D_5 receptors stimulate adenylate cyclase activity.

2. Protein and Gene Structure of the Dopamine Receptors

2.1. The Dopamine D2-Like Receptors

2.1.1. D_2 Receptors

The rat and murine D_{2S} and D_{2L} receptors are 415 and 444 residues in length *(4,12,34)*, with a predicted relative molecular mass *(M_r)* of 47,000 and 50,900, respectively. The bovine D_{2L} receptor (the only bovine D_2 receptor cDNA cloned) is also 444 amino acids long *(7)*. The respective human receptors are 414 and 443 amino acid residues long, one residue shorter than the rat receptors due to the absence of an isoleucine between Lys-331 and Asp-332 (numbered according to the human D_{2L}) within the C-terminal half of the third cytoplasmic loop (Fig. 1). The rat and murine proteins have 99% amino acid identity, whereas the human and bovine proteins each have ~95–96% amino acid identity with the rat sequence.

All the mammalian D_2 receptors have three consensus sites for asparagine-linked glycosylation, at Asn-5, Asn-17, and Asn-23. At least one of these sites is likely to be used, since D2-like receptors are heavily glycosylated (*see* Chapter 1). The receptors also share several potential sites of phosphorylation by protein kinase A (PKA). The optimal consensus sequence R-R/K-X-S^*/T^* *(113)* is present only once, at Ser-364 (human D_{2L} numbering). Interestingly, the location of this site, at the C-terminal-end of the third cytoplasmic loop, is similar to the location of Ser-261 of the human beta-2 adrenergic receptor. Phosphorylation of Ser-261 by PKA contributes to heterologous desensitization of beta-2 adrenergic receptors *(114)*. Other potential phosphorylation sites in the mammalian D_2 receptors have the less than optimal sequence R-X_{1-2}-S^*/T^*. These include Ser-147 and -148 in the second cytoplasmic loop, and Ser-229, -296, -and -354 in the third cytoplasmic loop.

Fig. 1. *(pp. 40–42)* The amino acid sequence of the dopamine receptors is shown using the standard single-letter abbreviations. GenBank accession numbers are provided in the text. Of the receptors mentioned in the text, those not included in the figure because of their similarity to other sequences are the murine D_4 receptor; the bovine and mouse D_2 receptors; and the opossum, rhesus, and porcine D_1 receptors. Only the long forms of alternatively spliced D_2 and D_3 receptors are shown, with the alternatively spliced region in italics. The first 101 amino acids of the *Drosophila* dopamine receptor are not shown. Residues that are conserved among all the D2-like or D1-like receptors (not necessarily including the *Drosophila* dopamine receptor) are boxed and shaded. Other residues that show significant conservation among the subtypes, or that may be sites of posttranslational modification, are boxed but not shaded. Figure is adapted from ref. *112a*.

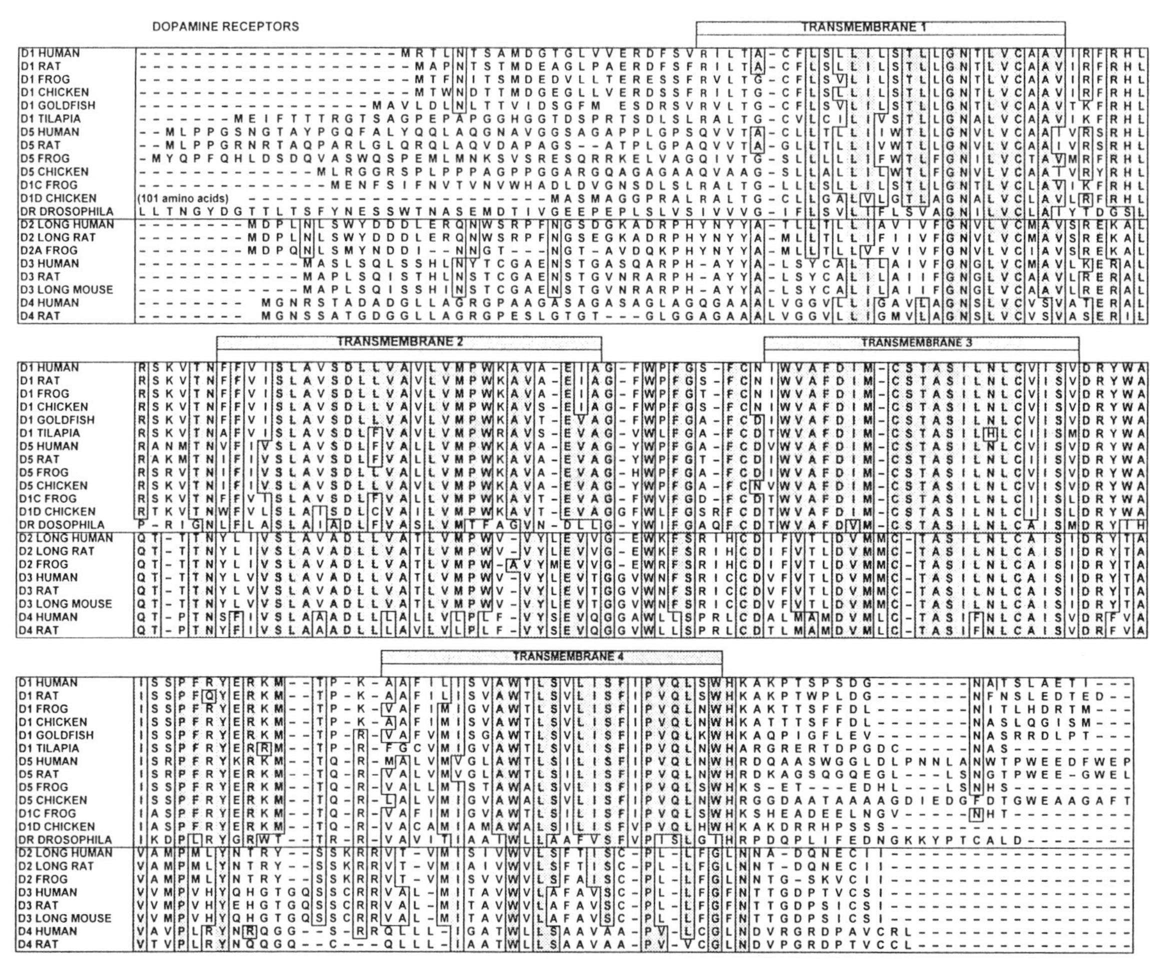
DOPAMINE RECEPTORS
TRANSMEMBRANE 1
D1 HUMAN -------------------MRTLNTSAMDGTGLVVERDFSVRILTA-CFLSLLILSTLLGNTLVCAAVIRFRHL
D1 RAT --------------------MAPNTSTMDEAGLPAERDFSFRILTA-CFLSLLILSTLLGNTLVCAAVIRFRHL
D1 FROG --------------------MTFNITSMDEDVLLTERESSFRVLTG-CFLSVLILSTLLGNTLVCAAVIRFRHL
D1 CHICKEN --------------------MTWNDTTMDGEGLLVERDSSFRILTG-CFLSLLILSTLLGNTLVCAAVIRFRHL
D1 GOLDFISH -----------------MAVLDLNLTTVIDSGFM ESDRSVRVLTG-CFLSVLILSTLLGNTLVCAAVTKFRHL
D1 TILAPIA -------MEIFTTTRGTSAGPEPAPGGHGGTDSPRTSDLSLRALTG-CVLCILIVSTLLGNALVCAAVIKFRHL
D5 HUMAN --MLPPGSNGTAYPGQFALYQQLAQGNAVGGSAGAPPLGPSQVVTA-CLLTLLIIWTLLGNVLVCAAIVRSRHL
D5 RAT --MLPPGRNRTAQPARLGLQRQLAQVDAPAGS--ATPLGPAQVVTA-GLLTLLIVWTLLGNVLVCAAIVRSRHL
D5 FROG -MYQPFQHLDSDQVASWQSPEMLMNKSVSRESQRRKELVAGQIVTG-SLLLLLIFWTLFGNILVCTAVMRFRHL
D5 CHICKEN ---------------MLRGGRSPLPPPAGPPGGARGQAGAGAAQVAAG-SLLALLILWTLFGNVLVCAATVRYRHL
D1C FROG ----------------MENFSIFNVTVNVWHADLDVGNSDLSLRALTG-LLLSLLILSTLLGNTLVCLAVIKFRHL
D1D CHICKEN (101 amino acids) ------------------MASMAGGPRALRALTG-CLLGALVLGTLAGNALVCLAVLRFRHL
DR DROSOPHILA LLTNGYDGTTLTSFYNESSWTNASEMDTIVGEEPEPLSLVSIVVVG-IFLSVLTFLSVAGNILVCLAIYTDGSL
D2 LONG HUMAN --------MDPLNLSWYDDDLERQNWSRPFNGSDGKADRPHYNYYA-TLLTLLIAVIVFGNVLVCMAVSREKAL
D2 LONG RAT --------MDPLNLSWYDDDLERQNWSRPFNGSEGKADRPHYNYYA-MLLTLLIFIIVFGNVLVCMAVSREKAL
D2A FROG --------MDPQNLSMYNDDI--NNGT---NGT-AVDQKPHYNYYA-MLLTLLVFVIVFGNVLVCIAVSREKAL
D3 HUMAN ------------MASLSQLSSHLNYTCGAENSTGASQARPH-AYYA-LSYCALILAIIFGNGLVCMAVLKERAL
D3 RAT ------------MAPLSQISTHLNSTCGAENSTGVNRARPH-AYYA-LSYCALILAIIFGNGLVCAAVLRERAL
D3 LONG MOUSE ------------MAPLSQISSHINSTCGAENSTGVNRARPH-AYYA-LSYCALILAIIFGNGLVCAAVLRERAL
D4 HUMAN -------- MGNRSTADADGLLAGRGPAAGASAGASAGLAGQGAAALVGGVLLIGAVLAGNSLVCVSVATERAL
D4 RAT -------- MGNSSATGDGGLLAGRGPESLGTGT---GLGGAGAAALVGGVLLIGMVLAGNSLVCVSVASERIL
TRANSMEMBRANE 2
TRANSMEMBRANE 3
D1 HUMAN RSKVTNFFVISLAVSDLLVAVLVMPWKAVA-EIAG-FWPFGS-FCNIWVAFDIM-CSTASILNLCVISVDRYWA
D1 RAT RSKVTNFFVISLAVSDLLVAVLVMPWKAVA-EIAG-FWPFGS-FCNIWVAFDIM-CSTASILNLCVISVDRYWA
D1 FROG RSKVTNFFVISLAVSDLLVAVLVMPWKAVA-EIAG-FWPFGT-FCNIWVAFDIM-CSTASILNLCVISVDRYWA
D1 CHICKEN RSKVTNFFVISLAVSDLLVAVLVMPWKAVS-EIAG-FWPFGS-FCNIWVAFDIM-CSTASILNLCVISVDRYWA
D1 GOLDFISH RSKVTNFFVISLAVSDLLVAVLVMPWKAVT-EVAG-FWPFGA-FCDIWVAFDIM-CSTASILNLCVISVDRYWA
D1 TILAPIA RSKVTNAFVISLAVSDLFVAVLVMPWRAVS-EVAG-VWLFGA-FCDTWVAFDIM-CSTASILHLCIISMDRYWA
D5 HUMAN RANMTNVFIVSLAVSDLFVALLVMPWKAVA-EVAG-YWPFGA-FCDVWVAFDIM-CSTASILNLCVISVDRYWA
D5 RAT RAKMTNIFIVSLAVSDLFVALLVMPWKAVA-EVAG-YWPFGT-FCDIWVAFDIM-CSTASILNLCIISVDRYWA
D5 FROG RSRVTNIFIVSLAVSDLLVALLVMPWKAVA-EVAG-HWPFGA-FCDIWVAFDIM-CSTASILNLCVISVDRYWA
D5 CHICKEN RSKVTNIFIVSLAVSDLLVALLVMPWKAVA-EVAG-YWPFGA-FCNVWVAFDIM-CSTASILNLCVISVDRYWA
D1C FROG RSKVTNFFVISLAVSDLFVAVLVMPWKAVT-EVAG-FWVFGD-FCDTWVAFDIM-CSTASILNLCIISVDRYWA
D1D CHICKEN RTKVTNWFVLSLAISDLCVAILVMPWKAVT-EVAGGFWLFGSRFCDTWVAFDIM-CSTASILNLCIISLDRYWA
DR DOSOPHILA P-RIGNLFLASLAIADLFVASLVMTFAGVN-DLLG-YWIFGAQFCDTWVAFDVM-CSTASILNLCAISMDRYIH
D2 LONG HUMAN QT-TTNYLIVSLAVADLLVATLVMPWV-VYLEVVG-EWKFSRIHCDIFVTLDVMMC-TASILNLCAISIDRYTA
D2 LONG RAT QT-TTNYLIVSLAVADLLVATLVMPWV-VYLEVVG-EWKFSRIHCDIFVTLDVMMC-TASILNLCAISIDRYTA
D2 FROG QT-TTNYLIVSLAVADLLVATLVMPW-AVYMEVVG-EWRFSRIHCDIFVTLDVMMC-TASILNLCAISIDRYTA
D3 HUMAN QT-TTNYLVVSLAVADLLVATLVMPWV-VYLEVTGGVWNFSRICCDVFVTLDVMMC-TASILNLCAISIDRYTA
D3 RAT QT-TTNYLVVSLAVADLLVATLVMPWV-VYLEVTGGVWNFSRICCDVFVTLDVMMC-TASILNLCAISIDRYTA
D3 LONG MOUSE QT-TTNYLVVSLAVADLLVATLVMPWV-VYLEVTGGVWNFSRICCDVFVTLDVMMC-TASILNLCAISIDRYTA
D4 HUMAN QT-PTNSFIVSLAAADLLLALLVLPLF-VYSEVQGGAWLLSPRLCDALMAMDVMLC-TASIFNLCAISVDRFVA
D4 RAT QT-PTNYFIVSLAAADLLLALVVLPLF-VYSEVQGGVWLLSPRLCDTLMAMDVMLC-TASIFNLCAISVDRFVA
TRANSMEMBRANE 4
D1 HUMAN ISSPFRYERKM--TP-K-AAFILISVAWTLSVLISFIPVQLSWHKAKPTSPSDG-------NATSLAETI---
D1 RAT ISSPFQYERKM--TP-K-AAFILISVAWTLSVLISFIPVQLSWHKAKPTWPLDG-------NFNSLEDTED--
D1 FROG ISSPFRYERKM--TP-K-VAFIMIGVAWTLSVLISFIPVQLSWHKAKTTSFFDL-------NITLHDRTM---
D1 CHICKEN ISSPFRYERKM--TP-K-AAFIMISVAWTLSVLISFIPVQLNWHKATTTSFFDL-------NASLQGISM---
D1 GOLDFISH ISSPFRYERKM--TP-R-VAFVMISGAWTLSVLISFIPVQLKWHKAQPIGFLEV-------NASRRDLPT---
D1 TILAPIA ISSPFRYERRM--TP-R-FGCVMIGVAWTLSVLISFIPVQLNWHARGRERTDPGDC------NAS-------
D5 HUMAN ISRPFRYKRKM--TQ-R-MALVMVGLAWTLSILISFIPVQLNWHRDQAASWGGLDLPNNLANWTPWEEDFWEP
D5 RAT ISRPFRYERKM--TQ-R-VALVMVGLAWTLSILISFIPVQLNWHRDKAGSQGQEGL---LSNGTPWEE-GWEL
D5 FROG ISSPFRYERKM--TQ-R-VALLMISTAWALSVLISFIPVQLSWHKS-ET---EDHL---LSNHS--------
D5 CHICKEN ISSPFRYERKM--TQ-R-LALVMIGVAWALSVLISFIPVQLNWHRGGDAATAAAAGDIEDGFDTGWEAAGAFT
D1C FROG IASPFRYERKM--TQ-R-VAFIMIGVAWTLSILISFIPVQLSWHKSHEADEELNGV-----NHT--------
D1D CHICKEN IASPFRYERKM--TQ-R-VACAMIAMAWALSILISFVPVQLHWHKAKDRRHPSSS---------------
DR DROSOPHILA IKDPLRYGRWT--TR-R-VAVITIAATWLLAAFVSFVPISLGTHRPDQPLIFEDNGKKYPTCALD-------
D2 LONG HUMAN VAMPMLYNTRY--SSKRRVT-VMISIVWVLSFTISC-PL-LFGLNNA-DQNECI---------------
D2 LONG RAT VAMPMLYNTRY--SSKRRVT-VMIAIVWVLSFTISC-PL-LFGLNNT-DQNECI---------------
D2 FROG VAMPMLYNTRY--SSKRRVT-VMISVVWVLSFAISC-PL-LFGLNNTG-SKVCII---------------
D3 HUMAN VVMPVHYQHGTGQSSCRRVAL-MITAVWVLAFAVSC-PL-LFGFNTTGDPTVCSI---------------
D3 RAT VVMPVHYEHGTGQSSCRRVAL-MITAVWVLAFAVSC-PL-LFGFNTTGDPSICSI---------------
D3 LONG MOUSE VVMPVHYQHGTGQSSCRRVAL-MITAVWVLAFAVSC-PL-LFGFNTTGDPSICSI---------------
D4 HUMAN VAVPLRYNRQGG--S-RRQLL-IGATWLLSAAVAA-PV-LCGLNDVRGRDPAVCRL---------------
D4 RAT VTVPLRYNQQGQ--C---QLLL-IAATWLLSAAVAA-PV-VCGLNDVPGRDPTVCCL---------------

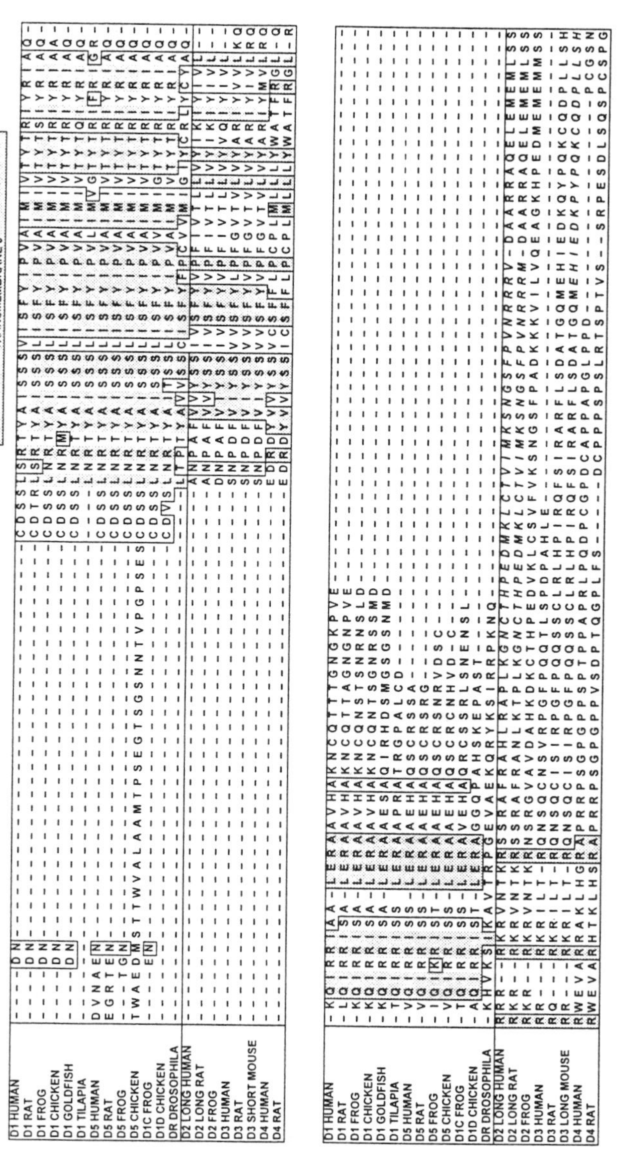

Fig. 1. *(continued)*

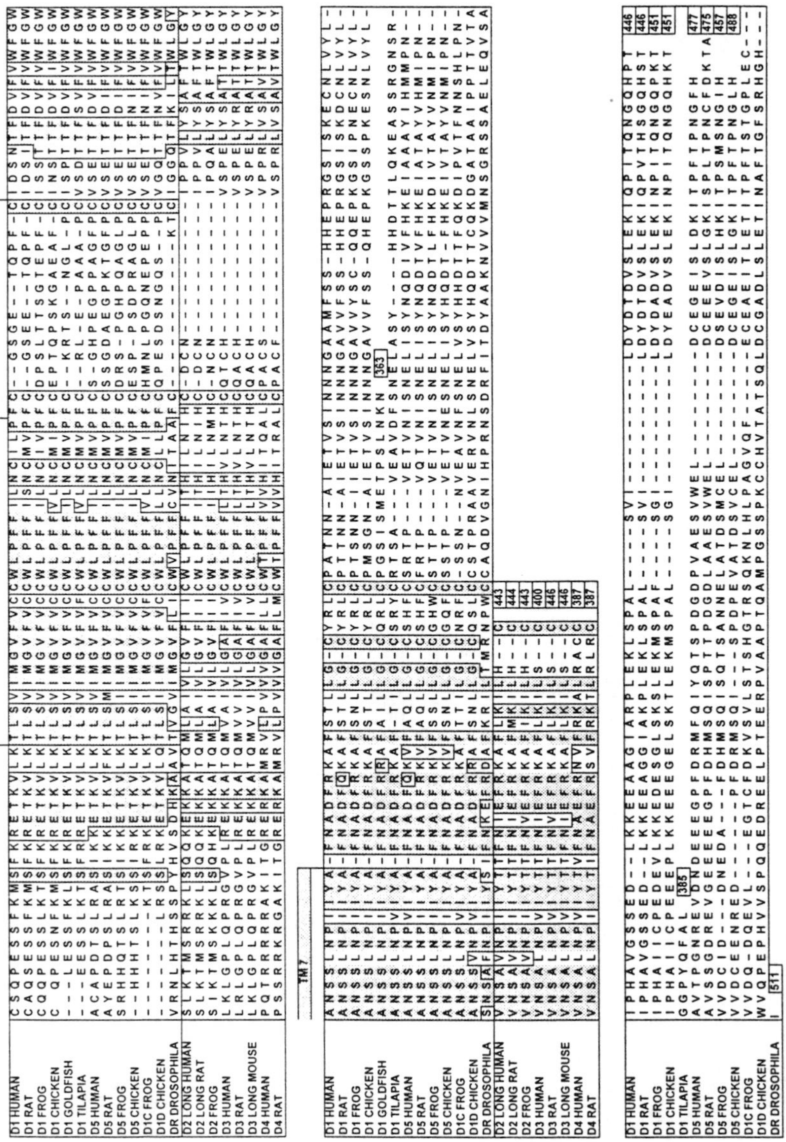

```
Regions: TRANSMEMBRANE 6 (cols 31-46); TRANSMEMBRANE 7 (cols 61-74)

D1 HUMAN        CSQPESSFKMSFKRETKVLKTLSVIMGVFVCCWLPFFILNCILPFC--GSGE--TQPF-CIDSNTFDVFVWFGW
D1 RAT          CAQSESSFKMSFKRETKVLKTLSVIMGVFVCCWLPFFISNCMVPFC--GSEE--TQPF-CIDSITFDVFVWFGW
D1 FROG         CQQPESSLKTSFKRETKVLKTLSVIMGVFVCCWLPFFILNCIVPFCDPSLTTSGTEPF-CISSTTFDVFVWFGW
D1 CHICKEN      C QPESNFKMSFKRETKVLKTLSVIMGVFVCCWLPFFVLNCMIPFCEPTQPSKGAEAF-CINSTTFDVFIWFGW
D1 GOLDFISH     ---LESSFKLSFKRETKVLKTLSVIMGVFVCCWLPFFILNCMVPFC--KRTS--NGL-PCISPTTFDVFVWFGW
D1 TILAPIA      ---EESSLKTSFRRETKVLKTLSVIMGVFVFCWLPFFVLNCMVPFC--RL-E-PAAA-PCVSDTTFSVFVWFGW
D5 HUMAN        ACAPDTSLRASIKKETKVLKTLSVIMGVFVCCWLPFFILNCMVPFCS-GHPEGPPAGFPCVSETTFDVFVWFGW
D5 RAT          AYEPDPSLRASIKKETKVFKTLSMIMGVFVCCWLPFFILNCMVPFCSSGDAEGPKTGFPCVSETTFDIFVWFGW
D5 FROG         SRHHQTSLRTSIKKETKVLKTLSIIMGVFVCCWLPFFILNCMVPFCDRS-PGHPQAGLPCVSETTFDIFVWFGW
D5 CHICKEN      --HHHTSLKSSIRKETKVLKTLSIIMGVFVCCWLPFFILNCMVPFCESP-PSDPRAGLPCVSETTFNIFVWFGW
D1C FROG       --------KTSFRKETKVLKTLSIIMGVFVFCWLPFFVLNCMIPFCHMNLPGQNEPEPPCVSETTFNIFVWFGW
D1D CHICKEN     -------LRSSLRKETKVLQTLSIIMGVFVCCWLPFFLLNCLLPFCQPESDSNGQS--PCVGQTTFNVFVWFGW
DR DROSOPHILA   VRNLHTHSSPYHVSDHKAAVTVGVIMGVFLICWVPFFCVNITAAFC-----------KTCIGGQTFKILTWLGY
D2 LONG HUMAN   SLKTMSRRKLSQQKEKKATQMLAIVLGVFIICWLPFFITHILNIHC-DCN----------IPPVLYSAFTWLGY
D2 LONG RAT     SLKTMSRRKLSQQKEKKATQMLAIVLGVFIICWLPFFITHILNIHC-DCN----------IPPVLYSAFTWLGY
D2 FROG         SIKTMSKKKLSQHKEKKATQMLAIVLGVFIICWLPFFIIHILNMHC-NCN----------IPQALYSAFTWLGY
D3 HUMAN        LKLGPLQPRGVPLREKKATQMVAIVLGAFIVCWLPFFLTHVLNTHCQTCH----------VSPELYSATTWLGY
D3 RAT          LRLGPLQPRGVPLREKKATQMVVIVLGAFIVCWLPFFLTHVLNTHCQACH----------VSPELYRATTWLGY
D3 LONG MOUSE   LKLGPLQPRGVPLREKKATQMVVIVLGAFIVCWLPFFLTHVLNTHCQACH----------VSPELYRATTWLGY
D4 HUMAN        PQTRRRRRAKITGRERKAMRVLPVVVGAFLLCWTPFFVVHITQALCPACS----------VPPRLVSAVTWLGY
D4 RAT          PSSRRKRGAKITGRERKAMRVLPVVVGAFLMCWTPFFVVHITRALCPACF----------VSPRLVSAVTWLGY
```

```
Region: TM 7 (cols 1-11)

D1 HUMAN        ANSSLNPIIYA-FNADFRKAFSTLLG-CYRLCPATNN-AIETVSINNNGAAMFSS-HHEPRGSISKECNLVYL-
D1 RAT          ANSSLNPIIYA-FNADFQKAFSTLLG-CYRLCPTTNN-AIETVSINNNGAVVFSS-HHEPRGSISKDCNLVYL-
D1 FROG         ANSSLNPIIYA-FNADFRKAFSNLLG-CYRLCPTSNN-IIETVSINNNGAVVYSC-QQEPKGSIPNECNLVYL-
D1 CHICKEN      ANSSLNPIIYA-FNADFRKAFSTLLG-CYRLCPMSGN-AIETVSINNNGAVVFSS-QHEPKGSSPKESNLVYL-
D1 GOLDFISH     ANSSLNPIIYA-FNADFRRAFAILLG-CQRLCPGSISMETPSLNKN 363
D1 TILAPIA      ANSSLNPVIYA-FNADFRKAF-TILG-CSRYCRTSA---VEAVDFSNELASY----HHDTTLQKEASSRGNSR-
D5 HUMAN        ANSSLNPVIYA-FNADFQKVFAQLLG-CSHFCSRTP---VETVNISNELISYNQDIVFHKEIAAAYIHMMPN--
D5 RAT          ANSSLNPIIYA-FNADFRKVFAQLLG-CSHFCFRTP---VQTVNISNELISYNQDTVFHKEIATAYVHMIPN--
D5 FROG         ANSSLNPIIYA-FNADFRKVFSSLLG-CGHWCSTTP---VETVNISNELISYNQDTLFHKDIVTAYVNMIPN--
D5 CHICKEN      ANSSLNPIIYA-FNADFRKVFSNLLG-CGQFCSSTP---VETVNESNELISYHQDT-FHKEIVTAYVNMIPN--
D1C FROG       ANSSLNPVIYA-FNADFRKAFTTILG-CNRFC-SSN--NVEAVNFSNELVSYHHDTTFQKDIPVTFNNSHLPN-
D1D CHICKEN     ANSSVNPVIYA-FNADFRRAFSNLLG-CQSLCCSTPRAAVERVNLSNELVSYHQDTTCQKDGATAAIPPTTVTA
DR DROSOPHILA   SNSAFNPIIYSIFNKEFRDAFKRILTMRNPWCCAQDVGNIHPRNSDRFITDYAAKNVVVMNSGRSSAELEQVSA
D2 LONG HUMAN   VNSAVNPIIYTTFNIEFRKAFLKILH--C 443
D2 LONG RAT     VNSAVNPIIYTTFNIEFRKAFMKILH--C 444
D2 FROG         VNSAVNPIIYTTFNVEFRKAFIKILH--C 443
D3 HUMAN        VNSALNPVIYTTFNIEFRKAFLKILS--C 400
D3 RAT          VNSALNPVIYTTFNVEFRKAFLKILS--C 446
D3 LONG MOUSE   VNSALNPVIYTTFNIEFRKAFLKILS--C 446
D4 HUMAN        VNSALNPVIYTVFNAEFRNVFRKALRACC 387
D4 RAT          VNSALNPIIYTIFNAEFRSVFRKTLRLRC 387
```

```
D1 HUMAN        IPHAVGSSED--LKKEEAAGIARPLEKLSPAL----SVI---------LDYDTDVSLEKIQPITQNGQHPT 446
D1 RAT          IPHAVGSSED--LKKEEAGGIAKPLEKLSPAL----SVI---------LDYDTDVSLEKIQPVTHSGQHST 446
D1 FROG         IPHAIICPEDEVLKKEDESGLSKSLEKMSPAF----SGI---------LDYDADVSLEKINPITQNGQPKT 451
D1 CHICKEN      IPHAIICPEEEPLKKEEEGELSKTLEKMSPAL----SGI---------LDYEADVSLEKINPITQNGQHKT 451
D1 TILAPIA      GGPYQFAL 385
D5 HUMAN        AVTPGNREVDNDEEEGPFDRMFQIYQTSPDGDPVAESVWEL--------DCEGEISLDKITPFTPNGFH 477
D5 RAT          AVSSGDREVGEEEEEGPFDHMSQISPTTPDGDLAAESVWEL--------DCEEEVSLGKISPLTPNCFDKTA 475
D5 FROG         VVDCID---DNEDA---FDHMSQISQTSANNELATDSMCEL--------DSEVDISLHKITPSMSNGIH 457
D5 CHICKEN      VVDCEENRED------PFDRMSQI---SPDPEVATDSVCEL--------DCEGEISLGKITPFTPNGLH 488
D1C FROG       VVDQ-DQEVL---EGTCFDKVSVLSTSHGTRSQKNLHLPAGVQF-----ECEAEITLETITPFTSTGPLEC---
D1D CHICKEN     WVQPEPHVVSPQQEDREELPTEERPVAAPTQAMPGSSPKCCHVTATSQLDCGADLSLETINAFTGFSRHGH---
DR DROSOPHILA   I 511
```

```
D1C FROG       LQLVADEDRHYTTKLY 465
D1D CHICKEN     HPIPEG 445
```

Fig. 1.

The *Xenopus* D_{2A} receptor is a 442-amino acid protein that corresponds to mammalian D_{2L} receptors. Both *Xenopus* D_{2A} and D_{2B} have sequences homologous to the mammalian alternatively spliced exon, but in the D_{2A} receptor this region is 33 amino acids long, rather than the 29 amino acid-long fragment found in *Xenopus* D_{2B} and the mammalian D_2 receptors. *Xenopus* D_{2A} and D_{2B} receptors have ~94% amino acid identity with each other, and ~75% identity with the mammalian D_2 receptors. The homology between *Xenopus* and mammalian D_2 receptors is lowest at the amino terminus, in the second extracellular loop, and in the third intracellular loop (Fig. 1). Among the TM regions, which are highly conserved, the lowest homology is in TM 1. When the *Xenopus* receptor is expressed and characterized pharmacologically, it may be possible to relate differences in function to these structural differences.

The *Xenopus* D_{2A} receptor shares with mammalian receptors the potential glycosylation sites at the amino terminus, at Asn-5, -15, and -18 of the *Xenopus* sequence *(36)*. The 5'-end of the D_{2B} cDNA has not been cloned. The two potential PKA phosphorylation sites in the second cytoplasmic loop of mammalian D_2 receptors are conserved in both the D_{2A} and D_{2B} receptors, at Ser-141 and -142, as are the sites in the third cytoplasmic loop corresponding to Ser-229 (Ser-223 in *Xenopus*) and Ser-354 (Ser-353). Interestingly, the optimal consensus sequence at Ser-364 of mammalian D_2 receptors is lost at Ser-363 of the *Xenopus* D_{2A} receptor owing to a change from Arg to Lys-360, but the *Xenopus* D_{2B} receptor maintains the optimal sequence R-K-X-S*.

The sequence of the alternatively spliced exon is well-conserved across species, being identical in rat and mouse, and differing between rat and human by only 1 of 29 amino acids. In this region, the *Xenopus* D_2 receptors differ from the mammalian D_{2L} by 11 of 29 amino acids (62% identity), but 8 of the 11 amino acid substitutions are conservative. Fryxell *(115)* observed that the high degree of conservation is consistent with an important function for the region.

In contrast to most G protein-coupled receptors, which lack introns or have an intron only within the 5'-untranslated region of the cDNA, the human dopamine D_2 receptor gene contains eight exons *(116)* and spans over 270 kb; notably, the first intron separates exon 1 from exon 2 by about 250 kb *(117)*. In Fig. 2, the structure of this gene is compared with that of other dopamine receptor genes in the D2 class. In both human and rodent dopamine receptor genes, exon 6 is alternatively spliced, yielding two D_2 receptor mRNAs that differ in size by 87 nucleotides *(4,8,12)*. All exon/intron junctions possess consensus splice sequences except for the rodent exon/intron 4 junction, in which the second position of the canonical GT dinucleotide has been replaced with a C *(13)*.

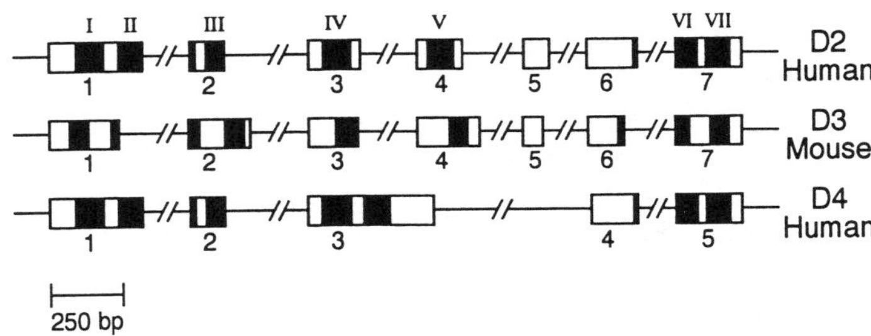

Fig. 2. Structures of the D2-like receptor gene family are shown. The coding regions of the D_2 and D_3 receptor genes comprise seven exons (although, as noted in Section 2.1.2., there is disagreement regarding the presence of exon four in the rat D_3 gene) whereas the D_4 receptor gene has five exons. Transmembrane domains 1–7 are indicated in black. The introns are of variable length.

Both the human and rodent D_2 receptor cDNAs are unusually rich in G + C content. This base composition is particularly striking in the 1 kb region upstream of exon 1, which is >70% G + C *(116)*. The human and rat genes contain several SP1-like elements in the presumptive promoter region *(116,118)*. Gandelman et al. *(116)* identified a putative TATA box at nucleotide –587 (relative to the translational start codon) in the human D_2 promoter, but showed that it is probably not used as a signal for transcription initiation and have posited the existence of additional exons 5' to this site. This speculation is buttressed by the inability of other laboratories to find such consensus promoter motifs as the TATA and CAAT boxes in the rat D_2 receptor promoter regions identified thus far *(118,119)*. Their data are also consistent with the propensity of receptor genes to possess TATA-less promoters (*119*, and references therein). The rat promoter region possesses two transcriptional start sites, one about 320 bp upstream of the 3' end of exon one *(118,119)* and a second, less robust site about 70 bp further 5' *(119)*. These distinct transcriptional start sites reflect the existence of two different promoters. Interestingly, whereas Valdenaire et al. *(119)* were able to show activity of the 1.1 kb *Eco*RI-*Xho*I promoter region in C6 glioma cells and in other cell lines that do not express endogenous D_2 receptors, Minowa et al. *(118)* were unable to induce expression of a reporter gene fused to this promoter in C6 cells. Deletion studies have revealed that the primary activator of the rat D_2 receptor gene is located between nucleotides –75 and –29 relative to the primary transcriptional start site *(118)*, whereas two distinct *cis*-acting negative modulators are found between –160 and –135, and between –116 and –76 *(120)*.

The human dopamine D_2 receptor gene has been localized to chromosome 11q22–q23 *(121)*. Numerous DNA sequence polymorphisms have been identified in the D_2 receptor gene, and are listed in Table 2. These polymorphisms, in particular the *Taq*I, *Eco*RI, and *Nco*I restriction fragment length polymorphisms (RFLPs), have been used to study linkage of the D_2 receptor with mental illness and substance abuse, with contradictory results (*see* Chapter 15).

2.1.2. D_3 Receptors

The rat D_3 receptor cDNA encodes a 446-amino acid protein with over 50% amino acid identity to the rat D_{2L} receptor. The sequence identity between the receptors is highest in TM 2 and TM 3, and lowest (within the TM regions) in TM 1 and TM 4. The human D_3 receptor has ~90% amino acid sequence identity with the rat receptor, but is 46 amino acids shorter (Fig. 1). This discrepancy in length primarily is the result of a stretch of amino acids present in the third cytoplasmic loop of the rat, but not the human, D_3 receptor, spanning an intronic site in the genomic sequence after Met-267 of the rat gene *(146)*. The lack of sequence homology begins after Gln-241, which is the site of an intron in the human gene *(147)*. The human and rat receptors have 10 and 57 amino acid residues, respectively, before sequence homology resumes at Leu-252 of the human and Leu-299 of the rat receptor. The murine D_3 receptor has 97% amino acid identity with the rat receptor. Interestingly, alternative splicing of the murine D_3 receptor gene yields two forms of the murine cDNA, both of which encode functional receptors *(56)*. The longer form (D_{3L}) has 446 amino acids like the rat receptor. D_{3S}, however, is shorter by a deletion of 21 amino acid residues after Met-267. Met-267 precedes intron 5 of the mouse and human genes *(147),* and intron 4 of the rat gene (*146*; Fig. 2).

The D_3 receptors cloned from various species all have two potential sites for N-linked glycosylation that are toward the amino terminus at Asn-12 and -19, and a third site in the second extracellular domain, Asn-97. They also share one potential site for phosphorylation by PKA, at Ser-348 of the rat receptor, toward the carboxyl-terminal end of the third cytoplasmic loop, with the sequence R-K-X-S*. Several sites within the third cytoplasmic loop have the less optimal sequence R-X_{1-2}-S*.

The murine and human dopamine D_3 receptor genes possess seven exons and six introns *(147),* in contrast to the rat gene, for which only five introns have been reported *(146, but see 148)*. Comparison of the murine D_3 and D_2 *(33)* genes reveals that the six introns occur at homologous positions in the two receptors. Introns 1, 2, and 4 of the D_4 receptor are located at positions corresponding to those of introns 1, 2, and 6 in the murine D_2 and D_3 genes. Mouse D_3 introns 1–3 are each significantly longer than their human and mouse D_2 counterparts, and introns 4–6 are of comparable size in all three genes.

Table 2
Polymorphisms and Chromosomal Locations of the Human Dopamine Receptor Genes

Receptor	Chromosome	Polymorphisms
D_1	5q35.1 *(122)*	RFLPs
		*Eco*RI *(122)*, *Taq*I *(123)*
		DNA sequence polymorphisms
		–48A → G, 90A → G, 222A → C *(124)*
D_2	11q22-q23 *(30,43)*	RFLPs
		*Taq*IA in intron preceding exon 8 *(121)*
		*Taq*IB in intron preceding first coding exon *(125)*
		*Taq*IC in intron 2, exon 3, or intron 3 *(126)*
		*Nco*I (3420C → T) in exon 6 *(127)*
		*Bcl*I in exon 2 or intron 2 *(126)*
		*Hinc*II in intron 7 or 8 *(126)*
		DNA sequence polymorphisms
		423G → A in exon 3, 3208G → T in intron between exons 5 and 6 *(127)*
		ser311 → cys *(128)*
		Intron 4, numbering relative to start of exon 5: -208A deleted, –175T → C *(129)*
		In exon 5: 764T → C *(129)*
		Intron 5, relative to start of exon 6: -46G deleted, -24G deleted, –9T → C *(129)*
		In exon 6: 939A → C, 957C → T, 1101A → G *(129)*
		PCR single-strand conformation polymorphism in 3'-untranslated region *(130)*
		Repeats
		TG microsatellite in intron preceding exon 3 *(125)*
D_3	3q13.3 *(131)*	RFLPs
		*Bal*I (9ser → gly) in exon 1 *(132)*
		*Pvu*II *(133)*

D_4	11p15.5 *(134,135)*	RFLPs
		*Sma*I PCR-RFLP in 5'-untranslated region *(136)*
		*Hinc*II *(137)*
		Repeats
		48-bp repeat in exon 3 *(86,94)*
		12-bp repeat in exon 1 *(138)*
		$(G)_n$-Mononucleotide polymorphism in intron 1 *(139)*
		Other
		Null allele (13 bp deletion) in exon 1 *(140)*
D_5	4 *(141)*	DNA sequence polymorphisms
	4p16.1 *(142)*	978T → C *(143)*
	4p15.1-15.3 *(144)*	
		Repeats
		$(CT/GT/GA)_n$ *(145)*

Examination of the sequences of the 5'- and 3'-untranslated regions flanking the coding region of the murine D_3 receptor cDNA indicates that they are only 75% homologous with the same rat D_3 receptor regions *(147)*. It appears, then, that the untranslated sequences abutting the D_3 coding region are less conserved among species than are these sequences in the mouse D_2 receptor, which show a 94 and 70% homology with the rat and human D_2 untranslated sequences, respectively *(33,116)*. The mouse D_3 untranslated regions are very long compared to those of the murine D_2 receptor, with the 5' untranslated sequence being approx 6.5 kb.

Like the dopamine D_2 receptor gene, the D_3 receptor gene undergoes alternative splicing when it is transcribed, although many more variants of the D_3 than of the D_2 gene have been discovered. Giros et al. *(146)* described a frameshift deletion in a rat D_3 receptor transcript which encodes only a 100-amino acid protein, and an in-frame 54-bp deletion in the sequence encoding TM 5. A similar deletion has been detected in a human dopamine D_3 receptor transcript: deletion of 113 bp encoding the end of the first extracellular loop and beginning of TM 3 introduces 19 novel amino acids followed by a stop codon *(54)*. The human D_3 receptor gene is, moreover, alternatively spliced in lymphocytes; a frameshift deletion of 143 bp in the sequence encoding TM 4 results in a predicted 138-amino acid protein containing only the TM 1–3 of the D_3 receptor *(149)*. Pagliusi et al. *(150)* reported an 84-bp insertion encoding 28 extra amino acids in the first extracellular loop. A "short" isoform of the murine D_3 receptor is generated by the deletion of 63 bp (encoding 21 amino acids) in the third cytoplasmic domain-encoding sequence of the receptor *(56)*. The 63 bp found in the "long" form, which was the form initially cloned, is not encoded by a distinct exon but rather is at the 5' end of the exon encoding the third cytoplasmic loop of the receptor protein, suggesting the use of an alternate internal splice acceptor sequence to generate the short form. The 425-amino acid murine D_{3S} is the only splice variant reported that encodes a functional receptor.

The gene encoding the human dopamine D_3 receptor resides on chromosome 3q13.3 *(131)*. *Bal*I *(132)* and *Pvu*II *(133)* RFLPs (Table 2) have been used to probe the possible involvement of the dopamine D_3 receptor in schizophrenia, bipolar disorder, and obsessive compulsive disorder, with generally negative results (*see* Chapter 15).

2.1.3. D_4 Receptors

The most striking structural feature of the D_4 receptor is the presence, in some species, of a 16-amino acid direct repeat in the third cytoplasmic loop of the receptor. In humans, there are at least 19 different repeat units (i.e., 19 different nucleotide sequences, encoding for 10 different amino acid repeats) present in 2–10 copies. Because the order and number of copies of the repeats can also vary, the potential number of haplotypes is quite large. To date, 27 haplotypes encoding 20 different protein sequences have been identified in humans *(81,94)*. The first copy of the repeated sequence always has the α sequence PAPRLPQDPCGPDCAP, and the last copy (in all but the $D_{4.2}$ and $D_{4.3}$ variants) has the ζ/ξ amino acid sequence PAPGLPPDPCGSNCAP. The most common allele is $D_{4.4}$αβθζ *(94)*. The 48-bp repeat, with intra- and interspecies variations in both the sequences and number of copies of the repeats, is also found in the D_4 receptor of nonhuman primates *(151,152)*, but not in the rat D_4 receptor *(80,81)*.

The human $D_{4.4}$ receptor is 419 amino acid residues long, with homology of ~40% to either D_2 or D_3 receptors *(78)*. The rat D_4 receptor, like the human $D_{4.2}$, has 387 amino acids *(81)*. The two species have unusually low homology for mammalian species variants, with 73% amino acid identity. Much of the difference is in the first extracellular domain and the third cytoplasmic loop, with the sequence identity in the TM regions (~93%) approaching that between rat and human D_2 receptors *(80)*. The murine D_4 receptor has 80 and 95% amino acid sequence identity with the human and rat receptors, respectively.

The D_4 receptors share one potential site for N-linked glycosylation, at Asn-3, and one site with the consensus sequence for phosphorylation by PKA, R-R-X-S*, at Ser-234, toward the amino terminal end of the third cytoplasmic loop. The rat D_4 receptor differs from the human receptor by the presence of two additional potential PKA sites, with the less optimal sequence R-X_{1-2}-S*, at Ser-262 and -271.

The human dopamine D_4 receptor gene is contained within 5 kb of DNA *(78)*. It possesses five exons, with introns 1, 2, and 4 corresponding to introns 1, 2, and 6 of the dopamine D_2 receptor gene. Intron 3 lacks canonical splice donor and acceptor sites. The rat D_4 receptor gene, which shares 77% nucleic acid sequence homology with the coding region of the human gene, has one less exon than does the human gene and comprises only about 3.5 kb *(80)*. Unlike the human intron 3, the rat intron 3 is bounded by conventional splice donor and acceptor sequence.

The human D_4 receptor gene is localized to chromosome 11p15.5 *(134,135)*. To date, none of the allelic variants of the 48-bp repeat has been shown to be associated with or linked to bipolar disorder, schizophrenia, or responsiveness to clozapine (*143,153,154*, *see* Chapter 15). However, Catalano et al. *(138)* identified a polymorphic 12-bp repeat in the D_4 receptor gene and found significant differences in frequencies of the two alleles of this polymorphism between patients suffering from delusional disorder and controls.

2.2. The Dopamine D1-Like Receptors

2.2.1. D_1 Receptors

All the mammalian D_1 receptors are 446-amino acid proteins, with M_r ~49,000. The human D_1 and D_{2L} receptors have ~44% overall amino acid identity, whereas the human and rat D_1 receptors are 92% identical *(41)*. The opossum D_1 receptor is ~87% identical to either the rat or human receptor. Not surprisingly, the rhesus and human proteins are virtually identical, differing by only two amino acid residues *(46)*.

The mammalian D_1 receptors have two potential sites for N-linked glycosylation: Asn-5 at the amino terminus (Asn-4 in the rat and opossum), and Asn-175 in the third extracellular domain (Asn-174 in the rat and opossum). At least one of these sites is used, since glycosylation of D_1 receptors has been demonstrated by several groups (*see* Chapter 1). Several potential sites for phosphorylation by PKA are also shared, including Thr-136 in the second cytoplasmic loop (Thr-135 in rat and opossum), within the sequence R-K-X-T*, and Thr-268, at the C-terminal end of the third cytoplasmic loop, within the less commonly phosphorylated sequence R-X-T*. Interestingly, the optimal sequence R-R-X-S* is present in the rat and opossum D_1 receptors at Ser-229 toward the amino-terminal end of the third cytoplasmic loop, where the primate D_1 receptors have an alanine residue (Fig. 1). The mammalian D_1 receptors also have multiple serine and threonine residues in the long cytoplasmic tail that are potential sites of phosphorylation by G protein-coupled receptor kinases.

The goldfish D_1 receptor has ~75% amino acid identity with the rat or human receptors. The two potential sites for N-linked glycosylation are conserved, at Asn-7 and Asn-176. In addition, all the potential PKA phosphorylation sites in the rat and opossum D_1 receptors are conserved, at Thr-137, Ser-231, and Thr-266. An interesting feature of the goldfish D_1 receptor is that it is truncated relative to other D_1 receptors. At 363 amino acids, the protein lacks 80 amino acids that are present in the cytoplasmic C-terminal tail of the mammalian receptors. Functional differences that might be attributed to this truncation have yet to be identified. The sequence for a D1-like receptor from another fish, *T. mossambica,* has also been deposited in the GenBank database (Fig. 1). Unlike the goldfish D_1 receptor, which is more closely related to the human D_1 receptor (91% identity within the TM regions) than to the D_5 receptor (85% identity within the TM regions), the tilapia D_1 receptor is equally homologous to the D_1 and D_5 receptors within the TM regions (~80% identity), and is moderately more closely related to the frog D_{1C} receptor (85% identity within the TM regions). The tilapia D_1 receptor lacks a potential site for glycosylation at the amino terminus, but shares with most of the other D1-like receptors the potential glycosylation site in the third extracellular domain, and the three potential sites for PKA.

The frog and chicken D_1 (D_{1A}) receptors are 451 amino acids long, have 80% amino acid sequence identity with the human D_1 receptor *(50,51),* and share potential glycosylation sites (Asn-4 and -174) and PKA phosphorylation sites (Thr-135, -267, and Ser-228) with the rat, opossum, and goldfish D_1 receptors. The frog D_{1C} receptor is 465 amino acids long with ~54% amino acid identity to the human D_1 or D_5 receptors, and possesses all the potential glycosylation and PKA sites found in the frog D_1 receptor. In addition, there is a third potential site for N-linked glycosylation, at Asn-3, and many additional potential phosphorylation sites with the sequence R-X_{1-2}-S^* (e.g., Ser-230, -242, and -243, -246). The chicken D_{1D} receptor is 445 amino acids long, with ~60% amino acid identity to each of the subfamilies of D_1 receptors (D_1, D_5, and frog D_{1C}). The D_{1D} receptor lacks N-linked glycosylation sites, but retains the three consensus PKA sites found in rat and opossum D_1 receptors.

The *Drosophila* D1-like receptor is a 511-amino acid protein with only ~25% amino acid identity with the human D_1 receptor. Within the TM regions, however, the identity between the two receptors increases to 53%, which is similar to the identity between *Drosophila* D_1 and human D_5 receptors (51%), but higher than the sequence conservation between the TM regions of the *Drosophila* D1-like receptor and those of other receptors including the human beta-2 adrenergic and D_2 receptors and the *Drosophila* tyramine and $5HT_1$ receptors *(49).*

Based on the predicted sequence and hydropathy profile of the *Drosophila* D_1 receptor, the most striking structural feature that differentiates it from mammalian D1-like receptors is the long, 142-amino acid, first extracellular domain. Within this long amino-terminal region, the *Drosophila* D_1 receptor has two short hydrophobic segments, at residues 10–23 and 24–37 *(49).* Similar hydrophobic segments in other *Drosophila* receptors have been proposed to serve as internal signal peptides. There are five potential sites for N-linked glycosylation in this amino terminal region, at Asn-53, -63, -74, -117, and -123. Although the potentially glycosylated asparagine residue in the third extracellular domain of other D_1 receptors is conserved at Asn-295, in the *Drosophila* D_1 receptor it is not a potential site of glycosylation. There are four potential sites of phosphorylation by PKA, one in the third cytoplasmic loop (Ser-361) and three in the C-terminal tail (Thr-462, Ser-481, and Ser-502), but all have the less than optimal sequence R-X_{1-2}-S^*/T^*.

The approximately 4 kb, rat dopamine D_1 receptor gene is virtually intronless, consisting of two exons separated by a 115-bp intron *(155).* The intron is in the 5'-untranslated region of the D_1 receptor mRNA, so that the coding sequence is contained in its entirety within exon 2, whereas exon 1 comprises only 313 bp of the D_1 mRNA 5'-untranslated region. The 5'-flanking sequence of the rat D_1 receptor gene, like that of the D_2 receptor gene, does not possess TATA and CAAT consensus sequences, but has a high G + C content and Sp1 binding sites *(155)*. The human D_1 receptor gene 5'-flanking region appears to be similarly organized, although the intron in the 5'-untranslated region has been estimated to be 116 bp (rather than the rat's 115 bp) *(156)*. The transcription start site of the rat gene is at –864 relative to the start of translation *(155)*, whereas the human gene has multiple transcription start sites between –1061 and –1040 *(156)*. Both rat and human D_1 receptor gene promoters possess a cAMP-response element, as well as binding sites for the transcription factors AP1 and AP2; the rat promoter exhibits, in addition, a glucocorticoid response element sequence. Although the D_1 promoter is a housekeeping type of promoter, it seems to exert tissue-specific control of expression *(155,156)*. It has two activator regions, located at –1154 to –1137 and –1197 to –1154, whereas the core promoter is located 3' of bp –1102 *(157)*.

The human dopamine D_1 receptor gene resides on the long arm of chromosome 5, at q35.1 *(122)*. The *Eco*RI and *Taq*I RFLPs *(122,123)* and three DNA sequence polymorphisms *(124)* that have been reported (Table 2) have not revealed linkage or association of the dopamine D_1 receptor gene with schizophrenia or manic depressive illness (*see* Chapter 15).

2.2.2. D_5 Receptors

The human D_5 receptor is a 477-amino acid protein with ~60% amino acid identity with the D_1 receptor, increasing to ~80% in the TM regions, and only ~30% identity with D_2 or D_3 receptors. Homology with the D_1 receptor is particularly high in TM 2, 3, 5, 6, and 7. The rat (475 amino acids) and human D_5 receptors have ~83% amino acid sequence identity *(105,108)*, which is low for homologs of these species, although higher than the homology between the rat and human D_4 receptors. Within the TM regions, the sequence identity increases to ~95%. The mammalian D_5 receptors have a potential site for N-linked glycosylation, at Asn-7. Both rat (Asn-194) and human (Asn-199) receptors have a second potential site in the third extracellular domain. Two potential sites for phosphorylation by PKA are present at Thr-153 in the second cytoplasmic loop and Ser-260 in the third cytoplasmic loop (Thr-151 and Ser-254 in the rat). These two sites are conserved in all the D1-like receptors identified to date, including the nonmammalian D1-like receptors, except that the primate D_1 receptors have an alanine (Ala-229) in place of the potentially phosphorylated serine residue conserved in the other receptors, and the *Drosophila* D1-like receptor lacks both sites (Fig. 1). Also in the third cytoplasmic loop, the rat and human proteins have one and two additional sites, respectively, with the sequence R-X-S*.

The chicken D_5 receptor (488 amino acids) has 66% amino acid identity with the mammalian D_5 receptors, increasing to 90% within the TM regions *(51)*. One characteristic of the rat and human (but not the frog) D_5 receptors is a third extracellular domain that is long relative to that of the D_1 receptors. This characteristic is particularly pronounced in the chicken D5 receptor, in which the loop contains 65 amino acids. The frog D_5 receptor has 457 amino acids and 66% identity with the human D_5 receptor, but only 55% identity with the human D_1 receptor. The relatively low sequence identity for homologs of D_5 receptors among all the species, compared to species conservation of D_1 and D_2 receptors, suggests an accelerated rate of evolutionary mutation of D_5 receptors *(50,115)*.

The frog D_5 receptor has a potential glycosylation site at the amino terminus (Asn-24), and the site in the third extracellular domain present in other D1-like receptors is also present in the frog. The chicken D_5 receptor lacks a site in the amino terminus, but has a site in the third extracellular domain. As noted, the two potential sites for phosphorylation by PKA present in the second and third cytoplasmic loops of most other D1-like receptors are conserved in the frog and chicken receptors, although substitution of a lysine for arginine produces the less optimal sequence K-R-X-S* in the third cytoplasmic loop of the frog D_5 receptor.

The rat and human dopamine D_5 receptors, as with the D_1 receptor, contain no intron within their coding sequence *(103,105)*. Two human dopamine D_5 receptor pseudogenes have been identified *(104)*. The functional gene has been localized to chromosome 4 *(141)*, specifically to its short arm. Its subchromosomal localization has been reported to be 4p16.3, 4p16.1, and 4p15.1–15.3 (refs. *144*, *142*, and *105*, respectively). Two polymorphisms in the functional gene have been reported (Table 2), a polymorphic microsatellite *(145)* and a T to C transition *(158)*.

3. Structure–Function Analysis of the Dopamine Receptors

Two approaches have been taken towards the evaluation of the structural determinants of dopamine receptor function by in vitro mutagenesis. In one approach, single amino acids are mutated (sometimes with several mutations combined in one receptor), on the assumption that the selective loss of a function resulting from the mutation of an amino acid reflects a contribution of that amino acid to the function. In the second approach, chimeric receptors are assembled using parts of two receptors that differ in defined characteristics. This approach can be useful for identifying the structural basis of functions that involve the action of multiple contiguous amino acid residues, or

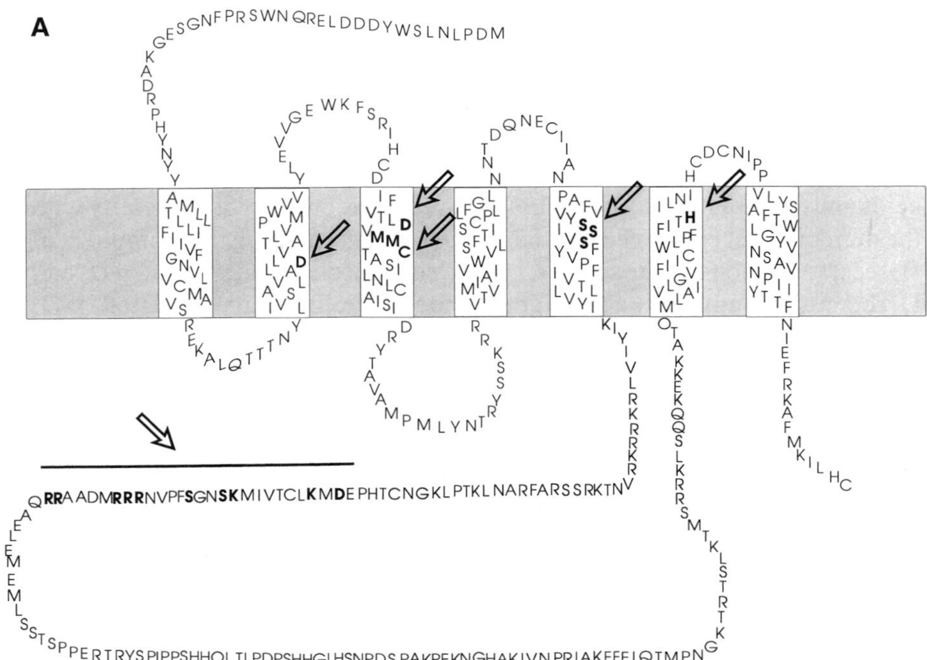

Fig. 3. Predicted structures of the dopamine receptors are shown. Amino acid residues that have been mutated in D_1 or D_2 receptors are indicated by arrows. **(A)** Sequence of the rat D_{2L} receptor.

useful for the preliminary identification of regions that might contain single residues that underlie a difference between the two "parent" receptors. One strength of this second approach is that it is often possible to assess the same property in reciprocal chimeras to verify that a domain associated with a loss of function when it is removed from one parent also causes a gain of function when added to the other parent.

For in vitro mutagenesis studies in which single amino acid residues are altered, a variety of considerations go into the choice of target residues (Fig. 3). It can be valuable to assess the function of residues that are conserved among some subgroup of receptors (e.g., the three serine residues conserved in TM 5 of catecholamine receptors) or the 12 residues that are conserved among all the dopamine receptors, but that are not shared by the beta-2 adrenergic receptor. It is also informative to evaluate residues that have been implicated in the function of related receptors. For example, mutagenesis studies contributed greatly to the development of a model of catecholamine binding to the

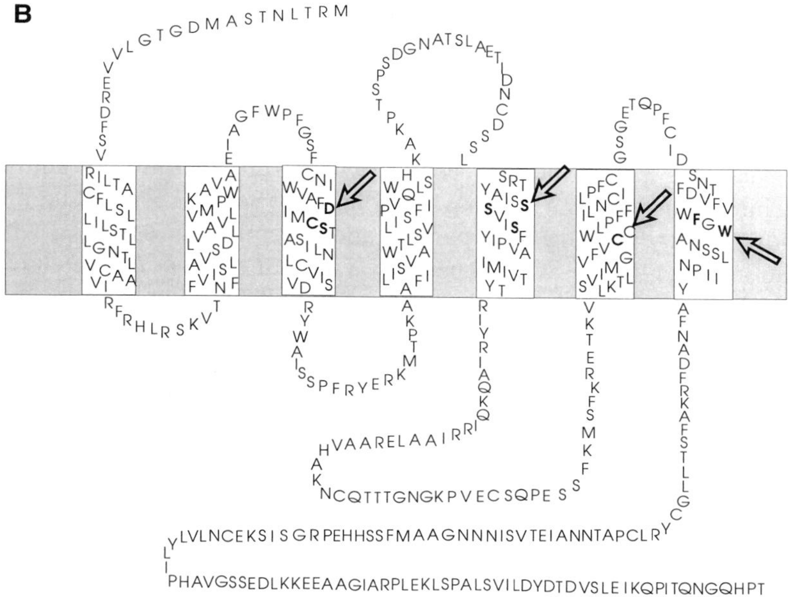

Fig. 3. *(continued)* **(B)** Sequence of the human D1 receptor.

beta-2 adrenergic receptor, a model that has been liberally extended to other catecholamine receptors as they were cloned. Other mutagenesis studies have, in turn, evaluated the validity of applying this model to dopamine receptors. Yet another consideration is to use in vitro mutagenesis to evaluate residues that are implicated by three-dimensional receptor modeling studies in the binding of ligands. Several amino acid residues are repeatedly identified in these models, some of which were reviewed in Dahl (*159*; *see also 160–162*). As three-dimensional models become increasingly sophisticated, we expect that the testing of these models will be an increasingly important, and useful, motivation for in vitro mutagenesis.

Considerable evidence indicates that three residues contribute greatly to the binding of catecholamines to the beta-2 adrenergic receptors. Asp-113, in TM 3, is proposed to participate in an ionic interaction with the positively charged amine of beta adrenergic receptor ligands, whereas Ser-204 and Ser-207 in TM 5 are postulated to interact with the catecholamine *meta-* and *para-*hydroxyl groups, respectively *(163)*. Several groups have assessed the contribution of these residues to the interactions of ligands with dopamine receptors. Similarly, data from studies of beta adrenergic and other receptors indicate that the third cytoplasmic loop of G protein-coupled receptors is the primary determinant of selective coupling to G proteins, although all the intracellular loops contribute to efficiency of coupling *(164)*.

3.1. Conserved Serine Residues in TM 5

Three groups have mutated two or three of the residues Ser-193, -194, and -197 in TM 5 of the D_2 receptor, with results that are not entirely in agreement. Cox et al. *(165)*, using D_{2S}, found that Ser-193 contributes most to the low-affinity binding of dopamine and several other agonists, whereas mutation of Ser-197 or Ser-194 has only a modest effect or no effect, respectively, on the low-affinity binding of dopamine. On the other hand, the binding of some agonists such as *N*-propylnorapomorphine (NPA) and quinpirole is sensitive to either mutation of Ser-194 or Ser-193. Furthermore, the contribution of each serine varies according to the particular ligand that is being tested, and the potency of some antagonists is also reduced by these mutations, findings consistent with other work on these residues *(166,167)*. Mansour et al. *(166)* created mutants of D_{2L} in which either Ser-194, Ser-197, or both were replaced with alanine, then assessed the high-affinity binding of agonists. Ser-197 was found to be more important than Ser-194 in the binding of five of the six agonists tested, including NPA and quinpirole. The binding of the agonist [^{3}H]N-0437, however, is unaffected by the mutations. On the other hand, simultaneous mutation of Ser-194 and -197 reduces the affinity of [^{3}H]N-0437 to the extent that specific binding is no longer detectable. Woodward et al. *(167)* mutated Ser-193, -194, or -197 in D_{2L} and assessed both low- and high-affinity binding of agonists. Low-affinity binding of dopamine is affected most by mutation of Ser-193, and to a lesser extent by mutation of Ser-194. The high-affinity binding of several agonists is abolished by mutation of Ser-197.

There are several areas of agreement among these studies. First, the observed effect of a particular mutation can vary depending on the drug that is being tested. The ligand-dependent effect of each mutation emphasizes the need to test many drugs when evaluating mutant receptors. Second, mutation of these serine residues has modest effects on the binding of some antagonists, in addition to more marked effects on the binding of many agonists. Third, mutation of Ser-193 causes the greatest decrease in the low-affinity binding of dopamine. Finally, Ser-197 is most important for the high-affinity binding of several agonists, including dopamine.

We have also assessed the ability of agonists to inhibit adenylate cyclase activity via the mutant D_{2S} receptors. We found that although mutation of Ser-194 has no effect on the binding of dopamine or *para*-tyramine, it abolishes inhibition of adenylate cyclase by these agonists, whereas mutation of the other serine residues has no effect on efficacy *(165)*. These results seem to be in conflict with the importance of Ser-197 for high-affinity (G protein-coupled) binding of dopamine to D_{2L} *(166)*, but may reflect differential G protein selection by D_{2S} and D_{2L} receptors. Interestingly, the efficacy of a third agonist, *meta*-tyramine, is unaltered by mutation of Ser-194 or the other serine residues. This indicated to us that *meta*-tyramine and dopamine activate the D_{2S} receptor via different mechanisms. In preliminary studies, we have evaluated this further by testing additional agonists.

Our results suggest that there are multiple independent structural determinants of efficacy that contribute differentially to inhibition of adenylate cyclase activity depending on the agonist being tested *(168)*. For example, mutation of Ser-193 selectively reduced the efficacy of quinpirole and apomorphine, whereas mutation of Ser-197 reduced the D_{2S} receptor efficacy of the D1-selective agonists SKF 38393 and fenoldepam. We are still determining the effect that receptor density could have on the apparent efficacy of agonists at these mutant receptors, but our data suggest that some of these effects, such as the loss of efficacy of dopamine after mutation of Ser-194, cannot be overcome by increasing the receptor density, whereas other effects, such as the loss of efficacy after mutation of Ser-197, are graded effects that may not be manifested if there is a sufficient receptor density.

Mutation of the corresponding residues of the dopamine D_1 receptor, Ser-198, -199, and -202, has been described in two studies. As observed for D2 receptors, Pollock et al. *(169)* determined that the potency of antagonists and agonists is decreased by mutation of the serine residues, with the effect of a particular mutation varying by drug. Mutation of Ser-198 (corresponding to D_2 Ser-193) profoundly decreases the potency of [^{3}H]SCH 23390, so that radioligand binding is undetectable, making it difficult to assess the potency of other ligands for that mutant. For the remaining two serine residues, the potency for low-affinity binding of dopamine is reduced dramatically after mutation of Ser-202 (corresponding to the D_2 Ser-197), and also greatly decreased by mutation of Ser-199. On the other hand, the potencies of the benzazepine agonist SKF 38393 and the benzazepine antagonist SCH 23390 are decreased more by mutation of Ser-199 than Ser-202. Dopamine and SKF 38393 stimulate adenylate cyclase activity via all of the mutant receptors, although the efficacy of SKF 38393 appears to be reduced more than that of dopamine. As noted by the authors, the apparent selective loss of efficacy of SKF 38393 could reflect the existence of a receptor reserve for dopamine that is not present for SKF 38393, which is a partial agonist. Tomic et al. *(170)* constructed a receptor in which both Ser-199 and Ser-202 were mutated, and determined that the affinity of the mutant for dopamine was moderately reduced and the affinity for [^{3}H]SCH 23390 was unchanged. These results appear to be in conflict with the much greater effect that mutating either of the residues singly has on the binding of dopamine and [^{3}H]SCH 23390 *(169)*.

3.2. Conserved Aspartate Residues

Whereas mutation of conserved serine residues yields complicated results that differ, sometimes markedly, from the model developed for the beta-2 adrenergic receptor, the effects of mutation of conserved aspartate residues in TM 2 and TM 3 are more straightforward. As described for Asp-113 of the beta-2 adrenergic receptor, Asp-114 in TM 3 of D_2 receptors is critical for the binding of ligands, so that mutation to either asparagine or glycine abolishes detectable radioligand binding *(166,171)*. The more conservative substitution of a glutamate for Asp-114 has less severe consequences on the potency of ligand, primarily decreasing the potency of at least one substituted benzamide antagonist and several agonists without altering the ability of the receptors to inhibit adenylate cyclase *(171)*. These data are consistent with a model in which an ionic interaction between this residue and the amine present in all dopamine receptor ligands is essential for binding.

The mutation of Asp-80 in TM 2 of D_{2S} has less severe consequences on the potency of ligands, but produces a receptor that is resistant to conformational changes induced by a variety of influences, including sodium ions, pH, and the binding of agonists *(171,172)*. Thus, ligand binding to the mutant receptor is insensitive to sodium and less sensitive to pH (*see* Chapter 1, for further discussion of regulation of dopamine receptors by ions). Furthermore, agonists do not induce a high-affinity, GTP-sensitive receptor conformation, and the receptor loses the ability to inhibit adenylate cyclase. A similar role for the corresponding residue Asp-70 in the regulation of D_1 receptors by sodium ions has been described *(170)*. Although some have interpreted the loss of affinity of these mutants for agonists as being consistent with a direct contribution of this residue to binding *(159,173)*, our interpretation is that the residue does not interact directly with ligands. Rather, this residue, perhaps through an intramolecular interaction with a residue in TM 7 *(44)*, contributes to the maintenance of a conformation in which the receptor is able to respond to stimuli (binding of agonists, sodium, G proteins, etc.) with an appropriate conformational change *(171)*. As discussed by Zhou et al. *(44)*, other arguments against a direct role of this aspartate residue in ligand binding include its extremely high degree of conservation among G protein-coupled receptors, its position (which is deeper within the membrane than other residues thought to form the binding pocket), and the finding that loss of binding owing to mutation of the corresponding residue (Asn-87) in GnRH receptors can be reversed by a reciprocal mutation in TM 7.

3.3. Other Amino Acid Residues

Mansour et al. *(166)* evaluated the function of two methionine residues in the D_{2L} receptor (Fig. 3). Met-116 is conserved among the dopamine receptors, but is not present in the adrenergic receptors, and Met-117 is shared by D_2 and D_3 receptors (Fig. 1). Mutation of these residues singly or together has no effect on the affinity of the receptors for several agonist and antagonist ligands.

Molecular modeling *(174)* and studies of the pH-dependence of ligand binding *(175)* suggest that His-394 of the rat D_{2L}, present in TM 6 of all the dopamine receptor subtypes, might be involved in the binding of some ligands. Naylor and colleagues *(176)* found that replacing the histidine residue with a leucine selectively alters the affinity of the receptors for some substituted benzamide derivatives. Depending on the nature of the substituents in the five and four positions on the aromatic ring of a particular substituted benzamide, the mutation either increases, decreases, or has no effect on the potency of the drug. It was suggested that a hydrogen bond between a 5-sulphone/sulphonamide substituent and His-394 favors binding, whereas either a 5-chloro or 4-amino substituent interacts unfavorably with His-394. Mutation of His-394 does not decrease, and may increase, the pH- or sodium-dependence of the binding of substituted benzamides.

Considerable evidence points toward the involvement of cysteine residues in forming the dopamine-binding pocket of D_2 receptors (*see* Chapter 1). Javitch et al. *(177)* confirmed this using polar, sulfhydryl-specific, methanethiosulfonate (MTS) derivatives. Positively charged MTS derivatives irreversibly inhibit radioligand binding to D_{2L}, an effect diminished by D_2 receptor agonists and antagonists. By individually converting seven cysteine residues to serine, it was determined that MTS-induced loss of radioligand binding is caused by formation of a disulfide bond between the reagent and Cys-118, a cysteine residue in TM 3 that is shared by the D2-like receptors and several other biogenic amine receptors. Interestingly, Cys-118 does not appear to participate directly in the binding of [^{3}H]YM-09151-2 to D_2 receptors, since neither mutation of Cys-118 to serine nor reaction with methyl-MTS, which changes the cysteine to a methionine-like side chain, affects the binding of the radioligand. The MTS-induced inhibition of binding is apparently not owing to the modification of Cys-118 per se, but to the presence of covalently bound MTS blocking access to the ligand binding site.

In a related approach, Javitch et al. *(178)* identified residues that are accessible in the D_2 receptor binding pocket by constructing 22 mutant receptors in each of which one of the residues in and flanking TM 3 was converted to cysteine. A cysteine residue (and, therefore, the residue replaced by cysteine) was assumed to face the binding cavity if reaction with MTS reagents irreversibly inhibited radioligand binding to the corresponding mutant receptor, and if the inhibition of binding could be prevented by prior occupancy with the D2 receptor ligand sulpiride. According to these criteria, 10 of the 22 residues are exposed to the binding pocket: Asp-108, Ile-109,

Phe-110, Val-111, Val-115, Cys-118, Ser-121, Ile-122, Leu-125, and Ser-129. If the α-helix is assumed to begin with Phe-110, so that Asp-108 and Ile-109 precede the helix, the remaining residues are within a 140° arc that includes Asp-114.

To investigate the structural basis of the differing abilities of D_{2S} and D_{2L} to inhibit adenylate cyclase in JEG3 cells, Guiramand et al. *(179)* mutated several residues within the 29-amino acid insert of D_{2L} (Fig. 3A). Substitution of valine for basic residues (lysines) within or adjacent to the insert decreases the expression (B_{max}) of D_{2L}. Two other mutants, one with Asp-249 replaced with valine and one with Ser-259 and -262 replaced with alanine, displayed an increased ability to inhibit adenylate cyclase, similar to that of D_{2S}, as reflected by a lower EC_{50} for dopamine, suggesting that Asp-249, Ser-259, and Ser-262 contribute to the lesser efficiency of coupling of D_{2L} to the G proteins present in JEG3 cells.

Tomic et al. *(170)* constructed a receptor with two mutations of amino acid residues close to Asp-103 in TM 3 of the D_1 receptor (corresponding to D_2 Asp-114), Cys-106, and Ser-107. Ser-107 is in the same position as Cys-118 of the D_2 receptor, four residues away from the conserved aspartic acid. The double mutation causes a modest decrease in affinity for [^{3}H]SCH 23390 and dopamine. Other mutations evaluated were Phe-319 → Leu, which has no effect on the binding of [^{3}H]SCH 23390 or dopamine, Trp-321 → Tyr, and Cys-283 → Val, both of which modestly decrease the binding of [^{3}H]SCH 23390 but not dopamine. Thus, several of these amino acid residues could participate in forming the binding pocket for [^{3}H]SCH 23390 or dopamine.

3.4. Chimeric Dopamine Receptors

3.4.1. Chimeric D_1/D_2 Receptors

A D_1/D_2 chimeric receptor that has D_1 receptor sequence up to the amino-terminal end of TM 6 (thus including the D_1 third cytoplasmic loop) and D_2 sequence from that junction to the C-terminus, mediates stimulation of adenylate cyclase activity by dopamine and the D1 agonist SKF 38393. Interestingly, the D2 agonist quinpirole, which has virtually no efficacy at the D_1 receptor, is efficacious at this D_1/D_2 chimera *(180)*. These results agree with the identification of the third cytoplasmic loop as the primary determinant of the specificity of coupling to G proteins. The chimeric receptor has a higher affinity than does the D_1 receptor for quinpirole, but a lower affinity for the D1 ligands SCH 23390 and SKF 38393. Interestingly, the chimeric receptor has decreased affinity for the nonselective agonist dopamine, compared to the D_1 or the D_2 receptor.

We constructed a number of D_1/D_2 chimeric receptors, with one of them (CH3) virtually identical to the chimeric receptor described. Our results with this chimera *(181)* were very similar to those of MacKenzie et al. *(180)*. Based on our analysis of these chimeras, we formulated several hypotheses that we are testing by construction and characterization of additional chimeric receptors. First, none of the chimeric receptors that had D_2 sequence at the amino terminus and D_1 sequence at the carboxy terminus were detectable by radioligand binding. Our hypothesis, which is supported by preliminary data obtained from novel chimeric receptors *(182)*, is that TM 1 or TM 2 from the D_2 receptor includes one or more amino acid residues that are adjacent to and sterically incompatible with residues in TM 7 of the D_1 receptor *(44)*. Second, radioligand binding data suggest that TM 7 is a particularly important determinant of the selective affinity of the D_1 receptor for two benzazepine derivatives, SCH 23390 and SKF 38393, and a determinant of the selective affinity of the D_2 receptor for quinpirole. On the other hand, several D2-selective or nonselective drugs have lower potencies at all the chimeras than at D_1 or D_2 receptors, which could be the result of nonspecific structural changes associated with expression of the artificial receptors. Third, confirming the results of MacKenzie et al. *(180)*, TM 6 appears to be an important determinant of selective efficacy, since the D2 receptor agonists quinpirole *(181)* and bromocriptine *(182)* gain the ability to stimulate adenylate cyclase when TM 6 is from the D_2 receptor. Fourth, the third cytoplasmic loop of the D_1 receptor may be sufficient, or at least necessary, for coupling to G_s and stimulation of adenylate cyclase, but both the second and third cytoplasmic loops of the D_2 receptor are needed for even minimal inhibition of adenylate cyclase *(181)*.

3.4.2. Chimeric D_2/D_3 Receptors

Two characteristics of the D_3 receptor have stimulated considerable interest in the analysis of D_2/D_3 receptors. First, the D_3 receptor displays GTP-resistant, high-affinity binding of many agonists. Second, the D_3 receptor has much less efficacy than the D_2 receptor for inhibition of adenylate cyclase. In many systems, it is not possible to demonstrate any inhibition of enzyme activity by the D_3 receptor. This makes the D_3 receptor nearly a neutral background on which to test D_2 receptor domains for the ability to confer coupling to this pathway, whereas the close relationship between the D_2 and D_3 receptors makes it less likely that nonspecific structural distortion of chimeric receptors will influence experimental outcomes.

McAllister et al. *(183)* constructed a chimeric receptor consisting of the human D_3 receptor, except for the third cytoplasmic loop, which was from the human D_2 receptor. The junctions between D_2 and D_3 sequence were approximately at the border between the TM and cytoplasmic domains. The chimera retains the high affinity for agonists typical of the D_3 receptor, but is still unable to inhibit adenylate cyclase, suggesting that the D_2 third cytoplasmic loop is not sufficient to confer this property, and that the D_3 third cytoplasmic loop is not needed to obtain high affinity for agonists. Robinson et al. *(184)* characterized a set of four D_2/D_3 receptor chimeras. Two consisted of D_2 receptor sequence except for the D_3 second or third cytoplasmic loops, and the other two were reciprocal chimeras in which the D_2 loops were put into the D_3 receptor. The junctions between D_2 and D_3 sequence were in the middle of the TM regions. Addition of the D_2 third cytoplasmic loop to the D_3 receptor moderately decreases affinity for the agonists dopamine and quinpirole, whereas addition of the D_3 third cytoplasmic loop to the D_2 receptor moderately increases affinity for the agonists. Agonist affinity is not altered by exchange of the second intracellular loops. Neither the second nor the third intracellular loop appears to contribute to the selective affinity of D_2 and D_3 receptors for antagonists.

In a preliminary report, reciprocal D_2/D_3 chimeric receptors were characterized in which sequence up to the middle of the third cytoplasmic loop was from one parent and the remainder of the sequence was from the other *(185)*. The affinities of the chimeras for the D_3-selective agonist 7-OH-DPAT were intermediate between the affinities of the wild-type receptors, but each chimera was more similar to the TM 1–5 donor than the TM 6–7 donor. Both chimeric receptors had a modest ability to inhibit adenylate cyclase activity in CHO-K1 cells. An additional report described a D_3/D_2 chimeric receptor that consisted of all D_3 receptor sequences expect for the D_2 third cytoplasmic loop *(186)*. For this particular chimera, the high affinity for agonists that is characteristic of the D_3 receptor was unaltered, but the D_3/D_2 chimeric receptor gained the ability to inhibit adenylate cyclase in CHO cells.

4. Future Directions

The importance of the dopamine system in normal and abnormal human behavior has stimulated considerable work aimed at characterizing the D_1–D_5 subtypes. Several broad lines of investigation appear to be particularly significant for future consolidation of the dopamine receptor field and for application of basic biological knowledge of the receptors to clinical medicine. The development of agonists and antagonists that are selective for each of the subtypes is critical, but far from trivial. The D_1 and D_5 receptors are so similar in sequence and pharmacological profile that developing selective drugs is certain to be difficult, and the same is true for the D_2 and D_3 receptors. Drug design should be aided by the development of more sophisticated molecular models of the receptors. With respect to receptor modeling, it seems likely that an iterative process will unfold, in which in vitro mutagenesis is used to confirm the importance of receptor amino acid residues that are identified by molecular modeling of the receptors, and the results of mutagenesis studies are used in turn to refine the molecular models. At the same time, the ability to generate large quantities of recombinant receptors may lead to the purification and crystallization of the receptors for physical determination of their three-dimensional structures, which would greatly increase the accuracy of computer simulations in which ligands are docked in a receptor.

A second crucial goal, one that would be made easier with more selective agonists and antagonists, is to identify behavioral correlates of stimulation or blockade of each dopamine receptor subtype. Considering the complicated interactions already known to exist between D1-like and D2-like receptors (*see* Chapter 7) this is likely to be a challenging task. One line of investigation that will play an increasingly prominent role in achieving this goal is the use of genetic intervention, including antisense inhibition of expression and gene "knockouts," to determine the effect of the loss of a receptor subtype (*see* Chapter 14).

The obvious question, but a question without a definite answer, is whether there are more subtypes of mammalian dopamine receptors still to be discovered. The 4 years that have passed since the D_4 and D_5 receptors were cloned suggests that the answer to this question is "no." On the other hand, the cloning of novel D1-like receptors from the chicken and frog that are distinct from the respective species homologs of the mammalian D_1 and D_5 receptors suggests that additional D1-like receptors may also be present in mammals. Similarly, the failure of any of the recombinant D1-like receptors to activate phospholipase C, a capability that has been demonstrated for some endogenous D1 receptors, leads to speculation that at least one mammalian D1-like receptor remains to be discovered.

Acknowledgments

This work is supported by the Veterans Affairs Merit Review and Career Scientist Programs, and by Public Health Service grants MH45372 and HD24236. We thank Dr. Philip Seeman for a file with alignments of some of the dopamine receptors shown in Fig. 1, and Laura Kozell for additional assistance with the figures.

References

1. Dixon, R. A. F., Kobilka, B. K., Strader, D. J., Benovic, J. L., Dohlman, H. G., Frielle, T., et al. (1986) Cloning of the gene and cDNA for mammalian β-adrenergic receptor and homology with rhodopsin. *Nature* **321,** 75–79.

2. Yarden, Y., Rodriguez, H., Wong, S. K.-F., Brandt, D. R., May, D. C., Burnier, J., Harkins, R. N., Chen, E. Y., Ramachandran, J., Ullrich, A., and Ross, E. M. (1986) The avian β-adrenergic receptor: Primary structure and membrane topology. *Proc. Natl. Acad. Sci. USA* **83,** 6795–6799.
3. Kubo, T., Fukuda, K., Mikami, A., Maeda, A., Takahashi, H., Mishina, M., Haga, T., Haga, K., Ichiyama, A., Kangawa, K., Kojima, M., Matsuo, H., Hirose, T., and Numa, S. (1986) Cloning, sequencing and expression of complementary DNA encoding the muscarinic acetylcholine receptor. *Nature* **323,** 411–416.
4. Bunzow, J. R., Van Tol, H. H. M., Grandy, D. K., Albert, P., Salon, J., Christie, M., Machida, C. A., Neve, K. A., and Civelli, O. (1988) Cloning and expression of a rat D_2 dopamine receptor cDNA. *Nature* **336,** 783–787.
5. Albert, P. R., Neve, K. A., Bunzow, J. R., and Civelli O. (1990) Coupling of a cloned rat dopamine-D_2 receptor to inhibition of adenylate cyclase and prolactin secretion. *J. Biol. Chem.* **265,** 2098–2104.
6. Neve, K. A., Henningsen, R. A., Bunzow, J. R., and Civelli, O. (1989) Functional characterization of a rat dopamine D-2 receptor cDNA expressed in a mammalian cell line. *Mol. Pharmacol.* **36,** 446–451.
7. Chio, C. L., Hess, G. F., Graham, R. S., and Huff, R. M. (1990) A second molecular form of D_2 dopamine receptor in rat and bovine caudate nucleus. *Nature* **343,** 266–269.
8. Dal Toso, R., Sommer, B., Ewert, M., Herb, A., Pritchett, D. B., Bach, A., Shivers, B. D., and Seeburg, P. H. (1989) The dopamine D_2 receptor: two molecular forms generated by alternative splicing. *EMBO J.* **8,** 4025–4034.
9. Giros, B., Sokoloff, P., Martres, M.-P., Riou, J.-F., Emorine, L. J., and Schwartz, J.-C. (1989) Alternative splicing directs the expression of two D_2 dopamine receptor isoforms. *Nature* **342,** 923–926.
10. Grandy, D. K., Marchionni, M. A., Makam, H., Stofko, R. E., Alfano, M., Frothingham, L., Fischer, J. B., Burke-Howie, K. J., Bunzow, J. R., Server, A. C., and Civelli, O. (1989) Cloning of the cDNA and gene for a human D_2 dopamine receptor. *Proc. Natl. Acad. Sci. USA* **86,** 9762–9766.
11. Miller, J. C., Wang, Y., and Filer, D. (1990) Identification by sequence analysis of a second rat brain cDNA encoding the dopamine (D2) receptor. *Biochem. Biophys. Res. Commun.* **166,** 109–112.
12. Monsma, F. J., McVittie, L. D., Gerfen, C. R., Mahan, L. C., and Sibley, D. R. (1989) Multiple D_2 dopamine receptors produced by alternative RNA splicing. *Nature* **342,** 926–929.
13. O'Malley, K. L., Mack, K. J., Gandelman, K.-Y., and Todd, R. D. (1990) Organization and expression of the rat $D2_A$ receptor gene: identification of alternative transcripts and a variant donor splice site. *Biochemistry* **29,** 1367–1371.
14. Rao, D. D., McKelvy, J., Kebabian, J., and MacKenzie, R. G. (1990) Two forms of the rat D_2 dopamine receptor as revealed by the polymerase chain reaction. *FEBS Lett.* **263,** 18–22.
15. Selbie, L. A., Hayes, G., and Shine, J. (1989) The major dopamine D2 receptor: molecular analysis of the human $D2_A$ subtype. *DNA* **8,** 683–689.
16. Neve, K. A., Neve, R. L., Fidel, S., Janowsky, A., and Higgins, G. A. (1991) Increased abundance of alternatively spliced forms of D-2 receptor mRNA after denervation. *Proc. Natl. Acad. Sci. USA* **88,** 2802–2806.

17. Snyder, L. A., Roberts, J. L., and Sealfon, S. C. (1991) Distribution of dopamine D_2 receptor mRNA splice variants in the rat by solution hybridization/protection assay. *Neurosci. Lett.* **122,** 37–40.
18. Leysen, J. E., Gommeren, W., Mertens, J., Luyten, W. H. M. L., Pauwels, P. J., Ewert, M., and Seeburg, P. (1993) Comparison of in vitro binding properties of a series of dopamine antagonists and agonists for cloned human dopamine D_{2S} and D_{2L} receptors and for D_2 receptors in rat striatal and mesolimbic tissues, using [^{125}I] 2'-iodospiperone. *Psychopharmacology* **110,** 27–36.
19. Castro, S. W. and Strange, P. G. (1993) Differences in the ligand binding properties of the short and long versions of the D2 dopamine receptor. *J. Neurochem.* **60,** 372–375.
20. Malmberg, Å., Jackson, D. M., Eriksson, A., and Mohell, N. (1993) Unique binding characteristics of antipsychotic agents interacting with human dopamine D_{2A}, D_{2B}, and D_3 receptors. *Mol. Pharmacol.* **43,** 749–754.
21. Castro, S. W. and Strange, P. G. (1993) Coupling of D_2 and D_3 dopamine receptors to G-proteins. *FEBS Lett.* **315,** 223–226.
22. Falardeau, P. (1994) Functional distinctions of dopamine D_2long and D_2short receptors, in *Dopamine Receptors and Transporters* (Niznik, H. B., ed.), Dekker, New York, pp. 323–342.
23. Montmayeur, J. P., Guiramand, J., and Borrelli, E. (1993) Preferential coupling between dopamine D2 receptors and G-proteins. *Mol. Endocrinol.* **7,** 161–170.
24. Hayes, G., Biden, T. J., Selbie, L. A., and Shine, J. (1992) Structural subtypes of the dopamine D2 receptor are functionally distinct: expression of the cloned $D2_A$ and $D2_B$ subtypes in a heterologous cell line. *Mol. Endocrinol.* **6,** 920–926.
25. Montmayeur, J.-P. and Borrelli, E. (1991) Transcription mediated by a cAMP-responsive promoter element is reduced upon activation of dopamine D_2 receptors. *Proc. Natl. Acad. Sci. USA* **88,** 3135–3139.
26. Starr, S., Kozell, L. B., and Neve, K. A. (1995) Drug-induced proliferation of dopamine D2 receptors on cultured cells. *J. Neurochem.* **65,** 569–577.
27. Zhang, L.-J., Lachowicz, J. E., and Sibley, D. R. (1994) The D_{2S} and D_{2L} dopamine receptor isoforms are differentially regulated in Chinese hamster ovary cells. *Mol. Pharmacol.* **45,** 878–889.
28. Liu, Y. F., Jakobs, K. H., Rasenick, M. M., and Albert, P. R. (1994) G protein specificity in receptor-effector coupling. Analysis of the roles of G_o and G_i2 in GH4C1 pituitary cells. *J. Biol. Chem.* **269,** 13,880–13,886.
29. Senogles, S. E. (1994) The D2 dopamine receptor isoforms signal through distinct $G_{i\alpha}$ proteins to inhibit adenylyl cyclase. A study with site-directed mutant $G_{i\alpha}$ proteins. *J. Biol. Chem.* **269,** 23,120–23,127.
30. Castellano, M. A., Liu, L.-X., Monsma, F. J., Jr., Sibley, D. R., Kapatos, G., and Chiodo, L. A. (1993) Transfected D_2 short dopamine receptors inhibit voltage-dependent potassium current in neuroblastoma x glioma hybrid (NG108-15) cells. *Mol. Pharmacol.* **44,** 649–656.
31. Liu, L.-X., Monsma, F. J., Sibley, D. R., and Chiodo, L. A. (1994) D_{2S} and D_{2L} receptors couple to K^+ currents in NG108-15 cells via different signal transduction pathways. *Soc. Neurosci. Abstr.* **20,** 523.
32. Stormann, T. M., Gdula, D. C., Weiner, D. M., and Brann, M. R. (1990) Molecular cloning and expression of a dopamine D2 receptor from human retina. *Mol. Pharmacol.* **37,** 1–6.

33. Mack, K. J., Todd, R. D., and O'Malley, K. L. (1991) The mouse dopamine $D2_A$ receptor gene: sequence homology with the rat and human genes and expression of alternative transcripts. *J. Neurochem.* **57,** 795–801.
34. Montmayeur, J. P., Bausero, P., Amlaiky, N., Maroteaux, L., Hen, R., and Borrelli, E. (1991) Differential expression of the mouse D_2 dopamine receptor isoforms. *FEBS Lett.* **278,** 239–243.
35. Martens, G. J. M., Molhuizen, H. O. F., Gröneveld, D., and Roubos, E. W. (1991) Cloning and sequence analysis of brain cDNA encoding a *Xenopus* D_2 dopamine receptor. *FEBS Lett.* **281,** 85–89.
36. Martens, G. J. M., Groenen, P. M. A., Gröneveld, D., and Van Riel, M. C. H. M. (1993) Expression of the *Xenopus* D_2 dopamine receptor: tissue-specific regulation and two transcriptionally active genes but no evidence for alternative splicing. *Eur. J. Biochem.* **213,** 1349–1354.
37. O'Dowd, B. F., Nguyen, T., Tirpak, A., Jarvie, K. R., Israel, Y., Seeman, P., and Niznik, H. B. (1990) Cloning of two additional catecholamine receptors from rat brain. *FEBS Lett.* **262,** 8–12.
38. Dearry, A., Gingrich, J. A., Falardeau, P., Fremeau, R. T., Bates, M. D., and Caron, M. G. (1990) Molecular cloning and expression of the gene for a human D_1 dopamine receptor. *Nature* **347,** 72–75.
39. Monsma, F. J., Mahan, L. C., McVittie, L. D., Gerfen, C. R., and Sibley, D. R. (1990) Molecular cloning and expression of a D_1 dopamine receptor linked to adenylyl cyclase activation. *Proc. Natl. Acad. Sci. USA* **87,** 6723–6727.
40. Sunahara, R. K., Niznik, H. B., Weiner, D. M., Stormann, T. M., Brann, M. R., Kennedy, J. L., Gelernter, J. E., Rozmahel, R., Yang, Y., Israel, Y., Seeman, P., and O'Dowd, B. F. (1990) Human dopamine D_1 receptor encoded by an intronless gene on chromosome 5. *Nature* **347,** 80–83.
41. Zhou, Q.-Y., Grandy, D. K., Thambi, L., Kushner, J. A., Van Tol, H. H. M., Cone, R., Pribnow, D., Salon, J., Bunzow, J. R., and Civelli, O. (1990) Cloning and expression of human and rat D_1 dopamine receptors. *Nature* **347,** 76–80.
42. Kebabian, J. W. and Calne, D. B. (1979) Multiple receptors for dopamine. *Nature (Lond.)* **277,** 93–96.
43. Spano, P. F., Govoni, S., and Trabucchi, M. (1978) Studies on the pharmacological properties of dopamine receptors in various areas of the central nervous system. *Adv. Biochem. Psychopharm.* **19,** 155–165.
44. Zhou, W., Flanagan, C., Ballesteros, J. A., Konvicka, K., Davidson, J. S., Weinstein, H., Millar, R. P., and Sealfon, S. C. (1994) A reciprocal mutation supports helix 2 and helix 7 proximity in the gonadotropin-releasing hormone receptor. *Mol. Pharmacol.* **45,** 165–170.
45. Nash, S. R., Godinot, N., and Caron, M. G. (1993) Cloning and characterization of the opossum kidney cell D1 dopamine receptor: expression of identical D1A and D1B dopamine receptor mRNAs in opossum kidney and brain. *Mol. Pharmacol.* **44,** 918–925.
46. Machida, C. A., Searles, R. P., Nipper, V., Brown, J. A., Kozell, L. B., and Neve, K. A. (1992) Molecular cloning and expression of the rhesus macaque D1 dopamine receptor gene. *Mol. Pharmacol.* **41,** 652–659.
47. Grenader, A. C., O'Rourke, D. A., and Healy, D. P. (1995) Cloning of the porcine D_{1A} dopamine receptor gene expressed in renal epithelial LLC-PK_1 cells. *Am. J. Physiol.* **268,** F423–F434.

48. Frail, D. E., Manelli, A. M., Witte, D. G., Lin, C. W., Steffey, M. E., and MacKenzie, R. G. (1993) Cloning and characterization of a truncated dopamine D1 receptor from goldfish retina: stimulation of cyclic AMP production and calcium mobilization. *Mol. Pharmacol.* **44,** 1113–1118.
49. Gotzes, F., Balfanz, S., and Baumann, A. (1994) Primary structure and functional characterization of a *Drosophila* dopamine receptor with high homology to human $D_{1/5}$ receptors. *Receptors Channels* **2,** 131–141.
50. Sugamori, K. S., Demchyshyn, L. L., Chung, M., and Niznik, H. B. (1994) D_{1A}, D_{1B}, and D_{1C} dopamine receptors from *Xenopus laevis*. *Proc. Natl. Acad. Sci. USA* **91,** 10,536–10,540.
51. Demchyshyn, L. L., Sugamori, K. S., Lee, F. J. S., Hamadanizadeh, S. A., and Niznik, H. B. (1995) The dopamine D1D receptor. Cloning and characterization of three pharmacologically distinct D1-like receptors from *Gallus domesticus*. *J. Biol. Chem.* **270,** 4005–4012.
52. Sokoloff, P., Giros, B., Martres, M.-P., Bouthenet, M.-L., and Schwartz, J.-C. (1990) Molecular cloning and characterization of a novel dopamine receptor (D_3) as a target for neuroleptics. *Nature* **347,** 146–151.
53. Bouthenet, M.-L., Souil, E., Martres, M.-P., Sokoloff, P., Giros, B., and Schwartz, J.-C. (1991) Localization of dopamine D_3 receptor mRNA in the rat brain using in situ hybridization histochemistry: comparison with dopamine D_2 receptor mRNA. *Brain Res.* **564,** 203–219.
54. Snyder, L. A., Roberts, J. L., and Sealfon, S. C. (1991) Alternative transcripts of the rat and human dopamine D3 receptor. *Biochem. Biophys. Res. Commun.* **180,** 1031–1035.
55. Sokoloff, P., Andrieux, M., Besançon, R., Pilon, C., Martres, M.-P., Giros, B., and Schwartz, J.-C. (1992) Pharmacology of human dopamine D_3 receptor expressed in a mammalian cell line: comparison with D_2 receptor. *Eur. J. Pharmacol.–Mol. Pharmacol.* **225,** 331–337.
56. Fishburn, C. S., Belleli, D., David, C., Carmon, S., and Fuchs, S. (1993) A novel short isoform of the D_3 dopamine receptor generated by alternative splicing in the third cytoplasmic loop. *J. Biol. Chem.* **268,** 5872–5878.
57. Landwehrmeyer, B., Mengod, G., and Palacios, J. M. (1993) Differential visualization of dopamine D_2 and D_3 receptor sites in rat brain. A comparative study using in situ hybridization histochemistry and ligand binding autoradiography. *Eur. J. Neurosci.* **5,** 145–153.
58. Levant, B., Grigoriadis, D. E., and DeSouza, E. B. (1993) [^{3}H]Quinpirole binding to putative D_2 and D_3 dopamine receptors in rat brain and pituitary gland: a quantitative autoradiographic study. *J. Pharmacol. Exp. Ther.* **264,** 991–1001.
59. Murray, A. M., Ryoo, H., and Joyce, J. N. (1992) Visualization of dopamine D_3-like receptors in human brain with [^{125}I]epidepride. *Eur. J. Pharmacol-Mol. Pharmacol.* **227,** 443–445.
60. Freedman, S. B., Patel, S., Marwood, R., Emms, F., Seabrook, G. R., Knowles, M. R., and McAllister, G. (1994) Expression and pharmacological characterization of the human D_3 dopamine receptor. *J. Pharmacol. Exp. Ther.* **268,** 417–426.
61. Kula, N. S., Baldessarini, R. J., Kebabian, J. W., and Neumeyer, J. L. (1994) *S*-(+)-aporphines are not selective for human D_3 dopamine receptors. *Cell. Mol. Neurobiol.* **14,** 185–192.

62. MacKenzie, R. G., VanLeeuwen, D., Pugsley, T. A., Shih, Y.-H., Demattos, S., Tang, L., Todd, R. D., and O'Malley, K. L. (1994) Characterization of the human dopamine D_3 receptor expressed in transfected cell lines. *Eur. J. Pharmacol.-Mol. Pharmacol.* **266,** 79–85.
63. Tang, L., Todd, R. D., Heller, A., and O'Malley, K. L. (1994) Pharmacological and functional characterization of D_2, D_3 and D_4 dopamine receptors in fibroblast and dopaminergic cell lines. *J. Pharmacol. Exp. Ther.* **268,** 495–502.
64. Van der Zee, P., Koger, H. S., Gootjes, J., and Hespe, W. (1980) Aryl 1,4-dialk(en)ylpiperazines as selective and very potent inhibitors of dopamine uptake. *Eur. J. Med. Chem.* **15,** 363–370.
65. Watts, V. J., Lawler, C. P., Knoerzer, T., Mayleben, M. A., Neve, K. A., Nichols, D. E., and Mailman, R. B. (1993) Hexahydrobenzo[a]phenanthridines: novel dopamine D_3 receptor ligands. *Eur. J. Pharmacol.* **239,** 271–273.
66. Chumpradit, S., Kung, M.-P., and Kung, H. F. (1993) Synthesis and optical resolution of (*R*)- and (S)-*trans*-7-hydroxy-2-[*N*-propyl-*N*-(3'-iodo-2'-propenyl)amino]tetralin: a new D3 dopamine receptor ligand. *J. Med. Chem.* **36,** 4308–4312.
67. Levant, B. and DeSouza, E. B. (1993) Differential pharmacological profile of striatal and cerebellar dopamine receptors labeled by [^{3}H]quinpirole: identification of a discrete population of putative D_3 receptors. *Synapse* **14,** 90–95.
68. Lévesque, D., Diaz, J., Pilon, C., Martres, M. P., Giros, B., Souil, E., Schott, D., Morgat, J. L., Schwartz, J. C., and Sokoloff, P. (1992) Identification, characterization, and localization of the dopamine D_3 receptor in rat brain using 7-[^{3}H]hydroxy-*N,N*-di-*n*-propyl-2-aminotetralin. *Proc. Natl. Acad. Sci. USA* **89,** 8155–8159.
69. Burris, K. D., Filtz, T. M., Chumpradit, S., Kung, M.-P., Foulon, C., Hensler, J. G., Kung, H. F., and Molinoff, P. B. (1994) Characterization of [^{125}I](*R*)-*trans*-7-hydroxy-2-[*N*-propyl-*N*-(3'-iodo-2'-propenyl)amino]tetralin binding to dopamine D3 receptors in rat olfactory tubercle. *J. Pharmacol. Exp. Ther.* **268,** 935–942.
70. Gonzalez, A. M. and Sibley, D. R. (1995) [^{3}H]7-OH-DPAT is capable of labeling dopamine D_2 as well as D_3 receptors. *Eur. J. Pharmacol.* **272,** R1–R3.
71. Schoemaker, H. (1993) [^{3}H]7-OH-DPAT labels both dopamine D_3 receptors and σ sites in the bovine caudate nucleus. *Eur. J. Pharmacol.* **242,** R1,R2.
72. Chio, C. L., Lajiness, M. E., and Huff, R. M. (1994) Activation of heterologously expressed D3 dopamine receptors: comparison with D2 dopamine receptors. *Mol. Pharmacol.* **45,** 51–60.
73. Cox, B. A., Rosser, M. P., Kozlowski, M. R., Duwe, K. M., Neve, R. L., and Neve, K. A. (1995) Regulation and functional characterization of a rat recombinant dopamine D3 receptor. *Synapse* **21,** 1–9.
74. Pilon, C., Lévesque, D., Dimitriadou, V., Griffon, N., Martres, M.-P., Schwartz, J.-C., and Sokoloff, P. (1994) Functional coupling of the human dopamine D_3 receptor in a transfected NG 108-15 neuroblastoma-glioma hybrid cell line. *Eur. J. Pharmacol.-Mol. Pharmacol.* **268,** 129–139.
75. Diaz, J., Lévesque, D., Griffon, N., Lammers, C. H., Martres, M.-P., Sokoloff, P., and Schwartz, J.-C. (1994) Opposing roles for dopamine D_2 and D_3 receptors on neurotensin mRNA expression in nucleus accumbens. *Eur. J. Neurosci.* **6,** 1384–1387.

76. Seabrook, G. R., Kemp, J. A., Freedman, S. B., Patel, S., Sinclair, H. A., and McAllister, G. (1994) Functional expression of human D_3 dopamine receptors in differentiated neuroblastoma X glioma NG108-15 cells. *Br. J. Pharmacol.* **111,** 391–393.
77. Swarzenski, B. C., Tang, L., Oh, Y. J., O'Malley, K. L., and Todd, R. D. (1994) Morphogenic potentials of D_2, D_3, and D_4 dopamine receptors revealed in transfected neuronal cell lines. *Proc. Natl. Acad. Sci. USA* **91,** 649–653.
78. Van Tol, H. H. M., Bunzow, J. R., Guan, H.-C., Sunahara, R. K., Seeman, P., Niznik, H. B., and Civelli, O. (1991) Cloning of the gene for a human dopamine D_4 receptor with high affinity for the antipsychotic clozapine. *Nature* **350,** 610–614.
79. Matsumoto, M., Hidaka, K., Tada, S., Tasaki, Y., and Yamaguchi, T. (1995) Full-length cDNA cloning and distribution of human dopamine D4 receptor. *Mol. Brain Res.* **29,** 157–162.
80. O'Malley, K. L., Harmon, S., Tang, L., and Todd, R. D. (1992) The rat dopamine D_4 receptor: sequence, gene structure, and demonstration of expression in the cardiovascular system. *N. Biol.* **4,** 137–146.
81. Asghari, V., Schoots, O., Van Kats, S., Ohara, K., Jovanovic, V., Guan, H.-C., Bunzow, J. R., Petronis, A., and Van Tol, H. H. M. (1994) Dopamine D4 receptor repeat: analysis of different native and mutant forms of the human and rat genes. *Mol. Pharmacol.* **46,** 364–373.
82. Fishburn, C. S., Carmon, S., and Fuchs, S. (1995) Molecular cloning and characterisation of the gene encoding the murine D_4 dopamine receptor. *FEBS Lett.* **361,** 215–219.
83. Chio, C. L., Drong, R. F., Riley, D. T., Gill, G. S., Slightom, J. L., and Huff, R. M. (1994) D4 dopamine receptor-mediated signaling events determined in transfected Chinese hamster ovary cells. *J. Biol. Chem.* **269,** 11,813–11,819.
84. McHale, M., Coldwell, M. C., Herrity, N., Boyfield, I., Winn, F. M., Ball, S., Cook, T., Robinson, J. H., and Gloger, I. S. (1994) Expression and functional characterisation of a synthetic version of the human D_4 dopamine receptor in a stable human cell line. *FEBS Lett.* **345,** 147–150.
85. Mills, A., Allet, B., Bernard, A., Chabert, C., Brandt, E., Cavegn, C., Chollet, A., and Kawashima, E. (1993) Expression and characterization of human D4 dopamine receptors in baculovirus-infected insect cells. *FEBS Lett.* **320,** 130–134.
86. Van Tol, H. H. M., Wu, C. M., Guan, H.-C., Ohara, K., Bunzow, J. R., Civelli, O., Kennedy, J., Seeman, P., Niznik, H. B., and Jovanovic, V. (1992) Multiple dopamine D4 receptor variants in the human population. *Nature* **358,** 149–152.
87. Kane, J., Honigfeld, G., Singer, J., and Meltzer, H. (1988) Clozapine for the treatment-resistant schizophrenic. *Arch. Gen. Psychiat.* **45,** 789–796.
88. Casey, D. E. (1989) Clozapine: neuroleptic-induced EPS and tardive dyskinesia. *Psychopharmacology* **99,** S47–S53.
89. Coward, D. M. (1992) General pharmacology of clozapine. *Br. J. Psychiatry* **160,** 5–11.
90. Meltzer, H. Y., Matsubara, S., and Lee, J. C. (1989) Classification of typical and atypical antipsychotic drugs on the basis of dopamine D-1, D-2 and serotonin$_2$ pK_i values. *J. Pharmacol. Exper. Ther.* **251,** 238–246.
91. Monsma, F. J., Shen, Y., Ward, R. P., Hamblin, M. W., and Sibley, D. R. (1993) Cloning and expression of a novel serotonin receptor with high affinity for tricyclic psychotropic drugs. *Mol. Pharmacol.* **43,** 320–327.

92. Shen, Y., Monsma, F. J., Jr., Metcalf, M. A., Jose, P. A., Hamblin, M. W., and Sibley, D. R. (1993) Molecular cloning and expression of a 5-hydroxytryptamine$_7$ serotonin receptor subtype. *J. Biol. Chem.* **268,** 18,200–18,204.
93. Lawson, C. F., Mortimore, R. A., Schlachter, S. K., and Smith, M. W. (1994) Pharmacology of a human dopamine D_4 receptor expressed in HEK293 cells. *Methods Find. Exp. Clin. Pharmacol.* **16,** 303–307.
94. Lichter, J. B., Barr, C. L., Kennedy, J. L., Van Tol, H. H. M., Kidd, K. K., and Livak, K. J. (1993) A hypervariable segment in the human dopamine receptor D_4 (DRD4) gene. *Hum. Mol. Genet.* **6,** 767–773.
95. Asghari, V., Sanyal, S., Buchwaldt, S., Paterson, A., Jovanovic, V., and Van Tol, H. H. M. (1995) Modulation of intracellular cyclic AMP levels by different human dopamine D4 receptor variants. *J. Neurochem.* **65,** 1157–1165.
96. Murray, A. M., Hyde, T. M., Knable, M. B., Herman, M. M., Bigelow, L. B., Carter, J. M., Weinberger, D. R., and Kleinman, J. E. (1995) Distribution of putative D4 dopamine receptors in postmortem striatum from patients with schizophrenia. *J. Neurosci.* **15,** 2186–2191.
97. Seeman, P., Guan, H.-C., and Van Tol, H. H. M. (1993) Dopamine D4 receptors elevated in schizophrenia. *Nature* **365,** 441–445.
98. Sumiyoshi, T., Stockmeier, C. A., Overholser, J. C., Thompson, P. A., and Meltzer, H. Y. (1994) Dopamine D_4 receptors and effects of guanine nucleotides on [^{3}H]raclopride binding in postmortem caudate nucleus of subjects with schizophrenia or major depression. *Brain Res.* **681,** 109–116.
99. Cohen, A. I., Todd, R. D., Harmon, S., and O'Malley, K. L. (1992) Photoreceptors of mouse retinas possess D_4 receptors coupled to adenylate cyclase. *Proc. Natl. Acad. Sci. USA* **89,** 12,093–12,097.
100. Zawilska, J. B., Derbiszewska, T., and Nowak, J. Z. (1994) Clozapine and other neuroleptic drugs antagonize the light-evoked suppression of melatonin biosynthesis in chick retina: involvement of the D4-like dopamine receptor. *J. Neural Transm.* **97,** 107–117.
101. Zawilska, J. B. and Nowak, J. Z. (1994) Does D_4 dopamine receptor mediate the inhibitory effect of light on melatonin biosynthesis in chick retina. *Neurosci. Lett.* **166,** 203–206.
102. Chabert, C., Cavegn, C., Bernard, A., and Mills, A. (1994) Characterization of the functional activity of dopamine ligands at human recombinant dopamine D_4 receptors. *J. Neurochem.* **63,** 62–65.
103. Sunahara, R. K., Guan, H.-C., O'Dowd, B. F., Seeman, P., Laurier, L. G., Ng, G., George, S. R., Torchia, J., Van Tol, H. H. M., and Niznik, H. B. (1991) Cloning of the gene for a human dopamine D_5 receptor with higher affinity for dopamine than D_1. *Nature* **350,** 614–619.
104. Grandy, D. K., Zhang, Y., Bouvier, C., Zhou, Q.-Y., Johnson, R. A., Allen, L., Buck, K., Bunzow, J. R., Salon, J., and Civelli, O. (1991) Multiple human D_5 receptor genes: a functional receptor and two pseudogenes. *Proc. Natl. Acad. Sci. USA* **88,** 9175–9179.
105. Tiberi, M., Jarvie, K. R., Silvia, C., Falardeau, P., Gingrich, J. A., Godinot, N., Bertrand, L., Yang-Feng, T. L., Fremeau, R. T., and Caron, M. G. (1991) Cloning, molecular characterization, and chromosomal assignment of a gene encoding a second D_1 dopamine receptor subtype: differential expression pattern in rat brain compared with the D_{1A} receptor. *Proc. Natl. Acad. Sci. USA* **88,** 7491–7495.

106. Weinshank, R. L., Adham, N., Macchi, M., Olsen, M. A., Branchek, T. A., and Hartig, P. R. (1991) Molecular cloning and characterization of a high affinity dopamine receptor ($D_{1\beta}$) and its pseudogene. *J. Biol. Chem.* **266,** 22,427–22,435.
107. Pedersen, U. B., Norby, B., Jensen, A. A., Schiødt, M., Hansen, A., Suhr-Jessen, P., Scheideler, M., Thastrup, O., and Andersen, P. H. (1994) Characteristics of stably expressed human dopamine D_{1a} and D_{1b} receptors: atypical behavior of the dopamine D_{1b} receptor. *Eur. J. Pharmacol.-Mol. Pharmacol.* **267,** 85–93.
108. Jarvie, K. R., Tiberi, M., Silvia, C., Gingrich, J. A., and Caron, M. G. (1993) Molecular cloning, stable expression and desensitization of the human dopamine D1B/D5 receptor. *J. Recept. Res.* **13,** 573–590.
109. Laurier, L. G., O'Dowd, B. F., and George, S. R. (1994) Heterogeneous tissue-specific transcription of dopamine receptor subtype messenger RNA in rat brain. *Mol. Brain Res.* **25,** 344–350.
110. Boyson, S. J., McGonigle, P., and Molinoff, P. B. (1986) Quantitative autoradiographic localization of the D_1 and D_2 subtypes of dopamine receptors in rat brain. *J. Neurosci.* **6,** 3177–3188.
111. Tiberi, M. and Caron, M. G. (1994) High agonist-independent activity is a distinguishing feature of the dopamine D1B receptor subtype. *J. Biol. Chem.* **269,** 27,925–27,931.
112. Pei, G., Samama, P., Lohse, M., Wang, M., Codina, J., and Lefkowitz, R. J. (1994) A constitutively active mutant β_2-adrenergic receptor is constitutively desensitized and phosphorylated. *Proc. Natl. Acad. Sci. USA* **91,** 2699–2702.
112a. Seeman, P. (1992) Dopamine receptor sequences: therapeutic levels of neuroleptics occupy D2 receptors, clozapine occupies D4. *Neuropsychopharmacology* **7,** 261–284.
113. Kennelly, P. J. and Krebs, E. G. (1991) Consensus sequences as substrate specificity determinants for protein kinases and protein phosphatases. *J. Biol. Chem.* **266,** 15,555–15,558.
114. Bouvier, M., Collins, S., O'Dowd, B. F., Campbell, P. T., De Blasi, A., Kobilka, B. K., MacGregor, C., Irons, G. P., Caron, M. G., and Lefkowitz, R. J. (1989) Two distinct pathways for cAMP-mediated down-regulation of the β_2-adrenergic receptor: phosphorylation of the receptor and regulation of its mRNA level. *J. Biol. Chem.* **264,** 16,786–16,792.
115. Fryxell, K. J. (1994) The evolution of the dopamine receptor gene family, in *Dopamine Receptors and Transporters* (Niznik, H. B., ed.), Dekker, New York, pp. 237–260.
116. Gandelman, K.-Y., Harmon, S., Todd, R. D., and O'Malley, K. L. (1991) Analysis of the structure and expression of the human dopamine $D2_A$ receptor gene. *J. Neurochem.* **56,** 1024–1029.
117. Eubanks, J. H., Djabali, M., Selleri, L., Grandy, D. K., Civelli, O., McElligott, D. L., and Evans, G. A. (1992) Structure and linkage of the D2 dopamine receptor and neural cell adhesion molecule genes on human chromosome 11q23. *Genomics* **14,** 1010–1018.
118. Minowa, T., Minowa, M. T., and Mouradian, M. M. (1992) Analysis of the promoter region of the rat D_2 dopamine receptor gene. *Biochemistry* **31,** 8389–8396.
119. Valdenaire, O., Vernier, P., Maus, M., Dumas Milne Edwards, J.-B., and Mallet, J. (1994) Transcription of the rat dopamine-D2-receptor gene from two promoters. *Eur. J. Biochem.* **220,** 577–584.

120. Minowa, T., Minowa, M. T., and Mouradian, M. M. (1994) Negative modulator of the rat D_2 dopamine receptor gene. *J. Biol. Chem.* **269,** 11,656–11,662.
121. Grandy, D. K., Litt, M., Allen, L., Bunzow, J. R., Marchionni, M., Makam, H., Reed, L., Magenis, R. E., and Civelli, O. (1989) The human dopamine D2 receptor gene is located on chromosome 11 at q22-q23 and identifies a TaqI RFLP. *Am. J. Hum. Genet.* **45,** 778–785.
122. Grandy, D. K., Zhou, Q.-Y., Allen, L., Litt, R., Magenis, R. E., Civelli, O., and Litt, M. (1990) A human D_1 dopamine receptor gene is located on chromosome 5 at q35.1 and identifies an EcoRI RFLP. *Am. J. Hum. Genet.* **47,** 828–834.
123. Litt, M., al-Dhalimy, M., Zhou, Q.-Y., Grandy, D., and Civelli, O. (1991) A TaqI RFLP at the DRD1 locus. *Nucleic Acids Res.* **19,** 3161.
124. Ohara, K., Ulpian, C., Seeman, P., Sunahara, R. K., Van Tol, H. H. M., and Niznik, H. B. (1993) Schizophrenia: dopamine D_1 receptor sequence is normal, but has DNA polymorphisms. *Neuropsychopharmacology* **8,** 131–135.
125. Hauge, X. Y., Grandy, D. K., Eubanks, J. H., Evans, G. A., Civelli, O., and Litt, M. (1991) Detection and characterization of additional DNA polymorphisms in the dopamine D2 receptor gene. *Genomics* **10,** 527–530.
126. Suarez, B. K., Parsian, A., Hampe, C. L., Todd, R. D., Reich, T., and Cloninger, C. R. (1994) Linkage disequilibria at the D_2 dopamine receptor locus (DRD2) in alcoholics and controls. *Genomics* **19,** 12–20.
127. Sarkar, G., Kapelner, S., Grandy, D. K., Marchionni, M., Civelli, O., Sobell, J., Heston, L., and Sommer, S. S. (1991) Direct sequencing of the dopamine D_2 receptor (DRD2) in schizophrenics reveals three polymorphisms but no structural change in the receptor. *Genomics* **11,** 8–14.
128. Itokawa, M., Arinami, T., Futamura, N., Hamaguchi, H., and Toru, M. (1993) A structural polymorphism of human dopamine D2 receptor, D2($Ser^{311}\rightarrow Cys$). *Biochem. Biophys. Res. Commun.* **196,** 1369–1375.
129. Seeman, P., Ohara, K., Ulpain, C., Seeman, M. V., Jellinger, K., Van Tol, H. H. M., and Niznik, H. B. (1993) Schizophrenia: normal sequence in the dopamine D_2 receptor region that couples to G-proteins: DNA polymorphisms in D_2. *Neuropsychopharmacology* **8,** 137–142.
130. Bolos, A. M., Dean, M., Lucas-Derse, S., Ramsburg, M., Brown, G. L., and Goldman, D. (1990) Population and pedigree studies reveal a lack of association between the dopamine D2 receptor gene and alcoholism. *JAMA* **264,** 3156–3160.
131. Le Coniat, M., Sokoloff, P., Hillion, J., Martres, M.-P., Giros, B., Pilon, C., Schwartz, J.-C., and Berger, R. (1991) Chromosomal localization of the human D_3 dopamine receptor gene. *Hum. Genet.* **87,** 618–620.
132. Lannfelt, L., Sokoloff, P., Martres, M.-P., Pilon, C., Giros, B., Jönnson, E., Sedval, G., and Schwartz, J. C. (1992) Amino acid substitution in the dopamine D3 receptor as a useful polymorphism for investigating psychiatric disorders. *Psychiat. Genet.* **2,** 249–256.
133. Sabaté, O., Campion, D., d'Amato, T., Martres, M. P., Sokoloff, P., Giros, G., et al. (1994) Failure to find evidence for linkage or association between the dopamine D_3 receptor gene and schizophrenia. *Am. J. Psychiatry* **151,** 107–111.
134. Gelernter, J., Kennedy, J. L., Van Tol, H. H. M., Civelli, O., and Kidd, K. K. (1992) The D4 dopamine receptor (DRD4) maps to distal 11p close to HRAS. *Genomics* **13,** 208–210.

135. Petronis, A., Van Tol, H. H. M., Lichter, J. B., Livak, K. J., and Kennedy, J. L. (1993) The D4 dopamine receptor gene maps on 11p proximal to HRAS. *Genomics* **18,** 161–163.
136. Petronis, A., Van Tol, H. H. M., and Kennedy, J. L. (1994) A *Sma*I PCR-RFLP in the 5' noncoding region of the human D4 dopamine receptor gene (DRD4). *Hum. Hered.* **44,** 58–60.
137. Kennedy, J. L., Sidenberg, D. G., Van Tol, H. H. M., and Kidd, K. K. (1991) A HincII RFLP in the human D4 dopamine receptor locus (DRD4). *Nucleic Acids Res.* **19,** 5801.
138. Catalano, M., Nobile, M., Novelli, E., Nöthen, M. M., and Smeraldi, E. (1993) Distribution of a novel mutation in the first exon of the human dopamine D4 receptor gene in psychotic patients. *Biol. Psychiat.* **34,** 459–464.
139. Petronis, A., O'Hara, K., Barr, C. L., Kennedy, J. L., and Van Tol, H. H. M. (1994) $(G)_n$-Mononucleotide polymorphism in the human D4 dopamine receptor (DRD4) gene. *Hum. Genet.* **93,** 719.
140. Nöthen, M. M., Cichon, S., Hemmer, S., Hebebrand, J., Remschmidt, H., Lehmkuhl, G., Poustka, F., Schmidt, M., Catalano, M., Fimmers, R., Körner, J., Rietschel, M., and Propping, P. (1994) Human dopamine D4 receptor gene: frequent occurrence of a null allele and observation of homozygosity. *Hum. Mol. Genet.* **3,** 2207–2212.
141. Polymeropoulos, M. H., Xiao, H., and Merril, C. R. (1991) The human D5 dopamine receptor (DRD5) maps on chromosome 4. *Genomics* **11,** 777,778.
142. Grandy, D. K., Allen, L. J., Yuan, Z., Magenis, R. E., and Civelli, O. (1992) Chromosomal localization of three human D5 dopamine receptor genes. *Genomics* **13,** 968–973.
143. Sommer, S. S., Lind, T. J., Heston, L. L., and Sobell, J. L. (1993) Dopamine D_4 receptor variants in unrelated schizophrenic cases and controls. *Am. J. Med. Genet.* **48,** 90–93.
144. Eubanks, J. H., Altherr, M., Wagner-McPherson, C., McPherson, J. D., Wasmuth, J. J., and Evans, G. A. (1992) Localization of the D5 dopamine receptor gene to human chromosome 4p15.1–p15.3, centromeric to the Huntington's disease locus. *Genomics* **12,** 510–516.
145. Sherrington, R., Mankoo, B., Attwood, J., Kalsi, G., Curtis, D., Buetow, K., Povey, S., and Gurling, H. (1993) Cloning of the human dopamine D5 receptor gene and identification of a highly polymorphic microsatellite for the DRD5 locus that shows tight linkage to the chromosome 4p reference marker RAF1P1. *Genomics* **18,** 423–425.
146. Giros, B., Martres, M.-P., Pilon, C., Sokoloff, P., and Schwartz, J.-C. (1991) Shorter variants of the D_3 dopamine receptor produced through various patterns of alternative splicing. *Biochem. Biophys. Res. Commun.* **176,** 1584–1592.
147. Park, B.-H., Fishburn, C. S., Carmon, S., Accili, D., and Fuchs, S. (1995) Structural organization of the murine D_3 dopamine receptor gene. *J. Neurochem.* **64,** 482–486.
148. Fu, D., Skryabin, B., Brosius, J., and Robakis, N. K. (1993) Detection of an additional intron in the dopamine D_3 receptor gene. *Soc. Neurosci. Abstr.* **19,** 735.
149. Nagai, Y., Ueno, S., Saeki, Y., Soga, F., and Yanagihara, T. (1993) Expression of the D3 dopamine receptor gene and a novel variant transcript generated by alternative splicing in human peripheral blood lymphocytes. *Biochem. Biophys. Res. Commun.* **194,** 368–374.

150. Pagliusi, S., Chollet-Daemerius, A., Losberger, C., Mills, A., and Kawashima, E. (1993) Characterization of a novel exon within the D3 receptor gene giving rise to an mRNA isoform expressed in rat brain. *Biochem. Biophys. Res. Commun.* **194,** 465–471.
151. Livak, K. J., Rogers, J., and Lichter, J. B. (1995) Variability of dopamine D_4 receptor (DRD4) gene sequence within and among nonhuman primate species. *Proc. Natl. Acad. Sci. USA* **92,** 427–431.
152. Matsumoto, M., Hidaka, K., Tada, S., Tasaki, Y., and Yamaguchi, T. (1995) Polymorphic tandem repeats in dopamine D4 receptor are spread over primate species. *Biochem. Biophys. Res. Commun.* **207,** 467–475.
153. Macciardi, F., Petronis, A., Van Tol, H. H. M., Marino, C., Cavallini, M. C., Smeraldi, E., and Kennedy, J. L. (1994) Analysis of the D_4 dopamine receptor gene variant in an Italian schizophrenia kindred. *Arch. Gen. Psychiatry* **51,** 288–293.
154. Shaikh, S., Collier, D., Kerwin, R. W., Pilowsky, L. S., Gill, M., Xu, W.-M., and Thornton, A. (1993) Dopamine D4 receptor subtypes and response to clozapine. *Lancet* **341,** 116.
155. Zhou, Q.-Y., Li, C., and Civelli, O. (1992) Characterization of gene organization and promoter region of the rat dopamine D1 receptor gene. *J. Neurochem.* **59,** 1875–1883.
156. Minowa, M. T., Minowa, T., Monsma, F. J., Sibley, D. R., and Mouradian, M. M. (1992) Characterization of the 5' flanking region of the human D_{1A} dopamine receptor gene. *Proc. Natl. Acad. Sci. USA* **89,** 3045–3049.
157. Minowa, M. T., Minowa, T., and Mouradian, M. M. (1993) Activator region analysis of the human D_{1A} dopamine receptor gene. *J. Biol. Chem.* **268,** 23,544–23,551.
158. Sommer, S. S., Sobell, J. L., and Heston, L. L. (1993) A common exonic polymorphism in the human D_5 dopamine receptor gene. *Hum. Genet.* **92,** 633,634.
159. Dahl, S. G. and Edvardsen, Ø. (1994) Molecular modeling of dopamine receptors, in *Dopamine Receptors and Transporters* (Niznik, H. B., ed.), Dekker, New York, pp. 265–282.
160. Malmberg, Å., Nordvall, G., Johansson, A. M., Mohell, N., and Hacksell, U. (1994) Molecular basis for the binding of 2-aminotetralins to human dopamine D_{2A} and D_3 receptors. *Mol. Pharmacol.* **46,** 299–312.
161. Moereels, H. and Leysen, J. E. (1993) Novel computational model for the interaction of dopamine with the D_2 receptor. *Receptors Channels* **1,** 89–97.
162. Teeter, M. M., Froimowitz, M., Stec, B., and DuRand, C. J. (1994) Homology modeling of the dopamine D_2 receptor and its testing by docking of agonists and tricyclic antagonists. *J. Med. Chem.* **37,** 2874–2888.
163. Strader, C. D., Sigal, I. S., and Dixon, R. A. F. (1989) Structural basis of β-adrenergic receptor function. *FASEB J.* **3,** 1825–1832.
164. Probst, W. C., Snyder, L. A., Schuster, D. I., Brosius, J., and Sealfon, S. C. (1992) Sequence alignment of the G-protein coupled receptor superfamily. *DNA Cell Biol.* **11,** 1–20.
165. Cox, B. A., Henningsen, R. A., Spanoyannis, A., Neve, R. L., and Neve, K. A. (1992) Contributions of conserved serine residues to the interactions of ligands with dopamine D2 receptors. *J. Neurochem.* **59,** 627–635.

166. Mansour, A., Meng, F., Meador-Woodruff, J. H., Taylor, L. P., Civelli, O., and Akil, H. (1992) Site-directed mutagenesis of the human dopamine D_2 receptor. *Eur. J. Pharmacol-Mol. Pharmacol.* **227,** 205–214.
167. Woodward, R., Coley, C., Daniell, S., Naylor, L. H., and Strange, P. G. (1996) Investigation of the role of conserved serine residues in the long form of the rat D dopamine receptor using site-directed mutagenesis. *J. Neurochem.* **66,** 394–402.^{2}
168. Neve, K. A and Wiens, B. L. (1995) Four ways of being an agonist: multiple sequence determinants of efficacy at D2 dopamine receptors. *Biochem. Soc. Trans.* **23,** 112–116.
169. Pollock, N. J., Manelli, A. M., Hutchins, C. W., Steffey, M. E., MacKenzie, R. G., and Frail, D. E. (1992) Serine mutations in transmembrane V of the dopamine D1 receptor affect ligand interactions and receptor activation. *J. Biol. Chem.* **267,** 17,780–17,786.
170. Tomic, M., Seeman, P., George, S. R., and O'Dowd, B. F. (1993) Dopamine D1 receptor mutagenesis: role of amino acids in agonist and antagonist binding. *Biochem. Biophys. Res. Commun.* **191,** 1020–1027.
171. Neve, K. A., Henningsen, R. A., Kozell, L. B., Cox, B. A. and Neve, R. L. (1994) Structure–function analysis of dopamine-D_2 receptors by in vitro mutagenesis, in *Dopamine Receptors and Transporters* (Niznik, H. B., ed.), Dekker, New York, pp. 299–321.
172. Neve, K. A., Cox, B. A., Henningsen, R. A., Spanoyannis, A., and Neve, R. L. (1991) Pivotal role for aspartate-80 in the regulation of D2 receptor affinity for drugs and inhibition of adenylyl cyclase. *Mol. Pharmacol.* **39,** 733–739.
173. Hutchins, C. (1994) Three-dimensional models of the D1 and D2 dopamine receptors. *Endocr. J.* **2,** 7–23.
174. Livingstone, C. D., Strange, P. G., and Naylor, L. H. (1992) Molecular modelling of D_2-like dopamine receptors. *Biochem. J.* **287,** 277–282.
175. Williamson, R. A. and Strange, P. G. (1990) Evidence for the importance of a carboxyl group in the binding of ligands to the D2 dopamine receptor. *J. Neurochem.* **55,** 1357–1365.
176. Woodward, R., Daniell, S. J., Strange, P. G., and Naylor, L. H. (1994) Structural studies on D_2 dopamine receptors: Mutation of a histidine residue specifically affects the binding of a subgroup of substituted benzamide drugs. *J. Neurochem.* **62,** 1664–1669.
177. Javitch, J. A., Li, X., Kaback, J., and Karlin, A. (1994) A cysteine residue in the third membrane-spanning segment of the human D_2 dopamine receptor is exposed in the binding-site crevice. *Proc. Natl. Acad. Sci. USA* **91,** 10,355–10,359.
178. Javitch, J. A., Fu, D., Chen, J., and Karlin, A. (1995) Mapping the binding-site crevice of the dopamine D2 receptor by the substituted-cysteine accessibility method. *Neuron* **14,** 825–831.
179. Guiramand, J., Montmayeur, J.-P., Ceraline, J., Bhatia, M., and Borrelli, E. (1995) Alternative splicing of the dopamine D2 receptor directs specificity of coupling to G-proteins. *J. Biol. Chem.* **270,** 7354–7358.
180. MacKenzie, R. G., Steffey, M. E., Manelli, A. M., Pollock, N. J., and Frail, D. E. (1993) A D1/D2 chimeric dopamine receptor mediates a D1 response to a D2-selective agonist. *FEBS Lett.* **323,** 59–62.
181. Kozell, L. B., Machida, C. A., Neve, R. L., and Neve, K. A. (1994) Chimeric D1/D2 dopamine receptors: distinct determinants of selective efficacy, potency, and signal transduction. *J. Biol. Chem.* **269,** 30,299–30,306.

182. Kozell, L. B. and Neve, K. A. (1995) Determinants of dopamine receptor function. *Soc. Neurosci. Abstr.* **21,** 621.
183. McAllister, G., Knowles, M. R., Patel, S., Marwood, R., Emms, F., Seabrook, G. R., Graziano, M., Borkowski, D., Hey, P. J., and Freedman, S. B. (1993) Characterisation of a chimeric hD_3/D_2 dopamine receptor expressed in CHO cells. *FEBS Lett.* **324,** 81–86.
184. Robinson, S. W., Jarvie, K. R., and Caron, M. G. (1994) High affinity agonist binding to the dopamine D_3 receptor: chimeric receptors delineate a role for intracellular domains. *Mol. Pharmacol.* **46,** 352–356.
185. Lachowicz, J. E. and Sibley, D. R. (1994) Dopamine-mediated inhibition of adenylyl cyclase by chimeric D_2/D_3 receptors with partial D2 third cytoplasmic loops. *Soc. Neurosci. Abstr.* **20,** 522.
186. Van Leeuwen, D. H., Eisenstein, J., O'Malley, K. L., and MacKenzie, R. G. (1995) Characterization of a chimeric human dopamine D3/D2 receptor that is functionally coupled to adenylyl cyclase in CHO cells. *Mol. Pharmacol.* **48,** 344–351.

CHAPTER 3

Distribution of Dopamine Receptors

Marjorie A. Ariano

1. Introduction

Dopamine neurotransmission has a very broad clinical relevance because of its alteration in several prevalent psychomotor disturbances (*see* Chapter 15 for more details). The most frequently occurring dopaminergic dysfunction is associated with Parkinson's disease, caused by the progressive loss of the dopaminergic neurons of the substantia nigra *(1)*. Although the etiology of affective psychoses, addictions to cocaine or alcohol, and hypertension are less well understood, a large body of clinical and experimental observations document that changes occur in dopamine mechanisms in these disorders *(2,3)*. One of the prevalent treatments for dopaminergic dysfunctions is the use of exogenous dopamine replacement/attenuation therapies in modern neurology and psychiatry to activate/block dopamine receptors. This type of intervention is prescribed whether or not the dopaminergic system has a direct etiology for the disorder, for example, Huntington's disease *(4)*, schizophrenia *(5)*, addiction *(6)*, and stress *(7)* in addition to the well-established regimen of L-DOPA precursor treatment for amelioration of the symptoms associated with Parkinson's disease.

The simple replacement of dopamine in these numerous neurologic disorders has not proved to be a therapeutic panacea. Side effects frequently develop and may become quite debilitating; tardive dyskinesia perhaps representing the most extreme example. As a consequence, alternative treatments specifically designed to lessen side effect development and restore normal dopamine function are still sought. The investigations to elucidate alternative methods to restore dopamine in the nervous system and the periphery encompass all aspects of the dopamine neuron, from genetic regulation in the cell soma to augmentation of transmitter synthesis by stimulation of the biosyn-

The Dopamine Receptors Eds.: K. A. Neve and R. L. Neve
Humana Press Inc., Totowa, NJ

thetic enzymes, attenuation of the catabolism of dopamine, or modification of the dopaminergic transporter at the terminal fields.

Additionally, dopaminergic effects can be modified at the level of dopamine receptors on target cells. This is the rationale for the intensive research into the mechanism of action of the dopamine receptors, and the impetus to determine precise regional and cellular expression patterns throughout the body, but especially within the brain. In this chapter, I present an overview of the different forms of the dopamine receptor and their detection using different experimental methods, and catalog their distribution patterns in the nervous system and body tissues.

2. Historical Perspective and Classification

Dopaminergic effects may be mediated by two principal receptor types: receptors distributed on nondopamine neurons, which are referred to as postsynaptic receptors, and receptors located on dopaminergic neurons, termed autoreceptors. In addition to these two types of receptors segregated by their cellular location, the more numerous postsynaptic receptors can be classed further into two subfamilies based on functional and pharmacological criteria (*8*; *see* Chapter 1 for more details). This scheme recognizes a D1* subfamily which binds substituted benzazepines, is predominantly postsynaptic at the dopamine terminal, and whose responses are transduced by elevations in target cell cyclic adenosine 3',5'-monophosphate (cAMP). Early attempts to determine the distribution of the D1 receptor subfamily capitalized on this biochemical change in the target cell, and measured increases in the cAMP second messenger following activation of this dopamine receptor *(9)* or selective lesions of the striatonigral pathway *(10–12)*. Although indirect, this approach did corroborate the biochemical studies showing cAMP elevations in identified neurons in the central and peripheral nervous system after dopamine application *(13,14)*. Subsequent studies of the D1 receptor subfamily distribution used the substituted benzazepines SKF 38393 and SCH 23390, which became available as radiolabeled ligands in mid-1980 *(15–18)*.

The D2 receptor subfamily is defined by the use of numerous transduction systems (Ca^{2+} mobilization, inositol phosphate metabolism, adenylate cyclase inhibition) and by the fact that these receptors are antagonized by neuroleptic drugs. The regional and cellular localization of this receptor type could not be determined experimentally until radiolabeled neuroleptics became available in mid-1980 *(19)*, since manipulation of target cell transduc-

*As detailed in the Preface, the terms D1 or D1-like and D2 or D2-like are used to refer to the subfamilies of DA receptors, or used when the genetic subtype (D_1 or D_5; D_2, D_3, or D_4) is uncertain.

tion systems did not produce consistent changes that could be detected morphologically. In addition to some postsynaptic dopamine receptors, the less numerous presynaptic autoreceptors exhibit D2 pharmacological profiles *(20,21)*. The function of the autoreceptor is self-regulatory, providing a negative feedback response to increased levels of dopamine at the synapse *(22,23)*. Stimulation of autoreceptors on somata located in the substantia nigra for example decreases their firing rate *(24)*. The reader is referred to Chapter 8 for further discussion of autoreceptors.

In the past 6 years, molecular biologists *(25–30)* have cloned five different dopamine receptors (*see* refs. *31,32* for reviews, and Chapter 2). The multiple clones of the dopamine receptors are divided between the broad D1 and D2 pharmacological subfamilies. These include the D_{1A} and D_{1B} subtypes in the rat or their corresponding D_1 and D_5 homologs in humans *(29,33)*, which are classed into the D1 subfamily. (The D_1/D_5 notation is used in this chapter.) The D2 pharmacologic subfamily exhibits numerous forms, which include the D_2, D_3, and D_4 subtypes. The D_2 and D_3 subtypes exhibit functional RNA-splice variants *(34,35)*. At present, specific ligands have not been developed to distinguish between the different cloned forms in each of the two dopamine receptor subfamilies (*see* Chapter 4 for more details). This has led to the development of alternative experimental strategies to elucidate the distribution and physiological role of each of these cloned forms (*see also* Chapters 9 and 14).

Morphological investigations using ligand binding support a heterogeneous localization pattern for the dopamine receptor subtypes. These data argue for an important role for dopamine in psychomotor activity on the basis of regional receptor distribution patterns *(36–41)*. The messenger RNA (mRNA) encoding these receptors can be detected using *in situ* hybridization histochemistry, which demonstrates widespread patterns of expression of the dopamine receptors throughout the central nervous system, pituitary, retina, and kidney. The dopamine receptor mRNA distribution patterns in the brain supports a role for dopamine in cognition, reward mechanisms, neuroendocrine regulation, and movement patterns *(42–46)*. The reader is referred to Chapter 16 for more information.

3. Localization Methods for Receptor Analysis

The methods to distinguish dopamine receptor distributions within the body are limited morphologically to:

1. Visualization of the two receptor subfamilies by binding to the ligand recognition site;
2. Detection of the mRNA coding the individual clones; or
3. Resolution of the translated receptor proteins using subtype-selective antipeptide antisera and immunohistochemistry.

These techniques historically have relied on the use of radiolabeled analyses of autoradiograms (*see* Section 4.1.) for regional localization; however, recent studies have employed fluorescent labeling (*see* Section 4.2.) to achieve detection at cellular resolution and to colocalize more than one dopamine receptor biosynthetic component. The methods used for receptor analysis are schematized in Fig. 1.

The detection of mRNA abundance using *in situ* hybridization histochemistry for the cloned dopamine receptor subtypes does not show whether the transcript is translated into the encoded protein. Nor does this experimental approach demonstrate the complete pattern of cellular receptor distribution within the dendritic arbors or terminal field areas; it will only detect somata mRNA levels. Additionally, the abundance of an mRNA transcript for a specific dopamine receptor subtype does not always reflect the level of viable ligand binding sites *(40,47–52)*. This mismatch of mRNA level to protein distribution may reflect differential receptor and transcript stability and mRNA translation rates.

An alternative strategy to *in situ* hybridization that has been employed recently to distinguish the cloned dopamine receptor forms is the detection of the native proteins using subtype-selective antipeptide antibodies *(53–58)*. This experimental paradigm enables visualization of individual cloned forms of the native proteins throughout the entire neuron. The sites used as antigens are chosen on the basis of their uniqueness to each receptor clone, as well as their immunogenicity. Thus, antipeptide antisera generated against the native receptor proteins avoid the ligand recognition site and transmembrane-spanning regions which are highly conserved among the five dopamine receptor subtypes. An additional advantage to targeting antibodies to epitopes outside of the ligand recognition site is that application of the antisera to localize the cellular distribution of the receptor does not block the ligand recognition site of the molecule *(40,53,59)*. A caveat of this method however is that receptors which are detected may not be physiologically active, as the antigenic sites distinguished by the antisera may also be present on receptor proteins being synthesized, catabolized, or in the process of transport *(60)*.

The ultimate measure of receptor distribution is elucidation of the ligand recognition site by binding analysis. This is a direct assessment of the pharmacologically responsive receptor in the tissue under study. However, drugs that discriminate among the protein products of different clones within the dopamine D1 and D2 receptor subfamilies have not yet been developed. Thus, determination of the role of the different receptor subtypes in dopaminergic neurotransmission is problematic. Attempts have been made to use one or more ligands to bind to receptor subtypes that have a higher affinity followed by radiolabeled assessment of the subtype in question with a proto-

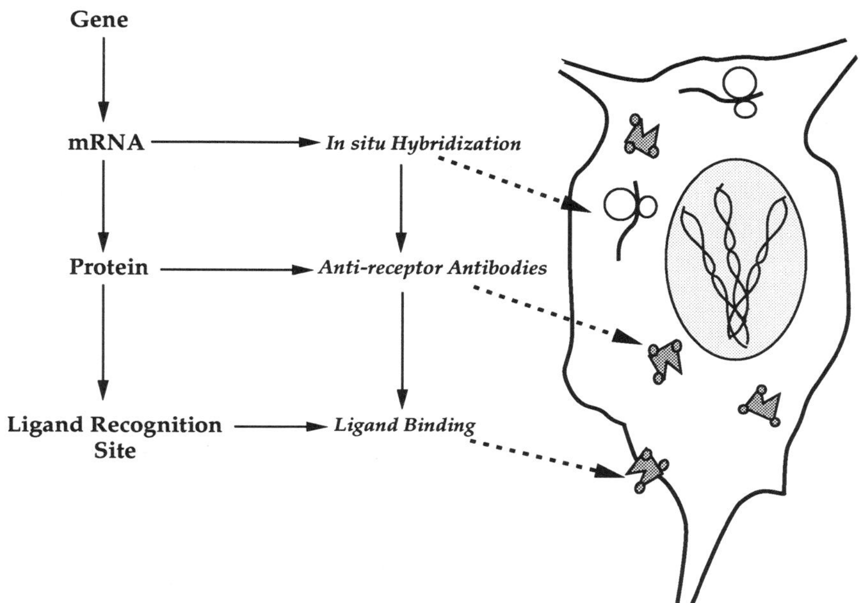

Fig. 1. The different experimental methods that can be employed to elucidate dopamine receptors in various tissues are drawn. Studies may examine regulation of the gene by measuring mRNA transcript abundance via *in situ* hybridization histochemistry. This approach will label cell bodies with sufficiently high levels of message to be resolved using autoradiography by short oligonucleotide probes, full length riboprobes (Section 4.2.), or incorporation of fluorescent bases into a cDNA/ mRNA heteroduplex (*see* Section 5.3.). To determine if the encoded protein is translated from the mRNA template, investigations have analyzed the content of the native protein by standard immunohistochemical methods with antireceptor antibodies. This approach expands the visualization of receptor locale to the entire cell and its processes, and can elucidate low levels of protein due to the added sensitivity offered by the antigen–antibody paradigm (*see* Section 5.2.). In an effort to establish functional receptor binding sites, studies have detected viable ligand recognition sites using in vitro autoradiography (*see* Section 4.1.) or fluorescently coupled probes (*see* Section 5.1.). These three morphological techniques visualize distinctly different aspects of the receptor biosynthetic pathway, and are localized to unique cellular compartments. Each method has its experimental advantages and limitations.

typical D1 or D2 subfamily ligand, so-called "subtractive binding." This type of study is fraught with inconsistencies, and generally is avoided for the in vitro morphologic assays, which employ higher concentrations of ligand than isolated membranes because of diffusion barriers encountered in the intact tissue section.

All of the morphological resolution techniques have limitations when used in isolation. Concurrent application of more than one of these methods is exacting for the tissue section, whereas some paradigms are not compatible for use together (e.g., mRNA detection and simultaneous binding studies). Recent work has employed combinations that determine receptor biosynthesis to establish the distribution of the different cloned subtypes (cf. ref. *61*). These simultaneous detection methods have illustrated occasional mismatches of transcript and binding sites *(49,50)*, and these are noted in subsequent sections of this chapter.

4. Autoradiographic Detection

Employing radioisotopes to determine the distribution of either dopamine receptor binding or their unique mRNAs has been the experimental detection method of choice until quite recently. In ligand binding investigations, the experimental paradigm necessitates performing total and nonspecific ligand binding incubations, since specific binding can only be estimated by visual subtraction of nonspecific binding from the total binding condition in the in vitro situation. The radiolabel is detected using emulsions, either in an X-ray sensitive film clamped on top of the glass slide, or in a nuclear tract monolayer applied directly over the tissue section, or in a coated overlying coverslip which is clamped to the glass slide. The resolution is excellent for regional, quantitative analysis of receptor binding levels, since radiolabeled standards can be made using known concentrations of ligand and processed concurrently with the experimental tissue slice. Cellular detection of the signal is somewhat limited however, as the isotopic particles emitted from the immobilized ligand produce a "cone" of exposed halide grains in the emulsions which spreads out over the cellular ligand binding sites. This hot spot of radioactivity continues to grow as the time of exposure is increased in order to detect lower binding levels, because the isotope continues to emit α-, β-, or γ-particles, depending on the label employed in the analysis. A signal produced by cellular processes exhibiting receptor binding cannot be resolved from the background level of exposed silver grains *(63)*. The use of a tritiated label can be problematic owing to quenching of the isotopic emission by fats, such as myelinated fibers in different regions of the central nervous system. Thus, experiments are performed sometimes using the highly energetic iodinated isotopes which are not compatible with cellular resolution of receptor binding sites *(64)*. Nonetheless, sufficient information can be derived from radiolabeled ligand binding studies with autoradiographic detection such that regional and cellular distribution of viable ligand recognition sites for the two pharmacologically defined families of dopamine receptors has been achieved.

Autoradiographic localization of mRNA abundance levels is the standard used in *in situ* hybridization studies. The isotopes are typically sulfur or phosphate, with some use of tritium and iodine. The same caveats as discussed for ligand binding analysis with reference to production of a cone of exposed silver grains over the transcript of interest, myelinated structures, and high energy isotopes are cogent to the *in situ* hybridization method. Nonetheless, excellent quantitative regional data on the distribution of the different dopamine receptor clones has been obtained. More recent experimental paradigms have employed colored detection systems with avidin–biotin complexes, and/or digoxigenin linked to immunohistochemical procedures. However, there is concern that the sensitivity of these modifications limits the detection of some transcripts. My laboratory has developed a fluorescent adaptation of *in situ* transcription *(52,65)* in order to combine analysis of message abundance and protein expression, which is discussed in Section 5.3.

4.1. Ligand Binding Sites

The distribution of D1 and D2 receptor binding sites in the brain was reported in a large number of investigations throughout the 1980s. Most studies used the prototypical antagonist SCH 23390, to assess D1 binding sites whereas numerous ligands were used to establish D2 localization patterns (e.g., sulpiride, spiperone, raclopride, eticlopride, domperidone). I confine my remarks to a few of these numerous works *(36,38,39)* which reported on the receptor distributions in the extended neuraxis of the rat in their presentations. Their findings, in addition to those reported by others, have been compiled into summary binding distribution patterns in Figs. 2 and 3 for the forebrain and hindbrain regions of the rat. The interested reader is referred to a comparative analysis of D1 and D2 receptor binding in the basal ganglia of some other animal species *(67)*.

These investigations demonstrate that the D1 receptor is detected throughout the forebrain, with the highest concentrations of binding sites in the brain found in the caudate-putamen complex, nucleus accumbens, olfactory tubercle, medial substantia nigra, and entopeduncular nucleus (i.e., the rodent homolog of the internal segment of the globus pallidus). D1 binding in the cortex exhibits a rostral to caudal attenuation *(66)* with the highest binding levels detected in the prefrontal cortex and lowest concentrations found in the occipital cortex. A bilaminar pattern of D1 receptor binding is present in most cytoarchitectonic areas of the neocortex, with the most abundant labeling visible in supragranular layers 1, 2, and 3a, and infragranular layers 5 and 6. The choroid plexus and retina, in the region of the horizontal cells *(68,69)*, also display robust binding for the D1 dopamine receptor. D1 receptor binding is present in greater density than D2 receptors in all regions examined except

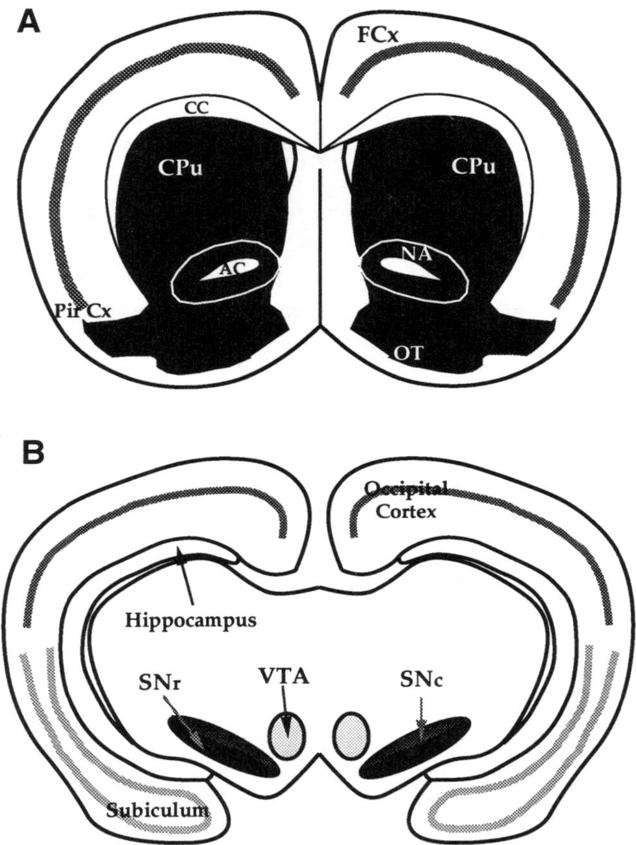

Fig. 2. D1 binding in rat brain: **(A)** coronal section of the rat forebrain; **(B)** coronal section of the rat hindbrain. Black is the highest level of binding, whereas gray depicts moderate to low levels. FCx, frontal cortex; CC, corpus callosum; CPu, caudate-putamen; NA, nucleus accumbens; AC, anterior commissure; Pir Cx, piriform cortex; OT, olfactory tubercle; VTA, ventral tegmental area; SNr, substantia nigra pars reticulata; SNc, substantia nigra pars compacta. Adapted from refs. *36, 37, 39, 41, 50,* and *66.*

the olfactory nerve layer, the ventral tegmental area *(39),* and the pituitary (*see* Fig. 4). Little specific binding for the D1 or D2 dopamine receptors is found within the superior cervical ganglion of the sympathetic nervous system *(62)*; however, D1 receptor binding can be detected in peripheral vasculature *(73,74)* and in capillaries of the central and peripheral nervous system *(53,75)*. Distributions in humans reflect the animal data *(76,77)*.

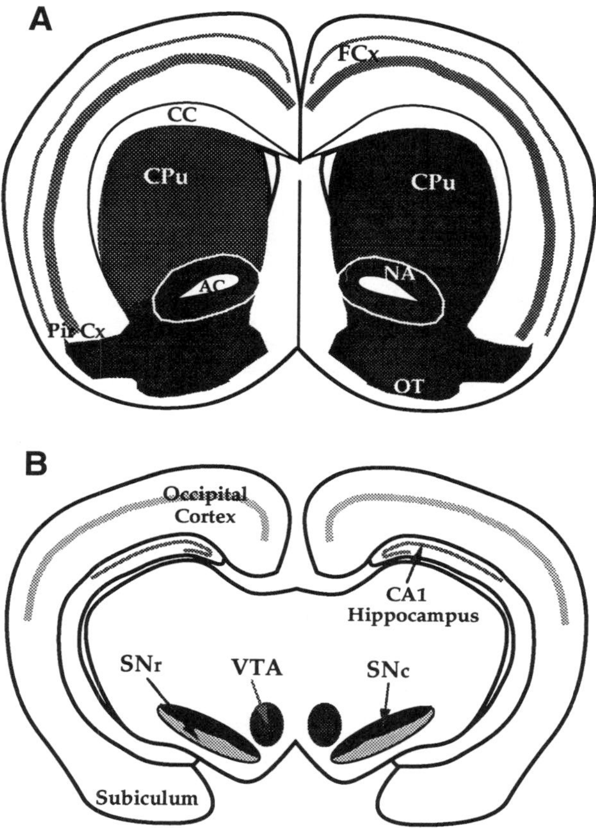

Fig. 3. D2 receptor binding in rat brain: **(A)** coronal section of the rat forebrain; **(B)** coronal section of the rat hindbrain. Black is the highest level of binding, whereas gray depicts moderate to low levels. *See* Fig. 1 for abbreviations. Adapted from refs. *37, 38, 41, 49,* and *66.*

Regions with the highest levels of D2 receptor binding sites are the caudate-putamen, nucleus accumbens, olfactory tubercle, substantia nigra compacta, ventral tegmental area, lateral septal area, central amygdaloid nucleus, both colliculi, islands of Calleja, and the molecular layer of the hippocampus. Lower binding densities for this receptor are detected in the thalamus, pons, and dorsal horn of the spinal cord *(38)*. A low density of D2 dopamine receptor binding was found in all layers of all cortical areas examined in the primate *(66)*. The highest density of binding was consistently detected in layer 5, corresponding to the principal projection lamina of the neocortex. The pituitary gland exhibits D2 receptor binding, which is robust

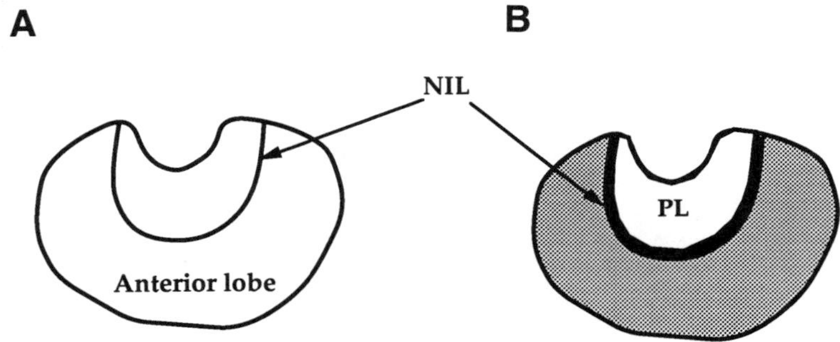

Fig. 4. Binding of both dopamine receptors in the pituitary gland. A horizontal section is illustrated, with black as the highest detected levels of binding. **(A)** D1 receptor binding is nonexistent in the pituitary gland. **(B)** D2 receptor binding is intense in the NIL, with about one third of the lactotrophs in the anterior lobe exhibiting binding sites. NIL, neurointermediate lobe; PL, posterior lobe. Adapted from refs. *37, 38, 50, 70, 71,* and *72*.

within the intermediate lobe, and localized to about one third of the lactotrophs within the anterior lobe *(70–72)*. D2 binding sites also have been measured in the adrenal medulla *(78)* and the retina *(68,79,80)*. The distribution of D2 dopamine receptor binding in humans reflects the data obtained in animal experiments.

With the reports of different genes encoding dopamine receptor subtypes comprising the pharmacologic families, the ligand binding data must be reevaluated *(31)*. The ultimate distribution patterns and the physiological relevance of each of the clones awaits development of subtype-specific ligands.

4.2. Transcript Abundance

4.2.1. D_1 mRNA Distribution Patterns

Investigations in the late 1980s to early 1990s demonstrated that the highest abundance for the rat or human D_1 receptor mRNA is in the striatal complex, which includes the caudate-putamen, nucleus accumbens, and olfactory tubercle *(33,50,81,82)* (*see* Fig. 2). Medium diameter neurons of the striatum have an intense signal for D_1 receptor transcripts, with the occasional large-diameter striatal neurons producing a moderate hybridization signal *(83–86)*. A very good correspondence of mRNA was observed with the binding sites for the D1 receptor subfamily in the rostral neocortex, amygdala, and retina. However, marked discrepancies between receptor binding and the mRNA transcripts were detected in the entopeduncular nucleus, subthalamic nucleus, substantia nigra pars reticulata, hippocampus, and cerebellum *(50)*.

The mismatches are most likely caused by transport of the receptors from the site of protein synthesis into terminal fields or distal dendritic trees of neurons. An example illustrates this point in that D_1 receptor binding sites are quite dense in the substantia nigra although no mRNA can be detected. The results of an elegant series of combined binding and *in situ* hybridization analysis studies performed on serial tissue sections, in combination with experimental lesions in the striatum, confirmed that the D_1 receptor is indeed translated within the striatonigral perikarya, then anterogradely transported to the terminal field within the substantia nigra pars reticulata. Moreover, neurons in the dentate gyrus and the granular layer of the cerebellum contain D_1 receptor mRNA, but completely transport the protein to the molecular layer of the hippocampus *(50)*. This type of investigation provides clues to the receptor biosynthesis in different brain regions. Few investigations have examined the distribution of this receptor mRNA in other body tissues, with the exception of the retina *(68)*.

4.2.2. D_2 mRNA Distribution Patterns

The D_2 cDNA clone represented the first dopamine receptor gene to be identified, by Civelli and colleagues in 1988 *(25)*. It was rapidly determined that two forms of the receptor were produced as RNA splice variants of one another *(35)*. In general, the longer form of the receptor, noted as D_{2L}, is more abundant in the nervous system. The interested reader is referred to Chapter 2 for more details. In general, the abundance of D_2 transcripts in the brain and pituitary parallels the distribution of D2 receptor binding sites *(49)*, with high expression in the caudate-putamen complex (in medium- and large-sized neurons) *(44,84–86)*, nucleus accumbens, olfactory tubercle, globus pallidus, substantia nigra pars compacta, and ventral tegmental area. The pituitary distribution of the mRNA is very intense within the neurointermediate lobe, and approximately one third of the cells in the anterior lobe of the gland *(70)* (*see* Figs. 3 and 4). However, discrepancies could be detected, using a combination of sequential *in situ* hybridization analysis and in vitro receptor binding, as just described for the D_1 receptor *(49)*. A lack of correspondence was detected in the olfactory bulb, neocortex, hippocampus, and zona incerta. These apparent mismatches probably reflect the fact that the protein is located on fibers and other cellular processes, and is transported away from the site of synthesis on mRNA templates within the perikarya. Another series of lesion studies demonstrated that chemical interruption of the nigrostriatal pathway at the median forebrain bundle produced a complete loss of D_2 receptor mRNA in the substantia nigra and ventral tegmental area, with a compensatory increase in this clone in the denervated striatum *(49)*. The experimental finding regarding elevated mRNA levels in the dopamine-depleted striatum is reflected also as additional supersensitivity in D_2 receptor binding in this same nucleus following 6-hydroxydopamine lesions. Human D_2 receptor mRNA abundance levels are homologous to those detected in animal studies.

4.2.3. D_3 mRNA Distribution Patterns

The D_3 receptor was cloned after the D_1 subtype, and demonstrated that a much more complex organization was present for the dopamine system. The initial report *(27)* demonstrated a high abundance for the transcript in limbic areas of the brain such as the ventral striatal complex consisting of the nucleus accumbens, olfactory tubercle, ventral pallidum, ventral tegmental area, and islands of Calleja. These investigators suggested that the D_3 receptor mediated cognitive functions associated with dopaminergic neurotransmission, as opposed to motoric behaviors. Subsequent studies using more sensitive assays such as full-length riboprobes and reverse transcription-polymerase chain reaction (RT-PCR) *(86)* have detected a broader distribution pattern than that originally reported. D_3 mRNA also is distributed in the medium-sized cells of the striatum, large neurons of the basal forebrain (including the ventral pallidum), substantia nigra, frontal cortex, hippocampus, archicerebellum, and mammillary bodies in the rat. There are slight species differences in the monkey and human. No D_3 mRNA is detected within the pituitary gland, much like the results presented for the D_1 receptor subtype. Since few available ligands can discriminate effectively between the members of the D2 subfamily of dopamine receptors, the alignment of D_3 transcripts and the functional binding sites awaits development of better reagents.

4.2.4. D_4 mRNA Distribution Patterns

PCR and low stringency screening were used to isolate this clone in early 1992 *(26)*. The characterization of this receptor subtype demonstrated that it had a high affinity for the atypical antipsychotic drug clozapine, had an extremely high expression pattern within the cardiovascular system, and exhibited a restricted distribution within the brain. The mRNA has been detected in the hypothalamus, thalamus, olfactory bulb, and frontal cortex. More recent data have demonstrated the presence of D_4 mRNA in the normal rodent striatum (*86*, unpublished data). There has been some suggestion that the D_4 receptor in humans mediates dopaminergic involvement in schizophrenia and depression based on its high affinity for clozapine, but these data need further confirmation using a pharmacologically specific ligand which will not also bind to other congeners of the D2 subfamily of dopamine receptors.

4.2.5. D_5 mRNA Distribution Patterns

Characterization of the expression pattern of an alternative D1-like receptor *(29)* using tritiated oligonucleotide primers demonstrated a very different mRNA distribution in the brain. This gene expressed little or no mRNA in the basal ganglia or frontal cortex, although use of RT-PCR (J. Surmeier, personal communication) or full-length riboprobes *(85)* has demonstrated low levels of the transcript throughout the striatum and cortex in rat and monkey. Both medium-diameter and large-sized (putatively cholinergic) striatal neurons exhibit a strong hybridization signal. High levels are present within the hippocampus, parafasicular nucleus of the thalamus (a striatal afferent site), the mammillary bodies, and the anterior pretectal nuclei. These nervous system regions have little or no levels of D1 binding *(29,36,45)*, and therefore a novel function for the D_5 receptor is suggested by these data.

5. Fluorescence Detection

Using fluorescence offers a number of experimental advantages:

1. Two different components (e.g., both D_1 and D_2 receptor binding sites, binding sites combined with protein, or protein detected with message), may be visualized simultaneously;
2. Cellular resolution is possible with reactive processes seen readily; and
3. Rapid analysis is possible, since the fluorescent signal does not need to be exposed to an emulsion for finite times to amplify the binding or hybridization signal.

The limitations imposed by the experimental application of fluorescent probes are that:

1. The signal is lost with time (fading) and during the experimental analysis (quenching), so that tissues and slides cannot be archived;
2. No amplification of the fluorochrome is possible, and signal-to-noise ratios become critical;
3. Quantitation is rarely possible; and
4. Low-magnification regional analysis is not compatible with the fluorescence-capturing capabilities of immersion microscope lenses.

I have employed fluorescence methods extensively in my laboratory, initially to avoid the prohibitive expense of radioactive waste disposal, but more importantly, to examine multiple aspects of dopamine receptor biosynthesis in identified projection neurons of the brain. These methods encompass ligand binding with fluoroprobes, immunocytochemistry with antireceptor antibodies, and nascent cDNA synthesis from the receptor mRNA template (fluorescent *in situ* transcription [FIST]) for the dopaminergic system. The fluorescent reagents achieve cellular resolution, which is my major research focus, and exhibit regional distribution patterns analogous to those previously reported using radiolabeled detection methods. The reader is referred to Figs. 2–4 for the regional expression patterns in the rodent brain.

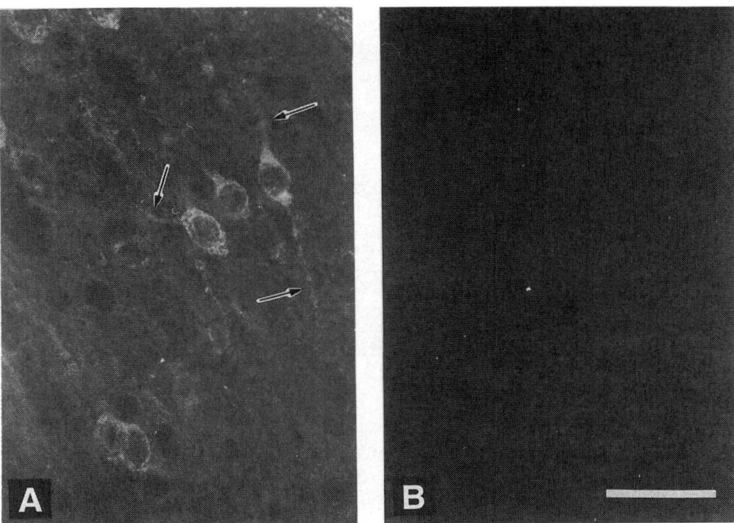

Fig. 5. D2 subfamily binding using fluorescently labeled N-aminophenyl-spiperone (NAPS) in the ventral striatal complex. Calibration bar, 50 μm **(A)** Rhodamine-NAPS was applied at 25 n*M* for 1 h at room temperature to a 10 μm thick, fresh frozen rat tissue slice and incubated in a dark, moist environment (e.g., total binding experiment). D2 binding was detected using narrow bandpass dichroic filters and epifluorescence. Arrows illustrate binding along numerous processes emanating from reactive perikarya. **(B)** Rhodamine-NAPS binding was challenged with 10 μ*M* unlabeled spiperone, applied 15-min prior to the fluoroprobe, and then in combination with labeled NAPS for the 1-h incubation period (e.g., nonspecific binding experiment). Photographic development and printing were identical to those used to obtain the enlargement in panel A. Thus, the fluorescent signal can be directly compared between the two photomicrographs. Specific binding of the D2 ligand is defined as the visual subtraction of B from A. In these experimental conditions, specific binding is nearly equivalent to the fluorescent staining visible in the total binding incubation.

5.1. Ligand Binding Sites

Fluorescently labeled D1 and D2 subfamily antagonists were developed in collaboration with Molecular Probes (Eugene, OR) to detect binding sites in vitro *(37,87,88)*. The specificity of the ligand is altered by this coupling process *(89)*; however, the fluoroprobes allow cellular resolution to be achieved in the tissue slice, and resolve stained cellular processes (Fig. 5). As in any ligand binding study, specific binding can be estimated only by the visual subtraction of nonspecific binding from the total binding experiment; under the experimental conditions we use routinely, total incubations of the ligand are nearly equivalent to specific binding.

Binding sites are particularly prevalent within the striatal complex *(37,87,90)*, and can be resolved on identified striatonigral neurons *(91)*, nucleus accumbens neurons *(40)*, primary cultures of the hypothalamus *(92)*, and cultured accumbens and ventral tegmental neurons *(93)*. The fluoroprobes distinguish cellular binding of D1 receptors in the substantia nigra pars reticulata, outer segments of the photoreceptors in the retina, postganglionic neurons in the superior cervical ganglion, and adrenal medullary cells *(37,94)*. Several researchers *(41,87)* have demonstrated cortical binding of the D1 receptors in large pyramidal neurons of lamina 3 and 5, as well as in smaller diameter cells of the deepest layers. No D1 binding could be detected within the pituitary gland.

Both antagonist and agonist fluoroprobes have been developed for the D2 receptor subfamily *(37,87,88–90)*. Cellular binding is prevalent on medium-sized retrogradely identified striatonigral *(91)* and accumbens-ventral tegmental area (VTA) neurons *(61)*, and on the occasional large interneuron of these regions. Cortical D2 binding sites also are visible *(40,41)* on pyramidal and nonpyramidal neurons and their processes. Binding has been detected within subregions of the nucleus accumbens, lateral septum, olfactory tubercle, piriform cortex, CA1 of the hippocampus, neurointermediate lobe cells and processes, and on lactotrophs of the anterior portion of the pituitary gland with high specificity. In an attempt to distinguish subtypes within the D2 dopamine receptor subfamily, fluoroprobe binding has been combined with immunofluorescent detection of the native protein using subtype-selective antireceptor antibodies *(40)*.

5.2. Cloned Native Proteins

Antireceptor antisera have been generated to unique, nonconserved polypeptide fragments derived from the published clone sequences of the D_1, D_{2S}, D_{2L}, D_3, D_4, and D_5 receptors (*53,54,56,57,* unpublished data). Multiple antisera have been produced to both intracellular and extracellular epitopes to minimize the potential for false-positive localization of homologous peptide sequences in unknown protein(s) with an enriched neuronal distribution. The recognition of the peptide antigen by the antisera is determined using immunoblots or enzyme-linked immunosorbent assay, whereas the specificity for the appropriate receptor subtype has employed stably transfected Chinese hamster ovary cells. The staining reaction also is examined using preimmune sera, derived from the rabbit (or immunologic host) prior to the inoculation regimen. These investigations have determined a robust expression of the D_1, D_{2S}, D_{2L}, D_3, D_4, and D_5 receptors within striatonigral neurons (*53,95,* unpublished observations). D_2 receptor protein expression can be seen in all three neuronal populations of the striatum (e.g., the medium spiny projection neurons), and the medium aspiny and large aspiny interneuron groups

(60). Subcellular distribution patterns for the D_2 protein show that it is widely expressed within the neuronal cell body and its primary dendritic processes and is transported anterogradely in the axon to termination fields *(56,60)*. Presynaptic detection of the D_2 receptor protein is seen, corresponding to the autoreceptor of the nigrostriatal pathway *(58,60)*.

D_2 and D_3 receptors can be identified within corticospinal neurons, detected following encroachment of the retrograde label into the pyramidal tract in the midbrain (during injection of the SNr), in large pyramidal-shaped neurons of layers 3 and 5 of the neocortex *(40,53)*, in the densely packed neurons of the olfactory tubercle *(59)*, in accumbens-VTA neurons and their processes *(61)*, and in the substantia nigra compacta *(58,59)*. D_2 protein is prevalent within the lateral septum and shell region of the nucleus accumbens *(40)*, as well as in the photoreceptor inner and outer segments, and the outer and inner plexiform layers of the retina of many species *(80)*.

D_3 receptor protein is found within the striatal complex, but is more robust in ventral regions, within the ventral pallidum, the striatopallidal neurons and fibers terminating in the globus pallidus, the core region of the nucleus accumbens, sensorimotor and parietal cortices, the hippocampus and subiculum *(40,53)*. D_4 receptors have been elucidated within the neurons of layers 2 through 6 of neocortex and in the dorsal striatum in low numbers of medium-diameter neurons (unpublished data).

In an attempt to determine functional receptor distribution patterns for the D_2 and D_3 receptors, we have combined the fluorescent visualization of subtype selective antipeptide antisera for the respective cloned D_2 and D_3 forms of the dopamine receptors with ligand binding using the D_2 antagonist fluoroprobe *(40)*. The double-labeling paradigm enabled direct cellular comparisons of functional binding in defined receptor clones and provided a mechanism to contrast the regional and cellular distribution patterns for the D_2 and D_3 subtypes. We determined that heterogeneities occurred in the cellular expression of functional D_2 and D_3 receptors in forebrain dopaminoceptive areas such that D_3 receptors are more related to basal ganglia and structures involved with motoric behavior, whereas D_2 is associated with regions involved with cognitive/affective functions. The only area of overlap between the functional D_2 and D_3 receptors is detected within the dorsal striatum, a more motorically related portion of this nucleus *(96)*. This work provides a strong caution regarding receptor functions; mRNA abundance levels should not be used in isolation to predict the functional role of the receptor. Posttranscriptional and posttranslational regulation processes may alter the final protein levels of a dopamine receptor in ways which are not fully understood at the present. Further, the binding affinity for a specific receptor subtype may not necessarily reflect the mRNA transcript abundance in an individual cell.

5.3. Transcript Abundance

The cellular localization of mRNA has been adapted to a fluorescent detection method in my laboratory through modification of the very sensitive cDNA *in situ* transcription technique *(65)*. The FIST technique allows direct examination of dopamine receptor mRNA transcripts (or any other pertinent transcript) by specific hybridization of a short (36–39 mer) cDNA oligonucleotide primer. The annealed primer is enzymatically extended from the 3' end, using reverse transcriptase and incorporation of a fluorescently labeled dNTP. In our experiments, we have used rhodamine-dUTP and occasionally, fluorescein-dUTP (Amersham, Arlington Heights, IL). The synthesis of the nascent cDNA/mRNA heteroduplex identifies cells with the appropriate transcript after 2 h of reverse transcriptase extension, and the experiment can then be visualized using standard fluorescence microscopy. We have used moderate stringency levels in performing hybridization overnight at room temperature, and consequently the tissue can be processed further for other components of receptor biosynthesis such as protein levels. Studies have examined the FIST detection of all five cloned dopamine receptor subtypes and some excitatory amino acid receptor subunits *(52)*. Negative experimental controls include: omission of the primer, omission of the reverse transcriptase, and omission of the fluorescent dUTP. No signal is detected in any of these conditions. Positive experimental controls include: substitution of oligo $d(T)_{36}$, which labels all poly A-tailed mRNA in every cell undergoing protein synthesis; substitution of a missense primer, with only three nucleotide base order rearrangements; and using different primers, to different portions of the mRNA sequence, for each dopamine receptor subtype. The fluorescent signal is abolished completely following missense primer hybridization, and all primers for an individual receptor subtype yield homologous mRNA fluorescence patterns within the experimental tissues.

FIST increases the cellular hybridization signal of low abundance transcripts (like the dopamine receptors) because the quantity of fluorescence incorporated into the nascent cDNA is dependent on the number of fluorescently labeled dNTP species used. A minimum of 25% of the nucleotide bases would account for the cellular reaction with a single fluorescent dNTP base, as used in the experimental paradigm. The option of using two, three, or completely labeled dNTP in the enzymatic extension and incorporation step is theoretically possible, but not necessary. Sensitivity also may be increased by varying the length of time the enzymatic extension of the oligonucleotide primer is allowed to proceed. The entire FIST paradigm is completed within 24 h with readily visible signals (*see* Figs. 6A and 7A). The stringency conditions used in the hybridization minimize tissue damage. Figures 6 and

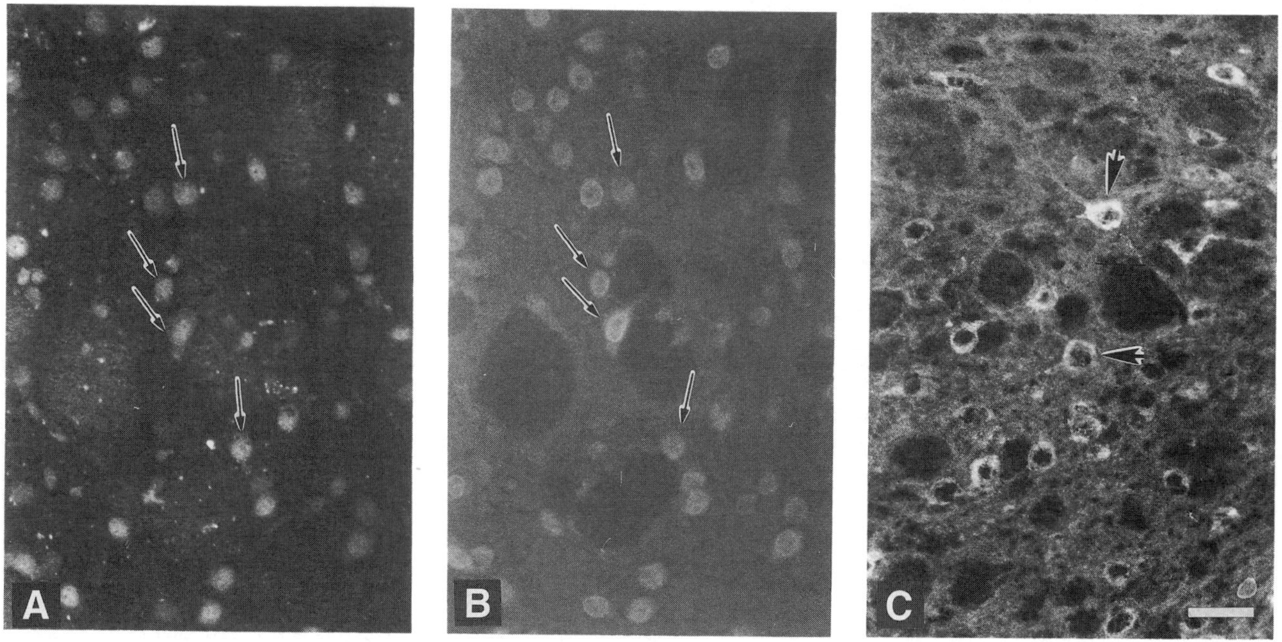

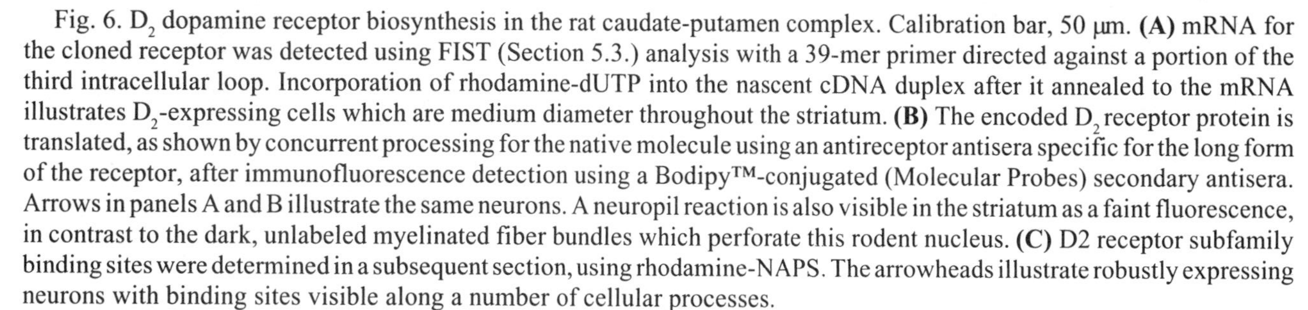

Fig. 6. D_2 dopamine receptor biosynthesis in the rat caudate-putamen complex. Calibration bar, 50 μm. **(A)** mRNA for the cloned receptor was detected using FIST (Section 5.3.) analysis with a 39-mer primer directed against a portion of the third intracellular loop. Incorporation of rhodamine-dUTP into the nascent cDNA duplex after it annealed to the mRNA illustrates D_2-expressing cells which are medium diameter throughout the striatum. **(B)** The encoded D_2 receptor protein is translated, as shown by concurrent processing for the native molecule using an antireceptor antisera specific for the long form of the receptor, after immunofluorescence detection using a Bodipy™-conjugated (Molecular Probes) secondary antisera. Arrows in panels A and B illustrate the same neurons. A neuropil reaction is also visible in the striatum as a faint fluorescence, in contrast to the dark, unlabeled myelinated fiber bundles which perforate this rodent nucleus. **(C)** D2 receptor subfamily binding sites were determined in a subsequent section, using rhodamine-NAPS. The arrowheads illustrate robustly expressing neurons with binding sites visible along a number of cellular processes.

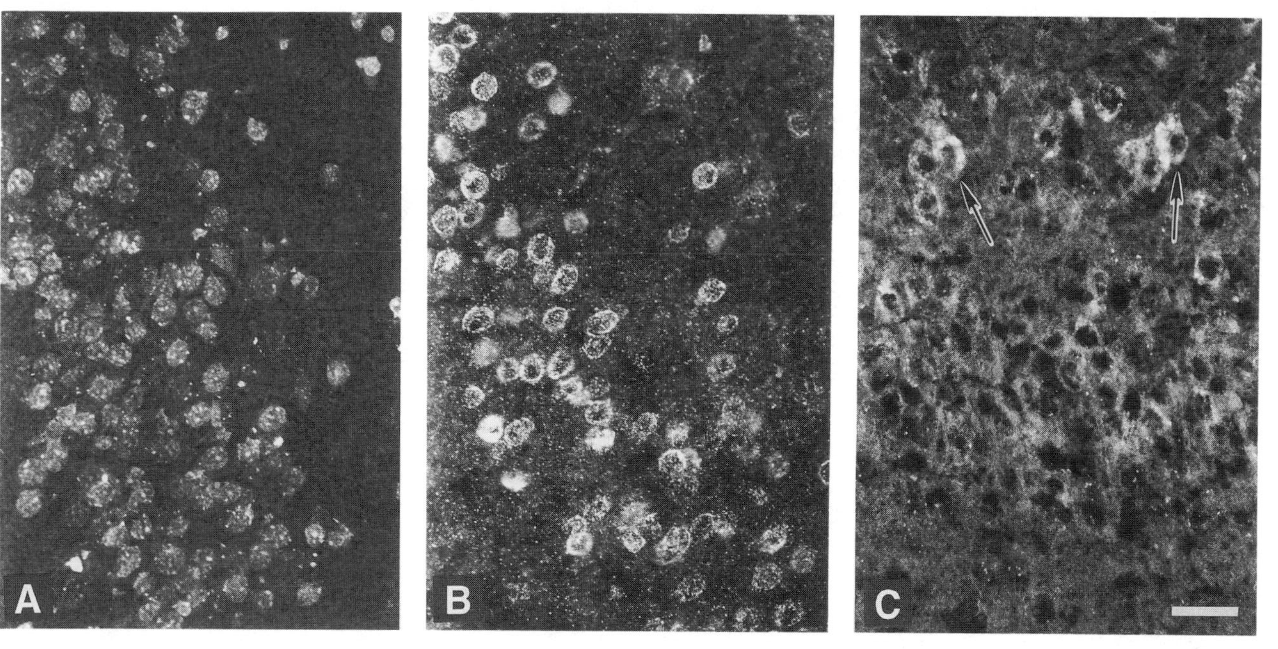

Fig. 7. D_3 dopamine receptor biosynthesis in the rat olfactory tubercle. Calibration bar, 50 µm. **(A)** mRNA for the cloned dopamine receptor was visualized by FIST after annealing a 39-mer cDNA primer to a unique sequence in the third cytoplasmic loop. Rhodamine-dUTP was incorporated enzymatically into the growing heteroduplex and identifies the large, densely packed pyramidal neurons of this region robustly expressing the D_3 receptor subtype. **(B)** A sequential section shows the protein is translated and can be detected using a selective antireceptor antisera, directed against a peptide sequence within the third extracellular loop of the native receptor. Immunofluorescence is intense within the thin rim of cytoplasm surrounding the unstained cellular nucleus of the pyramidal neurons. **(C)** A third section was processed for the presence of D2 receptor subfamily binding sites and shows high levels of rhodamine-NAPS binding in homologous large pyramidal cells to those seen in panels A and B. Arrows illustrate particularly intense binding in large neurons of the olfactory tubercle.

7 provide examples of these types of experiments using the D_2 and D_3 clones for illustration.

Visualization of nascent cDNA synthesis using FIST demonstrates fluorescence staining in the cytoplasm of reactive cells, heterogeneously spread throughout the soma (Figs. 6A and 7A). In the striatum, medium-sized cells express FIST-detected mRNA for the D_1, D_2, D_3 and occasionally D_4 transcripts but low to no levels of D_5. This correlates with autoradiographically detected mRNA abundance levels *(29,49,83,86),* and protein expression profiles *(40,53,59),* as well as cellular ligand binding sites *(87,91).* The medium-sized cells (~20 μm) expressing D_1 and D_2 in the striatum were evenly distributed throughout the dorsal regions of the nucleus. D_3-containing neurons were more prevalent in lateral and ventral portions of the striatal complex and not as numerous as D_1- and D_2-expressing striatal somata. Examination of the frontal cortex, nucleus accumbens, and olfactory tubercle has been assessed, and the results are homologous to the regional autoradiographic analyses of transcripts reported by other investigators.

We have retrogradely identified the striatonigral pathway using infusions of latex microspheres, coupled to different fluorochromes (Molecular Probes). In this experimental paradigm, FIST mRNA can be detected for D_1, D_2, and D_3 dopamine receptor clones; D_1, D_2, and D_3 protein staining using antireceptor antibodies; and viable ligand recognition sites for the D_1 and D_2 subfamilies using selective antagonist fluoroprobes in this output pathway. Thus, the utilization of fluorescence methods allows visualization of multiple aspects of the receptor biosynthetic process. The limitations of fluorescence (principally its stability) are far outweighed by the benefits provided by increased sensitivity and dual (or triple) detection.

6. Summary and Conclusions

The dopamine receptors are widespread throughout the brain, retina, cardiovascular system, and kidney. The cloning of five different forms of these receptors, classified into two broad pharmacological families, argues for a much more complex mediation of the effects of this neurotransmitter than was initially proposed. New experimental strategies have elucidated the heterogeneous distribution patterns for the different clones, based on detection of their unique mRNA transcripts, identification of the encoded protein via antireceptor antibodies, and determination of viable ligand recognition sites using selective ligand probes. However, additional reagents need to be designed to elucidate the physiological relevance of the identified receptor forms within tissues and within discrete cells. This field has made enormous progress in the past 5 years and greatly expanded our knowledge of receptor (co)expression patterns and dopaminergic neurobiology; its future holds great promise and the anticipation of solving receptor-mediated disorders of the nervous and cardiovascular systems.

Acknowledgments

This work was supported in part by United States Public Health Service (USPHS) grants NS 23079 and NS 32277. The technical help of Eric Larson and Kurt Noblett is acknowledged most gratefully.

References

1. Ehringer, J. and Hornykiewicz, O. (1960) Verteilung von adrenalin und dopamin (3-hydroxytyramin)im gtehirn des menschen und i hr verhalten bei erkran-kungen des extrapyramidalen systems. *Wien Klin. Wschr.* **38,** 1236–1239.
2. Carlsson, A. (1987) Perspectives on the discovery of central monoaminergic neurotransmission. *Annu. Rev. Neurosci.* **10,** 19–40.
3. Kalivas, P. and Nemeroff, C. R. (1988) The mesocorticolimbic system. *Ann. NY Acad. Sci.* **537,** 1–537.
4. DeLong, M. (1990) Primate models of involvement disorders of basal ganglia origin. *TINS* **13,** 281–285.
5. Reynolds, G. P. (1992) Developments in the drug treatment of schizophrenia. *TIPS* **13,** 116–121.
6. Rossetti, Z. L., Honaiden, Y., and Gessa, G. L. (1992) Marked inhibition of mesolimbic dopamine release: a common feature of ethanol, morphine, cocaine and amphetamine abstinence in rats. *Eur. J. Pharmacol.* **221,** 227–234.
7. Cenci, M. A., Kalén, P., Mandel, R. J., and Bjorklund, A. (1992) Regional differences in the regulation of dopamine and noradrenaline release in medial frontal cortex, nucleus accumbens and caudate-putamen: a microdialysis study in the rat. *Brain Res.* **581,** 217–228.
8. Kebabian, J. W. and Calne, D. B. (1979) Multiple receptors for dopamine. *Nature* **277,** 93–96.
9. Ariano, M. A., Briggs, C. A., and McAfee, D. A. (1982) Cellular localization of cyclic nucleotide changes in rat superior cervical ganglion. *Cell Mol. Neurobiol.* **2,** 143–156.
10. Ariano, M. A. and Adinolfi, A. M. (1977) Cyclic nucleotide phosphodiesterase: subcellular localization in caudate following selective interruption of striatal afferents. *Exp. Neurol.* **57,** 426–433.
11. Ariano, M. A., Butcher, L. L., and Appleman, M. M. (1980) Cyclic nucleotides in rat caudate-putamen complex: histochemical characterization and effects of deafferentation and kainic acid infusion. *Neuroscience* **5,** 1269–1276.
12. Spano, P. F., Trabucchi, M., and DiChiara, G. (1977) Localization of nigral dopamine-sensitive adenylate cyclase on neurons originating from the corpus striatum. *Science* **196,** 1343–1345.
13. Briggs, C. A., Whiting, C., Ariano, M. A., and McAfee, D. A. (1982) Cyclic nucleotide metabolism in the sympathetic ganglion. *Cell Mol. Neurobiol.* **2,** 129–142.

14. Ariano, M. A. and Ufkes, S. K. (1983) Cyclic nucleotide distribution within rat striatonigral neurons. *Neuroscience* **9,** 23–29.
15. Ariano, M. A. and Kenny, S. L. (1 985) Neurotransmitter receptor autoradiography in immunohistochemically identified neurons. *J. Neurosci. Methods* **15,** 49–61.
16. Billard, W., Ruperto, V., Crosby, G., Iorio, L. C., and Barnett, A. (1984) Characterization of the binding of ^{3}H-SCH 23390, a selective D-1 receptor antagonist ligand, in rat striatum. *Life Sci.* **35,** 1885–1893.
17. Iorio, L. C., Barnett, A., Leitz, F. H., Houser, V. P., and Korduba, C. (1983) SCH 23390, a potential benzazepine antipsychotic with unique interactions on dopamine systems. *J. Pharmacol. Exp. Ther.* **226,** 462–468.
18. Scatton, B. and Dubois, A. (1985) Autoradiographic localization of D_1 dopamine receptors in the rat brain with ^{3}H-SKF 38393. *Eur. J. Pharmacol.* **111,** 145,146.
19. Jastrow, T. R., Richfield, E., and Gnegy, M. E. (1984) Quantitative autoradiography of ^{3}H-sulpiride binding sites in rat brain. *Neurosci. Lett.* **51,** 47–53.
20. Drukarch, B. and Stoof, J. C. (1990) D_2 dopamine autoreceptor selective drugs: do they really exist? *Life Sci.* **47,** 361–376.
21. Plantje, J. F., Hansen, H. A., Daus, F. J., and Stoof, J. C. (1984) The effects of SCH 23390, YM 09151-2, (+)- and (–)-3-PPP and some classical neuroleptics on D-1 and D-2 receptors in rat neostriatum in vitro. *Eur. J. Pharmacol.* **105,** 73–83.
22. Wolf, M. E. and Roth, R. H. (1987) Dopamine autoreceptors, in *Receptor Biochemistry and Methodology* (Venter, J. C. and Harrison, L. C., eds.), Wiley, New York, pp. 45–96.
23. Wolf, M. E. and Roth, R. H. (1990) Autoreceptor regulation of dopamine synthesis. *Ann. N.Y. Acad. Sci.* **604,** 323–343.
24. Skirboll, L. R., Grace, A. A., and Bunney, B. S. (1979) Dopamine auto- and postsynaptic receptors: electrophysiological evidence for differential sensitivity to dopamine agonists. *Science* **206,** 80–82.
25. Bunzow, J. R., Van Tol, H. H. M., Grandy, D. K., Albert, P., Salon, J., Christy, M., Machida, C. A., Neve, K., and Civelli, O. (1988) Cloning and expression of a rat D_2 dopamine receptor cDNA. *Nature* **336,** 22–29.
26. O'Malley, K. L., Harmon, S., Tang, L., and Todd, R. D. (1992) The rat dopamine D_4 receptor: sequence, gene structure, and demonstration of expression in the cardiovascular system. *N. Biol.* **4,** 137–146.
27. Sokoloff, P., Giros, B., Martres, M.-P., Bouthenet, M.-L., and Schwartz, J.-C. (1990) Molecular cloning and characterization of a novel dopamine receptor (D3) as a target for neuroleptics. *Nature* **347,** 146–151.
28. Sunahara, R. K., Guan, H.-C., O'Dowd, B. F., Seeman, P., Laurier, L. G., Ng, G., George, S. R., Torchia, J., Van Tol, H. H. M., and Niznik, H. B. (1991) Human dopamine D1 receptor encoded by an intronless gene on chromosome 5. *Nature* **350,** 614–619.
29. Tiberi, M., Jarvie, K. R., Silvia, C., Falardeau, P., Gingrich, J. A., Godinot, N., Bertrand, L., Yang-Feng, T. L., Fremeau, R. P., Jr., and Caron, M. G. (1991) Cloning, molecular characterization, and chromosomal assignment of a gene encoding a second D_1 dopamine receptor subtype: differential expression pattern in rat brain compared with the D_{1A} receptor. *Proc. Natl. Acad. Sci. USA* **88,** 7491–7495.

30. Van Tol, H. H. M., Bunzow, J. R., Guan, H.-C., Sunahara, R. K., Seeman, P., Niznik, H. B., and Civelli, O. (1991) Cloning of the gene for a human D_4 receptor with high affinity for the antipsychotic clozapine. *Nature* **350,** 610–614.
31. Andersen, P. H., Gingrich, J. A., Bates, M. D., Dearry, A., Falardeau, P., Senogles, S. E., and Caron, M. G. (1990) Dopamine receptor subtypes: beyond the D_1/D_2 classification. *Trends Pharmacol. Sci.* **11,** 231–236.
32. Sibley, D. R. and Monsma, F. J., Jr. (1992) Molecular biology of dopamine receptors. *TIPS* **13,** 61–69.
33. Monsma, F. J., Jr., Mahan, L. C., McVittie, L. D., Gerfen, C. R., and Sibley, D. R. (1990) Molecular cloning and expression of a D_1 dopamine receptor linked to adenylyl cyclase activation. *Proc. Natl. Acad. Sci. USA* **87,** 6723–6727.
34. Fishburn, C. S., Belleli, D., David, C., Carmon, S., and Fuchs, S. (1993) A novel short isoform of the D_3 receptor generated by alternative splicing in the third cytoplasmic loop. *J. Biol. Chem.* **268,** 5872–5878.
35. Monsma, F. J., Jr., McVittie, L. D., Gerfen, C. R., Mahan, L. C., and Sibley, D. R. (1989) Multiple D2 dopamine receptors produced by alternative RNA splicing. *Nature* **342,** 926–929.
36. Altar, C. A. and Marien, M. R. (1986) Picomolar affinity of ^{125}I-SCH23982 for D_1 receptors in brain demonstrated with digital subtraction autoradiography. *J. Neurosci.* **7,** 213–222.
37. Ariano, M. A., Kang, H. C., Haugland, R. P., and Sibley, D. R. (1991) Multiple fluorescent ligands for dopamine receptors. II. Visualization in neural tissues. *Brain Res.* **547,** 208–222.
38. Bouthenet, M.-L., Martres, M.-P., Sales, N., and Schwartz, J.-C. (1987) A detailed mapping of dopamine D-2 receptors in rat central nervous system by autoradiography with ^{125}I-iodosulpride. *Neuroscience* **20,** 117–155.
39. Boyson, S. J., McGonigle, P., and Molinoff, P. (1986) Quantitative autoradiographic localization of the D-1 and D-2 subtypes of dopamine receptors in rat brain. *J. Neurosci.* **6,** 3177–3188.
40. Larson, E. R. and Ariano, M. A. (1995) D_3 and D_2 dopamine receptors: visualization of cellular expression patterns in motor and limbic structures. *Synapse* **20,** 325–337.
41. Vincent, S. L., Khan, Y., and Benes, F. M. (1993) Cellular distribution of dopamine D_1 and D_2 receptors in rat medial prefrontal cortex. *J. Neurosci.* **13,** 2557–2564.
42. Bouthenet, M. L., Souil, M.-P., Sokoloff, P., Giros, B., and Schwartz, J.-C. (1991) Localization of dopamine D_3 receptor mRNA in the rat brain using in situ hybridization histochemistry: comparison with D_2 receptor mRNA. *Brain Res.* **564,** 203–219.
43. Fremeau, R. T., Jr., Duncan, G. E., Fornaretto, M.-G., Dearry, A., Gingrich, J. A., Breese, G. R., and Caron, M. G. (1991) Localization of D_1 dopamine receptor mRNA in brain supports a role in cognitive, affective, and neuroendocrine aspects of dopaminergic neurotransmission. *Proc. Natl. Acad. Sci. USA* **88,** 3772–3776.
44. Le Moine, C. and Bloch, B. (1991) Rat striatal and mesencephalic neurons contain the long form of the D_2 receptor mRNA. *Mol. Brain Res.* **10,** 283–289.
45. Meador-Woodruff, J. H., Mansour, A., Grandy, D. K., Damask, S. P., Civelli, O., and Watson, S. J., Jr. (1992) Distribution of D_5 receptor mRNA in rat brain. *Neurosci. Lett.* **145,** 209–212.
46. Meador-Woodruff, J. H., Mansour, A., Van Tol, H. H. M., Watson, S. J., Jr., and Civelli, O. (1989) Distribution of D_2 dopamine receptor mRNA in rat brain. *Proc. Natl. Acad. Sci. USA* **86,** 7625–7628.

47. Large, C. H. and Stubbs, C. M. (1994) The dopamine D_3 receptor: Chinese hamsters or Chinese whispers? *TIPS* **15,** 46,47.
48. Liu, J.-C., Cox, R. F., Greif, G. J., Freedman, J. E., and Waszczak, B. L. (1994) The putative dopamine D_3 receptor agonist 7-OH-DPAT: lack of mesolimbic selectivity. *Eur. J. Pharmacol.* **264,** 265–269.
49. Mansour, A., Meador-Woodruff, J., Bunzow, J. R., Civelli, O., Akil, H., and Watson, S. J., Jr. (1990) Localization of dopamine D_2 receptor mRNA and D_1 and D_2 receptor binding in the rat brain and pituitary: an in situ hybridization-receptor autoradiographic analysis. *J. Neurosci.* **10,** 2587–2600.
50. Mansour, A., Meador-Woodruff, J., Zhou, Q., Civelli, O., Akil, H., and Watson, S. J., Jr. (1991) A comparison of D_1 receptor binding and mRNA in rat brain using receptor autoradiography and in situ hybridization histochemistry techniques. *Neuroscience* **45,** 359–377.
51. Qin, Z. H., Chen, J. F., and Weiss, B. (1994) Lesions of mouse striatum induced by 6-hydroxydopamine differentially alter the density, rate of synthesis, and level of gene expression of D_1 and D_2 dopamine receptors. *J. Neurochem.* **62,** 411–420.
52. Noblett, K. L. and Ariano, M. A. (1996) Co-expression of receptor mRNA and protein: striatal dopamine and excitatory amino acid subtypes. *J. Neurosci. Methods* **66,** 61–66.
53. Ariano, M. A. and Sibley, D. R. (1994) Dopamine receptor distribution in the rat CNS: elucidation using anti-peptide antisera directed against D_{1A} and D_3 subtypes. *Brain Res.* **648,** 95–110.
54. David, C., Ewert, M., Seeburg, P. H., and Fuchs, S. (1991) Antipeptide antibodies differentiate between long and short isoforms of the D_2 receptor. *Biochem. Biophys. Res. Commun.* **179,** 824–829.
55. Huang, Q., Zhou, D., Chase, K., Gusella, J. F., Aronin, N., and DiFiglia, M. (1992) Immunohistochemical localization of the D_1 dopamine receptor in rat brain reveals its axonal transport, pre- and postsynaptic localization, and prevalence in the basal ganglia, limbic system, and thalamic reticular nucleus. *Proc. Natl. Acad. Sci. USA* **89,** 11,988–11,992.
56. Levey, A. I., Hersch, S. M., Rye, D. B., Sunahara, R. K., Niznik, H. B., Kitt, C. A., Price, D. L., Maggio, R., Brann, M. R., and Ciliax, B. J. (1993) Localization of D_1 and D_2 dopamine receptors in brain with subtype-specific antibodies. *Proc. Natl. Acad. Sci. USA* **90,** 8861–8865.
57. McVittie, L. D., Ariano, M. A., and Sibley, D. R. (1991) Immunocytochemical localization of the D_2 dopamine receptor in the rat striatum using anti-peptide antibodies. *Proc. Natl. Acad. Sci. USA* **88,** 1441–1445.
58. Sesack, S. R., Aoki, C. J., and Pickel, V. M. (1994) Ultrastructural localization of D_2 receptor-like immunoreactivity in midbrain dopamine neurons and their striatal targets. *J. Neurosci.* **14,** 88–108.
59. Ariano, M. A., Fisher, R. S., Smyk-Randall, E., Sibley, D. R., and Levine, M. S. (1993) D_2 dopamine receptor distribution in the CNS using subtype specific anti-peptide antisera. *Brain Res.* **609,** 71–80.
60. Fisher, R. S., Levine, M. S., Sibley, D. R., and Ariano, M. A. (1994) D_2 dopamine receptor protein location: Golgi impregnation gold toned and ultrastructural analysis of the rat neostriatum. *J. Neurosci. Res.* **38,** 551–564.

61. Ariano, M. A., Larson, E. R., and Noblett, K. L. (1995) Cellular dopamine receptor subtype localization, in *Cellular and Molecular Mechanisms of Striatal Function* (Ariano, M. A. and Surmeier, D. J., eds.), Landes, Austin, TX, pp. 59–70.
62. Ariano, M. A. (1987) Comparison of dopamine binding sites in the rat superior cervical ganglion and caudate nucleus. *Brain Res.* **421,** 245–254.
63. Ariano, M. A. (1988) Striatal D_1 dopamine receptor distribution following chemical lesion of the nigrostriatal pathway. *Brain Res.* **443,** 204–214.
64. Ariano, M. A. (1989) Long-term changes in striatal D_1 dopamine receptor distribution after dopaminergic deafferentation. *Neuroscience* **32,** 203–212.
65. Tecotte, L. H., Barchas, J. D., and Eberwine, J . H. (1988) In situ transcription: specific synthesis of complementary DNA in fixed tissue sections. *Science* **240,** 1661–1664.
66. Lidow, M. S., Goldman-Rakic, P. S., Gallager, D. W., and Rakic, P. (1991) Distribution of dopamine receptors in the primate cerebral cortex: quantitative autoradiographic analysis using ^{3}H-raclopride, ^{3}H-spiperone, and ^{3}H-SCH-23390. *Neuroscience* **40,** 657–671.
67. Richfield, E. K., Young, A. B., and Penney, J. B. (1987) Comparative distribution of dopamine D_1 and D_2 receptors in the basal ganglia of turtles, pigeons, rats, cats and monkeys. *J. Comp. Neurol.* **262,** 446–463.
68. Dearry, A., Edelman, J., Miller, S., and Burnside, B. (1990) Dopamine induces light-adaptive retinomotor movements in bullfrog cones via D_2 receptors and in retinal pigment epithelium via D_1 receptors. *J. Neurochem.* **54,** 1367–1378.
69. Makman, M. H. and Dvorkin, B. (1986) Binding sites of ^{3}H-SCH-23390 in retina: properties and possible relationship to dopamine D_1 receptor mediating stimulation of adenylate cyclase. *Mol. Brain Res.* **1,** 261–270.
70. Chronwall, B. M., Dickerson, D. S., Huerter, B. S., Sibley, D. R., and Millington, W. R. (1994) Regulation of heterogeneity in D_2 dopamine receptor gene expression among individual melanotropes in the rat pituitary intermediate lobe. *Mol. Cell. Neurosci.* **5,** 35–45.
71. De Sousa, E. B. (1986) Serotonin and dopamine receptors in the rat pituitary gland: autoradiographic identification, characterization, and localization. *Endocrinology* **119,** 1534–1542.
72. Levant, B., Grigoriadis, D. E., and DeSouza, E. B. (1993) ^{3}H-Quinpirole binding to putative D_2 and D_3 dopamine receptors in rat brain and pituitary gland: a quantitative autoradiographic study. *J. Pharmacol. Exp. Ther.* **264,** 991–1001.
73. Alkadhi, K. A., Sabouni, M. J., Ansar, A. F., and Lokhandwala, M. F. (1986) Activation of DA_1 receptors by dopamine or fenoldopam increases cyclic AMP levels in the renal artery but not in the superior cervical ganglion of the rat. *J. Pharmacol. Exp. Ther.* **238,** 547–553.
74. Lokhandwala, M. F. and Steenberg, M. L. (1984) Evaluation of the effects of SKF 82526 and LY 17155 on presynaptic (DA_2) and postsynaptic (DA_1) dopamine receptors in rat kidney. *J. Auton. Pharmacol.* **4,** 273–277.
75. O'Connell, D. P., Botkin, S. J., Ramos, S. I., Sibley, D. R., Ariano, M. A., Felder, R. A., and Carey, R. M. (1995) Localization of the D_{1A} receptor protein in the rat kidney. *Am. J. Physiol.* **268,** F1185–F1197.
76. De Keyser, J., Claeys, A., De Backer, J. P., Ebinger, G., Rocis, F., and Vauquelin, G. (1988) Autoradiographic localization of D_1 and D_2 dopamine receptors in the human brain. *Neurosci. Lett.* **91,** 142–147.

77. Seeman, P. (1987) Dopamine receptors in human brain diseases, in *Dopamine Receptors* (Creese, I. and Fraser, C. M., eds.), Liss, New York, pp. 233–245.
78. Artalejo, C. R., Garcia, A. G., Montiel, C., and Sanchez, G. P. (1985) A dopaminergic receptor modulates catecholamine release from the cat adrenal gland. *J. Physiol.* **362,** 359–368.
79. Brann, M. R. and Young, W. S., III (1986) Dopamine receptors are located on rods in bovine retina. *Neurosci. Lett.* **69,** 221–226.
80. Wagner, H.-J., Luo, B. G., Ariano, M. A., Sibley, D. R., and Stell, W. K. (1993) Localization of D_2 dopamine receptors in vertebrate retinae with anti-peptide antibodies. *J. Comp. Neurol.* **331,** 469–481.
81. Dearry, A., Gingrich, J. A., Falardeau, P., Fremeau, R. T., Jr., Bates, M. D., and Caron, M. G. (1990) Molecular cloning and expression of the gene for a human D_1 dopamine receptor. *Nature* **347,** 72–76.
82. Meador-Woodruff, J. H., Mansour, A., Healy, D. J., Kuehnt R., Zhou, Q.-Y., Bunzow, J. R., Akil, H., Civelli, O., and Watson, S. J., Jr. (1991) Comparisons of the distributions of D_1 and D_2 dopamine receptor mRNAs in rat brain. *Neuropsychopharmacology* **5,** 231–242.
83. Le Moine, C., Normand, E., and Bloch, B. (1991) Phenotypical characterization of the rat striatal neuron expression the D1 dopamine receptor gene. *Proc. Natl. Acad. Sci. USA* **88,** 4205–4209.
84. Lester, J., Fink, J. S., Aronin, N., and DiFiglia, M. (1992) Co-localization of dopamine D_1 and D_2 receptor mRNAs in striatal neurons. *Brain Res.* **621,** 106–110.
85. Rappoport, M. S., Sealfon, S. C., Prikhozhan, A., Huntley, G. N., and Morrison, J. H. (1993) Heterogeneous distribution of D_1, D_2, and D_5 receptor mRNAs in monkey striatum. *Brain Res.* **616,** 242–250.
86. Surmeier, D. J., Eberwine, J. H., Wilson, C. J., Stefani, A., and Kitai, S. T. (1992) Dopamine receptor subtypes co-localize in rat: striatonigral neurons. *Proc. Natl. Acad. Sci. USA* **89,** 10,178–10,182.
87. Ariano, M. A., Monsma, F. J., Jr., Barton, A. C., Kang, H. C., Haugland, R. P., and Sibley, D. R. (1989) Direct visualization and cellular localization of D_1 and D_2 dopamine receptors in rat forebrain by use of fluorescent ligands. *Proc. Natl. Acad. Sci. USA* **86,** 9570–9575.
88. Monsma, F. J., Jr., Barton, A. C., Kang, H. C., Brassard, D. L., Haugland, R. P., and Sibley, D. R. (1989) Molecular characterization of novel fluorescent ligands with high affinity for D_1 and D_2 dopaminergic receptors. *J. Neurochem.* **52,** 1641–1644.
89. Barton, A. C., Kang, H. C., Rinaudo, M. S., Monsma, F. J., Jr., Stewart-Fram, R. M., Macinko, J. A., Jr., Haugland, R. P., Ariano, M. A., and Sibley, D. R. (1991) Multiple fluorescent ligands for dopamine receptors. I. Pharmacological characterization and receptor selectivity. *Brain Res.* **547,** 199–207.
90. Madras, B. K., Canfield, D. R., Pfaelzer, C., Vittimberga, F. J., Jr., DiFiglia, M., Aronin, N., Bakthavachalam, V., Baindur, N., and Neumeyer, J. L. (1990) Fluorescent and biotin probes for dopamine receptors: D_1 and D_2 receptor affinity and selectivity. *Mol. Pharmacol.* **37,** 833–839.
91. Larson, E. R. and Ariano, M. A. (1994) Dopamine receptor binding on identified striatonigral neurons. *Neurosci. Lett.* **172,** 101–106.
92. Mathiasen, J. R., Larson, E. R., Ariano, M. A., and Sladek, C. D. (1996) Neurophysin expression is stimulated by dopamine D_1 agonist in dispersed hypothalamic cultures. *Am. J. Physiol.* **39,** R404–R412.

93. Rayport, S. and Sulzer, D. (1995) Visualization of antipsychotic drug binding to living mesolimbic neurons reveals D_2 receptor, acidotropic and lipophilic components. *J. Neurochem.* **65,** 691–703.
94. Artalejo, C. R., Ariano, M. A., Perlman, R. L., and Fox, A. P. (1990) Activation of facilitation calcium channels in chromaffin cells by D_1 dopamine receptors through a cAMP/protein kinase A-dependent mechanism. *Nature* **348,** 239–242.
95. Gerfen, C. R. (1992) The neostriatal mosaic: multiple levels of compartmental organization. *TINS* **15,** 133–139.

CHAPTER 4

Molecular Drug Design and Dopamine Receptors

Richard B. Mailman, David E. Nichols, and Alexander Tropsha

1. Introduction

The purpose of this chapter is to discuss the approaches we and others have used for the rational design of novel ligands for dopamine receptors. It is impossible in one chapter to provide both a detailed review of the technologies used for molecular drug design and their application to dopamine receptor drug design. Our goal, therefore, is to communicate a summary of the general strategies that are used for molecular drug design, a summary of current issues in the field and their development, and some examples from our own research with dopamine receptors that highlight the application of such methods.

There are two strategies by which computational methods are applied to molecular drug design. "Ligand-based" drug design is an inferential, but predictive process that deduces information about a receptor or other macromolecular drug targets on the basis of interactions of the receptor with known drugs. Thus, quantitative structure–activity relationship (QSAR) studies on a series of compounds (both active and inactive) can define the steric and electronic factors that allow a drug to bind to its target site. QSAR studies can lead to the design of bioisosteric molecules that are based on existing lead molecules, or in the case of enzymes, transition state mimics that can block key catalytic reactions.

On the other hand, a second approach termed "structure-based" drug design requires more detailed information about the biophysical characteristics of the receptive protein. Based on knowledge about the chemistry, structure, and function of the target macromolecule, one can theoretically design ligands that have both high binding affinity and selectivity for this "receptor."

The Dopamine Receptors Eds.: K. A. Neve and R. L. Neve
Humana Press Inc., Totowa, NJ

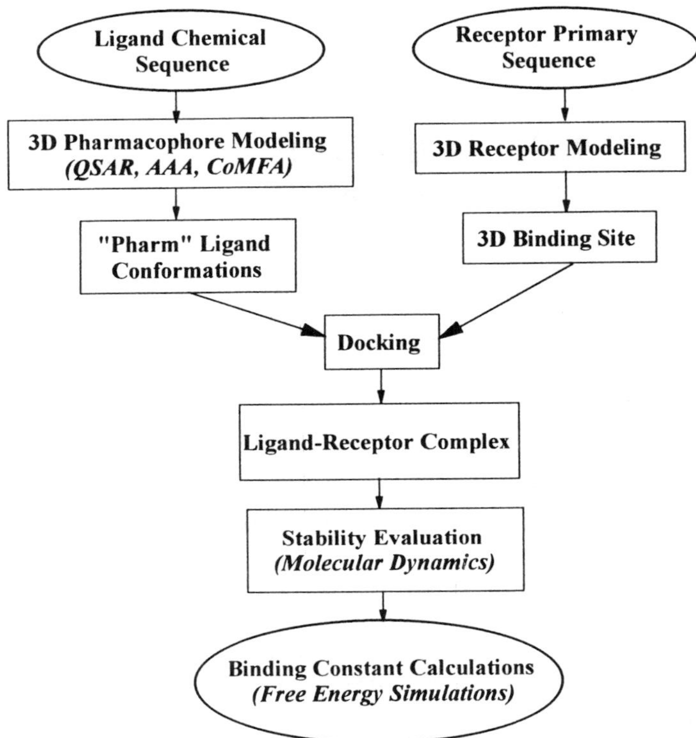

Fig. 1. The flow chart of theoretical modeling drug–receptor interaction and relation to experiment. The objects of theoretical investigation are in rectangles, and the experimental information is in the ovals.

Although this latter strategy may be both inherently more powerful and more intellectually appealing, it is often limited by a dearth of necessary structural information, as well as the demands such an approach makes on available computer hardware and software. At present, the interactive use of both approaches seems obligatory to understand the binding and activation of complex molecules like the dopamine receptors.

Experimental and theoretical methods of analysis of drug–receptor interaction should complement and enrich each other. We can schematically represent the drug–receptor interaction and its experimental findings as shown in Fig. 1. Molecular modeling uses experimental observations as input data to programs. Thus, ligand chemical structure and biological activity (e.g., binding constants) serve as experimental input in ligand-based computer-assisted drug design approaches such as QSARs, the active analog approach (AAA), and comparative molecular field analysis (CoMFA; *see* Section 2.).

Recent successes in deciphering primary sequences of many G protein-coupled receptors (GPCRs), and the development of several computerized protein-modeling tools, have enabled researchers to postulate three-dimensional models of these receptors (*see* Section 3.). These models incorporate three-dimensional, real (but, at this point, not yet realistic) receptor binding sites (i.e., constructed from real residues as opposed to a negative image of the pharmacophore) based solely on the primary structure of receptors. Finally, independent development of a specific computer modeling approach called "docking" (which allows one to bring together the ligand and the receptor, and "dock" the former into the latter) has completed the circle of computer representation of the natural process of ligand–receptor interaction (i.e., a pharmacophore-based, conformationally constrained ligand is brought into the vicinity of a modeled receptor binding site via a docking routine). This whole computerized process, together with some modeling tools applicable to evaluation and refinement of the putative ligand–receptor model, is summarized in Fig. 1. An important point illustrated here is the interaction of experiment (ovals in Fig. 1) and calculation (rectangles in Fig. 1). In our view, a model should not be considered "real" until it can predict both ligand affinity and functional behavior.

2. Ligand-Based Receptor Modeling and Drug Design

Ligand-based design has for much of the past decade had a significant impact on the design and synthesis of novel agents. This technology is now a routine part of rational drug design. As noted in Section 1., this approach involves the formulation of a hypothetical model based on known ligands for the recognition site (e.g., receptor) of interest. A key aspect involves the definition of a "pharmacophore"—the chemical features (both steric and electronic) that are believed to influence both ligand affinity for the macromolecule, and subsequent effects on functional activity. The pharmacophore model then can be used to design new ligands that are predicted to have higher affinity, greater selectivity, and/or cause different functional changes than existing drugs. This approach requires considerable data on the binding and activity of a series of ligands.

The application of computational techniques in drug design began with the development and use of the QSAR approach. This method correlates biological activity (or binding constants if available) of known compounds with their measurable or calculable physicochemical and structural parameters (e.g., the Hansch equation). Although not directly involved with receptor modeling *per se,* this avenue generated many useful and, at times, predictive QSAR equations, some of which led to documented drug discoveries *(1)*.

The advent of faster computers and the development of new algorithms led, in the late 1970s and early 1980s, to the development and use of three-dimensional negative receptor image modeling tools such as the AAA. First proposed by Marshall et al. *(2)*, this approach lets the researcher infer the size, the shape, and some physicochemical parameters of the receptor binding site by modeling receptor ligands. The key assumption of this approach is that all the active receptor ligands adopt, in the vicinity of the binding site, conformations that present to the receptor common, pharmacophoric functional groups. Thus, the receptor is thought of as a negative image of the generalized pharmacophore that incorporates all structural and physicochemical features of all active analogs overlapped in their pharmacophoric conformations. The latter are found in a course of conformational searches, starting from the most rigid active analogs and proceeding through more flexible compounds. Conformational restraints imposed by more rigid compounds with respect to internal geometry of common functional groups are used to facilitate the search for more flexible compounds. The AAA, or similar approaches, already has been successfully applied to negative-image modeling of practically all major GPCRs, yielding important leads for rational drug design. Most importantly, these models now can be incorporated into common databases providing the source of ligand structures for ligand–receptor docking studies. Furthermore, the same database can be used to search for potential specific activity or crossreactivity of independently designed or synthesized ligands by comparing them with known three-dimensional pharmacophores.

CoMFA is one of the most recent developments in the area of ligand-based drug design. This approach combines traditional QSAR analysis and three-dimensional ligand alignment central to AAA into a powerful three-dimensional QSAR tool. CoMFA correlates three-dimensional electrostatic and van der Waals fields around sample ligands typically overlapped in their pharmacophoric conformations with their biological activity. This approach, first proposed only a few years ago *(3)*, has been already successfully applied to many classes of ligands.

It is important not to underestimate the significance of ligand-based receptor modeling tools in the general process of computerized drug design. Ligand-based receptor modeling approaches are well developed and incorporated in every major molecular modeling package (e.g., SYBYL, QUANTA/CHARMm, INSIGHT/DISCOVER). The power of these approaches has always been in their ability not only to explain the (relative) activity of known receptor ligands based solely on their chemical structures but, in many cases, to predict accurately the activity of *de novo* designed drugs prior to their synthesis and experimental evaluation. Most importantly, these approaches provide limitations on the very large conformational space available to many structurally flexible ligands, focusing on conformations most likely to occur in the vicinity of the receptor binding site. The knowledge of these pharmacophoric conformations is crucial to the development of receptor models, particularly when the structure of the receptor is not known, as in the case of GPCRs. These conformations are used to ensure that the receptor model incorporates binding site complementarity to known active receptor ligands in their pharmacophoric conformation. Moreover, a receptor model tested on known ligands can be used to design new active molecules using structure-based design methods. Thus, it appears crucial to combine ligand-based receptor modeling with protein-based receptor modeling since together they provide realistic spatial limitations on the receptor binding site.

Although the ligand-based approach is of clear use in the search for new ligands and drugs, this strategy is best applied when incorporated into an iterative, interdisciplinary approach, because the modeling itself is based on the quality of the biological data used for quantitative structure–activity analysis. For example, one of the current issues with all receptor systems is the nature of the structural features of the ligand that may affect the functional changes resulting from ligand "docking" (i.e., is a ligand an agonist, antagonist, partial agonist, etc.). Ligand-based approaches have resulted in new drugs causing different functional effects, and this in turn, has led to hypotheses about the mechanisms of efficacy. New drugs provide important tools to investigate the allosteric changes resulting from ligand–receptor interaction. In most systems, such ideas have been an abstract concept, but during the next decade they should become well-understood (*see* Section 4.). One of the goals of new software is to perform automated and wide-ranging searches for new ligands that may have structurally different backbones. The development of a truly useful pharmacophore model requires an extensive database of ligands with different structures. The validity of the model is then tested with new biological data (e.g., molecular modifications or cloning), which in turn, will lead to refinements of the pharmacophoric model.

2.1. Ligand-Based Drug Design for Dopamine Receptors

It is important to review the history of dopamine receptor classification from the historical perspective of scientists interested in molecular drug design. Toward the end of the 1970s, most scientists in the field had recognized that there was no such thing as "the dopamine receptor." It had become clear that the dopamine receptor linked to stimulation of cyclic adenosine-3',5'-monophosphate (cAMP) in neural tissue *(4)* was not the site at which many actions of drugs then available took place. The D1/D2* terminology used today *(5)* reflects a differentiation that was widely recognized (e.g., *see* ref. *6*) by the time of its publication. More recently, molecular cloning studies have demonstrated that there are at least five genes in mammalian brain that code for dopamine receptors. Two of these genes code for products (D_1, D_5) that are similar to the traditional pharmacological D1 receptor. The other three genes code for several forms of D2 receptors that have been termed D_2, D_3, and D_4, of which there also are several splice variants. Although current research is beginning to make progress in differentiating the pharmacophore for these isoforms, there are still many questions remaining in differentiating drugs with selectivity for the subfamilies (i.e., D1 vs D2).

2.1.1. Foundation for Pharmacophoric Models of Dopamine Receptor Ligands

Ligand-based design of dopaminergic ligands has benefited greatly from the fact that a large number of dopamine-like agents actually incorporate elements of the dopamine structure. This is in contrast to many other classes of therapeutic agents, where a synthetic agonist often bears little resemblance to the natural neurotransmitter. In some cases this analogy is not obvious, as for example with the dopaminergic ergolines, but on closer examination the resemblance becomes more evident. And, although D2 classes of dopaminergic drugs perhaps diverge most widely from the basic dopamine structure, it is particularly fortunate that agonists for the D1 dopamine receptor generally incorporate a very obvious dopamine fragment, even to the requirement (at least to date) that they possess a catechol moiety to be full agonists.

A powerful approach in ligand-based drug design that we have employed for many years is the development of rigid analogs. If one can identify the pharmacophoric elements, it is often possible to incorporate them into a rigid framework. It is unfortunately the case that if a rigid analog is inactive, one typically cannot know which of the molecular features has led to loss of activity. Rigid analogs, however, if appropriately designed, are often powerful probes of receptor geometry. In addition, not only is it believed that they provide a good approximation of the binding conformation of the flexible template, but good rigid analogs typically have higher affinity for the receptor because entropic factors that are important in binding of the flexible ligand are less relevant.

There were several key players involved in the development of rigid agonists for dopamine D2 receptors. These efforts all led to a variety of sub-

*As detailed in the Preface, the terms D1 or D1-like and D2 or D2-like are used to refer to the subfamilies of DA receptors, or used when the genetic subtype (D_1 or D_5; D_2, D_3, or D_4) is uncertain.

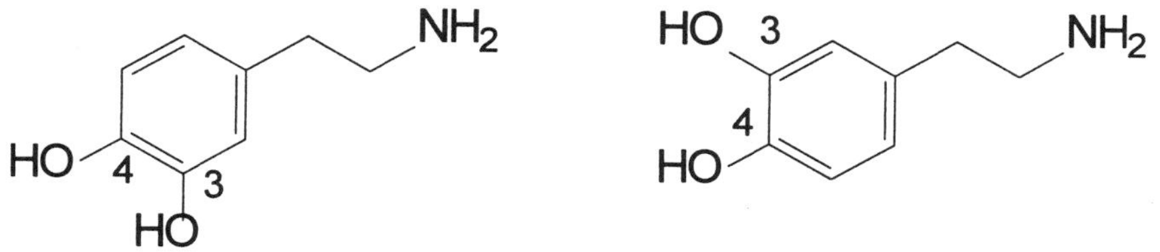

Fig. 2. The structure of dopamine represented as conformational extremes. The one designated as the alpha rotamer is on the left, and the beta rotamer is on the right *(7)*.

stituted 2-aminotetralins or benzoquinolines. Nearly all of them were evaluated prior to the recognition of subtypes of dopamine receptors, so we unfortunately know little about their structure–activity relationships at the various D2 receptors (D_2, D_3, and D_4). We can summarize the structure–activity relationships of virtually all of these molecules by describing them as rigid analogs containing a dopamine moiety locked into what Cannon *(7)* first defined as the "alpha" rotameric conformation, as shown on the left in Fig. 2.

It has been shown clearly that in rigid analogs incorporating a dopamine fragment in this conformation, the hydroxyl corresponding to the 3-position of dopamine (the "meta" position) is the most critical for receptor affinity and activation. Furthermore, although dopamine itself has relatively low affinity for D2 receptors, *N,N*-dipropyldopamine is much more potent. This feature carries through into rigid analogs, where tertiary amines, particularly with an *N*-n-propyl group, prove to be most active.

When chiral analogs are prepared, the stereochemistry in the side chain is also important for biological activity. McDermed et al. *(8)* first resolved a series of 2-aminotetralins into their enantiomers, and found that the more active enantiomers had opposite stereochemistry, depending on whether the molecule incorporated an alpha- or beta-rotameric dopamine element. This led to a general model for dopamine receptors that provided the foundation for more focused efforts that led ultimately to the design of high-affinity D1-dopamine selective full agonists like dihydrexidine. The elements of this receptor model are shown in Figs. 3 and 4. Figure 3 presents the hypothetical receptor with a dopamine alpha rotameric moiety bound, whereas Fig. 4 shows how the same receptor features were also proposed to accommodate a molecule incorporating a beta rotameric dopamine moiety.

It is important to note that at the time when these general models were proposed, it was unclear to most scientists that there was a clear differentiation between D1 and D2 receptors, let alone the subtypes within each subfamily. Thus, these models blur the important distinctions between the D2- and D1-like receptors. Although Goldberg et al. *(9)* first proposed that an accessory bind-

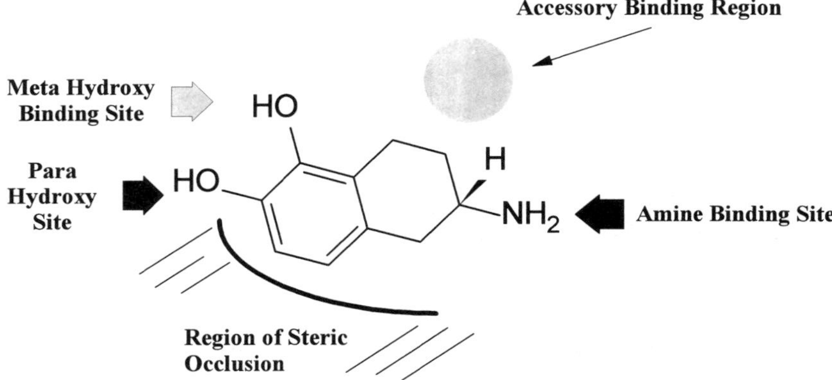

Fig. 3. This schematic receptor model is based on a proposal modified from McDermed et al. *(8)*. This figure shows the interaction with an alpha-rotameric dopamine moiety. The stereochemistry of the depicted more active enantiomer is designated as S.

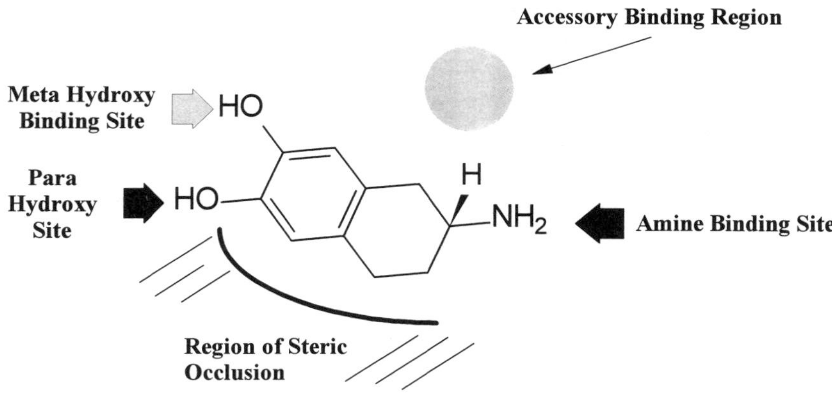

Fig. 4. This schematic receptor model is based on a proposal modified from McDermed et al. *(8)*. This figure shows the interaction with a beta-rotameric dopamine moiety. The stereochemistry of the depicted more active enantiomer is designated as R.

ing region might exist on the receptor to accommodate the nonhydroxylated phenyl ring of apomorphine (now known to be a nonselective D1/D2 ligand), it was not clear until more recently that this could be a distinguishing feature for the development of subtype-selective ligands.

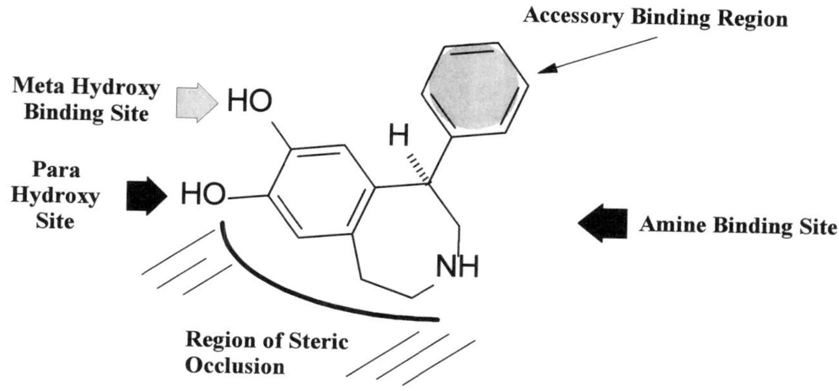

Fig. 5. Schematic receptor with the active enantiomer of SKF 38393 in place.

When SKF 38393 was discovered to be a D1-selective ligand *(10)*, the importance of exploiting the accessory binding region to develop selective D1 ligands began to be realized. This was the basis for the working hypothesis that a "beta-phenyldopamine" moiety might serve as the pharmacophore for D1-selective ligands. This idea is presented schematically in Fig. 5, in which SKF 38393 is placed into the working receptor model.

The most salient problem with the two-dimensional receptor model is the fact that the side chain of SKF 38393 is not transoid, as in most other rigid analogs, but is more cisoid, and the azepine ring has some degree of conformational flexibility. These features are not easily accounted for in the model. It should also be noticed that SKF 38393 has been oriented into the receptor as a beta rotamer of dopamine. This appears to be another structural feature that distinguishes D1- from D2-like ligands.

Dopamine ligands also were known to lose their affinity for the D1 receptor if the amine nitrogen was alkylated. Whereas D2 ligands have optimal affinity when the amino group is tertiary, and particularly when one alkyl group is an n-propyl, this structural modification dramatically decreases D1 affinity (and in agonists, decreases efficacy). This led to the working hypothesis that large alkyl groups might occupy an equatorial orientation, forcing interaction between the protonated amine lone pair and an anionic site in the receptor (possibly a conserved aspartate in TM 3) along a vector perpendicular to the plane of the molecule (Fig. 6). At the D2 receptor, this orientation may be optimal, whereas at the D1 receptor the protonated amine lone pair is forced to approach the anionic site aligned with a vector lying roughly in the plane of the molecule (Fig. 7). Such data provided the basis for more focused development of ligands for D1-like receptors.

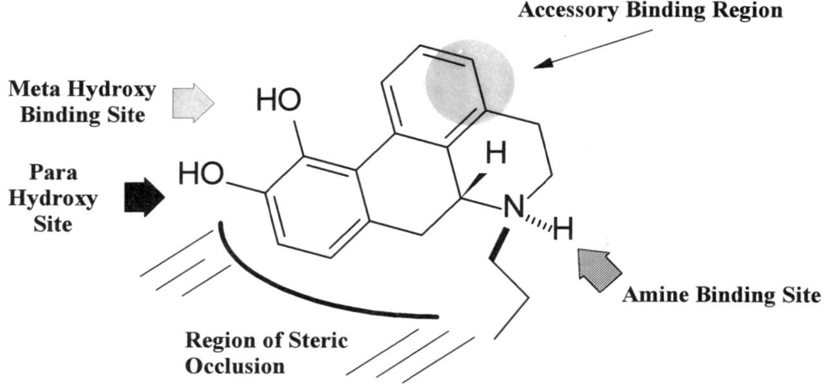

Fig. 6. *N*-n-propyl-norapomorphine is shown in place in a hypothetical D2-like receptor.

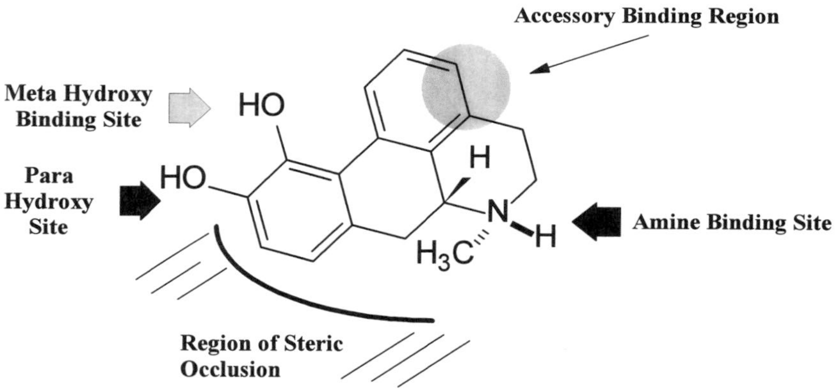

Fig. 7. Apomorphine is shown in place in a hypothetical D1-like receptor.

2.1.2. Ligand-Based Drug Design for the D1 Receptor

Whereas the models discussed in the previous section served as a focus for many medicinal chemists, by the mid-1980s neuropharmacological advances made these matters of more than heuristic interest to a variety of neuroscientists. Following the demonstration of the critical role of the D1 receptor in brain function *(11–13)*, a variety of converging lines of evidence made it clear that the pharmacological armamentarium to characterize the significance of this receptor was woefully lacking, both for basic investigation of the function of this receptor, and for potential clinical uses. Although the

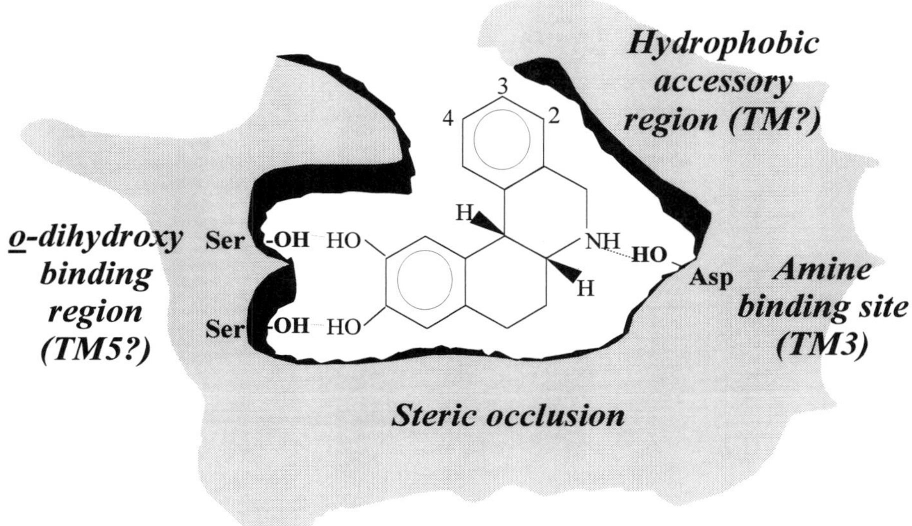

Fig. 8. Model of ligand interaction with the D1 receptor. The molecule shown "docked" is dihydrexidine, the first high affinity full D1 agonist.

selective antagonist SCH 23390 had been reported in 1983 *(14)*, the available agonists were also of the same structural class (i.e., phenyltetrahydrobenzazepines), and most were either of partial intrinsic activity in rat striatum *(10,15)* or had other pharmacological limitations *(16,17)*. It was for this reason that we began to focus on the atomic and molecular factors involved in ligand recognition by the D1 receptor, and the mechanisms by which a ligand becomes a full or partial agonist, or an antagonist. Coincidentally, it was during this same period that the combination of improved software and affordable hardware allowed the first widespread use of computer-assisted molecular drug design. Our group had the goal of the development of new and improved ligands that either had specific functional characteristics (e.g., being a full agonist), and/or were selective for molecular subtypes of the dopamine receptors (i.e., D_1 vs D_5). Because we have had some success in achieving the first of these goals, we discuss this in more detail.

The next conceptual advance from the model shown in Fig. 7 is the somewhat more detailed model shown in Fig. 8 (*see* ref. *18*). Essentially, this model proposes that several factors influence the affinity of a ligand for the D1 receptor, and the ability of a ligand to act as an agonist. An intrinsic part of this model is the hydrophobic accessory region. One hypothesis inherent in

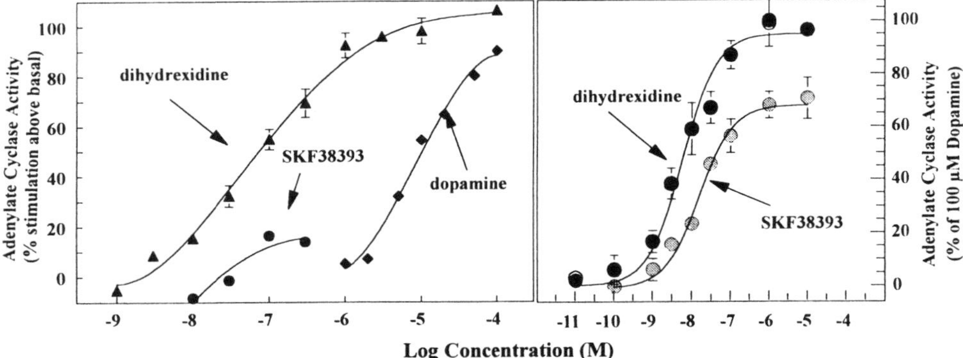

Fig. 9. Dihydrexidine is a full agonist. The left-hand frame illustrates results using rat striatal homogenates, whereas the right-hand frame is an experiment using C6 cells transfected with the monkey D_1 receptor. Note that dihydrexidine causes functional activation (i.e., increased cAMP synthesis) equal to dopamine in both preparations, whereas SKF 38393 is only partially effective. The relatively greater intrinsic activity of SKF 38393 in the C6 cells vs rat striatum reflects the greater receptor reserve in this particular preparation (*see* ref. *25*).

this model is that a key determinant of whether a ligand is a partial agonist or antagonist, rather than a full agonist, is the interaction of the ligand with this hydrophobic accessory region. In addition, the orientation of the dopamine backbone, elucidated from ligand-based modeling of D1 antagonists and analogs, suggested that the size and shape of the hydrophobic group and its spatial orientation were of importance *(19)*. Based on this scheme, one of our foci has been on modeling the agonist pharmacophore, with special concern for functional groups that may be required for activation of the D1 receptor.

These efforts resulted in the synthesis of dihydrexidine, the first full D1 agonist. Figure 9 illustrates the differences between dihydrexidine and the partial agonist SKF 38393. Not only does dihydrexidine have significant affinity for the D1 receptor ($K_{0.5}$ = 10 n*M*), but unlike available D1 agonists, it is as effective as dopamine in activating the D1 receptor *(15,21)*. This drug appears to have full intrinsic activity in that it has efficacy equal to dopamine in every preparation tested, including those with little or no receptor reserve *(22–25)*. Dihydrexidine is also bioavailable to brain after parenteral administration. These data provided the basis for using dihydrexidine as a tool to study brain function. It is particularly noteworthy that full agonists have, in several paradigms, been reported to cause functional effects in the intact organism not seen with agonists of less than full intrinsic activity *(26–28)*. The mechanism(s) for such effects are unclear, but make these types of studies of some importance.

2.1.3. The D1 Agonist Pharmacophore

Recently, we have used computer-aided conformational analysis to refine the agonist pharmacophore for D1 dopamine receptor recognition and activation that led to the synthesis of dihydrexidine *(29)*. These modeling efforts relied on dihydrexidine as a structural template for determining molecular geometry because it is not only a high-affinity full agonist, but it has limited conformational flexibility relative to other more flexible agonists. Using the AAA to pharmacophore building *(2)*, conformational analysis and molecular mechanics calculations were used to determine the lowest energy conformation of the active analogs (i.e., full agonists), as well as the conformations of each compound that displayed a common pharmacophoric geometry. We hypothesized that dihydrexidine and other full agonists may share a D1 pharmacophore made up of two hydroxy groups, the nitrogen atom (~7 Å from the oxygen of the meta-hydroxyl) and the accessory ring system characterized by the angle between its plane and that of the catechol ring (except for dopamine and A77636). For all full agonists studied (dihydrexidine, SKF 89626, SKF 82958, A70108, A77636, and dopamine) the energy difference between the lowest energy conformer and those that displayed a common pharmacophore geometry was relatively small (<5 kcal/mol). The pharmacophoric conformations of the full agonists were also used to infer the shape of the receptor binding site. Based on the union of the van der Waals density maps of the active analogs, the excluded receptor volume was calculated. Various inactive analogs (partial agonists with D1 $K_{0.5} > 300$ n*M*) subsequently were used to define the receptor essential volume (i.e., sterically intolerable receptor regions). These volumes, together with the pharmacophore results, were integrated into a three-dimensional model estimating the D1 receptor active site topography (Fig. 10).

2.2. The Utility of Three-Dimensional Models

As is clear from the historical review presented earlier, chemical intuition has been a keen component of molecular drug design for a century. What is now equally clear is that computer-assisted drug design is not only useful, but can accelerate drug design exponentially. Two examples from our own recent experience may illustrate this point. The first illustrates that the model is adequate, at least for current compounds. Dihydrexidine (10,11-dihydroxy-hexahydrobenzo[*a*]phenanthridine) can exist in four stereoisomers. Early on, it was shown that the *trans*-, and not the *cis*-, compounds were the active species, but because of the difficulty of resolving the enantiomers, it was unclear whether the 6aR, 12bS-, or 6aS, 12bR-isomer would be active. From the modeling, it was quite clear that the former should be the active species, and this was confirmed by the eventual separation and testing of the enanti-

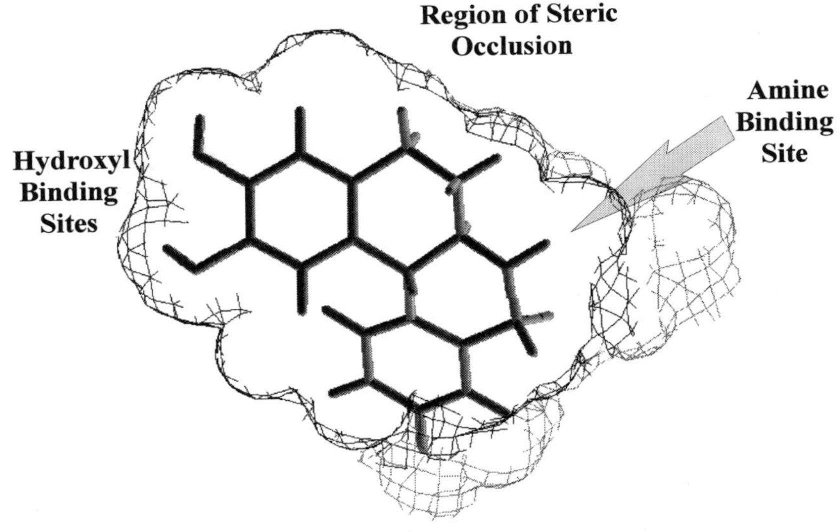

Fig. 10. Excluded volume for the D1 agonist pharmacophore. The mesh volume shown by the black lines is a cross-section of the excluded volume representing the receptor binding pocket. Dihydrexidine is shown in the receptor pocket. The gray mesh represents the receptor essential volume of inactive analogs. The hydroxyl binding, amine binding, and accessory regions are labeled, as is the steric region originally described by McDermed et al. *(8)*.

omers *(30)*. Of course, we would agree that a model that could not even predict its own template is valueless, and its utility is in predicting the activity of compounds *a priori*.

The second example is an interesting anecdote. In considering novel structures that would meet the criteria of our model, one exciting candidate was 8,9-dihydroxy-2,3,7,11b-tetrahydro-1*H*-naph[1,2,3-*de*]isoquinoline, a molecule that contains a rigid β-phenyl dopamine pharmacophore fitting elegantly into the model shown in Fig. 10. When the first small amount of the final product of the initial synthesis of this molecule was tested, we were shocked to find it had no D1 affinity! Yet, so confident were we in the foundation of this work, that we nicknamed this first product "*uncertamine*" and hypothesized that, despite its apparently correct molecular formula, it was not, in fact, the expected material. Subsequently, the correct compound was synthesized by a different route. Biological testing demonstrated that this latter material had high D1 affinity (almost identical to dihydrexidine), and also was a full D1 agonist. "*Uncertamine,*" as predicted by the molecular modeling, was structurally different and had been produced by an unexpected quirk of the chemistry.

2.3. Current Issues in Ligand-Based Design

Although for primary amines and secondary amines the accessory phenyl ring of molecules such as SKF 38393 and dihydrexidine seems to confer D1-like receptor selectivity, when an *N*-n-propyl group is placed on the basic nitrogen of dihydrexidine, the molecule becomes a potent D2-like receptor ligand. This suggests that some basic reorientation of the ligand in the receptor must occur when the nitrogen atom is alkylated. If a conserved aspartate in helix three binds to the basic nitrogen atom, it seems quite possible that it could interact with the protonated amine either from an equatorial or axial direction, with respect to the plane of the ligand molecule. These different binding approaches would clearly necessitate a reorientation of the ligand within the recognition domain of the receptor, and the accessory phenyl (or other hydrophobic group) might find complementarity in the D2 receptor, whereas it was not tolerated within the binding cavity of the D1 receptor. It will be interesting to see how easily structure-based design can accommodate these apparently divergent binding orientations, and indeed, the ability to accommodate them will of necessity be a key test of the validity of any dopamine receptor models.

3. Structure-Based Drug Design

3.1. Introduction

Two factors provide formidable challenges to structure-based drug design. The three-dimensional organization of the target macromolecule (e.g., receptor or transduction protein) must be understood, yet the necessary data are often unavailable, of inadequate resolution, or of analogous molecules rather than the one of specific interest. In addition, the computational algorithms and molecular graphics techniques necessary to solve such problems are themselves undergoing active development and modification. In the face of these challenges, however, is the plethora of new data on biological macromolecules resulting from the widespread application of molecular biological techniques. To use these data requires overcoming several important theoretical hurdles.

An important one is termed the "protein folding problem;" this is sometimes seen as the Holy Grail of computational methods. It has been known for decades that, in vitro, a denatured protein can often, under appropriate conditions, spontaneously refold into a correct and functioning three-dimensional structure *(31)*. These data suggest that the three-dimensional structure of the protein somehow is encoded solely in the amino acid sequence (second genetic code). Clearly, a computational solution to the protein folding problem would instantly (at least given enough computer hardware) convert the huge genomic database into a gold mine of structural information that could be used for many purposes. This information would encompass not only rational drug design, but also the fundamental understanding of biomolecular mechanisms, and new approaches to gene therapy and protein engineering.

The prediction of three-dimensional structure from sequence information has been approached using existing structural data, and also by attempting *de novo* folding via physical simulation and the exploration of a large conformational energy landscape. Methods utilizing existing structural knowledge include secondary and tertiary structure prediction based on local sequence information—helix, sheet, turn propensity matrices, and homology model building. Homology modeling has probably been the most widely used, and successful, at least in the area of drug design in which, in some instances, the three-dimensional structure of the target receptor may be unknown, but a highly related protein structure is known. Computer programs exist to substitute the appropriate sequence into the known structure, and then energy minimize the resulting structure to give a reasonable model for the target. Although this approach has been successfully used, it is still fraught with difficulties, and is limited to the few situations where homologous structures are known.

There has been an assumption that when a target receptor structure was well-characterized it would then be relatively easy to calculate the binding and energetics of particular ligands to the receptor. Although determination of receptor structure was thought to be key to structure-based drug design, recent evidence has shown that a structural model of the receptor may be necessary, but not sufficient, for effective structure-based design. It has, for instance, been exceedingly difficult to correlate experimental binding constants of ligands to calculated energy values. One aspect of this relates to the accurate and complete treatment of solvation and desolvation of both ligands and receptors. An especially important problem concerns the structural changes in the receptor that may occur on binding the ligands (induced fit). The current methods, although useful, are limited, and new techniques are needed to accurately represent changes in solvation and in local structure when ligands bind to receptors.

The issue of docking is of particular relevance to this approach. In terms of receptors, this involves first the docking of ligands to receptors, and then the docking of one protein with others (protein–protein interaction). The problem of representing flexibility in the receptor in addition to the ligand is still largely unsolved. One possible approach is to identify regions of limited motion within the receptor site and represent those side chains as pseudoligand atoms, constraining all but their torsional degrees of freedom.

Another important technique that is used to predict the interaction of ligands with receptors involves the scanning of small molecule three-dimensional databases to identify structures that match a "shape cast" of the receptor site. The goal of programs such as DOCK is to find new classes of compounds as potential drug leads for a well-defined target site. Since the search is based principally on the shape of the receptor site, and the structures of the small molecules as they are defined in the database, it may overlook promising compounds if the bound conformation differs from those represented in the database. Also, changes in the shape of the receptor site on complexation cannot be accounted for in the current methods. Nonetheless, the technique has been successfully used to suggest new classes of compounds as possible leads for AIDS and cancer therapies.

3.2. *Automated* de Novo *Drug Design*

One of the long-term goals of structure-based drug development is the automatic design of candidate compounds based on the characteristics of the target receptor. Many of the methods described represent steps in that direction. Receptor mapping and pharmacophoric analyses require data from existing lead compounds. Automated docking techniques require the identification of a candidate ligand. Shape-based searching techniques can automatically identify candidates from a preexisting database. Recently, new techniques have been developed that can algorithmically design trial ligands directly into a structurally characterized receptor site. Two approaches to the problem have been called "outside-in" and "inside-out." The "outside-in" approach will usually start with the identification of favorable sites for interaction within the receptor binding site. These sites, which may be hydrogen bond acceptor or donor sites, charge sites, or hydrophobic pockets, determine the placement of complementary molecular fragments. The construction algorithm then must find a reasonable molecular framework to link these fragments together into a candidate ligand. The "inside-out" approach starts with an initial fragment, or seed, placed in the binding site. Atoms or fragments are then iteratively attached to the seed fragment and to each other to build up a molecule that will reach the favorable receptor sites. A potential weakness of the "inside-out" approach is that it may not be possible to reach the interaction sites from a given seed position with the available building blocks. On the other hand, the "outside-in" approach may not always be able to connect the fragments together in a reasonable way.

3.3. *An Integrated Approach to Modeling GPCRs*

Before approaching the problem of modeling GPCRs, it is important to define the requirements for a viable receptor model. In our view, such a three-dimensional model must satisfy several criteria as follows:

1. It must be a reasonable protein model from the general protein architecture viewpoint. In the case of GPCRs, which are believed to be α-helical in their transmembrane regions, side chain conformations should be, in most cases, close to their average rotamers found in known α-helices, and the side chains from adjacent helices should be well packed against each other.
2. It must contain binding site complementarity to known receptor ligands in their optimized (pharmacophoric) conformations. This is obtained by applying traditional, pharmacophore modeling tools to known receptor ligands.
3. It must provide at least qualitative explanation(s) for different binding affinities of receptor ligands. The calculated relative binding constants of receptor ligands must follow the experimental pattern.
4. It must distinguish, on a structural basis, among agonists, antagonists, and partial agonists of the same receptor.
5. It must be receptor-specific (i.e., explain specificity of known receptor ligands toward the given receptor as opposed to receptors of other ligands).
6. As a highly desirable property, it should explain the signal transduction mechanism coupling ligand binding with receptor activation (e.g., conformational change, activation of transmembrane ion current, etc.).

We would argue that the receptor model that satisfies all these criteria will be a true, realistic model. We suggest, however, that the first two, and to some extent the third criteria, are absolutely necessary features of a reasonable receptor model, whereas the last three criteria may provide a means to select the most viable model from among several possible candidate models that will pass the initial criteria.

In order to satisfy these criteria, the tentative three-dimensional receptor model should be built on the basis of protein homology modeling tools. The model must incorporate a putative binding site (i.e., a surface cavity in the receptor structure complementary to known receptor ligands). Since, in most cases, the receptor ligands are conformationally flexible molecules, it is generally impossible to select *a priori* the conformation of the ligand that would be complementary to the binding site. This is a crucial choice since it will dictate the specific shape, size, and physicochemical properties of the postulated binding site. Ligand-based receptor modeling tools (e.g., AAA) generate this crucial information about the conformations of ligands in their receptor-bound form. Once the putative receptor model is generated, and the most probable ligand conformation is selected, the (preferably, interactive) docking of the ligand into the receptor must be performed to complete the ligand-bound receptor model. Further, since docking is a molecular mechanics-based routine and the docked model may be trapped in a local minimum, the stability of the proposed model should be evaluated using molecular simulation techniques such as molecular dynamics.

At this point, only two types of experimental data are used for receptor modeling: the ligand chemical structure (from which one generates the three-dimensional pharmacophore conformation) and the primary receptor sequence (which is used to generate the three-dimensional receptor model). The most important property of the receptors, however, is their ability to discriminate ligands on the basis of their chemical structure; we quantify this as ligand binding constants. The ability to reproduce binding constants, or, at least their relative order, is the most sensitive test of any putative receptor model. The technique that, in principle, can provide such data is the free energy simulations/thermodynamic cycle approach, a method that relies on full atomistic representation of both ligand and receptor in the molecular model *(32)*. This approach is very robust, and requires substantial computational resources that limit its practical applicability for drug design. In several reported cases, however, it was able to reproduce accurately the experimental binding constants *(33,34)*.

Three-dimensional receptor modeling represents a major molecular modeling effort incorporating the best experience accumulated so far in different areas of molecular modeling/computational chemistry. The development of an adequate receptor model requires integration of all existing knowledge about structure–activity relationships of known ligands, primary sequence of the receptor, results of mutational experiments, and structural data on homologous proteins such as rhodopsin.

3.4. Structure-Based Drug Design: Dopamine and Other GPCRs

In the absence of any experimental information on the three-dimensional structure of dopamine and other GPCRs, predictions from primary sequence remain the only source of generating the receptor structure. Earlier attempts to predict the structure of the GPCRs were limited to very general seven transmembrane cylinder models based on the easily identifiable stretches of largely hydrophobic amino acid residues in the primary sequence of these receptors. Recent determination of a middle-resolution crystallographic structure of bacteriorhodopsin *(35)*, a seven-helix transmembrane protein with very limited sequence homology to GPCRs, generated numerous attempts to model the GPCRs at atomic resolution. The theoretical basis for these attempts was the proposed structural similarity between bacteriorhodopsin and the GPCRs, despite, again, the extremely low sequence homology between the receptors and bacteriorhodopsin (6–11%; *36*); it is important to note that in the general case of protein homology modeling, these considerations alone would exclude bacteriorhodopsin as a possible candidate template structure (e.g., ref. *20*). Yet, the compelling interests and needs of investigators have led to several such models during the last few years (*see* ref. *37* for a recent review), including ones for dopamine receptors *(38–42)*. Most of the studies used the helix arrangement of bacteriorhodopsin (BR) as a template for model building; one group, modeling the $5HT_2$ receptor *(43)*, has proposed a different assembly of helices. In the majority of these studies, only the transmembrane portion of the receptor was modeled (however, *see* ref. *44*); this is justified by the fact that only the transmembrane fragment of BR was determined experimentally *(35)*, and that the binding site is presumed to be buried in the interior of the receptor in a location approximately analogous to the retinal moiety in BR. Recently, a low-resolution projection map of bovine rhodopsin was published. In contrast to bacteriorhodopsin, rhodopsin shares modest sequence homology with most of the GPCRs, and is itself a GPCR *(45)*. This map shows basically the same arrangement of seven transmembrane helices as in bacteriorhodopsin, except that the orientation and mutual tilt of the helices is somewhat different. These data provide additional evidence that GPCRs share a seven transmembrane helix motif and similar arrangements of the helices.

It is important to realize that the accurate prediction of the three-dimensional structure of GPCRs is an extremely difficult task, and with the present relatively meager knowledge of how to predict protein structure from sequence in general, it is almost impossible to claim that anyone has devised an accurate model of any GPCR. Nevertheless, modeling receptor structure is still a very useful excursion that, in general, allows the researcher to attempt to think about the receptor protein in three dimensions and to relate available experimental information from protein mutagenesis and ligand-binding studies to the receptor three-dimensional organization. The purpose of this section is to review briefly the general approaches to receptor modeling, and then to discuss in more detail the available models for dopamine receptors.

3.4.1. GPCR Homology Model Building

The template-based protein homology model-building process involves three major steps:

1. Sequence alignment of the target (modeled) protein with the template;
2. Target structure generation based on template modification; and
3. Final model analysis and refinement.

In general, proteins with sequence homology to the target protein of <30% are not selected as templates *(20)*, whereas the sequence homology between bacteriorhodopsin and any of the GPCRs is in the range of 6–11%

(36). As a result, traditional sequence alignment methods are not capable of generating an unambiguous alignment. For the alignment of GPCRs and bacteriorhodopsin, a different approach is used. The alignment of the target (GPCR) to the template (BR) is usually deduced from the alignment of the hydrophobic α-helical regions.

Once the alignment of a GPCR to the bacteriorhodopsin template has been decided, the next step is to assign the GPCR residues to the BR structure template. This means that a residue that is aligned against a particular residue of bacteriorhodopsin must assume the position of that residue in the three-dimensional structure of BR. This assignment is usually done in one of three ways. In the first, the most simple approach, the side chains of BR residues in the three-dimensional structure are mutated into corresponding residues of a GPCR (e.g., *41*). Thus, a chimeric protein is generated that has the backbone of BR and the side chains of the target protein. This protein is then subjected to structure refinement using molecular mechanics and/or molecular dynamics simulations to relieve bad steric contacts. This approach generates a structure in a straightforward manner that can be easily automated for practically any GPCR. One technical problem that arises on model building with this approach concerns the proline residues of both the template and target proteins. The prolines introduce a kink in the structure of α-helices that obviously influences the general organization of the transmembrane helices. When a proline residue of the template is mutated into some other residue, the kink stays in place since the modification only influences the side chain composition. For the same reason, the kink is not introduced automatically when a residue of the template is changed into a proline. Molecular mechanics optimization of the structure subsequent to initial model building eventually will correct the geometry of the modeled protein. Intrinsically poor initial geometry of the model, however, brings unnecessary complications to the process of model building and refinement.

To circumvent the proline problem, a slightly different approach to template-based model building was proposed. In this approach, the helices of the target protein are built separately and then superimposed onto the corresponding helices of bacteriorhodopsin (e.g., *39,46*). The structure alignment of helices is done in such a way that the residues of the target are superimposed with the positionally equivalent residues of the template (the position equivalency is based on the primary sequence alignment). The structure generated with this approach still requires further refinement and fine tuning with respect to helix orientation and register.

A different approach was employed in several recent publications *(47–49)*. Here, the model generation starts from multiple sequence alignment of homologous GPCRs such as, for example, sequences of cationic amine receptors. Particular care is taken to align residues conserved across all the receptors; it is suggested that the conserved residues play a structural role and are responsible for maintaining similar three-dimensional contacts in all the receptors. Based on this alignment, the boundaries of each helix are determined. The helices are then built individually, similar to the second structure generation method just described. The structure of bacteriorhodopsin, or, more recently, of bovine rhodopsin *(45)* is used as a general structural motif with respect to helix arrangement. The actual helix orientation and register is selected based on several considerations; thus, the conserved residues in the spatially adjacent helices must make contacts with each other, and no polar residues found in postulated transmembrane domains should be exposed to the lipid phase but should rather face the interior of the receptor and make contacts with either each other or the backbone.

3.4.2. Active Site Modeling

Early molecular modeling studies of GPCRs (e.g., ref. *42*) have been primarily concerned with finding reasonable orientations, register, and amino acid contacts between seven transmembrane helices. However, the major feature of each particular receptor that distinguishes it from other receptors is the ability to bind selectively ligands specific to that receptor. In early studies a single ligand, most often a native neuromediator (e.g., dopamine) was docked into the postulated active site, mainly to demonstrate that it could be reasonably well accommodated within the active site. However, it has been recognized *(41,50)* that accurate receptor modeling can only be realized if the homology model building is combined with the results of ligand-based pharmacophore receptor modeling (as discussed in Section 2.1. for dopaminergic ligands). Thus, the active site in the proposed model must be able to accommodate all known active receptor ligands in their pharmacophoric conformation.

Several authors have attempted modeling of various dopaminergic receptors *(38–42)*. Trumpp-Kallmeyer et al. *(42)* developed a simple D_2 receptor model along with several models for other GPCRs and suggested that the active site is formed between helices 3, 4, 5, and 6 in the neighborhood of Asp-308 (i.e., the eighth residue in the third helix near the top of transmembrane helix 3). This residue is believed to be responsible for interaction with the cationic portion of ligands. The authors have shown that dopamine can be easily accommodated in the active site of the model such that the cationic head of the ligand interacts with the Asp-308, and the rest of the ligand binds into a hydrophobic pocket formed by the adjacent aromatic residues in helices 4, 5, and 6.

A similar approach was taken by Livingstone et al. *(39)*. The authors modeled D_2, D_3, and D_4 receptors, and showed that the proposed models may account for affinity but not receptor specificity of dopaminergic ligands. The latter problem represents a challenge for molecular modelers; its resolution requires more sensitive construction of the active sites of different receptor subtypes, taking into account differences in their primary sequences. Hutchins *(38)* proposed somewhat different models of ligand binding to D_1 and D_2 dopamine receptors based on the analysis of primary sequence similarity, site-directed mutagenesis data, and structure–activity relationships for several receptor ligands. Those models also are able to accommodate a variety of dopaminergic ligands, and can explain the existing structure–activity relationships.

Detailed molecular models of D_2 and D_3 dopamine receptors were developed by Malmberg et al. *(40)*. The authors showed that several 2-aminotetralins can be docked into the active site of both receptor models so that their protonated nitrogen interacts with the Asp-308, the aromatic ring of the ligands makes face-to-edge interaction with a phenylalanine of TM 6, and the hydroxyl groups of the ligands interact with a Ser residue in TM 5. They also suggested that antagonists have a different mode of binding than agonists. Thus, the model is able to explain qualitatively the existing structure–activity relationships of known aminotetralins. This model agrees with other models of dopamine receptors in terms of active site residues involved in interaction with the ligands. Regarding the adequacy of model building discussed earlier, a major criterion should be the ability to dock structurally diverse ligands. Thus, a major limitation of the work of Malmberg et al. *(40)* is that it is focused only on one class of compounds, the aminotetralins. Prior to the submission of their work, our group had reported that one analog of dihydrexidine (4-methyl-*N*-n-propyl dihydrexidine [4-MNPrDHX]) had approximately the same D_3/D_2 selectivity as 7-OH-DPAT *(51)*. Because both 4-MNPrDHX and dihydrexidine are semirigid ligands having markedly different D_2/D_3 selectivity, it would seem they provide an important test of any receptor model purporting to predict the D_2 and D_3 active sites.

In the majority of papers referenced, the active site residues are guessed from the alignment of the receptor and bacteriorhodopsin helices. An interesting and promising approach to more accurate mapping of the active site residues was recently proposed by Javitch et al. *(52,53)*. The approach, termed the substituted-cysteine accessibility method, allows the identification of residues exposed in the binding site crevice of the receptor. Using this method, the authors first showed that a cysteine residue in the middle of the third transmembrane segment of the D_2 receptor is exposed in the active site *(53)*. Later, by substituting cysteines for several residues in various transmembrane domains, they were able to identify 10 residues that are exposed in the active site. These experiments provide very important constraints for the refinement of the alignment and mutual orientation of the receptor helices on model building.

The modeling of GPCRs remains an exceptionally challenging task. The most advanced models obtained so far have been able to combine and explain various pieces of information about these receptors coming from homology model building, site-directed mutagenesis, and pharmacophore modeling. The real utility of these models for drug design, however, can only be assessed if they possess the power to predict structures of novel receptor ligands of high affinity and predictable functional characteristics. So far, no such studies have been reported in the literature. Despite this fact, the growing body of experimental information on receptors and their ligands, combined with rapid developments in the area of database searching and receptor-based drug design, promise that the quality of the models ultimately will be improved to afford such new leads.

4. Conclusions and Future Goals

Our work has focused on the D1 class of dopamine receptors. By using the traditional tools of neuropharmacology, medicinal chemistry, biochemistry, and molecular biology, combined with computational chemistry, we have been able to understand aspects of the function of these receptors, and to use these data for the design of novel drugs with both clinical and research utility. Although the goals of our work have been focused on dopamine receptors, these directions are aimed at fundamental issues that are common to understanding receptors and, more generally, other macromolecular targets.

Ideally, we would like workable models of all of the dopamine receptors, models sufficient to design and predict the effects of new ligands (e.g., ligand-receptor docking). Although this is still years away, the present tools permit exciting advances of the types illustrated herein. It may be of use to note some of the questions and ideas that are likely to be of importance in the next decade. One obvious and traditional goal is the need for new drugs with selectivity for one specific molecular receptor subtype—the proverbial pharmacological "magic bullet." In this regard, there have been several recent advances regarding new dopamine receptor ligands. In addition to our work with D1 ligands, scientists at Abbott Pharmaceuticals (Chicago, IL) have achieved similar results with the family of isochromans (e.g., ref. *54*). Also as we discussed earlier, there have been several interesting leads toward the development of D_3-selective ligands *(51,55),* and it is anticipated that new selective agonists and antagonists will emerge for all of the dopamine receptor subtypes in the near future, even for subtypes of great homology (e.g., the D_1 vs D_5 receptors).

As noted, the focus in drug design has been for drugs with specific types of action at a single receptor. In fact, it may be that a "magic shotgun" also will be of clinical use in many cases. Such drugs would target two or more receptors simultaneously, with the goal of obtaining a therapeutic benefit not possible with action at only one receptor. (As an example, a drug that is a combined D1 agonist and D2 antagonist might be effective against both positive and negative symptoms of schizophrenia and have few neurological side effects.)

An interesting hypothesis has arisen from our pharmacological characterization of dihydrexidine. The receptor models discussed had led us to expect that this drug would have high D1, but low D2, affinity. In fact, we were surprised to find the drug was only 10-fold D1 selective. In pursuing what was initially expected to be a routine functional characterization of these properties, we found that dihydrexidine, as well as some analogs like *N*-n-propyldihydrexidine (NPrDHX), had a totally unique and distinct D2 receptor functional profile. They act as agonists only at postsynaptic D2-mediated functions, not at presynaptic functions, although they bind equally well to both sites. Some of the postsynaptic functions (mediated primarily by receptors located on nondopaminergic cells) include causing D2-mediated inhibition of cAMP efflux in rat striatal slices, and D2-mediated increases in potassium conductance in pituitary lactotroph cells. In vivo, both dihydrexidine and NPrDHX act as D2 receptor agonists to inhibit prolactin secretion by activating D2 receptors present on pituitary cells. In marked contrast, there is a surprising lack of effect at D2 receptor-mediated functional effects of dihydrexidine and analogs associated with presynaptic action. These hexahydrobenzo[a]phenanthridine analogs neither activate those D2 receptors controlling dopamine synthesis or release, nor do they inhibit firing of dopamine neurons. Finally, the effects of these atypical agonists are owing to neither actions at nondopamine receptors, nor to these drugs acting as indirect dopamine agonists.

Such data raise the possibility that drugs may be designed that are functionally selective for the same receptor subtype. This raises the second concept that the functional effects of receptor occupation may depend on the cellular milieu of the receptor. One consequence of this is that a drug may be an agonist in one brain region, yet have antagonist effects in another. This may lead to new ways of tailoring drug action. It is known that antagonism of presynaptic receptors can cause effects that are undesired when one wishes to achieve an effect by antagonizing postsynaptic receptors. Thus, if a drug has opposing functional effects at presynaptic vs postsynaptic receptors, significant therapeutic utility might result. As these examples illustrate, recent basic advances in the understanding of synaptic function have and will continue to have an impact on the design of novel psychotherapeutic agents.

There is little doubt that the next decade of neuropsychopharmacology will see a level of understanding of the nervous system, and availability of new therapeutic modalities, unforeseen a decade ago. These advances will depend on the use of integrated approaches in which disciplines such as molecular biology, cell biology, and computational chemistry/molecular modeling are coupled with modern psychopharmacological techniques. The ability to use such interdisciplinary approaches will be key to the advances of the next decade.

Acknowledgments

This work was supported by PHS grants MH40537 and MH42705, Center Grants HD03310 and MH33127, and a research grant from Hoechst Marion Roussel/Hoechst Cellanese Corporation.

References

1. Boyd, D. B. (1990) Successes of computer-assisted molecular design, in *Reviews in Computational Chemistry, vol. 1* (Lipkowitz, K. B. and Boyd, D. B., eds.), VCH Publishers, New York, pp. 355–371.
2. Marshall, G. R., Barry, C. D., Bosshard, H. E., Dammkoehler, R. A., and Dunn, D. A. (1979) The conformational parameter in drug design: the active analog approach. *ACS Symposium Series* **112,** 205–226.
3. Cramer, R. D., III, Patterson, D. E., and Bunce, J. D. (1988) Comparative molecular field analysis (CoMFA). 1. Effect of shape on binding of steroids to carrier proteins. *J. Am. Chem. Soc.* **110,** 5959–5967.
4. Clement-Cormier, Y. C., Kebabian, J. W., Petzold, G. L., and Greengard, P. (1974) Dopamine-sensitive adenylate cyclase in mammalian brain: a possible site of action of antipsychotic drugs. *Proc. Natl. Acad. Sci. USA* **71,** 1113–1117.
5. Kebabian, J. W. and Calne, D. B. (1979) Multiple receptors for dopamine. *Nature* **277,** 93–96.
6. Garau, L., Govoni, S., Stefanini, E., Trabucchi, M., and Spano, P. F. (1978) Dopamine receptors: pharmacological and anatomical evidence indicate two distinct dopamine receptor populations are present in rat striatum. *Life Sci.* **23,** 1745–1750.
7. Cannon, J. G. (1975) Chemistry of dopaminergic agonists, in *Advances in Neurology, vol. 9* (Calne, D. B., Chase, T. N., and Barbeau, A., eds.), Raven, New York, pp. 177–183.
8. McDermed, J. D., Freeman, H. S., and Ferris, R. M. (1979) Enantioselectivity in the binding of (+)- and (–)-2-amino-6,7-dihydroxy-1,2,3,4-tetrahydronaphthalene and related agonists to dopamine receptors, in *Catecholamines: Basic and Clinical Frontiers,* vol. I (Usdin, E., Kopin, I., and Barchas, J., eds.), Pergamon, New York, pp. 568–570.
9. Goldberg, L. I., Kohli, J. D., Kotake, A. N., and Volkman, P. H. (1978) Characteristics of the vascular dopamine receptor: comparison with other receptors. *Fed. Proc.* **37,** 2396–2402.
10. Setler, P. E., Sarau, H. M., Zirkle, C. L., and Saunders, H. L. (1978) The central effects of a novel dopamine agonist. *Eur. J. Pharmacol.* **50,** 419–430.

11. Christensen, A. V., Arnt, J., Hyttel, J., Larsen, J. J., and Svendsen, O. (1984) Pharmacological effects of a specific dopamine D-1 antagonist SCH23390 in comparison with neuroleptics. *Life Sci.* **34,** 1529–1540.
12. Clark, D. and White, F. J. (1987) Review: D_1 dopamine receptor—the search for a function: a critical evaluation of the D_1/D_2 dopamine receptor classification and its functional implications. *Synapse* **1,** 347–388.
13. Mailman, R. B., Schulz, D. W., Lewis, M. H., Staples, L., Rollema, H., and DeHaven, D. L. (1984) SCH-23390: a selective D_1 dopamine antagonist with potent D_2 behavioral actions. *Eur. J. Pharmacol.* **101,** 159,160.
14. Iorio, L. C., Barnett, A., Leitz, F. H., Houser, V. P., and Korduba, C. A. (1983) SCH23390, a potential benzazepine antipsychotic with unique interactions on dopaminergic systems. *J. Pharmacol. Exp. Ther.* **226,** 462–468.
15. Lovenberg, T. W., Brewster, W. K., Mottola, D. M., Lee, R. C., Riggs, R. M., Nichols, D. E., Lewis, M. H., and Mailman, R. B. (1989) Dihydrexidine, a novel selective high potency full D_1 dopamine receptor agonist. *Eur. J. Pharmacol.* **166,** 111–113.
16. Andersen, P. H., Nielsen, E. B., Scheel-Kruger, J., Jansen, J. A., and Hohlweg, R. (1987) Thienopyridine derivatives identified as the first selective, full efficacy, dopamine D_1 receptor agonists. *Eur. J. Pharmacol.* **137,** 291,292.
17. Truex, L. L., Foreman, M. M., Riggs, R. M., and Nichols, D. E. (1985) Effects of modifications of the 4-(3,4-dihydroxyphenyl)-1,2,3,4-tetrahydroisoquinoline structure on dopamine sensitive rat retinal adenylate cyclase activity. *Soc. Neurosci. Abstr.* **11,** 315.
18. Nichols, D. E. (1983) The development of novel dopamine agonists. *ACS Symposium* **224,** 201–218.
19. Charifson, P. S., Bowen, J. P., Wyrick, S. D., Hoffman, A. J., Cory, M., McPhail, A. T., and Mailman, R. B. (1989) Conformational analysis and molecular modeling of 1-phenyl-, 4-phenyl-, and 1-benzyl-1,2,3,4-tetrahydroisoquinolines as D_1 dopamine receptor ligands. *J. Med. Chem.* **32,** 2050–2058.
20. Chothia, C. and Lesk, A. M. (1986) The relation between the divergence of sequence and structure in proteins. *EMBO J.* **5,** 823–826.
21. Brewster, W. K., Nichols, D. E., Riggs, R. M., Mottola, D. M., Lovenberg, T. W., Lewis, M. H., and Mailman, R. B. (1990) Trans-10,11-dihydroxy-5,6,6a,7,8,12b-hexahydrobenzo[*a*]phen-anthridine: a highly potent selective dopamine D_1 full agonist. *J. Med. Chem.* **33,** 1756–1764.
22. Gilmore, J. H., Watts, V. J., Lawler, C. P., Noll, E. P., Nichols, D. E., and Mailman, R. B. (1995) "Full" dopamine D_1 agonists in human caudate: biochemical properties and therapeutic implications. *Neuropharmacology* **34,** 481–488.
23. Lovenberg, T. W., Roth, R. H., Nichols, D. E., and Mailman, R. B. (1991) D_1 dopamine receptors of NS20Y neuroblastoma cells are functionally similar to rat striatal D_1 receptors. *J. Neurochem.* **57,** 1563–1569.
24. Watts, V. J., Lawler, C. P., Gilmore, J. H., Southerland, S. B, Nichols, D. E., and Mailman, R. B. (1993) Efficacy at D_1 dopamine receptors in primates and rodents: comparison of full (dihydrexidine) and partial (SKF38393) efficacy dopamine agonists. *Eur. J. Pharmacol.* **242,** 165–172.
25. Watts, V. J., Lawler, C. P., Gonzales, A. J., Zhou, Q.-Y., Civelli, O., Nichols, D. E., and Mailman, R. B. (1995) Efficacy of D_1 dopamine receptor agonists: the role of spare receptors. *Synapse* **21,** 177–187.

26. Arnsten, A. F., Cai, J. X., Murphy, B. L., and Goldman-Rakic, P. S. (1994) Dopamine D_1 receptor mechanisms in the cognitive performance of young adult and aged monkeys. *Psychopharmacology (Berl.)* **116,** 143–151.
27. Schneider, J. S., Sun, Z.-Q., and Roeltgen, D. P. (1994) Effects of dihydrexidine, a full dopamine D-1 receptor agonist, on delayed response performance in chronic low dose MPTP-treated monkeys. *Brain Res.* **663,** 140–144.
28. Taylor, J. R., Lawrence, M. S., Redmond, D. E., Jr., Elsworth, J. D., Roth, R. H., Nichols, D. E., and Mailman, R. B. (1991) Dihydrexidine, a full dopamine D_1 agonist, reduces MPTP-induced Parkinsonism in African green monkeys. *Eur. J. Pharmacol.* **199,** 387,388.
29. Mottola, D. M., Laiter, S., Watts, V. J., Tropsha, A., Wyrick, S. D., Nichols, D. E., and Mailman, R. B. (1996) Conformational analysis of D_1 dopamine receptor agonists: pharmacophore assessment and receptor mapping. *J. Med. Chem.* **39,** 285–296.
30. Knoerzer, T. A., Nichols, D. E., Brewster, W. K., Watts, V. J., Mottola, D. M., and Mailman, R. B. (1994) Dopaminergic benzo[*a*]phenanthridines: resolution and pharmacological evaluation of the enantiomers of dihydrexidine, the full efficacy D_1 dopamine receptor agonist. *J. Med. Chem.* **37,** 2453–2460.
31. Anfinsen, C. B. (1973) Principles that govern the folding of protein chains. *Science* **181,** 223–230.
32. Tembe, T. L. and McCammon, J. A. (1984) Ligand–receptor interactions. *Comput. Chem.* **8,** 281–283.
33. Hirono, S. and Kollman, P. A. (1990) Calculation of the relative binding free energy of 2'-GMP and 2'-AMP to ribonuclease T1 using molecular dynamics/free energy perturbation approaches. *J. Mol. Biol.* **212,** 197–209.
34. Tropsha, A. and Hermans, J. (1992) Application of free energy simulations to the binding of a transition-state-analogue inhibitor to HIV protease. *Protein Engineering* **51,** 29–31.
35. Henderson, R., Baldwin, J. M., Ceska, T. A., Zemlin, F., Beckman, E., and Downing, K. H. (1990) Model for the structure of bacteriorhodopsin based on high-resolution electron cryo-microscopy. *J. Mol. Biol.* **213,** 899–929.
36. Hibert, M. F., Trumpp-Kallmeyer, S., Bruinvels, A., and Hoflack, J. (1991) Three-dimensional models of neurotransmitter G-binding protein-coupled receptors. *Mol. Pharmacol.* **40,** 8–15.
37. Donnelly, D. and Findlay, J. B. C. (1994) Seven-helix receptors: structure and modeling. *Curr. Opinion Struct. Biol.* **4,** 582–589.
38. Hutchins, C. (1994) Three-dimensional models of the D_1 and D_2 dopamine receptors. *Endocrine J.* **2,** 7–23.
39. Livingstone, C. D., Strange, P. G., and Naylor, L. H. (1992) Molecular modeling of D_2-like dopamine receptors. *Biochem. J.* **287,** 277–282.
40. Malmberg, A., Nordvall, G., Johansson, A. M., Mohell, N., and Hacksell, U. (1994) Molecular basis for the binding of 2-aminotetralins to human dopamine D_{2A} and D_3 receptors. *Mol. Pharmacol.* **46,** 299–312.
41. Teeter, M. M., Froimowitz, M., Stec, B., and DuRand, C. J. (1994) Homology modeling of the dopamine D_2 receptor and its testing by docking of agonists and tricyclic antagonists. *J. Med. Chem.* **37,** 2874–2888.
42. Trumpp-Kallmeyer, S., Hoflack, J., Bruinvels, A., and Hibert, M. (1992) Modeling of G-protein-coupled receptors application to dopamine, adrenaline, serotonin, acetylcholine, and mammalian opsin receptors. *J. Med. Chem.* **35,** 3448–3462.

43. Pardo, L., Ballesteros, J. A., Osman, R., and Weinstein, H. (1992) On the use of the transmembrane domain of bacteriorhodopsin as a template for modeling the three dimensional structure of guanine nucleotide binding of regulatory protein-coupled receptors. *Proc. Natl. Acad. Sci. USA* **89,** 4009–4012.
44. Maloney Huss, K. and Lybrand, T. P. (1992) Three-dimensional structure for the beta$_2$ adrenergic receptor protein based on computer modeling studies. *J. Mol. Biol.* **225,** 859–871.
45. Schertler, G. F. X., Villa, C., and Henderson, R. (1993) Projection structure of rhodopsin. *Nature* **362,** 770–772.
46. Dahl, S. G., Edvardsen, Ø., and Sylte, I. (1991) Molecular dynamics of dopamine at the D$_2$ receptor. *Proc. Natl. Acad. Sci. USA* **88,** 8111–8115.
47. Baldwin, J. M. (1993) The probable arrangement of the helices in G protein-coupled receptors. *EMBO J.* **12,** 1693–1703.
48. Jones, D. T., Taylor, W. R., and Thornton, J. M. (1994) A model recognition approach to the prediction of α-helical membrane protein structure and topology. *Biochemistry* **33,** 3038–3049.
49. Jones, D. T., Taylor, W. R., and Thornton, J. M. (1994) A mutation data matrix for transmembrane proteins. *FEBS Lett.* **139,** 269–275.
50. Nordvall, G. and Hacksell, U. (1993) Binding site modeling of the muscarinic m$_1$ receptor: a combination of homology based and indirect approaches. *J. Med. Chem.* **36,** 967–976.
51. Watts, V. J., Lawler, C. P., Knoerzer, T., Mayleben, M. A., Neve, K. A., Nichols, D. E., and Mailman, R. B. (1993) Hexahydrobenzo[a]phenanthridines: ligands with high affinity and selectivity for D$_3$ dopamine receptors. *Eur. J. Pharmacol.* **239,** 271–273.
52. Javitch, J. A., Fu, D., Chen, J., and Karlin, A. (1995) Mapping the binding site crevice of the dopamine D$_2$ receptor by the substituted-cysteine accessibility method. *Neuron* **14,** 825–831.
53. Javitch, J. A., Li, X., Kaback, J., and Karlin, A. (1994) A cysteine residue in the third membrane-spanning segment of the human D$_2$ dopamine receptor is exposed in the binding-site crevice. *Proc. Natl. Acad. Sci. USA* **91,** 10,355–10,359.
54. DeNinno, M. P., Schoenleber, R., Perner, R. J., Lijewski, L., Asin, K. E., Britton, D. R., MacKenzie, R., and Kebabian, J. W. (1991) Synthesis and dopaminergic activity of 3-substituted 1-(aminomethyl)-3,4-dihydro-5,6-dihydroxy-1H-2-benzopyrans: characterization of an auxiliary binding region in the D$_1$ receptor. *J. Med. Chem.* **34,** 2561–2569.
55. Lévesque, D., Diaz, J., Pilon, C., Martres, M.-P., Giros, B., Souil, E., Schott, D., Morgat, J.-L., Schwartz, J.-C., and Sokoloff, P. (1992) Identification, characterization, and localization of the dopamine D$_3$ receptor in rat brain using 7-[^{3}H]hydroxy-N,N-di-n-propyl-2-aminotetralin. *Proc. Natl. Acad. Sci. USA* **89,** 8155–8159.

Part II

Biochemical Mechanisms of Receptor Action

Chapter 5

Interactions of Dopamine Receptors with G Proteins

Susan W. Robinson and Marc G. Caron

1. Introduction

Dopamine is an important neurotransmitter, playing roles in motor control, emotion and affect, neuroendocrine regulation, and regulation of sodium uptake in the kidney, among other functions. The receptors for dopamine are members of the superfamily of G protein-coupled receptors (GPCRs). To date, five mammalian dopamine receptors have been cloned *(1,2)* and other subtypes may await discovery. These receptors have been characterized with regard to their pharmacology and modulation of second messengers in a variety of cell types and in vivo. However, less well studied has been the intermediate step between ligand binding and second messenger regulation (i.e., interaction of the receptor with G proteins). This aspect of dopamine receptor function may be quite significant in understanding the actions of dopamine in various tissues or regions of the brain which may contain different complements of receptors and G proteins. In this chapter, we discuss the evidence for which structural features of dopamine receptors are responsible for interaction with G proteins, as well as studies examining interactions of various subtypes of dopamine receptors with specific G proteins.

2. Structural Determinants of Coupling to G Proteins

Since the advent of molecular cloning of GPCRs, a large body of work has accumulated regarding the structure and function of these receptors. Extensive mutagenesis, particularly of the adrenergic receptors, has elucidated at least in a general way, those regions of receptors that are involved in

The Dopamine Receptors Eds.: K. A. Neve and R. L. Neve
Humana Press Inc., Totowa, NJ

ligand binding and coupling to G proteins *(3,4)*. The third intracellular loop (IL3) of these receptors has been shown to be essential for the interaction between receptors and G proteins in most GPCRs. Many studies utilizing chimeric receptors have shown that this region can determine the second messenger pathway to which a receptor is coupled. For example, a chimeric beta-2 adrenergic receptor containing the IL3 from the alpha-1 adrenergic receptor activated phosphatidylinositol (PI) hydrolysis, which is the signaling pathway of the alpha-1 adrenergic receptor *(5)*. Similarly, an alpha-2 adrenergic receptor with the IL3 from the beta-2 adrenergic receptor was able to stimulate rather than inhibit cAMP formation *(6)*. There is evidence from several receptors that the membrane proximal regions of the IL3 are most important for G protein coupling. Deletion of eight residues in the N-terminal part of the beta-2 adrenergic receptor IL3 abolished activation of adenylate cyclase *(7)*. Also, a peptide from the C-terminal region of the alpha-2 adrenergic receptor IL3 could mimic the receptor and bind G_i *(8)*. Chimeric muscarinic receptors also suggest the importance of a small region at the N-terminal portion of the IL3 for G protein coupling *(9)*. However, mutant rhodopsin with a deletion of a region in the middle of the IL3 could bind, but not activate transducin *(10)*, suggesting the importance of the entire IL3 in this receptor.

Mutagenesis experiments also support a role for the C-terminal tail of some receptors in interactions with G proteins. Certain point mutations in the beta-2 adrenergic receptor C-terminal tail reduce the ability of the receptor to activate adenylate cyclase *(11)*. Also, peptides from the C-terminal tail of rhodopsin block the ability of the receptor to stimulate transducin GTPase activity *(12)*. There is also evidence for the role of the first (IL1) or second (IL2) intracellular loops in receptor–G protein coupling. A peptide from the alpha-2 adrenergic receptor IL2 inhibited agonist binding to receptor *(8)*, suggesting that it might interfere with receptor–G protein interactions. Similarly, a peptide from the N-formyl peptide receptor IL2 blocked antibody recognition of $G_{i\alpha}$, ADP-ribosylation of $G_{i\alpha}$, and high affinity agonist binding to the receptor *(13)*. In addition, a peptide encoding the IL2 of the $5HT_{1A}$ receptor could inhibit adenylate cyclase activity and stimulate GTPγS binding to purified G_i/G_o proteins, suggesting that it could mimic receptor interactions with G proteins *(14)*. Mutations in the IL2 of rhodopsin resulted in a receptor that could bind transducin, but failed to activate it *(10)*. Furthermore, beta-1/m1 receptor chimeras suggest an interaction between IL2 and IL3 may play a role in specificity of coupling to G proteins in these receptors *(15)*. Finally, O'Dowd et al. *(11)* found that a point mutation in the IL1 of the beta-2 adrenergic receptor could substantially reduce stimulation of cAMP, suggesting that this region of the receptor might also contact G proteins. Although not a comprehensive review of the literature, all of this evidence supports the notion that in different receptors, there may be different regions that are involved in G protein coupling, and also suggests that multiple receptor regions may be necessary for productive interactions with G proteins.

2.1. Peptide Studies

One approach to the elucidation of the role of specific receptor domains in coupling to G proteins has been the use of synthetic peptides. Peptides may have a variety of effects on receptor–G protein coupling. They may inhibit agonist binding to the receptor or block receptor modulation of second messengers, suggesting that they might prevent the interaction between receptor and G protein. Peptides may also stimulate G protein GTPase activity or the rate of binding of guanine nucleotide analogs to G proteins, suggesting that they might mimic the receptor interaction with the G protein. A number of studies have utilized peptides from regions of the dopamine receptors to determine which regions might be important for interactions with G proteins.

2.1.1. D_1 Receptors*

Peptides from various regions of the rat D_1 receptor have been tested for the ability to compete with purified D_1 receptor for binding of purified G_s. Peptides encompassing the entire IL2, the N-terminal part of the IL3, or the membrane proximal region of the C-terminal tail were all able to block G_s binding to purified D_1 receptor *(16)*. These peptides all had similar potency and efficacy in their inhibition of G_s binding to the receptor. This suggests that like the beta-2 adrenergic receptor and rhodopsin, multiple regions of the D_1 receptor may be involved in interactions with G proteins. In contrast to these results, Voss et al. *(17)* found that peptides from either the N- or C-terminal part of the human D_1 IL3 did not specifically block dopamine-stimulated cAMP accumulation in 293 cell membranes. Further, these peptides were not able to stimulate GTPase activity in 293 cell membranes. This suggests that these peptides were not able to interfere with coupling of the D_1 receptor to G proteins in membranes, although a peptide containing a very similar region from the rat D_1 IL3 could block interactions between purified rat D_1 receptor and G_s *(16)*.

An approach that allows the study of the effects of a peptide in an intact cell has recently been utilized by Luttrell et al. *(18)*. The IL3 of various receptors can be expressed in a cell line using a minigene construct and the effect of the peptide on receptor coupling to second messenger systems can be examined. Expression of a minigene encoding the entire IL3 from the human

*As detailed in the Preface, the terms D1 or D1-like and D2 or D2-like are used to refer to the subfamilies of DA receptors, or used when the genetic subtype (D_1 or D_5; D_2, D_3, or D_4) is uncertain.

D_1 receptor reduced D_1-stimulated cAMP accumulation in 293 cells at receptor levels <2 pmol/mg membrane protein *(18)*, suggesting that in an intact system, the IL3 is important for G protein coupling. At higher levels of receptor expression, no inhibition was observed. This might suggest that the peptide competes with the receptor for binding to the G protein and therefore that higher receptor levels can overcome the inhibition. The D_1 IL3 was unable to block alpha-1B adrenergic receptor stimulation of PI hydrolysis, suggesting that its effect was specific to G_s *(18)*. In further studies using this approach, Hawes et al. *(19)* demonstrated that the ability of various IL3 minigenes to block receptor signaling is likely to be through binding of the peptide to the appropriate G protein, disrupting receptor–G protein coupling.

2.1.2. D_2 Receptor

Synthetic peptides have also been used as probes for regions of the D_2 receptor that interact with G proteins. Peptides from either the N- or C-terminal region of the IL3 of the human D_2 long (D_{2L}) receptor could block dopamine-mediated inhibition of prostaglandin E_1 (PGE_1)-stimulated cAMP accumulation in 293 membranes in a dose-dependent fashion *(20)*. These peptides also increased GTPase activity in 293 membranes, suggesting a direct interaction with G proteins. Voss et al. *(17)* also observed that a peptide from the N-terminal IL3 of the human D_{2L} could block dopamine inhibition of PGE_1-stimulated cAMP in 293 cell membranes. This peptide could also increase GTPase activity in a pertussis toxin (PTX) sensitive manner in 293 cell membranes and increase the rate of GTPγS binding to purified G_i/G_o proteins, suggesting a direct peptide–G protein interaction. In direct contrast to the results of Malek et al. *(20)*, in this study a peptide from the C-terminal part of the D_2 IL3 could not block inhibition of cAMP accumulation by the D_2 receptor, nor could it stimulate GTPase activity in a purified preparation of G_i/G_o proteins *(17)*.

These studies on dopamine receptors clearly point out the difficulties involved in the use of synthetic peptides to identify regions of a receptor involved in G-protein contact. In the case of both D_1 and D_2 receptors, there are contradictory reports in the literature about the effects of very similar or even identical peptides. The results obtained may be dependent on whether the effect of a peptide is examined using purified components, a membrane preparation, or intact cells. Another factor may be the conformation of the particular peptide. For example, a larger peptide may be able to adopt a conformation more similar to the case in the native receptor, whereas a smaller peptide may be unable to have the appropriate conformation. In addition, in any particular system, it may be necessary to have peptides from more than one region of the receptor to observe any effect. For example, using the minigene system, expression of either the N- or C-terminal part of the alpha-1B adrenergic receptor has no effect on the coupling of the alpha-1B adrenergic receptor to PI hydrolysis, however when both peptides are expressed simultaneously, they are able to have an effect similar to that when the entire IL3 is expressed *(18)*. Thus, peptide studies are one tool to examine structure–function relationships and can suggest which regions of the receptor are important, although the lack of an effect may not necessarily mean a lack of involvement in coupling. However, this technique can help point the way for further study using other approaches.

2.2. Receptor Chimeras

Another approach to understanding the structure–function relationship of GPCRs has been to make receptor chimeras. Chimeras between less closely related receptors may help to define general characteristics of GPCRs, whereas chimeras between closely related receptors can help dissect specific regions responsible for more subtle variations in pharmacology and coupling between receptor subtypes. This approach has been used extensively, particularly with adrenergic and muscarinic receptor subtypes (reviewed in refs. *3* and *4*). Recent studies have applied this method to the dopamine receptor family as well.

Several D_1/D_2 receptor chimeras have been constructed. MacKenzie et al. *(21)* made a human D_1 receptor that contained rat D_2 sequence from the beginning of transmembrane domain (TM) 6 to the end of the receptor. This receptor had decreased affinity for dopamine and D_1-selective compounds and increased affinity for the D_2-selective compound quinpirole as compared to the wild-type D_1 receptor. This chimera had the ability to stimulate cAMP accumulation in response to quinpirole. This suggests the importance of the IL3 in determining coupling to G proteins, as this chimera contained the D_1 IL3 *(21)*. Kozell et al. *(22)* also constructed a series of chimeras consisting of macaque D_1 receptor containing increasing amounts of rat D_2 short (D_{2S}) sequence. Chimeras that contain the IL3 of the D_1 receptor stimulate cAMP formation. A chimera with the IL3 from the D_2 receptor, but the IL2 from the D_1 receptor was unable to stimulate or inhibit cAMP accumulation, whereas a chimera that contained both the IL2 and IL3 of the D_2 receptor was able to inhibit cAMP formation *(22)*. These data strongly suggest that multiple receptor contacts may be required for coupling of the D_2 receptor to G_i proteins. This study did not include a chimera containing the D_2 IL2 and the D_1 IL3, which would allow the determination of whether the D_1 also requires both the IL2 and IL3 to couple to $G_{s,}$ or if the IL3 alone would be sufficient for this interaction.

Recently, some D_2/D_3 chimeras have also been made. A human D_3 receptor containing the IL3 of the human D_2 receptor was constructed and expressed in

Chinese hamster ovary (CHO) cells. In this cell type, the wild-type D_2 receptor can inhibit cAMP accumulation and also potentiate A23187-stimulated arachidonic acid release. However, the D_3/D_2 chimera was unable to couple to either of these pathways *(23)*. This suggests that the D_2 IL3 is not sufficient for coupling the D_2 receptor to second messengers, supporting the results of Kozell et al. *(22)*.

Our lab has also constructed a series of chimeric D_2/D_3 receptors. These chimeras have various intracellular regions switched between the human D_{2S} and human D_3 receptors *(24)*. We have found that although the IL3 appears to influence agonist affinity at the D_2 and D_3 receptors *(24)*, the IL2 seems to be a more important determinant of the ability to couple efficiently to inhibition of adenylate cyclase. Whereas all chimeras are able to couple to inhibition of adenylate cyclase, those containing the IL2 from the D_2 receptor couple with highest potency, similar to the wild-type D_2 receptor *(24a)*. This is consistent with the results of Kozell et al. *(22)*, suggesting that the D_2 IL2 is important for inhibition of adenylate cyclase. There is evidence from other GPCRs as well that the IL2 may be more important in coupling to G proteins than has previously been appreciated, as discussed *(8,10,13,14)*.

2.3. Other Approaches

2.3.1. Receptor Antibodies

Another approach to identification of receptor domains that interact with G proteins is to examine the effect of receptor antibodies on coupling of the receptor to G proteins. The measure of coupling that is frequently used in these experiments is the ability of an antibody to shift agonist binding to the receptor to a low affinity state, implying uncoupling of the receptor from G proteins. Several antibodies developed against the IL3 of D_2 receptors have been shown to decrease high affinity binding of dopamine. Antibodies to the IL3 of both the long and short forms of the rat D_2 receptor were made against peptides which excluded the conserved N- and C-terminal regions of this loop *(25)*. The antisera against both D_{2S} and D_{2L} were able to decrease agonist binding to D_2 receptors in solubilized rat caudate to the same extent as GTP. An antibody developed against amino acids 301–315 of the bovine D_2 receptor could decrease the affinity of dopamine binding to bovine striatal membranes *(26)*. This antibody recognizes a region that is internal to the IL3 and that is distal to the 29 amino acid insert found in D_{2L}. Similarly, an antibody against a region common to both D_{2L} and D_{2S} that is distal to the D_{2L} insert was able to inhibit high affinity binding of agonist to D_2 receptors in rat caudate, although an antibody against amino acids specific to the D_{2L} was ineffective in this assay *(27)*. Another antibody made against a region of the bovine D_2 IL3 that is not proximal to the membrane also could block high affinity agonist binding in the bovine caudate *(28)*. In contrast to evidence from peptide studies on the dopamine receptors, these antibody studies imply that internal regions of the IL3 may also be important for the interaction with G proteins. However, one possibility that must be considered is that the antibody could sterically hinder interaction of G proteins with portions of the receptor that are adjacent to the membrane, thus making the assignment of function to specific sequences difficult.

2.3.2. Site-Directed Mutagenesis

Site-directed mutagenesis of both D_1 and D_2 receptors has also suggested the importance of residues in the TM domains for coupling the receptor to G proteins. Mutation of serine residues in TM 5 of the human D_1 receptor results in a decrease in the ability of the receptor to stimulate accumulation of cAMP *(29)*. Similarly, mutation of an aspartic acid in TM 2 of the D_{2S} receptor results in a receptor that is unable to inhibit isoproterenol-stimulated cAMP *(30)*. All of these mutations are within the TM domains of the receptors, and therefore presumably cannot interact directly with G proteins. Importantly, although the serine mutations in the D_1 receptor do decrease the affinity of the receptor for agonist *(29)*, the mutation of the TM 2 aspartate in the D_2 receptor did not result in a substantial change in agonist affinity *(30)*. Thus, these studies point out the importance of a correct overall conformation of the receptor that is capable of coupling to G proteins.

3. Coupling of Dopamine Receptors in Cell Lines

Many studies have examined the properties of dopamine receptors in various tissues. However, especially in brain, it is difficult to work with a region that expresses only one of the dopamine receptor subtypes. With the molecular cloning of the dopamine receptors, it has become possible to study the properties of individual receptors in cultured cell lines. This has made it possible to examine the pharmacological specificity of ligands for various receptor subtypes and also to begin to explore the modulation of second messengers by these receptors. However, one drawback to heterologous expression of dopamine receptors in cultured cells is that these cells come from a variety of sources and may not express the appropriate G proteins or effectors that are physiologically relevant for the receptor. For this reason, data obtained from the same receptor in different cell lines may be different, or even contradictory, depending on the complement of signaling molecules expressed by the cells. Thus, although a dopamine receptor may couple to a particular signaling pathway in a cultured cell line, it must be remembered that this may not be the case in vivo, in which the same G proteins or effectors may not be expressed together with the receptor.

3.1. D1-Like Receptors

Molecular biology techniques have identified two mammalian D1-like receptors, D_1 and D_5. These receptors share a high degree of homology and have similar pharmacological profiles *(2)*. In all cell lines tested to date, these receptors stimulate formation of cAMP in response to agonist. In fact, recent evidence indicates that the D_5 receptor can constitutively activate adenylate cyclase, as well as causing a further increase in cAMP in response to agonist *(31)*. In addition to these receptors, two new vertebrate D1-like receptors have recently been cloned. They have been designated D_{1C} and D_{1D}, cloned from *Xenopus laevis* and chicken, respectively *(32,33)*, and both also stimulate adenylate cyclase.

In some cases, coupling of D1-like receptors to changes in intracellular Ca^{2+} concentrations has also been reported. In GH_4C_1 cells, the human D_1 receptor stimulates adenylate cyclase and also increases the opening of L-type Ca^{2+} channels, possibly through the activation of protein kinase A *(34)*. When the same receptor is expressed in Ltk^- cells, it stimulates adenylate cyclase and also stimulates PI hydrolysis *(34)*. Also, when either the human D_1 receptor or its goldfish homolog is expressed in 293 cells, both stimulate adenylate cyclase and increase intracellular Ca^{2+} *(35)*. This increase in Ca^{2+} levels may be owing to the activation of phospholipase C. Evidence for this comes from catfish horizontal cells which undergo neurite retraction in response to D_1 agonists *(36)*. This effect is mimicked by phorbol esters and diacylglycerol analogs, but not cAMP, suggesting that it occurs via activation of phospholipase C by the D_1 receptor. However, although both human D_1 and D_5 receptors stimulate adenylate cyclase in baby hamster kidney cells, neither receptor causes any change in intracellular Ca^{2+} levels *(37)*. In addition, neither of the newly cloned D_1-like receptors were able to consistently stimulate PI hydrolysis in COS-7 cells *(32,33)*. These discrepancies are likely the result of the system in which the receptors are expressed.

Despite the contradictory results concerning stimulation of PI turnover in cultured cells, there is some evidence that in brain membranes a D1-like receptor stimulates PI hydrolysis. D1 agonists stimulate formation of inositol phosphate (IP) in rat striatal slices *(38)*. There has been some suggestion that this increase in PI turnover is caused by an as yet unidentified D1-like receptor. In the amygdala, where there are D1 binding sites but no detectable D1-stimulated cAMP formation, the D1 agonist SKF 38393 stimulates a large increase in IP formation *(38)*. In addition, in rat striatal membranes there is no correlation between either the efficacy or potency of D1 agonists to stimulate adenylate cyclase or IP formation *(39)*. These data suggest the possibility that there may be a new D1 receptor subtype that couples to

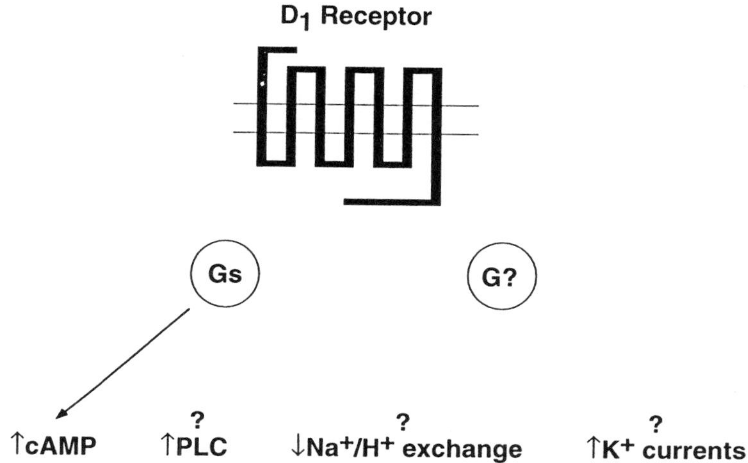

Fig. 1. Signaling pathways of the D1-like receptors. Circles, G protein α subunits; Arrows, known coupling of receptor/G protein/second messenger; G?, known or novel G protein α subunits with which the receptor may interact; ?, second messenger pathways with which the receptor might couple, but for which conflicting evidence exists in the literature.

stimulation of PI hydrolysis. However, there is evidence that the D_1 receptor can couple to this pathway in certain cultured cell types *(34,36),* so the PI hydrolysis that has been observed in brain may be mediated through this receptor subtype.

Other effects of D1-like receptors have been observed in peripheral tissues. In renal brush border membrane vesicles a D1 agonist can cause the inhibition of the amiloride sensitive Na^+/H^+ exchanger *(40,41)*. However, there is conflicting evidence as to whether this effect is mediated through changes in cAMP levels *(40)* or is cAMP independent *(41)*. Also, D1 agonists cause an increase in K^+ efflux in chick retina, a tissue in which dopamine does not stimulate cAMP or IP formation, suggesting that this might be a direct effect of a D1-like receptor *(42)*. A summary of the possible coupling of D1-like receptors to second messenger pathways is illustrated in Fig. 1.

3.2. D2-Like Receptors

To date, three D2-like receptors have been identified by cloning: D_2, D_3, and D_4. These have been grouped together on the basis of pharmacological and structural similarities. Because of the clinical importance of this group of receptors as targets for neuroleptic drugs, much effort has gone into understanding their signal transduction properties.

3.2.1. D_2 Receptor

The D_2 receptor has been shown to be coupled to $G_{i/o}$ proteins, and like other such receptors, seems to be coupled to multiple signaling pathways. In addition, there are two isoforms of this receptor, D_{2S} and D_{2L}, which are the result of alternative splicing of a single gene *(2)*. These forms of the D_2 receptor have slightly different coupling properties, as is discussed in this section and Section 4.1.

Initially, the D_2 receptor was defined as being coupled to the inhibition of adenylate cyclase, in contrast to the D_1 receptor which stimulates adenylate cyclase *(43)*. When this receptor was cloned, it was found that it inhibited the formation of cAMP in transfected cell lines in a PTX-sensitive manner. Studies that have examined the properties of both the D_{2S} and D_{2L} receptors have found that the D_{2S} receptor causes a greater maximal inhibition of cAMP and requires a lower dose of agonist to give half-maximal inhibition than the D_{2L} receptor *(44–46)*. These observations have been made in both JEG-3 cells *(45,46)* and CHO cells *(44)*.

Similar to other receptors that are coupled to PTX-sensitive G proteins, the D_2 receptor can influence a number of second messengers in addition to cAMP. In CHO cells, D_2 receptors are able to potentiate the release of arachidonic acid which is stimulated by ATP or calcium ionophores *(47–49)*. This effect is not a result of changes in cAMP levels caused by the receptor, however it is not clear whether or not this effect is mediated by a PTX-sensitive G protein. Piomelli et al. *(49)* showed that the potentiation of A23187-stimulated arachidonic acid release was blocked by PTX and Felder et al. *(47)* also observed PTX-sensitivity of potentiation of ATP-stimulated arachidonic acid release. However, Kanterman et al. *(48)* reported that potentiation of arachidonic acid release stimulated by A23187 was not sensitive to PTX. These conflicting observations are difficult to explain, as all experiments were performed in the same cell type.

The D_2 receptor also influences intracellular Ca^{2+} levels in several cell types. In both CHO and Ltk^- cells, stimulation of the D_2 receptor leads to an increase in intracellular Ca^{2+} *(44,50,51)*. In both cell lines, this increase is PTX-sensitive, and in the Ltk^- cells, it has been shown to be the result of a stimulation of PI hydrolysis *(51)*. The D_2 receptor has also recently been shown to stimulate PI turnover in CCL1.3 cells *(52,53)*, although this effect is not observed in CHO *(54)* or MN9D cells *(53)* expressing the D_2 receptor. In contrast, in some cell types, the D_2 receptor has an inhibitory effect on Ca^{2+}. In pituitary melanotrophs, which endogenously express D_2 receptors, dopamine inhibits calcium currents *(55)*. In GH_4C_1 cells, D_2 receptors also mediate a PTX-sensitive decrease in intracellular Ca^{2+} levels *(51,56,57)*. D_2 receptors also inhibit high-threshold Ca^{2+} currents in NG108-15 cells *(58)*.

Dopamine D_2 receptors also have been found to modify K^+ currents in several cell types. In pituitary cells, D_2 receptor agonists increase K^+ currents *(59,60)*. In pituitary lactotrophs, the D_2 receptor has been shown to stimulate voltage-activated K^+ currents *(61,62)*. However, in the neuroblastoma *x* glioma hybrid cell line, NG108-15, a transfected D_2 receptor inhibits voltage-dependent K^+ currents in a manner that is not sensitive to PTX *(63)*. In this case, the alteration in potassium current appears to be dependent on an increase in intracellular calcium levels, as this effect was mimicked by thapsigargin and blocked by ryanodine, which promote or inhibit increases in intracellular Ca^{2+}, respectively *(63)*. These differences in K^+ current response to D_2 receptors may be caused by differences in the cell type used for the experiments. It has also been shown recently that the D_2 receptor is able to inhibit both basal- and potassium-evoked dopamine release from the mesencephalic cell line MN9D *(64)*. This effect seems to be mediated through activation of K^+ channels in a PTX-sensitive manner by the D_2 receptor *(64)*.

Other effects of the D_2 receptor include stimulation of an amiloride-sensitive Na^+/H^+ exchanger in C_6 glioma cells and Ltk^- cells *(65)*. This effect is PTX insensitive in both cell lines *(65)*. Ganz et al. *(66)* reported that the D_2 receptor inhibits Na^+/H^+ exchange in primary cultures of rat anterior pituitary cells, through a PTX-insensitive manner. In contrast, a PTX-sensitive stimulation of extracellular acidification via the Na^+/H^+ exchanger by D_2 receptors has been observed in CHO cells *(67)*. Therefore, it may be that the D_2 receptor can have opposing effects on a single second messenger via different mechanisms in a variety of cell types.

D_2 receptors have also been observed to inhibit cell growth. When expressed in GH_4C_1 cells, this receptor inhibits [^{3}H]thymidine uptake, a measure of cell growth *(68,69)*. This effect has been reported to be both PTX-sensitive *(68)* and PTX-insensitive *(69)*. However, another recent paper reports stimulation of [^{3}H]thymidine uptake by the D_2 receptor in CHO cells *(70)*. This effect is PTX-sensitive and appears to be mediated by an increase in tyrosine phosphorylation in the cell *(70)*. Additionally, the D_2 receptor stimulates neurite outgrowth in the mesencephalic cell line MN9D *(71)*. This suggests that D_2 receptors may either stimulate or inhibit cell growth, possibly dependent on cell type, expression level, state of the cell, or other factors.

3.2.2. D_3 Receptor

The D_3 receptor has considerable structural similarity to the D_2 receptor. The human D_2 and D_3 receptors are 46% homologous overall, and 78% identical in the TM domains, which have been implicated in ligand binding *(72)*. Given this degree of homology, it might be expected that these receptors would exhibit similar coupling to G proteins and second messenger systems.

However, this has been difficult to show, despite the expression of the D_3 receptor in numerous cell culture lines.

One indicator of coupling to G proteins is the ability of guanine nucleotides to modulate agonist binding at a receptor. There have been conflicting reports as to whether the D_3 receptor exhibits a shift to low affinity agonist binding in the presence of GTP. When the rat D_3 receptor was originally cloned, it was reported that GppNHp did not alter dopamine binding to the receptor in CHO cells *(73)*. Similarly, no modulation of agonist binding in response to GTP has been reported when the D_3 receptor was expressed in MN9D *(58)*, CHO *(54)*, Sf9 *(74)*, or 293 cells *(75)*. In addition, GppNHp did not affect the binding of dopamine or the putatively D_3-selective agonist 7-OH-DPAT in rat olfactory tubercle membranes *(76)*. In contrast, guanine nucleotides have been reported to reduce agonist affinity at D_3 receptors in CHO *(52,67,77)*, GH_4C_1 *(78)*, and NG108-15 cells *(79)*. The data suggest that coupling of D_3 receptors to G proteins is dependent on the cell type utilized and perhaps on the characteristics of individual clonal cell lines, as in some CHO cell lines, the D_3 receptor appears to couple to G proteins, yet in others it does not. Alternatively, the D_3 receptor may have unique agonist binding characteristics that make it resistant to the effects of guanine nucleotides. It is important to note however, that in some cases coupling of the D_3 receptor to a second messenger pathway was observed in the absence of GTP modulation of agonist binding.

Recently, some evidence has accumulated that the D_3 receptor is negatively coupled to adenylate cyclase. Inhibition of adenylate cyclase by D_3 receptors has been shown in CHO 10001 *(67)*, 293 *(75)*, NG108-15 cells *(54)*, and cultured *Xenopus* melanophores *(80)*. In all cases, the inhibition of adenylate cyclase by the D_3 receptor was much weaker than the inhibition caused by the D_2 receptor in the same cell lines. Interestingly, in both 293 cells and NG108-15 cells, there was no modulation of agonist binding in the presence of GTP, despite the ability of the D_3 receptor to inhibit adenylate cyclase in these cell lines *(54,75)*. The inhibition of adenylate cyclase by the D_3 receptor is dependent on the cell line, as this receptor had no effect on cAMP levels in GH_4C_1 *(78)*, MN9D *(53)*, SK-N-MC *(52)*, CHO-K1 *(52)*, NG108-15 *(79)*, or CCL1.3 cells *(52,53)*. At this time, there is no obvious explanation for the ability of the D_3 receptor to inhibit adenylate cyclase in some cell lines, but not others, and also the conflicting observations of coupling in the same cell type by different groups.

The ability of the D_3 receptor to couple to other second messenger pathways that are modulated by the D_2 receptor has also been examined. Unlike the D_2 receptor, no potentiation of ATP-stimulated arachidonic acid release by the D_3 receptor has been observed in CHO-K1 *(52)* or GH_4C_1 cells *(78)*. In contrast, Freedman et al. *(54)* reported a small increase in arachidonic acid release in a CHO-D_3 cell line. As in the case of inhibition of adenylate cyclase, this response was much weaker than that mediated by the D_2 receptor in the same cell line *(54)*. Interestingly, in this cell line, despite the weak coupling to arachidonic acid release, agonist binding to the D_3 receptor was not modulated by guanine nucleotides.

In some cell lines, the D_2 receptor can stimulate production of inositol phosphates (*see* Section 3.2.1.). However, in no cell line tested has the D_3 receptor had this effect, including CCL1.3 *(52,53)*, GH_4C_1 *(78)*, MN9D *(53)*, CHO *(54)*, NG108-15 *(54,79)*, or rat 1 fibroblast cells *(54)*.

In addition to inhibition of adenylate cyclase, the D_2 receptor is also able to modulate some ion channels (*see* Section 3.2.1.). In differentiated NG108-15 cells, the D_3 receptor was able to cause an inhibition of high-threshold Ca^{2+} currents that was PTX-sensitive *(81)*. However no inhibition of Ca^{2+} currents by the D_3 receptor was observed in GH_4C_1 cells *(57)*. No effect of D_3 receptors on voltage-dependent K^+ currents was observed in GH_4C_1 cells *(78)*. Similar to the D_2 receptor, the D_3 receptor can inhibit dopamine release in MN9D cells, apparently by a mechanism involving the activation of potassium channels *(64)*. Interestingly, the D_3 receptor causes a greater inhibition of dopamine release than the D_2 receptor. This is the first report of a signaling pathway that has a greater modification by the D_3 receptor, rather than the D_2 receptor.

It has also been observed that D_2 receptors can couple to an amiloride-sensitive Na^+/H^+ exchanger in some cells *(65–67)*. The D_3 receptor also has been observed to stimulate extracellular acidification via this exchanger in a PTX-sensitive manner in CHO 10001 cells, although to a much lesser extent than the D_2 receptor in the same cell line *(67)*. It has also recently been observed that the D_3 receptor is able to stimulate [^{3}H]thymidine uptake in both CHO 10001 *(67)* and NG108-15 cells *(79)*. In both cell lines, this effect was abolished by PTX *(67,79)*. Finally, when D_3 receptors were expressed in MN9D cells, quinpirole stimulated neurite outgrowth from these cells *(71)*. These results suggest that the D_3 may have similar mitogenic effects to the D_2 receptor in some cell lines.

Among the family of dopamine receptors, the elucidation of the coupling of the D_3 receptor to second messenger pathways has been the most problematic. Several possibilities might explain why this has been the case. It may be that the cell lines that have been used to study the D_3 receptor are lacking a necessary G protein subunit and/or effector molecule for coupling of this receptor. Another possibility is that this receptor simply couples to G proteins with a much lower efficiency than the D_2 receptor. Alternatively, the D_3 receptor might function as a signal integrator, mediating crosstalk between various signaling pathways and does not dramatically alter any pathway on its own. Continuing study of the D_3 receptor will be necessary to clarify its signaling properties and its role in vivo.

3.2.3. D_4 Receptor

The D_4 receptor, like the D_3 receptor, was originally identified by molecular cloning based on its homology to the D_2 receptor. The signal transduction pathways of the D_4 receptor, like those of the D_3 receptor, have not been immediately obvious. It has been reported that in mouse retina, which expresses both D_2 and D_4 receptors, an inhibition of cAMP levels with a pharmacology consistent with D_4 receptors was observed *(82)*. Aside from this, there has been little in the literature concerning D_4 receptor signal transduction until recently. When the D_4 receptor was expressed in Sf9 *(83)* or COS-7 cells *(84)* an effect of guanine nucleotides on agonist binding affinity was observed, suggesting that the receptor was capable of coupling to G proteins. Recently, inhibition of adenylate cyclase by the D_4 receptor has been reported in 293 *(75,85)*, MN9D *(53)*, and CHO cells *(86)*. However, the D_4 receptor did not inhibit adenylate cyclase in CCL1.3 cells, although the D_2 receptor did *(53)*, once again suggesting that coupling of D_2-like receptors is dependent on cell line.

The D_4 receptor may be coupled also to several other second messengers that are modulated by the D_2 receptor. In CHO cells, the D_4 receptor potentiated ATP-stimulated arachidonic acid release in a PTX-sensitive manner *(86)*. Also in this cell line, the D_4 receptor stimulated extracellular acidification through the amiloride-sensitive Na^+/H^+ exchanger *(86)*. Similar to that of the D_2 and D_3 receptors in CHO cells, the effect on extracellular acidification was blocked by PTX *(86)*. The D_4 receptor did not stimulate PI hydrolysis in CCL1.3 or MN9D cells, although the D_2 receptor had this effect in CCL1.3 cells *(53)*. Like the D_2 receptor, the D_4 receptor inhibited Ca^{2+} currents in GH_4C_1 cells *(57)* and stimulated neurite outgrowth in MN9D cells *(71)*. Thus, it seems that in general, the D_4 receptor can couple to most of the same second messenger pathways as the D_2 receptor, although like both the D_2 and D_3 receptors, this appears to be highly dependent on the cell type in which the receptor is expressed.

4. Interactions with Specific G Proteins

As discussed, both D1- and D2-like receptors have the ability to couple to multiple signal transduction pathways depending on the cell type in which the receptor is expressed. Since there have been 20 different α subunits identified by molecular cloning, as well as at least five β and six γ subunits *(87)*, it seems likely that dopamine receptors may be able to interact with various G protein heterotrimers with different efficiencies to allow coupling to various second messenger pathways. Several approaches have been taken to dissect the coupling of dopamine receptors to second messenger systems via specific G protein α subunits. These include determination of the G protein content of cell lines in which a receptor modifies certain second messengers, use of various mutated or chimeric G proteins, and knockout of individual α subunits using antibodies or antisense techniques. In addition, it has recently come to be appreciated that $\beta\gamma$ subunits are also capable of signaling and that this pathway is utilized by dopamine receptors.

4.1. G Proteins Expressed in Cell Types

Many experiments aimed at determining which second messenger pathways are modulated by cloned receptors are carried out in cell lines in culture. These cells come from a variety of origins and each expresses its own complement of signaling molecules, including G protein subunits. As a result, by determining which G proteins are expressed in a given cell line, in some cases we may be able to infer which G protein couples a receptor to a particular second messenger. These studies have focused on D_2 receptors, since they couple to numerous signaling pathways, whereas the D1-like receptors cloned to date are primarily coupled to stimulation of cAMP formation in cell culture systems.

It has been observed that in some cases, the D_{2S} isoform can couple more efficiently to inhibition of cAMP than the D_{2L} receptor *(44–46)*. Montmayeur et al. *(46)* observed this phenomenon in the JEG-3 cell line which contains G_{i1} and G_{i3}, but not G_{i2}. However, in NCB20 cells that express all three G_i subtypes, D_{2S} and D_{2L} are coupled to inhibition of cAMP formation similarly *(46)*. This suggests that the D_{2L} isoform is able to couple better to G_{i2} than to G_{i1} or G_{i3} and also suggests that D_{2S} interacts more efficiently with G_{i1} or G_{i3}. In fact, coexpression of D_{2L} and G_{i2} in JEG-3 cells abolishes the differences observed between the two receptor isoforms, confirming this hypothesis *(46)*. Similar types of observations have been made when D_{2L} or D_4 receptors are expressed in CCL1.3 or MN9D cells. Both receptors are able to inhibit cAMP in MN9D cells, but only the D_{2L} receptor can inhibit cAMP in CCL1.3 cells *(53)*. Both cell types express G_{i2} but not G_{i1} or G_{i3}. It is likely that the D_{2L} receptor is coupled to $G_{i\alpha2}$, however the D_4 receptor does not seem to be able to interact with this α subunit and it is unclear from these results which G protein is utilized by the D_4 receptor to inhibit adenylate cyclase in MN9D cells.

In the pituitary tumor-derived GH_3 cell line, expression of D_2 receptors can be induced by treatment with epidermal growth factor (EGF) *(88)*. These D_2 receptors do not inhibit cAMP formation and are coupled to inhibition of prolactin release and activation of voltage-dependent K^+ currents apparently via cAMP-independent mechanisms. EGF treatment of GH_3 cells also caused a large increase in the expression of $G_{i\alpha 3}$, although G_{i1} and/or G_{i2} expression was unchanged, suggesting that the D_2 receptors may be coupled to cAMP-independent pathways via G_{i3} *(88)*. This observation has also been made in the related GH_4C_1 cell line transfected with D_{2S} receptors. In these cells, D_{2S} is able to inhibit cAMP formation and modify prolactin release and K^+ currents through both cAMP-dependent and cAMP-independent mechanisms *(89)*. When this cell line is treated with EGF, a large increase in G_{i3} levels is observed, yet $G_{i1,2}$ levels are unchanged and the D_{2S} receptor is no longer coupled to cAMP inhibition or other cAMP-dependent pathways *(89)*. These results suggest that D_2 receptors may couple to inhibition of cAMP through G_{i1} or G_{i2}, and couple to other cAMP-independent signaling pathways by G_{i3} in these cell lines.

Examination of G protein levels in pituitary tumors has also suggested which G proteins might be utilized by D_2 receptors to couple to various effectors. Two pituitary tumors, 7315a and MtTW15, and a pituitary adenoma were found to have altered levels of a number of Gα subunits. Each of these tumors had decreased levels of $G_i\alpha$ and $G_o\alpha$ subunits *(90)*. By examining the ability of D_2 receptors to inhibit cAMP levels and prolactin release in these cells, Bouvier et al. *(90)* suggested that $G_{i\alpha 3}$ couples D_2 receptors to inhibition of basal cAMP levels, and that $G_{i\alpha 1}$ mediates the inhibition of stimulated cAMP and/or prolactin levels. Whether some of the apparent differences in the coupling of D_2 receptors to various G_α subunits might be explained by different complements of βγ subunits remains to be determined.

4.2. Mutant G Proteins

Several different types of mutant G proteins have been used to explore the specificity of coupling of the D_2 receptor to G proteins. One type of G protein mutation that has been used in many systems is alteration of a conserved glutamine residue in α subunits that when mutated blocks the intrinsic GTPase activity and thus creates a constitutively active α subunit. In GH_4C_1 cells, the D_{2S} receptor can decrease transcription from the prolactin promoter. Lew et al. *(91)* showed that constitutively active G_{i2} or G_o α subunits could mimic this effect, suggesting that D_2 receptors are able to influence prolactin transcription via signal transduction pathways coupled to either of these G proteins. In addition, in these cells, activated $G_{i\alpha 2}$ inhibited basal cAMP levels *(91)*, further confirming previous results suggesting that D_2 receptors can couple to adenylate cyclase via this G protein.

In order to clarify further the coupling of the D_{2S} and D_{2L} receptor isoforms to specific G_i subtypes, Senogles *(92)* constructed a mutant of each $G_{i\alpha}$ subunit that cannot be modified by PTX. GH_4C_1 cells expressing D_{2S} or D_{2L}

receptors were transfected with a mutant $G_{i\alpha}$ and treated with PTX. This inactivates all endogenous G_i subtypes and therefore the D_2 receptor can signal only through the transfected $G_{i\alpha}$ subunit. This approach has revealed that neither D_2 subtype inhibits cAMP formation through G_{i1}. However, in this cell type, the D_{2S} inhibits adenylate cyclase exclusively via G_{i2}, whereas D_{2L} affects this pathway through G_{i3} *(92)*. Interestingly, this result is in direct contrast to that of Montmayeur et al. *(46)* who found that in JEG-3 cells, D_{2L} inhibits cAMP accumulation primarily via G_{i2}. This again points out the problems of utilizing cultured cell lines, which obviously are not equivalent in all respects.

Considerable evidence points to the C-terminal portion of G protein α subunits as the site of interaction with receptors *(93)*. Chimeric G proteins have been constructed which substitute C-terminal sequences of various α subunits. These constructs provide further information about which G proteins can interact with D_2 receptors. For example, it has been demonstrated that D_2 receptors can interact with $G_{i\alpha2}$. $G_{q\alpha}$ containing four C-terminal amino acids from $G_{i\alpha2}$ allows D_2 to stimulate PI hydrolysis in 293 cells *(94)*, a pathway to which the D_2 receptor is not normally coupled in these cells. Similar experiments showed that D_2 could activate PI hydrolysis through a G_o/G_q chimera *(94)*. It has also been shown that the D_2 receptor can interact with the PTX-insensitive $G_{z\alpha}$ using this chimeric approach. $G_{q\alpha}$ containing C-terminal $G_{z\alpha}$ sequence allowed the D_2 receptor to stimulate PI hydrolysis in 293 cells *(94)*, whereas a similar chimera consisting of $G_{13\alpha}$ with C-terminal $G_{z\alpha}$ sequence could couple the D_2 receptor to an increased rate of extracellular acidification in 293 cells *(95)*. Additional evidence of D_2 receptor coupling to $G_{z\alpha}$ comes from the observation that when it is cotransfected with $G_{z\alpha}$ in 293 cells, the D_2 receptor can inhibit cAMP accumulation in a PTX-insensitive manner *(96)*.

4.3. Other Approaches

Several approaches other than those already discussed may be used to determine directly with which G proteins a receptor can interact. For example, antibodies to specific G protein α subunits may disrupt signaling of a receptor and thus indicate an interaction between receptor and G protein. These experiments further support the notion that D2-like receptors are capable of interacting with multiple G proteins. Antibodies raised against $G_{i\alpha}$ and $G_{o\alpha}$ are capable of eliminating high affinity agonist binding to D2-like receptors in bovine striatal membranes *(97)*. Further evidence supporting interaction of D_2 receptors with both G_i and G_o comes from studies in pituitary lactotroph cells. In pituitary cells, D_2 receptors are able to decrease voltage-activated Ca^{2+} currents and also stimulate voltage-activated K^+ currents *(62)*. Lledo et al. *(62)*

loaded lactotrophs with G_i or G_o antibodies and then examined the effect of dopamine on potassium and calcium currents. An antibody against $G_{o\alpha}$ completely abolished inhibition of Ca^{2+} currents by dopamine, whereas antibodies against $G_{i\alpha 1,2}$ and $G_{i\alpha 3}$ were ineffective *(62)*. An antibody against $G_{i\alpha 3}$ blocked dopamine-stimulated K^+ currents, whereas anti-$G_{o\alpha}$ and anti-$G_{i\alpha 1,2}$ were not effective *(62)*. These results show that the D_2 receptor is capable of coupling to multiple G proteins, resulting in a variety of signals generated by the same receptor.

Recently, the use of antisense oligonucleotides to eliminate the expression of a specific protein has also been used to study coupling of D_2 receptors to G proteins. Confirming the previous data, in pituitary lactotrophs, antisense against the $G_{o\alpha}$ subunit abolished the dopamine inhibition of Ca^{2+} currents, which was unaffected by antisense to any of the G_i subtypes *(61)*. Dialysis with $G_{i\alpha 3}$ antisense oligonucleotides inhibited the ability of dopamine to stimulate K^+ currents in these cells whereas antisense $G_{o\alpha}$ or $G_{i\alpha 1}$ did not significantly block this stimulation *(61)*.

Similar techniques have been used in GH_4C_1 cells to examine the coupling of D_2 receptors to Ca^{2+} currents and cAMP levels. $G_{o\alpha}$ antisense oligonucleotides completely blocked the ability of the D_{2S} receptor to inhibit Bay K 8644-induced Ca^{2+} currents; however, this antisense only partially blocked the inhibition by D_{2L} receptors *(56)*. In addition, $G_{i\alpha 2}$ antisense eliminated the ability of both long and short receptor isoforms to inhibit basal cAMP accumulation, but it blocked the inhibition of stimulated cAMP formation only by the D_{2L}, but not the D_{2S} receptor *(56)*. These results indicate that D_2 receptors are capable of interacting with at least $G_{o\alpha}$, $G_{i\alpha 2}$, and $G_{i\alpha 3}$. In addition, it confirms that the short and long receptor isoforms have slightly different signaling properties at the level of G protein interaction.

The ability of dopamine receptors to couple to G proteins has been examined not only with studies that interfere with receptor–G protein interaction, but also with experiments that measure the ability of a receptor to influence guanine nucleotide binding properties of G proteins. By co-reconstituting purified receptors and G proteins, it is possible to examine the specificity of coupling between the two components. Senogles et al. *(98)* purified D2 receptors from bovine anterior pituitary that, when reconstituted with $G_{i\alpha 2}$, resulted in agonist stimulation of the GTPase activity of this protein. The GTPase activity of $G_{i\alpha 1}$ and $G_{i\alpha 3}$ were also stimulated by agonist, but at a lower receptor:G protein ratio and with a lower maximal stimulation, whereas $G_{o\alpha}$ GTPase activity was not stimulated at all *(98)*. Furthermore, D2 receptor agonists stimulated the rate of GTPγS binding and GDP release at $G_{i2\alpha}$ to a greater extent than at $G_{i\alpha 1}$ or $G_{i\alpha 3}$, and had no effect on these parameters of Go_α *(98)*. These results are in contrast to

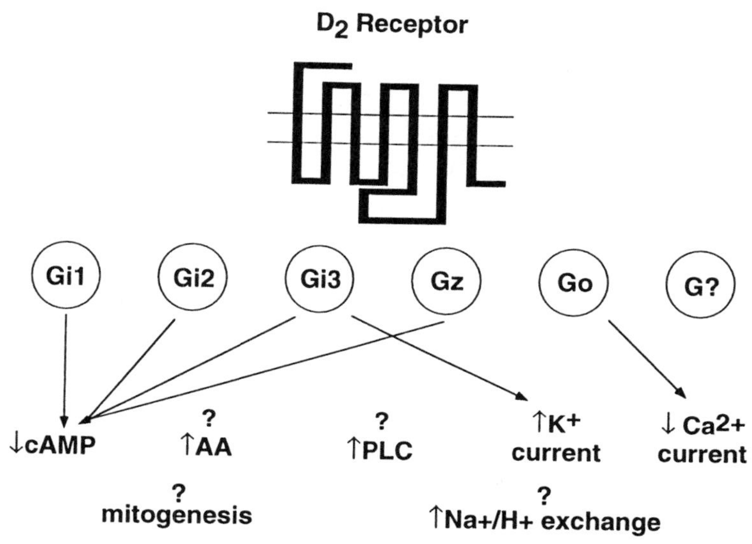

Fig. 2. Signaling pathways of the D_2 receptor. Circles, G protein α subunits; Arrows, coupling of receptor/G protein/second messenger which have been previously demonstrated; G?, any other G protein α subunit with which this receptor might couple; ?, second messenger pathways which the D_2 receptor has been shown to couple to in some cell lines, but not in others.

results in intact cells of pituitary origin in which coupling of D_2 receptors to $G_{o\alpha}$ has been demonstrated *(56,61,62)*. However, it is important to note that in these experiments, it is not known what proportion of D_{2S} and D_{2L} receptors were present.

Another method to detect receptor–G protein interactions is by the utilization of radiolabeled guanine nucleotides. In rat striatal membranes it is possible to measure the binding of [α^{32}P]GTP to α subunits of G_s and $G_{i/o}$ *(99)*. In striatal membranes, dopamine stimulated the binding of [α^{32}P]GTP to $G_{s\alpha}$ and $G_{i\alpha}$ *(99)*, consistent with the presence of both D1 and D2 receptors in this region of the brain. Dopamine did not stimulate GTP binding to $G_{o\alpha}$, supporting the result of Senogles et al. *(98)*, who showed that bovine anterior pituitary D_2 receptors do not interact with $G_{o\alpha}$. However, this is in contrast to the results of studies which decreased α subunit expression in pituitary cells and showed that the D_2 receptors couple to Ca^{2+} currents via $G_{o\alpha}$ *(56,61,62)*. Whether there are different conditions which alter the coupling of the D_2 receptor to $G_{o\alpha}$ remains to be determined. Figure 2 summarizes our current knowledge of the coupling of D_2 receptors to second messengers via G protein α subunits.

4.4. Role of $\beta\gamma$ Subunits

Most efforts to understand dopamine receptor–G protein coupling have been aimed at distinguishing which α subunits couple the receptors to various second messengers. However, recently it has come to be appreciated that $\beta\gamma$ subunits as well as α subunits of G proteins can modulate second messenger levels *(100)*. This is likely to be an important signaling pathway for many receptors, and recent work has shown that this may be the case for D_2 receptors. $\beta\gamma$ subunits modulate different isoforms of adenylate cyclase in different ways. It was shown recently that $\beta\gamma$ subunits in conjunction with activated $G_{s\alpha}$ stimulate the activity of adenylate cyclase type II *(101,102)*. Lustig et al. *(103)* demonstrated that D_2 receptors can stimulate adenylate cyclase type II via this mechanism in 293 cells. In addition, the activity of mitogen-activated protein (MAP) kinase is also stimulated by $\beta\gamma$ subunits *(104)*. Faure et al. *(104)* showed that in 293 cells, D_2 receptors are capable of activating MAP kinase by this mechanism. Thus it is important when discussing dopamine receptors and G proteins to consider that $\beta\gamma$ as well as α subunits are important players. In addition, if $\beta\gamma$ subunits can provide specificity of coupling, as they appear to with certain receptors *(105,106)*, the apparent selectivity displayed for α subunits in various cell types might be the result of differing $\beta\gamma$ complements in these cells.

5. Other Issues

Despite the volume of work that has accumulated regarding the coupling of dopamine receptors to G proteins, many unanswered or unexplored questions remain. For example, little work has been done concerning coupling of D1-like receptors to G proteins other than G_s. There is some evidence that members of this subfamily of dopamine receptors are able to stimulate PI hydrolysis, but it remains to be seen whether this occurs through D_1 coupling to a G_q-like G protein or via a new D1-like receptor subtype, or both. Another possibility is that the known D1-like receptors may stimulate PI turnover through G_s, as it was recently shown that knockout of this G protein by antisense could prevent stimulation of phospholipase C by the thyrotropin-releasing hormone receptor *(107)*.

Another relatively unexplored aspect of D1-like receptor coupling is the possibility that these receptors may stimulate adenylate cyclase through G_{olf}. It has recently been shown by immunoblotting and immunohistochemistry that G_{olf} is expressed in the substantia nigra, caudate putamen, and nucleus accumbens, perhaps at higher levels than G_s *(108)*. This coexpression of D_1 receptors and G_{olf} raises the possibility that this G protein could be at least partially responsible for the increase in cAMP mediated by the D_1 receptor in these brain regions.

Although there has been substantially more work done on the interaction of G proteins with D_2 receptors, there are still many unresolved questions. One important consideration is that although it is possible to show interactions between D_2 receptors and specific G proteins when they are coexpressed in cell culture lines, this does not prove their interaction in vivo. Such interactions, of course, depend on the coexpression of the receptor and G protein in vivo and may also depend on other factors that we have yet to understand.

One factor may be the actions of protein kinase C on either receptor or G protein. For example, phorbol ester treatment reduces the ability of D_{2S} and D_{2L} isoforms to alter intracellular Ca^{2+} levels in Ltk⁻ cells, yet having no effect on inhibition of cAMP levels by these receptors *(50)*. Furthermore, phorbol ester seems to alter the ability of D_2 receptors to modulate arachidonic acid and cAMP levels in CHO cells. Treatment of cells expressing D_2 receptors with phorbol ester increased the potentiation of arachidonic acid release and decreased the inhibition of adenylate cyclase evoked by dopamine *(109)*. As this shows, we still have much to learn about the ability of D_2 receptors to couple to different G proteins and second messenger pathways under various conditions.

Finally, the second messenger coupling of the D_3 and D_4 receptors is just beginning to be understood. Although it appears that these receptors may couple to many of the same second messenger systems as the D_2 receptors, there appear to be differences in the ability of these receptors to couple to these pathways. This might suggest that they couple to different G proteins, or couple to the same G proteins but with lower efficiency than the D_2 receptor. Thus, there is still much work to be done regarding these subtypes of receptors.

6. Summary

Since the cloning of five mammalian dopamine receptors, it has been found that the coupling of these receptors is more complicated than the original scheme of stimulation or inhibition of adenylate cyclase as put forward by Kebabian and Calne *(43)*. It has been shown that the D1-like receptors may stimulate PI hydrolysis, inhibit a Na^+/H^+ exchanger, and increase K^+ efflux in some tissues in addition to the stimulation of adenylate cyclase. The D2-like receptors have been shown to modulate a variety of second messengers including cAMP, intracellular Ca^{2+} levels, K^+ currents, and arachidonic acid. Furthermore, these receptors also seem to stimulate a Na^+/H^+ exchanger and affect cell growth.

Alterations in dopaminergic pathways have been implicated in several disease states including schizophrenia, Parkinson's disease, and Huntington's chorea *(110)*. Dopamine receptor antagonists are utilized in the treatment of schizophrenia, whereas dopamine receptor agonists are used to treat Parkinson's disease and also prolactinomas. A clearer understanding of the alterations that occur in these pathophysiologies, as well as the coupling of dopamine receptors, particularly the D2-like subtypes, to specific G proteins and second messengers, holds out the possibility of the development of therapeutic agents that could specifically modulate a particular signaling pathway that may be altered in the disease state.

The coupling of the dopamine receptors to specific second messengers through particular G proteins continues to be an area of active investigation. In addition to determining which G proteins can interact with each dopamine receptor, it is important to elucidate which G proteins do interact with the receptors in vivo. Such information may in the future allow the design of therapeutic agents that act at the level of G proteins with greater specificity than the currently utilized receptor ligands.

References

1. Civelli, O., Bunzow, J. R., and Grandy, D. K. (1993) Molecular diversity of the dopamine receptors. *Annu. Rev. Pharmacol. Toxicol.* **32,** 281–307.
2. Gingrich, J. A. and Caron, M. G. (1993) Recent advances in the molecular biology of dopamine receptors. *Annu. Rev. Neurosci.* **16,** 299–321.
3. Hedin, K. E., Duerson, K., and Clapham, D. E. (1993) Specificity of receptor–G protein interactions: searching for the structure behind the signal. *Cell Signal.* **5,** 505–518.
4. Savarese, T. M. and Fraser, C. M. (1992) *In vitro* mutagenesis and the search for structure–function relationships among G protein-coupled receptors. *Biochem. J.* **283,** 1–19.
5. Cotecchia, S., Exum, S., Caron, M. G., and Lefkowitz, R. J. (1990) Regions of the α_1-adrenergic receptor involved in coupling to phosphatidylinositol hydrolysis and enhanced sensitivity of biological function. *Proc. Natl. Acad. Sci. USA* **87,** 2896–2900.
6. Kobilka, B. K., Kobilka, T. S., Daniel, K., Regan, J. W., Caron, M. G., and Lefkowitz, R. J. (1988) Chimeric α_2-,β_2-adrenergic receptors: delineation of domains involved in effector coupling and ligand binding specificity. *Science* **240,** 1310–1316.
7. Strader, C. D., Dixon, R. A. F., Cheung, A. H., Candelore, M. R., Blake, A. D., and Sigal, I. S. (1987) Mutations that uncouple the β-adrenergic receptor from G_s and increase agonist affinity. *J. Biol. Chem.* **262,** 16,439–16,443.
8. Dalman, H. M. and Neubig, R. R. (1991) Two peptides from the α_{2A}-adrenergic receptor alter receptor G protein coupling by distinct mechanisms. *J. Biol. Chem.* **266,** 11,025–11,029.
9. Wess, J., Bonner, T. I., Dörje, F., and Brann, M. R. (1990) Delineation of muscarinic receptor domains conferring selectivity of coupling to guanine nucleotide-binding proteins and second messengers. *Mol. Pharmacol.* **38,** 517–523.
10. Franke, R. R., König, B., Sakmar, T. P., Khorana, H. G., and Hofmann, K. P. (1990) Rhodopsin mutants that bind but fail to activate transducin. *Science* **250,** 123–125.

11. O'Dowd, B. F., Hnatowich, M., Regan, J. W., Leader, W. M., Caron, M. G., and Lefkowitz, R. J. (1988) Site-directed mutagenesis of the cytoplasmic domains of the human β_2-adrenergic receptor. Localization of regions involved in G protein–receptor coupling. *J. Biol. Chem.* **263,** 15,985–15,992.
12. Takemoto, D. J., Takemoto, L. J., Hansen, J., and Morrison, D. (1985) Regulation of retinal transducin by C-terminal peptides of rhodopsin. *Biochem. J.* **232,** 669–672.
13. Schreiber, R. E., Prossnitz, E. R., Ye, R. D., Cochrane, C. G., and Bokoch, G. M. (1994) Domains of the human neutrophil *N*-formyl peptide receptor involved in G protein coupling. Mapping with receptor-derived peptides. *J. Biol. Chem.* **269,** 326–331.
14. Varrault, A., Nguyen, D. L., McClue, S., Harris, B., Jouin, P., and Bockaert, J. (1994) 5-Hydroxytryptamine$_{1A}$ receptor synthetic peptides. Mechanisms of adenylyl cyclase inhibition. *J. Biol. Chem.* **269,** 16,720–16,725.
15. Wong, S. K.-F., Parker, E. M., and Ross, E. M. (1990) Chimeric muscarinic cholinergic: β-adrenergic receptors that activate G_s in response to muscarinic agonists. *J. Biol. Chem.* **265,** 6219–6224.
16. König, B. and Grätzel, M. (1994) Site of dopamine D_1 receptor binding to G_s protein mapped with synthetic peptides. *Biochim. Biophys. Acta* **1223,** 261–266.
17. Voss, T., Wallner, E., Czernilofsky, A. P., and Freissmuth, M. (1993) Amphipathic α-helical structure does not predict the ability of receptor-derived synthetic peptides to interact with guanine nucleotide-binding regulatory proteins. *J. Biol. Chem.* **268,** 4637–4642.
18. Luttrell, L. M., Ostrowski, J., Cotecchia, S., Kendall, H., and Lefkowitz, R. J. (1993) Antagonism of catecholamine receptor signaling by expression of cytoplasmic domains of the receptors. *Science* **259,** 1453–1457.
19. Hawes, B. E., Luttrell, L. M., Exum, S. T., and Lefkowitz, R. J. (1994) Inhibition of G protein-coupled receptor signaling by expression of cytoplasmic domains of the receptor. *J. Biol. Chem.* **269,** 15,776–15,785.
20. Malek, D., Münch, G., and Palm, D. (1993) Two sites in the third inner loop of the dopamine D_2 receptor are involved in functional G protein-mediated coupling to adenylate cyclase. *FEBS Lett.* **325,** 215–219.
21. MacKenzie, R. G., Steffey, M. E., Manelli, A. M., Pollock, N. J., and Frail, D. E. (1993) A D_1/D_2 chimeric dopamine receptor mediates a D_1 response to a D_2-selective agonist. *FEBS Lett.* **323,** 59–62.
22. Kozell, L. B., Machida, C. A., Neve, R. L., and Neve, K. A. (1994) Chimeric D_1/D_2 dopamine receptors. Distinct determinants of selective efficacy, potency, and signal transduction. *J. Biol. Chem.* **269,** 30,299–30,306.
23. McAllister, G., Knowles, M. R., Patel, S., Marwood, R., Emms, F., Seabrook, G. R., Graziano, M., Borkowski, D., Hey, P. J., and Freedman, S. B. (1993) Characterisation of a chimeric hD_3/D_2 dopamine receptor expressed in CHO cells. *FEBS Lett.* **324,** 81–86.
24. Robinson, S. W., Jarvie, K. R., and Caron, M. G. (1994) High affinity agonist binding to the dopamine D_3 receptor: chimeric receptors delineate a role for intracellular domains. *Mol. Pharmacol.* **46,** 352–356.

24a. Robinson, S. W. and Caron, M. G. (1996) Chimeric D_2/D_3 dopamine receptors efficiently inhibit adenylyl cyclase in HEK 293 cells. *J. Neurochem.*, in press.

25. Boundy, V. A., Luedtke, R. R., and Molinoff, P. B. (1993) Development of polyclonal anti-D_2 dopamine receptor antibodies to fusion proteins: inhibition of D_2 receptor–G protein interaction. *J. Neurochem.* **60,** 2181–2191.

26. Plug, M. J., Dijk, J., Maassen, A., and Möller, W. (1992) An anti-peptide antibody that recognizes the dopamine D_2 receptor from bovine striatum. *Eur. J. Biochem.* **206,** 123–130.
27. Boundy, V. A., Luedtke, R. R., Artymyshyn, R. P., Filtz, T. M., and Molinoff, P. B. (1993) Development of polyclonal anti-D_2 dopamine receptor antibodies using sequence-specific peptides. *Mol. Pharmacol.* **43,** 666–676.
28. Chazot, P. L., Doherty, A. J., and Strange, P. G. (1993) Antisera specific for D_2 dopamine receptors. *Biochem. J.* **289,** 789–794.
29. Pollock, N. J., Manelli, A. M., Hutchins, C. W., Steffey, M. E., MacKenzie, R. G., and Frail, D. E. (1992) Serine mutations in transmembrane V of the dopamine D_1 receptor affect ligand interactions and receptor activation. *J. Biol. Chem.* **267,** 17,780–17,786.
30. Neve, K. A., Cox, B. A., Henningsen, R. A., Spanoyannis, A., and Neve, R. L. (1991) Pivotal role for aspartate-80 in the regulation of dopamine D_2 receptor affinity for drugs and inhibition of adenylyl cyclase. *Mol. Pharmacol.* **39,** 733–739.
31. Tiberi, M. and Caron, M. G. (1994) High agonist-independent activity is a distinguishing feature of the dopamine D_{1B} receptor subtype. *J. Biol. Chem.* **269,** 27,925–27,931.
32. Demchyshyn, L. L., Sugamori, K. S., Lee, F. J. S., Hamadanizadeh, S. A., and Niznik, H. B. (1995) The dopamine D_{1D} receptor. Cloning and characterization of three pharmacologically distinct D_1-like receptors from *Gallus domesticus*. *J. Biol. Chem.* **270,** 4005–4012.
33. Sugamori, K. S., Demchyshyn, L. L., Chung, M., and Niznik, H. B. (1994) D_{1A}, D_{1B}, and D_{1C} dopamine receptors from *Xenopus laevis*. *Proc. Natl. Acad. Sci. USA* **91,** 10,536–10,540.
34. Liu, Y. F., Civelli, O., Zhou, Q.-Y., and Albert, P. R. (1992) Cholera toxin-sensitive 3'5'-cyclic adenosine monophosphate and calcium signals of the human dopamine-D_1 receptor: selective potentiation by protein kinase A. *Mol. Endocrinol.* **6,** 1815–1824.
35. Frail, D. E., Manelli, A. M., Witte, D. G., Lin, C. W., Steffey, M. E., and MacKenzie, R. G. (1993) Cloning and characterization of a truncated dopamine D_1 receptor from goldfish retina: stimulation of cyclic AMP production and calcium mobilization. *Mol. Pharmacol.* **44,** 1113–1118.
36. Rodrigues, P. S. and Dowling, J. E. (1990) Dopamine induces neurite retraction in retinal horizontal cells via diacylglycerol and protein kinase C. *Proc. Natl. Acad. Sci. USA* **87,** 9693–9697.
37. Pedersen, U. B., Norby, B., Jensen, A. A., Schiødt, M., Hansen, A., Suhr-Jessen, P., Scheideler, M., Thastrup, O., and Andersen, P. H. (1994) Characteristics of stably expressed human D_{1a} and D_{1b} receptors: atypical behavior of the dopamine D_{1b} receptor. *Eur. J. Pharmacol.* **267,** 85–93.
38. Undie, A. S. and Friedman, E. (1990) Stimulation of a dopamine D_1 receptor enhances inositol phosphates formation in rat brain. *J. Pharmacol. Exp. Ther.* **253,** 987–992.
39. Undie, A. S., Weinstock, J., Sarau, H. M., and Friedman, E. (1994) Evidence for a distinct D_1-like dopamine receptor that couples to activation of phosphoinositide metabolism in brain. *J. Neurochem.* **62,** 2045–2048.
40. Felder, C. C., Campbell, T., Albrecht, F., and Jose, P. A. (1990) Dopamine inhibits Na^+–H^+ exchanger activity in renal BBMV by stimulation of adenylate cyclase. *Am. J. Physiol.* **259,** F297–F303.

41. Felder, C. C., Albrecht, F. E., Campbell, T., Eisner, G. M., and Jose, P. A. (1993) cAMP-independent, G protein-linked inhibition of Na^+/H^+ exchange in renal brush border by D_1 dopamine agonists. *Am. J. Physiol.* **264,** F1032–F1037.
42. Laitinen, J. T. (1993) Dopamine stimulates K^+ efflux in the chick retina via D_1 receptors independently of adenylyl cyclase activation. *J. Neurochem.* **61,** 1461–1469.
43. Kebabian, J. W. and Calne, D. B. (1979) Multiple receptors for dopamine. *Nature* **277,** 93–96.
44. Hayes, G., Biden, T. J., Selbie, L. A., and Shine, J. (1992) Structural subtypes of the dopamine D_2 receptor are functionally distinct: expression of the cloned D_{2A} and D_{2B} subtypes in a heterologous cell line. *Mol. Endocrinol.* **6,** 920–926.
45. Montmayeur, J.-P. and Borrelli, E. (1991) Transcription mediated by a cAMP-responsive promoter element is reduced upon activation of dopamine D_2 receptors. *Proc. Natl. Acad. Sci. USA* **88,** 3135–3139.
46. Montmayeur, J.-P., Guiramand, J., and Borrelli, E. (1993) Preferential coupling between dopamine D_2 receptors and G-proteins. *Mol. Endocrinol.* **7,** 161–170.
47. Felder, C. C., Williams, H. L., and Axelrod, J. (1991) A transduction pathway associated with receptors coupled to the inhibitory guanine nucleotide binding protein G_i that amplifies ATP-mediated arachidonic acid release. *Proc. Natl. Acad. Sci. USA* **88,** 6477–6480.
48. Kanterman, R. Y., Mahan, L. C., Briley, E. M., Monsma, F. J., Sibley, D. R., Axelrod, J., and Felder, C. C. (1991) Transfected D_2 dopamine receptors mediate the potentiation of arachidonic acid release in Chinese hamster ovary cells. *Mol. Pharmacol.* **39,** 364–369.
49. Piomelli, D., Pilon, C., Giros, B., Sokoloff, P., Martres, M.-P., and Schwartz, J.-C. (1991) Dopamine activation of the arachidonic acid cascade as a basis for D_1/D_2 synergism. *Nature* **353,** 164–167.
50. Liu, Y. F., Civelli, O., Grandy, D. K., and Albert, P. R. (1992) Differential sensitivity of the short and long human dopamine D_2 receptor subtypes to protein kinase C. *J. Neurochem.* **59,** 2311–2317.
51. Vallar, L., Muca, C., Magni, M., Albert, P., Bunzow, J., Meldolesi, J., and Civelli, O. (1990) Differential coupling of dopaminergic D_2 receptors expressed in different cell types. Stimulation of phosphatidylinositol 4,5-bisphosphate hydrolysis in Ltk^- fibroblasts, hyperpolarization, and cytosolic-free Ca^{2+} concentration decrease in GH_4C_1 cells. *J. Biol. Chem.* **265,** 10,320–10,326.
52. MacKenzie, R. G., VanLeeuwen, D., Pugsley, T. A., Shih, Y.-H., Demattos, S., Tang, L., Todd, R. D., and O'Malley, K. L. (1994) Characterization of the human dopamine D_3 receptor expressed in transfected cell lines. *Eur. J. Pharmacol.* **266,** 79–85.
53. Tang, L., Todd, R. D., Heller, A., and O'Malley, K. L. (1994) Pharmacological and functional characterization of D_2, D_3 and D_4 dopamine receptors in fibroblast and dopaminergic cell lines. *J. Pharmacol. Exp. Ther.* **268,** 495–502.
54. Freedman, S. B., Patel, S., Marwood, R., Emms, F., Seabrook, G. R., Knowles, M. R., and McAllister, G. (1994) Expression and pharmacological characterization of the human D_3 dopamine receptor. *J. Pharmacol. Exp. Ther.* **268,** 417–426.
55. Nussinovitch, I. and Kleinhaus, A. L. (1992) Dopamine inhibits voltage-activated calcium channel currents in rat pars intermedia pituitary cells. *Brain Res.* **574,** 49–55.

56. Liu, Y. F., Jakobs, K. H., Rasenick, M. M., and Albert, P. R. (1994) G protein specificity in receptor–effector coupling. Analysis of the roles of G_o and G_{i2} in GH_4C_1 pituitary cells. *J. Biol. Chem.* **269,** 13,880–13,886.
57. Seabrook, G. R., Knowles, M., Brown, N., Myers, J., Sinclair, H., Patel, S., Freedman, S. B., and McAllister, G. (1994) Pharmacology of high-threshold calcium currents in GH_4C_1 pituitary cells and their regulation by activation of human D_2 and D_4 dopamine receptors. *Br. J. Pharmacol.* **112,** 728–734.
58. Seabrook, G. R., McAllister, G., Knowles, M. R., Myers, J., Sinclair, H., Patel, S., Freedman, S. B., and Kemp, J. A. (1994) Depression of high-threshold calcium currents by activation of human D_2 (short) dopamine receptors expressed in differentiated NG108-15 cells. *Br. J. Pharmacol.* **111,** 1061–1066.
59. Einhorn, L. C., Gregerson, K. A., and Oxford, G. S. (1991) D_2 dopamine receptor activation of potassium channels in identified rat lactotrophs: whole-cell and single-channel recording. *J. Neurosci.* **11,** 3727–3737.
60. Memo, M., Pizzi, M., Belloni, M., Benarese, M., and Spano, P. F. (1992) Activation of dopamine D_2 receptors linked to voltage-sensitive potassium channels reduces forskolin-induced cyclic AMP formation in rat pituitary cells. *J. Neurochem.* **59,** 1829–1835.
61. Baertschi, A. J., Audigier, Y., Lledo, P.-M., Israel, J.-M., Bockaert, J., and Vincent, J.-D. (1992) Dialysis of lactotropes with antisense oligonucleotides assigns guanine nucleotide binding protein subtypes to their channel effectors. *Mol. Endocrinol.* **6,** 2257–2265.
62. Lledo, P. M., Homburger, V., Bockaert, J., and Vincent, J.-D. (1992) Differential G protein-mediated coupling of D_2 dopamine receptors to K^+ and Ca^{2+} currents in rat anterior pituitary cells. *Neuron* **8,** 455–463.
63. Castellano, M. A., Liu, L.-X., Monsma, F. J., Sibley, D. R., Kapatos, G., and Chiodo, L. A. (1993) Transfected D_2 short dopamine receptors inhibit voltage-dependent potassium current in neuroblastoma x glioma hybrid (NG108-15) cells. *Mol. Pharmacol.* **44,** 649–656.
64. Tang, L., Todd, R. D., and O'Malley, K. L. (1994) Dopamine D_2 and D_3 receptors inhibit dopamine release. *J. Pharmacol. Exp. Ther.* **270,** 475–479.
65. Neve, K. A., Kozlowski, M. R., and Rosser, M. P. (1992) Dopamine D_2 receptor stimulation of Na^+/H^+ exchange assessed by quantification of extracellular acidification. *J. Biol. Chem.* **267,** 25,748–25,753.
66. Ganz, M. B., Pachter, J. A., and Barber, D. L. (1990) Multiple receptors coupled to adenylate cyclase regulate Na–H exchange independent of cAMP. *J. Biol. Chem.* **265,** 8989–8992.
67. Chio, C. L., Lajiness, M. E., and Huff, R. M. (1994) Activation of heterologously expressed D_3 dopamine receptors: comparison with D_2 dopamine receptors. *Mol. Pharmacol.* **45,** 51–60.
68. Florio, T., Pan, M.-G., Newman, B., Hershberger, R. E., Civelli, O., and Stork, P. J. S. (1992) Dopaminergic inhibition of DNA synthesis in pituitary tumor cells is associated with phosphotyrosine phosphatase activity. *J. Biol. Chem.* **267,** 24,169–24,172.
69. Senogles, S. E. (1994) The D_2 dopamine receptor mediates inhibition of growth in GH_4ZR_7 cells: involvement of protein kinase-Cε. *Endocrinology* **134,** 783–789.
70. Lajiness, M. E., Chio, C. L., and Huff, R. M. (1993) D_2 dopamine receptor stimulation of mitogenesis in transfected Chinese hamster ovary cells: relationship to

dopamine stimulation of tyrosine phosphorylations. *J. Pharmacol. Exp. Ther.* **267,** 1573–1581.

71. Swarzenski, B. C., Tang, L., Oh, Y. J., O'Malley, K. L., and Todd, R. D. (1994) Morphogenic potentials of D_2, D_3, and D_4 dopamine receptors revealed in transfected neuronal cell lines. *Proc. Natl. Acad. Sci. USA* **91,** 649–653.
72. Giros, B., Martres, M.-P., Sokoloff, P., and Schwartz, J.-C. (1990) Clonage du gène du récepteur dopaminergique D_3 humain et identification de son chromosome. *C. R. Acad. Sci.* **311,** 501–508.
73. Sokoloff, P., Giros, B., Martres, M.-P., Bouthenet, M.-L., and Schwartz, J.-C. (1990) Molecular cloning and characterization of a novel dopamine receptor (D_3) as a target for neuroleptics. *Nature* **347,** 146–151.
74. Boundy, V. A., Luedtke, R. R., Gallitano, A. L., Smith, J. E., Filtz, T. M., Kallen, R. G., and Molinoff, P. B. (1993) Expression and characterization of the rat D_3 dopamine receptor: pharmacologic properties and development of antibodies. *J. Pharmacol. Exp. Ther.* **264,** 1002–1011.
75. McAllister, G., Knowles, M. R., Ward-Booth, S. M., Sinclair, H. A., Patel, S., Marwood, R., Emms, F., Patel, S., Seabrook, G. R., and Freedman, S. B. (1995) Functional coupling of human D_2, D_3, and D_4 dopamine receptors in HEK293 cells. *J. Receptor Signal Trans. Res.* **15,** 267–281.
76. Lévesque, D., Diaz, J., Pilon, C., Martres, M.-P., Giros, B., Souil, E., Schott, D., Morgat, J.-L., Schwartz, J.-C., and Sokoloff, P. (1992) Identification, characterization, and localization of the dopamine D_3 receptor in rat brain using 7-[^{3}H]hydroxy-*N,N*-di-*n*-propyl-2-aminotetralin. *Proc. Natl. Acad. Sci. USA* **89,** 8155–8159.
77. Castro, S. W. and Strange, P. G. (1993) Coupling of D_2 and D_3 dopamine receptors to G-proteins. *FEBS Lett.* **315,** 223–226.
78. Seabrook, G. R., Patel, S., Marwood, R., Emms, F., Knowles, M. R., Freedman, S. B., and McAllister, G. (1992) Stable expression of human D_3 dopamine receptors in GH_4C_1 pituitary cells. *FEBS Lett.* **312,** 123–126.
79. Pilon, C., Lévesque, D., Dimitriadou, V., Griffon, N., Martres, M.-P., Schwartz, J.-C., and Sokoloff, P. (1994) Functional coupling of the human dopamine D_3 receptor in a transfected NG108-15 neuroblastoma-glioma hybrid cell line. *Eur. J. Pharmacol.* **268,** 129–139.
80. Potenza, M. N., Graminski, G. F., Schmauss, C., and Lerner, M. R. (1994) Functional expression and characterization of human D_2 and D_3 dopamine receptors. *J. Neurosci.* **14,** 1463–1476.
81. Seabrook, G. R., Kemp, J. A., Freedman, S. B., Patel, S., Sinclair, H. A., and McAllister, G. (1994) Functional expression of human D_3 dopamine receptors in differentiated neuroblastoma x glioma NG108-15 cells. *Br. J. Pharmacol.* **111,** 391–393.
82. Cohen, A. I., Todd, R. D., Harmon, S., and O'Malley, K. L. (1992) Photoreceptors of mouse retinas possess D_4 receptors coupled to adenylate cyclase. *Proc. Natl. Acad. Sci. USA* **89,** 12,093–12,097.
83. Mills, A., Allet, B., Bernard, A., Chabert, C., Brandt, E., Cavegn, C., Chollet, A., and Kawashima, E. (1993) Expression and characterization of human D_4 dopamine receptors in baculovirus-infected cells. *FEBS Lett.* **320,** 130–134.
84. Asghari, V., Schoots, O., Van Kats, S., Ohara, K., Jovanovic, V., Guan, H.-C., Bunzow, J. R., Petronis, A., and Van Tol, H. H. M. (1994) Dopamine D_4 receptor repeat: analysis of different native and mutant forms of the human and rat genes. *Mol. Pharmacol.* **46,** 364–373.

85. McHale, M., Coldwell, M. C., Herrity, N., Boyfield, I., Winn, F. M., Ball, S., Cook, T., Robinson, J. H., and Gloger, I. S. (1994) Expression and functional characterisation of a synthetic version of the human D_4 dopamine receptor in a stable human cell line. *FEBS Lett.* **345,** 147–150.
86. Chio, C. L., Drong, R. F., Riley, D. T., Gill, G. S., Slightom, J. L., and Huff, R. M. (1994) D_4 dopamine receptor-mediated signaling events determined in transfected Chinese hamster ovary cells. *J. Biol. Chem.* **269,** 11,813–11,819.
87. Neer, E. J. (1995) Heterotrimeric G proteins: organizers of transmembrane signals. *Cell* **80,** 249–257.
88. Missale, C., Boroni, F., Castelletti, L., Dal Toso, R., Gabellini, N., Sigala, S., and Spano, P. F. (1991) Lack of coupling of D-2 receptors to adenylate cyclase in GH-3 cells exposed to epidermal growth factor. Possible role of a differential expression of G_i protein subtypes. *J. Biol. Chem.* **266,** 23,392–23,398.
89. Missale, C., Boroni, F., Sigala, S., Castelletti, L., Falardeau, P., Dal Toso, R., Caron, M. G., and Spano, P. F. (1994) Epidermal growth factor promotes uncoupling from adenylyl cyclase of the rat D_{2s} receptor expressed in GH_4C_1 cells. *J. Neurochem.* **62,** 907–915.
90. Bouvier, C., Forget, H., Lagacé, G., Drews, R., Sinnett, D., Labuda, D., and Collu, R. (1991) G proteins in normal rat pituitaries and in prolactin-secreting rat pituitary tumors. *Mol. Cell Endocrinol.* **78,** 33–44.
91. Lew, A. M., Yao, H., and Elsholtz, H. P. (1994) $G_{i\alpha 2}$- and $G_{o\alpha}$-mediated signaling in the Pit-1-dependent inhibition of the prolactin gene promoter. Control of transcription by dopamine D_2 receptors. *J. Biol. Chem.* **269,** 12,007–12,013.
92. Senogles, S. E. (1994) The D_2 dopamine receptor isoforms signal through distinct Giα proteins to inhibit adenylyl cyclase. A study with site-directed mutant Giα proteins. *J. Biol. Chem.* **269,** 23,120–23,127.
93. Conklin, B. R. and Bourne, H. R. (1993) Structural elements of Gα subunits that interact with Gβγ, receptors and effectors. *Cell* **73,** 631–641.
94. Conklin, B. R., Farfel, Z., Lustig, K. D., Julius, D., and Bourne, H. R. (1993) Substitution of three amino acids switches receptor specificity of $G_{q\alpha}$ to that of $G_{i\alpha}$. *Nature* **363,** 274–276.
95. Voyno-Yasenetskaya, T., Conklin, B. R., Gilbert, R. L., Hooley, R., Bourne, H. R., and Barber, D. L. (1994) Gα13 stimulates Na–H exchange. *J. Biol. Chem.* **269,** 4721–4724.
96. Wong, Y. H., Conklin, B. R., and Bourne, H. R. (1992) G_z-mediated hormonal inhibition of cyclic AMP accumulation. *Science* **255,** 339–342.
97. Plug, M. J., Möller, W., and Dijk, J. (1992) Interactions between the dopamine D_2 receptor and GTP-binding proteins. *Biochem. Int.* **28,** 21–29.
98. Senogles, S. E., Spiegel, A. M., Padrell, E., Iyengar, R., and Caron, M. G. (1990) Specificity of receptor–G protein interactions. Discrimination of G_i subtypes by the D_2 dopamine receptor in a reconstituted system. *J. Biol. Chem.* **265,** 4507–4514.
99. Friedman, E., Butkerait, P., and Wang, H.-Y. (1993) Analysis of receptor-stimulated and basal guanine nucleotide binding to membrane G proteins by sodium dodecyl sulfate-polyacrylamide gel electrophoresis. *Anal. Biochem.* **214,** 171–178.
100. Clapham, D. E. and Neer, E. J. (1993) New roles for G-protein βγ-dimers in transmembrane signalling. *Nature* **365,** 403–406.
101. Federman, A. D., Conklin, B. R., Schrader, K. A., Reed, R. R., and Bourne, H. R. (1992) Hormonal stimulation of adenylyl cyclase through G_i-protein βγ subunits. *Nature* **356,** 159–161.

102. Tang, W.-J. and Gilman, A. G. (1991) Type-specific regulation of adenylyl cyclase by G protein βγ subunits. *Science* **254,** 1500–1503.
103. Lustig, K. D., Conklin, B. R., Herzmark, P., Taussig, R., and Bourne, H. R. (1993) Type II adenylylcyclase integrates coincident signals from G_s, G_i, and G_q. *J. Biol. Chem.* **268,** 13,900–13,905.
104. Faure, M., Voyno-Yasenetskaya, T. A., and Bourne, H. R. (1994) cAMP and βγ subunits of heterotrimeric G proteins stimulate the mitogen-activated protein kinase pathway in COS-7 cells. *J. Biol. Chem.* **269,** 7851–7854.
105. Kleuss, C., Scherübl, H., Hescheler, J., Schultz, G., and Wittig, B. (1992) Different β-subunits determine G-protein interaction with transmembrane receptors. *Nature* **358,** 424–426.
106. Kleuss, C., Scherübl, H., Hescheler, J., Schultz, G., and Wittig, B. (1993) Selectivity in signal transduction determined by γ subunits of heterotrimeric G proteins. *Science* **259,** 832–834.
107. de la Peña, P., del Camino, D., Pardo, L. A., Domínguez, P., and Barros, F. (1995) G_s couples thyrotropin-releasing hormone receptors expressed in *Xenopus* oocytes to phospholipase C. *J. Biol. Chem.* **270,** 3554–3559.
108. Hervé, D., Lévi-Strauss, M., Marey-Semper, I., Verney, C., Tassin, J.-P., Glowinski, J., and Girault, J.-A. (1993) G_{olf} and G_s in rat basal ganglia: possible involvement of G_{olf} in the coupling of dopamine D_1 receptor with adenylyl cyclase. *J. Neurosci.* **13,** 2237–2248.
109. DiMarzo, V., Vial, D., Sokoloff, P., Schwartz, J.-C., and Piomelli, D. (1993) Selection of alternative G_i-mediated signaling pathways at the dopamine D_2 receptor by protein kinase C. *J. Neurosci.* **13,** 4846–4853.
110. Seeman, P. (1987) Dopamine receptors in human brain diseases, in *Receptor Biochemistry and Methodology, vol. 8: Dopamine Receptors* (Creese, I. and Fraser, C. M., eds.), Liss, New York, pp. 233–245.

CHAPTER 6

Signaling Pathways Modulated by Dopamine Receptors

Rita M. Huff

1. Introduction

Knowledge of the intracellular signaling events altered by the binding of dopamine to its receptors is critical to the understanding of how dopamine and dopaminergic drugs elicit their actions. Activation of a signal transduction mechanism that can regulate ion channels and turn on second messenger systems is the first step toward a variety of responses in neurons, from immediate changes in neuron excitability to long-term modulatory processes. Dopamine is a neurotransmitter and, thus, affects the excitability of neurons, but it also regulates protein kinases and transcription factors through signal transduction cascades initiated by the receptors. The long-term adaptive responses resulting from dopamine receptor activation or blockade are undoubtedly important for the effects of psychotropic drugs *(1)*. Characterization of the initial events triggered by activation of dopamine receptors at a molecular level can contribute to an understanding of how changes in ion channel actuation, protein phosphorylation, and the genetic programs of the cells can occur.

1.1. G Protein-Mediated Signal Transduction

All the known dopamine receptor subtypes are guanine nucleotide-binding protein (G protein)-linked receptors. This has been determined from various properties of the proteins. The historical designation of dopamine receptors as G protein-linked comes from demonstrations of allosteric regulation of agonist binding to the receptors by guanine nucleotides. This is the hallmark of receptor coupling to G proteins. Additionally, pertussis toxin, an agent which ADP-ribosylates the G_i/G_o type of G proteins and prevents them from

The Dopamine Receptors Eds.: K. A. Neve and R. L. Neve
Humana Press Inc., Totowa, NJ

coupling to receptors, was found to interfere with some dopamine-mediated signaling events. Finally, for all dopamine receptor subtypes, and this is particularly pertinent for the newer subtypes first discovered by cloning, the predicted amino acid structures have the putative seven membrane-spanning regions common to members of the superfamily of G protein-linked receptors. Knowing that dopamine receptor subtypes interact with G proteins is a significant first step for understanding how these receptors transduce the information from extracellular dopamine to alteration of intracellular events. However, much remains unknown as the complexity of receptor, G protein, and effector interactions is being realized and unraveled in general.

Since the known dopamine receptor subtypes interact with G proteins, it is necessary to introduce the components of G protein-mediated signal transduction *(2)*. This mode of receptor signaling has an obligatory requirement for at least one of many types of G proteins to couple receptors to intracellular or membrane effector molecules. The G proteins are heterotrimeric proteins of α-subunits that bind guanine nucleotides, and β- and γ-subunits. The β- and γ-subunits are always associated with each other, but α-subunits dissociate from βγ on activation of the G protein. The dissociation of α-subunits from βγ-subunits is initiated by agonist-activated receptors that stimulate the exchange of GTP for GDP on the α-subunits. There is considerable diversity among the α-, β-, and γ-subunits. The type of α-subunit defines the G protein and is important for both specificity of coupling to receptors and for specificity for activation of effectors. The type of β- and γ-subunits associated with each α-subunit also has been shown to be important for specificity of receptor interactions and effector activation *(3)*. Thus, both GTP-bound α-subunits and/or free βγ-subunits can interact with a large variety of effector proteins including adenylate cyclases, phospholipases, ion channels, Na^+/H^+ exchangers, and as yet unidentified proteins that link G protein-mediated signaling to the growth factor-stimulated tyrosine kinase cascade *(2,4)*.

G protein-linked receptors have some, but not absolute, specificity for the kinds of G proteins with which they interact. Generally, receptors either couple to a G_s resulting in stimulation of adenylate cyclase and increased production of cAMP, couple to a G_i resulting in inhibition of adenylate cyclase, or couple to a G_q resulting in stimulation of phospholipase C-βs and hydrolysis of phosphatidylinositols. In studies with the receptors that inhibit adenylate cyclase, it has become clear that reduction of cAMP is not the sole means of signal transduction, and these G_i-linked receptors are often found to regulate additional second messenger systems independently of adenylate cyclase inhibition *(5)*. This divergence in signaling could result either from the receptor activating a single type of G protein, producing active α-subunits and free βγ-subunits to activate separate effectors, or by the receptor activating multiple types of G proteins *(6)*. Interactions between α and effector or $\beta\gamma$ and effector also are not strictly specific; in other words, a single type of α or $\beta\gamma$ can activate more than one type of effector, and effectors can be activated by more than one type of α or by both α and $\beta\gamma$.

The divergence and convergence of receptor–G protein–effector interactions creates the potential for extensive signal processing to occur between receptors and second messenger production. The expression of certain α-, β-, and γ-subunits is tissue specific *(2)*. There is an increasing diversity among effector molecules being revealed as well, particularly in how they are activated, and the expression of some of these is also tissue specific *(7,8)*. Effector activation by a single type of receptor can differ from cell to cell depending on the signal transduction components expressed in that cell, and output can also vary within a cell depending on the presence or absence of other signaling inputs. For example, there are at least eight mammalian adenylate cyclases which are activated by α_s, however, they are differentially regulated by α_i-subunits, $\beta\gamma$-subunits, and Ca^{2+} *(7–9)*. Type II adenylate cyclase, which is predominantly expressed in the nervous system, is maximally activated by α_s, the $\beta\gamma$-subunits of G_i, and/or a protein kinase C (PKC)-dependent event resulting from activation of G_q *(10)*. This and other adenylate cyclase types have been suggested to serve as coincidence detectors for simultaneous signals from multiple extracellular neurotransmitters *(10,11)*. With this in mind, it is easy to understand why the repertoire of signaling outputs of a single type of receptor is dependent on the cell; in particular the presence and stoichiometry of G proteins, effector isozymes, and downstream substrates for the second messengers will determine the net changes in signals which can occur. Understanding these concepts is important in evaluating the results observed with dopamine receptor subtype signaling studies.

1.2. Historical Perspectives of Dopamine Receptor Subtype Classification

Dopamine receptors were originally split into two classes based on pharmacological characteristics and signal transduction differences: the D1* receptors that stimulate adenylate cyclase, and the D2 receptors that do not *(12)*. Later it was determined that D2 receptors actually inhibit adenylate cyclase *(13)*. With the cloning of additional dopamine receptor subtypes, the first features known about the previously unrecognized and rarer subtypes were the amino acid sequence and predicted structure. From the structure, it

*As detailed in the Preface, the terms D1 or D1-like and D2 or D2-like are used to refer to the subfamilies of DA receptors, or used when the genetic subtype (D_1 or D_5; D_2, D_3, or D_4) is uncertain.

could be deduced that these receptors also require G proteins for signal transduction. Furthermore, the D_5 receptors like the cloned D_1 receptors have characteristics of receptors that stimulate adenylate cyclase, including relatively short third cytoplasmic loops and long intracellular carboxy termini. Similarly, D_3 and D_4 receptors share structural features of cloned D_2 receptors, including relatively long third cytoplasmic loops and short carboxy termini *(14)*. Before any functional studies were performed, these were the first clues as to the kinds of signal transduction events these additional dopamine receptor subtypes might mediate.

1.3. Systems Used for Studying Dopamine Receptor-Mediated Signal Transduction

Characterization of dopamine receptor-mediated signaling events has been carried out in two types of settings, preparations in which dopamine receptors are found endogenously and preparations naturally lacking dopamine receptors but in which a dopamine receptor subtype has been heterologously expressed. Analysis of dopamine signaling in each type of setting has both good and bad features.

1.3.1. Tissues with Endogenous Dopamine Receptors

The signaling events elicited by dopamine receptors have been studied in the tissues in which they are naturally found. Dopamine receptor subtypes are found on neurons, predominantly in dopamine-rich areas of the brain including the striatum, substantia nigra, and prefrontal cortex, and in the pituitary. Dopamine receptors are also found in peripheral tissues including the kidney, but this chapter is concerned only with dopamine signaling in neurons and endocrine cells. Since the drugs used to characterize these receptors do not fully discriminate between the D1-like (D_1 and D_5) or between the D2-like (D_2, D_3, and D_4) subtypes, the identity of the receptor mediating the effects is not always definite. Studies of D1-like receptors have been carried out principally in the brain area most enriched for D_1 receptors, the striatum. D2-like receptors are most abundant in the striatum but are also expressed in the pituitary where there are no D1-like receptors, making this a better tissue for studying D2-like signal transduction *(15)*. In some cases, detection of receptor-mediated signal transduction can be difficult in tissues of heterogeneous cell types such as the striatum and pituitary, since receptor-mediated changes in second messengers are diluted by the second messengers in cells without receptors. Effects of receptor activation on ion channels are more easily detected in mixed cell populations, but direct effects are best determined on dissociated cells. Unfortunately, cellular studies of dopamine receptor signal transduction have been impeded by the lack of cell lines which maintain expression of dopamine receptors.

1.3.2. Heterologous Systems

In recent years, molecular biological approaches have been used to introduce the DNA for dopamine receptor subtypes into cells that ordinarily do not express them or have lost expression on culturing in order to enable heterologous expression of the receptor proteins. Pure populations of cells that express the receptor protein can be obtained by using a selection strategy to eliminate untransfected cells. With heterologous expression systems, responses can be directly attributed to the presence of the receptor since receptor-positive cells can be compared to receptor-negative cells. For studies of dopamine receptor signaling, these have provided valuable new cell models for studying dopamine receptor signal transduction, but they have also revealed additional complexities. Some of the earliest studies of heterologous expression of D_2 receptors in two different types of host cells demonstrate this principal: The dopamine-mediated signaling events were different depending on the host cell *(16)*. Thus, it has become clear that one cannot blindly extrapolate the observations of dopamine receptor signaling in one cell type to all cell types. Heterologous expression systems reveal the potential for receptor–G protein–effector interactions which can then be explored in the more complex native environments of dopamine receptors.

One of the biggest hurdles in studies of heterologously expressed dopamine receptor-mediated signaling is finding a host cell that perfectly mimics the cells in which dopamine receptors are naturally expressed. As pointed out, the signaling output is a result of many G proteins and effectors which are expressed at varied levels in different types of cells and in different culture conditions. Clearly, the first essential components for dopamine receptor-mediated signaling are the G proteins. For D_1 and D_5 receptors, the important G proteins are undoubtedly G_s and possibly G_{olf}. G_{olf} is a G protein highly homologous to G_s, initially characterized as an olfaction-specific G protein because it links odorant receptors to adenylate cyclase. $G_{olf}\alpha$ mRNA, however, is found at much higher levels than $G_s\alpha$ mRNA in striatonigral neurons where D1 receptors are also localized, leading to speculation that D1 receptors use G_{olf} rather than G_s in these cells to stimulate adenylate cyclase *(17)*. For D2-like receptors, the important G proteins are primarily the pertussis toxin-sensitive proteins, which include G_{oA}, G_{oB}, G_{i1}, G_{i2}, and G_{i3}, and maybe also the pertussis toxin-insensitive proteins G_q and G_z.

2. D_1 and D_5 Receptor-Mediated Signaling Events

2.1. D1-Like Receptor Signaling in Native Tissues

D1-like receptors stimulate adenylate cyclase *(18)*. This results in increased concentrations of cAMP, which activates protein kinase A (PKA) to carry out phosphorylation of specific substrate proteins. One well-characterized phosphoprotein specifically enriched in D1 receptor-containing neurons that is phosphorylated in response to dopamine and cAMP is DARPP-32. DARPP-32 has been termed an intracellular third messenger for D1 receptor activation *(19)*. Phosphorylated DARPP-32 is an inhibitor of protein phosphatase-1 and is itself dephosphorylated by calcineurin, a Ca^{2+} dependent phosphatase *(20)*, providing a mechanism for antagonism between cAMP and Ca^{2+} signaling. In fact, neurotransmitters such as glutamate that increase the excitability of medium spiny neurons decrease DARPP-32 phosphorylation *(21)*, whereas GABA and dopamine, which decrease neuronal excitability, increase the phosphorylation of DARPP-32 *(22)*. Changes in the phosphorylation of this protein thus parallel the changes in electrical activity mediated by these three neurotransmitters which interact in the neostriatum.

Downstream effects resulting from D1-like receptor-stimulated cAMP levels include an inhibition of arachidonic acid production in primary cultures of rat striatal neurons *(23)*. Arachidonic acid and its metabolites can affect the opening of K^+ channels and inhibit Ca^{2+}-calmodulin-dependent protein kinase II activity *(24)* and may have a role as a retrograde messenger in long-term potentiation *(25)*.

A D1-like receptor has been shown to stimulate inositol phosphate's formation in rat striatum and other brain areas such as the amygdala *(26)*. It has been suggested that the D1-like dopamine receptor which mediates this effect is independent of the receptor which stimulates adenylate cyclase since regional brain distributions of these activities do not match. Pharmacological characterization of each D1-like response also indicates that they may involve different receptors *(27)*.

D1-like receptor expression has been discovered in a few clonal cell lines including SK-N-MC human neuroblastoma cells *(28)* and NS20Y mouse neuroblastoma cells *(29)*. D1-like receptors on NS20Y neuroblastoma cells stimulate adenylate cyclase *(30,31)*. As D1-like receptors are expressed in kidney, several kidney cell lines have been found which maintain expression of D1-like receptors including the opossum kidney (OK) cells *(32)*.

2.2. D_1 and D_5 Receptor Signaling in Heterologous Systems

Cloned human and rat D_1 receptors stimulate adenylate cyclase when expressed in a variety of host cells including human embryonic kidney (HEK) 293 cells, mouse thymidine kinase$^-$ fibroblast (Ltk$^-$) cells, transformed African green monkey kidney (COS-7) cells, and Chinese hamster ovary (CHO) cells *(33–36)*. Cloned D_5 receptors expressed in HEK 293, COS-7, or CHO cells also stimulate adenylate cyclase *(36–40)*.

Linkage of D_1 or D_5 receptors to activation of phospholipase C has not been consistently observed in various cell types. Neither D_1 or D_5 receptors in

CHO or baby hamster kidney cells (BHK) couple to phospholipase C *(36)*. However, dual coupling of cloned human D_1 receptors both to stimulation of cAMP production and intracellular Ca^{2+} increases via stimulation of phosphatidylinositol hydrolysis was shown in Ltk^- cells *(41)*. In GH_4C_1 cells (a clonal pituitary cell line), D_1 receptor activation also results in increases in intracellular Ca^{2+} via opening of L-type voltage dependent calcium channels, but this is dependent on the increased cAMP levels. In HEK 293 cells, as in GH_4C_1 cells, D_1 receptor activation increases both cAMP and intracellular Ca^{2+}, but again the latter response is dependent on cAMP and is not through actuation of phospholipase C *(42)*.

3. D_2, D_3, and D_4 Receptor-Mediated Signaling Events

3.1. D2-Like Receptor Signaling in Native Tissues

Central nervous system D2-like dopamine receptors have been characterized principally on two types of cells, hormone-secreting cells of the pituitary and neurons. Within the pituitary, D2-like dopamine receptors have been found on cells in the anterior pituitary and in the intermediate pituitary. In nervous tissue, D2-like receptors are found on neurons predominantly in dopamine-rich areas such as the striatum, substantia nigra, and frontal cortex.

3.1.1. D2-Like Receptor Signaling in Endocrine Tissues

Inhibition of prolactin secretion by dopamine in the anterior pituitary is a well-characterized physiological response mediated by D2-like dopamine receptors on lactotroph cells. Within the lactotroph cell, dopamine activates multiple second messenger pathways. These include inhibition of adenylate cyclase *(13)*, inhibition of inositol phosphate production *(43)*, and inhibition of arachidonic acid release through a cAMP-independent pathway *(44)*. Dopamine also regulates the opening and closing of ion channels including activation of a K^+ current *(45,46)* through opening of K^+ channels *(47)*, and inhibition of Ca^{2+} currents *(48)*. The inhibition of inositol phosphate levels by D2 receptors does not appear to be a direct inhibition of phospholipase C, but rather a result of inhibition of voltage-gated Ca^{2+} channels *(49)*. D2 receptors also inhibit amiloride-sensitive Na^+/H^+ exchange in pituitary lactotrophs independently of cAMP modulation *(50)*. Regulation of Na^+/H^+ exchange can have either an essential or permissive role in signal transduction pathways including secretion *(5,50)*. This response is unique among the D2 responses in that it is not sensitive to pertussis toxin *(50)*.

Melanotrophs are the cells of the intermediate lobe of the pituitary that secrete the pro-opiomelanocortin (POMC) derived peptides α-melanocyte-stimulating hormone, β-endorphin, and corticotrophin-like intermediate peptide. D2 receptor activation results in inhibition of release of these peptides as well as inhibition of the synthesis of POMC. In these cells, dopaminergic agonists inhibit adenylate cyclase *(51),* and also alter ion currents through cAMP-independent pathways. Dopamine increases K^+ conductance and inhibits more than one type of Ca^{2+} current *(52–56).*

The MMQ cell line which was derived from the 7315a rat pituitary tumor, is one of the few tumor cell lines which has maintained expression of endogenous D_2 receptors, albeit at quite low levels. In these cells, dopamine inhibits cAMP accumulation and stimulates a voltage-dependent K^+ current *(57,58).*

D_2 receptors in pituitary cells may be an example of a single receptor coupling to multiple types of G proteins. Two complementary studies have indicated this potential. In one, antisera specific for different $G_{i/o}$ α-subunits were microinjected into lactotrophs to determine the G protein requirements for dopamine responses *(59).* In a separate study, lactotrophs were dialyzed with antisense oligonucleotides to $G_{i/o}$ α-subunits and tested 2 d later for the effects of dopamine on ion currents *(60).* Both studies resulted in similar conclusions; α_o is required for D_2 receptor-mediated decreases in voltage-activated Ca^{2+} currents and α_{i3} is required for D_2 receptor-mediated increases in voltage-activated K^+ currents. These studies may indicate that D_2 receptors couple with both G_o and G_{i3}, however, the possibility remains that more than one type of D_2 receptor selectively couples to either G protein. D_2 receptors are found as either of two isoforms which differ by the presence (D_{2L}) or absence (D_{2S}) of 29 amino acids in the third cytoplasmic loop *(14).* Since mRNA for both D_{2L} and D_{2S} isoforms of D_2 receptors is found in pituitary cells, it remains possible that each D_2 isoform couples to a different G protein. Nevertheless, the G-protein antibody and antisense oligonucleotide studies indicate the requirement for G_o for dopaminergic effects on Ca^{2+} channels and for G_{i3} for dopaminergic effects on K^+ channels.

The clonal pituitary tumor cell lines known as GH_3 and GH_4C_1 cells do not express D_2 receptors unless treated with epidermal growth factor (EGF) *(61,62).* Following treatment with EGF, there is an induction of both D_{2L} and D_{2S} receptors, an increase in α_{i3} protein and a reduction in α_{i2} protein *(63).* In these treated cells, the dopamine agonist quinpirole causes an opening of K^+ channels but does not inhibit vasoactive intestinal peptide (VIP)-stimulated cAMP production. These findings provide additional support for the role of α_{i3} in D_2 receptor coupling to K^+ channels. One could also speculate that the loss of α_{i2} contributes to the failure of D_2 receptors to couple to cAMP inhibition, although many other phenotypic changes may be occurring in the EGF-treated cells.

3.1.2. D2-Like Receptor Signaling in Neurons

D2-like dopamine receptors mediate inhibition of adenylate cyclase in neurons. In primary cultures of dissected striatal neurons, D2-like receptor agonists inhibit VIP-stimulated cAMP accumulation, and the response is blocked by pretreatment with pertussis toxin *(64)*. Also, using dissected neurons from rat striatum, it was shown that D2-like receptors activate K^+ channels *(65)*. In the substantia nigra zona compacta and the ventral tegmental area also, D2-like receptor activation increases K^+ conductance *(66,67)*. It was first demonstrated in the substantia nigra and later in striatal neurons that quinpirole-induced hyperpolarization and activation of outward K^+ current is abolished by tolbutamide, a sulfonylurea which is a potent blocker of ATP-sensitive K^+ channels in pancreatic β-cells *(68–70)*. D2 receptor regulation of this channel appears to be direct through G proteins without the requirement for second messengers *(70)*.

In primary cultures of rat striatal neurons, D2 receptor activation potentiates arachidonic acid release stimulated by ATP or a Ca^{2+} ionophore *(23)*. Whether or not this response results from reduced cAMP levels or opening of K^+ channels by dopamine is not known. The dopamine receptor-mediated modulation of cAMP in photoreceptors of mouse retina has the pharmacological specificity of D_4 receptors *(71)*. This is the first and only demonstration of putative D_4 receptor-mediated signaling events in native tissues.

3.2. D_2 Receptor-Mediated Signaling in Heterologous Systems

3.2.1. D_2 Receptor Signaling in Endocrine Cells

Many studies of D_2 receptor-mediated signaling in heterologous systems were carried out in the clonal mammosomatotroph cells, GH_4C_1, which as mentioned earlier do not express D_2 receptors unless treated with EGF. These cells have been useful for determining the potential D_2 receptor-mediated signaling events in endocrine cells since the effects of dopamine on prolactin and growth hormone secretion can also be measured. Both D_{2S} and D_{2L} isoforms of D_2 receptors are expressed in lactotrophs and are induced by EGF treatment of GH_3 cells *(63)*. Using heterologous expression of each dopamine receptor isoform individually in receptor-negative GH_4C_1 cells, the signaling properties of D_{2S} and D_{2L} have been delineated. D_{2S} and D_{2L} receptors both activate several pathways similarly to what has been observed in lactotrophs, including inhibition of adenylate cyclase, activation of K^+ channels, and inhibition of voltage-sensitive Ca^{2+} channels *(16,72–75)*.

Although the D_{2L} and D_{2S} receptor isoforms activate the same effectors, studies in GH_4C_1 cells have revealed that they differ in G-protein requirements for effector activation. The results from two groups support differential G protein usage by the two isoforms, although the details of G-protein requirements are not in agreement. In one study, pertussis toxin-resistant Gα-subunits were made by mutating the cysteine normally ADP-ribosylated by pertussis toxin to a serine residue. The assumption is made that the cysteine to serine switch does not interfere with receptor coupling to Gα. The DNAs for these proteins were transfected into cells to reconstitute D_{2L} or D_{2S} signaling in cells treated with pertussis toxin to inactivate all endogenous $G_{o/i}$s *(74)*. Interestingly, Senogles *(74)* found that D_{2L} receptor-mediated inhibition of forskolin-stimulated cAMP requires α_{i3}, and D_{2S} receptor-mediated inhibition of forskolin-stimulated cAMP requires α_{i2}. Albert's group *(72)* used a different strategy to ask the same question. By blocking specific G protein α-subunit synthesis with antisense oligonucleotides, they determined which signals mediated by D_{2L} or D_{2S} were prevented. They found that in the absence of α_{i2}, D_{2L} receptors fail to inhibit VIP-stimulated cAMP, whereas in the absence of α_{i3}, D_{2S} receptors fail to inhibit VIP-stimulated cAMP. Both receptor isoforms require α_{i2} to inhibit basal cAMP *(73)*. Montmayeur et al. *(76)* also came to the conclusion that D_{2L} has a greater requirement for G_{i2} than does D_{2S} to inhibit cAMP accumulation in two other nonpituitary cells, JEG cells and NCB20 cells. All of these studies have indicated that there are differences in G-protein coupling between D_{2L} and D_{2S}. Interestingly, the extra sequence in the D_{2L} does not seem to be directly involved in G-protein coupling, at least the G protein which mediates cAMP inhibition, based on studies using synthetic peptides to different regions of the third intracytoplasmic loop (ic3). D_{2L}-mediated inhibition of cAMP is attenuated by two peptides, one identical to the amino acids at the N terminal portion of ic3 and the other identical to amino acids at the C terminal portion of ic3. However, a peptide identical to part of the insert sequence of D_{2L} has no effect on dopamine inhibition of cAMP *(77)*.

Unequivocal evidence that a single type of D_2 receptor couples to at least two and potentially more G proteins within a single cell was provided by the studies of Liu et al. *(73)*. In the transfected GH_4C_1 cells, D_{2S}-mediated inhibition of basal cAMP requires G_{i2}, D_{2S}-mediated inhibition of Ca^{2+} entry requires G_o, and D_{2S}-mediated inhibition of VIP-stimulated cAMP accumulation is not prevented by blocked synthesis of either α_{i2} or α_{i3}, implicating a third G protein.

Most of the signaling studies of D_{2L} and D_{2S} receptors in the pituitary-like cells have focused on cAMP, K^+ channels, and Ca^{2+}, and the impact of these on prolactin secretion. Recent studies of a different downstream result of D_2 receptor activation in pituitary cells, inhibition of cell division, has revealed that D_2 receptors elicit additional early signaling events. The dopamine agonist bromocriptine has been known for some time to be an antiproliferative agent for lactotrophs in culture and some pituitary tumors *(78,79)*. Dopamine was found to inhibit proliferation in GH_4C_1 cells transfected with D_{2S} receptors

(80,81). Interestingly, the dopamine agonist-mediated effects on D_{2S}-transfected GH_4C_1 cell division are correlated with a dopamine agonist-stimulated increase in phosphotyrosine phosphatase activity. This dopamine response is blocked by pertussis toxin *(80)*. Senogles *(81)* also studied dopamine-mediated inhibition of mitogenesis in D_{2S} transfected GH_4C_1 cells and found that pertussis toxin is ineffective at preventing the response. She provided evidence that dopamine-mediated inhibition of mitogenesis may be related to dopamine stimulation of an isoform of PKC, PKCε, via a pathway independent of phospholipase C. In these cells, diacylglycerol from a source other than phosphatidylinositol must be released by dopamine in order to activate PKC. These findings have underscored the multiplicity of signaling events elicited by D_2 receptors and indicate that not all of the effectors activated by dopamine have been identified.

3.2.2. D_2 Receptor Signaling in Other Host Cells

CHO cells and Ltk^- fibroblasts have been a popular host for G protein-linked receptors for reasons such as their genetic stability and the paucity of endogenous G protein-linked receptors despite the presence of many postreceptor signaling components. Other studies of dopamine receptor signaling events have been initiated in transfected C_6 glioma cells; NG108-15 cells, which are neuroblastoma-glioma cells; and MN9D cells, which are an immortalized line of neuroblastoma cells fused with embryonic mouse mesencephalic dopamine-producing neurons *(82)*. In all of these cell types, D_{2S} and/or D_{2L} receptors inhibit cAMP accumulation *(83–88)*. Whether or not D_2 receptors mediate changes in phosphatidylinositol hydrolysis is more variable depending on the host cell. D_{2S} in Ltk^- and D_{2L} in CCL1.3 fibroblasts, increase inositol phosphate levels and consequently increase intracellular Ca^{2+} through pertussis toxin-sensitive G proteins *(16,41,87)*. D_{2S} receptors in NG108-15 cells cause a paradoxical inhibition of voltage-dependent K^+ current. The response is not blocked by pertussis toxin but is prevented by buffering intracellular Ca^{2+} released from internal stores *(89)*. D_2-mediated increases in inositol phosphates do not occur in GH_4C_1 cells, CHO-K1 cells, CHO 10001 cells, or MN9D cells *(16,87,90,91)*. In CHO-K1 cells both D_{2L} and D_{2S} receptors increase intracellular Ca^{2+} through a pertussis toxin-sensitive mechanism *(85)*.

D_{2L} receptors in CHO-K1 and CHO 10001 cells potentiate ATP- or Ca^{2+} ionophore-stimulated arachidonic acid release *(90,92,93)*. This response occurs through a cAMP-independent but PKC-dependent pathway *(90,92)*. Since activation of phosphatidylinositol hydrolysis does not occur in these cells, activation of PKC occurs via an undetermined mechanism *(90,91)*.

D_{2L} and D_{2S} receptor activation in C_6 glioma, Ltk^- cells, and CHO 10001 cells increases the rate of extracellular acidification measured with a microphysiometer through activation of an amiloride-sensitive Na^+/H^+ exchanger *(86,94)*. In C_6 glioma and Ltk^- cells, the response is unaffected by treatments with pertussis toxin that prevent D_2-mediated inhibition of cAMP accumulation *(86)*. In CHO 10001 cells, pertussis toxin blocks the D_2-mediated increase in Na^+/H^+ exchange; however, as in C_6 and Ltk^- cells, the response is independent of changes in cAMP. Activation of D_{2L} receptors expressed in CHO 10048 cells which have a mutant PKA and cannot respond to cAMP, results in an equivalent stimulation of extracellular acidification as in the CHO 10001 cells *(95)*.

The effects of D_2 receptors on ion channels in the cell lines which have neuronal characteristics are just beginning to be studied. In cAMP-differentiated NG108-15 cells, D_{2S} receptors partially inhibit two types of high threshold Ca^{2+} channels, ω-conotoxin GVIA-sensitive and dihydropyridine-sensitive channels *(96)*. As in the pituitary cells, D_2 receptors may require G_o, and perhaps only one of the G_o isoforms to mediate this response. There is evidence to suggest that G_o couples other receptors to Ca^{2+} currents in differentiated NG108-15 cells *(97,98)*. The two isoforms of G_o, which are differentially spliced products of the same gene, differ in amino acid sequence at the carboxyl terminal, the region of the α-subunits which is important for coupling to receptors and effectors. cAMP-mediated differentiation of NG108-15 cells induces expression of G_{oA} α, thus altering the ratio of the two G_o isoforms *(99)*. An induction of G_{oA} relative to G_{oB} is also associated with neuronal differentiation *(100)*. It is tempting to speculate that D_{2S} receptor-mediated effects on Ca^{2+} channel inhibition (as well as the D_3 receptor-mediated effects discussed in Section 3.3.) may involve coupling of D_2 receptors to the G_{oA} isoform of G_o.

The MN9D cells, which synthesize, store, and release dopamine, have been used to study D_2 receptor-mediated effects on these processes. Agonists at D_{2L} receptors inhibit basal and depolarization-induced dopamine release through a pertussis toxin-sensitive mechanism (101). Quinpirole-mediated release is blocked by the K^+ channel blockers 4-aminopyridine and tetraethylammonium, suggesting that D_2 receptors open K^+ channels in these cells and that K^+ channel opening is important for D_2-mediated inhibition of dopamine release. These cells express α_{i2}, α_{oA}, and α_{oB}, but no detectable α_{i1} or α_{i3} *(87)*. Thus, although α_{i3} is important for D_2 receptor effects on K^+ channels in pituitary cells, in MN9D cells it is apparently not necessary.

In MN9D cells, D_{2L} receptor activation has a morphogenic effect resulting in changes in neurite numbers and branching *(102)*. D2-like receptor activation has been shown to increase neurite extension and arborization in a subset of cells in cultures of rat embryonic frontal cortical neurons *(103)*. The initial signaling events responsible for the morphogenic effects have not been elucidated, but nevertheless these findings reveal a potential role for dopaminergic signal transduction in development and possibly in synaptic remodeling in adult brain.

D_{2L} and D_{2S} receptors stimulate mitogenesis in transfected CHO 10001 cells *(91)*. This is in contrast to the inhibitory effects of D_2 receptor activation on cell division in pituitary cells *(80,81)*. However, as was the case for studies in pituitary cells, characterization of dopamine-stimulated mitogenesis in CHO 10001 cells reveals a novel signaling pathway used by D_{2L} and D_{2S} receptors. In CHO 10001 cells, dopamine-stimulated cell division is independent of phorbol 12-myristate-13-acetate (PMA)-sensitive PKC, activation of Na^+/H^+ exchange, changes in cAMP, or arachidonic acid production. Mitogenesis is blocked, however, by an inhibitor of tyrosine kinases, and D_2 receptor activation results in rapid stimulation of tyrosine phosphorylation *(91)*. Two of the tyrosine phosphorylated substrates were identified as the mitogen-activated protein kinases (MAPKs), erk1 and erk2 *(95,104)*. MAPKs are rapidly activated by growth factors and also by G protein-coupled receptors by a pathway that is not yet fully characterized *(4)*. Transfection of some cells with a mutated α_{i2} that is constitutively active, *gip2*, results in constitutively activated MAPK *(105)*. G protein-linked receptor activation of MAPK has also been shown to require free $\beta\gamma$-subunits *(106,107)*. Although MAPK activation is part of the signaling pathways used by proliferative agents in dividing cells, erk2 is most abundant in neurons in brain areas including neocortex, hippocampus, and striatum *(108)*. MAPK activation in differentiated cells such as neurons is likely to have many functions unrelated to cell division. Known substrates for these kinases include tyrosine hydroxylase, phospholipase A_2, and microtubule-associated protein-2 *(109–111)*. In PC12 cells, MAPK activation is part of the signaling pathway used by nerve growth factor (NGF) to initiate neuritogenesis *(112)*. It is tempting to speculate that the morphogenic effects mediated by D_2 receptors expressed in MN9D cells could result from activation of MAPK *(87)*.

3.3. D_3 Receptor-Mediated Signaling Events

In contrast to the multiplicity of signaling events elicited by D_2 receptors, many of the initial attempts to characterize D_3 receptor functions resulted in a failure to detect effects on cAMP levels or other second messengers in a variety of cell lines including CHO-K1, NG108, GH_4C_1, CCL1.3, and SK-N-MC neuroblastoma cells *(87,113–116)*. Subsequent investigations have now revealed an array of D_3 receptor-initiated signaling events, although generally the effects are less than that achieved by full D_2 receptor activation.

In CHO 10001 cells, D_3 receptors inhibit cAMP accumulation with a modest efficacy *(94)*. D_3 receptors potentiate ATP-stimulated arachidonic acid secretion in CHO-K1 cells, although again the effect is less than that seen with D_2 receptor activation *(92)*. D_3 receptors actuate amiloride-sensitive Na^+/H^+ exchange through a pertussis toxin-sensitive pathway in CHO 10001 *(94)* and a pertussis toxin-insensitive pathway in C_6 glioma cells *(117)*. The pertussis toxin-sensitivity of the D_3 responses parallel the sensitivity of the D_2 effects on Na^+/H^+ exchange in these cell lines *(86,94)*. D_3 receptors, like D_{2S} receptors, partially inhibit high-threshold calcium currents in differentiated NG108-15 cells, but not in undifferentiated cells *(118)*. This response is prevented by pertussis toxin treatment.

In the mesencephalic MN9D cell line, D_3 receptors, like D_2 receptors, inhibit basal and potassium-stimulated dopamine release *(101)*. In these cells, D_3 receptor activation also results in changes in neurite outgrowth *(102)*.

D_3 receptors stimulate mitogenesis in CHO10001 and NG108-15 cells *(94,119)*. The D_3 receptor-mediated mitogenesis in CHO10001 cells is pertussis toxin-sensitive, does not require PKC, and may also involve the activation of tyrosine kinases as was shown for D_2-mediated mitogenesis *(94)*. In the NG108-15 cells, D_3 receptor-mediated activation of c-*fos* transcription is also detected *(119)*.

Generally, in the instances where D_2 and D_3 receptor functions have been compared in the same host cell types, the full response elicited by D_3 receptor activation is less than that seen with full D_2 receptor activation. In CHO10001 cells, cAMP inhibition, increased Na^+/H^+ exchange, and stimulated mitogenesis is less by D_3 receptor activation than by D_2 receptor activation. These differences are not caused by differences in receptor levels *(95)*. Although it is not known which G proteins are necessary for these responses, it is probable that multiple types of G proteins are involved. D_3 receptor depression of high threshold Ca^{2+} currents is also less than the D_2 receptor effect in differentiated NG108-15 cells and there is evidence to suggest that a G_o is required for this response *(96,118)*. There are exceptions to this: D_3 and D_2 receptor activations have equivalent effects on Na^+/H^+ exchange in C_6 glioma cells *(117)*, and in MN9D cells D_3 receptor activation has a greater effect than D_2 receptors on dopamine release *(101)*. Overall, it appears as if D_3 receptor activation induces coupling of the receptors to many of the same effectors as D_2 receptors, but usually with less efficiency. Studies of beta-1 and beta-2 adrenergic receptor efficacy on adenylate cyclase stimulation revealed that beta-1 receptors are consistently less effective at G_s activation and this could not be explained by abundance of receptors and G proteins, but is an inherent property of the receptor protein *(120)*. It will be interesting to determine if the differences in magnitude of response between D_2 and D_3 receptors are a result of insufficient

G proteins in the heterologous systems or are an inherent property of the D_3 receptor which limits G-protein activation.

3.4. D_4 Receptor-Mediated Signaling Events

D_4 receptors were predicted to belong to the G_i-linked class of D2-like receptors based on the predicted structure from the amino acid sequence *(121)*. After the D_4 receptor was cloned, it was discovered that mouse retina has D_4 mRNA, and dopaminergic agonists were found to inhibit the photoreceptor light-sensitive pool of cAMP with a D_4-like pharmacological profile *(71)*. This is the only clearly D_4 receptor-mediated response identified in tissues where D_4 receptors are endogenously expressed.

Initially, there were several difficulties with the cloning of a D_4 cDNA and with the development of cell lines with stable expression of D_4 receptors: perhaps all owing to the unusually high G + C content of the human gene. More than one group made synthetic versions of D_4 genes which had lowered G + C content but an unaltered coding sequence *(122–124)*. The human D_4 receptor gene has a highly polymorphic region in the area coding for the putative cytoplasmic loop 3 which results in a variable number of tandem repeats of 16 amino acid segments *(125)*. In all cases, the synthetic genes were designed to encode the same protein as the two-repeat form of the D_4 receptor or $D_{4.2}$ *(122–124)*. The rat D_4 receptor gene does not have the polymorphisms and encodes a protein with only 50% amino acid homology to the human D_4 receptor in the third cytoplasmic loop *(126)*.

$D_{4.2}$ receptors inhibit cAMP accumulation in CHO-K1 cells, CHO 10001 cells, and HEK 293 cells *(95,123,124)*. In GH_4C_1 cells, $D_{4.2}$ receptor activation partially depresses high threshold dihydropyridine-sensitive calcium currents, as does D_2 but not D_3 receptor activation *(127)*. Rat D_4 receptors inhibit cAMP accumulation in MN9D cells but not in CCL1.3 cells *(87)*. On the other hand, in MN9D cells, D_2 and D_3 receptors do, but D_4 receptors do not, inhibit basal or potassium-stimulated dopamine release *(101)*. In the same cells, the D_4 receptors compared to D_2 and D_3 receptors had the greatest effect on increasing neurite length and arborization *(102)*.

$D_{4.2}$ receptors in CHO-K1 and CHO 10001 cells mediate all of the signaling events determined to be activated by D_2 and D_3 receptors in the CHO 10001 cells including potentiation of ATP-stimulated arachidonic acid release through a PKC-dependent pathway, stimulation of amiloride-sensitive Na^+/H^+ exchange, and in CHO 10001 cells, stimulation of mitogenesis *(95,123)*. All of these signaling events require pertussis toxin-sensitive G proteins. The maximal efficacy for full agonists at D_4-mediated responses is greater or at least as great as the maximal efficacy for full agonists at D_3 responses in the same cell type despite the lower levels of D_4 receptor expression *(95)*.

4. Summary

It is remarkable that 16 years after dopamine receptors were originally split into two groups—those that stimulate adenylate cyclase and those that inhibit adenylate cyclase—this broad categorization has endured. Now the number of subtypes has expanded to five, but nevertheless, the D_1 and D_5 receptors stimulate adenylate cyclase and the D_2, D_3, and D_4 receptors inhibit adenylate cyclase. What has become evident over the years is that the effectors regulated by these receptors include many more than just the adenylate cyclases (e.g., *see* Fig. 1). Further characterization of the multiplicity of signaling events elicited by each receptor will be an important pursuit in order to understand how each of the dopamine receptor subtypes regulates cell function.

Progress has been made in determining which G protein α-subunits are required for D_2 receptor activation of various effectors. At the least these include α_o, α_{i2}, and α_{i3}. On the other hand, the importance of βγ-subunits in dopamine receptor coupling has not yet been examined. Other studies have shown that the type of β- and γ-subunits associated with α plays a role in the specificity of receptor interactions. Muscarinic and somatostatin receptors both inhibit Ca^{2+} currents in GH_3 cells via α_o, however, each receptor requires different β-subunits and different γ-subunits to mediate this response *(128,129)*. The types of βγ-subunits released by receptor activation may impact on the types of additional effectors which are regulated *(3,11)*. Determining the specificity of coupling of each of the dopamine receptors with each of the possible αβγ combinations will be challenging, but could be beneficial for determining the signaling pathways used by each receptor.

Interesting results from two groups suggest that the D_2 receptor isoforms use different α-subunits to activate the same effectors *(73,74)*. Why this occurs remains a mystery. As the heterogeneity of the various effectors, including adenylate cyclases, phospholipases, and ion channels, is discovered, we may find that the D_2 isoforms do not activate the exact same effector molecules. As is being shown with some of the types of adenylate cyclases, the characteristics of each effector isoform may dictate how a dopamine signal is integrated with other signals being received by the same cell *(10,11)*.

Now that it is clear that dopamine receptors activate multiple effectors within the same cell, it will be important to determine how dopamine signaling through each of these pathways is regulated. Some studies already have touched on this issue. In Ltk^- cells, D_{2S}-mediated increases in Ca^{2+} (resulting from inositol trisphosphate generation) are blocked by acute activation of PKC, whereas D_{2S} effects on cAMP accumulation are unaffected *(130)*. In CHO cells, stimulation of PKC increases the efficacy of D_2 receptor-mediated arachidonic acid release and reduces the efficacy of D_2 receptor-mediated

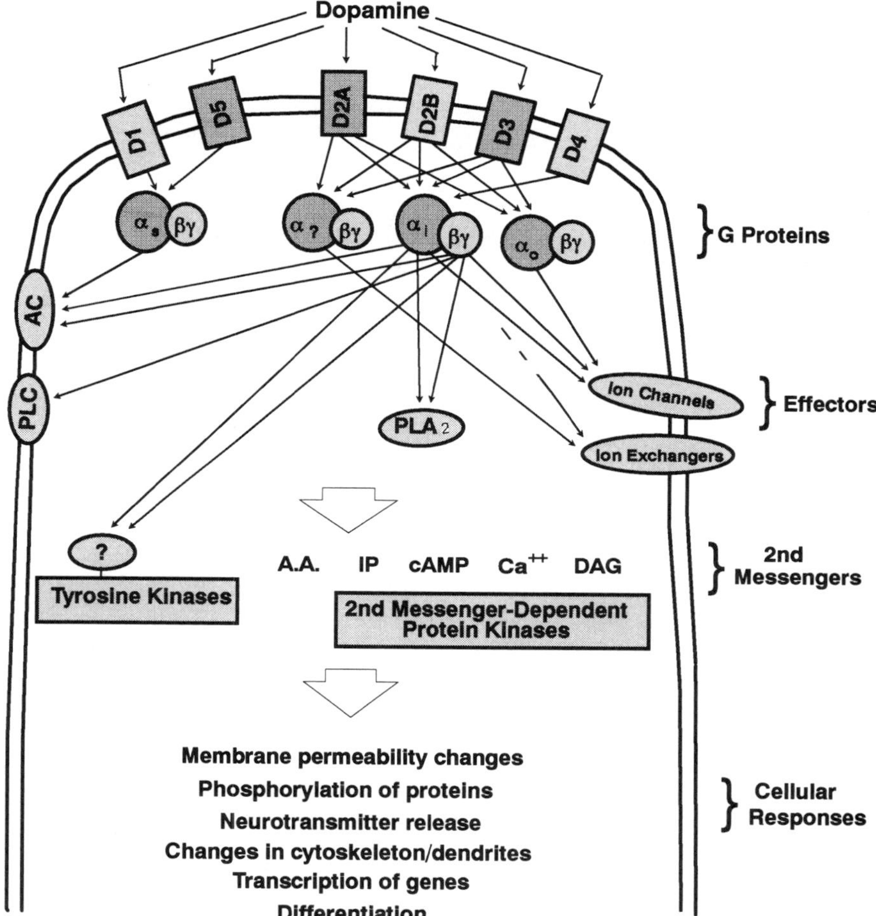

Fig. 1. Pathways for dopamine receptor-mediated cellular responses are shown. Dopamine receptor-mediated signal transduction occurs through several steps, originating with receptor activation, followed by G-protein activation, effector activation (or inhibition), and ultimately changes in cellular responses. The interactions of dopamine receptor subtypes with the indicated G proteins, G_s, G_i (including G_{i1-3}), and G_o as well as at least one additional pertussis toxin-insensitive G protein ($\alpha_?$) are indicated by the arrows. The potential ways in which G proteins can interact with effectors is indicated by the arrows emanating from the α- and $\beta\gamma$- subunits. Not shown are indirect pathways of effector activation which can occur through second messengers. The dotted arrow leading to ion exchangers is meant to indicate that there is no information regarding the identity of the G protein which mediates pertussis toxin-sensitive activation of Na^+/H^+ exchange. More details on the types of interaction are provided in the text. The diagram is not intended to imply that every dopamine receptor subtype is found in a single cell or that every possible interaction of receptors with G proteins and effecters occurs in a single cell.

cAMP accumulation *(131)*. These studies suggest the exciting possibility that a subset of signaling events mediated by dopamine receptors can be selectively dampened or enhanced by intracellular regulators.

With dopamine receptors activating multiple signal transduction pathways, it is possible that some agonists can differentially affect each of these pathways. If the domains of the receptor important for activating each type of G protein are different, not all agonist–receptor interactions may result in activation of the same kinds of G proteins. It may turn out that agonists can be found that induce a conformation of the receptor which is favorable for only a certain type of G-protein interaction, thus activating only a subset of receptor-mediated signaling events. An agonist at M1 muscarinic receptors has been described that selectively activates phosphatidylinositol hydrolysis and antagonizes other M1 agonists at cAMP inhibition *(132)*. Agonists such as this for dopamine receptor subtypes could result in signaling-specific therapeutic intervention. It may some day be possible, for example, to prevent D_2 receptor-mediated effects on gene transcription while mimicking D_2 receptor-mediated effects on neuron excitability.

References

1. Duman, R. S., Heninger, G. R., and Nestler, E. J. (1994) Molecular psychiatry, adaptations of receptor-coupled signal transduction pathways underlying stress- and drug-induced neural plasticity. *J. Nervous Mental Dis.* **182,** 692–700.
2. Hepler, J. R. and Gilman, A. G. (1992) G proteins. *Trends Biochem. Sci.* **17,** 383–387.
3. Sternweis, P. C. (1994) The active role of $\beta\gamma$ in signal transduction. *Curr. Opin. Cell Biol.* **6,** 198–203.
4. Lange-Carter, C. A., Pleiman, C. M., Gardner, A. M., Blumer, K. J., and Johnson, G. L. (1993) A divergence in the MAP kinase regulatory network defined by MEK kinase and raf. *Science* **260,** 315–319.
5. Limbird, L. (1988) Receptors linked to inhibition of adenylate cyclase: additional signalling mechanisms. *FASEB J.* **2,** 2686–2695.
6. Milligan, G. (1993) Mechanisms of multifunctional signalling by G protein-linked receptors. *Trends Pharmacol. Sci.* **14,** 239–244.
7. Iyenger, R. (1993) Molecular and functional diversity of mammalian G_s-stimulated adenylyl cyclases. *FASEB J.* **7,** 768–775.
8. Tang, W.-J. and Gilman, A. G. (1992) Adenylyl cyclases. *Cell* **70,** 869–872.
9. Taussig, R., Iniguez-Lluhi, J. A., and Gilman, A. G. (1993) Inhibition of adenylyl cyclase by $G_{i\alpha}$. *Science* **261,** 218–221.
10. Lustig, K. D., Conklin, B. R., Herzmark, P., Taussig, R., and Bourne, H. R. (1993) Type II adenylylcyclase integrates coincident signals from G_s, G_i, and G_q. *J. Biol. Chem.* **268,** 13,900–13,905.
11. Anholt, R. R. H. (1994) Signal integration in the nervous system: adenylate cyclases as molecular coincidence detectors. *Trends Neurosci.* **17,** 37–41.
12. Kebabian, J. W. and Calne, D. B. (1979) Multiple receptors for dopamine. *Nature* **277,** 93–96.

13. De Camilli, P., Macconi, D., and Spada, A. (1979) Dopamine inhibits adenylate cyclase in human prolactin-secreting pituitary adenomas. *Nature* **278,** 252–254.
14. Sibley, D. R. and Monsma, F. J., Jr. (1992) Molecular biology of dopamine receptors. *Trends Pharmacol. Sci.* **13,** 61–69.
15. Vallar, L. and Meldolesi, J. (1989) Mechanisms of signal transduction at the dopamine D_2 receptor. *Trends Pharmacol. Sci.* **10,** 74–77.
16. Vallar, L., Muca, C., Magni, M., Albert, P., Bunzow, J., Meldolesi, J., and Civelli, O. (1990) Differential coupling of dopaminergic D_2 receptors expressed in different cell types. *J. Biol. Chem.* **265,** 10,320–10,326.
17. Herve, D., Levi-Strauss, M., Marey-Semper, I., Verney, C., Tassin, J.-P., Glowinski, J., and Girault, J.-A. (1993) G_{olf} and G_s in rat basal ganglia: possible involvement of G_{olf} in the coupling of dopamine D_1 receptor with adenylyl cyclase. *J. Neurosci.* **13,** 2237–2248.
18. Kebabian, J. W., Petzold, G. L., and Greengard, P. (1972) Dopamine-sensitive adenylate cyclase in caudate nucleus of rat brain and its similarity to the "dopamine receptor." *Proc. Natl. Acad. Sci. USA* **69,** 2145–2149.
19. Hemmings, H. C., Jr., Walaas, S. I., Ouimet, C. C., and Greengard, P. (1987) Dopaminergic regulation of protein phosphorylation in the striatum: DARPP-32. *Trends Neurosci.* **10,** 377–383.
20. Hemmings, H. C., Jr., Greengard, P., Tung, H. Y. L., and Cohen, P. (1984) DARPP-32, a dopamine-regulated neuronal phosphoprotein, is a potent inhibitor of protein phosphatase-1. *Nature* **310,** 503–505.
21. Halpain, S., Girault, J.-A., and Greengard, P. (1990) Activation of NMDA receptors induces dephosphorylation of DARPP-32 in rat striatal slices. *Nature* **343,** 369–372.
22. Snyder, G. L., Fisone, G., and Greengard, P. (1994) Phosphorylation of DARPP-32 is regulated by GABA in rat striatum and substantia nigra. *J. Neurochem.* **63,** 1766–1771.
23. Schinelli, S., Paolillo, M., and Corona, G. L. (1994) Opposing actions of D_1- and D_2-receptors on arachidonic acid release and cyclic AMP production in striatal neurons. *J. Neurochem.* **62,** 944–949.
24. Piomelli, D. and Greengard, P. (1990) Lipoxygenase metabolites of arachidonic acid in neuronal transmembrane signalling. *Trends Pharmacol. Sci.* **11,** 367–373.
25. Fazeli, M. S. (1992) Synaptic plasticity: on the trail of the retrograde messenger. *Trends Neurosci.* **15,** 115–117.
26. Undie, A. S. and Friedman, E. (1990) Stimulation of a dopamine D1 receptor enhances inositol phosphates formation in rat brain. *J. Pharmacol. Exp. Ther.* **253,** 987–992.
27. Undie, A. S., Weinstock, J., Sarau, H. M., and Friedman, E. (1994) Evidence for a distinct D_1-like dopamine receptor that couples to activation of phosphoinositide metabolism in brain. *J. Neurochem.* **62,** 2045–2048.
28. Sidhu, A. and Fishman, P. H. (1990) Identification and characterization of functional D_1 dopamine receptors in a human neuroblastoma cell line. *Biochem. Biophys. Res. Commun.* **166,** 574–579.
29. Monsma F. J., Jr., Brassard, D. L., and Sibley, D. R. (1989) Identification and characterization of D_1 and D_2 dopamine receptors in cultured neuroblastoma and retinoblastoma clonal cell lines. *Brain Res.* **492,** 314–324.

30. Barton, A. C. and Sibley, D. R. (1990) Agonist-induced desensitization of D_1 dopamine receptors linked to adenylyl cyclase activity in cultured NS20Y neuroblastoma cells. *Mol. Pharmacol.* **38,** 531–541.
31. Lovenberg, T. W., Roth, R. H., Nichols, D. E., and Mailman, R. B. (1991) D_1 dopamine receptors of NS20Y neuroblastoma cells are functionally similar to rat striatal D_1 receptors. *J. Neurochem.* **57,** 1563–1569.
32. Nash, S. R., Godinot, N., and Caron, M. G. (1993) Cloning and characterization of the opossum kidney cell D1 dopamine receptor: expression of identical D1A and D1B dopamine receptor mRNAs in opossum kidney and brain. *Mol. Pharmacol.* **44,** 918–925.
33. Zhou, Q.-Y., Grandy, D. K., Thambi, L., Kushner, J. A., Van Tol, H. H. M., Cone, R., Pribnow, D., Salon, J., Bunzow, J. R., and Civelli, O. (1990) Cloning and expression of human and rat D_1 dopamine receptors. *Nature* **347,** 76–80.
34. Dearry, A., Gingrich, J. A., Falardeau, P., Fremeau, R. T., Jr., Bates, M. D., and Caron, M. G. (1990) Molecular cloning and expression of the gene for a human D_1 dopamine receptor. *Nature* **347,** 72–75.
35. Monsma, F. J., Jr., Mahan, L. C., McVittie, L. D., Gerfen, C. R., and Sibley, D. R. (1990) Molecular cloning and expression of a D_1 dopamine receptor linked to adenylyl cyclase activation. *Proc. Natl. Acad. Sci. USA* **87,** 6723–6727.
36. Pedersen, U. B., Norby, B., Jensen, A. A., Schiødt, M., Hansen, A., Suhr-Jessen, P., Scheideler, M., Thastrup, O., and Andersen, P. H. (1994) Characteristics of stably expressed human dopamine D_{1a} and D_{1b} receptors: atypical behavior of the dopamine D_{1b} receptor. *Eur. J. Pharmacol.* **267,** 85–93.
37. Tiberi, M., Jarvie, K. R., Silvie, C., Falardeau, P., Gingrich, J. A., Godinot, N., Bertrand, L., Yang-Feng, T. L., Fremeau, R. T., and Caron, M. G. (1991) Cloning, molecular characterization, and chromosomal assignment of a gene encoding a second D1 dopamine receptor subtype: differential expression pattern in rat brain compared with the D1A receptor. *Proc. Natl. Acad. Sci. USA* **88,** 7491–7495.
38. Weinshank, R. L., Adhan, N., Macchi, M., Olsen, M. A., Branchek, T. A., and Hartig, P. R. (1991). Molecular cloning and characterization of a high affinity dopamine receptor (D1 beta) and its pseudogene. *J. Biol. Chem.* **266,** 22,427–22,435.
39. Grandy, D. K., Zhang, Y., Bouvier, C., Zhou, Q.-Y., Johnson, R. A., Allen, L., Buck, K., Bunzow, J. R., Salon, J., and Civelli, O. (1991) Multiple human D5 dopamine receptor genes: a functional receptor and two pseudogenes. *Proc. Natl. Acad. Sci. USA* **88,** 9175–9179.
40. Sunahara, R. K., Guan, H. C., O'Dowd, B. F., Seeman, P., Laurier, L. G., Ng, G., George, S. R., Torchia, J., Van Tol, H. H. M., and Niznik, H. B. (1991) Cloning of the gene for a human dopamine D5 receptor with higher affinity for dopamine than D1. *Nature* **350,** 614–619.
41. Liu, Y. F., Civelli, O., Zhou, Q.-Y., and Albert, P. R. (1992) Cholera toxin-sensitive 3',5'-cyclic adenosine monophosphate and calcium signals of the human dopamine-D1 receptor: selective potentiation by protein kinase A. *Mol. Endocrinol.* **6,** 1815–1824.
42. Lin, C. W., Miller, T. R., Witter, D. G., Bianchi, B. R., Stashko, M., Manelli, A. M., and Frail, D. E. (1995) Characterization of cloned human dopamine D1 receptor-mediated calcium release in 293 cells. *Mol. Pharmacol.* **47,** 131–139.

43. Canonico, P. L., Valdenegro, C. A., and MacLeod, R. M. (1983) The inhibition of phosphatidylinositol turnover: a possible postreceptor mechanism for the prolactin secretion-inhibiting effect of dopamine. *Endocrinology* **113,** 7–14.
44. Canonico, P. L. (1989) D-2 dopamine receptor activation reduces free [^{3}H]arachidonate release induced by hypophysiotropic peptides in anterior pituitary cells. *Endocrinology* **125,** 1180–1186.
45. Israel, J. M., Kirk, C., and Vincent, J.-D. (1987) Electrophysiological responses to dopamine of rat hypophysial cells in lactotroph-enriched primary cultures. *J. Physiol.* **390,** 1–22.
46. Castelletti, L., Memo, M., Missale, C., Spano, P. F., and Valerio, A. (1989) Potassium channels involved in the transduction mechanism of dopamine D2 receptors in rat lactotrophs. *J. Physiol.* **410,** 251–265.
47. Einhorn, L. C., Gregerson, K. A., and Oxford, G. S. (1991) D_2 dopamine receptor activation of potassium channels in identified rat lactotrophs: whole-cell and single-channel recording. *J. Neurosci.* **11,** 3727–3737.
48. Lledo, P. M., Legendre, P., Israel, J. M., and Vincent, J.-D. (1990) Dopamine inhibits two characterized voltage-dependent calcium currents in identified rat lactotroph cells. *Endocrinology* **127,** 990–1001.
49. Vallar, L., Vincentini, L. M., and Meldolesi, J. (1988) Inhibition of inositol phosphate production is a late, Ca^{++}-dependent effect of D_2 dopaminergic receptor activation in rat lactotroph cells. *J. Biol. Chem.* **263,** 10,127–10,134.
50. Barber, D. L. (1991) Mechanisms of receptor-mediated regulation of Na-H exchange. *Cell. Signalling* **3,** 387–397.
51. Cote., T. E., Grewe, C. W., Tsurata, K., Stoof, J. C., Eskay, R. L., and Kebabian, J. W. (1982) D_2 dopamine receptor-mediated inhibition of adenylate cyclase activity in the intermediate lobe of the rat pituitary gland requires guanosine 5'-triphosphate. *Endocrinology* **110,** 812–819.
52. Williams, P. J., MacVicar, B. A., and Pittman, Q. J. (1989) A dopamine IPSP mediated by an increased potassium conductance. *Neuroscience* **31,** 673–681.
53. Williams, P. J., MacVicar, B. A., and Pittman, Q. J. (1990) Synaptic modulation by dopamine of calcium currents in rat pars intermedia. *J. Neurosci.* **10,** 757–763.
54. Stack, J. and Surprenant, A. (1991) Dopamine actions on calcium currents, potassium currents and hormone release in rat melanotrophs. *J. Physiol.* **439,** 37–58.
55. Keja, J. A., Stoof, J. C., and Kits, K. S. (1992) Dopamine D_2 receptor stimulation differentially affects voltage-activated calcium channels in rat pituitary melanotropic cells. *J. Physiol.* **450,** 409–435.
56. Nussinovitch, I. and Kleinhaus, A. L. (1992) Dopamine inhibits voltage-activated calcium channel currents in rat pars intermedia pituitary cells. *Brain Res.* **574,** 49–55.
57. Judd, A. M., Login, I. S., Kovacs, K., Ross, P. C., Spangelo, B. L., Jarvis, W. D., and MacLeod, R. M. (1988) Characterization of the MMQ cell, a prolactin-secreting clonal cell line that is responsive to dopamine. *Endocrinology* **123,** 2341–2350.
58. Login, I. S., Pancrazio, J. J., and Kim, Y. I. (1990) Dopamine enhances a voltage dependent transient K^+ current in the MMQ cell, a clonal pituitary line expressing functional D_2 dopamine receptors. *Brain Res.* **506,** 331–334.
59. Lledo, P. M., Homburger, V., Bockaert, J., and Vincent, J.-D. (1992) Differential G protein-mediated coupling of D_2 receptors to K^+ and Ca^{++} currents in rat anterior pituitary cells. *Neuron* **8,** 455–463.

60. Baertschi, A. J., Audigier, Y., Lledo, P.-M., Israel, J.-M., Bockaert, J., and Vincent, J. D. (1992) Dialysis of lactotropes with antisense oligonucleotides assigns guanine nucleotide binding proteins subtypes to their channel effectors. *Mol. Endocrinol.* **6,** 2257–2265.
61. Missale, C., Castelletti, L., Boroni, F., Memo, M., and Spano, P. (1991) Epidermal growth factor induces the functional expression of dopamine receptors in the GH3 cell line. *Endocrinology* **128,** 13–20.
62. Gardette, R., Rasolonjanahary, R., Kordon, C., and Enjalbert, A. (1994) Epidermal growth factor treatment induces D2 dopamine receptors functionally coupled to delayed outward potassium current (I_K) in GH4C1 clonal anterior pituitary cells. *Neuroendocrinology* **59,** 10–19.
63. Missale, C., Boroni, F., Castelletti, L., Dal Toso, R., Gabellini, N., Sigala, S., and Spano, P. F. (1991) Lack of coupling of D-2 receptors to adenylate cyclase in GH-3 cells exposed to epidermal growth factor. *J. Biol. Chem.* **266,** 23,392–23,398.
64. Weiss, S., Sebben, M., Garcia-Sainz, J. A., and Bockaert, J. (1985) D_2-dopamine receptor-mediated inhibition of cyclic AMP formation in striatal neurons in primary culture. *Mol. Pharmacol.* **27,** 595–599.
65. Freedman, J. E. and Weight, F. F. (1988) Single K^+ channels activated by D_2 receptors in acutely dissociated neurons from rat corpus striatum. *Proc. Natl. Acad. Sci. USA* **85,** 3618–3622.
66. Lacey, M. G., Mercuri, N. B., and North, R. A. (1987) Dopamine acts on D_2 receptors to increase potassium conductance in neurones of the rat substantia nigra zona compacta. *J. Physiol.* **392,** 397–416.
67. Momiyama, T., Todo, N., and Sasa, M. (1993) A mechanism underlying dopamine D_1 and D_2 receptor-mediated inhibition of dopaminergic neurones in the ventral tegmental area *in vitro*. *Br. J. Pharmacol.* **109,** 933–940.
68. Roeper, J., Hainsworth, A. H., and Ashcroft, F. M. (1990) Tolbutamide reverses membrane hyperpolarisation induced by activation of D_2 receptors and $GABA_B$ receptors in isolated substantia nigra neurones. *Eur. J. Physiol.* **416,** 473–475.
69. Lin, Y.-J., Greif, G. J., and Freedman, J. E. (1993) Multiple sulfonylurea-sensitive potassium channels: a novel subtype modulated by dopamine. *Mol. Pharmacol.* **44,** 907–910.
70. Freedman, J. E., Greif, G. J., and Lin, Y.-J. (1994) Dual modulation of a K^+ channel by dopamine receptors and by cellular metabolism in rat striatum. *Neuropsychopharmacology* **10,** P124–P126.
71. Cohen, A. I., Todd, R. D., Harmon, S., and O'Malley, K. L. (1992) Photoreceptors of mouse retinas possess D_4 receptors coupled to adenylate cyclase. *Proc. Natl. Acad. Sci. USA* **89,** 12,093–12,097.
72. Albert, P. R., Neve, K. A., Bunzow, J. R., and Civelli, O. (1990) Coupling of a cloned rat dopamine-D2 receptor to inhibition of adenylyl cyclase and prolactin secretion. *J. Biol. Chem.* **265,** 2098–2104.
73. Liu, Y. F., Jakobs, K. H., Rasenick, M. M., and Albert, P. R. (1994) G protein specificity in receptor-effector coupling. *J. Biol. Chem.* **269,** 13,880–13,886.
74. Senogles, S. E. (1994) The D2 dopamine receptor isoforms signal through distinct $G_{i\alpha}$ proteins to inhibit adenylyl cyclase. *J. Biol. Chem.* **269,** 23,120–23,127.
75. Burris, T. P. and Freeman, M. E. (1994) Comparison of the forms of the dopamine D_2 receptor expressed in GH_4C_1 cells. *Proc. Soc. Exptl. Biol. Med.* **205,** 226–235.

76. Montmayeur, J.-P., Guiramand, J., and Borrelli, E. (1993) Preferential coupling between dopamine D2 receptors and G-proteins. *Mol. Endocrinol.* **7,** 161–170.
77. Malek, D., Munch, G., and Palm, D. (1993) Two sites in the third inner loop of the dopamine D2 receptor are involved in functional G protein-mediated coupling to adenylate cyclase. *FEBS Lett.* **325,** 215–219.
78. McGregor, A. M., Scanlon, M. F., Hall, K., Cook, D., and Hall, R. (1979) Reduction in size of a pituitary tumor by bromocriptine therapy. *N. Engl. J. Med.* **300,** 291–293.
79. Lloyd, H. M., Meares, J. D., and Jacobi, J. (1975) Effects of oestrogen and bromocryptine on *in vivo* secretion and mitosis in prolactin cells. *Nature* **255,** 497,498.
80. Florio, T., Pan, M.-G., Newman, B., Hershberger, R. E., Civelli, O., and Stork, P. J. S. (1992) Dopaminergic inhibition of DNA synthesis in pituitary tumor cells is associated with phosphotyrosine phosphatase activity. *J. Biol. Chem.* **267,** 24,169–24,172.
81. Senogles, S. E. (1994) The D2 dopamine receptor mediates inhibition of growth in GH_4ZR_7 cells: involvement of protein kinase-Cε. *Endocrinology* **134,** 783–789.
82. Choi, H. K, Won, L. A., Kontur, P. J., Hammond, D. N., Fox, A. P., Wainer, B. H., Hoffman, P. C., and Heller, A. (1991) Immortalization of embryonic mesencephalic dopaminergic neurons by somatic cell fusion. *Brain Res.* **552,** 67–76.
83. Neve, K. A., Henningsen, R. A., Bunzow, J. R., and Civelli, O. (1989) Functional characterization of a rat dopamine D-2 receptor cDNA expressed in a mammalian cell line. *Mol. Pharmacol.* **36,** 446–451.
84. Bates, M. D., Senogles, S. E., Bunzow, J. R., Liggett, S. B., Civelli, O., and Caron, M. G. (1991) Regulation of responsiveness at D_2 dopamine receptors by receptor desensitization and adenylyl cyclase sensitization. *Mol. Pharmacol.* **39,** 55–63.
85. Hayes, G., Biden, T. J., Selbie, L. A., and Shine, J. (1992) Structural subtypes of the dopamine D2 receptor are functionally distinct: expression of the cloned $D2_A$ and $D2_B$ subtypes in a heterologous cell line. *Mol. Endocrinol.* **6,** 920–925.
86. Neve, K. A., Kozlowski, M. R., and Rosser, M. P. (1992) Dopamine D2 receptor stimulation of Na^+/H^+ exchange assessed by quantification of extracellular acidification. *J. Biol. Chem.* **267,** 25,748–25,753.
87. Tang, L., Todd, R. D., Heller, A., and O'Malley, K. L. (1994) Pharmacological and functional characterization of D_2, D_3, and D_4 dopamine receptors in fibroblast and dopaminergic cell lines. *J. Pharmacol Exp. Ther.* **268,** 495–502.
88. Zhang, L.-J., Lachowicz, J. E., and Sibley, D. R. (1994) The D_2 and D_{2L} dopamine receptor isoforms are differentially regulated in Chinese hamster ovary cells. *Mol. Pharmacol.* **45,** 878–889.
89. Castellano, M. A., Liu, L.-X., Monsma, F. J., Jr., Sibley, D. R., Kapatos, G., and Chiodo, L. A. (1993) Transfected D_2 short dopamine receptors inhibit voltage dependent potassium current in neuroblastoma x glioma hybrid (NG108-15) cells. *Mol. Pharmacol.* **44,** 649–656.
90. Kanterman, R. Y., Mahan, L. C., Briley, E. M., Monsma, F. J., Sibley, D. R., Axelrod, J., and Felder, C. C. (1991) Transfected D_2 dopamine receptors mediate the potentiation of arachidonic acid release in Chinese hamster ovary cells. *Mol. Pharmacol.* **39,** 364–369.
91. Lajiness, M. E., Chio, C. L., and Huff, R. M. (1993) D2 dopamine receptor stimulation of mitogenesis in transfected Chinese hamster ovary cells: relationship to dopamine stimulation of tyrosine phosphorylations. *J. Pharmacol. Exp. Ther.* **267,** 1573–1581.

92. Piomelli, D., Pilon, C., Giros, B., Sokoloff, P., Martres, M.-P., and Schwartz, J.-C. (1991) Dopamine activation of the arachidonic acid cascade as a basis for D1/D2 receptor synergism. *Nature* **353,** 164–167.
93. Lahti, R. A., Figur, L. M., Piercey, M. F., Ruppel, P. L., and Evans, D. L. (1992) Intrinsic activity determinations at the dopamine D2 guanine nucleotide-binding protein-coupled receptor: utilization of receptor state binding affinities. *Mol. Pharmacol.* **42,** 432–438.
94. Chio, C. L., Lajiness, M. E., and Huff, R. M. (1994) Activation of heterologously expressed D3 dopamine receptors: comparison with D2 dopamine receptors. *Mol. Pharmacol.* **45,** 51–60.
95. Lajiness, M. E., Chio, C. L., and Huff, R. M. (1995) Signaling mechanisms of D_2, D_3, and D_4 dopamine receptors determined in transfected cell lines. *Clin. Neuropharmacol.* **18,** 525–533.
96. Seabrook, G. R., McAllister, G., Knowles, M. R., Myers, J., Sinclair, H., Patel, S., Freedman, S. B., and Kemp, J. A. (1994) Depression of high-threshold calcium currents by activation of human D_2 (short) dopamine receptors expressed in differentiated NG108-15 cells. *Br. J. Pharmacol.* **111,** 1061–1066.
97. Hescheler, J., Rosenthal, W., Trautwein, W., and Schultz, G. (1987) The GTP-binding protein, G_o, regulates neuronal calcium channels. *Nature* **325,** 445–447.
98. McFadzean, I., Mullaney, I., Brown D. A., and Milligan, G. (1989) Antibodies to the GTP binding protein, G_o, antagonize noradrenaline-induced calcium current inhibition in NG108-15 hybrid cells. *Neuron* **3,** 177–182.
99. Mullaney, I. and Milligan, G. (1990) Identification of two distinct isoforms of the guanine nucleotide binding protein G_o in neuroblastoma x glioma hybrid cells: independent regulation during cyclic AMP-induced differentiation. *J. Neurochem.* **55,** 1890–1898.
100. Rouot, B., Charpentier, N., Chabbert, C., Carette, J., Zumbihl, R., Bockaert, J., and Homburger, V. (1992) Specific antibodies against G_o isoforms reveal the early expression of the $G_{o2\alpha}$ subunit and appearance of $G_{o1\alpha}$ during neuronal differentiation. *Mol. Pharmacol.* **41,** 273–280.
101. Tang, L., Todd, R. D., and O'Malley, K. L. (1994) Dopamine D_2 and D_3 receptors inhibit dopamine release. *J. Pharmacol. Exp. Ther.* **270,** 475–479.
102. Swarzenski, B. C., Tang, L., Oh, Y. J., O'Malley, K. L., and Todd, R. D. (1994) Morphogenic potentials of D_2, D_3, and D_4 dopamine receptors revealed in transfected neuronal cell lines. *Proc. Natl. Acad. Sci. USA* **91,** 649–653.
103. Todd, R. D., (1992) Neural development is regulated by classical neurotransmitters: dopamine D2 receptor stimulation enhances neurite outgrowth. *Biol. Psychiatry* **31,** 794–807.
104. Huff, R. M. and Lajiness, M. E. (1994) D2 dopamine receptors stimulate mitogen activated protein kinases. *J. Cell. Biochem.* **18B,** 220.
105. Gupta, S. K., Gallego, C., Johnson, G. L., and Heasley, L. E. (1992) MAP Kinase is constitutively activated in *gip2* and *src* transformed Rat 1a fibroblasts. *J. Biol. Chem.* **267,** 7987–7990.
106. Faure, M., Voyna-Yasenetskaya, T. A., and Bourne, H. R. (1994) cAMP and βγ subunits of heterotrimeric G proteins stimulate the mitogen-activated protein kinase pathway in COS-7 cells. *J. Biol. Chem.* **269,** 7851–7854.
107. Koch, W. J., Hawes, B. E., Allen, L. F., and Lefkowitz, R. J. (1994) Direct evidence that G_i-coupled receptor stimulation of mitogen-activated protein kinase is mediated by $G_{\beta\gamma}$ activation of $p21^{ras}$. *Proc. Natl. Acad. Sci. USA* **91,** 12,706–12,710.

108. Fiore, R. S., Bayer, V. E., Pelech, S. L., Posada, J., Cooper, J. A., and Baraban, J. M. (1993) p42 Mitogen-activated protein kinase in brain: prominent localization in neuronal cell bodies and dendrites. *Neuroscience* **55,** 463–472.
109. Haycock, J. W., Ahn, N. G., Cobb, M. H., and Krebs, E. G. (1992) ERK1 and ERK2, two microtubule-associated protein 2 kinases, mediate the phosphorylation of tyrosine hydroxylase at serine-31 *in situ*. *Proc. Natl. Acad. Sci. USA* **89,** 2365–2369.
110. Lin, L.-L., Wartmann, M., Lin, A. Y., Knopf, J. L., Seth, A., and Davis, R. J. (1993) $cPLA_2$ is phosphorylated and activated by MAP kinase. *Cell* **72,** 269–278.
111. Ray, L. B. and Sturgill, T. W. (1987) Rapid stimulation by insulin of a serine/threonine kinase in 3T3-L1 adipocytes that phosphorylates microtubule-associated protein 2 *in vitro*. *Proc. Natl. Acad. Sci. USA* **85,** 3753–3757.
112. Lloyd, E. D. and Wooten, M. W. (1992) $^{PP42/44}$MAP kinase is a component of the neurogenic pathway utilized by nerve growth factor in PC12 cells. *J. Neurochem.* **59,** 1099–1109.
113. Freedman, S. B., Patel, S., Marwood, R., Emms, F., Seabrook, G. R., Knowles, M. R., and McAllister, G. (1994) Expression and pharmacological characterization of the human D3 dopamine receptor. *J. Pharmacol. Exp. Ther.* **268,** 417–426.
114. Sokoloff, P., Giros, B., Martres, M. P., Bouthenet, M.-L., and Schwartz, J.-C. (1990) Molecular cloning and characterization of a novel dopamine receptor (D3) as a target for neuroleptics. *Nature* **347,** 146–151.
115. Seabrook, G. R., Patel, S., Marwood, R., Emms, F., Knowles, M. R., Freedman, S. B., and McAllister, G. (1992) Stable expression of human D_3 dopamine receptors in GH_4C_1 pituitary cells. *FEBS Lett.* **312,** 123–126.
116. MacKenzie, R. G., VanLeeuwen, D., Pugsley, T. A., Shih, Y.-H., Demattos, S., Tang, L., Todd, R. D., and O'Malley, K. L. (1994) Characterization of the human dopamine D_3 receptor expressed in transfected cell lines. *Eur. J. Pharmacol.* **266,** 79–85.
117. Cox, B. A., Rosser, M. P., Kozlowski, M. R., Duwe, K. M., Neve, R. L., and Neve, K. A. (1995) Regulation and functional characterization of a rat recombinant dopamine D3 receptor. *Synapse* **21,** 1–9.
118. Seabrook, G. R., Kemp, J. A., Freedman, S. B., Patel, S., Sinclair, H. A., and McAllister, G. (1994) Functional expression of human D_3 dopamine receptors in differentiated neuroblastoma x glioma NG108-15 cells. *Br. J. Pharmacol.* **111,** 391–393.
119. Pilon, C., Levesque, D., Dimitriadou, V., Griffon, N., Martres, M. P., Schwartz, J. C., and Sokoloff, P. (1994) Functional coupling of the human dopamine D_3 receptor in a transfected NG108-15 neuroblastoma-glioma hybrid cell line. *Eur. J. Pharmacol.* **268,** 129–139.
120. Levy, F. O., Zhu, X., Kaumann, A. J., and Birnbaumer, L. (1993) Efficacy of β_1-adrenergic receptors is lower than that of β_2-adrenergic receptors. *Proc. Natl. Acad. Sci. USA* **90,** 10,798–10,802.
121. Van Tol, H. H. M., Bunzow, J. R., Guan, H.-C., Sunahara, R. K., Seeman, P., Niznik, H. B., and Civelli, O. (1991) Cloning of the gene for a human dopamine D_4 receptor with high affinity for the antipsychotic clozapine. *Nature* **350,** 610–614.
122. Mills, A., Allet, B., Bernard, A., Chabert, C., Brandt, E., Cavegn, C., Chollet, A., and Kawashima, E. (1993) Expression and characterization of human D4 dopamine receptors in baculovirus-infected insect cells. *FEBS Lett.* **320,** 130–134.

123. Chio, C. L., Drong, R. F., Riley, D. T., Gill, G. S., Slightom, J. L., and Huff, R. M. (1994) D4 dopamine receptor-mediated signaling events determined in transfected Chinese hamster ovary cells. *J. Biol. Chem.* **269,** 11,813–11,819.
124. McHale, M., Coldwell, M. C., Herrity, N., Boyfield, I., Winn, F. M., Ball, S., Cook, T., Robinson, J. H., and Gloger, I. S. (1994) Expression and functional characterisation of a synthetic version of the human D_4 dopamine receptor in a stable human cell line. *FEBS Lett.* **345,** 147–150.
125. Van Tol., H. H. M., Wu, C. M., Guan, H.-C., Ohara, K., Bunzow, J. R., Civelli, O., Kennedy, J., Seeman, P., Niznik, H. B., and Jovanovic, V. (1992) Multiple dopamine D4 variants in the human population. *Nature* **358,** 149–152.
126. O'Malley, K. L., Harmon, S., Tang, L., and Todd, R. D. (1992) The rat dopamine D4 receptor: sequence, gene structure, and demonstration of expression in the cardiovascular system. *New Biologist* **4,** 137–146.
127. Seabrook, G. R., Knowles, M., Brown, N., Myers, J., Sinclair, H., Patel, S., Freedman, S. B., and McAllister, G. (1994) Pharmacology of high-threshold calcium currents in GH_4C_1 pituitary cells and their regulation by activation of human D_2 and D_4 dopamine receptors. *Br. J. Pharmacol.* **112,** 728–734.
128. Kleuss, C., Scherubl, H., Heschler, J., Schultz, G., and Wittig, B. (1992) Different β-subunits determine G-protein interaction with transmembrane receptors. *Nature* **358,** 424–426.
129. Kleuss, C., Scherubl, H., Hescheler, J., Schultz, G., and Wittig, B. (1993) Selectivity in signal transduction determined by γ-subunits of heterotrimeric G proteins. *Science* **259,** 832–834.
130. Liu, Y. F., Civelli, O., Grandy, D. K., and Albert, P. R. (1992) Differential sensitivity of the short and long human dopamine D_2 receptor subtypes to protein kinase C. *J. Neurochem.* **59,** 2311–2317.
131. Di Marzo, V., Vial, D., Sokoloff, P., Schwartz, J.-C., and Piomelli, D. (1993) Selection of alternative G_i-mediated signaling pathways at the dopamine D_2 receptor by protein kinase C. *J. Neurosci.* **13,** 4846–4853.
132. Gurwitz, D., Harin, R., Heldman, E., Fraser, C. M., Manor, D., and Fisher, A. (1994) Discrete activation of transduction pathways associated with acetylcholine m1 receptor by several muscarinic ligands. *Eur. J. Pharmacol.* **267,** 21–31.

CHAPTER 7

D1/D2 Dopamine Receptor Interactions in Basal Ganglia Functions

John F. Marshall, David N. Ruskin, and Gerald J. LaHoste

1. Historical Introduction

The classification by Kebabian and Calne *(1)* in 1979 of dopamine (DA) receptors into two subtypes (D1 and D2) was based on differential linkages to adenylate cyclase as well as on distinct pharmacological profiles. At the time that this classification was proposed, several lines of evidence pointed to a role for D2, but not D1, receptors in the behavioral functions mediated by dopamine.* For instance, in vivo administration of D2 antagonists was known to block the motor stimulant and rewarding properties of agents such as amphetamine or cocaine *(3,4)*. Administration of D2 agonists produces locomotor activity and stereotyped movements *(5)*. Also, the clinical potencies of a series of antipsychotic drugs were found to correlate with their affinities for D2 receptors *(6,7)*. In retrospect, it seems clear that the slowness to recognize the importance of D1 receptor agonism to dopamine's effects within basal ganglia occurred for two reasons. First, a lack of pharmacological agents specific for D1 receptors precluded researchers from isolating the functions of this receptor. Second, as we now appreciate, the influences of D1 receptors can be subtle to detect. The interactions between D1 and D2 receptors are profoundly changed by nigrostriatal injury; and, in intact animals, their interpretation depends on an appreciation of the role played by endogenous DA in providing tonic D1 agonism.

*Unless otherwise noted, the terms D1 and D2 will be used to refer to the two subfamilies of DA receptors constituted by the D_1 and D_5 subtypes on the one hand, and the D_2, D_3, and D_4 subtypes on the other *(2)*.

The Dopamine Receptors Eds.: K. A. Neve and R. L. Neve
Humana Press Inc., Totowa, NJ

Several key findings led to the early recognition that D1 receptors participate in control of basal gangliar motor functions. In 1978, Setler et al. *(8)* first established a role for D1 receptors in motor behaviors by showing that, in rats with unilateral nigrostriatal 6-hydroxydopamine (6-OHDA) injections, the D1 agonist SKF 38393 produces contralateral rotation comparable to that produced by the nonselective DA agonists apomorphine or L-dopa. Demonstration of a role for D1 receptors in the motor functions of intact rats occurred only later, however, as a result of several findings. As reviewed in Section 2.1., these included the observation that administration of a D1 antagonist alone can produce catalepsy and akinesia, as does a D2 antagonist *(9)*; and the findings that the locomotion and motor stereotypies elicited by the nonselective DA agonists apomorphine or amphetamine can be blocked by either a D1 or a D2 selective antagonist *(10,11)*. Two seminal electrophysiological studies showed the cooperative influences between D1 and D2 receptors in the control of striatal and pallidal firing rates *(12,13)*.

The history of understanding D1/D2 interactions also benefited from the authoritative review of the phenomenon in 1987 provided by Clark and White *(14)*. Much of the earlier literature is not revisited here. Instead, the focus of this review is on the manifestations of D1/ D2 interactions, and processes hypothesized to underlie them, that have emerged in the subsequent 9 years. For example, much recent evidence suggests that D1 and D2 receptors normally synergize in their influences on striatal neuropeptide expression as well as in their effects on immediate-early gene expression within striatal, pallidal, and cortical neurons. This chapter also contrasts the synergistic influences of D1 and D2 receptors in normal animals with the largely independent effects of these receptors in animals with nigrostriatal injury. We emphasize that the breakdown in normal synergistic interactions of D1 and D2 receptors, not an up-regulation in dopamine receptors or their mRNAs, appears responsible for the supersensitivity to dopamine agonists seen after nigrostriatal injury. We also consider how this D1/D2 independence occurs, noting the role of dopamine in maintaining normal synergistic interactions of D1 and D2 receptors. Finally, we consider several types of models of basal gangliar function that attempt to explain D1/D2 interactions.

2. Manifestations of D1/D2 Interactions: Normal Synergy

2.1. Motor Behaviors

For many central actions of DA, concomitant stimulation of D1 and D2 receptors is normally required, a phenomenon that we refer to as requisite

D1/D2 synergism. This type of receptor interaction represents an extreme form of synergism, the pharmacological principle whereby the effect of two combined drugs exceeds the algebraic sum of their separate effects. Other authors have referred to this as obligatory or qualitative synergism, or enabling. It is important to differentiate this type of synergism from the nonrequisite (or quantitative) type, as the two phenomena may be dissociable.

Requisite D1/D2 synergism is evident in the control of motor behavior. For example, the akinesia that results from acute DA depletion induced by reserpine can be reversed by combined but not separate administration of a selective D1 and a selective D2 agonist *(15)*. Furthermore, the locomotion and motor stereotypy elicited by the nonselective DA agonists apomorphine or amphetamine can be blocked by either a selective D1 or a selective D2 antagonist *(10,11)*. When selective D1 or D2 agonists are administered, motor stereotypy is observed only following combined D1/D2 stimulation (*see* Fig. 1A) *(12,17–22)*. This latter finding is only fully apparent, however, when care is taken to control for the influence of endogenous DA which can act at D1 or D2 receptors (*see,* e.g., refs. *18,12,22*). Basal levels of endogenous DA provide sufficient D1 tone to synergize with an exogenous D2 agonist. By contrast, basal endogenous DA does not provide sufficient D2 tone to synergize with an exogenous D1 agonist. Consequently, administration of a D1 agonist alone does not elicit motor stereotypy, whereas administration of a D2 agonist alone does. An exception to the rule that D1/D2 costimulation is required for behavioral effects involves grooming in rodents, which can be elicited by selective D1 agonists in a manner that is independent of D2 receptor stimulation *(22)*.

2.2. Electrophysiology

2.2.1. Striatum

The D1/D2 receptor synergism apparent for activation of motor behavior is also manifest in the electrophysiology of the basal ganglia. White, Wang, and colleagues utilized in vivo extracellular unit recording of nucleus accumbens septi and caudate-putamen neurons in anesthetized rats. These studies found that locally iontophoresed D1 or D2 agonists inhibit the glutamate-stimulated firing rate of striatal neurons. The inhibitory action of D2 agonists is greatly reduced by acute DA depletion with α-methyl-*p*-tyrosine (αMPT), which should decrease D1 receptor tone *(23)*. Furthermore, co-iontophoresis of D1 and D2 agonists produces a supra-additive inhibition of firing rate (nonrequisite synergism) *(13,24)*. This synergism also is seen with systemic injection of D1 and D2 agonists *(25)*. The D1/D2 synergism regarding striatal neuron activity with locally applied drugs provides a physiological basis for the description of this nucleus as a locus of behavioral

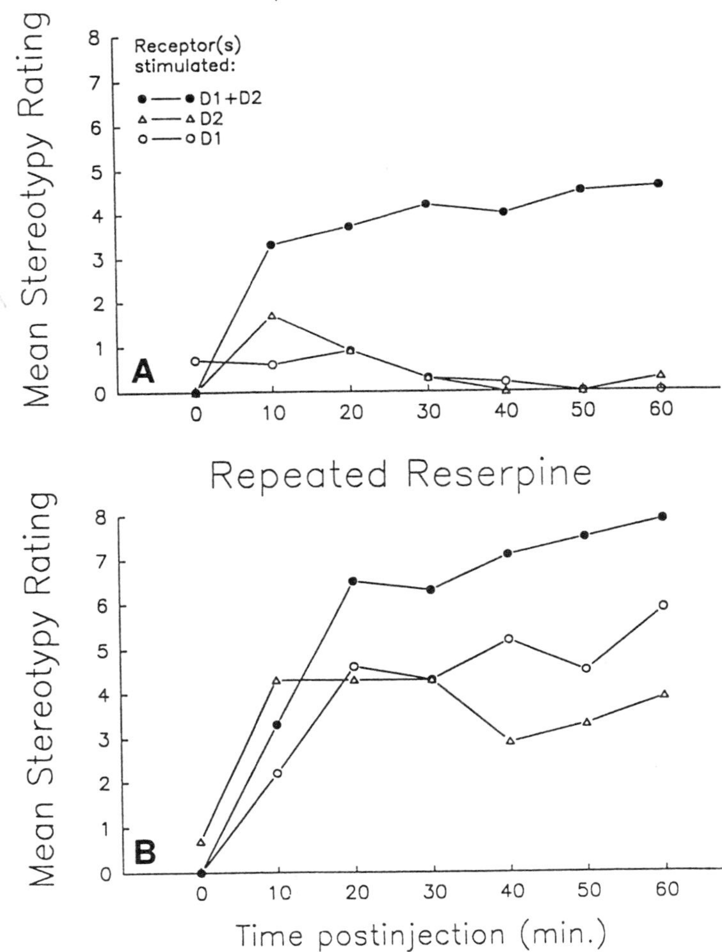

Fig. 1. **(A)** Stereotyped motor behaviors after administration of D1 and/or D2 agonists to normally sensitive rats or **(B)** to rats made supersensitive by reserpine administration (1 mg/kg, sc daily for 5 d). Stimulation of both D1 and D2 receptor subfamilies was achieved by administration of SKF 38393 (20 mg/kg, ip) and quinpirole (3 mg/kg, ip). Stimulation of D1 receptors alone was achieved by administration of SKF 38393 together with the D2 antagonist eticlopride (0.1 mg/kg, ip). Stimulation of D2 receptors alone was achieved by administration of quinpirole (3 mg/kg, ip) with the D1 antagonist SCH 23390 (0.1 mg/kg, sc). In normally sensitive rats, stereotyped behaviors are observed only in animals receiving combined D1 and D2 agonists, whereas in supersensitive rats, agonism of either receptor is sufficient to induce stereotyped behaviors. For behavioral ratings, 0, inactivity; 1, grooming; 2, locomotion; 3, sniffing directed upward; 4, sniffing with head down; 5, intense sniffing in a small circumscribed area; 6, intense sniffing in small area while climbing; 7, licking or gnawing of test apparatus; 8, self-directed licking or biting. From ref. *16* with permission.

D1/D2 synergism *(26)*. Electrophysiological manifestations of requisite synergism are not, however, in complete concordance with behavioral manifestations. Thus, in contrast to D2-mediated effects, D1-elicited inhibition of striatal neuron firing does not depend on D2 costimulation *(24)*. In addition, D2 autoreceptor effects on nigrostriatal and mesoaccumbens neurons, such as inhibition of firing rates and decreased tyrosine hydroxylase activity, appear not to depend on D1 receptors *(24)*.

The excitatory action of amphetamine on striatal neuron firing reflects D1/D2 interdependence. Peripherally injected amphetamine increases the firing rate of striatal neurons in awake, unrestrained rats, typically in neurons whose activity is related to movement *(27)*. This effect of amphetamine is completely blocked by either D1 or D2 antagonists, as is the behavioral activation owing to amphetamine. It appears contradictory that peripheral amphetamine causes striatal excitation in awake animals whereas locally applied DA agonists cause striatal inhibition in anesthetized animals. These differing effects are probably a consequence of the state of anesthesia, and not different routes of administration, since locally applied amphetamine causes striatal excitation in awake rats *(28)* and systemically applied DA agonists cause inhibition in anesthetized rats *(25)*.

Long-term depression represents another paradigm in which D1/D2 synergism in the control of striatal activity is evident. In vitro, tetanic stimulation of corticostriatal afferents induces a persistent depression of corticostriatal neurotransmission. The induction of this long-term depression can be blocked by bath application of either D1 or D2 antagonists *(29)*. These results show that D1 and D2 receptors can synergistically influence synaptic plasticity in the striatum in addition to normal striatal function. Since striatal long-term depression appears to be mediated by intracellular calcium *(30)*, D1 and D2 receptors may interact to regulate intracellular levels of this second messenger.

2.2.2. Globus Pallidus (GP)

Synergistic D1/D2 interactions have been reported for the single unit activity of the GP. In paralyzed rats, systemically injected D2 agonist increases the firing rate of GP neurons. This effect is potentiated by coinjection of D1 agonist and attenuated by acute DA depletion with αMPT (Fig. 2A) *(12,31)*. D1 agonists alone have no consistent effect. D1/D2 synergistic influences on the ventral pallidum have also been found, utilizing application of dopaminergic drugs directly into the nucleus accumbens septi *(33)*. Therefore, a dependence of D2 receptor actions on D1 receptor activity is apparent in the GP much as it is in the striatum, and indeed the D1/D2 interactions in the striatum may be causal to the interactions found in GP.

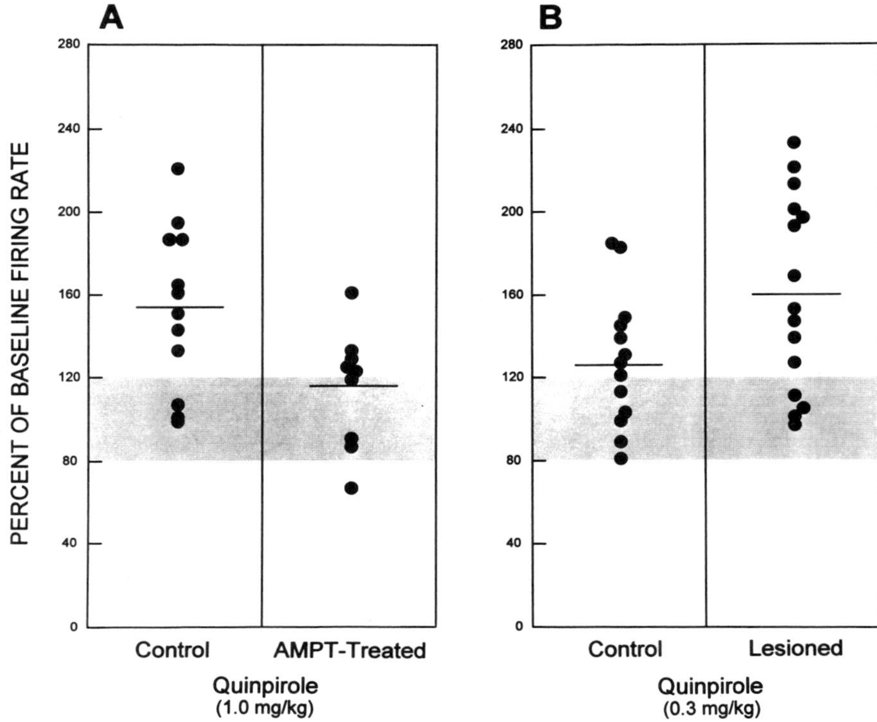

Fig. 2. Distributions of firing rate changes in individual GP neurons of gallamine-immobilized, locally anesthetized rats given the D2 agonist quinpirole. **(A)** The effect of quinpirole (1 mg/kg, iv) on extracellular unit activity, and its prevention in rats receiving αMPT pretreatment (300 mg/kg, ip 4 h and 200 mg/kg, ip 2 h prior to recording). **(B)** Comparison of the effect of a lower dose of quinpirole (0.3 mg/kg, iv) in control rats with its effect in rats given nigrostriatal 6-hydroxy-dopamine lesions 6–10 wk previously. In lesioned rats lacking sufficient endogenous dopamine tone to provide D1 receptor agonism, quinpirole is sufficient to increase GP firing rates. Adapted from refs. *31,32* with permission.

2.2.3. Substantia Nigra

Peripherally injected apomorphine has variable effects on the firing rates of neurons in the substantia nigra pars reticulata (SNr) of normal animals, with some units being excited, others being inhibited, and others showing no significant change *(34)*. Both the excitatory and inhibitory effects can be reversed by either D1 or D2 antagonists *(35)*, showing that these effects of apomorphine, although varying between cells, are the result of coactivation of D1 and D2 receptors.

2.3. 2-Deoxyglucose (2-DG)

The noncatalyzable glucose analog 2-DG is widely used as an ex vivo indicator of the metabolic activity of various tissues. In the nervous system, uptake of 2-DG is thought to reflect mainly the activity of axon terminals, rather than cell bodies *(36)*. Systemic administration of amphetamine or apomorphine increases 2-DG uptake in GP, subthalamic nucleus (STN), SNr, and entopeduncular nucleus (EPN). These metabolic effects are blocked by either D1 or D2 antagonists *(37,38)*. Hence, the control by DA of basal ganglia metabolism bears a striking similarity to that of motor activity. Striatal dopamine receptors may mediate these effects, since DA infused into the striatum or nucleus accumbens septi increases 2-DG incorporation in STN, SNr, and EPN *(39,40)*.

2.4. Immediate-Early Gene Expression

2.4.1. Striatum

Following systemic administration of the indirect DA agonist amphetamine, neurons within the calbindin-poor patch compartment of the striatum express nuclear immunoreactivity for Fos, the protein product of the c-*fos* gene, whereas those in the matrix compartment are relatively unresponsive *(41)*. This effect is completely blocked by the selective D1 antagonist SCH 23390, demonstrating the requirement for D1 stimulation *(41,42)*. D2 receptor involvement in this effect has been difficult to establish, since potent D2 antagonists themselves elicit striatal Fos, leading some authors to conclude that striatal Fos expression elicited by indirect DA agonists (amphetamine or cocaine) is purely a D1-mediated phenomenon. The issue is clarified by distinguishing the striatal neuron populations in which DA agonists or D2 antagonists stimulate Fos expression. D2 antagonist-induced Fos occurs primarily in striatopallidal neurons *(43)*, whereas DA agonist-mediated Fos occurs in striatonigral neurons *(44,45)*. When the striatonigral neurons are selectively labeled by retrograde transport of fluorogold from SNr, amphetamine- or cocaine-induced Fos in these neurons is fully antagonized by a potent D2 antagonist *(46)*. These findings are consistent with those obtained using directly acting selective D1 and D2 agonists. Combined, but not separate, administration of a selective D1 and a selective D2 agonist reveals clusters of neurons that are immunoreactive for Fos (Fig. 3) *(47)* and which appear to be largely localized to the striatal patch compartment. Thus, the DA-stimulated expression of the immediate-early gene c-*fos* depends on concomitant stimulation of D1 and D2 receptors.

2.4.2. GP

D1/D2 synergism is also manifest in DA-stimulated Fos expression of pallidal neurons. Systemic D1 agonist stimulation alone does not elicit

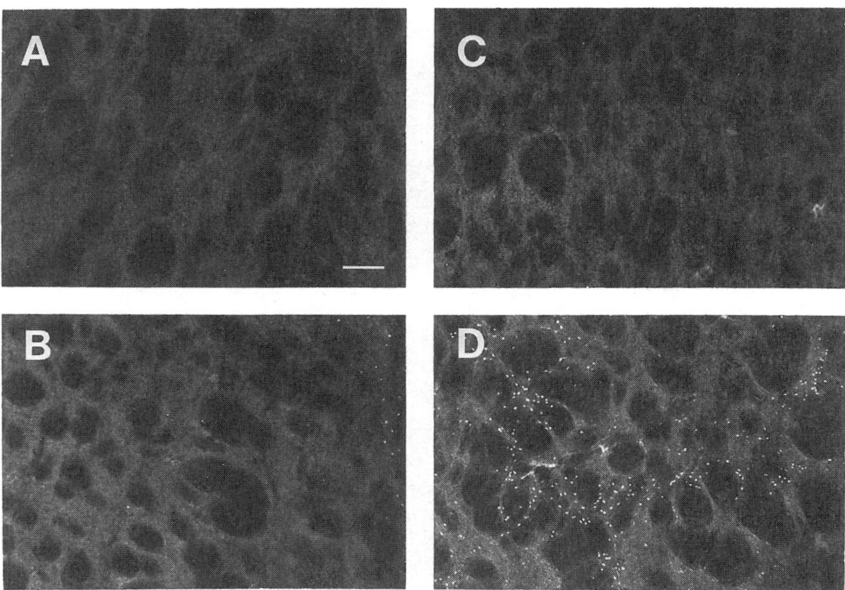

Fig. 3. Reverse-image photomicrographs of Fos-like immunoreactivity in striatum of rats that were treated with **(A)** saline, **(B)** SKF 38393 (20 mg/kg, ip), **(C)** quinpirole (3 mg/kg, ip), or **(D)** combined SKF 38393 and quinpirole. Very low levels of nuclear immunoreactivity are observed except after combined D1 and D2 agonist administration, in which clusters of intensely stained nuclei are seen. Bar, 0.1 mm. Adapted from ref. *47* with permission.

Fos expression in GP neurons; systemic D2 stimulation elicits a small amount of Fos expression, an effect that may be owing to synergistic actions of endogenous DA at D1 receptors. By contrast, combined D1/D2 stimulation produces pronounced Fos expression in GP *(48)*. These effects are mimicked by combined intrastriatal administration of D1 and D2 agonists (Ruskin and Marshall, unpublished observations), suggesting that pallidal Fos expression results from inhibition of inhibitory striatopallidal neurons.

2.4.3. Cerebral Cortex

Following systemic DA agonist administration, cerebral cortical neurons show pronounced nuclear Fos expression with a specific laminar and areal distribution. The greatest Fos expression occurs in primary somatosensory cortex (Par1); however, secondary somatosensory areas (Par2) and frontal cortex (Fr1) also show Fos expression, whereas cingulate neurons (Cg2) do not *(49)*. In all areas, the most pronounced expression occurs in layer IV, but other layers also show high levels of Fos expression. These effects occur following combined but not separate administration of D1 and D2 agonists or following amphetamine. Further, the amphetamine effects are blocked by either D1 or D2 antagonists *(50)*. Thus, cerebral cortical Fos expression displays the characteristics of requisite D1/D2 synergism.

2.5. Neuropeptide Regulation

Striatonigral neurons express the neuropeptides dynorphin (DYN) and substance P (SP) in addition to the classical transmitter γ-aminobutyric acid (GABA) *(51,52)*. Dopaminergic regulation of these peptides and their prepropeptide mRNA has remarkable similarities to the dopaminergic regulation of immediate-early genes in striatonigral cells, although chronic rather than acute treatments are typically used for peptide regulation studies (*see also*, Chapter 10). Chronic treatment with selective D1 or D2 agonists has no effect on striatal preprodynorphin mRNA or DYN peptide levels in normal animals or in the intact hemisphere of unilaterally 6-OHDA lesioned animals *(53–55)*; however, chronic administration of mixed D1/D2 agonists or indirect DA agonists effectively increases preprodynorphin mRNA levels *(56,57)*. Acute injection of indirect DA agonists also raises levels of striatal preprodynorphin mRNA *(58)*. The preprodynorphin mRNA response to chronic or acute dopaminergic stimulation is substantially larger in patches compared to the matrix *(56,58)*, mimicking the compartmental selectivity of the immediate-early gene response to mixed or indirect DA agonists.

The regulation of SP is similar to that of DYN. Chronic D1 or D2 agonist administration has no effect on preprotachykinin mRNA (the mRNA precursor to SP) in the intact striatum *(54)*, but levels of this mRNA species are increased by chronic mixed or indirect agonist treatment *(56,57)*, an effect that appears to happen in both patches and matrix. Acute injection of methamphetamine also increases striatal preprotachykinin mRNA, and this action is blocked by either D1 or D2 antagonists *(59)*.

These studies demonstrate that there is a requisite D1/D2 receptor synergism concerning control of the striatonigral peptides DYN and SP, such that both receptor subtypes must be activated for increases in peptide or prepropeptide mRNA to occur. Therefore, DA receptor subtypes synergistically increase the peptide expression in striatonigral cells much as they synergistically increase the immediate-early gene expression in these neurons.

3. Manifestations of D1/D2 Interactions: Independence After Nigrostriatal Injury or Reserpine Treatment

Injury to the dopaminergic nigrostriatal projection produces a state of supersensitivity to directly acting DA agonists which is evident in the animals' behavioral reactions as well as in basal gangliar electrophysiological, metabolic, biochemical, and genomic responses to these treatments. For example, bilateral 6-OHDA-induced nigrostriatal injury results in supersensitivity to the motor stereotypy-inducing effects of apomorphine *(60)*, whereas rats with unilateral nigrostriatal injury rotate contralaterally when given this DA agonist *(61)*. Quantification reveals that nigrostriatal injury increases sensitivity to apomorphine 10–40-fold *(62,63)*. The injury-induced changes are mimicked by treatment with the monoamine-depleting agent, reserpine *(64)*.

For many years, investigations concerning the basis for this supersensitivity centered on the role of striatal DA receptor up-regulation *(65–67)*. Although nigrostriatal injury in rats increases the density of striatal D2 binding sites *(65)*, as well as mRNA for D_2 receptors *(54,66,68)*, the D_2 up-regulation and sensitivity changes seen after nigrostriatal damage can be dissociated (reviewed in ref. *69*). Injury-induced supersensitivity also does not appear to be secondary to increased striatal D1-like or D_3 receptors *(70,71)*. An important alternative explanation for the dopaminergic supersensitivity after nigrostriatal injury has its basis in observations that the requisite synergistic influences of D1 and D2 receptors break down after nigrostriatal injury. It is our hypothesis that this breakdown in D1/D2 synergism is responsible for the supersensitivity of lesioned animals to DA agonists and may contribute to the emergence of dyskinetic reactions in Parkinsonian patients treated with L-dopa. In the remainder of this section, we review the evidence for D1/D2 independence after nigrostriatal injury (or reserpine treatment). In Section 4. we consider how this D1/D2 independence occurs as well as its role in supersensitivity to DA agonists.

3.1. Motor Behaviors/Rotation

Following destruction of central dopaminergic neurons with 6-OHDA, or prolonged depletion of monoamines with reserpine, the requisite D1/D2 synergism breaks down. Functions that previously required concomitant D1/D2 stimulation can be mediated by independent stimulation of either D1 or D2 receptors. For example, in sharp contrast to neurologically intact rats, rats with central 6-OHDA lesions display hyperactivity and motor stereotypy in response to selective D1 or D2 agonists administered alone, effects which can be blocked by pretreatment with the homotypic, but not heterotypic, antagonist *(72,73)*. These effects of 6-OHDA are mimicked by repeated daily systemic injections of the monoamine-depleting agent reserpine (Fig. 1B) *(16,64)*. In addition, rotation elicited by direct DA receptor stimulation in rats with unilateral 6-OHDA lesions can be elicited by either D1 or D2 agonists and can only be blocked by the homotypic antagonist *(74,75)*.

3.2. Electrophysiology

3.2.1. Striatum

After a nigrostriatal lesion, locally iontophoresed D1 or D2 agonists more potently inhibit the firing of striatal neurons, demonstrating a lesion-induced supersensitivity of both DA receptor subtypes *(76)*. In addition, after nigrostriatal injury, D2 receptor agonism no longer requires concomitant D1 receptor stimulation for its functional effects. Therefore, as for stimulation of motor behavior, a 6-OHDA lesion breaks down the requisite D1/D2 synergism concerning striatal firing rate. Such independence after nigrostriatal injury is not seen with all electrophysiological measures, however. Long-term depression in striatal slices still depends on activation of both D1 and D2 receptors after nigrostriatal lesion, just as in normal tissue *(29)*.

3.2.2. GP

Nigrostriatal lesion causes the D2 receptors modulating GP unit firing rates to become independent. In normal animals the excitatory effects of D2 agonists on pallidal neurons depend on D1 receptor activation by endogenous DA, but after a 6-OHDA lesion D2 agonists excite GP neurons even in the absence of D1 activation (Fig. 2B) *(32)*. D1 receptor actions are also changed after nigrostriatal 6-OHDA injections. Some GP cells in lesioned rats show large D1-mediated increases in firing, yet others show large decreases, effects that are not typically seen in control animals due to any dopaminergic treatment *(32)*. Therefore, after nigrostriatal lesion, D1 and D2 receptors become independent in increasing the firing rate of GP neurons. In addition, a novel inhibitory action of D1 receptors becomes apparent for some units.

3.2.3. SNr

D1 and D2 agonists have small and variable effects on SNr neurons in normal animals *(35)*, but after nigrostriatal lesion, high doses of D1 and D2 agonists effectively inhibit these cells *(77)*. Also, in the lesioned state, doses of D1 and D2 agonists low enough to have no effect alone produce a robust inhibition when combined *(77)*. Therefore, in the 6-OHDA-lesioned state, D1 and D2 receptors can independently influence the firing rate of cells in this nucleus, although a nonrequisite synergism is still apparent.

3.3. 2-DG

After nigrostriatal injury or reserpine administration, DA agonists alter 2-DG incorporation in several basal ganglia nuclei in a manner suggesting D1/D2 independence. For example, after nigrostriatal 6-OHDA injections, either D1 or D2 agonists alone increase STN 2-DG uptake *(78,79)*. Also, in the SNr and EPN, the magnitude of 2-DG responses to D1 agonists changes markedly after nigrostriatal injury or reserpine administration. In rats made supersensitive by such pretreatments, D1 agonists increase 2-DG uptake by three- or fourfold in SNr and EPN *(79,80)*, a greater effect than that of amphetamine in intact animals. These results have strong parallels with the studies of SNr electrophysiology discussed in Section 3.2.3. Specifically, after a 6-OHDA lesion, D1 agonists produce large increases in SNr 2-DG uptake and large decreases in SNr firing rate, suggesting that under these conditions D1 receptor activation excites the inhibitory GABAergic striatonigral projection.

3.4. Immediate-Early Gene Expression

3.4.1. Striatum

D1/D2 interactions in striatal immediate-early gene expression are also affected by 6-OHDA or by protracted reserpine treatments. D1 receptor stimulation induces Fos independently of D2 receptors, following nigrostriatal destruction *(81)* or following repeated reserpine administration (Fig. 4) *(47)*. Furthermore, under these supersensitive conditions, D1 agonists induce Fos in both the patch and matrix compartments, rather than in the patch-selective pattern seen in intact animals. Although the compartmentalization of the Fos response changes after 6-OHDA or repeated reserpine, DA agonists still induce Fos in the striatonigral population of cells, just as in the normal state *(43,44)*. Therefore, 6-OHDA or subchronic reserpine treatment allows D1 receptors to independently activate the striatonigral direct pathway, whereas in the normal state concomitant D2 receptor activity is needed to enable this D1 effect. It should be noted that after nigrostriatal lesion, D2 activation causes a low, ineffective dose of D1 agonist to become effective in inducing striatal Fos *(82)*, so that there may still be a nonrequisite D1/D2 synergism in the supersensitive state even though requisite synergism is gone.

3.4.2. GP

Twenty-four hours after a single dose of reserpine, D2 stimulation alone elicits pronounced Fos expression in GP neurons, in contrast to control animals *(83)*. Similar effects of D2 agonist are seen in the ipsilateral pallidum of rats sustaining prior unilateral nigrostriatal 6-OHDA injections *(48)*. In 6-OHDA-injected rats, D1 stimulation alone also elicits pallidal Fos expression *(83a)*, an effect that may be related to the increased firing rates seen in some pallidal neurons owing to D1 agonists after 6-OHDA *(32)*.

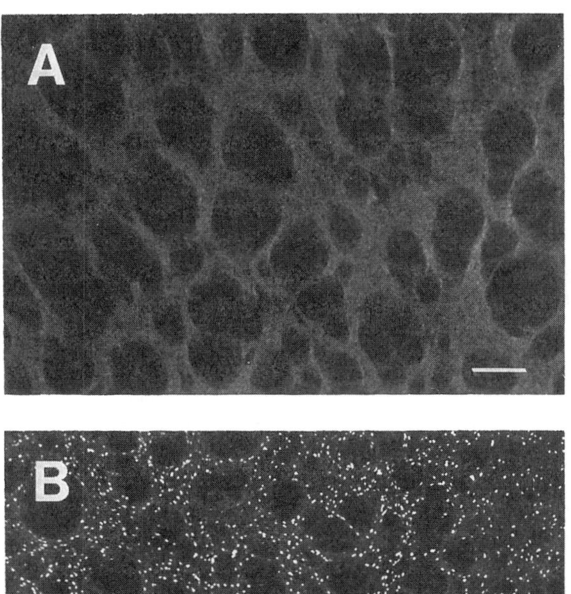

Fig. 4. Reverse-image photomicrographs of Fos-like immunoreactivity in striatum of rats injected daily with reserpine (1 mg/kg, sc) for 5 d. After the last reserpine treatment rats were given **(A)** saline or **(B)** SKF 38393 (20 mg/kg, ip). In contrast to normally sensitive rats, SKF 38393 administration is sufficient to induce pronounced nuclear Fos immunoreactivity in reserpine-pretreated animals. Bar, 0.1 mm. Adapted from ref. *47* with permission.

3.4.3. Cerebral Cortex

Following repeated reserpine treatment or 6-OHDA-induced nigrostriatal injury, there is a breakdown in requisite D1/D2 synergism with respect to cortical Fos expression. Under these conditions, separate stimulation of D1 or D2 receptors elicits Fos in cerebral cortex, whereas D1/D2 costimulation is normally required for this effect *(50)*. The areal and laminar distribution of Fos expression following separate D1 or D2 stimulation in reserpine- or 6-OHDA-treated rats is similar to that which occurs in normal rats following combined D1/D2 stimulation (*see* Section 2.4.3.).

3.5. Neuropeptide Regulation

Striatal preprotachykinin mRNA levels are decreased after a nigrostriatal lesion. Chronic D1 agonist treatment will reverse this decrease *(54)*. Chronic D1 agonist treatment also increases striatal levels of DYN peptide and preprodynorphin mRNA after nigrostriatal lesion *(53–55)*. This contrasts with findings in normal rats, where concomitant D1/D2 stimulation is necessary to increase levels of either of these neuropeptides. Chronic treatment with D2 agonist is without effect on either peptide in supersensitive tissue as it is in normosensitive tissue. Therefore, after a 6-OHDA lesion, D1 receptors can independently increase expression of neuropeptides as well as expression of immediate-early genes in striatonigral cells. The lack of effect of D2 agonists on both neuropeptide and immediate-early gene expression in these cells in 6-OHDA-lesioned tissue is further evidence of the importance of D1 receptors in controlling the striatonigral neurons of the direct pathway in the supersensitive state.

4. Processes Leading to D1/D2 Independence

4.1. Role of Dopamine Receptor Regulation in D1/D2 Independence

The two treatments that result in a breakdown in requisite synergism, 6-OHDA and reserpine treatment, both deplete DA, an effect that ultimately gives rise to proliferation of several striatal D2 receptors. Thus, changes in D1 and/or D2 receptor density might underlie the changes in receptor interaction. However, increasing D1 or D2 density alone or together via chronic daily antagonist treatment was shown not to alter the normal pattern of requisite D1/ D2 synergism with respect to motor stereotypy *(16)*. Furthermore, profound changes in D1/ D2 synergism induced by reserpine can be observed long before any changes in D2 receptor density, which do not begin to manifest themselves until 7 d after the initiating stimulus and are not maximal for 3 wk. By contrast, a breakdown in requisite D1/D2 synergism can be seen within 48 h after reserpine treatment *(84)* or earlier *(80,83)*. Thus, there is a double dissociation between DA receptor density and the state of D1/D2 synergism: Altering receptor density does not change the state of synergism, and a breakdown in synergism can be achieved without changes in receptor density *(16)*.

4.2. Does Independence Result from a Loss of DA Receptor Activity *per se*?

Although the data described in Section 4.1. indicate that receptor up-regulation cannot be the cause of the breakdown in requisite D1/D2 synergism, they do not indicate what is the cause. Perhaps a more continuous blockade of the action of synaptic DA might engage a different repertoire of intracellular responses than those elicited by single daily antagonist injections, and thereby influence the state of D1/D2 synergism. Since depletion of DA by 6-OHDA or reserpine can elicit a breakdown in requisite D1/D2 synergism within 48 h, closer examination of dopaminergic actions within this time period might help to identify the neural stimuli that govern the state of D1/D2 synergism and thereby help to elucidate the mechanism(s) by which this receptor interaction occurs. Using this strategy, it was found that interaction of synaptic DA with D2 receptors plays a major role in determining whether these receptors can function independently of D1 receptors. Continuous treatment over 48 h with reserpine, 6-OHDA, the tyrosine hydroxylase inhibitor αMPT, or the potent D2 antagonist eticlopride each liberated D2 receptors from the regulatory influence of D1 receptors *(84)*. The common action of these diverse treatments is that they all reduce the ability of DA to interact with its D2 receptors. By contrast, D1 receptors were liberated from the controlling influence of D2 receptors only by reserpine or 6-OHDA treatments. Other treatments that reduced the ability of DA to interact with D1 (and D2) receptors over 48 h did not have this effect, including combined treatment with αMPT and high doses of the potent D1 antagonist SCH 23390. This suggests that D1 receptor independence is governed either by a non-dopaminergic factor (e.g., a nigrostriatal terminal product sensitive to reserpine) or a non-D1, non-D2 action of DA. These findings also indicate that D1/D2 interdependence can be broken down unidirectionally such that D2 receptors can become independent of D1 receptors without the converse effect. A recent study *(85)* confirmed these findings with respect to D2 independence, except that chronic blockade of D1 receptors with SCH 23390 induces D2 receptors to become independent. Thus, there is not complete agreement on the stimuli that elicit D2 independence.

4.3. Role of D2 Receptor Liberation in D2 Sensitivity

The breakdown in requisite D1/D2 synergism following 6-OHDA or repeated reserpine treatments is accompanied by a profound increase in agonist sensitivity with respect to motor behavior (e.g., ref. *16*), electrophysiological phenomena *(76),* and Fos expression *(47)*. The magnitude of this supersensitivity exceeds the slightly enhanced potency of D2 agonists that follows D2 receptor up-regulation elicited by chronic D2 antagonist treatment *(16,86)*. It is interesting, and perhaps highly important, to note that D2 agonist sensitivity following 6-OHDA or reserpine treatments is very similar to that observed when the D2 agonists is administered to neurologically intact rats concomitantly with a maximally stimulating dose of a D1 agonist (Fig. 5) *(16)*.

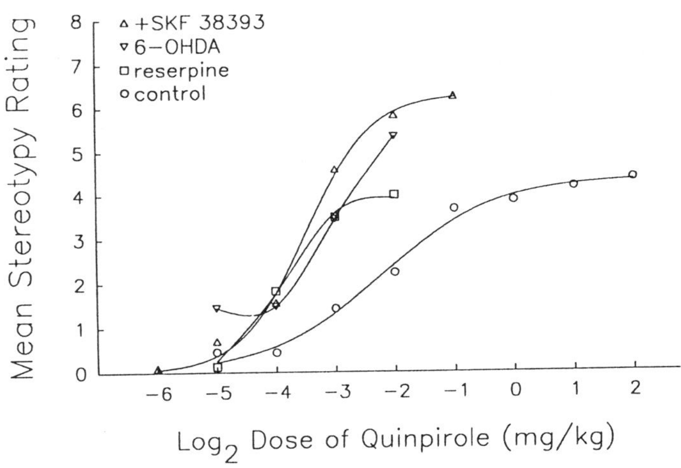

Fig. 5. Behavioral sensitivity to the D2 agonist quinpirole under conditions of normal sensitivity and supersensitivity. Quinpirole dose–response curves are depicted for otherwise untreated animals (controls), rats that received concomitant administration of SKF 38393 (20 mg/kg, ip) (+ SKF 38393), rats given bilateral nigrostriatal 6-hydroxydopamine lesions 4 d previously (6-OHDA), or animals given daily reserpine treatments (1 mg/kg, sc daily) starting 4 d previously (reserpine). Coadministration of SKF 38393 increases the sensitivity of rats to quinpirole, such that their dose–response curves closely match those of supersensitive animals. Note that scaling of abscissa is $\log_2$. Adapted from ref. *16* with permission.

This concordance leads to the suggestion that maximal D2 sensitivity is observed when D2 receptors are liberated from the regulatory influence of D1 receptors, either through a breakdown in synergism or by maximally stimulating D1 receptors.

5. Models of D1/D2 Interactions in Basal Ganglia

The D1/D2 synergism observed in intact rats, and its breakdown after denervation or prolonged DA depletion, needs to be considered in light of our present knowledge concerning the neuronal substrates for D1- and D2-dependent processes in intact animals. The current evidence is reviewed with respect to three possible models by which D1- and D2-mediated events can interact in the control of basal ganglia efferents. First, D1/D2 synergism may occur at the level of the single striatal neuron, in which case the D1 and D2 proteins must be expressed within the same neuron. Second, D1/D2 synergism may occur through local synaptic interactions of distinct D1- and

D2-expressing populations of striatal neurons. Third, this synergy may occur as a property of the circuitry of striatal efferents. Specifically, the direct and indirect striatal efferent pathways exert converging influences on common target structures (SNr and EPN [internal pallidum]), which may be controlled by D1 and D2 receptors, respectively (reviewed in refs. *54,87–89*). These possibilities are not mutually exclusive, and more than one substrate for D1/D2 synergism may underlie the range of behavioral and physiological interactions observed. Evidence pertaining to these three models is now considered in greater detail.

5.1. Striatal Neuron Colocalization

The possibility that D1/D2 synergism occurs at the level of individual striatal neurons is supported by several lines of evidence, including reports of extensive overlap of D1 and D2 mRNAs in striatal medium spiny neurons *(90–92),* extensive localization of D2 receptor immunoreactivity within striatonigral cell bodies *(93,94),* synergistic effects of D1 and D2 agonists with respect to Na^+/K^+ adenosine triphosphatase (ATPase) in dissociated striatal neurons *(95),* and the influence of D1 receptor ligands on D2 receptor binding in striatal tissue homogenates *(96).* If D1 and D2 receptors are colocalized within striatonigral and/or striatopallidal projection neurons, then the behavioral and neuronal manifestations of D1/D2 synergy may reflect the requirement for concurrent occupation of both receptor types in the control of these efferent pathways. Whereas D1/D2 receptor colocalization may provide a basis for the synergistic effects of these receptors observed in vivo, the opposing influences that D1 and D2 receptor subfamilies exert on several second messengers, including cyclic adenosine 5'-monophosphate (cAMP) *(97),* inositol phosphates *(98,99),* and arachidonic acid *(100),* suggest differently.

5.2. Local Synaptic Interactions

Alternatively, the interactions between D1 and D2 receptors may be a property of the synaptic interactions of neurons within striatum. Such interactions may occur through the recurrent collaterals of medium spiny neurons. As reviewed by Groves *(101),* inhibitory collateral interactions between populations of striatal neurons can profoundly influence the physiology of the striatum. The striatum contains at least three types of GABAergic neurons: the striatonigral and striatopallidal medium spiny neurons and the medium aspiny neurons that contain parvalbumin (reviewed in ref. *102*). The axon collaterals of these neurons that ramify within the striatum provide a neurobiological basis for D1/D2 synergism, even if the D1- and D2-expressing striatal neuron populations are largely independent *(54,103,104).* For example, if facilitatory influences of DA are mediated by way of D1 receptors on striatonigral neurons, whereas inhibitory dopaminergic influences are mediated via D2 receptors on striatopallidal cells, then inhibitory collateral interactions between these two classes of medium spiny neurons could account for synergistic D1/D2 effects on the activity of each type of neuron. However, recent evidence against inhibitory collaterals between medium spiny cells comes from electrophysiological studies in which antidromic activation of striatal projection neurons did not result in inhibitory postsynaptic potentials (IPSPs) in neighboring spiny neurons *(105)*. Instead, the medium aspiny GABAergic interneurons may underlie the IPSP component of the excitatory postsynaptic potential (EPSP)–IPSP sequence recorded in striatal spiny cells following stimulation of excitatory afferents *(106)*. The aspiny GABAergic neuron has been found to form synaptic specializations with medium spiny projection cells and itself is contacted by nigrostriatal boutons *(107,108)*.

Another neuron type that could contribute to D1/D2 interactions within striatum is the large aspiny intrinsic acetylcholine (ACh) neuron *(109)*. These neurons contain D2 but not D1 receptor mRNA *(110,111)*, and their processes often extend up to 1 mm from the soma. The ACh-containing neurons both receive contacts from, and distribute axon terminals to, medium spiny neurons *(112,113)*, strategically positioning them to mediate interactions between different populations of medium spiny cells. Their release of ACh in vivo is modulated by $GABA_A$ receptors *(114)* as well as dopamine receptors *(115,116)*.

5.3. Converging Projections of Medium Spiny Cells

The third possibility is that the interaction between D1 and D2 receptors is a property of the converging circuitry of the two basal ganglia efferent pathways. The two distinct striatal GABA-containing efferent pathways terminate primarily in either the GP or in the SNr and EPN. Whereas neurons of the striatopallidal pathway express enkephalin, those of the striatonigral/striato-entopeduncular pathway express SP and DYN *(51,52)*. The influences of these two striatal efferent pathways converge in the SNr and EPN, the major output structures of the basal ganglia *(117)*. The striatonigral/striato-entopeduncular pathway directly inhibits the neurons of the SNr/EPN, whereas the striatopallidal pathway exerts an indirect facilitatory influence on the SNr/EPN, which is achieved through disinhibiting neurons of subthalamic nucleus (reviewed in refs. *87–89*). The synergistic influences of D1 and D2 agonism on motor function can be explained by the common influences exerted by each on the SNr/EPN through these two parallel paths, even if the D1- and D2-expressing striatal neuron populations are largely independent *(54,103,104)*. This model predicts that D1-mediated excitation of the SP/DYN-containing efferents will directly inhibit the SNr/EPN, whereas D2-mediated inhibition of striatopallidal enkephalin-containing efferents will indirectly inhibit the same output structures.

Although a convergence of two independent efferent pathways may explain some features of the D1/D2 synergism reviewed in earlier sections, it cannot account for all of the effects. For example, co-iontophoresis of D1 and D2 agonists in striatum produces a greater inhibition of cell firing rate than does either agonist alone *(13,24)*, indicating a local interaction of D1 and D2 receptor influences on neuronal activity. Similarly, co-injection of D1 and D2 agonists into striatum produces greater stereotyped behavior than does either agonist alone *(26)*. Also, D1 and D2 agonists co-injected into nucleus accumbens septi synergize in their influences on ventral pallidal cell firing *(33)*.

6. Conclusions and Future Directions

1. Many findings now point to the conclusion that the motor influences of dopamine, and the neural processes of the basal ganglia, depend on the synergistic influences of this transmitter on D1 and D2 subfamilies of receptors. D1/D2 synergy applies not only to the effects of exogenously administered DA agonists; the efficacy of both D1 and D2 antagonists in retarding movement indicates that the behavioral effects of endogenously released dopamine also require interaction with both subfamilies of receptor.
2. Requisite D1/D2 synergism is manifest in numerous measures of basal ganglia function, including striatal neuropeptide expression and striatal and pallidal cell firing and immediate-early gene expression. Less well characterized are the interactions of D1 and D2 receptors in controlling cellular functions of nuclei on the output side of the basal ganglia (SNr and EPN).
3. Attempts to model the D1/D2 interactions within the basal ganglia are at an early stage of development. Resolution of issues concerning the extent of colocalization of D1 and D2 receptors within particular classes of striatal neurons is clearly important to this effort, as is an assessment of the functional significance of colocalization to the synergistic interactions observed in vivo. An important consideration in the development of such models is the dramatic breakdown in requisite D1/D2 synergism that occurs after extensive nigrostriatal injury or reserpine treatment. The rapid time course over which requisite D1/D2 synergism is replaced by D1/D2 independence after these treatments constrains the mechanisms hypothesized to underlie the D1/D2 interactions.
4. Extrastriatal dopamine receptor influences need to be considered more fully in developing models for D1/D2 interactions. The influence of dopamine receptors in the SNr, subthalamic nucleus, and pallidum on basal ganglia may bear on the interpretation of experiments in which dopamine agonists or antagonists are administered systemically. Further,

a role for extrastriatal dopamine receptors in the physiological influences of endogenously released dopamine warrants further consideration.

5. Further understanding of D1/D2 interactions within the basal ganglia provides a special opportunity for neuroscientists to integrate knowledge across several levels of analysis (molecular biology of receptors, anatomy, synaptic physiology, and behavior) to provide a coherent account of how dopamine shapes the outputs of these nuclei.

References

1. Kebabian, J. W. and Calne, D. B. (1979) Multiple receptors for dopamine. *Nature* **277,** 93–96.
2. Sibley, D. R. and Monsma, F. J., Jr. (1992) Molecular biology of dopamine receptors. *Trends Pharmacol. Sci.* **13,** 61–69.
3. Pijnenburg, A. J. J., Honig, W. M. M., and Rossum, J. M. V. (1975) Inhibition of d-amphetamine-induced locomotor activity by injection of haloperidol into the nucleus accumbens of the rat. *Psychopharmacologia* **41,** 87–95.
4. Yokel, R. A. and Wise, R. A. (1976) Attenuation of intravenous amphetamine reinforcement by central dopamine blockade in rats. *Psychopharmacology* **48,** 311–318.
5. Pijnenburg, A. J. J., Woodruff, G. N., and Rossum, J. M. V. (1973) Ergometrine induced locomotor activity following intracerebral injection into the nucleus accumbens. *Brain Res.* **59,** 289–302.
6. Creese, I., Burt, D. R., and Snyder, S. H. (1976) Dopamine receptor binding predicts clinical and pharmacological potencies of antischizophrenic drugs. *Science* **192,** 481–483.
7. Seeman, P., Lee, T., Chau-Wong, M., and Wong, K. (1976) Antipsychotic drug doses and neuroleptic/dopamine receptors. *Nature* **261,** 717–719.
8. Setler, P. E., Sarau, H. M., Zirkle, C. L., and Saunders, H. L. (1978) The central effects of a novel dopamine agonist. *Eur. J. Pharmacol.* **50,** 419–430.
9. Iorio, L. C., Barnett, A., Billard, W., and Gold, E. H. (1986) Benzazepines: structure–activity relationships between D_1 receptor blockade and selected pharmacological effects, in *Neurobiology of Central D_1-Dopamine Receptors* (Breese, G. R. and Creese, I., eds.), Plenum, New York, pp. 1–14.
10. Lewis, M. H., Widerlov, E., Knight, D. L, Kilts, C. D., and Mailman, R. B. (1983) N-oxides of phenothiazine antipsychotics: effects on in vivo and in vitro estimates of dopaminergic functions. *J. Pharmacol. Exp. Ther.* **225,** 539–545.
11. Mailman, R. B., Schultz, D. W., Lewis, M. H., Staples, L., Rollema, H., and Dehaven, D. L. (1984) SCH 23390: a selective D_1 dopamine antagonist with potent D_2 behavioral actions. *Eur. J. Pharmacol.* **101,** 159,160.
12. Walters, J. R., Bergstrom, D. A., Carlson, J. H., Chase, T. N., and Braun, A. R. (1987) D_1 dopamine receptor activation required for postsynaptic expression of D_2 agonist effects. *Science* **236,** 719–722.
13. White, F. J. and Wang, R. Y. (1986) Electrophysiological evidence for the existence of both D-1 and D-2 dopamine receptors in the rat nucleus accumbens. *J. Neurosci.* **6,** 274–280.

14. Clark, D. and White, F. J. (1987) Review: D1 dopamine receptor—the search for a function: a critical evaluation of the D1/D2 dopamine receptor classification and its functional implications. *Synapse* **1,** 347–388.
15. Gershanik, O., Heikkila, R. E., and Duvoisin, R. C. (1983) Behavioral correlations of dopamine receptor activation. *Neurology* **33,** 1489–1492.
16. LaHoste, G. J. and Marshall, J. F. (1992) Dopamine supersensitivity and D_1/D_2 synergism are unrelated to changes in striatal receptor density. *Synapse* **12,** 14–26.
17. Arnt, J., Hyttel, J., and Perregaard, J. (1987) Dopamine D-1 receptor agonists combined with the selective D-2 agonist quinpirole facilitate the expression of oral stereotyped behavior in rats. *Eur. J. Pharmacol.* **133,** 137–145.
18. Braun, A. R. and Chase, T. N. (1986) Obligatory D-1/D-2 receptor interaction in the generation of dopamine agonist related behaviors. *Eur. J. Pharmacol.* **131,** 301–306.
19. Jackson, D. M. and Hashizume, M. (1986) Bromocriptine induces marked locomotor stimulation in dopamine-depleted mice when D-1 dopamine receptors are stimulated with SKF38393. *Psychopharmacology* **90,** 147–149.
20. Mashurano, M. and Waddington, J. L. (1986) Stereotyped behaviour in response to the selective D-2 dopamine receptor agonist RU 24213 is enhanced by pretreatment with the selective agonist SK&F 38393. *Neuropharmacology* **25,** 947–949.
21. Meller, E., Bordi, F., and Bohmaker, K. (1988) Enhancement by the D1 dopamine agonist SKF 38393 of specific components of stereotypy elicited by the D2 agonists LY 171555 and RU 24213. *Life Sci.* **42,** 2561–2567.
22. White, F. J., Bednarz, L. M., Wachtel, S. R., Hjorth, S., and Brooderson, R. J. (1988) Is stimulation of both D1 and D2 receptors necessary for the expression of dopamine-mediated behaviors? *Pharmacol. Biochem. Behav.* **30,** 189–193.
23. White, F. J. (1987) D-1 dopamine receptor stimulation enables the inhibition of nucleus accumbens neurons by a D-2 receptor agonist. *Eur. J. Pharmacol.* **135,** 101–105.
24. Wachtel, S. R., Hu, X.-T., Galloway, M. P., and White, F. J. (1989) D1 dopamine receptor stimulation enables the postsynaptic, but not autoreceptor, effects of D2 dopamine agonists in nigrostriatal and mesoaccumbens dopamine systems. *Synapse* **4,** 327–346.
25. Hu, X.-T. and Wang, R. Y. (1988) Comparison of effects of D-1 and D-2 dopamine receptor agonists on neurons in the rat caudate putamen: an electrophysiological study. *J. Neurosci.* **8,** 4340–4348.
26. Bordi, F. and Meller, E. (1989) Enhanced behavioral stereotypies elicited by intrastriatal injection of D_1 and D_2 agonists in intact rats. *Brain Res.* **504,** 276–283.
27. Rosa-Kenig, A., Puotz, J. K., and Rebec, G. V. (1993) The involvement of D_1 and D_2 dopamine receptors in amphetamine-induced changes in striatal unit activity in behaving rats. *Brain Res.* **619,** 347–351.
28. Wang, Z. and Rebec, G. V. (1993) Neuronal and behavioral correlates of intrastriatal infusions of amphetamine in freely moving rats. *Brain Res.* **627,** 79–88.
29. Calabresi, P., Maj, R., Mercuri, N. B., and Bernardi, G. (1992) Coactivation of D_1 and D_2 dopamine receptors is required for long-term synaptic depression in the striatum. *Neurosci. Lett.* **142,** 95–99.
30. Calabresi, P., Pisani, A., Mercuri, N. B., and Bernardi, G. (1994) Post-receptor mechanisms underlying striatal long-term depression. *J. Neurosci.* **14,** 4871–4881.

31. Carlson, J. H., Bergstrom, D. A., Demo, S. D., and Walters, J. R. (1988) Acute reduction of dopamine levels alters responses of basal ganglia neurons to selective D-1 and D-2 dopamine receptor stimulation. *Eur. J. Pharmacol.* **152,** 289–300.
32. Carlson, J. H., Bergstrom, D. A., Demo, S. D., and Walters, J. R. (1990) Nigrostriatal lesion alters neurophysiological responses to selective and nonselective D-1 and D-2 dopamine agonists in rat globus pallidus. *Synapse* **5,** 83–93.
33. Yang, C. R. and Mogenson, G. J. (1989) Ventral pallidal neuronal responses to dopamine receptor stimulation in the nucleus accumbens. *Brain Res.* **489,** 237–246.
34. Waszczak, B. L., Lee, E. K., Ferraro, T., Hare, T. A., and Walters, J. R. (1984) Single unit responses of substantia nigra pars reticulata neurons to apomorphine: effects of striatal lesions and anesthesia. *Brain Res.* **306,** 307–318.
35. Walters, J. R., Kelland, M. D., Huang, K.-X., and Bergstrom, D. A. (1992) Effects of dopamine agonists on neuronal activity in the basal ganglia of normal and dopamine-denervated rats, in *Progress in Parkinson's Disease—2* (Hefti, F. and Weiner, W. J., eds.), Futura Publishing, Mount Kisco, NY, pp. 229–257.
36. Kadekaro, M., Crane, A. M., and Sokoloff, L. (1985) Differential effects of electrical stimulation of sciatic nerve on metabolic activity in spinal cord and dorsal root ganglion in the rat. *Proc. Natl. Acad. Sci. USA* **82,** 6010–6013.
37. Brown, L. L. and Wolfson, L. I. (1978) Apomorphine increases glucose utilization in the substantia nigra, subthalamic nucleus and corpus striatum of rat. *Brain Res.* **140,** 188–193.
38. Trugman, J. M. and James, C. L. (1993) D1 dopamine agonist and antagonist effects on regional cerebral glucose utilization in rats with intact dopaminergic innervation. *Brain Res.* **607,** 270–274.
39. Brown, L. L. and Wolfson, L. I. (1983) A dopamine-sensitive striatal efferent system mapped with [^{14}C]deoxyglucose in the rat. *Brain Res.* **261,** 213–299.
40. Patel, S., Slater, P., and Crossman, A. R. (1985) A lesioning and 2-deoxyglucose study of the hyperactivity produced by an intra-accumbens dopamine agonist. *Naunyn-Schmiedeberg's Arch. Pharmacol.* **331,** 334–340.
41. Graybiel, A. M., Moratalla, R., and Robertson, H. A. (1990) Amphetamine and cocaine induce drug-specific activation of the c-*fos* gene in striosome-matrix compartments and limbic subdivisions of the striatum. *Proc. Natl. Acad. Sci. USA* **87,** 6912–6916.
42. Young, S. T., Porrino, L. J., and Iadarola, M. J. (1991) Cocaine induces striatal c-Fos-immunoreactive proteins via dopaminergic D_1 receptors. *Proc. Natl. Acad. Sci. USA* **88,** 1291–1295.
43. Robertson, G. S., Vincent, S. R., and Fibiger, H. C. (1992) D_1 and D_2 dopamine receptors differentially regulate c-*fos* expression in striatonigral and striatopallidal neurons. *Neuroscience* **49,** 285–296.
44. Cenci, M. A., Campbell, K., Wictorin, K., and Bjorklund, A. (1992) Striatal c-*fos* induction by cocaine or apomorphine occurs preferentially in output neurons projecting to the substantia nigra. *Eur. J. Neurosci.* **41,** 376–380.
45. Johansson, B., Lindstrom, K., and Fredholm, B. B. (1994) Differences in the regional and cellular localization of c-*fos* messenger RNA induced by amphetamine, cocaine and caffeine in the rat. *Neuroscience* **59,** 837–849.
46. Ruskin, D. N. and Marshall, J. F. (1994) Amphetamine-and cocaine-induced Fos in the rat striatum depends on D_2 dopamine receptor activation. *Synapse* **18,** 233–240.

47. LaHoste, G. J., Yu, J., and Marshall, J. F. (1993) Striatal Fos expression is indicative of dopamine D1/D2 synergism and receptor supersensitivity. *Proc Natl. Acad. Sci. USA* **90,** 7451–7455.
48. Marshall, J. F., Cole, B. N., and LaHoste, G. J. (1993) Dopamine D_2 receptor control of pallidal fos expression: comparisons between intact and 6-hydroxydopamine-treated hemispheres. *Brain Res.* **632,** 308–313.
49. Zilles, K. (1985) *The Cortex of the Rat,* Springer-Verlag, New York.
50. LaHoste, G. J., Ruskin, D. N., and Marshall, J. F. (1996) Cerebrocortical Fos expression following dopaminergic stimulation: D1/D2 synergism and its breakdown. *Brain Res.,* in press.
51. Gerfen, C. R. and Young, W. S., III (1988) Distribution of striatonigral and striatopallidal peptidergic neurons in both patch and matrix compartments: an in situ hybridization histochemistry and fluorescent retrograde tracing study. *Brain Res.* **460,** 161–167.
52. Reiner, A. and Anderson, K. D. (1990) The patterns of neurotransmitter and neuropeptide co-occurrence among striatal projections neurons: conclusions based on recent findings. *Brain Res. Rev.* **15,** 251–265.
53. Engber, T. M., Boldry, R. C., Kuo, S., and Chase, T. N. (1992) Dopaminergic modulation of striatal neuropeptides: differential effects of D_1 and D_2 receptor stimulation on somatostatin, neuropeptide Y, neurotensin, dynorphin and enkephalin. *Brain Res.* **581,** 261–268.
54. Gerfen, C. R., Engber, T. M., Mahan, L. C., Susel, Z., Chase, T. N., Monsma, F. J., Jr., and Sibley, D. R. (1990) D_1 and D_2 dopamine receptor-regulated gene expression of striatonigral and striatopallidal neurons. *Science* **250,** 1429–1432.
55. Jiang, H.-K., McGinty, J. F., and Hong, J. S. (1990) Differential modulation of striatonigral dynorphin and enkephalin by dopamine receptor subtypes. *Brain Res.* **507,** 57–64.
56. Gerfen, C. R., McGinty, J. F., and Young, W. S., III (1991) Dopamine differentially regulates dynorphin, substance P, and enkephalin expression in striatal neurons: in situ hybridization histochemical analysis. *J. Neurosci.* **11,** 1016–1031.
57. Steiner, H. and Gerfen, C. R. (1993) Cocaine-induced c-*fos* messenger RNA is inversely related to dynorphin expression in striatum. *J. Neurosci.* **13,** 5066–5081.
58. Smith, A. J. W. and McGinty, J. F. (1994) Acute amphetamine or methamphetamine alters opioid peptide mRNA expression in rat striatum. *Mol. Brain Res.* **21,** 359–362.
59. Haverstick, D. M., Rubenstein, A., and Bannon, M. J. (1989) Striatal tachykinin gene expression regulated by interaction of D-1 and D-2 dopamine receptors. *J. Pharmacol. Exp. Ther.* **248,** 858–862.
60. Schoenfeld, R. and Uretsky, N. (1972) Altered response to apomorphine in 6-hydroxydopamine-treated rats. *Eur. J. Pharmacol.* **19,** 115–118.
61. Ungerstedt, U. (1971) Postsynaptic supersensitivity after 6-hydroxydopamine induced degeneration of the nigro-striatal dopamine system. *Acta Physiol. Scand.* **367,** 68–91.
62. Mandel, R. J. and Randall, P. K. (1985) Quantification of lesion-induced dopaminergic supersensitivity using the rotational model in the mouse. *Brain Res.* **330,** 358–363.
63. Marshall, J. F. and Ungerstedt, U. (1977) Supersensitivity to apomorphine following destruction of the ascending dopamine neurons: quantification using the rotational model. *Eur. J. Pharmacol.* **41,** 361–367.

64. Arnt, J. (1985) Behavioural stimulation is induced by separate dopamine D-1 and D-2 receptor sites in reserpine-pretreated but not normal rats. *Eur. J. Pharmacol.* **113,** 79–88.
65. Creese, I., Burt, D. R., and Snyder, S. H. (1977) Dopamine receptor binding enhancement accompanies lesion-induced behavioral supersensitivity. *Science* **197,** 596–598.
66. Neve, K. A., Neve, R. L., Fidel, S., Janowsky, A., and Higgins, G. A. (1991) Increased abundance of alternatively spliced forms of D2 dopamine receptor mRNA after denervation. *Proc. Natl. Acad. Sci. USA* **88,** 2802–2806.
67. Staunton, D. A., Wolfe, B. B., Groves, P. M., and Molinoff, P. B. (1981) Dopamine receptor changes following destruction of the nigrostriatal pathway: lack of a relationship to rotational behavior. *Brain Res.* **211,** 315–327.
68. Brene, S., Lindefors, N., Herrera-Marschitz, M., and Persson, H. (1990) Expression of dopamine D2 receptor and choline acetyltransferase mRNA in the dopamine deafferented rat caudate-putamen. *Exp. Brain Res.* **83,** 96–104.
69. LaHoste, G. J. and Marshall, J. F. (1996) Dopamine receptor interactions in the brain, in *CNS Neurotransmitters and Neuromodulators: Dopamine* (Stone, T. W., ed.), CRC Press, Boca Raton, FL, in press.
70. Fornaretto, M. G., Caccia, C., Caron, M. G., and Fariello, R. G. (1993) Dopamine receptor status after unilateral nigral 6-OHDA lesions. Autoradiographic and in situ hybridization study. *Mol. Chem. Neuropathol.* **19,** 147–162.
71. Marshall, J. F., Navarette, R., and Joyce, J. N. (1989) Decreased striatal D1 binding density following mesotelencephalic 6-hydroxydopamine injections: an autoradiographic analysis. *Brain Res.* **493,** 247–257.
72. Arnt, J. (1985) Hyperactivity induced by stimulation of separate dopamine D-1 and D-2 receptors in rats with bilateral 6-OHDA lesions. *Life Sci.* **37,** 717–723.
73. Breese, G. R. and Mueller, R. A. (1985) SCH-23390 antagonism of a D-2 dopamine agonist depends upon catecholaminergic neurons. *Eur. J. Pharmacol.* **113,** 109–114.
74. Arnt, J. and Hyttel, J. (1984) Differential inhibition by dopamine D-1 and D-2 antagonists of circling behavior induced by dopamine agonists in rats with unilateral 6-hydroxydopamine lesions. *Eur. J. Pharmacol.* **102,** 349–354.
75. Arnt, J. and Hyttel, J. (1985) Differential involvement of dopamine D-1 and D-2 receptors in the circling behavior induced by apomorphine, SK&F 38393, pergolide and LY 171555. *Psychopharmacology* **85,** 346–352.
76. Hu, X.-T., Wachtel, S. R., Galloway, M. P., and White, F. J. (1990) Lesions of the nigrostriatal dopamine projection increase the inhibitory effects of D_1 and D_2 dopamine agonists on caudate-putamen neurons and relieve D_2 receptors from the necessity of D_1 receptor stimulation. *J. Neurosci.* **10,** 2318–2329.
77. Weick, B. G. and Walters, J. R. (1987) Effects of D1 and D2 dopamine receptor stimulation on the activity of substantia nigra pars reticulata neurons in 6-hydroxydopamine lesioned rats: D_1/D_2 coactivation induces potentiated responses. *Brain Res.* **405,** 234–246.
78. Engber, T. M., Anderson, J. J., Boldry, R. C., Papa, S. M., Kuo, S., and Chase, T. N. (1994) Excitatory amino acid receptor antagonists modify regional cerebral metabolic responses to levodopa in 6-hydroxydopamine-lesioned rats. *Neuroscience* **59,** 389–399.

79. Trugman, J. M. and Wooten, G. F. (1987) Selective D_1 and D_2 dopamine agonists differentially alter basal ganglia glucose utilization in rats with unilateral 6-hydroxydopamine substantia nigra lesions. *J. Neurosci.* **7,** 2927–2935.
80. Trugman, J. M. and James, C. L. (1992) Rapid development of dopaminergic supersensitivity in reserpine-treated rats demonstrated with ^{14}C-2-deoxyglucose autoradiography. *J. Neurosci.* **12,** 2875–2879.
81. Robertson, H. A., Peterson, M. R., Murphy, K., and Robertson, G. S. (1989) D_1-dopamine receptor agonists selectively activate striatal c-*fos* independent of rotational behavior. *Brain Res.* **503,** 346–349.
82. Paul, M. L., Graybiel, A. M., David, J.-C., and Robertson, H. A. (1992) D1-like and D2-like dopamine receptors synergistically activate rotation and c-*fos* expression in the dopamine-depleted striatum in a rat model of Parkinson's disease. *J. Neurosci.* **12,** 3729–3742.
83. LaHoste, G. J. and Marshall, J. F. (1994) Rapid development of D_1 and D_2 dopamine receptor supersensitivity as indicated by striatal and pallidal Fos expression. *Neurosci. Lett.* **179,** 153–156.

83a. Ruskin, D. R. and Marshall, J. F. (1995) D_1 dopamine receptors influence Fos immunoreactivity in the globus pallidus and subthalamic nucleus of intact and nigrostriatal-lesioned rats. *Brain Res.* **703,** 156–164.

84. LaHoste, G. J. and Marshall, J. F. (1993) The role of dopamine in the maintenance and breakdown of D_1/D_2 synergism. *Brain Res.* **611,** 108–116.
85. Hu, X.-T. and White, F. J. (1994) Loss of D_1/D_2 dopamine receptor synergisms following repeated administration of D_1 or D_2 receptor selective antagonists: electrophysiological and behavioral studies. *Synapse* **17,** 43–61.
86. LaHoste, G. J. and Marshall, J. F. (1991) Chronic eticlopride and dopamine denervation induce equal nonadditive increases in striatal D_2 receptor density: autoradiographic evidence against the dual mechanism hypothesis. *Neuroscience* **41,** 473–781.
87. Albin, R. L., Young, A. B., and Penney, J. B. (1989) The functional anatomy of basal ganglia disorders. *Trends Neurosci.* **12,** 366–375.
88. Alexander, G. E. and Crutcher, M. D. (1990) Functional architecture of basal ganglia circuits: neural substrates of parallel processing. *Trends Neurosci.* **13,** 266–270.
89. Scheel-Kruger, J. (1986) Dopamine–GABA interactions: evidence that GABA transmits, modulates and mediates dopaminergic functions in the basal ganglia and the limbic system. *Acta Neurol. Scand. Suppl.* **107,** 1–54.
90. Lester, J., Fink, S., Aronin, N., and DiFiglia, M. (1993) Colocalization of D_1 and D_2 dopamine receptor mRNAs in striatal neurons. *Brain Res.* **621,** 106–110.
91. Meador-Woodruff, J. H., Mansour, A., Healy, D. J., Kuehn, R., Zhou, Q. Y., Bunzow, J. R., Akil, H., Civelli, O., and Watson, S. J. (1991) Comparisons of the distributions of D_1 and D_2 dopamine receptor mRNAs in rat brain. *Neuropsychopharmacology* **5,** 231–242.
92. Surmeier, D. J., Eberwine, J., Wilson, C. J., Cao, Y., Stefani, A., and Kitai, S. T. (1992) Dopamine receptor subtypes colocalize in rat striatonigral neurons. *Proc. Natl. Acad. Sci. USA* **89,** 10,178–10,182.
93. Ariano, M. A., Stromski, C. J., Smyk-Randall, E. M., and Sibley, D. R. (1992) D2 dopamine receptor localization on striatonigral neurons. *Neurosci. Lett.* **144,** 215–220.

94. Larson, E. R. and Ariano, M. A. (1994) Dopamine receptor binding on identified striatonigral neurons. *Neurosci. Lett.* **172,** 101–106.
95. Bertorello, A. M., Hopfield, J. F., Aperia, A., and Greengard, P. (1990) Inhibition by dopamine of (Na^+ + K^+)ATPase activity in neostriatal neurons through D1 and D2 dopamine receptor synergism. *Nature* **347,** 386–388.
96. Seeman, P., Sunahara, R. K., and Niznik, H. B. (1994) Receptor–receptor link in membranes revealed by ligand competition: example for dopamine D1 and D2 receptors. *Synapse* **17,** 62–64.
97. Onali, P., Olianas, M. C., and Gessa, G. L. (1984) Selective blockade of dopamine D-1 receptors by SCH 23390 discloses striatal dopamine D-2 receptors mediating the inhibition of adenylate cyclase. *Eur. J. Pharmacol.* **99,** 127,128.
98. Pizzi, M., Prada, M. D., Valerio, A., Memo, M., Spano, P. F., and Haefely, W. E. (1988) Dopamine D2 receptor stimulation inhibits inositol phosphate generating system in rat striatal slices. *Brain Res.* **456,** 235–240.
99. Undie, A. S. and Friedman, E. (1990) Stimulation of a dopamine D1 receptor enhances inositol phosphates formation in rat brain. *J. Pharmacol. Exp. Ther.* **253,** 987–992.
100. Schinelli, S., Paolillo, M., and Corona, G. L. (1994) Opposing actions of D_1- and D_2-dopamine receptors on arachidonic acid release and cyclic AMP production in striatal neurons. *J. Neurochem.* **62,** 944–949.
101. Groves, P. M. (1983) A theory of the functional organization of the neostriatum and the neostriatal control of voluntary movement. *Brain Res. Rev.* **5,** 109–132.
102. Parent, A. and Hazrati, L.-N. (1995) Functional anatomy of the basal ganglia. I. The cortico-basal ganglia-thalamo-cortical loop. *Brain Res. Rev.* **20,** 91–127.
103. Le Moine, C., Normand, E., and Bloch, B. (1991) Phenotypical characterization of the rat striatal neurons expressing the D_1 dopamine receptor gene. *Proc. Natl. Acad. Sci. USA* **88,** 4205–4209.
104. Le Moine, C., Normand, E., Guitteny, A. F., Fouque, B., Teoule, R., and Bloch, B. (1990) Dopamine receptor gene expression by enkephalin neurons in rat forebrain. *Proc. Natl. Acad. Sci. USA* **87,** 230–234.
105. Jaeger, D., Kita, H., and Wilson, C. J. (1994) Surround inhibition among projection neurons is weak or nonexistent in the rat neostriatum. *J. Neurophysiol.* **72,** 2555–2558.
106. Wilson, C. J., Kita, H., and Kawaguchi, Y. (1989) GABAergic interneurons, rather than spiny cell axon collaterals, are responsible for the IPSP responses to afferent stimulation in neostriatal spiny neurons. *Soc. Neurosci. Abst.* **15,** 901.
107. Bennett, B. D. and Bolam, J. P. (1994) Synaptic input and output of parvalbumin-immunoreactive neurons in the neostriatum of the rat. *Neuroscience* **62,** 707–719.
108. Kita, H. (1993) GABAergic circuits of the striatum. *Prog. Brain Res.* **99,** 51–72.
109. Di Chiara, G. and Morelli, M. (1993) Dopamine–acetylcholine–glutamate interactions in the striatum—a working hypothesis, in *Advances in Neurology* (Narabayashi, H., Nagatsu, T., Yanagisawa, N., and Mizuno, Y., eds.), Raven, New York, pp. 102–106.
110. Le Moine, C., Tison, F., and Bloch, B. (1990) D2 dopamine receptor gene expression by cholinergic neurons in the rat striatum. *Neurosci. Lett.* **117,** 248–252.
111. Weiner, D. M., Levey, A. I., Sunahara, R. K., Niznik, H. B., O'Dowd, B. F., Seeman, P., and Brann, M. R. (1991) D_1 and D_2 dopamine receptor mRNA in rat brain. *Proc. Natl. Acad. Sci. USA* **88,** 1859–1863.

112. Bolam, J. P. and Izzo, P. N. (1988) The postsynaptic targets of substance P-immunoreactive terminals in the rat neostriatum with particular reference to identified spiny striatonigral neurons. *Exp. Brain Res.* **70,** 361–377.
113. Izzo, P. N. and Bolam, J. P. (1988) Cholinergic synaptic input to different parts of spiny striatonigral neurons in the rat. *J. Comp. Neurol.* **269,** 219–234.
114. De Boer, P. and Westerink, B. H. (1994) GABAergic modulation of striatal cholinergic interneurons: an in vivo microdialysis study. *J. Neurochem.* **62,** 70–75.
115. Damsma, G., Robertson, G. S., Tham, C. S., and Fibiger, H. C. (1991) Dopaminergic regulation of striatal acetylcholine release: importance of D1 and N-methyl-D-aspartate receptors. *J. Pharmacol. Exp. Ther.* **259,** 1064–1072.
116. De Boer, P., Abercrombie, E. D., Heeringa, M., and Westerink, B. H. C. (1993) Differential effect of systemic administration of bromocriptine and L-DOPA on the release of acetylcholine from striatum of intact and 6-OHDA-treated rats. *Brain Res.* **608,** 198–203.
117. Nauta, W. J. H. and Domesick, V. B. (1984) Afferent and efferent relationships of the basal ganglia, in *Functions of the Basal Ganglia* (Evered, D. and O'Connor, M., eds.), Pitman, Newark, NJ, pp. 3–29.

Part III

Dopamine Receptors and Function

CHAPTER 8

Dopamine Autoreceptor Pharmacology and Function

Recent Insights

John D. Elsworth and Robert H. Roth

1. Historical Perspective

The potential importance of local feedback mechanisms in regulating dopaminergic activity was first noted by Farnebo and Hamberger *(1)*, who observed that dopamine (DA) agonists were effective in attenuating the stimulus-evoked release of [^{3}H]DA from striatal slices by an interaction with presynaptic receptors. When it became appreciated that catecholamine neurons, in addition to possessing receptors on their nerve terminals, appear to have receptors distributed over other parts of the neuron, such as the soma, dendrites, and preterminal axons, the term presynaptic receptors became inappropriate as a description for all these receptors. Carlsson *(2)* suggested that autoreceptor was a more appropriate term to describe them, as the sensitivity of these catecholamine receptors to the neurons' own transmitter seemed more significant than their location at the synapse.

The term autoreceptor achieved rapid acceptance and stimulated pharmacological research in the succeeding years, resulting in the detection and description of autoreceptors on neurons in many chemically defined neuronal systems. The presence of autoreceptors on some neurons may turn out to be of pharmacological interest only, for they may rarely, if ever, encounter effective concentrations of appropriate endogenous agonists in vivo. Other autoreceptors, however, in addition to their pharmacological responsiveness, may play a very important physiological role in neuronal homeostasis. This certainly appears to be the case for DA autoreceptors. It has become clear that

The Dopamine Receptors Eds.: K. A. Neve and R. L. Neve
Humana Press Inc., Totowa, NJ

DA autoreceptors provide an important homeostatic mechanism by which the DA neuron can regulate cellular functions such as neurotransmitter release, synthesis, and impulse flow.

Since there have been comprehensive reviews on DA autoreceptors in the past *(3–5),* this chapter focuses primarily on advances that have been achieved in the past 5 years or so. However, in order to provide some perspective for this chapter, a brief background is given in the remainder of this section.

As noted, functionally relevant DA autoreceptors are conceptualized as being located on soma, dendrites, and terminals of most DA neurons. Three types of DA autoreceptor have been classified according to the function they modulate (i.e., impulse flow, synthesis rate, and release rate). Stimulation of DA autoreceptors in the somatodendritic region slows the firing rate of DA neurons, whereas stimulation of autoreceptors located on DA nerve terminals results in an inhibition of DA synthesis and release. There is also evidence for the existence of release-modulating autoreceptors on dendrites of DA neurons in the substantia nigra *(6),* although it is not clear how or whether these interact with impulse-modulating autoreceptors. Somatodendritic autoreceptors may also regulate DA release and synthesis in the terminal regions by changing impulse flow, and there is recent direct evidence to support the notion that somatodendritic and nerve terminal autoreceptors work in concert to exert feedback regulatory effects on dopaminergic transmission *(7,8)*. In general, all DA autoreceptors can be classified as D2-like* DA receptors. Whereas all DA neurons possess release-modulating autoreceptors, the distribution of impulse-modulating and synthesis-modulating autoreceptors is restricted. DA neurons that project to the prefrontal and cingulate cortices, as well as those projecting to some nuclei of the amygdala, appear to have either a greatly diminished number of these receptors or to lack them entirely. The absence or diminished numbers of these important modulatory receptors on meso-prefrontal and mesocingulate DA neurons, and perhaps also mesoamygdaloid neurons, appears in part to be responsible for some of the distinctive characteristics of these neurons when compared to other mesotelencephalic DA neurons that possess autoreceptors (nigrostriatal, mesolimbic, and meso-piriform systems). For example, mesoprefrontal and mesocingulate DA neurons exhibit a faster firing rate, more bursting, a more rapid turnover of transmitter, and a diminished biochemical responsiveness to DA agonists and antagonists, than the nigrostriatal, mesolimbic, and mesopiriform DA neurons.

*As detailed in the Preface, the terms D1 or D1-like and D2 or D2-like are used to refer to the subfamilies of DA receptors, or used when the genetic subtype (D_1 or D_5; D_2, D_3, or D_4) is uncertain.

Following chronic administration of antipsychotic drugs, tolerance develops to the metabolite-elevating effects of these agents in midbrain DA systems possessing impulse-modulating and nerve terminal synthesis-modulating autoreceptors, but not in systems lacking such autoreceptors. Transmitter synthesis is also more readily influenced by altered availability of DA precursor, tyrosine, in midbrain DA neurons lacking impulse-modulating and synthesis-modulating autoreceptors. There are data to suggest distinct differences in the autoreceptor regulation of hypothalamic DA neurons. Thus, in contrast to the incertohypothalamic DA neurons, tuberoinfundibular DA neurons appear to lack synthesis-modulating autoreceptors. Only a subset of the tuberohypophysial DA neurons appear to have synthesis-modulating autoreceptors.

2. Methodology

Several different techniques have been employed to detect and study DA autoreceptors. These include biochemical (e.g., γ-butyrolactone [GBL] model, evoked DA release), electrophysiological (e.g., DA cell firing), and behavioral (e.g., locomotor activity) methods, which have been reviewed previously *(4,5)*. There have been several exciting methodological advances in the past few years that have furthered the understanding of DA autoreceptor function.

One novel procedure is the use of fast cyclic voltammetry. This approach has been used to measure drug-induced changes in stimulated DA release from brain slices *(9)*. The main advantage of the technique is that it permits real-time resolution of released endogenous DA. Thus, it has been found that DA overflow measured in striatal slices in response to the second pulse of a 10 Hz train is only 58% of that released by the first pulse, indicating that autoinhibition occurs within 100 ms *(10)*. McElvain and Schenk *(11)* used a rotating disk electrode voltammetric technique to examine the dynamics of potassium-stimulated release of endogenous DA from striatal suspensions. They found that low concentrations of haloperidol, used to block DA autoreceptors, result in an increase in the amplitude of DA release, but not the duration or initial rate of the release.

Sesack et al. *(12)* used light and electron microscopic techniques to visualize the occurrence of D_2 receptor-like immunoreactivity and tyrosine-hydroxylase (TH) immunogold labeling in the mesostriatal DA system. They found that D_2 receptor-like protein appears to be located strategically to subserve autoreceptor functions on dendrites of DA neurons in the midbrain, and on presynaptic dopaminergic axon terminals in the striatum.

Another approach for evaluating autoreceptor function has been to reduce the number of D_2 receptors expressed in midbrain DA neurons by intracerebral infusion of an antisense oligodeoxynucleotide directed against the D_2 receptor mRNA. For example, evidence of a role for nigrostriatal DA autoreceptors in the motor actions of cocaine was obtained by this technique *(13)*.

As mentioned in more detail in Sections 5. and 6., the *in situ* hybridization technique has allowed the identification of regions in which the mRNA for DA receptors is present. This approach has indicated, for example, that DA neurons in the primate ventral tegmental area express little or no mRNA for D_2 or D_3 receptors, which tends to suggest DA D2-like autoreceptors are not expressed in any abundance by these neurons.

Investigations aimed at determining the role of the D_3 DA receptor in autoregulatory events have often been hampered by the lack of specific D_3 agonists and antagonists. However, recent data indicate that the relative D_2/D_3 selectivity of DA agonists may be markedly altered in vitro by conducting the assay in the presence or absence of guanosine-5'-triphosphate (GTP) *(14–16)*. Thus, in the presence of GTP, there is conversion of the D_2 receptor to its low-affinity state, and a rightward shift in the competition curve for the displacement of a D_2 antagonist by a D_2 agonist. However, there is only a small GTP-induced shift for the D_3 receptor. In the presence of GTP, 7-OH-DPAT has a greater than 1000-fold selectivity for the D_3, compared to the D_2, receptor.

Waters et al. *(17–19)* recently suggested that the relative change in dialysate DA and dihydroxyphenylacetic acid (DOPAC) elicited by systemic administration of an antagonist may be a neurochemical fingerprint that distinguishes the population of DA receptors that is recruited by the drug. This was based on the different DA/DOPAC ratios obtained with the autoreceptor preferring antagonists, (+)-AJ76 and (+)-UH232, compared with haloperidol or raclopride. If this holds true for other drugs, this may be a useful refinement of the in vivo microdialysis technique to assess drugs with an autoreceptor-preferring profile.

In another dialysis study, Robertson et al. *(20)* measured the striatal dialysate concentrations of both DA and acetylcholine following administration of B-HT 920 and SND 919. They found that these putative autoreceptor agonists reduce interstitial concentrations of DA only, in contrast to quinpirole which decreases the level of both transmitters. This may suggest that the relative effect of drugs on extracellular levels of DA and acetylcholine in the striatum may be reflective of autoreceptor selectivity.

3. Ontogeny

There have now been several studies on the development of DA autoreceptors in rat brain. Besides documenting the onset of autoregulation of DA neurons, these investigations have revealed some interesting shifts in the receptor subtypes that subserve DA autoreceptor function in several brain regions during development.

Two recent behavioral studies were designed to pinpoint the emergence of DA autoreceptor function during development. Lin and Walters *(21)* monitored the decrease in spontaneous stereotypy elicited by two putative DA autoreceptor-selective agonists (SND 919, PD 128483), whereas Van Hartesveldt et al. *(22)* employed the appearance of quinpirole-induced locomotor suppression as an index of DA autoreceptor function. Both studies suggested that D2 autoreceptor-mediated behaviors do not appear until after postnatal d 20 (P20). These data are consistent with the early study of Shalaby and Spear *(23)* who found that low doses of apomorphine fail to suppress behavior in rats younger than P28.

Tison et al. *(24)* determined that the mRNA for the D_2 receptor is present in the rat mesencephalon at embryonic d 14 (E14), suggesting that DA autoreceptor function mediated by D_2 receptors could commence soon after E14. In agreement with this, De Vries et al. *(25)* demonstrated that the functional striatal release-modulating DA D2-like autoreceptors in the rat are present at E17. This was based on D2 agonist-induced inhibition of electrically evoked [^{3}H]DA release from superfused striatal slices. However, whereas D1-stimulated adenylate cyclase activity is also present in striatum at E17, D2-mediated inhibition of D1-stimulated adenylate cyclase activity is not observed until P17. Thus, these data also indicate that there is a marked difference between the development of striatal release-modulating DA D2 autoreceptors and postsynaptic striatal D2 receptors.

Andersen and Gazzara *(26,27)* assessed the ontogeny of DA autoreceptors using in vivo microdialysis. These investigators measured decreases in both the basal and potassium-stimulated extracellular striatal level of DA in response to systemically administered apomorphine. In further studies these investigators *(28)* tested the ontogeny of the decrease in the potassium-stimulated extracellular striatal DA level elicited by direct application of quinpirole, and the reversal of this effect by (–)-sulpiride. These in vivo approaches indicate the presence of functional release-modulating DA autoreceptors at the earliest time point examined (P5).

Teicher et al. *(29)* examined the ontogeny of synthesis-modulating DA autoreceptors in striatum and prefrontal cortex; this latter region appears to be devoid of this type of autoreceptor in the adult rat. As in adults, quinpirole (a D2 receptor agonist) is able to attenuate DA synthesis in the striatum, but not prefrontal cortex, of P15 and P22 rats, as assessed by DOPA accumulation in the GBL model. This indicates that functional synthesis-modulating DA D2-like autoreceptors are in place in the striatum by P15. However, in developing rats (P15 and P22), but not adults, DA synthesis is inhibited by D1 agonists in both the striatum and prefrontal cortex. This leads to the interesting conclusion that synthesis-modulating D1-like DA autoreceptors emerge transiently in the developing brain. In the adult, autoregulation of DA synthesis is assumed by D2-like receptors in striatum, and in prefrontal cortex this ability appears to be lost.

A more recent study by Andersen and Teicher *(30)* explored regulation of DA synthesis during development by the D_3 agonist, (+)-7-OH-DPAT using the rat GBL model. Between P10 and P30, (+)-7-OH-DPAT inhibits the GBL elevation in DA synthesis in striatum, nucleus accumbens, and prefrontal cortex. However, in older rats (P40) GBL does not increase DA synthesis in prefrontal cortex. These results suggest that D_3 synthesis-modulating autoreceptors exist in rat brain; although the validity of this depends on the selectivity of (+)-7-OH-DPAT for the D_3 receptor. It is not clear whether the putative D_3 synthesis-modulating autoreceptors coexist with D_2 synthesis-modulating autoreceptors, or whether there is a transition from D_3 to D_2 autoreceptor. This study also confirms the transitory expression of synthesis-modulating autoreceptor function in prefrontal cortex.

An electrophysiological study on somatodendritic DA autoreceptors *(31)* revealed that, although rat nigrostriatal DA neurons at P14 are less sensitive to iv apomorphine or quinpirole than in the adult, this difference is not apparent when the drugs are given iontophoretically.

Taken together, these studies on the development of release-modulating DA autoreceptors in the nigrostriatal system suggest these autoreceptors are in place and can function by E17. Synthesis-modulating autoreceptors have been detected at P10. Synthesis-modulating D1 receptors may transiently emerge in the prefrontal cortex during development in the rat; if a similar postnatal alteration in regulation of DA neurons innervating prefrontal occurs in humans, it is conceivable that an abnormality of this transition could underlie some types of psychosis. Behavioral and electrophysiological responses to systemically administered drugs with affinity for DA autoreceptors appear not to resemble those in the adult until after the second or third postnatal week, possibly because expression of these effects requires the recruitment of other systems that lag behind the DA autoreceptor in terms of ontogeny. It is conceivable that the pattern of maturity of DA autoreceptors is involved in the inability to sensitize rats to psychostimulants unless the drug administration is commenced after P21 *(32),* bearing in mind the suggested role of DA autoreceptors in psychostimulant sensitization (*see* Section 8.2.).

4. Peptide Cotransmitters and Autoreceptor Function

Although once it was believed that a particular neuron could only possess a single transmitter, it now appears that it is common for a neuron to contain cotransmitters. In the case of some dopaminergic neurons, there is particularly compelling evidence for the neuronal coexistence of DA with the peptides neurotensin (NT) and cholecystokinin (CCK). In such neurons, in addition to DA-sensitive autoreceptors, there appear to be peptide-sensitive autoreceptors that can affect the release of DA. The presence of transmitter peptides in DA neurons, and the location of DA and peptide receptors both pre- and postsynaptic to the DA neurons, provides a framework for many possible interactions and modes of regulation. Recent investigations have furnished some interesting insights into the dynamics between DA, its cotransmitters, and autoreceptors.

4.1. Neurotensin

There are strong biochemical, behavioral, physiological, and anatomical data to indicate a close link between DA neurons and nondopaminergic NT neurons in brain *(33–36)*. In addition, NT is colocalized with DA in the majority of ventral tegmental area DA neurons that project to the prefrontal cortex *(37–39)*. Thus, virtually all TH-positive neurons in the rat prefrontal cortex also contain NT-like immunoreactivity (NT-LI), and all NT-LI neurons appear to be TH-positive. Furthermore, NT and DA are partially colocalized in the same intraneuronal storage compartment in the rat prefrontal cortex *(40,41)*. In the prefrontal cortex, there are few, if any, intrinsic DA- or NT-containing neurons, and this situation has been exploited to study the presynaptic release and interactions of DA and NT in the absence of postsynaptic cells containing DA or NT, which would confound data interpretation *(42–44)*. These studies have shown that the release of both DA and NT, as assessed by extracellular levels measured by in vivo microdialysis, is increased by stimulation of the median forebrain bundle, which activates mesocortical axons. However, at a frequency that is similar to the basal firing rate of some mesocortical DA neurons (2.5 Hz) DA, but not NT, levels are increased. Interestingly, as the stimulation frequency is increased, the ratio of DA release to NT release decreases exponentially. Mesencephalic DA neurons are known to alternate between a single spike firing mode and a burst firing mode *(45)*, and when mesocortical axons are stimulated to reproduce this burst firing pattern, a greater release of both DA and NT is obtained. A key finding in these studies was that pharmacological blockade or activation of DA autoreceptors has opposite effects on the release of DA and NT. Thus, stimulation of DA autoreceptors with low-dose apomorphine or the autoreceptor agonist, EMD-23448, results in a decrease in DA release and an increase in NT release in prefrontal cortex. Moreover, blockade of DA autoreceptors with (–)-sulpiride increases the release of DA and decreases NT release.

There is also evidence that NT can modulate the release of DA. In slice preparations, activation of presynaptic NT receptors (NT autoreceptors) by NT enhances depolarization-induced DA release *(46)*. The net neurotransmission induced by dopaminergic neurons innervating the prefrontal cortex may be influenced further by the distribution and density of postsynaptic NT and DA receptors on cortical follower cells. In the prefrontal cortex, DA predominantly generates inhibitory postsynaptic potentials, and NT generates excitation of follower cells *(47,48)*. Overall, these data demonstrate that depending on the physiological or pharmacological situation, DA and NT release from mesocortical neurons can be altered either differentially or in unison, and the functional effects these neurons exert will depend on the relative output of DA and NT (Fig. 1). As DA neurons that innervate the rat prefrontal cortex lack synthesis-modulating autoreceptors, the colocalization and corelease of NT in these neurons may provide an alternate means of regulating DA function. A dysregulation of the DA/NT interaction may conceivably occur in the schizophrenic brain. In support of this hypothesis is the evidence that DA neurons projecting to the prefrontal cortex may be associated with cognitive and affective function, together with links between NT and the action of antipsychotic drugs *(49–51)*.

It is interesting to compare the corelease of NT and DA with other neurons that colocalize a peptide and a nonpeptide. Electrical stimulation of motor and sympathetic neurons that contain peptide and nonpeptide cotransmitters also induces predominant release of the nonpeptide at low frequencies and preferential release of the peptide at higher frequencies *(52–55)*. However, in some aspects the described relationship between release of DA and NT differs from other colocalized neurotransmitters (Fig. 2). For instance, in some systems, burst stimulation produces a greater enhancement in peptide, relative to nonpeptide, release *(56)*.

DA neurons that innervate other regions, such as the striatum or nucleus accumbens, also appear to possess presynaptic NT receptors that can modulate DA release. However, they are not considered in detail here because there is not good evidence that NT is actually colocalized in mesostriatal DA neurons, and thus these NT receptors do not fulfill our extended definition of autoreceptors. The presynaptic receptors on mesostriatal DA neurons may be responsive to NT released from intrinsic NT neurons that are present in the striatum and nucleus accumbens.

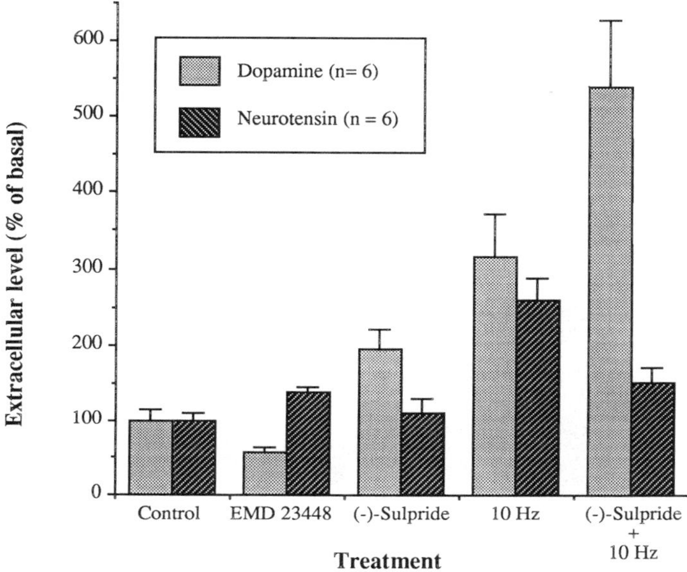

Fig. 1. DA and NT release from rat prefrontal cortex are illustrated. Studies were performed in chloral hydrate-anesthetized male rats. Each microdialysis sample from the prefrontal cortex was assayed for both DA (HPLC-ECD) and NT (RIA) concentrations. Drugs were administered through the dialysis probe. Application of the DA autoreceptor agonist, EMD 23448, resulted in a decrease in DA and an increase in NT extracellular fluid levels. The DA autoreceptor antagonist, (-)-sulpiride, increased DA concentration, and did not alter the level of NT. Electrical stimulation of the median forebrain bundle at 10 Hz enhanced the release of both DA and NT. At lower frequencies (e.g., 2.5 Hz) DA release was elevated, and NT release was unaffected (not shown here). When (-)-sulpiride was administered in conjunction with 10 Hz stimulation of the median forebrain bundle, an increase in DA and a decrease in NT release was observed, compared to 10 Hz stimulation alone. Thus, DA and NT can be altered in the same direction or in different directions, depending on the stimulus, such as neuronal firing frequency and DA autoreceptor occupation. For further details *see* refs. *43* and *44*.

DA neurons in which NT is costored originate in the ventral tegmental area. In this region and the neighboring substantia nigra, there are dense plexuses of NT-LI axon terminals that are in register with the cell bodies and dendrites of dopaminergic neurons. In addition, a large proportion of the dopaminergic neurons in these regions possess somatodendritic NT receptors *(33)*. Thus, it is possible that the neurons in which DA and NT coexist have NT autoreceptors at the cell body level, in addition to the terminal level. However, NT receptors located on the soma

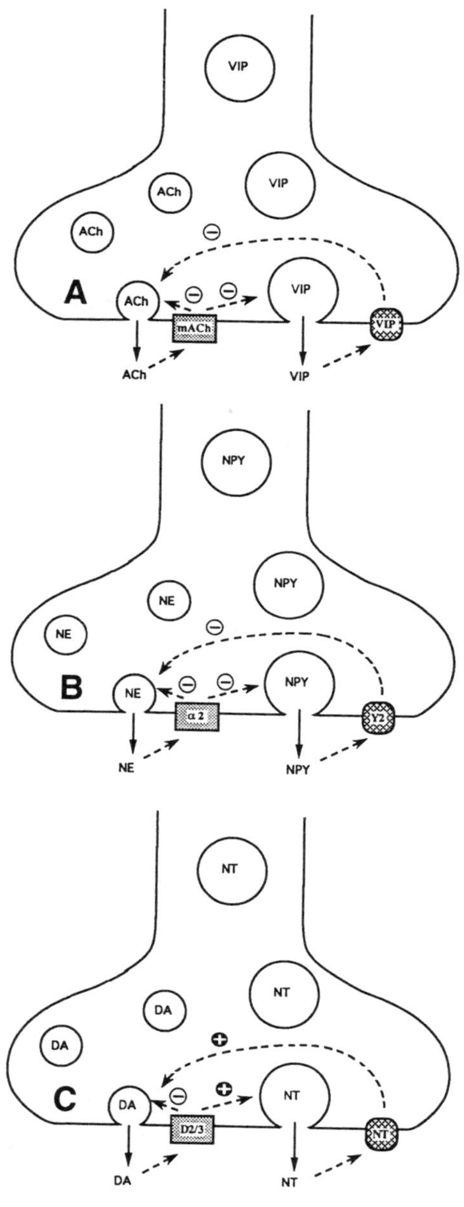

VIP
ACh
VIP
ACh
A
ACh
VIP
mACh
VIP
ACh
VIP
NPY
NE
NPY
NE
B
NE
NPY
α 2
Y2
NE
NPY
NT
DA
NT
DA
C
DA
NT
D2/3
NT
DA
NT

Fig. 2.

and dendrites of DA cells may function as postsynaptic receptors that respond to NT released from midbrain afferents. There have been several interesting electrophysiological studies on the interaction between DA and NT in the ventral tegmental area and substantia nigra. Although these studies cannot distinguish the activity of DA neurons that contain NT from those that do not, we briefly discuss the findings because of their potential importance to DA autoreceptor function.

In addition to a direct excitatory effect of NT on DA neurons in the rat midbrain, microiontophoretically applied NT at lower doses exerts an interesting modulation of the DA cell activity. Shi and Bunney *(35)* found that NT consistently attenuates the inhibition of basal activity induced by either DA or the DA agonist, quinpirole. This indicates that NT can antagonize the inhibitory effect that DA exerts by interacting with DA impulse-modulating autoreceptors. The mechanism of this interaction is not clear at present. However, there are some possible clues in the literature. In striatal tissue, NT decreases the affinity of DA agonists, but not antagonists, for the D2 receptor by a G protein-independent mechanism *(57,58)*. The studies of Shi and Bunney *(36)* suggest that in midbrain DA neurons, the modulatory effect of NT involves intracellular cyclic AMP and protein kinase A.

Whereas the peptide, NT, can influence the function of mesocortical DA neurons in which it is colocalized in several ways, most of these neurons also contain another peptide, CCK, which itself can interact with DA neuron function and potentially provide another level of neuronal regulation.

4.2. Cholecystokinin

CCK coexists with DA in some ventral tegmental area neurons that project to regions such as the posterior medial nucleus accumbens, medial

Fig. 2. *(previous page)* Autoreceptor regulation of classical transmitters and costored peptides in cholinergic, noradrenergic and dopaminergic neurons is shown. In panel A, acetylcholine (ACh) is shown colocalized in a neuron with vasoactive intestinal polypeptide (VIP); this represents the situation that occurs in certain neurons in the rat and cat submandibular gland, and rat cerebral cortex. Released ACh interacts with its muscarinic acetylcholine (mACh) autoreceptor to inhibit further ACh release, and also to inhibit VIP release. In addition, VIP that is released exerts a negative influence on ACh release. In panel B, parallel interactions are shown to occur between norepinephrine (NE) and neuropeptide Y (NPY), representing certain neurons in the rat and mouse vas deferens and pig splenic nerve. In panel C, DA is shown colocalized with NT; this situation occurs in certain neurons projecting to the rat prefrontal cortex. Released DA interacts with its autoreceptor to inhibit further DA release, but also to enhance NT release. In addition, NT that is released exerts a positive influence on DA release. Abbreviations: Y2, NPY receptor type Y2; $\alpha 2$, noradrenergic receptor type alpha 2.

olfactory tubercle, medial septum, and prefrontal cortex. These neurons comprise only a small contingent of all CCK neurons; in fact, CCK appears to be one of the most abundant neuropeptides in mammalian brain *(59,60)*. Although many studies on the reciprocal effects of CCK and DA release have been performed, the extensive distribution of CCK neurons in brain makes it difficult to ascribe alterations to CCK autoreceptors located on DA neurons, particularly in the in vivo situation. Nevertheless, we briefly mention some of these data to illustrate the diverse processes of self-regulation that can exist in DA neurons.

In vitro release studies have shown that CCK acting at CCK-A receptors in posterior medial nucleus accumbens potentiates both resting and potassium-stimulated DA release *(61)*. Acting at CCK-B receptors, CCK inhibits DA release from the anterior nucleus accumbens, a region in which DA and CCK are not colocalized to a high degree. In posterior medial nucleus accumbens, low-doses of a D2 DA agonist inhibit potassium-stimulated CCK-LI release; higher doses potentiate CCK-LI release. This is an unusual dose–response effect, but it could provide a means to maintain homeostasis of nucleus accumbens activity *(62),* since CCK and DA have opposite electrophysiological effects on nucleus accumbens neurons *(63)*.

Other electrophysiological studies *(64–66)* have shown that CCK increases the firing rate of DA neurons in ventral tegmental area and substantia nigra by a CCK-A receptor mechanism. Interestingly though, CCK given in combination with a DA agonist, such as apomorphine or quinpirole, potentiates the inhibitory effects of DA on firing rate of ventral tegmental area and substantia nigra DA neurons, an effect which seems to be mediated by CCK-B receptors or a subtype of CCK-A receptors. However, CCK microinjected into the ventral tegmental area has been reported to have no effect on DA release measured by microdialysis in the nucleus accumbens *(67)*. These latter data may seem at odds with the electrophysiological data suggesting that CCK augments the firing rate of ventral tegmental DA neurons, yet it is possible that the population of CCK-activated DA neurons in the ventral tegmental area are not those that terminate in the nucleus accumbens, sampled in the study of Laitinen et al. *(67)*.

Although at first a discussion of NT and CCK receptor function may seem tangential to the subject of DA autoreceptors, we hope that this section has justified its inclusion, since these receptors are, after all, present on DA neurons and their activation appears to modulate potently the function of the DA neuron on which they are located. These studies have not only provided an interesting perspective on the intricacies that exist in the regulation of DA neuron function, they also open up avenues of investigation that may lead to new strategies to manipulate dopaminergic function therapeutically.

5. Species Differences

In this section the evidence for the existence and function of DA autoreceptors in primates is reviewed, for the majority of the data on DA autoreceptors in the literature have derived from studies with rodents. This is obviously important, since the assumption underlying most neuroscience investigations in rodents is that the data generated will be relevant to primates, such as humans.

Two groups have provided direct evidence for DA autoreceptors in human brain. Hetey et al. *(68)* reported that DA inhibits potassium-evoked [^{3}H]DA release from synaptosomes prepared from cryopreserved human nucleus accumbens. In samples from schizophrenic patients, who were without neuroleptic medication for at least a year before death, release-modulating DA autoreceptors in the nucleus accumbens appear to be supersensitive. Fedele et al. *(69)* demonstrated that quinpirole inhibits electrically evoked release of [^{3}H]DA from slices prepared from fresh specimens of human cortex. These investigators took the precaution of preventing uptake of [^{3}H]DA release into non-DA monoamine terminals, and also showed that the effect of quinpirole is blocked by the DA antagonist, (–)-sulpiride.

Meador-Woodruff et al. *(70)* recently used the presence of mRNA for the D_2 DA receptor to infer the presence of DA autoreceptors in human and nonhuman primates. Whereas the substantia nigra in monkeys and humans contain mRNA for the D_2 DA receptor, this signal is lacking in the ventral tegmental area. Hurd et al. *(71)* also noted the low or absent expression of D_2 mRNA in human ventral tegmental area and retrorubral field. This result suggests that in primates, DA neurons arising from the ventral tegmental area to innervate limbic and cortical regions may be devoid of, or express far less, DA autoreceptors than the projections originating in the substantia nigra. In addition, mRNA for D_3 DA receptor is present in the human substantia nigra, but not in ventral tegmental area *(70)*. Although ventral tegmental area DA neurons that impinge on the prefrontal cortex in the rat appear to have a deficiency in impulse- and synthesis-modulating DA autoreceptors, one would still have predicted the presence of mRNA for D_2 DA receptors in primate ventral tegmental area, based on evidence for the presence of release-modulating D2-like DA autoreceptors in prefrontal cortex DA neurons in the rat. These data suggest that in primates and rodents the mesocorticolimbic DA systems may be differentially autoregulated.

An in vivo approach to determining the presence of DA autoreceptors has been implemented in the monkey *(72)*. A GBL-induced increase in short-term accumulation (15 min) of DOPA was used as a marker for the presence of synthesis-modulating DA autoreceptors after inhibition of DOPA decarboxylase. The results suggest that such autoreceptors are present in the caudate and putamen (especially in the caudal subregions) and nucleus accumbens; no evidence for synthesis-modulating DA autoreceptors was found using this approach in cortical regions (Fig. 3).

The data from these experiments indicate that relying totally on rodent data to dictate new directions for development of treatment for human dopaminergic disorders may be a poor strategy. The discussed studies *(68,69,72)* suggest that at least some primate DA neurons possess release- and/or synthesis-modulating DA autoreceptors in striatum, nucleus accumbens, and cortex. These data are not inconsistent with those of Meador-Woodruff et al. *(70),* who found no evidence that DA neurons in the primate ventral tegmental area synthesize D_2 receptors, as it is possible that the DA autoreceptors detected by Elsworth et al. *(72),* Hetey et al. *(68),* and Fedele et al. *(69)* derive from DA neurons arising in the substantia nigra, or alternatively, are a different subtype of DA receptor. It is clear that more work is necessary in this important field.

Section 4. reviewed the potential importance of peptide cotransmitters on DA autoreceptor function. It is important to note that there is a marked species difference in the extent of neuronal colocalization between DA and NT and CCK *(39,73–75)*. The rodent cerebral cortex has three types of DA and NT afferents with different regional and laminar distributions: separate DA, separate NT, and the largest group, mixed DA/NT. In human and nonhuman primates, DA innervation of the cerebral cortex is expanded, and no DA/NT colocalization is apparent. In addition, there are species differences in the coexpression of DA and CCK *(76,77)*. An mRNA for CCK is not detectable in most samples of human substantia nigra. Low amounts of CCK mRNA are present in ventral tegmental area neurons, partially colocalized with DA. A recent development in this field was that although CCK mRNA is in the substantia nigra of a minority of normal controls, it is expressed in the substantia nigra from the majority of schizophrenia patients. Although the potentially confounding influence of antipsychotic drug treatment is a concern, separate studies in other species that do not have detectable levels of CCK mRNA suggest that exposure to antipsychotic drugs does not induce CCK mRNA.

Further studies are now needed to clarify whether the lack of coexistence of NT and CCK with DA in primate brain is a permanent phenotype, or whether the colocalized peptide may be transiently expressed with DA in certain neuronal populations. In addition, it will be necessary to determine whether, in the primate brain, NT and CCK released from neurons that do not contain DA still can exert regulatory influences over nearby DA neurons.

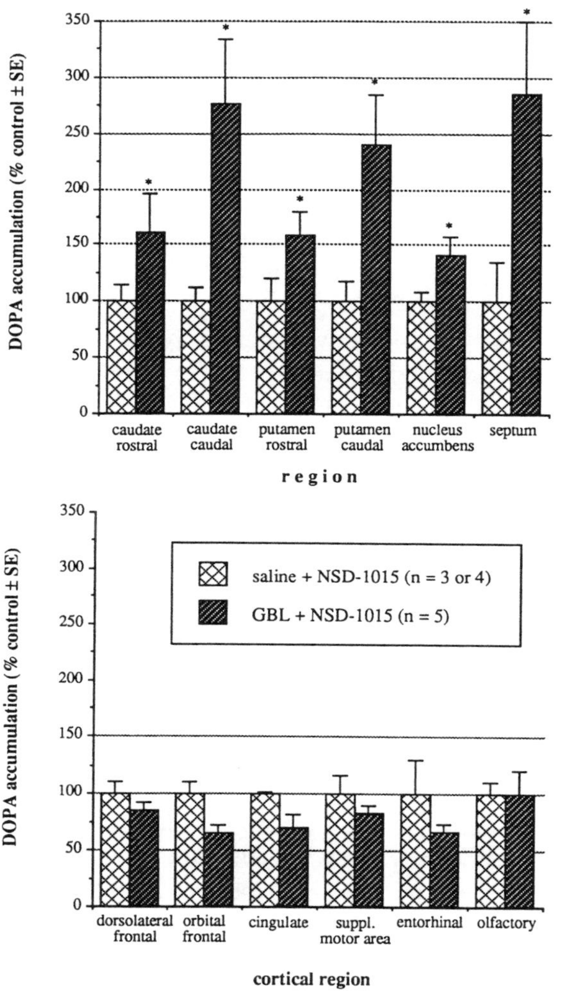

Fig. 3. Distribution of autoreceptors in primate brain. The GBL model was applied to the primate, as follows. Under light pentobarbital anesthesia, nine male African green monkeys were treated with γ-butyrolactone (GBL; 400 mg/kg iv) to interrupt impulse flow in DA neurons, or saline 20 min before sacrifice. Five min later, all monkeys received m-hydroxybenzylhydrazine dihydrochloride (NSD 1015; 100 mg/kg iv), a DOPA decarboxylase inhibitor. Under deep pentobarbital anesthesia, the monkeys were sacrificed. Brain regions were dissected and frozen until assay for DOPA concentration, by HPLC-ECD. These data show evidence for synthesis-modulating autoreceptors in caudate nucleus, putamen, nucleus accumbens, and septum, but not in any of the examined cortical regions.

6. D_3 DA Receptor—An Autoreceptor?

Five genes encoding DA receptors have been identified. The D_3 and D_4 receptors are considered D2-like receptors, based on their homology with the D_2 receptor. However, D_3 and D_4 receptors have a more limited distribution in brain than D_2 receptors, being expressed predominantly in regions of the limbic system. The localization of the mRNA for the D_3 receptor in the cell bodies of midbrain DA neurons led to the suggestion that it may function as an autoreceptor. This was strongly supported by the demonstration that the signal for mRNA for the D_3 DA receptor in the substantia nigra and ventral tegmental area is sensitive to local application of 6-hydroxydopamine *(78)*. However, it should be noted that the distribution of mRNA for the D_2 and the D_3 receptor within subregions of the substantia nigra and ventral tegmental area is different *(15,79)*. Particularly noteworthy is the relative predominance of mRNA for the D_2 receptor in medial substantia nigra pars compacta, and the higher density of mRNA for the D_3 receptor in lateral substantia nigra pars compacta. The fact that the D_3 receptor has the highest affinity for DA among the receptor subtypes identified so far also may be relevant to the proposed role of the D_3 receptor as an autoreceptor, for a high sensitivity to DA would appear to be a requisite feature for a DA autoreceptor. In addition to the midbrain, mRNA of the D_3 receptor is detected in telencephalic regions, particularly those innervated by A10 (ventral tegmental area) cell group, such as the shell of the nucleus accumbens, islands of Calleja, and bed nucleus of the stria terminalis, indicating that it also may act as a postsynaptic receptor *(79,80)*. The presence of the D_3 receptor in limbic regions of the brain, together with the fact that most antipsychotics have a similar affinity for D_3 receptor as for the D_2 receptor, invites the speculation that a disturbance in the expression or function of the D_3 receptor may occur in schizophrenic subjects.

The suggestion has been put forward that there may be distinct autoreceptors that regulate DA release and synthesis *(81,82)*. In addition, it has been speculated that the D_3 autoreceptor may preferentially regulate DA release *(19,83,84)* although the data in the literature regarding this are far from consistent. Thus, the evidence for D_3 regulation of synthesis, release, and impulse flow is considered separately, even though a change in one autoreceptor-controlled process (e.g., firing rate) does not preclude secondary changes in other neuronal events that can be regulated separately by autoreceptors (e.g., release).

6.1. D_3 Regulation of DA Synthesis

Using 7-OH-DPAT as a preferential D_3 agonist, Booth et al. *(85)* and Aretha et al. *(86)* provided evidence that the D_3 receptor can regulate DA synthesis. Aretha et al. *(86)* found that 7-OH-DPAT, at a dose of 0.03 mg/kg, reversed GBL activation of DOPA accumulation, in the presence of decarboxylase inhibition, in striatum and nucleus accumbens in vivo, an effect that was completely prevented by the D_3-preferring antagonist, (+)-UH-232. In parallel studies in vitro, 7-OH-DPAT inhibited DOPA accumulation in striatal slices in the presence of a decarboxylase inhibitor; this index of synthesis-modulating autoreceptors was also sensitive to (+)-UH-232. Although these studies rely on the specificity of D_3-selective drugs, they do provide supporting evidence for the presence of presynaptic D_3 autoreceptors that can exert control over the rate of DA synthesis. The findings of Meller et al. *(87)* are consistent with this conclusion; they reported that the potency of three DA agonists (quinpirole, quinelorane, and apomorphine) in reversing DOPA accumulation in the GBL model is correlated with their affinities for the D_3, but not the D_2 receptor. However, very recent studies with D_3 receptor-deficient mutant mice have revealed similar DOPA accumulation in the GBL model in D_3 mutants and wild-type strains, arguing against a critical role for D_3 regulation of synthesis *(87a)*.

6.2. D_3 Regulation of DA Release

Two groups *(19,83)* have used the relative changes in extracellular DA and DOPAC levels following systemic administration of DA antagonists to infer selective changes in DA release or metabolism, respectively. It has been suggested that a closer relationship exists between dialysate DOPAC levels and DA synthesis than between dialysate DOPAC levels and DA release *(88)*. However, it is not clear at this point whether DA agonist- or antagonist-induced changes in DOPAC dialysate levels can be related to an interaction of the agent with synthesis-modulating autoreceptors. Waters et al. *(19)* showed that the D_3- and autoreceptor-preferring antagonists, (+)-AJ76 and (+)-UH232, increase extracellular DA more than DOPAC in the awake rat, whereas for classical antagonists, such as haloperidol, the reverse is observed. Gainetdinov et al. *(83)* confirmed this finding, and extended it by testing a larger range of drugs. This enabled the relationship between the relative affinity of each drug for D_3 and D_2 receptors to be correlated with their relative potency in increasing DA and DOPAC in striatal dialysates in the conscious rat. A significant correlation was found, indicating that the greater the affinity for the D_3 receptor, the more a drug tends to elevate DA relative to DOPAC in the extracellular fluid. Gainetdinov et al. *(84)* also presented pilot data to suggest that the putative D_3-selective agonist 7-OH-DPAT given systemically at low doses (at least 0.025 mg/kg, which was stated not to affect the D_2 receptor function) decreases the extracellular level of DA, but not DOPAC, in the striatum. Gobert et al. *(89)* also reported preliminary data that 7-OH-

DPAT lowers the extracellular level of DA in the nucleus accumbens and striatum, and that this effect is reversed by the novel D_3-selective antagonist, S14297, but not by its less active stereoisomer.

Another approach to distinguish the autoreceptor function of the different D2-like receptors that largely sidesteps the issue of drug specificity has been employed recently *(90,91)*. In these studies D_2, D_3 and D_4 receptors were expressed separately in a mouse mesencephalic cell line that synthesizes and releases DA. In this cell line, stimulation of D_2 and D_3 receptors, but not D_4 receptors, inhibits DA release. Other preliminary data from this group *(90)* suggest that quinpirole is able to inhibit potassium-stimulated TH activity in cells expressing the D_2, but not the D_3, receptor. Together these data from dialysis and cell culture studies point to a role for the D_2 receptor in regulation of both synthesis and release. The D_3 receptor may function predominantly to control DA release. However, Gifford and Johnson *(92)* found that the IC_{50} values for a series of DA blockers to antagonize the quinpirole-induced inhibition of electrically evoked [^{3}H]DA release from striatal slices correlates better with the drugs' potency at cloned DA D_2 receptors than with their potency at cloned D_3 or D_4 receptors. This latter piece of information suggests that terminal DA autoreceptors show greatest similarity with DA D_2 receptors.

6.3. D_3 Regulation of Dopaminergic Firing Rate

Lejeune et al. *(93)* showed that the firing rate of ventral tegmental neurons in vivo is inhibited by iv administration of 7-OH-DPAT, and that this effect is antagonized stereospecifically by the D_3 antagonist, S14297. Another recent report *(94)* showed that there is a significant correlation between the potency of 11 DA agonists to inhibit the firing of substantia nigra DA neurons in vivo and their in vitro binding affinities at DA D_3, but not D_{2L}, receptors expressed in Chinese hamster ovary cells. Both these studies are consistent with the existence of D_3 dendritic autoreceptors that can regulate the firing rate of midbrain DA neurons.

6.4. D_3 as a Postsynaptic DA Receptor

Although Sections 6.2. and 6.3. have dealt with the possible role of the D_3 receptor as an autoreceptor, it should be remembered that mRNA for the D_3 receptor is also found in neurons that receive dopaminergic innervation, indicating that the D_3 receptor serves also as a postsynaptic receptor. In fact, Svensson et al. *(95)* argued that this is the principal function of the D_3 receptor. These investigators observed that D_3-preferring agonists ((+)-7-OH-DPAT and pramipexole) reduce locomotion in actively exploring rats, and importantly, that this effect occurs at doses that do not affect in vivo indices of DA synthesis or release. Furthermore, whereas D_2-preferring antagonists reduce locomotor activity in the rat, the D_3-preferring antagonist, U99194, increases locomotor activity *(18)*. Thus, in contrast to the postsynaptic D_2 receptor, the postsynaptic D_3 receptor has been proposed to exert an inhibitory influence on psychomotor functions.

These sections also indicate that the D_3 receptor probably plays an important role in the postsynaptic mediation of DA released from terminals in limbic regions of the brain. In addition, the data that support the premise that the D_3 receptor operates as a presynaptic receptor which can regulate DA release and firing rate are also persuasive. It seems very likely, therefore, that the D_3 receptor functions both as a presynaptic and a postsynaptic DA receptor. Rigorous testing of this conclusion will be possible with the availability of more specific agonists and antagonists for the various DA receptor subtypes. The use of DA receptor subtype antisense oligonucleotides, which prevent receptor synthesis by binding to complementary mRNA regions, may provide another tool to investigate the roles of DA receptors *(96)*, although investigators will have to beware of nonspecific effects *(97)*. An intriguing question that may be tested in the future is the possibility that D_3 and D_2 autoreceptors are predominantly present on different populations of DA neurons, which may conceivably be distinguished by the presence of cotransmitters or modulators. Future studies should be directed at further examining the possibility that D_3 and D_2 receptors may be responsible for different autoregulatory events (e.g., release or synthesis modulation). It will also be very instructive to examine further the comparative signal transduction mechanisms that the D_2 and D_3 autoreceptors utilize *(98)*. Although it may seem that the field has become quite complicated with the identification of various DA receptor subtypes, other developments are soon likely to muddy (or clarify) the water further. In fact, the expression of alternatively spliced forms of the D_3 receptor is one such new finding *(99,100)*.

7. Signal Transduction

This section is devoted to studies that have addressed the sequence of events that occur between interaction of DA, or a dopaminergic agonist, with a DA autoreceptor and subsequent inhibition of either membrane excitability, DA synthesis, or DA release.

7.1. Impulse-Regulating Autoreceptors

Significant progress has been made in the last few years in deciphering the ionic currents that exist in the DA neuron and how they are altered by somatodendritic autoreceptor stimulation; the results are summarized in this section.

Potassium conductance in mesencephalic DA neurons is mediated by several different potassium currents. Stimulation of DA somatodendritic autoreceptors is known to increase at least three distinct potassium currents. The coupling of the D_2 DA receptor to two of these currents, the transient A current (IA) and the delayed rectifier current (IK), is dependent on a G_o signal transduction pathway *(101)*.

Two molecular forms of the DA D_2 receptor are generated by alternative splicing; these isoforms are known as D_{2short} (D_{2S}) and D_{2long} (D_{2L}). Although these two variants coexist in all regions and species so far examined, when individually expressed in a cell line, it has been shown that D_{2S} couples to potassium channels by means of a pertussis toxin-insensitive mechanism (i.e., not through G_o or G_i) and D_{2L} by means of a pertussis toxin-sensitive mechanism *(102,103)*. Recent work using a cell line expressing the DA D_3 receptor has indicated that this receptor couples to potassium currents in a pertussis toxin-sensitive, G_o protein-dependent, manner *(104)*. Thus, it is possible that the two forms of D_2 receptor and the D_3 receptor may modulate potassium currents in a similar way, but by utilizing different transduction mechanisms.

Three different inward calcium currents have been identified in the cell bodies of mesencephalic DA neurons. Two of these (IL and IN, but not IT) appear to mediate autoreceptor-regulated impulse flow, as they are reduced by stimulation of the autoreceptor. The coupling of these currents involves a pertussis-sensitive G protein *(105)*.

In addition, recent studies on cultured mesencephalic DA neurons suggests that DA autoreceptors modulate tetrodotoxin-sensitive sodium currents. Stimulation of somatodendritic DA autoreceptors was found to increase the peak amplitude of the voltage-dependent sodium current (INa) *(106)*.

7.2. Release-Modulating Autoreceptors

Early studies on autoreceptor signal transduction indicated that D2 receptor modulation of evoked DA release from the striatum is not related to inhibition of adenylate cyclase *(107,108)*. More recent data have bolstered the case for the involvement of a potassium channel link in the mechanism underlying autoreceptor regulation of DA release in the striatum. In striatal slices, the inhibition of evoked DA overflow caused by a D2 agonist, N-0437, is blocked by potassium channel blockers (4-aminopyridine, tetraetylammonium) in a dose-dependent manner *(109)*, suggesting that D2 agonists acting at DA autoreceptors may increase potassium conductance and hyperpolarize the presynaptic terminal. However, in an in vivo study it was found that local application of quinine, a potassium channel blocker, prevented the increase in extracellular DA levels in the striatum elicited by a D2 DA receptor antagonist, but did not block the decrease in DA levels elicited by a D2 DA agonist *(110)*. The interpretation of these data, however, is complicated by the likely contribution of both pre- and postsynaptic D2 DA receptors to the overall effect, and the possibility that only a subpopulation of the receptors may be linked to quinine-sensitive potassium channels.

7.3. Synthesis-Modulating Autoreceptors

Several studies have shown that presynaptic inhibition of DA synthesis by DA autoreceptors is achieved by modulation of the activity of TH *(4)*. An increase in K_m of the pterin cofactor for TH develops in striatal preparations in vitro on exposure to a DA agonist *(111)*. More recently, the molecular mechanisms underlying the changes in kinetic state of TH have been investigated.

TH is a substrate for protein phosphorylation, and the proportion of TH in the phosphorylated form appears to be a critical determinant of its activity *(112,113)*. Studies using rat striatal slices have indicated that the reduction of TH activity induced by stimulation of DA autoreceptors is achieved by a reduction in the rate of TH phosphorylation *(114)*. Since phosphorylation of TH can potentially be catalyzed by several protein kinases, there have been investigations designed to determine the second messenger system that is most important in this control. There is now good evidence that a cyclic AMP (cAMP)-dependent protein kinase may be involved in the phosphorylation of TH *(115–117)*. Thus, forskolin (an activator of adenylate cyclase) stimulates TH activity in striatal preparations, and this stimulation may be inhibited by activation of DA autoreceptors. Furthermore, if the adenylate cyclase system is circumvented by directly stimulating cAMP-dependent protein kinase with dibutryl cAMP, then striatal TH activity is no longer inhibited by DA agonists. It is also conceivable that the activity of Ca^{2+}-dependent protein kinases is able to govern the extent of phosphorylation of TH. However, DA agonist-induced inhibition of TH activity in striatal synaptosomes is undiminished when the assay is performed in calcium-free buffer. In addition, eliciting the entry of calcium into synaptosomes with a calcium ionophore does not prevent DA agonist-induced inhibition of TH activity *(118)*. Although these data argue against a role for calcium-dependent protein kinases in DA autoreceptor control of TH activity, some recent data have indirectly supported the hypothesis. Thus, Tissari and Lillgäls *(119)*, using striatal synaptosomes from reserpine-treated rats, reported that the activation of TH activity evoked by reserpine is additive with that produced by the addition of dibutyrl cAMP to the assay. The authors argued that this observation, together with the knowledge that reserpine induces an increase in impulse activity, and consequently, an increase in free intraneuronal calcium concentration, favors the interpretation that phosphorylation of TH is mediated by calcium-dependent protein kinases.

Other support for this hypothesis was provided by Fujiwara et al. *(120)*, who found that a D2 agonist inhibits the intracellular calcium mobilization elicited by electrical stimulation of striatal slices. These data argue strongly for DA autoreceptor modulation of cAMP-dependent protein kinases that are involved in the phosphorylation-dephosphorylation of TH. However, a role for the calcium-dependent protein kinases cannot be ruled out.

It is clear that great progress has been made toward identifying the signal transduction pathways utilized by DA autoreceptors. It seems likely that autoreceptors modulating different events utilize different second messenger systems. This may be the result of different signal transduction pathways linked to the same receptor at various sites on the neuron. Also it is possible that subtypes of DA autoreceptor (e.g., D_{2L}, D_{2S}, D_3), may utilize distinct transduction pathways. Subpopulations of DA neurons may conceivably be associated with a distinct subtype(s) of D2-like autoreceptor.

8. Changes in Autoreceptor Sensitivity

Much of the impetus for research into DA autoreceptors is the hope that the information obtained may contribute to the design of new autoreceptor-selective drugs to treat disorders that may involve a dysfunctional dopaminergic system. Thus, a key consideration is whether the targeted autoreceptor population retains its original response on repeated exposures to a drug or other challenge.

8.1. Supersensitivity of Autoreceptors

Early evidence indicated that supersensitivity of synthesis-modulating autoreceptors may be involved in the development of biochemical tolerance to the metabolite-elevating effects observed following chronic administration of antipsychotic drugs. DA neurons that project to the prefrontal and cingulate cortices appear to lack synthesis-modulating autoreceptors, and these regions are relatively resistant to the development of biochemical tolerance to antipsychotic drugs. On the other hand, nigrostriatal, mesolimbic, and mesopiriform DA neurons, which possess synthesis-modulating autoreceptors, readily develop tolerance during antipsychotic drug treatment *(121–123)*. Furthermore, there is a parallel time course for acquisition of tolerance and acquisition of autoreceptor supersensitivity *(124)*.

Some recent data suggest that continuous exposure of rats to cocaine results in supersensitivity of release-modulating DA autoreceptors *(125)*. This dosing regimen produces tolerance to the effects of subsequent cocaine administration. The evidence for autoreceptor supersensitivity is an enhanced inhibition of behavior in response to an autoreceptor-selective dose of apomorphine in rats continuously treated with cocaine for 14 d and tested 7 d later. The absence of any change in mRNA levels for the D_2 receptor in striatum indicates that the effects are not the result of an alteration in postsynaptic D_2 receptors. In addition, the same treatment schedule decreases electrically stimulated DA release from striatal slices in the absence of sulpiride, and increases electrically stimulated DA release in the presence of sulpiride *(125)*. These data are consistent with the hypothesis that tolerance induced by continuous cocaine is caused by supersensitivity of release-modulating DA autoreceptors, presumably arising as a consequence of persistently elevated synaptic levels of DA induced by inhibition of DA uptake by cocaine.

8.2. Subsensitivity of Autoreceptors

In animals, intermittent pre-exposure to psychostimulants, such as cocaine or amphetamine, enhances the reinforcing or locomotor effects of a subsequent exposure to the drug, a phenomenon known as sensitization. In humans, sensitization has been suggested to be involved in cocaine addiction and/or psychosis *(126,127)*. DA autoreceptor subsensitivity or down-regulation has recently attracted much interest in connection with the mechanism of sensitization.

Extracellular recordings from slices of rat ventral tegmental area suggested that bath-applied amphetamine induces a subsensitivity of somatodendritic DA autoreceptors, which starts as early as 10 min after exposure to amphetamine *(128)*. In addition, there is good evidence that chronic psychostimulant treatment renders impulse-regulating DA autoreceptors of A9 (substantia nigra) or A10 neurons subsensitive. Thus, following repeated amphetamine or cocaine administration, the rate-suppressant effects of systemically or iontophoretically administered DA agonists are attenuated in A9 and A10 DA neurons *(129–132)*. Since subsensitivity of impulse-regulating DA autoreceptors results in enhanced DA cell activity, these findings originally appeared to provide a reasonable explanation for the enhanced mesostriatal DA synaptic levels and augmented locomotor behavior that occurs in response to a challenge dose of a psychostimulant in sensitized animals. It appears, though, that impulse-regulating autoreceptors located on DA neurons in the mesencephalon are only transiently (3–4 d) subsensitive following the last injection of the drug. Thus, subsequent studies revealed that the sensitivity of DA cells to the inhibitory effects of a DA agonist, such as apomorphine, returns to normal at a time when rats still display behavioral sensitization. It may be that the subsensitivity of impulse-regulating DA autoreceptors plays an important role in the establishment, but not the maintenance of behavioral sensitization *(133)*.

In addition, studies of electrically evoked endogenous DA release from rat striatal slices in the presence of DA agonists or antagonists indicated that pretreatment of rats with methamphetamine produces subsensitivity of terminal release-modulating DA autoreceptors, examined 9 d after the last dose of methamphetamine *(134)*. Chronic treatment of another abused drug, nicotine, also leads to subsensitivity of release-modulating DA autoreceptors; this was demonstrated using evoked release of [^{3}H]DA from striatal slices 18 h after the last injection *(136)*. There is also evidence that repeated cocaine treatment alters the sensitivity of release-modulating autoreceptors. Yi and Johnson *(136)* showed that Ca^{2+}-evoked release of [^{3}H]DA from slices prepared from either the striatum or nucleus accumbens of female rats treated chronically with cocaine was resistant to the effect of a D2 DA agonist. Tissue from these rats was collected 30 min after a challenge injection of cocaine, given 7 d following the last sensitizing dose. However, other studies using cocaine are not in agreement with these data, reporting no alteration of release-modulating DA autoreceptors in rats treated repeatedly with cocaine. Fitzgerald and Reid *(137)* found that the inhibitory effect of pergolide on electrically evoked [^{3}H]DA release from striatal slices was not altered by 14 d of cocaine treatment, when examined 1 d after the last dose. Gifford and Johnson *(138)* found no change in the potency of quinpirole to inhibit electrically induced release of [^{3}H]DA from slices of either striatum or nucleus accumbens, examined 7 d after 9 d of cocaine treatment. King et al. *(125)* treated rats intermittently with cocaine for 14 d and, when tested 7 d later, found no alteration in the behavioral response to apomorphine, and no changes in electrically evoked release of endogenous DA from striatal slices. In fact, Dwoskin et al. *(139)* observed that autoreceptors regulating electrically evoked [^{3}H]DA release in striatal slices were supersensitive to pergolide when examined 1 d after 8 or 14 d treatment with cocaine. Because these somewhat conflicting studies on the sensitivity of terminal release-modulating autoreceptors following intermittent cocaine employed different drug treatment and washout schedules, and different methods of eliciting DA release, these factors are probably important determinants for inducing or detecting changes in release-modulating DA autoreceptors.

In striatum of rats behaviorally sensitized to cocaine, a challenge injection of cocaine has been reported to result in less inhibitory effect on DOPA accumulation in vivo in the presence of a decarboxylase inhibitor than in previously untreated rats, which may suggest that in cocaine sensitization there is reduced sensitivity of receptors regulating DA synthesis *(140)*. In this study, though, the effect of cocaine on DOPA accumulation was also attenuated in sensitized rats in the median eminence *(140)*, a region that is thought to be relatively devoid of DA synthesis-modulating autoreceptors. Thus, it is possible that the effects observed are caused by, at least in part, cocaine-induced alterations at sites other than the DA synthesis-modulating autoreceptor. In fact, acute cocaine inhibits DOPA accumulation in vivo in the absence of GBL, but not after GBL treatment, indicating that the inhibition of synthesis in this situation is secondary to elevation of synaptic levels of DA and stimulation of release-modulating autoreceptors *(141)*. Thus, the effects of chronic cocaine on DOPA accumulation following decarboxylase inhibition without GBL treatment *(140)* may be due to changes in sensitivity of release-modulating autoreceptors, or to a postsynaptic adaption to cocaine.

Jeziorski and White *(142)* found that repeated treatment with autoreceptor-selective doses of apomorphine results in only a slightly subsensitive response of ventral tegmental area DA neurons to the inhibitory effects of DA on firing rate. Further investigation of this effect, using N-ethoxycarbonyl-2-ethoxy-1,2-dihydroquinoline (EEDQ) (see Section 9.), indicated that the existence of "spare" somatodendritic DA autoreceptors in the ventral tegmental area may have masked a significant autoreceptor down-regulation in response to daily low-dose apomorphine treatment *(142)*. Thus, there is ample evidence to support the theory that impulse-regulating and terminal release-regulating DA autoreceptors respond with changes in sensitivity to prolonged exposure to DA, DA agonists, or DA antagonists. However, the precise conditions necessary for inducing such changes is not clear in all cases.

9. Autoreceptor-Selective Drugs

Over the past few years, the number of agents that apparently possess selectivity for the DA autoreceptor has grown considerably. The impetus for this research is the therapeutic potential of autoreceptor-selective drugs. Table 1 lists many of the putative autoreceptor-selective agonists and antagonists that have been reported on in the past few years. The scope of this chapter does not permit an in-depth comparison of the compounds, but detailed discussions of their properties are provided in the cited references. It should be realized, though, that the weight of evidence supporting the autoreceptor selectivity is not the same for all these agents. Particularly when the classification depends mainly on an in vivo study, the possibility of confounding effects on postsynaptic DA receptors, or non-DA receptors, should be borne in mind.

9.1. Are DA Autoreceptors Different from DA D2-Like Postsynaptic Receptors?

Although it seemed likely at one time that the greater sensitivity of DA autoreceptors, compared with postsynaptic DA receptors, may be due to different subtypes of D2 receptor located at each site, there is no more convincing evidence for this now than there was before the discovery of the D_{2L}, D_{2S}, D_3

Table 1
Putative Dopamine Autoreceptor-Selective Agonists and Antagonists

Drug	Refs.
Agonist	
B-HT 920 (talipexole)	*143*
DPAI	*144*
EMD 23448, EMD 49980 (roxindol)	*145*
HW-165	*146*
OPC-4392	*147*
PD 118717	*148*
PD 119819	*149*
PD 143188	*150*
PD 128483	*151*
PD 118440	*152*
3-PPP	*153*
Ro 41-9067	*154*
SDZ 208-911, SDZ 208-912	*155*
SND 919 (pramipexole)	*156*
TDHL	*157*
TL-99	*158*
U-86170F, U-68553B	*159,160*
Antagonist	
(+)-AJ76	*161*
(+)-UH232	*161*
CGP 25454A	*162*
S-(–)-OSU6162	*163*
(–)-DS121	*164*

and D_4 members of the D2 family. Although there has been speculation that the D_3 receptor may be an autoreceptor, it does not appear to be exclusively an autoreceptor (*see* Section 6.). There is no strong evidence that the recognition site of the autoreceptor and the postsynaptic receptor are actually different *(3)*. Presently, one of the other theories outlined in this section, or a combination of them, seems the best explanation for the preferential action of some drugs at autoreceptors in vivo.

Preliminary evidence was presented at one time to suggest that D2 DA autoreceptors operate in the high-affinity agonist state, as opposed to DA postsynaptic receptors that operate in the low-affinity state *(165)*. Although this is an attractive hypothesis, the evidence supporting it is somewhat indirect, being based on correlations between the potency of a series of drugs at displacing [^{3}H]spiperone from D2 receptors in the high- or low-affinity state, and the potency of the drugs to inhibit either [^{3}H]DA release (presynaptic) or acetylcholine release (postsynaptic). This hypothesis does not explain the relatively greater presynaptic potency of some compounds.

Carlsson *(166)* put forward a hypothesis that accounted for the differential sensitivity of autoreceptors and postsynaptic DA receptors based on the previous agonist occupancy of the receptor. According to this proposition, the two classes of receptor are derived from a homogeneous population; however, the receptor, or part of the receptor complex, is envisaged to adapt itself to varying agonist occupancy by a slow conformational change that influences the subsequent responsiveness of the receptor to an agonist. Thus, with this hypothesis, the same receptor and agonist will show varying intrinsic activity depending on the state of the receptor. A postsynaptic receptor is assumed to have relatively high agonist occupancy and low sensitivity to agonists. An autoreceptor is assumed to be adapted to low receptor occupancy, since it is thought to be principally located outside the synaptic cleft, and to have high sensitivity to agonists. This would explain some of the unusual properties of the enantiomers of 3-PPP, and account for the supersensitivity of denervated DA postsynaptic receptors. However, supersensitivity of DA postsynaptic receptors, which follows denervation or chronic antagonist treatment, has been firmly associated with an increased density of D_2 receptors postsynaptically. Although this hypothesis does not rule out the synthesis or recruitment of new receptor molecules (to increase density), changes in receptor responsiveness are posited to be the main explanation for supersensitivity.

A later hypothesis *(167)* depended on the extent of current agonist occupancy rather than previous occupancy. More specifically, the greater in vivo potency of agonists at autoreceptors compared to postsynaptic receptors was speculated to be due to the presence of a larger receptor reserve at autoreceptor sites. Some experimental evidence to support the concept was obtained. The irreversible DA receptor antagonist EEDQ, which inactivates DA receptor binding sites, was used to determine the extent of receptor reserve for DA autoreceptors regulating TH activity in vivo, using the rat GBL model. By quantifying the shift in potency for various DA agonists in response to different doses of EEDQ, these investigators were able to derive the relationship between receptor occupancy and response. This showed that for a full agonist such as N-propylnorapomorphine, the DA autoreceptor reserve was approximately 70% (i.e., a maximal response was obtained with about 30% occupancy). The autoreceptor selective drugs, the 3-PPP isomers and EMD 23,448, were shown to be partial agonists, requiring a much larger proportion of receptor occupancy than N-propylnorapomorphine to achieve a given response. This theory has been supported by later studies employing EEDQ which indicated that there is a large receptor reserve for autoreceptor regulation of both electrically evoked release of [^{3}H]DA from striatal slices, and rate of firing of nigral DA neurons *(142,168,169)*. A subsequent study by Cox and Waszczak *(170)* indicated that the receptor reserve was greater in the A9 than the A10 DA neurons. In contrast, Meller et al. *(171)* found no evidence for receptor reserve at the postsynaptic DA receptor, as judged by changes in striatal acetylcholine concentration in response to N-propylnorapomorphine following EEDQ treatment. Thus, there appears to be a differential receptor reserve at pre- and postsynaptic receptor sites. This hypothesis provides an explanation for the greater potency of full agonists at the autoreceptor; that is, at the autoreceptor there is a receptor reserve (and an efficient coupling of receptor occupancy and response). For partial agonists (autoreceptor-selective drugs) the presynaptic receptor reserve accounts for their full or nearly full activity at this site, and the lack of a postsynaptic receptor reserve explains their relative lack of activity, or even antagonistic properties, at this site.

However, in contrast to the in vivo situation, Bohmaker et al. *(172)* found that there is no receptor reserve for autoreceptors regulating dopamine agonist inhibition of TH in an in vitro system. Although the explanation for the discrepancy is not clear, it is consistent with other data *(3)* that indicate no difference in potency of DA agonists when tested in vitro in assays of either autoreceptor or postsynaptic function; yet in assays of in vivo electrophysiology and behavior, autoreceptors seem more sensitive to the effects of D2 agonist than postsynaptic DA receptors. Furthermore, use of the in vitro DA release model may not reliably predict the apparent in vivo autoreceptor selectivity of D2 agonists *(173)*. Possibly, the multiple levels of regulation that exist for the DA neuron in vivo can combine, in certain situations, to provide an apparently altered sensitivity to D2 agonists.

Autoreceptors differ from postsynaptic D2 DA receptors with respect to their interactions with D1 DA receptors. It is now well established that D1 and D2 DA receptors act synergistically in many behavioral and electrophysiological models to elicit augmented responses. Stimulation of D1 receptors appears to play an enabling role, that is D1 receptor occupation is necessary, in addition to D2 receptor stimulation, to achieve full postsynaptic functional effects *(174)*. Interestingly, this does not appear to be the case for DA D2-like autoreceptors where synergy with D1 receptors is not observed *(175)*. Drukarch and Stoof *(3)* postulated that the presence of D1 DA receptors postsynaptic to the DA terminal, and the absence of D1 DA receptors on DA terminals may be an important factor in the preferential autoreceptor action of certain drugs. Thus, it is conceivable that low doses of D2-selective drugs would reduce release of DA by an action at release-modulating autoreceptors, so that the reduction in endogenous DA lessens the postsynaptic D1 receptor stimulation and prevents D1 receptors enabling the postsynaptic D2 effect.

Another possible contributing explanation to the differential sensitivity of autoreceptors and postsynaptic DA receptors to D2 agonists is that at these two sites the receptor may be associated with dissimilar signal transduction pathways which have different amplification factors.

There is no evidence that the autoreceptor and the postsynaptic receptor are different entities. Both D_2 and D_3 receptors appear to function as autoreceptors and postsynaptic receptors. Although some D_3-preferring drugs are thought to be autoreceptor preferring, it is doubtful that affinity for the D_3 receptor will necessarily dictate autoreceptor selectivity. Thus, although various hypotheses have attempted to explain why autoreceptors appear more sensitive to D2-like agonists than postsynaptic receptors, there is not a complete understanding of the phenomenon. It appears, at present, that the development of partial agonists at D2-like receptors offers the best chance to selectively manipulate DA autoreceptors.

9.2. Therapeutic Potential of Autoreceptor-Selective Drugs

The DA hypothesis of schizophrenia in its simplest form would predict a DA autoreceptor agonist to be a useful treatment of the disorder by decreasing the activity of most DA neurons. In fact, supersensitivity of release-modulating autoreceptors in schizophrenia has been suggested, based on analysis of postmortem tissue *(68)*. Clinical use of N-propylnorapomorphine for the treatment of schizophrenia was encouraging *(176)*. However, open clinical trials with DA autoreceptor-selective drugs have not resulted in consistent improvements in positive psychotic symptoms *(177,178)*. Interestingly though, minor to moderate improvements have been observed in the negative symptoms, which have been speculated to result from a DA deficit, rather than a DA excess. A possible mechanism for this effect may be stimulation of a subpopulation of supersensitive postsynaptic receptors, for autoreceptor selective drugs have been shown to have stimulatory effects in DA-depleted animals *(179)*. Svensson et al. *(180)* recently suggested that treatment of schizophrenia patients with the combination of a partial DA receptor agonist and a classical neuroleptic may minimize the appearance of extrapyramidal side-effects and hyperprolactinemia.

DA autoreceptor-selective drugs may also have utility in the treatment of major depression. In open clinical trials the autoreceptor agonist, roxindole, has been shown to be as effective as standard antidepressants *(177)*.

Theoretically, administration of a selective autoreceptor antagonist could be helpful in DA deficiency states, such as Parkinson's disease. The metabolic activity of surviving nigrostriatal DA neurons is increased in Parkinson's disease, presumably to compensate for the loss. Richard and Bennett *(181)* recently showed that, in rodents with lesions to the nigrostriatal DA system, a DA antagonist can further increase the metabolic activity of the remaining neurons. This suggests that even in moderately advanced Parkinson's disease, DA synthesis and release could be further elevated by an autoreceptor antagonist. Conversely, it has been suggested that some autoreceptor agonists may be of value in Parkinson's disease, as stimulatory effects of pramipexole were observed in the 1-methyl-4-phenyl-1,2,3,6-tetrahydropyridine (MPTP)-treated monkey, where there are likely to be supersensitive postsynaptic DA receptors in the striatum *(179)*.

Even though at first it may seem that it would be counterproductive to treat psychostimulant abuse with drugs that augment DA function, some autoreceptor DA antagonists have been proposed to treat cocaine and amphetamine abuse. (+)-AJ 76 and (+)-UH 232 are preferential DA autoreceptor antagonists with weak postsynaptic antagonistic effects. These drugs are able to block certain behavioral and electrophysiological effects of cocaine. It has been suggested that they may be useful in the treatment of cocaine abuse, as they may act as self-limiting indirect dopaminergic agonists that produce weak stimulant effects, without producing the anhedonia associated with administration of potent postsynaptic DA receptor antagonists *(182–184)*. Similar therapeutic hopes have been expressed for the autoreceptor antagonist, (–)-DS121, which does not itself seem to possess abuse liability, as the drug does not support self-administration behavior in rats when substituted for cocaine *(185)*. Autoreceptor agonists such as PD128483 or similar drugs have been suggested as a possible treatment for cocaine abuse, since PD128483 substitutes for cocaine as a discriminative stimulus, but is not self-administered *(186)*.

Although available DA autoreceptor drugs have not yet become an accepted treatment for a central nervous system disorder, there are encouraging signs that, in the near future, this may be accomplished. It seems that rather than there being a clear distinction between autoreceptor- and postsynaptic receptor-selective agonists and antagonists, a gradation of these properties is possible. Thus, it may be possible to tailor chemically the properties of DA receptor ligand to suit the relative pre- and postsynaptic modifications necessary for the treatment of a particular disorder.

10. Conclusion

This review has brought together advances that have been made in understanding the function and importance of DA autoreceptors. In the next few years we can look forward to the unraveling of currently unresolved issues, such as the possible subtypes of DA autoreceptors, and the signal transduction mechanisms involved in autoreceptor-mediated events. A major thrust of research in this field is likely to be toward development of drugs that are selective agonists or antagonists at DA autoreceptors. It will be particularly interesting to see whether the colocalization of DA with neuropeptides in certain groups of DA neurons can be exploited. Thus, it may soon be possible to manipulate the function of DA neurons by the use of a drug, or a combination of drugs, that are designed to alter the dynamics between DA and its costored neuropeptide(s). Such a strategy may have profound therapeutic significance, bearing in mind the number of disorders that have, or are speculated to have, a dysregulated dopaminergic component.

Acknowledgment

This work received support from MH-14092, NS 24032, Axion Research Foundation and St. Kitts Biomedical Research Foundation.

References

1. Farnebo, L.-O. and Hamberger, B. (1971) Drug-induced changes in the release of ^{3}H-monoamines from field stimulated rat brain slices. *Acta Physiol. Scand.* **371(Suppl),** 35–44.
2. Carlsson, A. (1975) Dopaminergic autoreceptors, in *Chemical Tools in Catecholamine Research II: Regulations of Catecholamine Turnover* (Almgren, O., Carlsson, A., and Engel, J., eds.), North-Holland, Amsterdam, pp. 219–225.
3. Drukarch, B. and Stoof, J. C. (1990) D-2 dopamine autoreceptor selective drugs: do they really exist? *Life Sci.* **47,** 361–376.
4. Wolf, M. E. and Roth, R. H. (1987) Dopamine autoreceptors, in *Receptor Biochemistry Methodology: Dopamine Receptors* (Creese, I. and Fraser, C. M., eds.), Liss, New York, pp. 45–96.
5. Wolf, M. E. and Roth, R. H. (1990) Autoreceptor regulation of dopamine synthesis. *Ann. NY Acad. Sci.* **604,** 323–343.
6. Westerink, B. H. C., de Boer, P., Santiago, M., and De Vris, J. B. (1994) Do nerve terminals and cell bodies of nigrostriatal dopaminergic neurons of the rat contain similar receptors? *Neurosci. Lett.* **167,** 109–112.
7. Santiago, M. and Westerink, B. H. C. (1991) The regulation of dopamine release from nigrostriatal neurons in conscious rats: the role of somatodendritic autoreceptors. *Eur. J. Pharmacol.* **204,** 79–85.
8. Westerink, B. H. C., Santiago, M., and De Vries, J. B. (1992) In vivo evidence for a concordant response of terminal and dendritic dopamine release during intranigral infusion of drugs. *Naunyn-Schmiedeberg's Arch. Pharmacol.* **346,** 637–643.
9. Bull, D. R., Palij, P., Sheehan, M. J., Millar, J., Stamford, J. A., Kruk, Z. L., and Humphrey, P. P. A. (1990) Application of fast cyclic voltammetry to measurement of electrically-evoked dopamine overflow from brain slices in vitro. *J. Neurosci. Methods* **32,** 37–44.
10. Kennedy, R. T., Jones, S. R., and Wightman, R. M. (1992) Dynamic observation of dopamine autoreceptor effects in rat striatal slices. *J. Neurochem.* **59,** 449–455.

11. McElvain, J. S. and Schenk, J. O. (1992) Blockade of dopamine autoreceptors by haloperidol and the apparent dynamics of potassium-stimulated endogenous release of dopamine from and reuptake into striatal suspensions in the rat. *Neuropharmacology* **31,** 649–659.
12. Sesack, S. R., Aoki, C., and Pickel, V. M. (1994) Ultrastructural localization of D_2 receptor-like immunoreactivity in midbrain dopamine neurons and their striatal targets. *J. Neurosci.* **14,** 88–106.
13. Silvia, C. P., King, G. R., Lee, T. H., Xue, Z. Y., Caron, M. G., and Ellinwood, E. H. (1994) Intranigral administration of D_2 dopamine receptor antisense oligodeoxynucleotides establishes a role for nigrostriatal D_2 autoreceptors in the motor actions of cocaine. *Mol. Pharmacol.* **46,** 51–57.
14. Freedman, S. B., Patel, S., Marwood, R., Emms, F., Seabrook, G. R., Knowles, M. R., and McAllister, G. (1994) Expression and pharmacological characterization of the human D_3 dopamine receptor. *J. Pharmacol. Exp. Ther.* **268,** 417–426.
15. Murray, A. M., Ryoo, H. L., Gurevich, E., and Joyce, J. N. (1994) Localization of dopamine D_3 receptors to mesolimbic and D_2 receptors to mesostriatal regions of human forebrain. *Proc. Natl. Acad. Sci. USA* **91,** 11,271–11,275.
16. Sokoloff, P., Andrieux, M., Besancon, R., Pilon, C., Martres, M.-P., Giros, B., and Schwartz, J.-C. (1992) Pharmacology of human dopamine D_3 receptor expressed in a mammalian cell line: comparison with D_2 receptor. *Eur. J. Pharmacol.* **225,** 331–337.
17. Waters, N., Hansson, L., Lofberg, L., and Carlsson, A. (1994) Intracerebral infusion of (+)-AJ76 and (+)-UH232: effects on dopamine release and metabolism in vivo. *Eur. J. Pharmacol.* **251,** 181–190.
18. Waters, N., Lofberg, L., Haadsma-Svensson, S., Svensson, K., Sonesson, C., and Carlsson, A. (1994) Differential effects of dopamine D2 and D3 receptor antagonists in regard to dopamine release, in vivo receptor displacement and behaviour. *J. Neural Transmiss.* **98,** 39–55.
19. Waters, N., Lagerkvist, S., Lofberg, L., Piercey, M., and Carlsson, A. (1993) The dopamine D_3 receptor and autoreceptor preferring antagonists (+)-AJ76 and (+)-UH232; a microdialysis study. *Eur. J. Pharmacol.* **242,** 151–163.
20. Robertson, G. S., Tham, C. S., Wilson, C., Jakubovic, A., and Fibiger, H. C. (1993) In vivo comparisons of the effects of quinpirole and the putative presynaptic dopaminergic agonists B-HT 920 and SND 919 on striatal dopamine and acetylcholine release. *J. Pharmacol. Exp. Ther.* **264,** 1344–1351.
21. Lin, M. Y. and Walters, D. E. (1994) The D2 autoreceptor agonists SND 919 and PD 128483 decrease stereotypy in developing rats. *Life Sci.* **54,** PL17–PL22.
22. Van Hartesveldt, C., Meyer, M. E., and Potter, T. J. (1994) Ontogeny of biphasic locomotor effects of quinpirole. *Pharmacol. Biochem. Behav.* **48,** 781–786.
23. Shalaby, I. and Spear, L. P. (1980) Psychopharmacological effects of low and high doses of apomorphine during ontogeny. *Eur. J. Pharmacol.* **67,** 451–459.
24. Tison, F., Normand, E., and Bloch, B. (1994) Prenatal ontogeny of D2 dopamine receptor and dopamine transporter gene expression in the rat mesencephalon. *Neurosci. Lett.* **166,** 48–50.
25. De Vries, T. J., Mulder, A. H., and Schoffelmeer, A. N. (1992) Differential ontogeny of functional dopamine and muscarinic receptors mediating presynaptic inhibition of neurotransmitter release and postsynaptic regulation of adenylate cyclase activity in rat striatum. *Brain Res.* **66,** 91–96.

26. Andersen, S. L. and Gazzara, R. A. (1993) The ontogeny of apomorphine-induced alterations of neostriatal dopamine release: effects on spontaneous release. *J. Neurochem.* **61,** 2247–2255.
27. Gazzara, R. A. and Andersen, S. L. (1994) The ontogeny of apomorphine-induced alterations of neostriatal dopamine release: effects on potassium-evoked release. *Neurochem. Res.* **19,** 339–345.
28. Andersen, S. L. and Gazzara, R. A. (1994) The development of D_2 autoreceptor-mediated modulation of K^+-evoked dopamine release in the neostriatum. *Brain Res.* **78,** 123–130.
29. Teicher, M. H., Gallitano, A. L., Gelbard, H. A., Evans, H. K., Marsh, E. R., Booth, R. G., and Baldessarini, R. J. (1991) Dopamine D_1 autoreceptor function: possible expression in developing rat prefrontal cortex and striatum. *Dev. Brain Res.* **63,** 229–235.
30. Andersen, S. L. and Teicher, M. H. (1995) Dopamine synthesis inhibition by (±)-7-OH-DPAT in striatum, accumbens, and prefrontal cortex in developing rats. *Soc. Neurosci. Abstr.* **21,** 1780.
31. Wang, L. and Pitts, D. K. (1995) Ontogeny of nigrostriatal dopamine neuron autoreceptors: iontophoretic studies. *J. Pharmacol. Exp. Ther.* **272,** 164–176.
32. Tsuchida, K., Ujike, H., Kanzaki, A., Fujiwara, Y., and Akiyama, K. (1994) Ontogeny of enhanced striatal dopamine release in rats with methamphetamine-induced behavioral sensitization. *Pharmacol. Biochem. Behav.* **47,** 161–169.
33. Beaudet, A. and Woulfe, J. (1992) Morphological substrate for neurotensin–dopamine interactions in the rat midbrain tegmentum. *Ann. NY Acad. Sci.* **668,** 173–185.
34. Deutch, A. Y. and Zahm, D. S. (1992) The current status of neurotensin–dopamine interactions. *Ann. NY Acad. Sci.* **668,** 232–252.
35. Shi, W.-X. and Bunney, B. S. (1992) Actions of neurotensin: a review of the electrophysiological studies. *Ann. NY Acad. Sci.* **668,** 129–145.
36. Shi, W.-X. and Bunney, B. S. (1992) Roles of intracellular cAMP and protein kinase A in the actions of dopamine and neurotensin on midbrain dopamine neurons. *J. Neurosci.* **12,** 2433–2438.
37. Hökfelt, T., Everitt, B. J., Theodorsson-Norheim, E., and Goldstein, M. (1984) Occurrence of neurotensin-like immunoreactivity in subpopulations of hypothalamic, mesencephalic, and medullary catecholamine neurons. *J. Comp. Neurol.* **222,** 543–559.
38. Seroogy, K. B., Ceccatelli, S., Schalling, M., Hökfelt, T., Frey, P., Dockray, G., Buchan, A., and Goldstein, M. (1988) A subpopulation of dopaminergic neurons in the rat ventral mesencephalon contain both neurotensin and cholecystokinin. *Brain Res.* **455,** 88–98.
39. Studler, J. M., Kitabgi, P., Tramu, G., Herve, D., Glowinski, J., and Tassin, J. P. (1988) Extensive colocalization of neurotensin with dopamine in rat meso-cortico-frontal dopaminergic neurons. *Neuropeptides* **11,** 95–100.
40. Bean, A. J., Adrian, T. E., Modlin, I. M., and Roth, R. H. (1989) Storage of dopamine and neurotensin in colocalized and non-colocalized neuronal populations. *J. Pharmacol. Exp. Ther.* **249,** 681–687.
41. During, M. J., Bean, A. J., and Roth, R. H. (1992) Effects of CNS stimulants on the in vivo release of the colocalized transmitters, dopamine and neurtensin, from rat prefrontal cortex. *Neurosci. Lett.* **140,** 129–133.

42. Bean, A. J., During, M. J., and Roth, R. H. (1989) Stimulation-induced release of coexistent transmitters in the prefrontal cortex: an in vivo microdialysis study of dopamine and neurotensin release. *J. Neurochem.* **53,** 655–657.
43. Bean, A. J., During, M. J., and Roth, R. H. (1990) Effects of dopamine autoreceptor stimulation on the release of colocalized transmitters: in vivo release of dopamine and neurotensin from rat prefrontal cortex. *Neurosci. Lett.* **108,** 143–148.
44. Bean, A. J. and Roth, R. H. (1991) Extracellular dopamine and neurotensin in rat prefrontal cortex in vivo: effects of median forebrain bundle stimulation frequency, stimulation pattern, and dopamine autoreceptors. *J. Neurosci.* **11,** 2694–2702.
45. Grace, A. A. and Bunney, B. S. (1984) The control of firing pattern in nigral dopamine neurons: burst firing. *J. Neurosci.* **4,** 2877–2890.
46. Hétier, E., Boireau, A., Dubédat, P., and Blanchard, J. C. (1988) Neurotensin effects on evoked release of dopamine in slices from striatum, nucleus accumbens, and prefrontal cortex in rat. *Naunyn Schmiedeberg's Arch. Pharmacol.* **337,** 13–17.
47. Audinat, E., Hermel, J. M., and Crepel, F. (1989) Neurotensin-induced excitation of neurons of the rat's frontal cortex studied intracellularly in vitro. *Exp. Brain Res.* **78,** 358–368.
48. Sesack, S. R. and Bunney, B. S. (1989) Pharmacological characterization of the receptor mediating electrophysiological responses to dopamine in the rat medial prefrontal cortex: a microiontophoretic study. *J. Pharmacol. Exp. Ther.* **248,** 1323–1333.
49. Deutch, A. Y. and Roth, R. H. (1990) The determinants of stress-induced activation of the prefrontal cortical dopamine system. *Prog. Brain Res.* **85,** 367–402.
50. Nemeroff, C. B. (1986) The interaction of neurotensin with dopaminergic pathways in the central nervous system: basic neurobiology and implications for the pathogenesis and treatment of schizophrenia. *Psychoneuroendocrinology* **11,** 15–37.
51. Weinberger, D. R. (1987) Implications of normal brain development for the pathogenesis of schizophrenia. *Arch. Gen. Psychiatry* **44,** 660–669.
52. Iverfelt, K., Peterson, L. L., Brodin, E., Ogern, S.-O., and Bartfai, T. (1986) Serotonin type-2 receptor mediated regulation of substance P release in the ventral spinal cord and the effects of chronic antidepressant treatment. *Naunyn Schmiedeberg's Arch. Pharmacol.* **333,** 1–6.
53. Lundberg, J. M., Rudehill, A., Sollevi, A., Fried, G., and Wallin, G. (1989) Co-release of neuropeptide Y and noradrenaline from pig spleen in vivo: importance of subcellular storage, nerve impulse frequency and pattern, feedback regulation, and resupply by axonal transport. *Neuroscience* **28,** 475–486.
54. Whim, M. D. and Lloyd, P. E. (1989) Frequency dependent release of peptide cotransmitters from identified cholinergic motor neurons in Aplysia. *Proc. Natl. Acad. Sci. USA* **86,** 9034–9038.
55. Iverfelt, K., Serfozo, P., Diaz Arnesto, L., and Bartfai, T. (1989) Differential release of coexisting neurotransmitters: frequency dependence of the efflux of substance P, thyrotropin releasing hormone and [^{3}H]serotonin from tissue slices of rat ventral spinal cord. *Acta Physiol. Scand.* **137,** 63–71.
56. Bartfai, T., Iverfelt, K., Fisone, G., and Serfozo, P. (1988) Regulation of the release of coexisting neurotransmitters. *Annu. Rev. Pharmacol. Toxicol.* **28,** 285–310.
57. von Euler, G. (1991) Biochemical characterization of the intramembrane interaction between neurotensin and dopamine D_2 receptors in the rat brain. *Brain Res.* **561,** 93–98.

58. von Euler, G., van der Ploeg, I., Fredholm, B. B., and Fuxe, K. (1991) Neurotensin decreases the affinity of dopamine D_2 agonist binding by a G protein-independent mechanism. *J. Neurochem.* **56,** 178–183.
59. Crawley, J. N. and Corwin, R. L. (1994) Biological actions of cholecystokinin. *Peptides* **15,** 731–755.
60. Raiteri, M., Paudice, P., and Vallebuona, F. (1993) Release of cholecystokinin in the central nervous system. *Neurochem. Int.* **22,** 519–527.
61. Marshall, F. H., Barnes, S., Hughes, J., Woodruff, G. N., and Hunter, J. C. (1991) Cholecystokinin modulates the release of dopamine from the anterior and posterior nucleus accumbens by two different mechanisms. *J. Neurochem.* **56,** 917–922.
62. Martin, J. R., Beinfeld, M. C., and Wang, R. Y. (1986) Modulation of cholecystokinin release from posterior nucleus accumbens by D-2 dopamine receptor. *Brain Res.* **397,** 253–258.
63. White, F. J. and Wang, R. Y. (1985) Interactions of cholecystokinin and dopamine on nucleus accumbens neurons. *Brain Res.* **300,** 161–166.
64. Freeman, A. S. and Bunney, B. S. (1987) Activity of A9 and A10 dopaminergic neurons in unrestrained rats: further characterization and effects of apomorphine and cholecystokinin. *Brain Res.* **405,** 46–55.
65. Hommer, D. W., Stoner, G., Crawley, J. N., Paul, S. M., and Skirboll, L. R. (1986) Cholecystokinin-dopamine coexistence: electrophysiological actions corresponding to cholecystokinin receptor subtypes. *J. Neurosci.* **6,** 3039–3043.
66. Skirboll, L. R., Grace, A. A., Homer, D. W., Rehfeld, J., Goldstein, M., Hökfelt, T., and Bunney, B. S. (1981) Peptide-monoamine coexistence: studies of the actions of cholecystokinin-like peptide on the electrical activity of midbrain dopamine neurons. *J. Neurosci.* **6,** 2111–2124.
67. Laitinen, K., Crawley, J. N., Mefford, I. N., and De Witte, P. (1990) Neurotensin and cholecystokinin microinjected into the ventral tegmental area modulate microdialysate concentrations of dopamine and metabolites in the posterior nucleus accumbens. *Brain Res.* **523,** 342–346.
68. Hetey, L., Schwitzkowsky, R., Ott, T., and Barz, H. (1991) Diminished synaptosomal dopamine (DA) release and DA autoreceptor supersensitivity in schizophrenia. *J. Neural Transmiss.* **83,** 25–35.
69. Fedele, E., Andrioli, G. C., Ruelle, A., and Raiteri, M. (1993) Release-regulating dopamine autoreceptors in human cerebral cortex. *Br. J. Pharmacol.* **110,** 20–22.
70. Meador-Woodruff, J. H., Damask, S. P., and Watson, S. J. J. (1994) Differential expression of autoreceptors in the ascending dopamine systems of the human brain. *Proc. Natl. Acad. Sci. USA* **91,** 8297–8301.
71. Hurd, Y. L., Pristupa, Z. B., Herman, M. M., Niznik, H. B., and Kleinman, J. E. (1994) The dopamine transporter and dopamine D_2 receptor messenger RNAs are differentially expressed in limbic- and motor-related subpopulations of human mesencephalic neurons. *Neuroscience* **63,** 357–362.
72. Elsworth, J. D., Redmond, D. E. J., Jr., and Roth, R. H. (1991) Are dopamine synthesis-modulating autoreceptors present in primate brain? *Soc. Neurosci. Abstr.* **17,** 529.
73. Febvret, A., Berger, B., Gaspar, P., and Verney, C. (1991) Further indication that distinct dopaminergic subsets project to the rat cerebral cortex: lack of colocalization with neurotensin in the superficial dopaminergic fields of the anterior cingulate, motor, retrosplenial and visual cortices. *Brain Res.* **547,** 37–52.

74. Gaspar, P., Berger, B., and Febvret, A. (1990) Neurotensin innervation of the human cerebral cortex: lack of colocalization with catecholamines. *Brain Res.* **530,** 181–195.
75. Satoh, K. and Matsumura, H. (1990) Distribution of neurotensin-containing fibers in the frontal cortex of the macaque monkey. *J. Comp. Neurol.* **298,** 215–233.
76. Palacios, M., Savasta, M., and Mengod, G. (1989) Does cholecystokinin colocalize with dopamine in the human substantia nigra? *Brain Res.* **488,** 369–375.
77. Schalling, M., Friberg, K., Seroogy, K., Riederer, P., Bird, E., Schiffmann, S., et al. (1990) Analysis of expression of cholecystokinin in dopamine cells in the ventral mesencephalon of several species and in humans with schizophrenia. *Proc. Natl. Acad. Sci. USA* **87,** 8427–8431.
78. Sokoloff, P., Giros, B., Martres, M. P., Bouthenet, M. L., and Schwartz, J.-C. (1990) Molecular cloning and expression of a novel dopamine receptor (D_3) as a target for neuroleptics. *Nature* **347,** 146–151.
79. Schwartz, J. C., Levesque, D., Martres, M. P., and Sokoloff, P. (1993) Dopamine D_3 receptor: basic and clinical aspects. *Clin. Neuropharmacol.* **16,** 295–314.
80. Bouthenet, M. L., Souil, E., Martres, M. P., Sokoloff, P., Giros, B., and Schwartz, J. C. (1991) Localization of dopamine D_3 receptor mRNA in the rat brain using in situ hybridization: comparison with dopamine D_2 receptor mRNA. *Brain Res.* **564,** 203–219.
81. Arbilla, S. and Langer, S. Z. (1978) Stereoselectivity of presynaptic autoreceptors modulating dopamine release. *Eur. J. Pharmacol.* **76,** 345–351.
82. Galloway, M. P., Wolf, M. E., and Roth, R. H. (1986) Regulation of dopamine synthesis in the medial prefrontal cortex is mediated by release modulating autoreceptors: studies in vivo. *J. Pharmacol. Exp. Ther.* **236,** 689–698.
83. Gainetdinov, R. R., Grekhova, T. V., Sotnikova, T. D., and Rayevsky, K. S. (1994) Dopamine D_2 and D_3 receptor preferring antagonists differentially affect striatal dopamine release and metabolism in conscious rats. *Eur. J. Pharmacol.* **261,** 327–331.
84. Gainetdinov, R. R., Sotnikova, T. D., Grekhova, T. V., and Rayevsky, K. S. (1994) Selective dopamine D3 receptor activation by 7-OH-DPAT induces hypomotility and decreases striatal dopamine release but not metabolism in awake rats. *Soc. Neurosci. Abstr.* **20,** 1355.
85. Booth, R. G., Baldessarini, R. J., Marsh, E., and Owens, C. E. (1994) Actions of (±)-7-hydroxy-*N,N*-dipropylaminotetralin (7-OH-DPAT) on dopamine synthesis in limbic and extrapyramidal regions of rat brain. *Brain Res.* **662,** 283–288.
86. Aretha, C. W., Keegan, M., and Galloway, M. P. (1994) Effects of D_3 preferring ligands on the autoregulation of dopamine (DA) synthesis. *Soc. Neurosci. Abstr.* **20,** 284.
87. Meller, E., Bohmaker, K., Goldstein, M., and Basham, D. A. (1993) Evidence that striatal synthesis-inhibiting autoreceptors are dopamine D_3 receptors. *Eur. J. Pharmacol.* **249,** R5,R6.
87a. White, F. J. (1996) Synaptic regulation of mesocorticolimbic dopamine neurons. *Annu. Rev. Neurosci.* **19,** 405–436.
88. Zetterström, T., Sharp, T., Collin, A. K., and Ungerstedt, U. (1988) In vivo measurement of extracellular dopamine and DOPAC in striatum after various dopamine-releasing drugs: implications for the origin of extracellular DOPAC. *Eur. J. Pharmacol.* **148,** 327–334.

89. Gobert, A., Rivet, J.-M., Audinot, V., Peglion, J.-L., and Millan, M. J. (1994) Modulation of mesolimbic, mesocortical and nigrostriatal dopamine release and synthesis by dopamine D_3 autoreceptors: influence of the selective D_3 antagonist, S14297. *Soc. Neurosci. Abstr.* **20,** 1355.
90. O'Hara, C. M., O'Malley, K. L., and Todd, R. D. (1994) Autoreceptor regulation of dopamine synthesis in a mouse mesencephalic cell line. *Soc. Neurosci. Abstr.* **20,** 644.
91. Tang, L., Todd, R. D., and O'Malley, K. L. (1994) Dopamine D_2 and D_3 receptors inhibit dopamine release. *J. Pharmacol. Exp. Ther.* **270,** 475–479.
92. Gifford, A. N. and Johnson, K. M. (1993) A pharmacological analysis of (+)-AJ76 and (+)-UH232 at release regulating pre- and postsynaptic dopamine receptors. *Eur. J. Pharmacol.* **237,** 169–175.
93. Lejeune, F. and Millan, M. J. (1995) Activation of dopamine D_3 autoreceptors inhibits firing of ventral tegmental dopaminergic neurones in vivo. *Eur. J. Pharmacol.* **275,** R7–R9.
94. Kreiss, D. S., Bergstrom, D. A., Gonzalez, A. M., Huang, K.-X., Sibley, D. R., and Walters, J. R. (1995) Dopamine receptor agonist potencies for inhibition of cell firing correlate with dopamine D_3 receptor binding affinities. *Eur. J. Pharmacol.* **277,** 209–214.
95. Svensson, K., Carlsson, A., Huff, R. M., Kling-Petersen, T., and Waters, N. (1994) Behavioral and neurochemical data suggest functional differences between dopamine D_2 and D_3 receptors. *Eur. J. Pharmacol.* **263,** 235–243.
96. Zhang, M., Tarazi, F. I., and Creese, I. (1994) Antisense knockout of rat CNS dopamine D_3 receptors: behavioral effects. *Soc. Neurosci. Abstr.* **20,** 909.
97. Kreiss, D. S., Mouradian, M. M., and Walters, J. R. (1994) Effects on in vivo dopamine cell firing of dopamine D_2 and D_3 receptor antisense oligonucleotides infused into rat substantia nigra pars compacta. *Soc. Neurosci. Abstr.* **20,** 909.
98. MacKenzie, R. G., VanLeewen, D., Pugsley, T. A., Shih, Y. H., Demattos, S., Tang, L., Todd, R. D., and O'Malley, K. L. (1994) Characterization of the human dopamine D_3 receptor expressed in transfected cell lines. *Eur. J. Pharmacol.* **266,** 79–85.
99. Fishburn, C. S., Belleli, D., David, C., Carmon, S., and Fuchs, S. (1993) A novel short isoform of the D_3 dopamine receptor generated by alternative splicing in the third cytoplasmic loop. *J. Biol. Chem.* **268,** 5872–5878.
100. Schmauss, C., Levenson, R., Bergson, C., and Liu, K. (1994) Expression of a dopamine D_3-receptor-like protein (D_{3nf}) in human cortex. *Soc. Neurosci. Abstr.* **20,** 644.
101. Liu, L., Shen, R.-Y., Kapatos, G., and Chiodo, L. A. (1994) Dopamine neuron membrane physiology: characterization of the transient outward current (IA) and demonstration of a common signal transduction pathway for IA and IK. *Synapse* **17,** 230–240.
102. Castellano, M. A., Liu, L.-X., Monsma, F. J. J., Sibley, D. R., Kapatos, G., and Chiodo, L. A. (1993) Transfected D_2 short dopamine receptors inhibit voltage-dependent potassium current in neuroblastoma X glioma (NG108-15) cells. *Mol. Pharmacol.* **44,** 649–656.
103. Liu, L.-X., Monsma, F. J. J., Sibley, D. R., and Chiodo, L. A. (1993) Coupling of D_2-long receptor isoform to K^+ currents in neuroblastoma X glioma (NG108-15) cells. *Soc. Neurosci. Abstr.* **19,** 79.

104. Chiodo, L. A., Liu, L.-X., Monsma, F. J. J., and Sibley, D. R. (1993) Transfected D_3 dopamine receptors inhibit voltage-dependent potassium current in neuroblastoma X glioma hybrid (NG108-15) cells. *Soc. Neurosci. Abstr.* **19,** 79.
105. Liu, L.-X., Kapatos, G., and Chiodo, L. A. (1992) DA autoreceptor modulation of different calcium currents in DA neurons. *Soc. Neurosci. Abstr.* **18,** 1516.
106. Chiodo, L. A. and Liu, L.-X. (1994) DA autoreceptor stimulation increases the voltage dependent Na^+ current observed in identified DA neurons. *Soc. Neurosci. Abstr.* **20,** 522.
107. Bowyer, J. F. and Weiner, N. (1989) K^+ channel and adenylate cyclase involvement in regulation of Ca^{2+}-evoked release of [^{3}H]dopamine from synaptosomes. *J. Pharmacol. Exp. Ther.* **248,** 514–520.
108. Memo, M., Missale, C., Carruba, M. O., and Spano, P. F. (1986) D_2 dopamine receptors associated with inhibition of dopamine release from rat neostriatum are independent from cyclic AMP. *Neurosci. Lett.* **71,** 192–196.
109. Cass, W. A. and Zahniser, N. R. (1991) Potassium channel blockers inhibit D_2 dopamine, but not A1 adenosine, receptor-mediated inhibition of striatal dopamine release. *J. Neurochem.* **57,** 147–152.
110. Tanaka, T., Vincent, S. R., Nomikos, G. G., and Fibiger, H. C. (1992) Effect of quinine on autoreceptor-regulated dopamine release in the rat striatum. *J. Neurochemistry* **59,** 1640–1645.
111. El-Mestikawy, S., Glowinski, J., and Hamon, M. (1986) Presynaptic dopamine autoreceptors control tyrosine hydroxylase activation in depolarized striatal dopaminergic terminals. *J. Neurochem.* **46,** 12–22.
112. Goldstein, M. (1995) Long- and short-term regulation of tyrosine hydroxylase, in *Psychopharmacology: The Fourth Generation of Progress* (Bloom, F. E. and Kupfer, D. J., eds.), Raven, New York, pp. 189–196.
113. Haycock, J. (1987) Stimulation-dependent phosphorylation of tyrosine hydroxylase in rat corpus striatum. *Brain Res. Bull.* **19,** 619–622.
114. Salah, R. S., Kuhn, D. M., and Galloway, M. P. (1989) Dopamine autoreceptors modulate the phosphorylation of tyrosine hydroxylase in rat striatum. *J. Neurochemistry* **52,** 1517–1522.
115. El-Mestikawy, S. and Hamon, M. (1986) Is dopamine-induced inhibition of adenylate cyclase involved in the autoreceptor mediated negative control of tyrosine hydroxylase in striatal dopaminergic terminals? *J. Neurochem.* **47,** 1425–1433.
116. Onali, P. L. and Olianas, M. C. (1989) Involvement of adenylate cyclase inhibition in dopamine autoreceptor regulation of tyrosine hydroxylase in rat nucleus accumbens. *Neurosci. Lett.* **102,** 91–96.
117. Strait, K. A. and Kuczenski, R. (1986) Dopamine autoreceptor regulation of the kinetic state of striatal tyrosine hydroxylase. *Mol. Pharmacol.* **29,** 561–569.
118. Onali, P., Mosca, E., and Olianas, M. C. (1992) Presynaptic dopamine autoreceptors and second messengers controlling tyrosine hydroxylase activity in rat brain. *Neurochem. Int.* **20,** 89S–93S.
119. Tissari, A. H. and Lillgäls, M. S. (1993) Reduction of dopamine synthesis inhibition by dopamine autoreceptor activation in striatal synaptosomes with in vivo reserpine administration. *J. Neurochem.* **61,** 231–238.
120. Fujiwara, H., Kato, N., Shuntoh, H., and Tanaka, C. (1987) D_2-dopamine receptor-mediated inhibition of intracellular Ca^{2+} mobilization and release of acetylcholine from guinea-pig neostriatal slices. *Br. J. Pharmacol.* **91,** 287–297.

121. Bannon, M. J., Reinhard, J. F. J., Bunney, E. B., and Roth, R. H. (1982) Mesocortical dopamine neurons: unique response to antipsychotic drugs explained by absence of terminal autoreceptors. *Nature* **296,** 444–446.
122. Matsumoto, T., Uchimura, H., Hirano, M., Kim, J. S., Yokoo, H., Shimomura, M., Nakahara, T., Inoue, K., and Oomagari, K. (1983) Differential effects of acute and chronic administration of haloperidol on homovanillic acid levels in discrete dopaminergic areas of rat brain. *Eur. J. Pharmacol.* **89,** 27–33.
123. Scatton, B. (1977) Differential regional development of tolerance to increase dopamine turnover upon repeated neuroleptic administration. *Eur. J. Pharmacol.* **46,** 363–369.
124. Nowycky, M. and Roth, R. H. (1978) Dopaminergic neurons: role of presynaptic receptors in the regulation of transmitter biosynthesis. *Prog. Neuro-Psychopharmacol.* **2,** 139–158.
125. King, G. R., Ellinwood, E. H. J., Silvia, C., Joyner, C. M., Xue, Z., Caron, M. G., and Lee, T. H. (1994) Withdrawal from continuous or intermittent cocaine administration: changes in D_2 receptor function. *J. Pharmacol. Exp. Ther.* **269,** 743–749.
126. Post, R. M. (1977) Progressive changes in behavior and seizures following chronic cocaine administration: relationship to kindling and psychosis, in *Advances in Behavioral Biology: Cocaine and Other Stimulants* (Ellinwoood, E. H. and Kilbey, M. M., eds.), Plenum, New York, pp. 353–372.
127. Robinson, T. E. and Berridge, K. C. (1993) The neural basis of drug craving: an incentive-sensitization theory of addiction. *Brain Res. Rev.* **18,** 247–291.
128. Seutin, V., Verbanck, P., Massotte, L., and Dresse, A. (1991) Acute amphetamine-induced subsensitivity of A10 dopamine autoreceptors in vitro. *Brain Res.* **558,** 141–144.
129. Antelman, S. M. and Chiodo, L. A. (1981) Dopamine autoreceptor subsensitivity: a mechanism common to the treatment of depression and the induction of amphetamine psychosis. *Biol. Psychiatry* **16,** 717–727.
130. Henry, D. J., Margaret, A. G., and White, F. J. (1989) Electrophysiological effects of cocaine in the mesoaccumbens dopamine system: repeated administration. *J. Pharmacol. Exp. Ther.* **251,** 833–839.
131. Kamata, K. and Rebec, G. V. (1984) Long-term amphetamine treatment attenuates or reverses the depression of neuronal activity produced by dopamine agonists in the ventral tegmental area. *Life Sci.* **34,** 2419–2427.
132. White, F. J. and Wang, R. Y. (1984) Electrophysiological evidence for A10 dopamine autoreceptor subsensitivity following chronic D-amphetamine treatment. *Brain Res.* **309,** 283–292.
133. Ackerman, J. M. and White, F. J. (1990) A10 somatodendritic dopamine autoreceptor sensitivity following withdrawal from repeated cocaine treatment. *Neuroscience Lett.* **117,** 181–187.
134. Yamada, S., Yokoo, H., and Nishi, S. (1991) Changes in sensitivity of dopamine autoreceptors in rat striatum after subchronic treatment with methamphetamine. *Eur. J. Pharmacol.* **205,** 43–47.
135. Harsing, L. G. J., Sershen, H., and Lajtha, A. (1992) Dopamine efflux from striatum after chronic nicotine: evidence for autoreceptor desensitization. *J. Neurochem.* **59,** 48–54.
136. Yi, S.-J. and Johnson, K. M. (1990) Chronic cocaine treatment impairs the regulation of synaptosomal ^{3}H-DA release by D_2 autoreceptors. *Pharmacol. Biochem. Behav.* **36,** 457–461.

137. Fitzgerald, J. L. and Reid, J. J. (1991) Chronic cocaine treatment does not alter rat striatal D_2 autoreceptor sensitivity to pergolide. *Brain Res.* **541,** 327–333.
138. Gifford, A. N. and Johnson, K. M. (1992) Effect of chronic cocaine treatment on D_2 receptors regulating the release of dopamine and acetylcholine in the nucleus accumbens and striatum. *Pharmacol. Biochem. Behav.* **41,** 841–846.
139. Dwoskin, L. P., Peris, J., Yasuda, R. P., Philpott, K., and Zahniser, N. R. (1988) Repeated cocaine administration results in supersensitivity of striatal D2 dopamine autoreceptors to pergolide. *Life Sci.* **42,** 255–262.
140. Baumann, M. H. and Rothman, R. B. (1993) Effects of acute and chronic cocaine on the activity of tuberoinfundibular dopamine neurons in the rat. *Brain Res.* **608,** 175–179.
141. Galloway, M. P. (1990) Regulation of dopamine and serotonin synthesis by acute administration of cocaine. *Synapse* **6,** 63–72.
142. Jeziorski, M. and White, F. J. (1989) Dopamine agonists at repeated "autoreceptor-selective" doses: effects upon the sensitivity of A10 dopamine autoreceptors. *Synapse* **4,** 267–280.
143. Anden, N.-E., Golembiowska-Nikitin, K., and Thornström, U. (1982) Selective stimulation of dopamine and noradrenaline autoreceptors by B-HT920 and B-HT933, respectively. *Naunyn Schmiedeberg's Arch. Pharmacol.* **321,** 100–104.
144. Clemens, J. A., Fuller, R. W., Phebus, L. A., Smalstig, E. B., and Hynes, M. D. (1984) Stimulation of presynaptic dopamine autoreceptors by 4-(2-di-n-propyl-aminoethyl)indol (DPAI). *Life Sci.* **34,** 1015–1022.
145. Seyfried, C. A., Fuxe, K., Wolf, H.-P., and Agnati, L. F. (1982) Demonstration of a new type of dopamine receptor agonist: an indolyl-3-butylamine. Actions at intact versus supersensitive dopamine receptors in the rat forebrain. *Acta Physiol. Scand.* **116,** 465–468.
146. Hjorth, S., Svensson, K., Carlsson, A., Wikstrom, H., and Andersson, B. (1986) Central dopaminergic properties of HW-165 and its enantiomers; transoctahydro-benzo(f)quinoline cogeners of 3-PPP. *Naunyn-Schmiedeberg's Arch. Pharmacol.* **333,** 205–218.
147. Yasuda, Y., Kikuchi, T., Suzuki, S., Tsutsui, M., Yamada, K., and Hiyama, T. (1988) 7-[3-(4-[2,3-Dimethylphenyl]piperazinyl)propoxy]-2(1H)-quinoline (OPC-4392), a presynaptic dopamine autoreceptor agonist and postsynaptic D2 receptor antagonist. *Life Sci.* **42,** 1941–1954.
148. Pugsley, T. A., Christofferson, C. L., Corbin, A., DeWald, H. A., Demattos, S., Meltzer, L. T., et al. (1992) Pharmacological characterization of PD 118717, a putative piperazinyl benzopyranone dopamine autoreceptor agonist. *J. Pharmacol. Exp. Ther.* **263,** 1147–1158.
149. Jaen, J. C., Wise, L. D., Heffner, T. G., Pugsley, T. A., and Meltzer, L. T. (1992) Dopamine autoreceptor agonists as potential antipsychotics. 2. (Aminoalkoxy)-4H-1-benzopyran-4-ones. *J. Med. Chem.* **34,** 248–256.
150. Wright, J. L., Caprathe, B. W., Downing, D. M., Glase, S. A., Heffner, T. G., Jaen, J. C., et al. (1994) The discovery and structure–activity relationships of 1,2,3,6-tetrahydro-4-phenyl-1-[(arylcyclohexenyl)alkyl]pyridines. Dopamine autoreceptor agonists and potential antipsychotic agents. *J. Med. Chem.* **37,** 3523–3533.
151. Meltzer, L. T., Caprathe, B. W., Christoffersen, C. L., Corbin, A. E., Jaen, J. C., Ninteman, F. W., et al. (1993) Pharmacological profile of the dopamine partial agonist, (±)-PD 128483 and its enantiomers. *J. Pharmacol. Exp. Ther.* **266,** 1177–1189.

152. Caprathe, B. W., Jaen, J. C., Wise, L. D., Heffner, T. G., Pugsley, T. A., Meltzer, L. T., and Parvez, M. (1991) Dopamine autoreceptor agonists as potential antipsychotics. 3. 6-Propyl-4,5,5a,6,7,8-hexahydrothiazolo[4,5-f]quinolin-2-amine. *J. Med. Chem.* **34,** 2736–2746.
153. Hjorth, S., Carlsson, A., Wikström, H., Lindberg, P., Sanchez, D., Hacksell, U., Arvidsson, L.-E., Svensson, U., and Nilsson, J. L. G. (1981) 3-PPP, a new centrally acting DA-receptor agonist with selectivity for autoreceptors. *Life Sci.* **28,** 1225–1238.
154. Nisoli, E., Tonello, C., Imhof, R., Scherschlicht, R., da Prada, M., and Carruba, M. O. (1993) Neurochemical and behavioral evidence that Ro 41-9067 is a selective presynaptic dopamine receptor agonist. *J. Pharmacol. Exp. Ther.* **266,** 97–105.
155. Coward, D. M., Dixon, A. K., Urwyler, S., White, T. G., Enz, A., Karobath, M., and Shearman, G. (1990) Partial dopamine-agonistic and atypical neuroleptic properties of the amino-ergolines SDZ 208-911 and SDZ 208-912. *J. Pharmacol. Exp. Ther.* **252,** 279–285.
156. Mierau, J. and Bechtel, W. D. (1988) SND 919 inhibits dopamine release in vivo and in vitro. *Psychopharmacology* **96,** S338.
157. Kehr, W. (1984) Transdihydrolisuride, a partial dopamine receptor antagonist: effects on monoamine metabolism. *Eur. J. Pharmacol.* **97,** 111–119.
158. Goodale, D. B., Rusterholz, D. B., Long, J. P., Flynn, J. R., Walsh, B., Cannon, J. G., and Lee, T. (1980) Neurochemical and behavioral evidence for a selective presynaptic receptor agonist. *Science* **210,** 1141–1143.
159. Lahti, R. A., Evans, D. L., Figur, L. M., Huff, R. M., and Moon, M. W. (1993) Pre- and postsynaptic dopaminergic activities of U-86170F. *Naunyn-Schmiedeberg's Arch. Pharmacol.* **344,** 509–513.
160. Piercey, M. F., Broderick, P. A., Hoffmann, W. E., and Vogelsang, G. D. (1990) U-66444B and U-68553B, potent autoreceptor agonists at dopaminergic cell bodies and terminals. *J. Pharmacol. Exp. Ther.* **254,** 369–374.
161. Svensson, K., Johansson, A. M., Magnusson, T., and Carlsson, A. (1986) (+)-AJ 76 and (+)-UH 232: central stimulants acting as preferential dopamine autoreceptor antagonists. *Naunyn-Schmiedeberg's Arch. Pharmacol.* **334,** 234–245.
162. Bischoff, S., Baumann, P., Krauss, J., Maître, L., Vassout, A., Storni, A., and Chouinard, G. (1994) CGP 25454A, a novel and selective presynaptic dopamine autoreceptor antagonist. *Naunyn-Schmiedeberg's Arch. Pharmacol.* **350,** 230–238.
163. Sonesson, C., Lin, C.-H., Hansson, L., Waters, N., Svensson, K., Carlsson, A., Smith, M. W., and Wilkström, H. (1994) Substituted (S)-phenylpiperidines and rigid congeners as preferred dopamine autoreceptor antagonists: synthesis and structure-activity relationships. *J. Med. Chem.* **37,** 2735–2753.
164. Kling-Petersen, T., Ljung, E., Wollter, L., and Svensson, K. (1995) Effects of the dopamine D3- and autoreceptor preferring antagonist (-)-DS121 on locomotor activity, conditioned place preference and intracranial self-stimulation in the rat. *Behav. Pharmacol.* **6,** 107–115.
165. Seeman, P., George, S., and Watanabe, M. (1984) Presynaptic dopamine receptors operate in the high affinity state for dopamine. Postsynaptic ones work in low-affinity state. *Soc. Neurosci. Abstr.* **10,** 278.
166. Carlsson, A. (1983) Dopamine receptor agonists: intrinsic activity vs. state of receptor. *J. Neural Transmiss.* **57,** 309–315.

167. Meller, E., Bohmaker, K., Namba, Y., Friedhoff, A. J., and Goldstein, M. (1987) Relationship between receptor occupancy and response at striatal dopamine autoreceptors. *Mol. Pharmacol.* **31,** 592–598.
168. Cox, R. F. and Waszczak, B. L. (1989) Differences in dopamine receptor reserve for *N*-n-propylnorapomorphine enantiomers: single unit recording studies after partial inactivation of receptors by N-ethoxycarbonyl-2-ethoxy-1,2-dihydroquinoline. *Mol. Pharmacol.* **35,** 125–131.
169. Yokoo, H., Goldstein, M., and Meller, E. (1988) Receptor reserve at striatal dopamine receptors modulating the release of [^{3}H]dopamine. *Eur. J. Pharmacol.* **155,** 323–327.
170. Cox, R. F. and Waszczak, B. L. (1990) Irreversible receptor inactivation reveals differences in dopamine receptor reserve between A9 and A10 dopamine systems: an electrophysiological analysis. *Brain Res.* **534,** 273–282.
171. Meller, E., Enz, A., and Goldstein, M. (1988) Absence of receptor reserve at striatal dopamine receptors regulating cholinergic neuronal activity. *Eur. J. Pharmacol.* **155,** 151–154.
172. Bohmaker, K., Puza, T., Goldstein, M., and Meller, E. (1989) Absence of spare autoreceptors regulating dopamine agonist inhibition of tyrosine hydroxylation in slices of rat striatum. *J. Pharmacol. Exp. Ther.* **248,** 97–103.
173. Seyfried, C. A. (1988) Comparison of EMD 38362, (+)3-PPP, (–)3-PPP and BHT 920 in pre- and postsynaptic models for D2-activity: discrepancies between in vivo and in vitro results, in *Pharmacology and Functional Regulation of Dopaminergic Neurons* (Beart, P. M., Woodruff, G. N., and Jackson, D. M., eds.), MacMillan, London, pp. 187–190.
174. Clark, D. and White, F. J. (1987) D_1 dopamine receptor—the search for a function: a critical evaluation of the D_1/D_2 dopamine receptor classification and its functional implications. *Synapse* **1,** 347–388.
175. Wachtel, S. R., Hu, X. T., Galloway, M. P., and White, F. J. (1989) D_1 dopamine receptor stimulation enables the postsynaptic, but not autoreceptor, effects of D_2 dopamine agonists in nigrostriatal and mesoaccumbens dopamine systems. *Synapse* **4,** 327–346.
176. Tamminga, C. A., Gotta, M. D., Thaker, G. K., Alphs, L. D., and Foster, N. L. (1986) Dopamine agonist treatment of schizophrenia with N-propylnorapomorphine. *Arch. Gen. Psychiat.* **43,** 398–402.
177. Benkert, O., Grunder, G., and Wetzel, H. (1992) Dopamine autoreceptor agonists in the treatment of schizophrenia and major depression. *Pharmacopsychiatry* **25,** 254–260.
178. Wetzel, H., Hillert, A., Grunder, G., and Benkert, O. (1994) Roxindole, a dopamine autoreceptor agonist, in the treatment of positive and negative schizophrenic symptoms. *Am. J. Psychiat.* **151,** 1499–1502.
179. Mierau, J. and Schingnitz, G. (1992) Biochemical and pharmacological studies on pramipexole, a potent and selective dopamine D_2 receptor agonist. *Eur. J. Pharmacol.* **215,** 161–170.
180. Svensson, K., Eriksson, E., and Carlsson, A. (1993) Partial dopamine receptor agonists reverse behavioral, biochemical and neuroendocrine effects of neuroleptics in the rat: potential treatment of extrapyramidal side effects. *Neuropharmacology* **32,** 1037–1045.

181. Richard, M. G. and Bennett, J. P. J. (1994) Regulation by D2 dopamine receptors of in vivo dopamine synthesis in striata of rats and mice with experimental parkinsonism. *Exp. Neurol.* **129,** 57–63.
182. Kling-Petersen, T., Ljung, E., and Svensson, K. (1994) The preferential dopamine autoreceptor antagonist (+)-UH 232 antagonizes the positive reinforcing effects of cocaine and d-amphetamine in the ICSS paradigm. *Pharmacol. Biochem. Behav.* **49,** 345–351.
183. Piercey, M. F., Lum, J. T., Hoffmann, W. E., Carlsson, A., Ljung, E., and Svensson, K. (1992) Antagonism of cocaine's pharmacological effects by the stimulant dopaminergic antagonists, (+)-AJ 76 and (+)-UH 232. *Brain Res.* **588,** 217–222.
184. Richardson, N. R., Piercey, M. F., Svensson, K., Collins, R. J., Myers, J. E., and Roberts, D. C. (1993) Antagonism of cocaine self-administration by the preferential dopamine autoreceptor antagonist, (+)-AJ 76. *Brain Res.* **619,** 15–21.
185. Clark, D., Exner, M., Furmidge, L. J., Svensson, K., and Sonesson, C. (1995) Effects of the dopamine autoreceptor antagonist (–)-DS121 on the discriminative properties of d-amphetamine and cocaine. *Eur. J. Pharmacol.* **275,** 67–74.
186. Vanover, K. E. and Woolverton, W. L. (1994) Behavioral effects of the dopamine autoreceptor agonist PD 128483 alone in combination with cocaine. *J. Pharmacol. Exp. Ther.* **270,** 1049–1056.

CHAPTER 9

Electrophysiological Effects of Dopamine Receptor Stimulation

Johan Grenhoff and Steven W. Johnson

1. Introduction

The involvement of dopamine in multiple aspects of brain function has produced a great interest in dopaminergic pharmacology, and a large number of dopamine receptor ligands have been developed and tested. Dopamine receptors were initially divided into D1 and D2* subtypes on the basis of pharmacological and biochemical criteria *(1)*. D1 receptors mediate stimulation, and D2 receptors inhibition, of adenylate cyclase by dopamine, and certain compounds interact selectively with the two types of receptor *(2)*. In recent years, molecular cloning studies have demonstrated the presence of at least five genetically different dopamine receptor subtypes in the mammalian nervous system *(3,4)*. Since the cloned D_1 and D_5 receptors are highly homologous and pharmacologically similar, they can be viewed as members of the D1 subfamily. In parallel, the cloned D_2, D_3, and D_4 subtypes belong to the D2 subfamily. The pharmacologically characterized D1 and D2 receptors may involve different members of these two subfamilies. In functional studies, such as those reviewed here, the relative lack of selective compounds necessitates the continued use of the conservative D1/D2 classification, where D1 and D2 refer to the entire subfamily and not the cloned subtype. The only exception is the section on expression systems, where genetically characterized subtypes have been studied in isolation. Still, the expanded knowledge of dopamine receptor structures calls for a re-evaluation of previous concepts

*As detailed in the Preface, the terms D1 or D1-like and D2 or D2-like are used to refer to the subfamilies of DA receptors, or used when the genetic subtype (D_1 or D_5; D_2, D_3, or D_4) is uncertain.

The Dopamine Receptors Eds.: K. A. Neve and R. L. Neve
Humana Press Inc., Totowa, NJ

and an enhanced understanding of the five (or more) subtypes, which will stimulate the development of even more selective pharmacological tools and clinically useful drugs.

The functional role of dopamine receptors in regulating the electrical activity of nerve cells has never been clear. One reason for this is that dopamine, unlike some other neurotransmitters, such as acetylcholine, has no well-defined peripheral synapse that can serve as a general model for dopaminergic transmission, so the brain has to be approached directly. In the initial studies during the 1960s and 1970s, neuronal firing in various brain regions was recorded extracellularly in anesthetized or paralyzed animals, and dopamine was mostly applied locally by iontophoresis. This type of experiment suffers from several limitations. One is the undefined role of anesthesia, another that the recorded cells cannot be identified with certainty. Furthermore, the effective concentration of focally applied substances is unknown, equilibrium conditions are unattainable, and the applied substance may diffuse and affect not only the recorded cell, but also neighboring neurons. The last mechanism is illustrated by the excitatory effect of iontophoretically applied opioids on hippocampal pyramidal cells and mesencephalic dopamine neurons, which in both cases is caused by inhibition of inhibitory interneurons, as clarified by intracellular recording *(5,6)*. Although a method that does not distinguish between excitatory and inhibitory effects of receptor stimulation cannot provide clear evidence of receptor mechanisms, extracellular recordings may provide useful adjunctive information when interpreted in the light of more detailed results derived from intracellular or patch recordings.

Another problem with extracellular single-unit recording is that only firing, i.e., nerve activity produced by action potentials, can be studied, which led researchers to focus on the question, Is the action of dopamine inhibitory or excitatory? A natural subject for the study of this question was the striatum, the terminal region of the nigrostriatal dopamine pathway. The results from these studies were complicated and contradictory enough to prompt contemporaneous reviewers to arrive at opposite conclusions as to whether the action of dopamine in the striatum was mainly excitatory *(7)* or inhibitory *(8)*. Meanwhile, intracellular recording from neurons in brain slices led to a growing appreciation of how several types of ionic conductance can participate in the regulation of cell firing *(9)* and the conclusion that neurotransmitters can say more than "yes" or "no" (i.e., they can be modulatory rather than purely excitatory or inhibitory) *(10)*. These modulatory transmitter effects were first described in two-electrode voltage clamp experiments in large invertebrate neurons, but the development of the single-electrode voltage clamp, which allowed recording of membrane currents from smaller cells in brain slices, suggested that modulatory actions of neurotransmitters are widespread also in the mammalian central nervous system (CNS) *(11)*. The use of the powerful patch clamp methodology in central neurons has provided more examples, and also enabled detailed study of synaptic function in the brain *(12)*. In the case of dopamine, these techniques have revealed multiple actions on voltage-dependent currents, actions that can partly explain the complexity of dopamine action in the older literature.

Intracellular recording and patch clamp methodology in isolated preparations offer some distinct advantages in the characterization of the functional role of dopamine receptors. Intracellular recording of membrane potential with current clamp allows the study of synaptic events and pharmacological effects that do not directly trigger or inhibit action potentials, and provides information on membrane conductances underlying de- and hyperpolarizations. Voltage clamp with intracellular microelectrodes (intracellular recording) or patch pipets (whole-cell recording) brings the analysis further by isolating the ionic current affected by a drug or transmitter from the interference of other voltage-dependent currents. Single-channel recording with patch pipets enables direct observation of the ion channel mediating the currents. These techniques can evaluate whether or not functional receptors are present on a certain type of cell, whether the effect is pre- or postsynaptic, and if the effect is voltage-dependent. Recording electrodes, particularly patch pipets, can also be used to manipulate the internal milieu of the cell, thereby providing information about transduction mechanisms and additional modulatory pathways, whereas the external medium of the preparation can be completely controlled, allowing analysis of ionic mechanisms and pharmacological dose–effect relations under equilibrium conditions.

This chapter attempts to answer the question, What are the electrophysiological effects of dopamine receptor stimulation?, by providing a comprehensive review of the relevant literature. The review is organized according to the type of preparation studied rather than to receptor subtype or effector mechanism, because we did not want to reconcile clearly disparate results by force, but rather acknowledge the complexity and variation of the subject. Nevertheless, parallels are occasionally drawn, reflections are made along the way, and a few general conclusions are offered at the end.

2. Expression Systems

The description of five different dopamine receptor subtypes and their distribution and receptor binding properties has prompted investigation into the functional properties of each subtype. This type of study is difficult in the brain, since the subtypes are not strictly segregated between regions, and few compounds have any appreciable selectivity for a certain subtype. Furthermore, it is not clear to what extent selectivity in receptor binding experiments translates to selectivity on the functional level. This problem is exemplified by 7-OH-DPAT, a substance which is 100-fold selective for the D_3 vs the D_2 receptor in binding experiments *(13),* but appears functionally equipotent at the two receptors in cultured cells expressing D_2 or D_3 receptors *(14).*

So far, the only type of study that can unambiguously distinguish among the five subtypes is the use of cultured cells transfected with the DNA of one subtype. However, it is difficult to draw conclusions about effectors (e.g., ion channels) from such studies, since effector mechanisms vary according to the employed expression system. An example is the expressed D_2 receptor, which can couple to completely different second messenger systems in two different cell lines *(15).* The report that stimulation of protein kinase C can switch the biochemical signaling pathway of transfected D_2 receptors from inhibition of adenylate cyclase to release of arachidonic acid *(16)* adds further weight to the importance of the cellular transduction machinery of the expression system. Lack of effect of a transfected receptor in a cell line might also be dependent on the cell: The D_3 receptor apparently is not coupled to any investigated effectors in the rat pituitary GH_4C_1 cell line *(17),* but couples to calcium channels in the neuroblastoma × glioma cell line NG108-15 (*see* Section 2.2.). Still, bearing these caveats in mind, studies of receptor subtypes in expression systems provide unique information that can be related to transduction mechanisms of endogenous dopamine receptors.

2.1. D_1 and D_5 Receptors

To our knowledge, no electrophysiological studies have been published on genetically expressed D_1 or D_5 receptors. In most expression systems, cloned D_1 and D_5 receptors are coupled to stimulation of adenylate cyclase via a G protein of the G_s subtype *(4).* Human D_1 receptors transfected to GH_4C_1 cells induce a cyclic adenosine monophosphate (cAMP)-dependent calcium increase measured with calcium-sensitive fluorescence *(18).* The calcium increase is blocked by the dihydropyridine nifedipine, indicating that it is mediated via activation of L-type calcium channels, as previously described for other receptors positively coupled to adenylate cyclase, e.g., beta adrenergic receptors in heart *(19)* and hippocampus *(20).*

2.2. D2-Like Receptors

D_2 receptors are expressed in two forms produced by alternative splicing, the short D_{2S} and long D_{2L} forms. The two forms do not differ with regard to receptor binding or transduction mechanisms in expression studies *(3,4),* but could be coupled to different G proteins in natural systems. Both forms couple to opening of potassium channels when expressed in GH_4 pituitary cells *(21)*. In contrast, in NG108-15 cells, dopamine D_{2S} receptor stimulation has been reported to inhibit potassium current via a calcium-dependent, pertussis toxin-insensitive mechanism *(22)*. Stimulation of the human D_{2S} receptor in differentiated NG108-15 cells leads to inhibition of both N- and L-type voltage-activated calcium currents studied with whole-cell voltage clamp *(23)*. Dihydropyridine-sensitive (L-type) calcium currents are decreased by stimulation of human D_{2S} receptors in GH_4C_1 cells as well *(24)*. Suppression of such high-threshold calcium currents is frequently connected with inhibition of neurotransmitter release mediated by endogenous D_2 receptors.

Transfected D_3 receptors do not couple to any investigated effector mechanism in nondifferentiated NG108-15 cells *(25)*, but, in differentiated cells of the same type, they mediate partial inhibition of high-threshold calcium currents studied with whole-cell recording *(26)*. The effect is blocked by the D_2/D_3 antagonist sulpiride and by pretreatment with pertussis toxin, suggesting mediation by pertussis toxin-sensitive G proteins of the G_i and/or G_o subtype. If endogenous D_3 receptors are similarly coupled, they could be involved in presynaptic inhibition of transmitter release produced by pharmacologically defined D2 receptors.

Similar to D_{2S} receptors, human D_4 receptors expressed in GH_4C_1 cells mediate depression of high-threshold calcium currents by dopamine *(24)*, an effect mimicked by the agonist quinpirole. Thus, the many reported examples of presynaptic inhibition with D2-like pharmacology could involve receptors of the D_4 subtype, if they are coupled to calcium channels in natural systems as well.

3. Invertebrates

As mentioned in Section 1., powerful electrophysiological techniques that have only recently become available for the study of the mammalian CNS could be applied much earlier in certain invertebrate preparations, e.g., voltage clamp in large molluscan neurons. Although results from invertebrates cannot be generalized to mammalian neurons, they can offer parallels, suggestions, and experimental detail that are of value for the understanding of dopamine receptor electrophysiology in more complex systems.

In the marine mollusk *Aplysia,* inhibitory responses to dopamine produced by an increased potassium conductance have been described in pleural *(27)* and cerebral *(28,29)* ganglion cells. Excitatory responses to dopamine, on the other hand, are mediated by a different, uncharacterized receptor and an unknown ionic mechanism *(27)*. The opening of potassium channels, which underlies the inhibitory responses to dopamine, acetylcholine, or histamine in ganglion cells, appears to be an example of receptor convergence on the G protein level *(30)*. Dopamine also decreases calcium currents in cultured R15 neurons from young *Aplysia (31)*.

In giant neurons from the snail *Lymnaea stagnalis,* dopamine decreases a calcium current by an action unrelated to intracellular [Ca^{2+}] or cAMP *(32)*. Growth hormone-producing cells in *Lymnaea,* on the other hand, display two different responses to dopamine: a hyperpolarization which is mimicked by high concentrations of D2 agonists and blocked by sulpiride, and an increased excitability mimicked by the D1 agonist SKF 38393 and blocked by the D1 antagonist SCH 23390 *(33)*. The excitatory effect is also produced by treatments enhancing cyclic adenosine monophosphate (cAMP) levels, the classical transduction mechanism of D1 receptors. The sulpiride-sensitive hyperpolarization is caused by opening of potassium channels, and is inhibited by treatments that raise cAMP levels, suggesting mediation via inhibition of adenylate cyclase activity *(34)*. Further analysis reveals some similarities, but also clear discrepancies between the pharmacology of this response and mammalian D2 receptor-mediated responses *(35)*.

An increased potassium current produced by dopamine has also been described in patch clamp recordings from neurons in pedal ganglia of *Planorbarius corneus,* a freshwater mollusk. Dopamine, muscarinic, and metabotropic glutamate receptors all enhance the probability of potassium channel opening via a pertussis toxin-sensitive G protein *(36)*.

In single neurons of the snail *Helix aspersa,* dopamine-induced closing of potassium channels recorded with two-electrode voltage clamp correlates with stimulation of adenylate cyclase *(37,38)*. This mechanism is similar to the cAMP-dependent closing of a class of potassium channels, the so-called S-channel, by serotonin which produces presynaptic facilitation in *Aplysia (39)*. In addition, dopamine can decrease the action potential plateau phase in *Helix* neurons by depressing a calcium current *(38)*. Similar to the dopamine effect in *Lymnaea* giant neurons, the calcium current suppression in *Helix* is not mediated by cAMP, cyclic guanosine monophosphate (cGMP), or intracellular calcium *(38)*. Instead, it can be blocked by pertussis toxin and an antibody to the α subunit of the G protein G_o *(40)*. Accordingly, intracellular injection of G_o mimics the dopamine response *(40)*. Suppression of calcium channel activity via G_i and G_o proteins is a major transduction pathway for receptors that decrease release of transmitters and hormones *(41,42)*.

Dopamine has been advanced as a transmitter candidate in the salivary gland of the cockroach, where afferent nerve stimulation produces secretion and a hyperpolarizing postsynaptic potential, the so-called secretory potential, which is caused by an increased potassium conductance *(43)*. Exogenously applied dopamine hyperpolarizes the cells via an increased potassium conductance, and both the dopamine effect and the secretory potential can be blocked stereospecifically by the *cis,* but not the *trans,* isomer of flupenthixol, implicating dopamine receptors in the response *(44)*. In mammalian exocrine glands, a hyperpolarizing potential involved in the complex secretory response to transmitters and hormones is caused by opening of calcium-dependent potassium channels that are activated by release of calcium from intracellular stores *(45)*. Dopamine may act on a similar mechanism in the cockroach salivary gland, since the potassium conductance increase produced by dopamine is gradually decreased in calcium-free medium *(46)*.

In the central pattern generator circuit of the lobster stomatogastric ganglion, a well-characterized neural network, dopamine modifies the rhythmic motor pattern by several mechanisms, including a reduced efficacy of graded chemical transmission, possibly via decreased input resistance *(47),* and a suppression of I_A *(48),* a transient potassium current that serves to modulate rhythmic firing in neurons *(49)*.

4. Vertebrates

4.1. Peripheral Cells

4.1.1. Peripheral Neurons and Endocrine Cells

Dopamine has been ascribed a modulatory neurotransmitter role in autonomic ganglia. However, in electrophysiological studies, both pre- and postsynaptic inhibitory effects of dopamine on autonomic ganglion cells appear to be mediated by stimulation of alpha adrenergic receptors *(50–53)*. In cultured neurons from sympathetic ganglia and dorsal root ganglion of chick embryos, both dopamine and noradrenaline slow the activation of high-voltage (possibly N- or L-type) calcium currents recorded in the whole-cell configuration *(54)*. In addition, the two catecholamines depress the amplitude of low-voltage (T-type) calcium currents in the dorsal root ganglion cells, an action which is observed as a decreased probability of channel opening in membrane patches. Dopamine and noradrenaline appeared generally equipotent in this study, but the identity of the mediating receptors was not investigated. Another pharmacologically uncharacterized action of dopamine is suppression of L-type calcium current in glomus cells from rabbit carotid body studied with whole-cell recording *(55)*. Since dopamine is considered a possible transmitter substance in glomus cells, this effect could mediate autoinhibition of release.

Chromaffin cells of the adrenal medulla release different catecholamines, including dopamine, which may activate dopamine autoreceptors and thereby inhibit further release of catecholamines. Such a mechanism seems to be present in bovine chromaffin cells, where dopamine release is decreased by stimulation of D2 receptors *(56,57)*. The D2-mediated inhibition of dopamine release is paralleled by decreases of high voltage-activated calcium currents recorded with whole-cell voltage clamp *(56,57)*. Resting potassium conductance, on the other hand, is unaltered *(57),* indicating that D2 inhibition of hormone release in chromaffin cells is mediated by calcium channel inhibition rather than hyperpolarization caused by opening of potassium channels. D2 receptor stimulation does not affect voltage-dependent sodium current, but slightly depresses the maximal depolarization-activated potassium current, presumably a current of the delayed rectifier type *(57)*. In addition, the dopamine agonist apomorphine decreases the inward current produced by nicotine *(57)*. Adding the nonhydrolyzable guanosine diphosphate (GDP) analog GDP-β-S, which prevents G protein activation, to the patch pipet solution inhibits the depression of calcium current without affecting the D2-mediated decrease in nicotinic current *(57)*.

Whereas no D1 effects on calcium currents were observed in the studies on chromaffin cells described, another patch clamp study of cultured bovine chromaffin cells demonstrated a D1-mediated increase in calcium current without any evidence of D2 receptor responses *(58)*. The reason for this discrepancy is not clear. The D1-mediated increase specifically affects the so-called facilitation current, a dihydropyridine-sensitive (L-type) current which is activated by long depolarizations and high-frequency stimulation, and could therefore play a positive feedback role in situations of increased hormone demand. Dopamine D1 receptor-induced recruitment of facilitation channels is mediated via stimulation of adenylate cyclase and activation of protein kinase A *(58),* thus resembling the previously mentioned facilitation of calcium channels by beta adrenergic receptor activation in the heart. Another electrophysiological effect of dopamine observed in whole-cell recording of bovine chromaffin cells is a D2-dependent potentiation of a calcium-dependent potassium current of the big K (BK) type via pertussis toxin-sensitive G proteins *(59)*. This effect of dopamine could serve to limit bursting and other forms of high-frequency firing of action potentials.

In PC12 cells, a rat pheochromocytoma cell line that resembles chromaffin cells, dopamine enhances whole-cell currents produced by stimulation of adenosine triphosphate (ATP) receptors *(60,61)*. The receptor mediating the dopamine effect seems highly atypical, since the effect is mimicked by agonists and antagonists for both D1 and D2 receptors. Inclusion of the nonhydrolyzable ATP analog ATP-γ-S in the patch pipet reduces the effect, suggesting that facilitation by dopamine is inhibited by an ATP-dependent process *(61)*.

4.1.2. Pituitary

4.1.2.1. Anterior Lobe

Release of prolactin from lactotrophs of the anterior pituitary gland is controlled by dopamine-containing neurons in the hypothalamus. These tuberohypophyseal neurons, which form the cell group A12 in the classification of brain catecholamine neurons of Dahlström and Fuxe, innervate the median eminence, the pituitary stalk, and the posterior and intermediate lobes of the pituitary *(62)*. The nerve terminals in the median eminence release dopamine into the portal blood, through which dopamine reaches the lactotrophs in the anterior pituitary. Thus, dopamine acts like a circulating hormone to tonically inhibit prolactin secretion. This mechanism underlies the hyperprolactinemia caused by dopamine receptor-blocking antipsychotic drugs, and constitutes the basis for treatment of galactorrhea and prolactinoma with dopamine agonists such as bromocriptine.

In the original D1/D2 classification, dopamine receptors in the pituitary gland were considered prototypical for D2 receptors *(1),* and later studies have confirmed this concept. The D_3 and D_4 subtypes are expressed in very low levels in the pituitary, so most hypophyseal dopamine receptors appear to be of the genetically characterized D_2 subtype. This selective expression, together with the clearly defined physiological role of dopamine, make hormone-secreting pituitary cells an interesting model system for responses mediated by dopamine D_2 receptors.

Dopamine hyperpolarizes rat lactotrophs in culture by increasing a potassium conductance, a response that is mimicked by D_2 agonists and blocked by sulpiride *(63,64)*. Voltage clamp analysis shows that the potassium channel opened by dopamine is independent of voltage, is blocked by quinine but not by tetraethylammonium and 4-aminopyridine, and has a single-channel conductance of 40 pS *(63)*. The potassium channel opening produced by dopamine is reduced by addition of GDP-β-S to the patch pipet solution, irreversibly activated by the nonhydrolyzable guanosine triphosphate (GTP) analog GTP-γ-S, and abolished by pretreatment with pertussis toxin, indicating that dopamine acts via a G protein of the $G_{i/o}$ class *(65)*. The potassium channel opening is also absent when antibodies against the α subunit of the G protein G_{i3} are included in the whole-cell recording pipet *(66)*. The interaction between the G protein and the channel appears to be membrane-delimited, i.e., independent of cAMP or other intracellular second messengers, since dopamine is able to open the channel in excised membrane patches as well *(65)*. In addition to this action on a voltage-independent potassium current, the voltage-dependent potassium currents I_K (delayed rectifier) and I_A are also enhanced by dopamine in rat lactotrophs *(67)* and in MMQ cells, a prolactin-secreting pituitary cell line *(68)*.

Voltage-dependent whole-cell calcium currents in lactotrophs are suppressed by dopamine via different mechanisms: The duration of the transient T-current is decreased and the activation of the L-current is shifted in a depolarizing direction *(69)*. Both of these actions display D_2 pharmacology *(69),* are blocked by GDP-β-S, potentiated by GTP-γ-S, and abolished by pertussis toxin *(70)*. Antibodies to the α subunit of the G protein G_o reduce the calcium channel inhibition by dopamine, indicating that the two actions of dopamine—calcium channel inhibition and potassium channel opening—are mediated by two different pertussis toxin-sensitive G proteins *(66)*. Apart from this acute effect, prolonged D_2 receptor stimulation can also prevent the development of voltage-dependent calcium current in cultured lactotrophs, presumably by decreasing synthesis of channel protein *(71)*.

4.1.2.2. Intermediate Lobe

Dopamine D_2 receptors are present also in the intermediate lobe of the pituitary, where dopamine inhibits the secretion of pro-opiomelanocortin-derived peptides, including α-melanocyte-stimulating hormone and β-endorphin, from melanotrophs. Unlike the situation in the anterior lobe, the tubero-hypophyseal dopamine neurons make synaptic contact with their target cells in the intermediate lobe. The functional role of this synaptic connection was demonstrated by Williams et al. *(72)*, who showed that a potassium-mediated inhibitory synaptic potential mediated by D_2 receptors be recorded intracellularly in rat melanotroph cells following electrical stimulation of the pituitary stalk. Intracellular and whole-cell recording from cultured rat melanotrophs also demonstrate opening of potassium channels by dopamine and quinpirole *(73,74)*. In addition, calcium currents in melanotrophs are inhibited by D_2 receptors activated synaptically by electrical stimulation of the pituitary stalk *(75)*, as illustrated in Fig. 1. Calcium currents are suppressed by dopamine in whole-cell recordings from cultured rat melanotrophs as well *(73,74,76)*, although the relative effects on different types of calcium current are disputed. Both the increased potassium conductance and the decreased calcium conductance appear to be mediated by pertussis toxin-sensitive G proteins *(72–74)*. Interestingly, when the electrophysiological effects of quinpirole are compared with release of β-endorphin from melanotrophs, the potassium current enhancement appears strongly correlated with inhibition of secretion, whereas suppression of calcium current appears at considerably higher concentrations of quinpirole *(74)*. This indicates that dopamine receptor stimulation decreases hormone release via hyperpolarization, thereby inhibiting voltage-dependent activation of calcium channels, rather than by a direct modulation of calcium channels. The opposite mechanism, i.e., direct inhibition of voltage-dependent calcium current rather than via hyperpolarization, was inferred in the study of D2 inhibition of dopamine release from bovine chromaffin cells mentioned above *(57)*.

Peptide secretion from frog melanotrophs is also decreased by dopamine, which inhibits firing of action potentials and hyperpolarizes the cell by increasing a potassium conductance *(77)*. In patch clamp recordings,

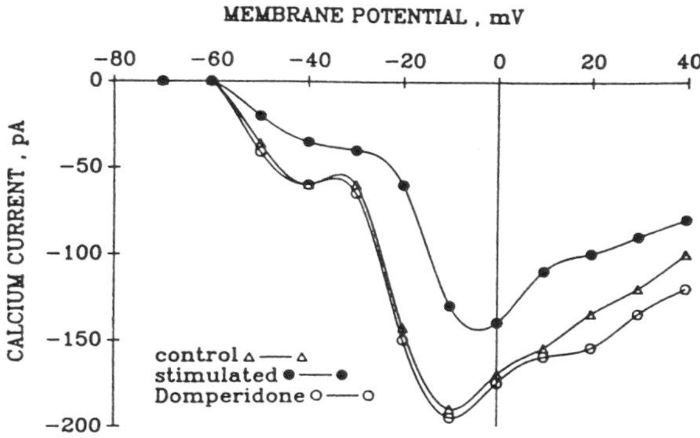

Fig. 1. Endogenously released dopamine inhibits voltage-activated calcium current by acting at the D_2 receptor in melanotrophs of the rat pars intermedia in vitro. Membrane currents were recorded with a microelectrode under single-electrode voltage-clamp. The D_2 receptor antagonist domperidone (1 μ*M*) blocked the inhibition of calcium current produced by dopamine released synaptically by electrical stimulation of the pituitary stalk. Currents were recorded in response to 10 mV incremental voltage steps from a holding potential of –90 mV. Voltage-activated sodium and potassium currents were blocked by QX-222, cesium acetate, and tetraethylammonium which were present in the pipet solution. Data were reproduced by permission from Williams et al. *(75)*.

dopamine increases the activity of two presumed potassium channels with unitary conductances of 100 and 30 pS, and increases a delayed rectifier-type whole-cell potassium current. In addition, an inactivating sodium current and two different voltage-activated calcium currents are suppressed by dopamine *(77)*. These actions of dopamine are mediated by D2 receptors coupled to pertussis toxin-sensitive G proteins *(78)*.

In the rat pituitary, synaptic responses to dopamine are not restricted to melanotrophs. A dopamine-dependent hyperpolarization also has been observed in glial (stellate) cells of the intermediate lobe. Although the membrane mechanism of the postsynaptic potential is unknown, it is blocked by D2 antagonists *(79)*.

4.2. Central Neurons

4.2.1. Retina

Dopamine is a transmitter in amacrine or interplexiform cells in the vertebrate retina *(62),* where it innervates horizontal and amacrine cells that regulate the size of the receptive field and mechanisms of lateral inhibition. The modulatory role of retinal dopamine has been described in detail in studies

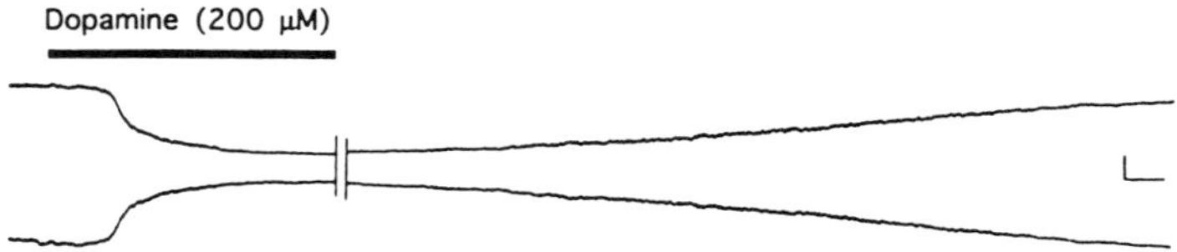

Fig. 2. Dopamine electrically uncouples white perch horizontal cells in primary cell culture. Traces represents currents recorded simultaneously from two electrically coupled cells recorded in the whole-cell configuration by patch electrodes. Each cell was voltage-clamped at a different potential (–30 mV in the cell in the upper trace, +10 mV in the cell in the lower trace), thus imposing a constant transjunctional voltage gradient between the two cells. Perfusion with dopamine, indicated by the solid bar, produces equal and opposite changes in the holding currents of the two clamp circuits, indicating decreased junctional conductance and uncoupling of the cells. Dopamine had no consistent effect on the nonjunctional resistance of horizontal cells. Calibration bars: 500 pA; 8 s. Data were reproduced by permission from McMahon et al. *(81)*.

of horizontal cells of teleost fish *(80)*. Horizontal cells are electrotonically coupled via gap junctions, and dopamine has been shown to decrease this coupling via D1 receptor stimulation of adenylate cyclase and protein kinase A, which presumably phosphorylates the gap junction channel *(80)* (*see* Fig. 2). The effect of dopamine, studied with noise analysis and single-channel recording in pairs of cultured horizontal cells, is to decrease the open time of gap junction channels without modifying their conductance *(81)*. In single horizontal cells, unitary hemi-gap junction currents have been observed that are inhibited by dopamine via protein kinase A activation *(82)*. Whereas D1 receptor stimulation leads to uncoupling of horizontal cells, presynaptic D2 receptors on interplexiform cells increase horizontal cell coupling by inhibiting dopamine release onto postsynaptic D1 receptors *(83)*.

In addition to modifying the conductance of gap junctions, dopamine also acts via D1 receptors and cAMP to enhance glutamate-induced depolarization of horizontal cells *(80)*. Dopamine increases the opening probability of single glutamate receptor channels without altering their conductance or the number of available channels *(84)*. This effect might be related to the recently described effect of dopamine on desensitization of glutamate responses in perch horizontal cells recorded in the whole-cell and outside-out configurations *(85)*. Very fast application of agonists like glutamate, quisqualate, or α-amino-3-hydroxy-5-methylisoxazole-4-propionic acid (AMPA) produces a transient current in these cells which rapidly desensitizes in the presence of agonist. Dopamine abolishes desensitization, thereby converting the transient responses to steady-state currents, presumably by phosphorylation of the glutamate receptor channel protein. The functional role of the dopamine-induced potentiation of glutamate responses is to reduce the light responsiveness of horizontal cells, thus modulating contrast sensitivity in the retina. Dopamine can also modulate responses of horizontal cells to their own neurotransmitter, GABA. In whole-cell recordings of catfish horizontal cells, the current induced by $GABA_C$ receptor activation is reduced by dopamine, via D1 receptors and stimulation of adenylate cyclase *(86)*.

In ganglion cells, the projection neurons of the retina, calcium currents are decreased by dopamine in whole-cell and perforated patch clamp recordings from isolated turtle ganglion cells *(87)*. Unlike the situation in many other cell types, this action is mediated by D1 receptors' activation of adenylate cyclase and protein kinase A. Dopamine suppresses both a transient and a sustained calcium current, which leads to decreased excitability of the ganglion cell and a shift from high-frequency to single spike firing of action potentials recorded in current clamp.

4.2.2. Spinal Cord and Medulla

Dopamine is coreleased with serotonin from neurons in the lamprey spinal cord, where both monoamines regulate firing in interneurons of the locomotor network, albeit by different mechanisms: Serotonin decreases a calcium-activated potassium conductance (gK_{Ca}) responsible for the afterhyperpolarization, whereas dopamine decreases calcium influx during the action potential, thereby decreasing gK_{Ca} indirectly *(88)*. The decrease of calcium current appears to be mediated by D2 receptors, since it was mimicked by quinpirole and blocked by the D2 antagonist eticlopride. The D1 agonist SKF 38393 was without effect.

Dopamine also modulates electrotonic and glutamatergic synaptic connections of the VIIIth cranial nerve to the Mauthner cell in the goldfish medulla. Dopamine, acting via D1 receptors, enhances both forms of transmission by stimulating adenylate cyclase and protein kinase A *(89)*. The action on electrotonic coupling is thus the opposite of the D1 effect in the retina, whereas the effect on glutamatergic transmission is similar.

4.2.3. Hypothalamus

Hypothalamic neurons in the supraoptic nucleus send their axons to the posterior pituitary, where they release oxytocin and vasopressin into the circulation. In intracellular recordings from rat hypothalamic explants, supraoptic neurons are depolarized by dopamine via D2 receptors. Voltage clamp analysis indicates that dopamine acts through an unusual mechanism: an increase of a calcium-dependent nonselective cation conductance *(90)*. This conductance has been shown to regulate firing patterns in other neurons via long-lasting depolarizations *(91)*.

4.2.4. Midbrain

The dopamine cells in the retrorubral nucleus, substantia nigra, and ventral tegmental area in the mesencephalon (cell groups A8–A10) innervate the striatum, the neocortex, and various limbic areas, thereby giving rise to the mesostriatal and mesocorticolimbic dopamine systems (also known as the nigrostriatal, mesolimbic, and mesocortical systems) *(62)*. These systems have been extensively studied since the 1960s and remain a focus of interest for research on the basis and treatment of Parkinson's disease, schizophrenia, and addictive disorders.

The first electrophysiological studies of mesencephalic dopamine neurons, made with extracellular recording in anesthetized rats, indicated the presence of inhibitory somatodendritic dopamine receptors on the neurons *(92)*. Such autoreceptors, exerting negative feedback control of dopamine neurotransmission, had previously been suggested on the basis of biochemical and behavioral data. In vivo recordings with microiontophoretic application of drugs suggested a D2 profile of the autoreceptors *(93)*. Prevention of the D2-mediated inhibition of firing by local injection of pertussis toxin implicates G proteins of the G_o/G_i type in the D2 response *(94)*. Intracellular recording with current and voltage clamp in rat brain slices have shown that these D2 receptors open potassium channels, which can also be opened by GABA acting on somatodendritic $GABA_B$ receptors *(95,96)*, as seen in Fig. 3. Focal electrical stimulation of the brain slice produces an inhibitory postsynaptic potential in dopamine neurons mediated via D2 receptors, suggesting feedback via dendrites or axon collaterals *(97)*. Tolbutamide, a sulfonylurea which blocks ATP-sensitive potassium channels (K_{ATP}), reportedly blocks potassium channels opened by dopamine in acutely dissociated dopamine neurons *(98)*. This interesting finding has not been confirmed in subsequent studies in brain slices *(99)*.

The recent expansion of the D2 subfamily has raised the issue of reclassification of the autoreceptors of mesencephalic dopamine neurons (*see* Chapter 8). Intravenous administration of the putative D_3 agonist 7-OH-DPAT (*see* Section 1.) inhibits the firing of extracellularly recorded dopamine neurons in anesthetized rats *(100)*. This effect is partially antagonized by a new compound, S 14297, which shows 20-fold selectivity for binding to D_3 vs D_2 receptors in expression systems *(101)*. In another study utilizing iv administration of drugs in anesthetized rats, the potency of different agonists in inhibiting dopamine cell firing showed stronger correlation for affinity to D_3 than at D_2 receptors in expression systems *(102)*. Although these results suggest the existence of functional D_3 receptors on dopamine neurons, some objections can be raised: Pharmacokinetic differences between drugs could complicate comparisons of iv doses and concentrations at receptors in the brain; differ-

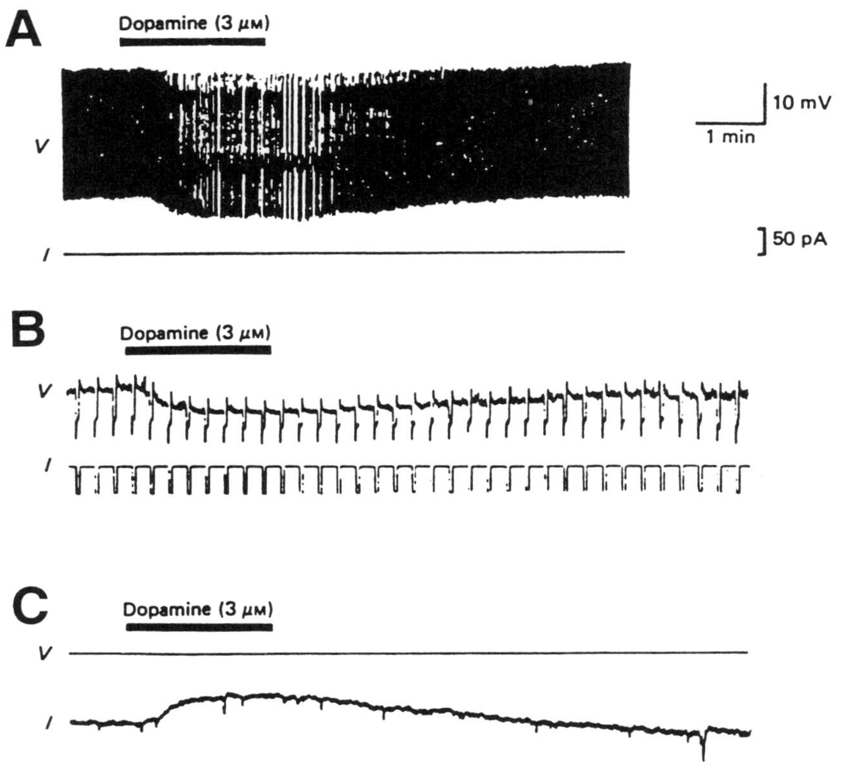

Fig. 3. Dopamine hyperpolarizes dopamine neurons by opening a potassium conductance. Neurons were recorded with microelectrodes in the substantia nigra brain slice. **(A)** dopamine slows spontaneous firing of action potentials. **(B)** Dopamine hyperpolarizes the membrane and reduces input resistance. **(C)** In voltage-clamp (–63 mV), dopamine evokes an outward current which was also shown to reverse direction at the expected equilibrium potential for potassium. These effects of dopamine were mimicked by quinpirole and blocked by domperidone, suggesting mediation by D2 receptors. Data were reproduced by permission from Lacey et al. *(95)*.

ences in binding affinity do not automatically translate to differences in potency, as observed with 7-OH-DPAT in expression systems *(14)*; and agonist potency is determined not only by affinity, but also by receptor reserve, transduction mechanisms, desensitization, etc. which make receptor classification by agonist potency inherently more problematic than classification by antagonist affinity *(103)*. In an intracellular recording study of dopamine neurons in rat brain slices, where drug concentration can be controlled, Schild analysis of antagonist affinity indicated that the inhibitory (hyperpolarizing) action of quinpirole and 7-OH-DPAT is more likely mediated by D_2 than by D_3 or D_4 receptors *(104)*. Definitive evidence on the issue of the functional role of D2 subtypes probably must await the development of more selective pharmacological tools or molecular biological methods.

Dopamine also enhances two voltage-dependent potassium conductances, I_A and I_K, studied with whole-cell recording from mesencephalic dopamine neurons in culture *(105)*. The effect is mimicked by quinpirole, unaffected by SKF 38393, and blocked by eticlopride and sulpiride, implicating D2 receptors in the response, similar to results from lactotrophs *(67)*. The hyperpolarization-activated cation current I_H, which is a characteristic of midbrain dopamine cells, is decreased by dopamine in whole-cell recordings in brain slices *(106)*. Quinpirole mimics and sulpiride blocks the response, implicating D2 receptors, and the occlusion of the effect by GTP-γ-S indicates a G protein-mediated mechanism.

No evidence for postsynaptic D1 receptors on midbrain dopamine neurons has been obtained, but presynaptic D1 receptors on GABA-containing afferents have been shown to enhance inhibitory synaptic potentials mediated via $GABA_B$ receptors in dopamine neurons studied with intracellular recording in guinea pig brain slices *(107)*. Forskolin mimicks the effect produced by D1 agonists such as SKF 38393 and SKF 82958, suggesting that D1 receptor activation potentiates GABA release by stimulation of adenylate cyclase.

4.2.5. Striatum

4.2.5.1. Dorsolateral Striatum (Caudate-Putamen)

Since the nigrostriatal pathway was the first dopamine-containing projection to be described, interest in the neurotransmitter role of dopamine initially focused on the action of dopamine in the caudate nucleus and putamen. The predominant electrophysiological technique for studies of brain neurotransmitter action in the 1960s was extracellular recording of action potentials from single neurons in anesthetized animals, with microiontophoretic application of drugs. Several studies with this technique demonstrated that iontophoretically applied dopamine mostly inhibits spontaneous and glutamate-induced firing of striatal neurons, although excitation was occasionally observed *(8)*. Experiments with electrical stimulation, on the other hand, often suggested an excitatory role for dopamine *(7)*. In studies using intracellular recording of striatal neurons in anesthetized animals, microiontophoretically applied dopamine frequently induced a depolarization that was exclusively accompanied by either excitation *(108)* or inhibition *(109,110)*. Inhibitory hyperpolarizations were also observed *(110)*. Despite considerable effort, the question of whether dopamine in the striatum was excitatory or inhibitory could not be satisfactorily resolved on the basis of these and similar studies.

There were several obvious reasons for this confusion, as mentioned in Section 1.:

1. Extracellular recording with microiontophoretic drug application cannot separate between direct and indirect effects, e.g., excitation and disinhibition;
2. In vivo recording does not permit control of the ionic environment, spontaneous synaptic activity cannot be controlled, and anesthesia may alter neuronal responsiveness;
3. Microiontophoresis delivers only approximate amounts of drug, the extracellular drug concentration is unknown, and equilibrium conditions are never reached;
4. Artifacts of microiontophoresis, both ionic and electrical, are common, and negative results might be the result of inadequate ejection of drug solution;
5. Pharmacological subtypes of dopamine receptors had not yet been described; and
6. The prevalence and importance of modulatory actions of neurotransmitters were not generally realized.

Intracellular recording from striatal neurons in rat brain slices by Calabresi and coworkers *(111)* showed that dopamine causes a direct postsynaptic inhibition of firing by an unexpected, voltage-dependent action. Resting membrane potential and input resistance are unaffected by dopamine, but depolarizing current becomes less effective in producing depolarization and firing of action potentials (*see* Fig. 4). This effect is caused by suppression of a voltage-dependent persistent sodium conductance which is also tetrodotoxin sensitive. SKF 38393 and a cAMP analog mimic the effect of dopamine, which is blocked by SCH 23390; these data implicate D1 receptors positively coupled to adenylate cyclase *(111,112)*. Amphetamine mimics the effect of dopamine, suggesting that the decreased membrane excitability can also be produced by endogenously released dopamine *(113)*. Tolerance to the D1 effect was observed in animals with dopamine-depleting 6-hydroxydopamine lesions *(112)*. Akaike et al. *(114)* also reported that high concentrations of dopamine (100–500 μ*M*) reduced membrane excitability in striatal neurons in the brain slice without a concomitant change of the membrane potential.

Dopaminergic modulation of voltage-dependent inactivating sodium currents has been investigated by Surmeier et al. *(115)* with whole-cell voltage clamp methodology in acutely dissociated neurons from rat striatum. The amplitude of the sodium current is decreased, and the inactivation voltage is shifted in a hyperpolarizing direction by dopamine, SKF 38393, or the cAMP

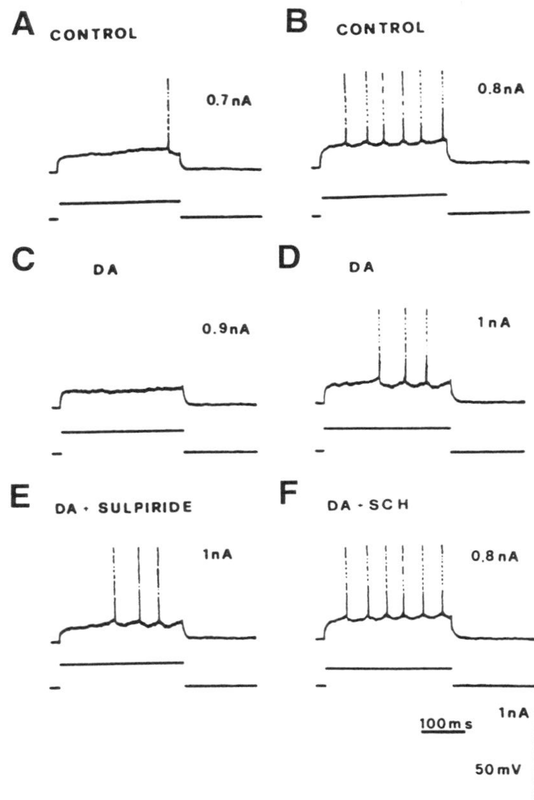

Fig. 4. Dopamine acts at D1 receptors to suppress the generation of action potentials in the striatal brain slice. In each figure, the upper trace is the voltage trace and the lower trace is the current trace. **(A)** A pulse of depolarizing current (0.7 nA) evokes a single action potential, whereas, in **(B)** a stronger depolarizing stimulus (0.8 nA) evokes a train of spikes. Dopamine (1 μ*M*) reduces the number of action potentials even when relatively strong depolarizing current pulses are applied (**C** and **D**). The D2 antagonist sulpiride (500 n*M*) has no effect on inhibition of spikes by dopamine **(E)**, whereas the D1 antagonist SCH 23390 (500 n*M*) completely blocks the inhibitory effect of dopamine, allowing the firing of a full train of action potentials **(F)**. All data were recorded with a microelectrode from the same cell. This effect of dopamine was mimicked by tetrodotoxin, but not by block of calcium channels, suggesting dopamine reduces conductance in voltage-gated sodium channels. Data were reproduced by permission from Calabresi et al. *(111)*.

analog 8-Br-cAMP, indicating that the action results from D1 receptor stimulation of adenylate cyclase. The effect on the cellular level is that smaller sodium currents are activated by depolarizing inputs during stimulation of D1 receptors, and the cell becomes less excitable. The D1-induced decrease of sodium current might involve phosphorylation of the channel protein by protein kinase A, a mechanism which has been characterized on the molecular level in brain membrane patches and cloned sodium channels *(116)*.

Voltage-dependent calcium channels can also be modulated by D1 receptors, as shown in a whole-cell recording study of acutely dissociated neurons from rat striatum *(117)*. Dopamine or D1 agonists such as SKF 38393 and SKF 82958 depress the amplitude of calcium current through N- and P-channels, defined pharmacologically by their sensitivity to ω-conotoxin and ω-agatoxin, respectively. The D1 effect is mimicked by cAMP analogs and the phosphodiesterase inhibitor isobutylmethylxanthine (IBMX), and it is inhibited by protein kinase A inhibitors. Presumably, D1 receptor stimulation of cAMP production activates protein kinase A, which phosphorylates protein phosphatase-1, which in turn dephosphorylates N- and P-channel proteins, leading to calcium current reduction. In the presence of phosphatase inhibitors, an additional effect of D1 agonists or cAMP analogs is observed in about half of the neurons: an increase in calcium current through L-type channels which is blocked by the protein kinase A inhibitor H-89. The D1-mediated enhancement of L-current in striatal neurons thus resembles the beta adrenergic enhancement of L-current in cardiac muscle cells, which is also produced via stimulation of adenylate cyclase and protein kinase A. The functional role of the divergent modulation of calcium currents by D1 receptors is ultimately determined by factors such as phosphorylation state of the channels, distribution of channel subtypes within the cell, and degree of synaptic influence from other transmitters. It is possible that the fast decrease of N- and P-currents serves to decrease the responsiveness of the cell to excitatory input on a short timescale, whereas the slower enhancement of L-current might play a role in long-term modification of synaptic responses, e.g., long-term depression.

Compared to D1 receptors, the effect of D2 receptor stimulation on striatal neurons is more controversial, and several different actions have been described. No D2 effects were observed by Calabresi et al. *(118)* in control slices. In slices from reserpine-treated rats, however, the D2 agonists bromocriptine and lisuride reduced membrane excitability without change in input resistance, similar to the effect of D1 receptor stimulation in slices from untreated rats. The uncovering of D2-mediated inhibition could be the result of increased receptor expression which has been observed in striatum following catecholamine depletion or chronic receptor blockade *(119–121)*.

D2-mediated modulation of voltage-dependent sodium current has been studied in patch clamp recordings from acutely dissociated striatal neurons *(115)*. In whole-cell recordings, the D2 agonists quinpirole and bromocriptine produce two different effects on the inactivating sodium current, both of which are blocked by sulpiride. The first is an increase in amplitude and a depolarizing shift of the inactivation curve, i.e., the opposite of the D1 effect on this current. This action, which should make the cell more excitable, appears to be mediated by a soluble second messenger, possibly decreased cAMP levels, since it can be evoked by bath-applied agonist in cell-attached recordings, where only the pipet solution is in contact with the extracellular membrane of the recorded patch. The second D2 effect is a hyperpolarizing shift of the inactivation curve without alteration of the maximum amplitude, leading to an effective decrease of depolarization-activated sodium current and a decreased excitability. It involves a membrane-delimited mechanism, since it is produced only when the agonist is present in the pipet solution in cell-attached recordings. The basis for this diversity of D2 effects is not clear. It could be owing to the presence of two different members of the D2 subfamily, e.g., D_2 and D_3, a concept that gained support when mRNA for different receptor subtypes was extracted from the recorded cell through the patch pipet *(115)*. D_1, D_2, and D_3 receptor mRNA are frequently found in the same cell, suggesting extensive colocalization of dopamine receptor subtypes *(122)*. However, the extent and importance of this phenomenon is controversial *(123),* since several types of evidence, including studies of mutant mice lacking the D1 receptor gene *(124),* point to a pronounced segregation of D_1 and D_2 receptors between different populations of striatal output neurons *(120)*. Another possible explanation of the diversity of D2 responses is that one receptor subtype produces two effects in the same cell type via two different G proteins, as suggested in a patch clamp study of D2 effects in lactotrophs *(66)*.

An excitatory action of dopamine was reported by Akaike et al. *(114)* in which dopamine (1 μ*M*) depolarized the membrane and increased spontaneous firing of action potentials recorded in neurons in current clamp in the rat striatal slice *(114)*. This response, which was blocked by domperidone but not SCH 23390, is difficult to reconcile with the inhibitory effects of D2 receptor stimulation described.

Patch clamp experiments in acutely dissociated neurons from rat striatum have revealed an additional action of dopamine, namely increased opening frequency of a potassium conductance of 85 pS *(125)*. The channel is opened by dopamine or quinpirole and is blocked by haloperidol and spiperone, but not by SCH 23390, suggesting D2 pharmacology. The dopamine-activated channel is dose-dependently blocked by quinine *(126),* similar to the D2-coupled channel in lactotrophs *(63),* but the unitary conductance is larger. An unusual property of the potassium channel opened by dopamine in striatal neurons is its blockade by the sulfonylurea drugs tolbutamide and glibenclamide *(127),* suggesting a similarity with pancreatic K_{ATP} channels. However, the order of potency between the two sulfonylureas is reversed in striatum compared to pancreas, and the striatal channel is not opened by diazoxide, a classical K_{ATP} channel opener. Still, it is interesting to note that the channel is opened by the NADH dehydrogenase inhibitor rotenone, indicating activation in situations of metabolic stress, where K_{ATP} channels have been suggested to play a protective role.

Voltage-dependent potassium currents in striatal neurons are also modulated by dopamine. A long-lasting A-like current evoked by membrane depolarization is selectively inhibited by dopamine acting at D1 receptors in whole-cell recordings from acutely dissociated cells from adult rats *(128)*. The D2 agonist quinpirole, on the other hand, increases the current slightly in a small proportion of cells. This long-lasting current, which resembles the D current described in hippocampal pyramidal cells *(129)*, is distinct from the A-current which inactivates much faster and is unaffected by dopamine. The D1-mediated decrease of the slowly inactivating A-current should facilitate the transition of striatal neurons from a hyperpolarized resting state to an active state, whereas the D2 stimulation of this current would stabilize the neurons in the resting state.

Iontophoretically applied dopamine reduces the membrane hyperpolarization that follows a train of action potentials evoked by a depolarizing current pulse, according to an intracellular study in rat striatal slices *(130)*. This effect of dopamine would be expected to facilitate burst firing of action potentials, yet is in apparent conflict with the simultaneously observed ability of dopamine to decrease membrane excitability. The combined net effect of dopamine is to raise the threshold for firing of action potentials, but to produce a stronger response once the threshold is reached, i.e., to promote burst firing over single spike firing. The identity of the receptor and the conductance producing the afterhyperpolarization decrease was not investigated in this study.

Dopamine might also modulate responses of striatal cells to other transmitters, e.g., excitatory amino acids. Cepeda et al. *(131)* reported that dopamine potentiates the membrane depolarization produced by microiontophoresis of N-methyl-D-aspartate (NMDA), and it reduces the amount of depolarization produced by agonists at non-NMDA receptors. The D1 agonist SKF 38393 also potentiates NMDA-induced depolarization, whereas the D2 agonist quinpirole reduces depolarizations produced both by NMDA and non-NMDA agonists. These modulatory actions of dopamine may be important in regulating the response of striatal neurons to synaptic inputs from glutamate-containing pathways arising from cerebral cortex and thalamus.

Dopamine transmission might also modulate excitatory transmission via presynaptic receptors. No presynaptic dopamine effect was observed by Calabresi et al. *(112,118,132)* in intracellular recordings from brain slices of control rats. However, presynaptic D2-mediated inhibition of corticostriatal excitatory postsynaptic potentials was observed in slices from rats treated chronically with haloperidol *(132)*, 6-hydroxydopamine *(112)*, or acutely with reserpine *(118)*. These findings might be linked to the functional supersensitivity of D2 receptors observed in behavioral experiments after dopamine depletion or treatment with neuroleptics.

Different forms of synaptic plasticity, including long-term potentiation (LTP) and long-term depression (LTD), have been advanced as neuronal mechanisms for memory and learning in the hippocampus and other brain regions. LTP is a long-lasting (days to weeks) enhancement of synaptic response after high-frequency stimulation of excitatory pathways, and requires stimulation of postsynaptic NMDA receptors for its induction *(133)*. LTD is a long-term decrease of synaptic response induced by other stimulation parameters, often involving low stimulation frequencies and concomitant stimulation of different afferent inputs *(134)*. In the rat striatum, where both LTP and LTD of excitatory transmission can be observed *(135)*, dopamine transmission seems to be critical for LTD. Induction of LTD by brief tetanic stimulation of the corticostriatal pathway is prevented by administration of either a D1 or a D2 antagonist and is absent in striatal slices from dopamine-depleted rats; it can be restored by dopamine or by a combination of D1 and D2 agonists *(136)*. On the behavioral plane, this observation points to a cellular basis for the involvement of dopamine in motor learning observed in recordings from the striatum of behaving animals *(137)*.

As in the retina, dopamine may also control the extent of electrotonic coupling between striatal neurons. Dye-coupling between striatal neurons is significantly increased after dopamine is depleted by surgical or chemically induced lesions *(138)*. The mechanism behind this action of dopamine presumably involves modulation of gap junction channels.

4.2.5.2. Ventral Striatum (Nucleus Accumbens)

The nucleus accumbens septi, olfactory tubercle, and islands of Calleja are prominent terminal regions for the midbrain dopamine neurons which make up the mesolimbic pathway *(62)*. Recent anatomical research has stressed the similarity between these limbic forebrain areas and the dorsal striatum (caudate-putamen), leading to their designation as the ventral striatum *(139)*. Although there are similarities between the dorsal and ventral striatum, there are also sufficient differences in transmitter distribution, organization of cell clusters, and projection areas to warrant occasional separate treatment. Whether there are fundamental similarities or differences in the electrophysiological consequences of dopamine receptor activation between the dorsal and ventral striatum remains to be clarified.

Early electrophysiological investigations in the accumbens were hampered by the same problems as in the dorsal striatum. In 1986, however,

Uchimura et al. *(140)* utilized intracellular recording in brain slices from guinea pig nucleus accumbens to show two different postsynaptic actions of dopamine: a hyperpolarization associated with decreased input resistance and a depolarization associated with increased input resistance. Both the hyperpolarization and the depolarization reverse direction around the reversal potential for potassium, suggesting that they are mediated by an increased and a decreased potassium conductance, respectively. Sulpiride is effective in antagonizing the depolarization, but not the hyperpolarization, leading to the suggestion that the depolarization is mediated by D2, and the hyperpolarization by D1, dopamine receptors. The hyperpolarization is mimicked by dibutyryl cAMP, further indicating D1 involvement *(140)*. These findings were confirmed by Higashi et al. *(141)* in accumbens slices from untreated rats and rats chronically treated with methamphetamine. Interestingly, the D1 response is 100-fold more sensitive, whereas the D2 response is 10-fold less sensitive in slices from methamphetamine-treated rats. The sensitivity to cAMP and forskolin is also enhanced after methamphetamine treatment, suggesting a possible sensitization of transduction mechanisms. Whereas calcium spikes are unaffected in control slices, they are suppressed by dopamine via a presumed D1 action in slices, from methamphetamine-treated rats *(141)*. In rat accumbens slices the dopamine-induced hyperpolarization and depolarization are mimicked by acute administration of cocaine, suggesting that the effects may be produced by release of endogenous dopamine *(142)*. The hyperpolarization and depolarization are blocked by SCH 23390 and sulpiride, respectively, confirming the results from guinea pig accumbens of D1-induced hyperpolarization and D2-induced depolarization.

Presynaptic modulation of synaptic inputs to nucleus accumbens neurons by dopamine receptors has been demonstrated by several investigators. In the study of Higashi et al. *(141)*, dopamine decreased the amplitude of excitatory postsynaptic potentials without affecting the response to exogenous glutamate; this D1-mediated action was encountered mainly in slices from methamphetamine-treated rats. D1 receptor-mediated inhibition of inhibitory as well as excitatory postsynaptic potentials was demonstrated by Pennartz et al. *(143)*. However, the nucleus accumbens has recently been divided into two main subregions, core and shell, on the basis of differences in projection patterns and neurochemistry *(144)*, and the ability of dopamine to presynaptically modulate synaptic transmission may differ in these subregions *(145)*. In an intracellular recording study of neurons in the core region, sulpiride increased the amplitude of excitatory postsynaptic potentials, which may reflect D2 dopaminergic tone in the slice *(146)*. This action of sulpiride was not seen in dopamine-depleted slices; however, quinpirole decreased the amplitude of excitatory postsynaptic potentials, perhaps owing to activation of supersensitive presynaptic D2 receptors as described previously in dorsal striatum *(112,118,132)*.

Dopaminergic regulation of electrotonic coupling is observed not only in dorsal, but also in ventral striatum *(147)*. In the core region, D1 receptor stimulation decreases coupling, similar to dopamine action in the retina. In the posterior shell region, on the other hand, stimulation of D2 receptors increases coupling.

4.2.6. Hippocampus

The dopaminergic innervation of the hippocampus is not prominent, but several articles describing the electrophysiological action of dopamine in this region have been published. In CA1 pyramidal cells recorded in guinea pig brain slices, dopamine increased the afterhyperpolarization that follows a train of spikes and produced a hyperpolarization that reverses around –90 mV: Both effects were attributed to an increased calcium-dependent potassium conductance *(148)*. These effects of dopamine were mimicked by cAMP and blocked by flupenthixol and chlorpromazine, suggesting mediation by D1-like receptors *(149)*. In contrast, Malenka and Nicoll *(150)* found that dopamine induced the opposite effect in both rat and guinea pig hippocampal slices: a decrease in afterhyperpolarization which was blocked by propranolol and timolol, indicating that the response was mediated by beta adrenergic receptors. In intracellular recordings in rat hippocampal slices by Stanzione et al. *(151)*, dopamine also decreased the afterhyperpolarization of pyramidal cells. In this study, however, the dopamine action was blocked by domperidone and not by timolol, indicating mediation by dopamine rather than beta adrenergic receptors. Furthermore, dopamine decreased the excitability of the cells in a manner similar to that described in the striatum *(111)*.

Dopamine transmission also appears to play a role in LTP induced by stimulation of the Schaffer collateral input to CA1 pyramidal neurons. Repeated tetanic stimulation induces a long-lasting LTP that can be dissected into an early and a late component. The late component, which develops about 1 h after LTP induction, is mediated by cAMP and protein kinase A and requires protein synthesis for its expression *(152)*. Dopamine and D1 agonists can induce the late, but not the early, component of LTP of field potentials in rat hippocampal slices *(153)*. The D1 action is mimicked and occluded by a cAMP analog, and prevented by SCH 23390. SCH 23390 alone selectively depresses the late component of LTP induced by tetanic stimulation, suggesting that synaptic activation of D1 receptors is critical for this long-lasting response.

4.2.7. Neocortex

The mesocortical dopamine system arises from cell bodies in the ventral tegmental area and terminates in various neocortical areas, primarily in the prefrontal cortex *(62)*. Since its discovery, it has attracted attention relating to its function in behavior and as a possible site of action for antipsychotic drugs *(154)*. Yet, the functional responses produced by dopamine receptors in cortical neurons are not well understood. Early studies employing extracellular recording and microiontophoresis suffered from the drawbacks described in the section on the striatum, and failed to give a consistent picture of dopaminergic actions.

In intracellular recordings from slices of rat prefrontal cortex, dopamine has small and variable actions on resting membrane properties and firing of pyramidal neurons *(155)*. However, the frequency of spontaneous inhibitory postsynaptic potentials mediated by $GABA_A$ receptors is consistently increased, suggesting that dopamine increases the firing of GABA-containing interneurons. This suggestion was confirmed by results with extracellular recording in slices of rat piriform cortex, where bath-applied dopamine increases action potential firing of interneurons via an uncharacterized receptor *(156)*.

Dopamine also modulates evoked synaptic responses in pyramidal cells. In the entorhinal cortex, excitatory postsynaptic potentials mediated by NMDA and non-NMDA receptors are markedly reduced in amplitude by dopamine via stimulation of D1 receptors *(157)*. Dopamine reduces NMDA and non-NMDA-mediated postsynaptic potentials in the prefrontal cortex as well, but in contrast to effects in the entorhinal cortex, dopamine also inhibits GABAergic inhibitory postsynaptic potentials *(158)*. In an intracellular study in slices of human cerebral cortex, Cepeda et al. *(159)* found that dopamine potentiates membrane depolarization produced by exogenous NMDA, whereas dopamine inhibits the depolarization produced by non-NMDA agonists. Thus, dopamine may modulate glutamatergic synaptic transmission by both pre- and postsynaptic mechanisms in cortical pyramidal neurons.

Synaptic plasticity in the form of both LTP and LTD have recently been observed in the rat prefrontal cortex slice. Dopamine counteracts the induction of LTP and favors the expression of LTD in intracellularly recorded pyramidal neurons *(160)*. The mechanism is unknown, but the effect is of great potential interest in view of the role of cortical dopamine in working memory and other aspects of cognitive function *(154,161)*.

5. Conclusions

Reviews on pharmacological receptors are often organized according to receptor subtype or effector mechanisms. We believe, however, that application of either organizational principle to the present subject would act like a Procrustean bed, stretching or cutting the results beyond recognition. Dopamine is involved in many aspects of health and disease, and the diversity of electrophysiological results is a natural consequence of its multifaceted function. Still, it might be worthwhile to return to the question posed in the Introduction: What are the electrophysiological consequences of dopamine receptor stimulation? Although no simple answer can be given, it is possible to discern some patterns in the reviewed literature.

The D1 and D2 receptor subfamilies clearly differ in their coupling to transduction mechanisms. D1 receptors seem to be strictly coupled to $G_{s/olf}$ and stimulation of adenylate cyclase, and the cloned D_1 and D_5 subtypes enhance cAMP production in all expression systems *(3,4)*. Since this signaling pathway leads to activation of protein kinase A and protein phosphorylation, the effects of adenylate cyclase stimulation will be determined by the expression of specific proteins which may differ between neurons. In some cases, stimulation of adenylate cyclase and protein kinase A clearly enhances transmitter release, e.g., in presynaptic facilitation in *Aplysia (39)* and at glutamatergic synapses in rat hippocampus *(162)*. Furthermore, D1 receptor stimulation of adenylate cyclase increases synaptic release of glutamate in goldfish Mauthner cells *(89)* and GABA in guinea pig substantia nigra *(107)*. The D1-mediated enhancement of calcium current in adrenal chromaffin cells *(58)* and striatal neurons *(117)*, which results from phosphorylation of L-type calcium channels, may also facilitate transmitter release. However, there are also many examples in which stimulation of D1 receptors produces results which are not consistent with a general function of D1 receptors to facilitate transmitter release. Thus, dopamine-induced reduction of calcium current in retinal ganglion cells *(87)* and in rat striatal neurons *(117)* would be expected to reduce transmitter release. In addition, D1 receptors decrease synaptic potentials produced by glutamate and GABA in cerebral cortex *(157,158)* and nucleus accumbens *(141,143)*.

The difficulty in generalizing between responses produced by D1 receptor stimulation is shared with other receptors which activate adenylate cyclase and protein kinase A. For example, beta adrenergic receptor stimulation in liver cells phosphorylates metabolic enzymes leading to increased lipolysis and glycogenolysis, whereas stimulation of beta receptors in the heart leads to phosphorylation of ion channel proteins, causing decreased opening of sodium channels and facilitation of calcium channels, thereby modulating heart rate and contractility. Accordingly, protein kinase A activation by D1 receptors produces different effects in different cells, such as closing of gap junction channels in retinal horizontal cells *(80)* and suppression of sodium current in striatal neurons *(115)*. Thus, the electrophysiological effect of D1 receptor stimulation cannot be generalized, since it is necessarily dependent on the identity of the channels and other proteins that are modulated by protein kinase A in the cell type under study.

In comparison with D1 receptors, the D2 receptor subfamily is more promiscuous (or versatile, depending on perspective) in its relations with G proteins which, in their turn, are even more promiscuous in their choice of effector proteins (adenylate cyclase inhibition, voltage-dependent calcium channels, etc.). In some cases, such as pituitary lactotrophs, one receptor type (D_2) might couple to two different G proteins affecting calcium and potassium channels, respectively *(66)*. There is also confusion surrounding the signaling pathways employed by the newly discovered D_3 and D_4 subtypes. Still, although the transduction machinery of D2 receptors is more variable than for D1 receptors, the repertoire of cellular responses studied with electrophysiological methods is smaller, consisting mostly of increased potassium and decreased calcium conductances. A general scheme is receptor activation of a pertussis toxin-sensitive G protein, G_i or G_o, which activates a voltage-independent potassium conductance and inhibits voltage-dependent calcium conductances via a membrane-delimited pathway, i.e., without participation by a cytosolic second messenger. Examples include D2 responses in pituitary lactotrophs *(63,65)* and melanotrophs *(73,74)*, calcium current decrease in *Helix* neurons *(38,40)*, and potassium channel opening in striatal neurons *(125)*. Naturally, there are exceptions to this scheme, like putative second messenger involvement in modulation of $I_{K(Ca)}$ in chromaffin cells *(59)* and D2-induced decrease of sodium current in striatal neurons *(115)*, as well as examples of modulation of other ion conductances, such as the voltage-dependent potassium conductances I_A and I_K *(67,128)* and nonselective cation conductances regulated by calcium *(90)* or voltage *(106)*. Still, the general scheme appears to be widespread and is probably predominant in presynaptic inhibition of hormone or transmitter release mediated by D2 dopamine receptors. Several other receptors, including subtypes of adrenergic, muscarinic, and opioid receptors, produce inhibition of nerve activity and transmitter release in a similar fashion, frequently converging on the G protein or channel level *(163–165)*.

What remains to be done in the electrophysiological pharmacology of dopamine receptors? As should be apparent from this review: plenty. Understanding of the functional role of receptor subtypes should be improved by the development of selective ligands. The present knowledge is largely based on results with the agonists SKF 38393 and quinpirole, and the antagonists SCH 23390 and sulpiride, which can distinguish between the D1 and D2 subfamilies, but not the subtypes within the subfamilies. More studies of cloned receptor subtypes in expression systems, including the use of chimeric receptors and site-directed mutagenesis, will enhance knowledge of transduction mechanisms and facilitate comparison with endogenous receptors in systems with receptor heterogeneity. Improved electrophysiological techniques may also help to resolve some of the inconsistencies in the present literature. Voltage clamp, single-channel recording, and careful distinction between pre- and postsynaptic effects would be useful in this regard. The possibility of separating between pre- and postsynaptic sites of action is a great advantage of modern electrophysiological methods in the study of receptor function. Progress is also dependent on another important development, which is more difficult to define in technical terms: the growing understanding of the complexity of biological systems, as well as their parallels and relations. Sometimes the main problem of a scientific issue lies not in finding the right answer, but in formulating the right question. Instead of asking whether dopamine is excitatory or inhibitory, or whether dopamine receptor subtypes generally act in synergy or antagonistically, actions and mechanisms need to be addressed, analyzed, and interpreted on the appropriate level: G protein, ion channel, cell, synapse, network, and so on, all the way up to the impetus for this unending amassment of knowledge: the treatment of human disease.

References

1. Kebabian, J. W. and Calne, D. B. (1979) Multiple receptors for dopamine. *Nature* **277,** 93–96.
2. Stoof, J. C. and Kebabian, J. W. (1984) Two dopamine receptors: biochemistry, physiology and pharmacology. *Life Sci.* **35,** 2281–2296.
3. Civelli, O., Bunzow, J. R., and Grandy, D. K. (1993) Molecular diversity of the dopamine receptors. *Annu. Rev. Pharmacol. Toxicol.* **32,** 281–307.
4. Gingrich, J. A. and Caron, M. G. (1993) Recent advances in the molecular biology of dopamine receptors. *Annu. Rev. Neurosci.* **16,** 299–321.
5. Johnson, S. W. and North, R. A. (1992) Opioids excite dopamine neurons by hyperpolarization of local interneurons. *J. Neurosci.* **12,** 483–488.
6. Madison, D. V. and Nicoll, R. A. (1988) Enkephalin hyperpolarizes interneurones in the rat hippocampus. *J. Physiol. (Lond.)* **398,** 123–130.
7. York, D. H. (1979) The neurophysiology of dopamine receptors, in *The Neurobiology of Dopamine* (Horn, A. S., Korf, J., and Westerink, B. H. C., eds.), Academic, London, pp. 395–415.
8. Moore, R. Y. and Bloom, F. E. (1978) Central catecholamine neuron systems: anatomy and physiology of the dopamine systems. *Annu. Rev. Neurosci.* **1,** 129–169.
9. Llinás, R. R. (1988) The intrinsic electrophysiological properties of mammalian neurons: insights into nervous system function. *Science* **242,** 1654–1664.
10. Nicoll, R. A. (1982) Neurotransmitters can say more than just "yes" or "no." *Trends Neurosci.* **5,** 369–374.
11. Nicoll, R. A., Malenka, R. C., and Kauer, J. A. (1990) Functional comparison of neurotransmitter receptor subtypes in mammalian central nervous system. *Physiol. Rev.* **70,** 513–565.
12. Sakmann, B. (1992) Elementary steps in synaptic transmission revealed by currents through single ion channels. *Science* **256,** 503–512.
13. Lévesque, D., Diaz, J., Pilon, C., Martres, M.-P., Giros, B., Souil, E., et al. (1992) Identification, characterization, and localization of the dopamine D_3 receptor in rat

brain using 7-(^{3}H)hydroxy-*N,N*-di-*n*-propyl-2-aminotetralin. *Proc. Natl. Acad. Sci. USA* **89,** 8155–8159.

14. Chio, C. L., Lajiness, M. E., and Huff, R. M. (1994) Activation of heterologously expressed D3 dopamine receptors: comparison with D2 dopamine receptors. *Mol. Pharmacol.* **45,** 51–60.
15. Vallar, L., Muca, C., Magni, M., Albert, P., Bunzow, J., Meldolesi, J., and Civelli, O. (1990) Differential coupling of dopaminergic D_2 receptors expressed in different cell types. *J. Biol. Chem.* **265,** 10,320–10,326.
16. Di Marzo, V., Vial, D., Sokoloff, P., Schwartz, J.-C., and Piomelli, D. (1993) Selection of alternative G_i-mediated signaling pathways at the dopamine D_2 receptor by protein kinase C. *J. Neurosci.* **13,** 4846–4853.
17. Seabrook, G. R., Patel, S., Marwood, R., Emms, F., Knowles, M. R., Freedman, S. B., and McAllister, G. (1992) Stable expression of human D_3 dopamine receptors in GH_4C_1 pituitary cells. *FEBS Lett.* **312,** 123–126.
18. Liu, Y. F., Civelli, O., Zhou, Q.-Y., and Albert, P. R. (1992) Cholera toxin-sensitive 3',5'-cyclic adenosine monophosphate and calcium signals of the human dopamine-D1 receptor: selective potentiation by protein kinase A. *Mol. Endocrinol.* **6,** 1815–1824.
19. Reuter, H. (1983) Calcium channel modulation by neurotransmitters, enzymes and drugs. *Nature* **301,** 569–574.
20. Gray, R. and Johnston, D. (1987) Noradrenaline and β-adrenoceptor agonists increase activity of voltage-dependent calcium channels in hippocampal neurons. *Nature* **327,** 620–622.
21. Einhorn, L. C., Falardeau, P., Caron, M., Civelli, O., and Oxford, G. S. (1990) Both isoforms of the D2 dopamine receptor couple to a G protein-activated K^+ channel when expressed in GH_4 cells. *Soc. Neurosci. Abstr.* **16,** 382(Abstract)
22. Castellano, M. A., Liu, L.-X., Monsma, F. J., Jr., Sibley, D. R., Kapatos, G., and Chiodo, L. A. (1993) Transfected D_2 short dopamine receptors inhibit voltage-dependent potassium current in neuroblastoma x glioma hybrid (NG108-15) cells. *Mol. Pharmacol.* **44,** 649–656.
23. Seabrook, G. R., McAllister, G., Knowles, M. R., Myers, J., Sinclair, H., Patel, S., Freedman, S. B., and Kemp, J. A. (1994) Depression of high-threshold calcium currents by activation of human D_2 (short) dopamine receptors expressed in differentiated NG108-15 cells. *Br. J. Pharmacol.* **111,** 1061–1066.
24. Seabrook, G. R., Knowles, M., Brown, N., Myers, J., Sinclair, H., Patel, S., Freedman, S. B., and McAllister, G. (1994) Pharmacology of high-threshold calcium currents in GH_4C_1 pituitary cells and their regulation by activation of human D_2 and D_4 dopamine receptors. *Br. J. Pharmacol.* **112,** 728–734.
25. Freedman, S. B., Patel, S., Marwood, R., Emms, F., Seabrook, G. R., Knowles, M. R., and McAllister, G. (1994) Expression and pharmacological characterization of the human D-3 dopamine receptor. *J. Pharmacol. Exp. Ther.* **268,** 417–426.
26. Seabrook, G. R., Kemp, J. A., Freedman, S. B., Patel, S., Sinclair, H. A., and McAllister, G. (1994) Functional expression of human D_3 dopamine receptors in differentiated neuroblastoma × glioma NG108-15 cells. *Br. J. Pharmacol.* **111,** 391–393.
27. Ascher, P. (1972) Inhibitory and excitatory effects of dopamine on *Aplysia* neurones. *J. Physiol. (Lond.)* **225,** 173–209.

28. Ascher, P. and Chesnoy-Marchais, D. (1982) Interactions between three slow potassium responses controlled by three distinct receptors in *Aplysia* neurones. *J. Physiol. (Lond.)* **324,** 67–92.
29. Gruol, D. L. and Weinreich, D. (1979) Two pharmacologically distinct histamine receptors mediating membrane hyperpolarization on identified neurons of *Aplysia californica. Brain Res.* **162,** 281–301.
30. Sasaki, K. and Sato, M. (1987) A single GTP-binding protein regulates K^+-channels coupled with dopamine, histamine and acetylcholine receptors. *Nature* **325,** 259–262.
31. Lotshaw, D. P. and Levitan, I. B. (1988) Reciprocal modulation of calcium current by serotonin and dopamine in the identified *Aplysia* neuron R15. *Brain Res.* **439,** 64–76.
32. Akopyan, A. R., Chemeris, N. K., and Iljin, V. I. (1985) Neurotransmitter-induced modulation of neuronal Ca current is not mediated by intracellular Ca^{2+} or cAMP. *Brain Res.* **326,** 145–148.
33. Stoof, J. C., De Vlieger, T. A., and Lodder, J. C. (1984) Opposing roles for D-1 and D-2 dopamine receptors in regulating the excitability of growth hormone-producing cells in the snail *Lymnaea stagnalis. Eur. J. Pharmacol.* **106,** 431–435.
34. De Vlieger, T. A., Lodder, J. C., Stoof, J. C., and Werkman, T. R. (1986) Dopamine receptor stimulation induces a potassium dependent hyperpolarizing response in growth hormone producing neuroendocrine cells of the gastropod mollusc *Lymnaea stagnalis. Comp. Biochem. Physiol.* **83C,** 429–433.
35. Werkman, T. R., Lodder, J. C., De Vlieger, T. A., and Stoof, J. C. (1987) Further pharmacological characterization of a D-2-like dopamine receptor on growth hormone producing cells in *Lymnaea stagnalis. Eur. J. Pharmacol.* **139,** 155–161.
36. Bolshakov, V. Y., Gapon, S. A., Katchman, A. N., and Magaznik, L. G. (1993) Activation of a common potassium channel in molluscan neurones by glutamate, dopamine and muscarinic agonist. *J. Physiol. (Lond.)* **468,** 11–33.
37. Deterre, P., Paupardin-Tritsch, D., Bockaert, J., and Gerschenfeld, H. M. (1982) cAMP-mediated decrease in K+ conductance evoked by serotonin and dopamine in the same neuron: a biochemical and physiological single-cell study. *Proc. Natl. Acad. Sci. USA* **79,** 7934–7938.
38. Paupardin-Tritsch, D., Colombaioni, L., Deterre, P., and Gerschenfeld, H. M. (1985) Two different mechanisms of calcium spike modulation by dopamine. *J. Neurosci.* **5,** 2522–2532.
39. Kandel, E. R. (1985) Cellular mechanisms of learning and the biological basis of individuality, in *Principles of Neural Science,* 2nd ed. (Kandel, E. R. and Schwartz, J. H., eds.), Elsevier, New York, pp. 816–833.
40. Harris-Warrick, R. M., Hammond, C., Paupardin-Tritsch, D., Homburger, V., Rouot, B., Bockaert, J., and Gerschenfeld, H. M. (1988) An α_{40} subunit of a GTP-binding protein immunologically related to G_o mediates a dopamine-induced decrease of Ca^{2+} current in snail neurons. *Neuron* **1,** 27–32.
41. Hille, B. (1994) Modulation of ion-channel function by G-protein-coupled receptors. *Trends Neurosci.* **17,** 531–536.
42. Rosenthal, W., Hescheler, J., Trautwein, W., and Schultz, G. (1988) Control of voltage-dependent Ca^{2+} channels by G protein-coupled receptors. *FASEB J.* **2,** 2784–2790.

43. Ginsborg, B. L., House, C. R., and Silinsky, E. M. (1974) Conductance changes associated with the secretory potential in the cockroach salivary gland. *J. Physiol. (Lond.)* **236,** 723–731.
44. House, C. R. and Ginsborg, B. L. (1976) Actions of a dopamine analogue and a neuroleptic at a neuroglandular synapse. *Nature* **261,** 332,333.
45. Petersen, O. H. (1992) Stimulus-secretion coupling: cytoplasmic calcium signals and the control of ion channels in exocrine acinar cells. *J. Physiol. (Lond.)* **448,** 1–51.
46. Ginsborg, B. L., House, C. R., and Mitchell, M. R. (1980) On the role of calcium in the electrical responses of cockroach salivary gland to dopamine. *J. Physiol. (Lond.)* **303,** 325–335.
47. Johnson, B. R. and Harris-Warrick, R. M. (1990) Aminergic modulation of graded synaptic transmission in the lobster stomatogastric ganglion. *J. Neurosci.* **10,** 2066–2076.
48. Harris-Warrick, R. M., Coniglio, L. M., Barazangi, N., Guckenheimer, J., and Gueron, S. (1995) Dopamine modulation of transient potassium current evokes phase shifts in a central pattern generator network. *J. Neurosci.* **15,** 342–358.
49. Rogawski, M. A. (1985) The A-current: how ubiquitous a feature of excitable cells is it? *Trends Neurosci.* **8,** 214–219.
50. Brown, D. A. and Caulfield, M. P. (1979) Hyperpolarizing "α_2"-adrenoceptors in rat sympathetic ganglia. *Br. J. Pharmacol.* **65,** 435–445.
51. Cole, A. E. and Shinnick-Gallagher, P. (1981) Comparison of the receptors mediating the catecholamine hyperpolarization and slow inhibitory postsynaptic potential in sympathetic ganglia. *J. Pharmacol. Exp. Ther.* **217,** 440–444.
52. Dun, N. and Nishi, S. (1974) Effects of dopamine on the superior cervical ganglion of the rabbit. *J. Physiol. (Lond.)* **239,** 155–164.
53. Willems, J. L., Buylaert, W. A., Lefebvre, R. A., and Bogaert, M. G. (1985) Neuronal dopamine receptors on autonomic ganglia and sympathetic nerves and dopamine receptors in the gastrointestinal system. *Pharmacol. Rev.* **37,** 165–216.
54. Marchetti, C., Carbone, E., and Lux, H. D. (1986) Effects of dopamine and noradrenaline on Ca channels of cultured sensory and sympathetic neurons of chick. *Pflugers Arch.* **406,** 104–111.
55. Benot, A. R. and Lopez-Barneo, J. (1990) Feedback inhibition of Ca^{2+} currents by dopamine in glomus cells of the carotid body. *Eur. J. Neurosci.* **2,** 809–812.
56. Bigornia, L., Allen, C. N., Jan, C.-R., Lyon, R. A., Titeler, M., and Schneider, A. S. (1990) D_2 dopamine receptors modulate calcium channel currents and catecholamine secretion in bovine adrenal chromaffin cells. *J. Pharmacol. Exp. Ther.* **252,** 586–592.
57. Sontag, J.-M., Sanderson, P., Klepper, M., Aunis, D., Takeda, K., and Bader, M.-F. (1990) Modulation of secretion by dopamine involves decreases in calcium and nicotinic currents in bovine chromaffin cells. *J. Physiol. (Lond.)* **427,** 495–517.
58. Artalejo, C. R., Ariano, M. A., Perlman, R. L., and Fox, A. P. (1990) Activation of facilitation calcium channels in chromaffin cells by D1 dopamine receptors through a cAMP/protein kinase A-dependent mechanism. *Nature* **348,** 239–242.
59. Twitchell, W. A. and Rane, S. G. (1994) Nucleotide-independent modulation of Ca^{2+}-dependent K^+ channel current by a μ-type opioid receptor. *Mol. Pharmacol.* **46,** 793–798.

60. Inoue, K., Nakazawa, K., Watano, T., Ohara-Imaizumi, M., Fujimori, K., and Takanaka, A. (1992) Dopamine receptor agonists and antagonists enhance ATP-activated currents. *Eur. J. Pharmacol.* **215,** 321–324.
61. Nakazawa, K., Watano, T., and Inoue, K. (1993) Mechanisms underlying facilitation by dopamine of ATP-activated currents in rat pheochromocytoma cells. *Pflugers Arch.* **422,** 458–464.
62. Björklund, A. and Lindvall, O. (1984) Dopamine-containing systems in the CNS, in *Handbook of Chemical Neuroanatomy, vol. 2* (Björklund, A. and Hökfelt, T., eds.), Elsevier, Amsterdam, pp. 55–122.
63. Einhorn, L. C., Gregerson, K. A., and Oxford, G. S. (1991) D_2 dopamine receptor activation of potassium channels in identified rat lactotrophs: whole-cell and single-channel recording. *J. Neurosci.* **11,** 3727–3737.
64. Israel, J. M., Kirk, C., and Vincent, J. D. (1987) Electrophysiological responses to dopamine of rat hypophysial cells in lactotroph-enriched primary cultures. *J. Physiol. (Lond.)* **390,** 1–22.
65. Einhorn, L. C. and Oxford, G. S. (1993) Guanine nucleotide binding proteins mediate D_2 dopamine receptor activation of a potassium channel in rat lactotrophs. *J. Physiol. (Lond.)* **462,** 563–578.
66. Lledo, P. M., Homburger, V., Bockaert, J., and Vincent, J.-D. (1992) Differential G protein-mediated coupling of D-2 dopamine receptors to K^+ and Ca^{2+} currents in rat anterior pituitary cells. *Neuron* **8,** 455–463.
67. Lledo, P.-M., Legendre, P., Zhang, J., Israel, J.-M., and Vincent, J.-D. (1990) Effects of dopamine on voltage-dependent potassium currents in identified rat lactotroph cells. *Neuroendocrinology* **52,** 545–555.
68. Login, I. S., Pancrazio, J. J., and Kim, Y. I. (1990) Dopamine enhances a voltage-dependent transient K^+ current in the MMQ cell, a clonal pituitary line expressing functional D_2 dopamine receptors. *Brain Res.* **506,** 331–334.
69. Lledo, P.-M., Legendre, P., Israel, J.-M., and Vincent, J.-D. (1990) Dopamine inhibits two characterized voltage-dependent calcium currents in identified rat lactotroph cells. *Endocrinology* **127,** 990–1001.
70. Lledo, P.-M., Israel, J.-M., and Vincent, J.-D. (1990) A guanine nucleotide protein mediates the inhibition of voltage-dependent calcium currents by dopamine in rat lactotrophs. *Brain Res.* **528,** 143–147.
71. Lledo, P.-M., Israel, J. M., and Vincent, J.-D. (1991) Chronic stimulation of D_2 dopamine receptors specifically inhibits calcium but not potassium currents in rat lactotrophs. *Brain Res.* **558,** 231–238.
72. Williams, P. J., MacVicar, B. A., and Pittman, Q. J. (1989) A dopaminergic inhibitory postsynaptic potential mediated by an increased potassium conductance. *Neuroscience* **31,** 673–681.
73. Keja, J. A., Stoof, J. C., and Kits, K. S. (1992) Dopamine D_2 receptor stimulation differentially affects voltage-activated calcium channels in rat pituitary melanotropic cells. *J. Physiol. (Lond.)* **450,** 409–435.
74. Stack, J. and Surprenant, A. (1991) Dopamine actions on calcium currents, potassium currents and hormone release in rat melanotrophs. *J. Physiol. (Lond.)* **439,** 37–58.
75. Williams, P. J., MacVicar, B. A., and Pittman, Q. J. (1990) Synaptic modulation by dopamine of calcium currents in rat pars intermedia. *J. Neurosci.* **10,** 757–763.
76. Nussinovitch, I. and Kleinhaus, A. L. (1992) Dopamine inhibits voltage-activated calcium channel currents in rat pars intermedia pituitary cells. *Brain Res.* **574,** 49–55.

77. Valentijn, J. A., Louiset, E., Vaudry, H., and Cazin, L. (1991) Dopamine-induced inhibition of action potentials in cultured frog pituitary melanotrophs is mediated through activation of potassium channels and inhibition of calcium and sodium channels. *Neuroscience* **42,** 29–39.
78. Valentijn, J. A., Louiset, E., Vaudry, H., and Cazin, L. (1991) Dopamine regulates the electrical activity of frog melanotrophs through a G protein-mediated mechanism. *Neuroscience* **44,** 85–95.
79. Mudrick-Donnon, L. A., Williams, P. J., Pittman, Q. J., and MacVicar, B. A. (1993) Postsynaptic potentials mediated by GABA and dopamine evoked in stellate glial cells of the pituitary pars intermedia. *J. Neurosci.* **13,** 4660–4668.
80. Dowling, J. E. (1991) Retinal neuromodulation: the role of dopamine. *Vis. Neurosci.* **7,** 87–97.
81. McMahon, D. G., Knapp, A. G., and Dowling, J. E. (1989) Horizontal cell gap junctions: single-channel conductance and modulation by dopamine. *Proc. Natl. Acad. Sci. USA* **86,** 7639–7643.
82. DeVries, S. H. and Schwartz, E. A. (1992) Hemi-gap-junction channels in solitary horizontal cells of the catfish retina. *J. Physiol. (Lond.)* **445,** 201–230.
83. Harsanyi, K. and Mangel, S. C. (1992) Activation of a D_2 receptor increases electrical coupling between retinal horizontal cells by inhibiting dopamine release. *Proc. Natl. Acad. Sci. USA* **89,** 9220–9224.
84. Knapp, A. G., Schmidt, K. F., and Dowling, J. E. (1990) Dopamine modulates the kinetics of ion channels gated by excitatory amino acids in retinal horizontal cells. *Proc. Natl. Acad. Sci. USA* **87,** 767–771.
85. Schmidt, K.-F., Kruse, M., and Hatt, H. (1994) Dopamine alters glutamate receptor desensitization in retinal horizontal cells of the perch *(Perca fluviatilis)*. *Proc. Natl. Acad. Sci. USA* **91,** 8288–8291.
86. Dong, C.-J. and Werblin, F. S. (1994) Dopamine modulation of $GABA_C$ receptor function in an isolated retinal neuron. *J. Neurophysiol.* **71,** 1258–1260.
87. Liu, Y. and Lasater, E. M. (1994) Calcium currents in turtle retinal ganglion cells. II. Dopamine modulation via a cyclic AMP-dependent mechanism. *J. Neurophysiol.* **71,** 743–752.
88. Schotland, J., Shupliakov, O., Wikström, M., Brodin, L., Srinivasan, M., You, Z.-B., Herrera-Marschitz, M., Zhang, W., Hökfelt, T., and Grillner, S. (1995) Control of lamprey locomotor neurons by colocalized monoamine transmitters. *Nature* **374,** 266–268.
89. Pereda, A., Triller, A., Korn, H., and Faber, D. S. (1992) Dopamine enhances both electrotonic coupling and chemical excitatory postsynaptic potentials at mixed synapses. *Proc. Natl. Acad. Sci. USA* **89,** 12,088–12,092.
90. Yang, C. R., Bourque, C. W., and Renaud, L. P. (1991) Dopamine D_2 receptor activation depolarizes rat supraoptic neurones in hypothalamic explants. *J. Physiol. (Lond.)* **443,** 405–419.
91. Partridge, L. D. and Swandulla, D. (1988) Calcium-activated non-specific cation channels. *Trends Neurosci.* **11,** 69–72.
92. Aghajanian, G. K. and Bunney, B. S. (1973) Central dopaminergic neurons: neurophysiological identification and responses to drugs, in *Frontiers in Catecholamine Research* (Usdin, E. and Snyder, S. H., eds.), Pergamon, New York, pp. 643–648.

93. White, F. J. and Wang, R. Y. (1984) Pharmacological characterization of dopamine autoreceptors in the rat ventral tegmental area: microiontophoretic studies. *J. Pharmacol. Exp. Ther.* **231,** 275–280.
94. Innis, R. B. and Aghajanian, G. K. (1987) Pertussis toxin blocks autoreceptor-mediated inhibition of dopaminergic neurons in rat substantia nigra. *Brain Res.* **411,** 139–143.
95. Lacey, M. G., Mercuri, N. B., and North, R. A. (1987) Dopamine acts on D-2 receptors to increase potassium conductance in neurones of the rat substantia nigra zona compacta. *J. Physiol. (Lond.)* **392,** 397–416.
96. Lacey, M. G., Mercuri, N. B., and North, R. A. (1988) On the potassium conductance increase activated by GABA-B and dopamine D-2 receptors in rat substantia nigra neurones. *J. Physiol. (Lond.)* **401,** 437–453.
97. Johnson, S. W. and North, R. A. (1992) Two types of neurone in the rat ventral tegmental area and their synaptic inputs. *J. Physiol. (Lond.)* **450,** 455–468.
98. Roeper, J., Hainsworth, A. H., and Ashcroft, F. M. (1990) Tolbutamide reverses membrane hyperpolarisation induced by activation of D_2 receptors and $GABA_B$ receptors in isolated substantia nigra neurones. *Pflugers Arch.* **416,** 473–475.
99. Hicks, G. A. and Henderson, G. (1992) Lack of evidence for coupling of the dopamine D2 receptor to an adenosine triphosphate-sensitive potassium (ATP-K^+) channel in dopaminergic neurones of the rat substantia nigra. *Neurosci. Lett.* **141,** 213–217.
100. Lejeune, F. and Millan, M. J. (1995) Activation of dopamine D_3 autoreceptors inhibits firing of ventral tegmental dopaminergic neurones in vivo. *Eur. J. Pharmacol.* **275,** R7–R9.
101. Millan, M. J., Audinot, V., Rivet, J.-M., Gobert, A., Vian, J., Prost, J.-F., Spedding, M., and Peglion, J.-L. (1994) S 14297, a selective ligand at cloned human dopamine D_3 receptors, blocks 7-OH-DPAT-induced hypothermia in rats. *Eur. J. Pharmacol.* **260,** R3–R5.
102. Kreiss, D. S., Bergstrom, D. A., Gonzalez, A. M., Huang, K.-X., Sibley, D. R., and Walters, J. R. (1995) Dopamine receptor agonist potencies for inhibition of cell firing correlate with dopamine D_3 receptor binding affinities. *Eur. J. Pharmacol.* **277,** 209–214.
103. Colquhoun, D. (1987) Affinity, efficacy, and receptor classification: is the classical theory still useful?, in *Perspectives on Receptor Classification* (Black J. W., Jenkinson D. H., and Gerskowitch V. P., eds.), Liss, New York, pp. 103–114.
104. Bowery, B., Rothwell, L. A., and Seabrook, G. R. (1994) Comparison between the pharmacology of dopamine receptors mediating the inhibition of cell firing in rat brain slices through the substantia nigra pars compacta and ventral tegmental area. *Br. J. Pharmacol.* **112,** 873–880.
105. Liu, L., Shen, R.-Y., Kapatos, G., and Chiodo, L. A. (1994) Dopamine neuron membrane physiology: characterization of the transient outward current (I_A) and demonstration of a common signal transduction pathway for I_A and I_K. *Synapse* **17,** 230–240.
106. Jiang, Z.-G., Pessia, M., and North, R. A. (1993) Dopamine and baclofen inhibit the hyperpolarization-activated cation current in rat ventral tegmental neurones. *J. Physiol. (Lond.)* **462,** 753–764.
107. Cameron, D. L. and Williams, J. T. (1993) Dopamine D1 receptors facilitate transmitter release. *Nature* **366,** 344–347.

108. Kitai, S. T., Sugimori, M., and Kocsis, J. D. (1976) Excitatory nature of dopamine in the nigro-caudate pathway. *Exp. Brain Res.* **24,** 351–363.
109. Bernardi, G., Marciani, M. G., Morocutti, C., Pavone, F., and Stanzione, P. (1978) The action of dopamine on rat caudate neurones intracellularly recorded. *Neurosci. Lett.* **8,** 235–240.
110. Herrling, P. L. and Hull, C. D. (1980) Iontophoretically applied dopamine depolarizes and hyperpolarizes the membrane of cat caudate neurons. *Brain Res.* **192,** 441–462.
111. Calabresi, P., Mercuri, N., Stanzione, P., Stefani, A., and Bernardi, G. (1987) Intracellular studies on the dopamine-induced firing inhibition of neostriatal neurons in vitro: evidence for D1 receptor involvement. *Neuroscience* **20,** 757–771.
112. Calabresi, P., Mercuri, N. B., Sancesario, G., and Bernardi, G. (1993) Electrophysiology of dopamine-denervated striatal neurons. *Brain* **116,** 433–452.
113. Calabresi, P., Benedetti, M., Mercuri, N. B., and Bernardi, G. (1988) Endogenous dopamine and dopaminergic agonists modulate synaptic excitation in neostriatum: intracellular studies from naive and catecholamine-depleted rats. *Neuroscience* **27,** 145–157.
114. Akaike, A., Ohno, Y., Sasa, M., and Takaori, S. (1987) Excitatory and inhibitory effects of dopamine on neuronal activity of the caudate nucleus neurons in vitro. *Brain Res.* **418,** 262–272.
115. Surmeier, D. J., Eberwine, J., Wilson, C. J., Cao, Y., Stefani, A., and Kitai, S. T. (1992) Dopamine receptor subtypes colocalize in rat striatonigral neurons. *Proc. Natl. Acad. Sci. USA* **89,** 10,178–10,182.
116. Li, M., West, J. W., Lai, Y., Scheuer, T., and Catterall, W. A. (1992) Functional modulation of brain sodium channels by cAMP-dependent phosphorylation. *Neuron* **8,** 1151–1159.
117. Surmeier, D. J., Bargas, J., Hemmings, H. C., Jr., Nairn, A. C., and Greengard, P. (1995) Modulation of calcium currents by a D_1 dopaminergic protein kinase/phosphatase cascade in rat neostriatal neurons. *Neuron* **14,** 385–397.
118. Calabresi, P., Benedetti, M., Mercuri, N. B., and Bernardi, G. (1988) Depletion of catecholamines reveals inhibitory effects of bromocriptine and lysuride on neostriatal neurones recorded intracellularly in vitro. *Neuropharmacology* **27,** 579–587.
119. Coirini, H., Schumacher, M., Angulo, J. A., and McEwen, B. S. (1990) Increase in striatal dopamine D_2 receptor mRNA after lesions or haloperidol treatment. *Eur. J. Pharmacol.* **186,** 369–371.
120. Gerfen, C. R., Engber, T. M., Mahan, L. C., Susel, Z., Chase, T. N., Monsma, F. J., Jr., and Sibley, D. R. (1990) D_1 and D_2 dopamine receptor-regulated gene expression of striatonigral and striatopallidal neurons. *Science* **250,** 1429–1432.
121. Le Moine, C., Normand, E., Guitteny, A. F., Fouque, B., Teoule, R., and Bloch, B. (1990) Dopamine receptor gene expression by enkephalin neurons in rat forebrain. *Proc. Natl. Acad. Sci. USA* **87,** 230–234.
122. Surmeier, D. J., Reiner, A., Levine, M. S., and Ariano, M. A. (1993) Are neostriatal dopamine receptors co-localized? *Trends Neurosci.* **16,** 299–305.
123. Gerfen, C. R., Keefe, K. A., Bloch, B., Le Moine, C., Surmeier, D. J., Reiner, A., Levine, M. S., and Ariano, M. A. (1994) Neostriatal dopamine receptors. *Trends Neurosci.* **17,** 2–5.

124. Drago, J., Gerfen, C. R., Lachowicz, J. E., Streiner, H., Hollon, T. R., Love, P. E., et al. (1994) Altered striatal function in a mutant mouse lacking D_{1A} dopamine receptors. *Proc. Natl. Acad. Sci. USA* **91,** 12,564–12,568.
125. Freedman, J. E. and Weight, F. F. (1988) Single K^+ channels activated by D_2 dopamine receptors in acutely dissociated neurons from rat corpus striatum. *Proc. Natl. Acad. Sci. USA* **85,** 3618–3622.
126. Freedman, J. E. and Weight, F. F. (1989) Quinine potently blocks single K^+ channels activated by dopamine D-2 receptors in rat corpus striatum neurons. *Eur. J. Pharmacol.* **164,** 341–346.
127. Lin, Y.-J., Greif, G. J., and Freedman, J. E. (1993) Multiple sulfonylurea-sensitive potassium channels: a novel subtype modulated by dopamine. *Mol. Pharmacol.* **44,** 907–910.
128. Kitai, S. T. and Surmeier, D. J. (1993) Cholinergic and dopaminergic modulation of potassium conductances in neostriatal neurons, in *Advances in Neurology, Vol. 60: Parkinson's Disease: From Basic Research to Treatment* (Narabayashi, H., Nagatsu, T., Yanagisawa, N., and Mizuno, Y., eds.), Raven, New York, pp. 40–52.
129. Storm, J. F. (1990) Potassium currents in hippocampal pyramidal cells, in *Understanding the Brain Through the Hippocampus* (Storm-Mathisen, J., Zimmer, J., and Ottersen, O. P., eds.), Elsevier, Amsterdam, pp. 161–187.
130. Rutherford, A., Garcia-Munoz, M., and Arbuthnott, G. W. (1988) An afterhyperpolarization recorded in striatal cells "in vitro": effect of dopamine administration. *Exp. Brain Res.* **71,** 399–405.
131. Cepeda, C., Buchwald, N. A., and Levine, M. S. (1993) Neuromodulatory actions of dopamine in the neostriatum are dependent upon the excitatory amino acid receptor subtypes activated. *Proc. Natl. Acad. Sci. USA* **90,** 9576–9580.
132. Calabresi, P., De Murtas, M., Mercuri, N. B., and Bernardi, G. (1992) Chronic neuroleptic treatment: D2 dopamine receptor supersensitivity and striatal glutamatergic transmission. *Ann. Neurol.* **31,** 366–373.
133. Bliss, T. V. P. and Collingridge, G. L. (1993) A synaptic model for memory: long-term potentiation in the hippocampus. *Nature* **361,** 31–39.
134. Linden, D. J. (1994) Long-term synaptic depression in the mammalian brain. *Neuron* **12,** 457–472.
135. Calabresi, P., Pisani, A., Mercuri, N. B., and Bernardi, G. (1992) Long-term potentiation in the striatum is unmasked by removing the voltage-dependent magnesium block of NMDA receptor channels. *Eur. J. Neurosci.* **4,** 929–935.
136. Calabresi, P., Maj, R., Mercuri, N. B., and Bernardi, G. (1992) Coactivation of D_1 and D_2 dopamine receptors is required for long-term synaptic depression in the striatum. *Neurosci. Lett.* **142,** 95–99.
137. Aosaki, T., Graybiel, A. M., and Kimura, M. (1994) Effect of the nigrostriatal dopamine system on acquired neural responses in the striatum of behaving monkeys. *Science* **265,** 412–415.
138. Cepeda, C., Walsh, J. P., Hull, C. D., Howard, S. G., Buchwald, N. A., and Levine, M. S. (1989) Dye-coupling in the neostriatum of the rat: I. Modulation by dopamine-depleting lesions. *Synapse* **4,** 229–237.
139. Heimer, L., Switzer, R. D., and Van Hoesen, G. W. (1982) Ventral striatum and ventral pallidum—components of the motor system? *Trends Neurosci.* **5,** 83–87.

140. Uchimura, N., Higashi, H., and Nishi, S. (1986) Hyperpolarizing and depolarizing actions of dopamine via D-1 and D-2 receptors on nucleus accumbens neurons. *Brain Res.* **375,** 368–372.
141. Higashi, H., Inanaga, K., Nishi, S., and Uchimura, N. (1989) Enhancement of dopamine actions on rat nucleus accumbens neurones *in vitro* after methamphetamine pre-treatment. *J. Physiol. (Lond.)* **408,** 587–603.
142. Uchimura, N. and North, R. A. (1990) Actions of cocaine on rat nucleus accumbens neurones *in vitro*. *Br. J. Pharmacol.* **99,** 736–740.
143. Pennartz, C. M. A., Dolleman-Van der Weel, M. J., Kitai, S. T., and Lopes da Silva, F. H. (1992) Presynaptic dopamine D1 receptors attenuate excitatory and inhibitory limbic inputs to the shell region of the rat nucleus accumbens studied in vitro. *J. Neurophysiol.* **67,** 1325–1334.
144. Heimer, L. and Alheid, G. F. (1991) Piecing together the puzzle of basal forebrain anatomy, in *The Basal Forebrain* (Napier, T. C., Kalivas, P. W., and Hanin, I., eds.), Plenum, New York, pp. 1–42.
145. Pennartz, C. M. A., Dolleman-Van der Weel, M., and Lopes da Silva, F. H. (1992) Differential membrane properties and dopamine effects in the shell and core of the rat nucleus accumbens studied in vitro. *Neurosci. Lett.* **136,** 109–112.
146. O'Donnell, P. and Grace, A. A. (1994) Tonic D_2-mediated attenuation of cortical excitation in nucleus accumbens neurons recorded in vitro. *Brain Res.* **634,** 105–112.
147. O'Donnell, P. and Grace, A. A. (1993) Dopaminergic modulation of dye coupling between neurons in the core and shell regions of the nucleus accumbens. *J. Neurosci.* **13,** 3456–3471.
148. Benardo, L. S. and Prince, D. A. (1982) Dopamine modulates a Ca^{2+}-activated potassium conductance in mammalian hippocampal pyramidal cells. *Nature* **297,** 76–79.
149. Benardo, L. S. and Prince, D. A. (1982) Dopamine action on hippocampal pyramidal cells. *J. Neurosci.* **2,** 415–423.
150. Malenka, R. C. and Nicoll, R. A. (1986) Dopamine decreases the calcium-activated afterhyperpolarization in hippocampal CA1 pyramidal cells. *Brain Res.* **379,** 210–215.
151. Stanzione, P., Calabresi, P., Mercuri, N., and Bernardi, G. (1984) Dopamine modulates CA1 hippocampal neurons by elevating the threshold for spike generation: an *in vitro* study. *Neuroscience* **13,** 1105–1116.
152. Frey, U., Huang, Y.-Y., and Kandel, E. R. (1993) Effects of cAMP simulate a late stage of LTP in hippocampal CA1 neurons. *Science* **260,** 1661–1664.
153. Huang, Y.-Y. and Kandel, E. R. (1995) D1/D5 receptor agonists induce a protein synthesis-dependent late potentiation in the CA1 region of the hippocampus. *Proc. Natl. Acad. Sci. USA* **92,** 2446–2450.
154. Le Moal, M. and Simon, H. (1991) Mesocorticolimbic dopaminergic network: functional and regulatory roles. *Physiol. Rev.* **71,** 155–234.
155. Penit-Soria, J., Audinat, E., and Crepel, F. (1987) Excitation of rat prefrontal cortical neurons by dopamine: an in vitro electrophysiological study. *Brain Res.* **425,** 263–274.
156. Gellman, R. L. and Aghajanian, G. K. (1993) Pyramidal cells in piriform cortex receive a convergence of inputs from monoamine activated GABAergic interneurons. *Brain Res.* **600,** 63–73.

157. Pralong, E. and Jones, R. S. G. (1993) Interactions of dopamine with glutamate- and GABA-mediated synaptic transmission in the rat entorhinal cortex *in vitro*. *Eur. J. Neurosci.* **5,** 760–767.
158. Law-Tho, D., Hirsch, J. C., and Crepel, F. (1994) Dopamine modulation of synaptic transmission in rat prefrontal cortex: an in vitro electrophysiological study. *Neurosci. Res.* **21,** 151–160.
159. Cepeda, C., Radisavljevic, Z., Peacock, W., Levine, M. S., and Buchwald, N. A. (1992) Differential modulation by dopamine of responses evoked by excitatory amino acids in human cortex. *Synapse* **11,** 330–341.
160. Law-Tho, D., Desce, J. M., and Crepel, F. (1995) Dopamine favours the emergence of long-term depression versus long-term potentiation in slices of rat prefrontal cortex. *Neurosci. Lett.* **188,** 125–128.
161. Sawaguchi, T. and Goldman-Rakic, P. S. (1991) D1 dopamine receptors in prefrontal cortex: involvement in working memory. *Science* **251,** 947–950.
162. Chavez-Noriega, L. E. and Stevens, C. F. (1994) Increased transmitter release at excitatory synapses produced by direct activation of adenylate cyclase in rat hippocampal slices. *J. Neurosci.* **14,** 310–317.
163. Nicoll, R. A. (1988) The coupling of neurotransmitter receptors to ion channels in the brain. *Science* **241,** 545–551.
164. North, R. A. (1989) Drug receptors and the inhibition of nerve cells. *Br. J. Pharmacol.* **98,** 13–28.
165. North, R. A. (1992) Opioid actions on membrane ion channels, in *Handbook of Experimental Pharmacology, Vol. 104/I: Opioids I* (Herz, A., Akil, H., and Simon, E. J., eds.), Springer-Verlag, Berlin, pp. 773–797.

CHAPTER 10

Dopamine Receptor Modulation of Gene Expression in the Brain

Monique R. Adams, Raymond P. Ward, and Daniel M. Dorsa

1. Introduction

Numerous changes in neuronal activity have been documented to be under the control of dopaminergic systems. Prominent among these are changes in transcription of neurotransmitter genes which are expressed in dopamine receptor positive neurons throughout the brain. The dopamine receptors are subdivided in two subfamilies: the D1-like (D_1, D_5) and the D2-like (D_2, D_3, D_4).* As described in Chapter 6, members of this class of receptors are either positively or negatively coupled to some of the most important signal transduction pathways involved in regulating gene transcription. Prominent among these is the adenylate cyclase–cyclic adenosine monophosphate (cAMP) system, in which the cascade of events leading to gene transcription has in large part been elucidated. Agonism of receptors positively or negatively coupled to adenylate cyclase alters the conversion of ATP to cAMP. Subsequent binding of this nucleotide to the regulatory subunit of protein kinase A (PKA) leads to activation of its catalytic domain which carries out phosphorylation of numerous neuronal proteins. Among these downstream, rapid signaling events is the phosphorylation of a protein termed the cAMP response element-binding protein or CREB. Phosphorylation on a serine residue in position 133 of this protein enhances its ability to promote gene transcription. CREB binds as a dimer to a canonical sequence of nucle-

*As detailed in the Preface, the terms D1 or D1-like and D2 or D2-like are used to refer to the subfamilies of DA receptors, or used when the genetic subtype (D_1 or D_5; D_2, D_3, or D_4) is uncertain.

The Dopamine Receptors Eds.: K. A. Neve and R. L. Neve
Humana Press Inc., Totowa, NJ

otides termed the cAMP response element (CRE) found in the upstream 5' flanking region of numerous genes. Interestingly, this element is found in the promoter regions of genes encoding other transcriptional activators such as members of the immediate-early gene family, including c-fos. Thus agonism or antagonism of any given dopamine receptor can (as is described in Section 3.) and does lead to induction of a variety of factors capable of influencing transcription of neurotransmitter genes, such as those encoding the neuropeptides enkephalin and neurotensin.

This raises several important questions about the role of dopamine receptors in regulating neurotransmitter gene expression, which we attempt to address in this chapter. Do dopamine receptors play a physiologically important role in transsynaptic regulation of gene expression? Can distinct effects associated with any one dopamine receptor be discerned within the context of normal brain circuitry? When a given receptor is expressed in different regions of the brain, are the downstream signaling and transcriptional responses associated with it the same in all locations? In the present chapter we summarize some of the data which address these issues.

The neuropharmacologic approach has been most commonly used to investigate the involvement of dopamine and other receptors in regulating gene transcription. Nearly all of the drugs used in such studies lack specificity for any one receptor. These studies describe changes in transcription factor and neurotransmitter gene expression following exposure of anatomically identified populations of neurons to drugs of varying specificities. For example, in this chapter, we frequently refer to the effects of indirect dopamine agonists such as amphetamine and cocaine. These are psychomotor stimulants that act to enhance dopaminergic transmission, but can be expected to affect other aminergic neurotransmitter systems as well. Cocaine primarily blocks reuptake of dopamine at the synapse *(1–3),* but also blocks the reuptake of serotonin *(2,4)* and norepinephrine *(4,5)*. Amphetamine appears to exert most or all of its effects in the central nervous system by releasing biogenic amines from their storage sites in the nerve terminals. However, amphetamine also promotes the release of serotonin from serotonergic nerve terminals *(6)*.

We also summarize the effects of agents which antagonize dopamine receptors; prominent among these are the antipsychotic agents. These compounds are widely used clinically; however their mechanisms of action are poorly understood. It takes several weeks for the manifestation of the full therapeutic effects of antipsychotic drugs to emerge. Based on this observation, it has been suggested that the interaction of antipsychotic agents with neurotransmitter receptors is merely the initial step in their actions and their therapeutic effects result from adaptive processes that may occur in response to chronic receptor occupancy. An important mechanism may involve the induction of cellular immediate-early genes which can mediate drug-induced neural plasticity.

There are two distinct classes of antipsychotic drugs termed typical and atypical agents. Typical antipsychotic drugs (e.g., haloperidol and fluphenazine) in addition to their therapeutic effects, are accompanied by a variety of acute and chronic motor side effects. Atypical antipsychotic drugs (e.g., clozapine and risperidone) induce a much lower incidence of the extrapyramidal motor side effects and have a clinical efficacy equal to typical agents. Although clearly an improvement in our ability to treat schizoprenia, the atypicals are not without unwanted side effects. Clozapine, for example, can induce a potentially lethal agranulocytosis.

As previously stated, all antipsychotic drugs bind to and antagonize dopamine receptors. However, these agents also exhibit a high affinity for other receptors. Typical antipsychotic agents, such as haloperidol, also bind to alpha-1 adrenergic and sigma receptors with a high affinity. Atypical antipsychotic agents have an increased potency as serotonergic, muscarinic, and adrenergic receptor antagonists *(7,8)*. The atypical antipsychotic clozapine has a high affinity for the dopamine D_4 receptor *(7,8),* however, it also binds with very high affinity to at least three subclasses of serotonin receptors, namely the serotonin 2A, serotonin 6, and serotonin 7 receptors *(9)*. The serotonin 6 and 2A receptors are likely to be coexpressed at least in striatal neurons—the model system in which the vast majority of dopaminergic neuropharmacologic studies have been carried out. With this complexity in mind, it is possible to view the drugs as probes of overall neuronal function within the context of a complex circuit, in which the dopamine receptors play an important but not exclusive role in regulating transcription of other neurotransmitter genes. Since the striatum is discussed throughout this chapter, we begin by briefly reminding the reader of its anatomy.

2. Striatal Molecular Neuroanatomy

The striatum has two major output systems (*see* Fig. 1 for a diagrammatic representation of the main inputs and outputs of the striatum and nucleus accumbens), and dopamine has been shown to affect gene expression in both of them. In one of these systems, the striatopallidal neurons project to the globus pallidus where they synapse with neurons that then project to the substantia nigra. The other output system consists of neurons which project directly from the striatum to the substantia nigra with no intermediate synapses *(10,11)*. Both of these pathways are GABAergic *(12),* but they also express different neuropeptide transmitters. The striatopallidal neurons

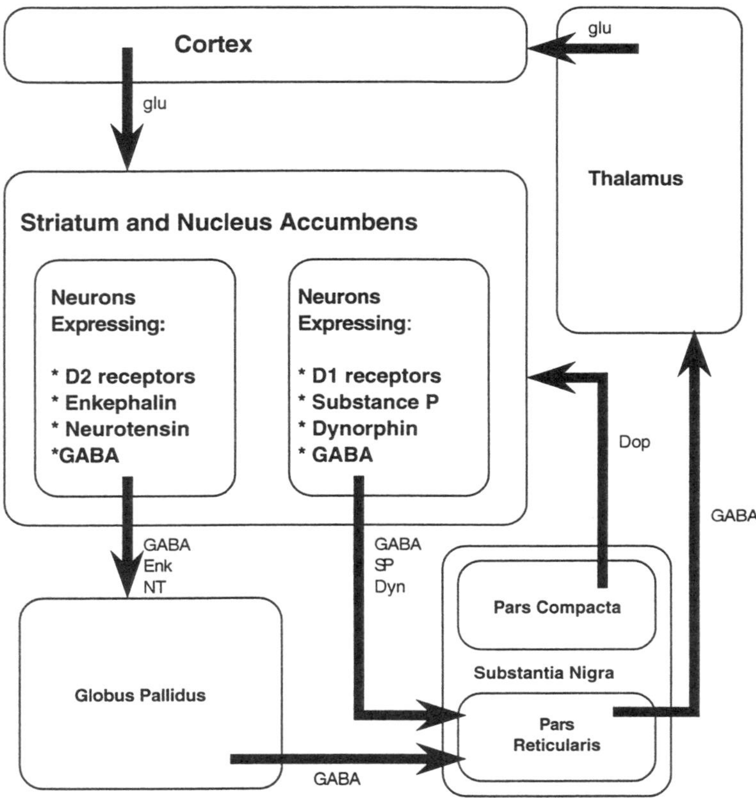

Fig. 1. Diagrammatic representation of major efferent and afferent connections of the basal ganglia. The largest inputs to striatum and nucleus accumbens are an excitatory glutamatergic input from cortex, and a dopaminergic input from the pars compacta of the substantia nigra. All striatal and accumbal outputs are GABAergic. Neurons that project to globus pallidus tend also to use enkephalin and neurotensin as coneurotransmitters, whereas neurons that project to the substantia nigra pars reticularis also use substance P and dynorphin as coneurotransmitters. Abbreviations: Dop, dopamine; Dyn, dynorphin; Enk, enkephalin; GABA, gamma aminobutyric acid; glu, glutamate; NT, neurotensin; SP, substance P.

coexpress enkephalin and neurotensin, whereas the striatonigral neurons contain substance P and dynorphin *(10,11,13,14)*. As described later in the chapter, expression of the genes encoding these peptides is under dopaminergic control. For example, proenkephalin *(15,16)* and proneurotensin mRNA *(17)* are increased following neuroleptic treatment. Substance P and dynorphin mRNA are elevated by dopamine agonists and decreased by dopamine depletion *(15,16,18–22)*. Combined *in situ* hybridization histochemistry and flourogold track-tracing suggest that the majority of D_2 receptor mRNA-containing neurons coexpress proenkephalin and proneurotensin mRNA, whereas D_1 receptor mRNA-containing neurons coexpress prodynorphin mRNAs *(19)*. Therefore, the majority of striatopallidal neurons express D_2 receptors, whereas striatonigral neurons express D_1 receptors. However, this general classification does not take into account the other members of the dopamine receptor family.

Most agents that interact with D_1 receptors have similar actions at D_5 receptors, and most agents which interact with D_2 receptors also bind D_3 and D_4 receptors. It is therefore likely that many of the studies which have attempted to determine whether the dopamine D_1 or D_2 receptor subtype is responsible for a given phenomenon may need to be re-examined with the use of more specific agents and markers. By far, the majority of the work which has examined dopaminergic effects has been in rodent basal ganglia, and this simplifies the problem somewhat, as the cells here have not been shown to express D_4 *(23)* or D_5 *(24)* receptors. Both D_3 mRNA and binding however have been shown in the basal ganglia *(25,26)*, with higher levels in the nucleus accumben than in the striatum. Because of the possibilities of uncharacterized receptors interacting with available probes, or actions of receptors which are present at levels too low to be currently detected, it is extremely difficult to determine with certainty the effect of any one receptor on downstream transcriptional responses. These molecular anatomical features of the system must be kept in mind in interpreting the data summarized in this chapter.

3. Immediate-Early Genes

Dopaminergic agents are known to affect the expression of several immediate-early genes at the level of both their mRNAs (Table 1, pp. 310–311) and proteins (Table 2, pp. 312–313). The immediate-early genes include two primary classes of proteins related to Fos and Jun. The Fos family includes, in addition to Fos, several known Fos-related proteins called Fos-related antigens (FRAs), including FRA-1, FRA-2, Fos B, and its truncated version, ΔFosB *(64–67)*. The proto-oncogene c-fos encodes Fos, a 55-kDa phosphoprotein *(68,69)*, whereas the related transcription factor FosB is encoded by a separate gene *(70)*. The Jun family of transcription factors consists of three genes, c-jun, junB, and junD, each of which codes for a protein of the same name *(71–73)*. The Fos family of proteins form heterodimers *(73)* and bind to the activating protein-1 (AP-1) response element as transcriptional activating factors. Jun proteins, which may form heterodimers (e.g., Fos/Jun)

Table 1
Dopaminergic Regulation of the Genes for Transcription Factors[a]

Drugs	c-fos mRNA	fosB mRNA	c-jun mRNA	junB mRNA	junD mRNA	zif268 mRNA
Typical antipsychotics	↑ in STR *(27–31)*	↑ in STR and NA (after 8 h) *(33)*	Rapid ↑ in STR *(31)* returns to baseline by 2 h *(33)*	↑ in STR and NA *(31)*	Rapid ↑ in STR and NA (1 mg/kg) *(34)* returns to baseline by 2 h *(33)*	↑ in STR *(30,33)* and NA *(33,35)*
Atypical antipsychotics	No effect in STR or NA at 3 mg/kg *(34,36)*; ↑ in LST; and NA at 30 mg/kg 2 h and after administration *(33)*; ↑ in PFC *(37)*		No effect in STR *(33,34)*	No effect in STR or NA at 3 mg and 30 min *(34,36)*; ↑ in LST at 30 mg/kg and 2 h *(38)*	No effect in STR or NA *(33)*	No effect in STR or NA at 3 mg/kg and 30 min *(34)* or at 30 mg/kg, 2 h *(33)*; ↑ in STR at 10 and 20 mg/kg, 30 min *(30)*
D1 agonist	↑ in MST, VST, and NA *(39)*; ↑ in CCTX *(40)*	No effect in STR *(39,41)*	No effect in STR *(39)*; little to no effect in CCTX *(40)*	↑ in MST and CCTX *(40)*	No effect in STR *(39)*	↑ in CCTX *(40)*
D2 agonist	No effect in STR *(42)*			No effect in STR *(43)*		No effect in CCTX *(40)*
Direct D1 and D2 agonists	↑ in STR *(43)*			↑ in STR *(43)*		↑ in STR *(43,44)*
Indirect agonist						
Amphetamine						
Acute	↑ in STR *(30,45)*, thalamus, BS, MB, HYP, and CTX *(45)*	↑ in STR CTX, CB, HIP, thalamus, and HYP *(45)*	↑ in CTX, CB, thalamus, STR, BS, MB, and HYP *(45)*	↑ in STR *(30,45)* CB, thalamus, BS, MB, and HYP *(45,46)*		↑ in STR *(30,45)*, CTX, CB, HIP, thalamus, HYP, BS, and MB *(45)*

Chronic	↓ in STR, thalamus, BS, MD, HYP, and CTX *(45)*	↓ in STR, CTX, CB, HIP, thalamus, and HYP *(45)*	↓ in CTX, CB, thalamus, STR, BS, MB, and HYP *(45)*	↓ in STR, thalamus, BS, MB, and HYP *(45)*		↑ in STR, CTX, CB, HIP, thalamus, HYP, BS, and MB *(45)*
Cocaine						
Acute	↑ in STR, NA *(47)*, ↑ in core of STR and DLST, CTX, and CB *(48–50)*	↑ in NA *(47)*	↑ in NA *(47)*; no effect in STR *(51)*	↑ in MST, DMST, ↑ in STR, CTX *(51)*, and NA *(47)*		↑ in STR and NA *(43,48,51)*
Chronic	↓ in NA *(47)*	↑ in NA but less than acute *(52)*	↓ in NA *(47)*	↑ in NA but less than acute *(52)*		↓ in NA *(47)*, ↓ basal exp. in STR and PFC *(53)*

[a]Abbreviations: BS, brainstem; CB, cerebellum; CCTX, cerebral cortex; CTX, cortex; DLST, dorsolateral striatum; DMST, dorsomedial striatum; HIP, hippocampus; HYP, hypothalamus; LST, lateral striatum; MB, midbrain; MST, medial striatum; NA, nucleus accumbens; PFC, prefrontal cortex; STR, striatum; VST, ventral striatum.

Table 2
Dopaminergic Regulation of the Immunoreactivity of Transcription Factors[a]

Drugs	c-Fos-IR	FRA-IR	FosB-IR	c-Jun-IR
Typical antipsychotics				
Acute	↑ in NA, DLST, septum, CC, and CeA *(32,33,53,54)*	↑ in DLST, DM shell of NA, and CC *(55,56)*	↑ in STR and NA after 8 h *(33,35)*	No effect in STR and NA 2 h after administration *(33)*
Chronic	↑ in DLST, VLST, and LS *(54)*			
Atypical antipsychotics				
Acute	↑ in NA, septum, PFC, MST, PVN, SON, and CeA *(33,54,58)*	↑ in MST, NA shell, and VS *(27,33)*	No effect in STR and NA *(33)*	No effect in STR, NA, and LS islands of Calleja *(33)*
Chronic	↑ in NA, LS, PVN, SON, and CeA *(54)*			
Direct D1 and D2 agonists	↑ in STR following 6-OHDA lesion *(56)*; ↑ in MST and DLST *(55)*	↑ in MST, DLST, NA, cortex, septum islands of Calleja *(55)* and LLH *(38)*		↑ in STR after 6-OHDA lesion *(56)*
Indirect agonist				
Amphetamine	↑ in MST and core of STR *(30,60)*	↑ in LLH *(38)*		
Cocaine				
Acute	↑ in STR with greatest ↑ in the core *(49,60–63)*		↑ in STR and CB *(61)*	
Chronic		↑ in FRA doublet *(35–37)* kDa *(52)*		

Drugs	JunB-IR	JunD-IR	PCREB-IR	zif268-IR
Typical antipsychotics				
Acute	↑ STR and NA *(33)*	No effect in STR and NA 2 h after administration *(33)*	↑ in STR *(57)* and NA (unpublished results)	↑ in STR and NA *(33)*
Atypical antipsychotics				
Acute	↑ in LST and VS *(33)*	No effect in STR and NA *(33)*		↑ in LST, NA, and VS *(33,35)*
D1 agonist			↑ in STR *(59)*	
Indirect agonist				
Amphetamine				
Acute	↑ in STR and CB *(61)*	↑ in STR and CB *(61)*	↑ in STR *(62)*	
Chronic				

[a]Abbreviations: CC, cingulate cortex; CeA, central amygdala; FRA, Fos-related antigen; LLH, lateral habenula; LS, lateral septum; PVN, periventricular nucleus; SON, supraoptic nucleus; VLST, ventrolateral striatum; VS, ventral septum. For more abbreviations, *see* Table 1 footnote.

as well as homodimers (e.g., Jun/Jun), and which bind DNA with varying affinities depending on the protein composition *(67,70,74),* and facilitate or repress the expression of genes containing AP-1 sequences. For example, the c-Jun homodimer positively regulates the transcription of its own gene *(75),* whereas c-Jun heterodimerized with JunB represses the transcription of c-jun *(76)*. Likewise, studies have also shown that a truncated form of FosB can inhibit the transcriptional activity of Fos and Jun proteins *(77)*. An understanding of the exact pattern of the immediate-early genes induced by dopaminergic intervention is therefore prerequisite to understanding dopamine's longer term transcriptional effects.

3.1. Dopaminergic Effects on Fos Family Members

Various dopamine D1 and D2 agonists and antagonists are known to affect the expression of Fos. Apomorphine is both a D1 and D2 agonist and appears to elicit effects on immediate-early genes only following 6-hydroxy-dopamine(6-OHDA)-induced dopamine depletion. 6-OHDA lesions alone increase the expression of c-Fos and 35 kDa FRA. This increase is further augmented in the presence of apomorphine *(56)*. In addition, apomorphine significantly induces AP-1 binding activity *(56)* as measured in gel shift assays. Morphological examination of the effects of apomorphine reveals that apomorphine leads to the induction of c-Fos-IR in the medial and lateral striatum, whereas FRA expression is induced in the nucleus accumbens, cortex, septum, and islands of Calleja *(55)*. Apomorphine also increases c-fos and fosB mRNA in the striatum *(43)*.

Using immunofluorescence, in dopamine-depleted striatum, the specific dopamine D1 receptor agonist SKF 38393 was shown to induce c-fos expression in medium-sized neurons that project to the substantia nigra pars reticulata *(42,78)*. Quinpirole, a D2 agonist, produces little or no effect on c-fos expression in the striatum. However, in the presence of SKF 38393, quinpirole augments c-fos expression in striosomes in the dorsolateral striatum *(42)*. A reciprocal relationship is seen in the globus pallidus, where quinpirole increases Fos expression, and the D1 agonist alone has no effect. When the D1 agonist is given in combination with quinpirole there is an augmentation of Fos expression *(42,79)*. Therefore, it appears that in certain situations a D1 agonist can synergize with agonists of the D2 family of receptors. Morphine acts at mu opiate receptors to release DA indirectly in the striatum. Morphine induces c-fos mRNA in the striatum and this effect is probably mediated by D1 receptors since its effects are blocked by pretreatment of selective D1 antagonists, SCH 23390 and SCH 39166 *(39)*. Morphine has no effect on fosB mRNA *(39)*. D1 agonists also increase c-fos expression in primary cultures prepared from rat cerebral cortex *(40)*.

3.2. Antipsychotic Agents and Fos Family Members

Haloperidol treatment leads to a rapid and transient induction of c-fos mRNA in the rat striatum *(27–31)* and nucleus accumbens *(27,28,30,32)*. This increase in mRNA is followed by an enhancement of c-Fos and FRA immunoreactivity in neurons of the striatum *(28,58)*, nucleus accumbens *(58)*, and lateral septum *(58)*. Haloperidol also induces AP-1 binding activity in the striatum *(30)*. Following chronic haloperidol treatment, c-fos mRNA seen with acute exposure returns to control levels *(80)*. A reduction in Fos-like immunoreactivity following chronic haloperidol treatment is also observed in the striatum, but not in the nucleus accumbens and lateral septum *(54)*.

Clozapine, an atypical antipsychotic drug, also affects c-fos mRNA and Fos protein expression, but the anatomical pattern of expression does not mimic that observed after treatment with typical neuroleptics. Both acute and chronic clozapine, but not haloperidol, treatment robustly induce c-fos mRNA in the prefrontal cortex. This is an effect that appears to be mediated by dopamine D_3 receptor blockade *(37)*. Clozapine administered acutely at low to moderate doses (3–20 mg/kg) does not affect the c-fos mRNA in the striatum or nucleus accumbens *(30,34,81)* but at higher doses (30 mg/kg) 2 h after clozapine administration, there is a slight increase in c-fos mRNA expression in the lateral striatum *(33)*. Detection of c-fos mRNA in the striatum in the latter study may be the result of cross-hybridization of the c-fos mRNA probe with mRNAs encoding other FRAs, as suggested by the investigators *(33)*. Similarly, clozapine evokes little or no increase in striatal Fos protein *(32,33,54)*. However, clozapine increases Fos-like immunoreactivity in limbic regions including the nucleus accumbens *(27,30,32,33,54)*, the lateral septum, and central amygdala *(54)*. Long-term treatment of clozapine reduces the acute response significantly in the lateral septum and central amygdala while maintaining its induction of Fos-like immunoreactivity in the nucleus accumbens *(54)*. Based on the observation that typical antipsychotic drugs increase c-fos in the dorsolateral striatum (a region thought to be associated with the regulation of movement) *(82–84)*, and atypical antipsychotic drugs fail to do so, it has been suggested that the induction of c-fos in this brain region may account for the extrapyramidal motor symptoms associated with the typical neuroleptics *(27,28,32,81)*. The ventral striatum and nucleus accumbens are more closely associated with the limbic system which suggests involvement in mood and affect. It is in this anatomic locus that antipsychotic drugs are believed to exert their therapeutic effects. Thus, the increase in the expression of Fos induced by both typical and atypical antipsychotic drugs in this region of the brain suggests that this system may represent a common neuronal pathway for the antipsychotic effects of these two classes of drugs.

From a functional perspective, these differences may account for some of the motor side effects seen with neuroleptics *(28)*.

3.3. *Effects of Psychomotor Stimulants on Fos Family Members*

Psychomotor stimulants produce behavioral effects that are opposite to those seen with antipsychotic drugs and, as might be predicted, have different effects on the induction of immediate-early genes. Amphetamine and cocaine are stimulant drugs that act on central monoaminergic neurons to produce both acute psychomotor activation and long-lasting behavioral effects, including addiction and a psychosis-like syndrome in humans. Like the typical antipsychotic haloperidol, amphetamine *(30,49)* and cocaine *(30,48,49, 51,85)* increase the expression of the c-fos gene in the striatum. Amphetamine effects on c-fos appear to be mediated by dopamine D1 receptors since its effects are blocked by selective D1 receptor antagonists but not by D2 receptor antagonists *(30,43,49)*. Cocaine appears to increase c-fos mRNA expression in all striatal areas, with its greatest effect in the dorsal central striatum *(51,85)*. The regional expression of c-fos mRNA by methamphetamine, an amphetamine derivative, appears to be the greatest in the medial region of the central core of the striatum *(86)*. Thus, although both classes of agents cause a transient increase in striatal c-fos expression, the psychomotor stimulants affect a distinctly different population of striatal neurons from those affected by the typical neuroleptics.

Cocaine increases AP-1 binding activity and c-fos protein in the striatum and nucleus accumbens *(47,52,63,87)*. Cocaine's effect on c-fos is dose-dependent, is apparent at 1 h, and maximal at 2 h *(63)*. Administration of SCH 23390, a D1 receptor antagonist, and sulpiride, a D2 receptor antagonist, prior to cocaine indicate a significant role for D1 but not for D2 receptors in mediating this effect *(63)*. Chronic treatment of rats with cocaine leads to long-term biochemical changes in the nucleus accumbens, a brain region implicated in mediating the reinforcing effects of cocaine and other drugs of abuse. Immediate-early genes and their protein products appear to play an important role in transducing extracellular stimuli into altered patterns of cellular gene expression and therefore into long-term changes in cellular function. Following chronic cocaine treatment, the levels of mRNA encoding c-fos, c-jun, fosB, jun B, and zif268 in the nucleus accumbens are reduced to control levels *(47)*. Similarly, the level of Fos-like immunoreactivity, which is enhanced by acute cocaine, is reduced to control levels in chronic cocaine-treated rats *(47)*. With acute cocaine treatment, AP-1 binding activity is increased within 1 h but returns to control levels within 8–12 h. However, with chronic cocaine treatment, AP-1 binding activity remains elevated 18 h after the last treatment, a time point at which c-fos and c-jun mRNA and Fos-like immunoreactivity have returned to control levels *(47)*. There are at least four FRAs, with molecular weights ranging from 35–37 kDa, that are selectively induced by chronic treatment and probably comprise the increase in AP-1 binding *(52)*. Based on this evidence, it appears that the persistent increase in AP-1 binding produced by chronic cocaine treatment could have long-term effects on gene expression, which may lead to some of the physiological and behavioral changes associated with cocaine addiction.

Induction of Fos by dopaminergic intervention is not limited to those brain areas expressing D1 and D2 receptors. In rats treated systemically with either amphetamine or apomorphine, an increase in Fos-like immunoreactivity can be detected in the lateral zone of the lateral habenula *(38)*. It has been postulated that the effects of dopamine agonist on the lateral region of the lateral habenula nucleus are mediated by alterations in striatal activity which are relayed to the habenula through the entopeduncular nucleus.

3.4. Dopaminergic Effects on Jun Family Members

As summarized in Tables 1 and 2, dopamine agonists, antipsychotic drugs, and psychomotor stimulants also produce variable effects on several Jun family members at the level of both mRNA and protein. Apomorphine and SKF 38393 cause a rapid, but transient, dose-dependent increase in junB mRNA levels in the striatum *(43)*. Consistent with their lack of effect on Fos levels, selective D2 agonists have no effect on junB mRNA levels in the striatum *(43)* or in the cerebral cortex *(40)*. This would suggest that apomorphine probably exerts its effects mainly through D1 receptors. This statement is further supported by the evidence that D1-selective antagonists block apomorphine's effect on immediate-early gene expression *(43)*, whereas D2-selective antagonists are without effect *(40,43)*. When the effects of D1 agonist were examined more carefully, it was discovered that junB mRNA is increased in the medial and ventral striatum and nucleus accumbens *(39)*. Dopamine D1-selective agonists have little to no effect on c-jun mRNA in the striatum *(39)* or cerebral cortex *(40)*. D1 agonists are without effect on junD expression in the striatum *(39)*.

3.5. Antipsychotic Drugs and Jun Family Members

Typical neuroleptics lead to a rapid and transient increase in c-jun and junD mRNA in the striatum *(31)* and junD mRNA in the nucleus accumbens *(31)* that returns to control levels by 2 h *(33)*. These agents also increase junB mRNA in the striatum and nucleus accumbens *(31)* and this expression persists for at least 2 h *(33)*. On the other hand, clozapine does not affect the expression of c-jun or junD mRNA *(33,34)*. Only at high doses (30 mg/kg) is clozapine able to elicit an increase in junB mRNA in the lateral striatum *(33)*.

Both haloperidol and clozapine increase the expression of JunB protein. Previous studies have shown that chronic administration of haloperidol leads to a decrease in protachykinin gene expression in striatonigral neurons *(88)*. The increase in JunB protein is of particular interest since it has been linked to the suppression rather than the enhancement of downstream gene transcription *(76,89)*. The ability of these drugs to induce JunB may underlie their ability to induce neuropeptide genes.

3.6. Psychomotor Stimulants on Jun Family Members

Psychomotor stimulants regulate Jun family members in a region- and time-dependent manner. Cocaine acutely induces c-jun and junB mRNA in the nucleus accumbens, whereas in the striatum junB and junD mRNA are elevated with no change in c-jun. Following chronic cocaine treatment in the nucleus accumbens, c-jun and junB mRNAs are reduced or completely eliminated *(47)*. This suggests that the induction of immediate-early genes seen with acute cocaine treatment may be desensitized following chronic treatment. It appears that the composition of AP-1 complex proteins induced with acute treatment differs from that induced with chronic treatment *(52)*. Amphetamine causes a rapid and transient induction in junB mRNA levels in the striatum that is abolished by striatal 6-OHDA lesion or by pretreatment with a specific D1 receptor antagonist but not by a specific D2 antagonist *(43)*. Amphetamine acutely increases c-jun and junB mRNA levels in the cortex, cerebellum, thalamus, striatum, hypothalamus, brainstem, and midbrain. Following chronic amphetamine treatment, c-jun and junB expression are not induced in any of the aforementioned brain regions *(45)*. Therefore, it can be assumed that the changes in transcription factor gene expression following acute and chronic amphetamine treatment will affect the expression of the neuropeptide genes coexpressed with the factors in neurons in the affected brain regions.

3.7. Effects on Other Transcription Factors

3.7.1. CREB

Dopaminergic agents not only have region-specific effects on Fos- and Jun-related proteins, but they also affect the expression and phosphorylation of other transcription factors. Most of the agents discussed so far appear to produce their effects mainly through D1-like receptor agonism or D2-like receptor antagonism. D1 receptors are positively coupled to adenylate cyclase, so dopamine binding leads to an increase in cAMP, subsequently activating cAMP-dependent PKA *(90)*. D1 receptor stimulation may also increase Ca^{2+} entry into neurons via L-type calcium channels, thereby activating one or more Ca^{2+}/calmodulin-dependent protein kinases (CaM kinases) *(91–94)*. Conversely, D2 receptors are negatively coupled to adenylate cyclase. Thus, D2 receptor antagonists remove this negative inhibition, leading to increases in cAMP and activation of PKA. PKA activates gene expression by phosphorylating a CREB at Serine 133 *(95),* switching this transcription factor from its inactive to its active state.

CREB is thought to be a primary mediator of cAMP-dependent gene transcription. The CREB protein binds to CREs found in the upstream promoter of many genes to regulate their expression. Preliminary studies in our laboratory have shown that SKF 82958, a selective D1 agonist, leads to a rapid but transient increase in CREB phosphorylation in the striatum and nucleus accumbens of organotypic cultures using an antiserum that specifically recognizes CREB when it is phosphorylated on Ser 133 *(96,97)*. It has been shown that in primary neurons prepared from striatum, dopamine increases the phosphorylated form of CREB (PCREB) and that this effect is blocked by a D1 receptor antagonist *(59)*. In rats, L-DOPA increases PCREB-immunoreactivity in the nuclei of medium-sized striatal neurons ipsilateral to an extensive lesion of the substantia nigra compacta region *(59)*. Again, this effect is blocked by pretreatment with a D1 receptor antagonist, but not with a D2 receptor antagonist *(59)*. These findings suggest that dopamine increases the levels of PCREB in striatal neurons by activating D1 receptors and that dopamine enhances the activity of PKA or CaM kinase, or both, in striatal neurons in vivo. Haloperidol has also been shown to increase the PCREB in the striatum *(57)* and nucleus accumbens (Fig. 2) after ip injection. Amphetamine induces PCREB in the rat striatum *(62)*. Amphetamine-induced CREB phosphorylation is blocked by pretreatment with a D1 receptor antagonist *(62)*. As discussed previously, all of these agents also increase c-fos expression. Since these agents also increase CREB phosphorylation, it is possible that the induction of PCREB precedes the induction of c-fos. The c-fos gene has both a CRE and a calcium response element (CaRE) in its upstream promoter. Both the CRE and the CaRE bind CREB. Further, intrastriatal injection of antisense oligonucleotides directed against CREB inhibits induction of c-fos by amphetamine *(62)*. Therefore CREB phosphorylation is likely to be an important step in linking D1 receptor agonism and D2 receptor antagonism to acute transcriptional responses as well as to long-term alterations in properties of striatal neurons.

3.7.2. Zif268

Zif268, also named NGFI-A *(98),* Krox-24 *(99),* or egr-1 *(100),* is a transcription factor which contains a DNA-binding domain consisting of three tandemly repeated zinc finger sequences *(98,100–102)*. This transcription factor also appears to be sensitive to dopaminergic manipulations. The D1 agonist

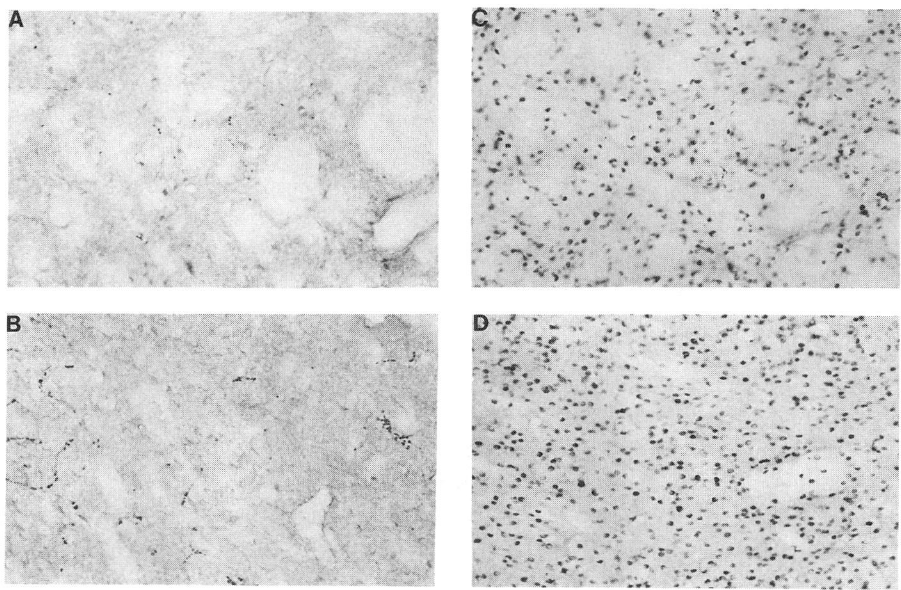

Fig. 2. Induction of phosphoCREB immunoreactivity in the dorsal lateral region of rat striatum and the nucleus accumbens by haloperiodol. **(A)** Photomicrograph (40×) of a section from the dorsolateral striatum and **(B)** the nucleus accumbens-stained with the PCREB antiserum 10 min after ip injection of saline. **(C)** A section taken at an identical striatal level as shown in A and **(D)** section taken at an identical nucleus accumbal level as shown in B, 10 min after haloperidol (2 mg/kg) injection stained with PCREB antiserum. Staining was restricted to cell nuclei throughout the striatum and nucleus accumbens.

SKF 38393 increases the expression of zif268 mRNA in primary neuron cultures prepared from rat cerebral cortex *(40)*. The effect of the selective D1 agonist is blocked by pretreatment with a D1-receptor selective antagonist *(40)*. Apomorphine alone has no effect on zif268, but following 6-OHDA lesion or reserpine treatment, apomorphine elicits a robust increase in zif268 mRNA in the striatum and olfactory tubercle *(103)*. This effect is blocked by pretreatment with the D1 antagonist SCH 23390 but not by haloperidol, suggesting that apomorphine exerts its effects via D1 receptor activation *(103)*. Haloperidol increases zif268 mRNA and protein in the striatum and nucleus accumbens *(30,31,33,34)*. This effect is specifically blocked by pretreatment with a D2 agonist, quinelorane *(31)*. Clozapine increases zif268 mRNA very rapidly and transiently *(30)*, which eventually leads to an increase in zif268-immunoreactivity in the lateral striatum, ventral striatum, and nucleus accumbens *(33)*. Amphetamine, like haloperidol, increases zif268 mRNA in the striatum *(30)*. This agent also increases zif268 expression in the cortex, cerebellum, thalamus, hypothalamus, brainstem, and midbrain *(45)*. However, chronic administration of amphetamine leads to a down-regulation of zif268 mRNA in all brain regions *(45)*. An acute dose of cocaine increases zif268 mRNA in the nucleus accumbens *(47)* and striatum *(48,104)*. This response appears to involve the dopamine system, since it is abolished by a selective D1 receptor antagonist or by 6-OHDA lesions *(104)*, but it appears that the serotonin system also plays a role *(48)*. Following chronic cocaine treatment, the expression of zif268 mRNA returns to control levels in the nucleus accumbens *(47)* and is even suppressed below basal in the striatum and overlying cortex *(53)*. There does appear to be regional heterogeneity, since this treatment spares zif268 mRNA levels in the olfactory tubercle and the pyramidal cell layer of the hippocampus *(53)*. It appears that ongoing synaptic activity is critical for maintaining basal zif268 expression *(53)*. Thus, it is possible that the reduction produced by chronic cocaine administration reflects a suppression of the types of synaptic activity important for driving basal zif268 expression.

4. Dopamine Receptor-Mediated Transcription of Neurotransmitter Genes

Dramatic and region-specific effects on expression and phosphorylation of transcriptionally active proteins can be elicited by either agonism or antagonism of dopamine receptors. In some regions of the brain (e.g., the striatum), region- and cell-type specific changes in neurotransmitter, primarily neuropeptide, gene expression also can be observed. The time course of these changes is longer than that typically seen for the transcription factors (days or weeks rather than hours), and the changes tend to be sustained for the duration of a given treatment, rather than rapidly increasing or decreasing in level as many of the transcription factors do. It is likely that the transcription factor changes are necessary intermediates to some of the neuropeptide changes, but exact relationships are only just beginning to be determined. This section focuses on neuropeptide genes encoding enkephalin, dynorphin, substance P, and neurotensin and oncogenes encoding enzymes responsible for the biosynthesis of specific neurotransmitters. Note that an excellent review regarding the function and regulation of these genes has been published recently *(105)*.

The responsiveness of these genes to the transcription factors just detailed can be expected to be determined partially by the structure of their respective promoters, although the experiments to link definitively particular response elements to observed changes in expression have not yet been

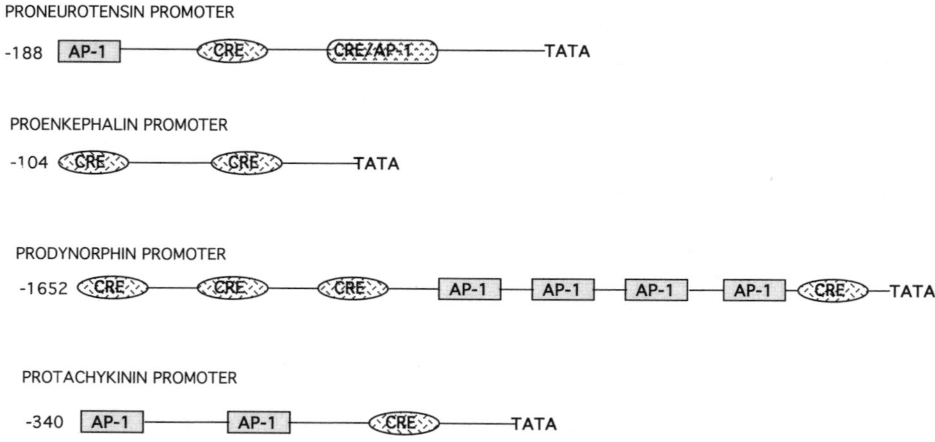

Fig. 3. Diagrammatic representation of the CRE and AP-1 enhancer elements present in the proneurotensin *(106)*, proenkephalin *(107)*, prodynorphin *(108,109)*, and protachykinin *(110,111)* promoters. Abbreviations: CRE, cAMP response element; AP-1, activating protein-1.

reported. Figure 3 summarizes the general topology of the promoter sequences which have been reported to be important in regulating transcription of their associated neuropeptide genes. Examination of this figure clearly shows that each gene contains at least one, if not several, copies of the elements which can support cAMP (phosphoCREB) and/or AP-1 (primarily fos-jun)-mediated transcriptional activation. In the following sections, we review the literature which examines changes in the mRNA or protein products of each of the genes as a result of dopaminergic manipulations comparable to those detailed previously for the transcription factors. These changes are summarized in Table 3.

4.1. Enkephalin

Enkephalins are found at high levels in the striatum, nucleus accumbens, and olfactory tubercle *(133)*. They are pentapeptides which are cleaved from a larger protein precursor, preproenkephalin, and among their many functions they are thought to modulate dopaminergic transmission. For example, they can increase dopamine turnover *(134)*. In the basal ganglia, enkephalin-containing neurons are a subset of the GABAergic projection neurons of the striatum and nucleus accumbens. They are fairly evenly spread throughout these structures, and have been shown to coexpress the dopamine D_2 receptor and project to the substantia nigra *(11)*. The enkephalins are therefore relevant to the discussion of regulation of dopaminergic neurotransmission since they

Table 3
Dopaminergic Regulation of Striatal Neuropeptides

Drugs	Enkephalin	Substance P	Dynorphin	Neurotensin
D1 agonists	Can increase expression under some conditions *(112–114)*	No increase, but provide a necessary basal tone for D2 interaction *(115)*	Increase, but only after 6-OHDA lesion *(116)*	Increase expression in some subregions *(117)*
D1 antagonists	Can decrease expression *(112–114)*	Mixed results *(118,119)*	Decrease, but only after apomorphine-induced increase *(120)*	Decrease expression in some subregions *(117)*
D2 agonists	Decrease expression *(112–114)*	Increase expression if basal D1 tone is present *(115)*	No effect *(116)*	Decrease expression *(121)*
D2 antagonists antipsychotics	Increase expression *(112–114,117,122,123)*	No effect or decrease expression *(88,118,119)*	No effect *(120)*	Increase expression *(121)*
Nonselective dopamine agonists	Little effect *(26)*	Increase expression *(20,124,125)*	Increase expression *(20,125–127)*	Increase expression *(86,124)*
Dopamine 6-OHDA lesion	Increases expression *(128,129)*	Decreases expression *(15,20)*	Decreases expression *(20,116)*	Decreases expression *(130)*
Receptor knockout		D1 knockout decreases expression *(131)*	D1 knockout decreases expression in patch *(132)*	
Overall influence of dopaminergic tone	Dopamine inhibits expression	Dopamine stimulates expression	Dopamine stimulates expression	More complex; both dopamine stimulation and inhibition can increase expression

are expressed in cells receiving dopaminergic input and which express dopamine receptors, and also because they make reciprocal connections which can have effects on the activity of dopaminergic cells themselves.

A basic observation, which has been verified using several different techniques, is that the removal of dopaminergic tone to the striatum results in an increase in preproenkephalin mRNA and protein. This has been demonstrated using selective lesions of dopaminergic neurons *(128,129)*, by depletion of dopaminergic stores *(135)*, as well as by the blockade of dopamine receptors *(122,123)*. This increase can then be reversed by the reintroduction of dopamine *(129)*. Starting with this basic observation, further studies have attempted to determine which dopamine receptor subtypes may be responsible for the changes, and the time course over which the effect can be seen.

Acute treatment (6 h) with D2 receptor antagonists (haloperidol or raclopride) increases preproenkephalin mRNA expression in both striatum and nucleus accumbens, whereas agonists (LY-171555) reduce it. In contrast, the D1-specific antagonist SCH 23390 reduces preproenkephalin expression, and the D1-specific agonist SKF 38393 has little or no effect *(112)*. A somewhat different pattern of effects can be observed by examining the protein products (Met5 and Leu5) of enkephalin gene transcription, rather than the preproenkephalin mRNA. Six hours after administration of a D1 or D2 agonist or antagonist, only the D2 antagonist (sulpiride) was found to have an effect in striatum, where it increases levels of both Met5 and Leu5. Different patterns of effects are seen in nucleus accumbens and substantia nigra, where D1 agonists and antagonists promote greater effects than their D2 counterparts *(117)*.

Studies which use longer treatment paradigms reinforce the idea that D1 agonism can increase, and D2 signaling can decrease, the level of enkephalin gene transcription. This has been shown both by direct treatment with D1 and D2 selective agents *(113)*, as well as by treatment with these agents after a dopaminergic lesion *(114)*. Of the two opposing influences, however, the inhibitory D2 influence is clearly the stronger, as can be seen by the overall increase in preproenkephalin gene expression levels when both inputs are lost. Recent examination of the time course of changes in preproenkephalin mRNA expression after 6-OHDA lesion revealed an increase which begins on d 1, reaches significance on d 3, and steadily continues upward throughout the 21-d period of the study *(136)*.

Developmental data are consistent with this reciprocal regulation. D1 receptors become able to couple to the stimulation of adenylate cyclase as early as embryonic d 17, whereas the D2 receptor family members do not exhibit the ability to inhibit cyclase activity until sometime in the second postnatal week *(137)*. Blockade of dopaminergic transmission during this time by administration of haloperidol results in a 40% decrease in striatal levels of preproenkephalin mRNA *(138),* as would be expected by the removal of an excitatory input. In normal development, the amounts of both preproenkephalin mRNA and enkephalin increase from the time of birth through the first month of life. If dopaminergic inputs are removed via 6-OHDA lesions on postnatal d 3, both the preproenkephalin mRNA and protein increase normally during the first 25 d, but are significantly elevated by d 35 *(139)*. The increase in enkephalin mRNA caused by neonatal dopaminergic lesion apparently is permanent, as it persists in the adult animal *(140)*.

Although this tonic inhibitory effect of dopamine on enkephalin expression is a well-described phenomenon, it is important to realize that there are many other factors that can influence enkephalin expression, especially in disease states. For example, examination of postmortem brain tissue from Parkinsonian patients showed that even when an 80% reduction of their dopamine content occurred, enkephalin content was reduced as well *(141)*.

4.2. Substance P

Substance P and neurokinin A are two neuropeptides which are cleaved from the same precursor, termed preprotachykinin *(142)*. Of these two, substance P is by far the better characterized. The highest levels of substance P are found in the hypothalamus and substantia nigra pars reticulata, whereas moderate levels of expression are found throughout the nucleus accumbens and striatum *(143)*. In contrast to enkephalin, which tends to be expressed homogeneously throughout the striatum, and to colocalize with the D_2 receptor in cells that project to the globus pallidus, substance P has a more focal striatal expression, and tends to colocalize with D_1 receptors in cells that project directly to the substantia nigra *(11,19)*.

A further contrast with the enkephalins is the response of substance P and neurokinin A to dopaminergic input. Dopaminergic inputs sustain the high level of striatal substance P expression, and removal of these inputs causes a reduction in substance P mRNA and immunoreactivity. These changes have been noted using both 6-OHDA dopaminergic lesions *(15,16,20)* and dopamine antagonists *(88)*. Furthermore, stimulation of the dopaminergic system via nonselective or indirect agonists such as apomorphine or methamphetamine can cause an increase in substance P expression *(20,124,125)*. The time course of this stimulation is faster than many of the other effects of dopamine on neuropeptide expression, occurring within a few hours of drug administration *(115,124)*.

The time course of the substance P response to dopaminergic denervation is somewhat faster than that of the enkephalin response. One day after 6-OHDA lesion, substance P mRNA declines by 30%, and remains near that level for up to 21 d *(136)*. Interestingly, this decrease on the first day following the lesion takes place even though the dopamine lesion has not yet completely taken effect, and striatal dopamine content is actually higher than control. It is not until d 2 following the lesion that striatal dopamine drops below control values. This pattern of a rapid reduction in substance P in response to dopaminergic neurotoxins is also seen in neonatal animals. Animals that receive a 6-OHDA lesion on postnatal d 3 already exhibit a significant reduction of substance P protein and mRNA by d 5, and both values remain significantly below control levels for up to 35 d *(139)*. A recently described mutant mouse which is missing the dopamine D_1 receptor shows reduced levels of substance P *(131)*. Examinations of the different cleavage products of preprotachykinin, substance P, and neurokinin A generally show that they are similarly regulated after dopamine blockade *(88,144)*.

Efforts to determine the role of the individual dopamine receptors have provided useful information, but have left some unanswered questions. Blockade of either D1 or D2 receptors is sufficient to inhibit the methamphetamine-induced up-regulation of substance P mRNA; and D2, but not D1 agonists can mimic the up-regulation. Furthermore, concurrent administration of a D1 antagonist is able to block the D2 agonist-induced up-regulation *(115)*. Taken together, these data indicate that D1 receptors provide an enabling basal tone that is necessary for D2 receptors to manifest their ability to enhance substance P levels. It is notable that D2 receptors seem to be important in this phenomenon, even though substance P is thought to colocalize in cells with the D_1 receptor, but not the D_2 *(145)*. Whether D2 agonists exert their transcriptional effects in merely a subpopulation of cells that express D_2 as well as D_1 and substance P, or indirectly via striatal interneurons, or other indirect means, remains to be determined.

It has been reported that antagonists of either of the D1- or D2-like receptors tend to cause reductions in substance P expression *(118)*. On the other hand, another group has reported that repeated administration of a D1 receptor antagonist actually promotes an increase in substance P immunoreactivity, whereas a D2 receptor antagonist has no effect *(119)*. Furthermore, when the two antagonists are given together, substance P immunoreactivity still increases, although to a lesser degree than after D1 antagonist alone. Whether D1 antagonists can cause both increases and decreases in striatal substance P expression, or whether there is some alternative explanation, remains to be determined.

Because transcriptional effects are indirect, requiring at least two or three intermediate signaling events between dopamine binding to its receptors and actual changes in transcription, it is to be expected that different combinations of these intermediate factors (e.g., G proteins, cyclases, or DNA-binding proteins) can lead to different transcriptional outputs, even given the same dopaminergic input. Studies which have examined these responses in other brain areas, and at different developmental time points, have shown this sometimes to be the case for substance P. In rats, a 6-OHDA lesion induced on postnatal d 3 or 6 of an animal's life has been reported to cause an up-regulation in substance P mRNA in both frontoparietal and cingulate cortex after 2 wk, while no change was found in the striatum. By adulthood, the substance P mRNA in cortical areas of these animals had returned to normal levels, and the expected decrease in striatal mRNA was observed *(146)*. Studies that have examined substance P in the substantia nigra have not been as extensive as those concentrating on the striatum and accumbens, but have generally found similar patterns of regulation, that is, general or D2-specific dopaminergic block produces a reduced expression, whereas D1-specific block has less effect *(147)*.

In the interpretation of all of these studies, it is important to remember that an elimination of dopaminergic inputs can have indirect as well as direct effects on target cells. For example, it has been shown that a neonatal 6-OHDA lesion results in a substantial increase in the amount of serotonin present in striatum *(148)*, and that serotonergic innervation also has a positive effect on the levels of substance P mRNA and peptide *(149)*.

4.3. Dynorphin

The neuropeptide dynorphin is a member of the endogenous opiate family, and its first five amino acids are identical to leu-enkephalin. Like other neuropeptides, dynorphin is cleaved from a larger precursor polypeptide termed prodynorphin. Medium to high levels of both the dynorphin peptide and prodynorphin mRNA are found in the striatum and nucleus accumbens *(14,150)*. Both in its distribution and responses to dopaminergic manipulation, however, it is much more similar to substance P than to enkephalin. Within the striatum, dynorphin expression is higher in patch than matrix, and it tends to colocalize with substance P in neurons that project to the substantia nigra rather than neurons that project to the globus pallidus *(11)*.

The expression of dynorphin, like that of substance P, is stimulated by dopamine input. This has been shown using dopamine agonists such as apomorphine or amphetamine *(125–127)*. Neurons within striatal patches exhibit a much greater increase (100%) in levels of dynorphin mRNA in response to dopamine agonists than do matrix (25%) dynorphin-expressing neurons *(20)*. Removal of dopaminergic input via 6-OHDA lesion results in a decrease in dynorphin striatal mRNA expression *(20)*. Recently, a mutant knock-out mouse lacking the D_1 receptor was described, and it also displays a reduced amount of dynorphin expression *(132)*. It was noted in both of these studies that the loss of dynorphin expression in striatal patches was greater than the loss throughout the rest of striatal matrix. These data, together with those showing dynorphin increases after apomorphine, demonstrate the increased sensitivity of patch dynorphin expression to dopamine regulation.

Curiously, agents specific for the dopamine receptor subtypes have shown little effect on dynorphin expression when administered alone. However, D1-specific agents are able to reverse the changes in dynorphin expression caused by generalized dopamine increases or decreases. The D1-specific antagonist SCH 23390 blocks apomorphine-induced dynorphin increases *(120),* whereas a D1-specific agonist attenuates 6-OHDA-induced dynorphin decreases *(116)*. In no case have D2 agonists or antagonists been shown to affect levels of dynorphin *(116,120)*.

Efforts to understand the intermediate signals leading to increased dynorphin expression have implicated the Fos family of transcription factors as key players. Treatment with D1 agonists following unilateral 6-OHDA lesion increases the levels of c-fos and of a 35 kDa FRA, but has no effect on levels of jun family members *(151)*. This result builds on a previous observation of a 35 kDa FRA which is colocalized in neurons with substance P and dynorphin, but not with enkephalin *(152)*. An examination of the time course of treatment with the indirect dopamine agonist cocaine revealed a unique inverse relationship between c-fos mRNA and dynorphin mRNA *(85)*. Areas that were low in basal dynorphin expression showed a strong c-fos response to acute cocaine treatment, whereas areas with high basal dynorphin expression did not. After 4 d of repeated cocaine administration, the striatal subregions which had originally showed a strong c-fos response exhibited enhanced dynorphin expression, but a diminished level of c-fos induction. This evidence is consistent with the idea that dopamine, perhaps via D1 receptors, increases c-fos, leading to increased transcription of the dynorphin gene, which perhaps then feeds back to inhibit the expression of c-fos. The existence of AP-1 sites in the dynorphin gene promoter, shown in Fig. 3, are consistent with this chain of events. Undoubtedly, this is an incomplete characterization of striatal dynorphin induction by dopamine and other intermediate signaling molecules are likely to play significant roles in this process.

4.4. Neurotensin

Neurotensin is a tridecapeptide originally isolated from bovine hypothalamus *(153)* and is expressed in various central nervous system locations including the striatum, nucleus accumbens, globus pallidus, amygdala, hypothalamus, and septum *(154,155)*. Unlike dynorphin, substance P, and enkephalin, which are expressed at fairly high levels, neurotensin is expressed only in specific subregions of the striatum in the basal state *(155,156),* and the general pattern of its expression is consistent with its presence in the enkephalin-D_2 receptor positive cells of the striatum. A variety of stimulatory and inhibitory dopaminergic interventions can lead to increased neurotensin expression, which has led to the description of distinct striatal subpopulations of neurotensin-containing neurons.

This and other laboratories have carried out a series of studies detailing the effects of antipsychotic drugs on anatomically distinct populations of neurons in the striatum of the rat brain. Several of these studies have shown that administration of antipsychotic drugs increases the number of neurotensin immunoreactive cells in the striatum of the rat *(157–159)*. The population of neurons that express neurotensin shows differential responses to acute and chronic treatment with various antipsychotic drugs *(160)*. Transcription of the neurotensin gene in the dorsolateral striatum is dramatically increased by acute treatment with typical antipsychotic drugs such as haloperidol, but not by atypical antipsychotic drugs such as clozapine. On the other hand, all antipsychotics tested (which included both typicals: haloperidol and fluphenazine, and atypicals: clozapine, remoxipride, and thioridazine) dramatically increase neurotensin mRNA levels in the shell region of the accumbens.

Both the cells of the accumbal shell and dorsolateral striatum show increased neurotensin mRNA following chronic haloperidol treatment, but tolerance is observed in the dorsolateral striatum response *(80)*. This tolerance shows an excellent correlation with the time course of induction of tolerance to haloperidol-induced catalepsy in rats. Interestingly, the accumbal response to haloperidol is sustained throughout the chronic exposure. Chronic haloperidol treatment also induces a late-onset, sustained increase in neurotensin mRNA in the ventrolateral striatum, a region thought to be involved in induction of oral movements which model tardive dyskinesia in rats *(80)*. Chronic clozapine treatment, on the other hand, elicits the sustained elevation of neurotensin mRNA only in the accumbal shell.

To summarize, the accumbal neurotensin neurons demonstrate responsiveness to all antipsychotic drugs so far tested and, in contrast to neurons of the dorsolateral striatum, sustain this enhancement of neurotensin expression after chronic treatment. These data lead us to hypothesize that these neurons may be an important neural target involved in the therapeutic effects of both classes of antipsychotic drugs. In contrast, changes in neurotensin gene expression in the dorsolateral striatum may be related to the acute extrapyramidal effects of typical drugs. Finally, the neurotensin response of the ventrolateral neurons could be involved in the tardive dyskinesia-like oral movements seen with chronic drug exposure.

Neurotensin gene expression is not only responsive to D2 receptor blockers, but also can be increased with dopaminergic agonists. Neurotensin mRNA expression is enhanced acutely after treatment with the indirect dopamine agonist methamphetamine *(86,124)*. This effect was more pronounced in caudal dorsal striatum than in rostral striatum *(124)*, and greater in dorsomedial than dorsolateral striatum *(86)*. The increase could be suppressed by coadministration of the D1 antagonist SCH 23390 but not the D2 antagonist sulpiride *(126,161)*, indicating the existence of a population of striatal cells that can increase neurotensin expression in response to D1 activation. This idea is corroborated by evidence showing a dose-dependent increase in striatal neurotensin immunoreactivity following acute treatment with the D1 agonist, and a dose-dependent decrease following D1 antagonist exposure *(117)*.

4.5. Possible Involvement of c-fos and CREB in Antipsychotic Drug-Induced Neurotensin Gene Expression

As described previously, antipsychotic drugs dramatically enhance c-fos gene expression in the striatum. The increased Fos could, along with Jun family members, interact with the promoter regions of numerous genes. Interestingly, the neurotensin gene has been shown to have a functional AP-1 consensus element (Fig. 3). Therefore, the changes in c-fos expression in response to antipsychotic drugs are likely to be involved in the neurotensin transcriptional response noted. Interestingly, Merchant *(162)* recently reported that antisense oligonucleotides that inhibit fos expression block the effect of haloperidol on dorsolateral striatum neurotensin mRNA.

CREB is another transcription factor that is undoubtedly involved in antipsychotic drug-induced transcription of the neurotensin gene. The neurotensin promoter possesses, in addition to its AP-1 site, CRE. Occupancy of this site is required for maximal activation of neurotensin gene transcription *(163)*. In addition, CREB can form dimers with c-fos, and can interact with AP-1 sites as well as CREs. As described previously, we and others have used antibodies specific for the phosphorylated form of CREB to observe changes in CREB phosphorylation in dorsolateral striatum and nucleus accumbens, which occur in the same region as the antipsychotic drug-sensitive neurotensin neurons just described. Since the transcription of the c-fos gene also relies on cAMP-dependent CREB activation, a cascade of events leading to increases in neurotensin expression can be constructed. That is, antipsychotic drugs bind to and antagonize D2-like receptors, which leads to an enhancement of cellular cAMP, activation of PKA, and subsequent phosphorylation of CREB. CREB not only acts on the neurotensin gene itself, but induces transcription of c-fos. Translation to c-Fos protein allows enhancement of AP-1 binding activity and binding to AP-1 or AP-1/CRE sites on the neurotensin promoter, increasing neurotensin gene transcription.

4.6. Neurotransmitter-Synthesizing Enzymes

Glutamate decarboxylase is the enzyme responsible for synthesizing gamma amino butyric acid (GABA) from its precursor glutamate. The inhibitory neurotransmitter GABA is the main transmitter used by the output neurons of the striatum and globus pallidus. Levels of glutamate decarboxylase mRNA seem to be negatively regulated by dopaminergic input, since both dopamine deafferentation *(140,164)*, and dopamine antagonists *(165)* cause increases in glutamate decarboxylase mRNA levels. 6-OHDA lesion has also been shown to cause an increase in the levels of glutamate decarboxylase activity *(166)*.

Tyrosine hydroxylase (TH) and DOPA decarboxylase, the enzymes responsible for the production of dopamine from its precursor amino acid tyrosine, also have been examined. Protein levels measured by immunocytochemistry, and mRNA levels measured by *in situ* hybridization, were increased substantially 8 h after treatment with the typical antipsychotic haloperidol *(167)*. On the other hand, a separate study found no effect on TH mRNA levels after 32 d of treatment with either haloperidol or loxapine *(168)*. By contrast, the same study showed that DOPA decarboxylase mRNA levels increased by nearly 200% in response to the same treatment protocol, indicating that perhaps this enzyme is more important than TH for long-term regulation of dopamine production.

5. Conclusion

In this chapter, we have attempted to provide an updated summary of the literature regarding dopamine receptor-mediated effects on downstream transcriptional responses. The preponderance of evidence suggests that D1- and D2-like receptors play critical roles in the regulation of both basal and stimulated levels of transcription of target genes in neurons of the striatum and other brain regions. It seems likely that the events leading to transcriptional responses make use of cAMP- and AP-1-mediated gene transcription. Clearly, however, when a given receptor is present in different brain regions, the downstream transcriptional responses with which it is associated can be dramatically different. This is, in part, a reflection of differing phenotypes of receptor-positive neurons, but is also a consequence of the other receptors which are coexpressed by these cells, with which the nonselective drugs we have discussed may also interact.

Although there is a large body of data concerning dopamine's effects on both immediate-early genes and neuropeptides, the links showing how the former are able to bring about transcription of the latter are just beginning to be forged. The next few years should see impressive gains in this field, which will yield a more complete understanding of both the normal physiological role of dopamine, and the mechanisms by which dopaminergic agents elicit their many clinical and behavioral manifestations.

References

1. Heikkila, R. E., Orlansky, O. H., and Cohen, G. (1975) Studies on the distinction between uptake inhibition and release of ^{3}H dopamine in rat brain slices. *Biochem. Pharmacol.* **24,** 847–852.
2. Koe, B. K. (1976) Molecular geometry of inhibitors of the uptake of catecholamines and serotonin in synaptosomal preparations of rat brain. *J. Pharmacol. Exp. Ther.* **199,** 649–661.
3. Moore, K. E., Chiueh, C. C., and Zeldes, G. (1977) Release of neurotransmitters from the brain in vivo by amphetamine methylphenidate and cocaine, in *Cocaine and Other Stimulants* (Ellinwood, E. H. and Kilbey, M. M., eds.), Plenum, New York, pp. 143–160.
4. Ross, S. F. and Renyi, A. L. (1967) Accumulation of tritiated 5-hydroxytryptamine in brain slices. *Life Sci.* **6,** 1407–1415.
5. Hertting, G., Axelrod, J., and Whitby, L. G. (1961) Effect of drugs on the uptake and metabolism of ^{3}H norepinephrine. *J. Pharmacol. Exp. Ther.* **134,** 146–153.
6. Weiner, N. (1972) Pharmacology of the central nervous system stimulants, in *Drug Abuse: Proceedings of the International Conference* (Zarafonetis, C. J. D., ed.), Lea & Febiger, Philadelphia, pp. 243–251.
7. Seeman, P., Guan, H. C., Civelli, O., Van Tol, H. H. M., Sunahara, R. K., and Niznik, H. B. (1992) *Eur. J. Pharmacol.* **227,** 139–146.
8. Seeman, P., Niznik, H. B., and Guan, H. C. (1990) Elevation of dopamine D_2 receptors in schizophrenia is underestimated by radioactive raclopride. *Arch. Gen. Psychiatry* **47,** 1170–1172.
9. Roth, B. L., Craigo, S. C., Choudhary, M. S., Uluer, A., Monsma, F. J., Jr., Shen, Y., Meltzer, H. Y., and Sibley, D. R. (1994) Binding of typical and atypical antipsychotic agents to 5-hydroxytryptamine-6 and 5-hydroxytryptamine-7 receptors. *J. Pharmacol. Exp. Ther.* **268,** 1403–1410.
10. Becksted, R. M. and Kersey, K. S. (1985) Immunohistochemical demonstration of differential substance P-, met enkephalin, and glutamic acid decarboxylase-containing cell and axon distributions in the corpus striatum of the cat. *J. Comp. Neurol.* **232,** 481–498.
11. Gerfen, C. R. and Young, W. S., 3rd. (1988) Distribution of striatonigral and striatopallidal peptidergic neurons in both patch and matrix compartments: an in situ hybridization histochemistry and fluorescent retrograde tracing study. *Brain Res.* **460,** 161–167.
12. Kita, H. and Kitai, S. T. (1988) Glutamate decarboxylase immunoreactive neurons in the rat neostriatum: their morphological types and populations. *Brain Res.* **447,** 346–352.
13. Brownstein, M. J., Mroz, E. A., Tappaz, M. L., and Leeman, S. E. (1977) On the origin of substance P and glutamic acid decorboxylase (GAD) in the substantia nigra. *Brain Res.* **135,** 315–323.
14. Vincent, S. R., Hokfelt, T., Christensson, I., and Terenius, L. (1982) Dynorphin-immunoreactive neurons in the central nervous system of the rat. *Neurosci. Lett.* **33,** 185–190.
15. Voorn, P., Roest, G., and Groenewegen, H. J. (1987) Increase of enkephalin and decrease of substance P immunoreactivity in the dorsal and ventral striatum of the rat after midbrain 6-hydroxydopamine lesions. *Brain Res.* **412,** 391–396.

16. Young, W. S. D., Bonner, T. I., and Brann, M. R. (1986) Mesencephalic dopamine neurons regulate the expression of neuropeptide mRNAs in the rat forebrain. *Proc. Natl. Acad. Sci. USA* **83,** 9827–9831.
17. Merchant, K. M., Dobie, D. J., and Dorsa, D. M. (1992) Expression of the proneurotensin gene in the rat brain and its regulation by antipsychotic drugs. *Ann. NY Acad. Sci.* **668,** 54–69.
18. Bannon, M. J., Elliot, P. J., and Bunney, E. B. (1987) Striatal tackykinin iosynthesis: regulation of mRNA and peptide levels by dopamine agonist and antagonist. *Mol. Brain Res.* **3,** 31–37.
19. Gerfen, C. R., Engber, T. M., Mahan, L. C., Susel, Z., Chase, T. N., Monsma, F. J. J., and Sibley, D. R. (1990) Dl and D2 dopamine receptor-regulate gene expression of striatonigral and striatopallidal neurons. *Science* **250,** 1429–1432.
20. Gerfen, C. R., McGinty, J. F., and Young, W. S., 3rd (1991) Dopamine differentially regulates dynorphin, substance P, and enkephalin expression in striatal neurons: in situ hybridization histochemical analysis. *J. Neurosci.* **11,** 1016–1031.
21. Hanson, G. R., Merchant, K. M., Letter, A. A., Bush, L., and Gibb, J. W. (1988) Characterization of methamphetamine effects on the striatal-nigral dynorphin system. *Eur. J. Pharmacol.* **144,** 245,246.
22. Li, S. J., Jiang, H. K., Stachowiak, M. S., Hudson, P. M., Owyang, V., and Nanry, K. (1990) Influence of nigrostriatal dopaminergic tone on the biosynthesis of dynorphin and enkephalin in rat striatum. *Mol. Brain Res.* **8,** 219–225.
23. O'Malley, K. L., Harmon, S., Tang, L., and Todd, R. D. (1992) The rat dopamine D4 receptor: sequence, gene structure, and demonstration of expression in the cardiovascular system. *New Biol.* **4,** 137–146.
24. Meador Woodruff, J. H., Mansour, A., Grandy, D. K., Damask, S. P., Civelli, O., and Watson, S. J., Jr. (1992) Distribution of D5 dopamine receptor mRNA in rat brain. *Neurosci. Lett.* **145,** 209–212.
25. Herroelen, L., De Backer, J. P., Wilczak, N., Flamez, A., Vauquelin, G., and De Keyser, J. (1994) Autoradiographic distribution of D3-type dopamine receptors in human brain using [3H]7-hydroxy-N,N-di-n-propyl-2-aminotetralin. *Brain Res.* **648,** 222–228.
26. Wallace, D. R. and Booze, R. M. (1995) Identification of D3 and sigma receptors in the rat striatum and nucleus accumbens using (+/-)-7-hydroxy-N,N-di-n-[3H]propyl-2-aminotetralin and carbetapentane. *J. Neurochem.* **64,** 700–710.
27. Deutch, A., Lee, M. C., and Iadarola, M. (1992) Regionally specific effects of atypical antipsychotic drugs on striatal Fos expression: the nucleus accumbens shell as a locus of antipsychotic action. *Mol. Cell. Neurosci.* **3,** 332–341.
28. Dragunow, M., Robertson, G. S., Faull, R. L., Robertson, H. A., and Jansen, K. (1990) D2 dopamine receptor antagonists induce fos and related proteins in rat striatal neurons. *Neuroscience* **37,** 287–294.
29. Miller, J. C. (1990) Induction of c-fos mRNA expression in rat striatum by neuroleptic drugs. *J. Neurochem.* **54,** 1453–1455.
30. Nguyen, T. V., Kosofsky, B. E., Birnbaum, R., Cohen, B. M., and Hyman, S. E. (1992) Differential expression of c-fos and zif268 in rat striatum after haloperidol, clozapine, and amphetamine. *Proc. Natl. Acad. Sci. USA* **89,** 4270–4274.
31. Rogue, P. and Vincendon, G. (1992) Dopamine D2 receptor antagonists induce immediate early genes in the rat striatum. *Brain Res. Bull.* **29,** 469–472.

32. Robertson, G. S. and Fibiger, H. C. (1992) Neuroleptics increase c-fos expression in the forebrain: contrasting effects of haloperidol and clozapine. *Neuroscience* **46,** 315–328.
33. MacGibbon, G. A., Lawlor, P. A., Bravo, R., and Dragunow, M. (1994) Clozapine and haloperidol produce a differential pattern of immediate early gene expression in rat caudate-putamen, nucleus accumbens, lateral septum and islands of Calleja. *Brain Res. Mol. Brain Res.* **23,** 21–32.
34. Simpson, C. S. and Morris, B. J. (1994) Haloperidol and fluphenazine induce junB gene expression in rat striatum and nucleus accumbens. *J. Neurochem.* **63,** 1955–1961.
35. Conneely, O. M., Power, R. F., and O'Malley, B. W. (1992) Regulation of gene expression by dopamine: implications in drug addiction. *NIDA Res. Monogr.* **126,** 84–97.
36. O'Donovan, M. C., Buckland, P. R., Spurlock, G., and McGuffin, P. (1992) Bidirectional changes in the levels of messenger RNAs encoding gamma-aminobutyric acid A receptor alpha subunits after flurazepam treatment. *Eur. J. Pharmacol.* **226,** 335–341.
37. Merchant, K. M., Figur, L. M., and Evans, D. L. (1995) Induction of c-fos mRNA in the rat medial prefrontal cortex by antipsychotic drugs: role of dopamine D2 and D3 receptors. *Corebral Cortex,* submitted.
38. Wirtshafter, D., Asin, K. E., and Pitzer, M. R. (1994) Dopamine agonists and stress produce different patterns of Fos-like immunoreactivity in the lateral habenula. *Brain Res.* **633,** 21–26.
39. Liu, J., Nickolenko, J., and Sharp, F. R. (1994) Morphine induces c-fos and junB in striatum and nucleus accumbens via D1 and N-methyl-D-aspartate receptors. *Proc. Natl. Acad. Sci. USA* **91,** 8537–8541.
40. Vaccarino, F. M., Hayward, M. D., Le, H. N., Hartigan, D. J., Duman, R. S., and Nestler, E. J. (1993) Induction of immediate early genes by cyclic AMP in primary culture of neurons from rat cerebral cortex. *Mol. Brain Res.* **19,** 76–82.
41. Matsunaga, T., Ohara, K., Natsukari, N., and Fujita, M. (1991) Dopamine D2-receptor mRNA level in rat striatum after chronic haloperidol treatment. *Neurosci. Res.* **12,** 440–445.
42. Paul, M. L., Graybiel, A. M., David, J. C., and Robertson, H. A. (1992) D1-like and D2-like dopamine receptors synergistically activate rotation and c-fos expression in the dopamine-depleted striatum in a rat model of Parkinson's disease. *J. Neurosci.* **12,** 3729–3742.
43. Cole, A. J., Bhat, R. V., Patt, C., Worley, P. F., and Baraban, J. M. (1992) D1 dopamine receptor activation of multiple transcription factor genes in rat striatum. *J. Neurochem.* **58,** 1420–1426.
44. Cadet, J. L., Zhu, S. M., and Angulo, J. A. (1992) Quantitative in situ hybridization evidence for differential regulation of proenkephalin and dopamine D2 receptor mRNA levels in the rat striatum: effects of unilateral intrastriatal injections of 6-hydroxydopamine. *Brain Res. Mol. Brain Res.* **12,** 59–67.
45. Persico, A. M., Schindler, C. W., O'Hara, B. F., Brannock, M. T., and Uhl, G. R. (1993) Brain transcription factor expression: effects of acute and chronic amphetamine and injection stress. *Brain Res. Mol. Brain Res.* **20,** 91–100.
46. Borgundvaag, B., Kudlow, J. E., Mueller, S. G., and George, S. R. (1992) Dopamine receptor activation inhibits estrogen-stimulated transforming growth factor-

alpha gene expression and growth in anterior pituitary, but not in uterus. *Endocrinology* **130,** 3453–3458.

47. Hope, B., Kosofsky, B., Hyman, S. E., and Nestler, E. J. (1992) Regulation of immediate early gene expression and AP-1 binding in the rat nucleus accumbens by chronic cocaine. *Proc. Natl. Acad. Sci. USA* **89,** 5764–5768.
48. Bhat, R. V. and Baraban, J. M. (1993) Activation of transcription factor genes in striatum by cocaine: role of both serotonin and dopamine systems. *J. Pharmacol. Exp. Ther.* **267,** 496–505.
49. Graybiel, A. M., Moratalla, R., and Robertson, H. A. (1990) Amphetamine and cocaine induce drug-specific activation of the c-fos gene in striosome-matrix compartments and limbic subdivisions of the striatum. *Proc. Natl. Acad. Sci. USA* **87,** 6912–6916.
50. Kosofsky, B. E., Genova, L. M., and Hyman, S. E. (1995) Postnatal age defines specificity of immediate early gene induction by cocaine in developing rat brain. *J. Comp. Neurol.* **351,** 27–40.
51. Moratalla, R., Vickers, E. A., Robertson, H. A., Cochran, B. H., and Graybiel, A. M. (1993) Coordinate expression of c-fos and jun B is induced in the rat striatum by cocaine. *J. Neurosci.* **13,** 423–433.
52. Hope, B. T., Nye, H. E., Kelz, M. B., Self, D. W., Iadarola, M. J., Nakabeppu, Y., Duman, R. S., and Nestler, E. J. (1994) Induction of a long-lasting AP-1 complex composed of altered Fos-like proteins in brain by chronic cocaine and other chronic treatments. *Neuron* **13,** 1235–1244.
53. Bhat, R. V., Cole, A. J., and Baraban, J. M. (1992) Chronic cocaine treatment suppresses basal expression of zif268 in rat forebrain: in situ hybridization studies. *J. Pharmacol. Exp. Ther.* **263,** 343–349.
54. Sebens, J. B., Koch, T., Ter Horst, G. J., and Korf, J. (1995) Differential Fos-protein induction in rat forebrain regions after acute and long-term haloperidol and clozapine treatment. *Eur. J. Pharmacol.* **273,** 175–182.
55. Dilts, R. P., Jr., Helton, T. E., and McGinty, J. F. (1993) Selective induction of Fos and FRA immunoreactivity within the mesolimbic and mesostriatal dopamine terminal fields. *Synapse* **13,** 251–263.
56. Pennypacker, K. R., Zhang, W. Q., Ye, H., and Hong, J. S. (1992) Apomorphine induction of Ap-1 DNA binding in the rat striatum after dopamine depletion. *Mol. Brain Res.* **15,** 151–155.
57. Konradi, C., Kobierski, L. A., Nguyen, T. V., Heckers, S., and Hyman, S. E. (1993) The cAMP-response-element-binding protein interacts, but Fos protein does not interact, with the proenkephalin enhancer in rat striatum. *Proc. Natl. Acad. Sci. USA* **90,** 7005–7009.
58. Robertson, G. S., Matsumura, H., and Fibiger, H. C. (1994) Induction patterns of Fos-like immunoreactivity in the forebrain as predictors of atypical antipsychotic activity. *J. Pharmacol. Exp. Ther.* **271,** 1058–1066.
59. Cole, D. G., Kobierski, L. A., Konradi, C., and Hyman, S. E. (1994) 6-Hydroxydopamine lesions of rat substantia nigra up-regulate dopamine-induced phosphorylation of the cAMP-response element-binding protein in striatal neurons. *Proc. Natl. Acad. Sci. USA* **91,** 9631–9635.
60. Berretta, S., Robertson, H. A., and Graybiel, A. M. (1992) Dopamine and glutamate agonist stimulate neuron-specific expression of fos-like protein in the striatum. *J. Neurophysiology* **68,** 767–777.

61. Cohen, B. M., Nguyen, T. V., and Hyman, S. E. (1991) Cocaine-induced changes in gene expression in rat brain. *NIDA Res. Monogr.* **105,** 175–181.
62. Konradi, C., Cole, R. L., Heckers, S., and Hyman, S. E. (1994) Amphetamine regulates gene expression in rat striatum via transcription factor CREB. *J. Neurosci.* **14,** 5623–5634.
63. Young, S. T., Porrino, L. J., and Iadarola, M. J. (1991) Cocaine induces striatal c-Fos-immunoreactive proteins via dopaminergic D1 receptors. *Proc. Natl. Acad. Sci. USA* **88,** 1291–1295.
64. Cohen, D. R. and Curran, T. (1988) fra-1: a serum-inducible, cellular immediate-early gene that encodes a fos-related antigen. *Mol. Cell Biol.* **8,** 2063–2069.
65. Dobrzanski, P., Noguchi, T., Kovary, K., Rizzo, C. A., Lazo, P. S., and Bravo, R. (1991) Both products of fosB gene, FosB and its short form, FosB/SF, are transcriptional activators in fibroblasts. *Mol. Cell. Biol.* **11,** 5470–5478.
66. Morgan, J. I. and Curran, T. (1991) Stimulus-transcription coupling in the nervous system: involvement of the inducible proto-oncogenes fos and jun. *Ann. Rev. Neurosci.* **14,** 421–451.
67. Nishina, H., Sato, H., Suzuki, T., Sato, M., and Iba, H. (1990) Isolation and characterisation of fra-2, an additional member of the fos gene family. *Proc. Natl. Acad. Sci. USA* **87,** 3619–3623.
68. Morgan, J. I. and Curran, T. (1989) Stimulus-transcription coupling neurons: role of cellular immediate early genes. *Trends Neurosci.* **12,** 459–462.
69. Sambucetti, L. C. and Curran, T. (1986) The fos protein is associated with DNA in isolated nuclei and binds to DNA cellulose. *Science* **234,** 1417–1419.
70. Zerial, M., Toschi, L., Ryseck, R.-P., Schuermann, M., Muller, R., and Brava, R. (1989) The product of a novel growth factor activated gene, fos B, interacts with JUN proteins enhancing their DNA binding activity. *EMBO J.* **8,** 805–813.
71. Hirai, S.-I., Ryseck, R.-P., Mechta, F., Brava, R., and Yaniv, M. (1989) Characterisation of junD: a new member of the jun protooncogene family. *EMBO J.* **8,** 1433–1439.
72. Ryder, K., Lau, L. F., and Nathans, D. (1988) Jun-D: a third member of the jun gene family. *Proc. Natl. Acad. Sci. USA* **86,** 1487–1491.
73. Vogt, P. K. and Bos, T. J. (1990) Jun: oncogene and transcription factor. *Adv. Cancer Res.* **55,** 1–35.
74. Nakabeppu, Y., Ryder, K., and Nathans, D. (1988) DNA binding activities of three murine Jun proteins: stimulation by Fos. *Cell* **55,** 907–915.
75. Angel, P., Hattori, K., Smeal, T., and Karin, M. (1988) The jun protooncogene is positively autoregulated by its own protein product, Jun/AP-1. *Cell* **55,** 875–885.
76. Chiu, R., Angel, P., and Karin, M. (1989) JunB differs in its biological properties from and is a negative regulator of c-Jun. *Cell* **59,** 979–986.
77. Nakabeuppu, Y. and Nathans, D. (1991) A naturally occurring truncated from of FosB that inhibits Fos/Jun transcriptional activity. *Cell* **64,** 751–759.
78. Robertson, G. S., Vincent, S. R., and Fibiger, H. C. (1990) Striatonigral projection neurons contain D1 dopamine receptor-activated c-fos. *Brain Res.* **523,** 288–290.
79. Marshall, J. F., Cole, B. N., and LaHoste, G. J. (1993) Dopamine D2 receptor control of pallidal fos expression: comparisons between intact and 6-hydroxy-dopamine-treated hemispheres. *Brain Res.* **632,** 308–313.
80. Merchant, K. M., Dobie, D. J., Filloux, F. M., Totzke, M., Aravagiri, M., and Dorsa, D. M. (1994) Effects of chronic haloperidol and clozapine treatment on neurotensin

and c-fos mRNA in rat neostriatal subregions. *J. Pharmacol. Exp. Ther.* **271,** 460–471.

81. Merchant, K. M. and Dorsa, D. M. (1993) Differential induction of neurotensin and c-fos gene expression by typical versus atypical antipsychotics. *Proc. Natl. Acad. Sci. USA* **90,** 3447–3451.
82. Carelli, R. M. and West, M. O. (1991) Representation of the body by single neurons in the dorsolateral striatum of the awake, unrestrained rat. *J. Comp. Neurol.* **309,** 231–249.
83. Pisa, M. (1988) Motor functions of the striatum of the rat: critical role of the lateral region in tongue and forelimb reaching. *Neuroscience* **24,** 453–463.
84. West, M. O., Michael, A. J., Knowles, S. E., Chapin, J. K., and Woodard, D. J. (1987) Striatal unit activity and the linkage between sensory and motor events, in *Basal ganglia and behavior: Sensory aspects of motor functioning* (Schneider, J. S. and Lidsky, T., eds.), Hans Huber Hogrefe, Toronto, Canada, pp. 27–35.
85. Steiner, H. and Gerfen, C. R. (1993) Cocaine-induced c-fos messenger RNA is inversely related to dynorphin expression in striatum. *J. Neurosci.* **13,** 5066–5081.
86. Merchant, K. M., Hanson, G. R., and Dorsa, D. M. (1994) Induction of neurotensin and c-fos mRNA in distinct subregions of rat neostriatum after acute methamphetamine: comparison with acute haloperidol effects. *J. Pharmacol. Exp. Ther.* **269,** 806–812.
87. Couceyro, P., Pollock, K. M., and Douglas, J. (1994) Cocaine differentially regulates activator protein-1 mRNA levels and DNA-binding complexes in the rat striatum and cerebellum. *Mol. Pharmacol.* **49,** 667–676.
88. Bannon, M. J., Lee, J. M., Giraud, P., Young, A., Affolter, H. U., and Bonner, T. I. (1986) Dopamine antagonist haloperidol decreases substance P, substance K, and preprotachykinin mRNAs in rat striatonigral neurons. *J. Biol. Chem.* **261,** 6640–6642.
89. Schutte, J., Viallet, J., Nau, M., Segal, S., Fedorko, J., and Minna, J. (1989) JunB inhibits and c-fos stimulates the transforming and transactivating activities of c-jun. *Cell* **59,** 987–997.
90. Kebabian, J. W. and Calane, D. B. (1979) Multiple receptors of dopamine. *Nature* **277,** 93–96.
91. Artalejo, C. R., Ariano, M. A., Perlaman, R. L., and Fox, A. P. (1990) Activation of facilitation calcium channels in chromaffin cells by D1 dopamine receptors through a cAMP/protein kinase A-dependent mechanism. *Nature* **348,** 239–242.
92. Bading, H., Ginty, D. D., and Greenburg, M. E. (1993) Regulation of gene expression in hippocampal neurons by distinct calcium signaling pathways. *Science* **260,** 181–186.
93. Hille, B. (1992) *Ionic Channels of Excitable Membranes,* Sinauer, Sutherland, MA.
94. Lerea, L. S., Butler, L. S., and McNamara, J. O. (1992) NMDA and non-NMDA receptor-mediated increase of c-fos mRNA in dentate gyrus neurons involves calcium influx via different routes. *J. Neurosci.* **12,** 2973–2981.
95. Gonzalez, G. A. and Montminy, M. R. (1989) Cyclic AMP stimulates somatostatin gene transcription by phosphorylation of CREB at serine 133. *Cell* **59,** 675–680.
96. Adams, M. R., Malouf, A., Unis, A., and Dorsa, D. M. (1994) Signal transduction effects of antipsychotic drugs *in vivo* and in organotypic cultures of rat brain striatum. *Society Neurosci. Abstracts* **18,** 223.

97. Ginty, D. D., Kornhauser, J. M., Thompson, M. A., Bading, H., Mayo, K. E., Takahashi, J. S., and Greenburg, M. E. (1993) Regulation of CREB phosphorylation in the suprachiasmatic nucleus by light and circadian clock. *Science* **260,** 238–241.
98. Milbrandt, J. (1987) A nerve growth factor-induced gene encodes a possible transcriptional regulatory factor. *Science* **238,** 797–799.
99. Chavrier, P., Zerial, M., Lemaire, P., Almendral, J., Bravo, R., and Charnay, P. (1988) A gene encoding a protein with zinc fingers if activated during G_0/G_1 transition in cultured cells. *EMBO J.* **7,** 29–35.
100. Sukhatme, V. P., Coa, X., Chang, L. C., Tsai-Morris, C., Stamenkovich, D., Ferreira, P. C. P., et al. (1988) A zinc finger-encoding gene coregulated with c-*fos* during growth and differentiation and after cellular depolarization. *Cell* **53,** 37–43.
101. Christy, B. A., Lau, L. F., and Nathans, D. (1988) A gene activated in mouse 3T3 cells by serum growth factors encodes a protein with "zinc fingers" sequences. *Proc. Natl. Acad. Sci. USA* **85,** 7857–7861.
102. Lemaire, P., Revelent, O., Bravo, R., and Charnay, P. (1988) Two mouse genes encoding potential transcription factors with identical DNA-binding domains are activated by growth factors in cultured cells. *Proc. Natl. Acad. Sci. USA* **85,** 4691–4695.
103. Bhat, R. V., Worley, P. F., Cole, A. J., and Baraban, J. M. (1992) Activation of the zinc finger encoding gene *krox-20* in adult rat brain: comparison with zif268. *Mol. Brain Res.* **13,** 263–266.
104. Bhat, R. V., Cole, A. J., and Baraban, J. M. (1992) Role of monoamine systems in activation of zif268 by cocaine. *J. Psychiatry Neurosci.* **17,** 94–102.
105. Angulo, J. A. and McEwen, B. S. (1994) Molecular aspects of neuropeptide regulation and function in the corpus striatum and nucleus accumbens. *Brain Res. Brain Res. Rev.* **19,** 1–28.
106. Kislauskis, E. and Dobner, P. R. (1990) Mutually dependent response elements in the cis-regulatory region of the neurotensin/neuromedin N gene integrate environmental stimuli in PC12 cells. *Neuron* **4,** 783–795.
107. MacArthur, L., Iacangelo, A. L., Hsu, C. M., and Eiden, L. E. (1992) Enkephalin biosynthesis is coupled to secretory activity via transcription of the proenkephalin A gene. *J. Physiol. Paris* **86,** 89–98.
108. Kaynard, A. H., McMurray, C. T., Douglass, J., Curry, T. E., Jr., and Melner, M. H. (1992) Regulation of prodynorphin gene expression in the ovary: distal DNA regulatory elements confer gonadotropin regulation of promoter activity. *Mol. Endocrinol.* **6,** 2244–2256.
109. Naranjo, J. R., Mellstrom, B., Achaval, M., and Sassone Corsi, P. (1991) Molecular pathways of pain: Fos/Jun-mediated activation of a noncanonical AP-1 site in the prodynorphin gene. *Neuron* **6,** 607–617.
110. Mendelson, S. C. and Quinn, J. P. (1993) Identification of potential regulatory elements within the rat preprotachykinin A promoter. *Biochem. Soc. Trans.* **21,** 372S.
111. Quinn, J. P., Morrison, C., McAllister, J., and Mendelson, S. (1993) Evolution of enhancer domains within the preprotachykinin promoter. *Biochem. Soc. Trans.* **21,** 371S.
112. Angulo, J. A. (1992) Involvement of dopamine D1 and D2 receptors in the regulation of proenkephalin mRNA abundance in the striatum and accumbens of the rat brain. *J. Neurochem.* **58,** 1104–1109.

113. Morris, B. J. and Hunt, S. P. (1991) Proenkephalin mRNA levels in rat striatum are increased and decreased, respectively, by selective D2 and D1 dopamine receptor antagonists. *Neurosci. Lett.* **125,** 201–204.
114. Pollack, A. E. and Wooten, G. F. (1992) Differential regulation of striatal preproenkephalin mRNA by D1 and D2 dopamine receptors. *Brain Res. Mol. Brain Res.* **12,** 111–119.
115. Haverstick, D. M., Rubenstein, A., and Bannon, M. J. (1989) Striatal tachykinin gene expression regulated by interaction of D-1 and D-2 dopamine receptors. *J. Pharmacol. Exp. Ther.* **248,** 858–862.
116. Engber, T. M., Boldry, R. C., Kuo, S., and Chase, T. N. (1992) Dopaminergic modulation of striatal neuropeptides: differential effects of D1 and D2 receptor stimulation on somatostatin, neuropeptide Y, neurotensin, dynorphin and enkephalin. *Brain Res.* **581,** 261–268.
117. Taylor, M. D., de Ceballos, M. L., Jenner, P., and Marsden, C. D. (1991) Acute effects of D-1 and D-2 dopamine receptor agonist and antagonist drugs on basal ganglia [Met5]- and [Leu5]-enkephalin and neurotensin content in the rat. *Biochem. Pharmacol.* **41,** 1385–1391.
118. Nylander, I. and Terenius, L. H. (1987) Dopamine receptors mediate alterations in striato-nigral dynorphin and substance P pathways. *Neuropharmacology* **26,** 1295–1302.
119. Oblin, A., Zivkovic, B., and Bartholini, G. (1987) Selective antagonists of dopamine receptor subtypes differentially affect substance P levels in the striatum and substantia nigra. *Brain Res.* **421,** 387–390.
120. Jiang, H. K., McGinty, J. F., and Hong, J. S. (1990) Differential modulation of striatonigral dynorphin and enkephalin by dopamine receptor subtypes. *Brain Res.* **507,** 57–64.
121. Singh, N. A., Bush, L. G., Gibb, J. W., and Hanson, G. R. (1992) Role of N-methyl-D-aspartate receptors in dopamine D1-, but not D2-, mediated changes in striatal and accumbens neurotensin systems. *Brain Res.* **571,** 260–264.
122. Hong, J. S., Yang, H. Y., Gillin, J. C., and Costa, E. (1980) Effects of long-term administration of antipsychotic drugs on enkephalinergic neurons. *Adv. Biochem. Psychopharmacol.* **24,** 223–232.
123. Tang, F., Costa, E., and Schwartz, J. P. (1983) Increase of proenkephalin mRNA and enkephalin content of rat striatum after daily injection of haloperidol for 2 to 3 weeks. *Proc. Natl. Acad. Sci. USA* **80,** 3841–3844.
124. Castel, M. N., Morino, P., Dagerlind, A., and Hokfelt, T. (1994) Up-regulation of neurotensin mRNA in the rat striatum after acute methamphetamine treatment. *Eur. J. Neurosci.* **6,** 646–656.
125. Li, S., Sivam, S. P., and Hong, J. S. (1986) Regulation of the concentration of dynorphin A1-8 in the striatonigral pathway by the dopaminergic system. *Brain Res.* **398,** 390–392.
126. Letter, A. A., Matsuda, L. A., Merchant, K. M., Gibb, J. W., and Hanson, G. R. (1987) Characterization of dopaminergic influence on striatal-nigral neurotensin systems. *Brain Res.* **422,** 200–203.
127. Li, S. J., Sivam, S. P., McGinty, J. F., Jiang, H. K., Douglass, J., Calavetta, L., and Hong, J. S. (1988) Regulation of the metabolism of striatal dynorphin by the dopaminergic system. *J. Pharmacol. Exp. Ther.* **246,** 403–408.

128. Angulo, J. A., Davis, L. G., Burkhart, B. A., and Christoph, G. R. (1986) Reduction of striatal dopaminergic neurotransmission elevates striatal proenkephalin mRNA. *Eur. J. Pharmacol.* **130,** 341–343.
129. Sirinathsinghji, D. J. and Dunnett, S. B. (1991) Increased proenkephalin mRNA levels in the rat neostriatum following lesion of the ipsilateral nigrostriatal dopamine pathway with 1-methyl-4-phenylpyridinium ion (MPP+): reversal by embryonic nigral dopamine grafts. *Brain Res. Mol. Brain Res.* **9,** 263–269.
130. Masuo, Y., P'Elaprat, D., Montagne, M. N., Scherman, D., and Rostene, W. (1990) Regulation of neurotensin-containing neurons in the rat striatum and substantia nigra. Effects of unilateral nigral lesion with 6-hydroxydopamine on neurotensin content and its binding site density. *Brain Res.* **510,** 203–210.
131. Drago, J., Gerfen, C. R., Lachowicz, J. E., Steiner, H., Hollon, T. R., Love, P. E., et al. (1994) Altered striatal function in a mutant mouse lacking D1A dopamine receptors. *Proc. Natl. Acad. Sci. USA* **91,** 12,564–12,568.
132. Xu, M., Moratalla, R., Gold, L. H., Hiroi, N., Koob, G. F., Graybiel, A. M., and Tonegawa, S. (1994) Dopamine D1 receptor mutant mice are deficient in striatal expression of dynorphin and in dopamine-mediated behavioral responses. *Cell* **79,** 729–742.
133. Hong, J. S., Yang, H.-Y. T., Fratta, W., and Costa, E. (1978) Rat striatal methionine-enkephalin content after chronic treatment with cataleptogenic and non-cataleptogenic drugs. *J. Pharmacol. Exp. Ther.* **205,** 141–147.
134. Kalivas, P. W. (1985) Interactions between neuropeptides and dopamine neurons in the ventromedial mesencephalon. *Neurosci. Biobehav. Rev.* **9,** 573–587.
135. Jaber, M., Fournier, M. C., and Bloch, B. (1992) Reserpine treatment stimulates enkephalin and D2 dopamine receptor gene expression in the rat striatum. *Brain Res. Mol. Brain Res.* **15,** 189–194.
136. Nisenbaum, L. K., Kitai, S. T., Crowley, W. R., and Gerfen, C. R. (1994) Temporal dissociation between changes in striatal enkephalin and substance P messenger RNAs following striatal dopamine depletion. *Neuroscience* **60,** 927–937.
137. De Vries, T. J., Mulder, A. H., and Schoffelmeer, A. N. (1992) Differential ontogeny of functional dopamine and muscarinic receptors mediating presynaptic inhibition of neurotransmitter release and postsynaptic regulation of adenylate cyclase activity in rat striatum. *Brain Res. Dev. Brain Res.* **66,** 91–96.
138. De Vries, T. J., Jonker, A. J., Voorn, P., Mulder, A. H., and Schoffelmeer, A. N. (1994) Adaptive changes in rat striatal preproenkephalin expression and dopamine-opioid interactions upon chronic haloperidol treatment during different developmental stages. *Brain Res. Dev. Brain Res.* **78,** 175–181.
139. Sivam, S. P., Krause, J. E., Breese, G. R., and Hong, J. S. (1991) Dopamine-dependent postnatal development of enkephalin and tachykinin neurons of rat basal ganglia. *J. Neurochem.* **56,** 1499–1508.
140. Soghomonian, J. J. (1993) Effects of neonatal 6-hydroxydopamine injections on glutamate decarboxylase, preproenkephalin and dopamine D2 receptor mRNAs in the adult rat striatum. *Brain Res.* **621,** 249–259.
141. Sivam, S. P. (1991) Dopamine dependent decrease in enkephalin and substance P levels in basal ganglia regions of postmortem Parkinsonian brains. *Neuropeptides* **18,** 201–207.
142. Krause, J. E., Chirgwin, J. M., Carter, M. S., Xu, Z. S., and Hershey, A. D. (1987) Three rat preprotachykinin mRNAs encode the neuropeptides substance P and neurokinin A. *Proc. Natl. Acad. Sci. USA* **84,** 881–885.

143. Brownstein, M. J., Mroz, E. A., Kizer, J. S., Palkovits, M., and Leeman, S. E. (1976) Regional distribution of substance P in the brain of the rat. *Brain Res.* **116,** 299–305.
144. Bannon, M. J., Haverstick, D. M., Shibata, K., and Poosch, M. S. (1991) Preprotachykinin gene expression in the forebrain: regulation by dopamine. *Ann. NY Acad. Sci.* **632,** 31–37.
145. Gerfen, C. R. (1992) The neostriatal mosaic: multiple levels of compartmental organization in the basal ganglia. *Annu. Rev. Neurosci.* **15,** 285–320.
146. Bren'e, S., Lindefors, N., and Persson, H. (1992) Midbrain dopamine neurons regulate preprotachykinin-A mRNA expression in the rat forebrain during development. *Brain Res. Mol. Brain Res.* **14,** 13–19.
147. Oblin, A. and Zivkovic, B. (1991) Tachykinins in the rat substantia nigra: effects of selective dopamine receptor antagonists. *Fundam. Clin. Pharmacol.* **5,** 129–138.
148. Kostrzewa, R. M., Gong, L., and Brus, R. (1993) Serotonin (5-HT) systems mediate dopamine (DA) receptor supersensitivity. *Acta Neurobiol. Exp. Warsz.* **53,** 31–41.
149. Walker, P. D., Ni, L., Riley, L. A., Jonakait, G. M., and Hart, R. P. (1991) Serotonin innervation affects SP biosynthesis in rat neostriatum. *Ann. NY Acad. Sci.* **632,** 485–487.
150. Civelli, O., Douglass, J., Goldstein, A., and Herbert, E. (1985) Sequence and expression of the rat prodynorphin gene. *Proc. Natl. Acad. Sci. USA* **82,** 4291–4295.
151. Bronstein, D. M., Ye, H., Pennypacker, K. R., Hudson, P. M., and Hong, J. S. (1994) Role of a 35 kDa fos-related antigen (FRA) in the long-term induction of striatal dynorphin expression in the 6-hydroxydopamine lesioned rat. *Brain Res. Mol. Brain Res.* **23,** 191–203.
152. Zhang, W. Q., Pennypacker, K. R., Ye, H., Merchenthaler, I. J., Grimes, L., Iadarola, M. J., and Hong, J. S. (1992) A 35 kDa Fos-related antigen is co-localized with substance P and dynorphin in striatal neurons. *Brain Res.* **577,** 312–317.
153. Carraway, R. and Leeman, S. E. (1979) The amino acid sequence of bovine hypothalamic substance P. Identity to substance P from colliculi and small intestine. *J. Biol. Chem.* **254,** 2944–2945.
154. Jennes, L., Stumpf, W. E., and Kalivas, P. W. (1982) Neurotensin: topographical distribution in rat brain by immunohistochemistry. *J. Comp. Neurol.* **210,** 211–224.
155. Zahm, D. S. and Heimer, L. (1988) Ventral striatopallidal parts of the basal ganglia in the rat: I. Neurochemical compartmentation as reflected by the distributions of neurotensin and substance P immunoreactivity. *J. Comp. Neurol.* **272,** 516–535.
156. Zoli, M., Cintra, A., Zini, I., Hersh, L. B., Gustafsson, J. A., Fuxe, K., and Agnati, L. F. (1990) Nerve cell clusters in dorsal striatum and nucleus accumbens of the male rat demonstrated by glucocorticoid receptor immunoreactivity. *J. Chem. Neuroanat.* **3,** 355–366.
157. Eggerman, K. W. and Zahm, D. S. (1988) Numbers of neurotensin-immunoreactive neurons selectively increased in rat ventral striatum following acute haloperidol administration. *Neuropeptides* **11,** 125–132.
158. Govoni, S., Hong, J. S., Yang, H. Y., and Costa, E. (1980) Increase of neurotensin content elicited by neuroleptics in nucleus accumbens. *J. Pharmacol. Exp. Ther.* **215,** 413–417.
159. Letter, A. A., Merchant, K., Gibb, J. W., and Hanson, G. R. (1987) Effect of methamphetamine on neurotensin concentrations in rat brain regions. *J. Pharmacol. Exp. Ther.* **241,** 443–447.

160. Merchant, K. M., Dobner, P. R., and Dorsa, D. M. (1992) Differential effects of haloperidol and clozapine on neurotensin gene transcription in rat neostriatum. *J. Neurosci.* **12,** 652–663.
161. Castel, M. N., Morino, P., Frey, P., Terenius, L., and Hokfelt, T. (1993) Immunohistochemical evidence for a neurotensin striatonigral pathway in the rat brain. *Neuroscience* **55,** 833–847.
162. Merchant, K. M. (1994) c-fos antisense oligonucleotide specifically attenuates haloperidol-induced increases in neurotensin/neuromedin N mRNA expression in rat dorsal striatum. *Mol. Cell. Neurosci.* **5,** 336–344.
163. Dobner, P. R., Kislauskis, E., and Bullock, B. P. (1992) Cooperative regulation of neurotensin/neuromedin N gene expression in PC12 cells involves AP-1 transcription factors. *Ann. NY Acad. Sci.* **668,** 17–29.
164. Vernier, P., Julien, J. F., Rataboul, P., Fourrier, O., Feuerstein, C., and Mallet, J. (1988) Similar time course changes in striatal levels of glutamic acid decarboxylase and proenkephalin mRNA following dopaminergic deafferentation in the rat. *J. Neurochem.* **51,** 1375–1380.
165. Chen, J. F. and Weiss, B. (1993) Irreversible blockade of D2 dopamine receptors by fluphenazine-N-mustard increases glutamic acid decarboxylase mRNA in rat striatum. *Neurosci. Lett.* **150,** 215–218.
166. Segovia, J., Tillakaratne, N. J., Whelan, K., Tobin, A. J., and Gale, K. (1990) Parallel increases in striatal glutamic acid decarboxylase activity and mRNA levels in rats with lesions of the nigrostriatal pathway. *Brain Res.* **529,** 345–348.
167. Stork, O., Hashimoto, T., and Obata, K. (1994) Haloperidol activates tyrosine hydroxylase gene-expression in the rat substantia nigra, pars reticulata. *Brain Res.* **633,** 213–222.
168. Buckland, P. R., O'Donovan, M. C., and McGuffin, P. (1992) Changes in dopa decarboxylase mRNA but not tyrosine hydroxylase mRNA levels in rat brain following antipsychotic treatment. *Psychopharmacology Berl.* **108,** 98–102.

CHAPTER 11

Dopamine Receptor-Mediated Gene Regulation in the Pituitary

James L. Roberts, Stuart C. Sealfon, and Jean Philippe Loeffler

1. Introduction

A common regulatory mechanism in the endocrine and neuroendocrine system is the coupling of secretion with the biosynthesis of a hormone. Factors that regulate hormone release also regulate its biosynthesis. This provides a cellular mechanism to ensure a ready supply of the protein or peptide. Thus, for any cell type in which secretion is regulated by dopamine via a dopamine receptor, it might be expected that dopamine would also have a significant effect on expression of the genes encoding the peptide/protein hormones themselves and/or those genes encoding the proteins involved in hormone production. Indeed, this linkage has been demonstrated in the first studies of regulation of prolactin and pro-opiomelanocortin (POMC) gene expression performed in model pituitary cell culture systems. Since these pituitary hormone-encoding genes were some of the first to be cloned in the early days of recombinant DNA technology, their regulation by dopamine has been well characterized but, surprisingly, some aspects of the underlying mechanisms have not been elucidated. In this chapter we discuss what is known about the expression of different dopamine receptor subtypes in the pituitary gland; their mechanisms of intracellular signaling; and processes by which the activation of these receptors interdicts the gene expression of two major pituitary peptide/protein hormone genes, prolactin and POMC.

The Dopamine Receptors Eds.: K. A. Neve and R. L. Neve
Humana Press Inc., Totowa, NJ

2. Dopamine Receptor Subtypes in the Pituitary

The predominant pituitary dopamine receptors have D2-like* pharmacological profiles. Using [^{3}H] SCH 23390 receptor autoradiography, minimal D1-like labeling is detected in the neural lobe of rat pituitary and no sites are present in the anterior or intermediate lobes *(1)*. In cultured pituitary cells, however, D1-like sites have been detected by autoradiography *(2)*. D2-like binding sites are densely distributed in the intermediate lobe and are present in scattered foci in the anterior lobe *(1,3)*. Not only are D2-like receptors expressed in melanocytes, but electrophysiological studies also have demonstrated the presence of these receptors in stellate glial cells of the intermediate lobe *(4)*. Among the cloned D2-like receptors, D_{2S}/D_{2L} and D_4 are expressed in the anterior pituitary *(5)*, and both of the D_2 mRNA splice variants are present in anterior and intermediate lobes of the rat (*6*; *see* Fig. 1). D_3 receptor mRNA is not detectable in RNA extracted from pituitary *(5,7)*. 7-OH-dipropylaminotetralin (DPAT) has been proposed as a relatively selective ligand for the dopamine D_3 receptor *(8)*, although its specificity is controversial (*see* refs. *9* and *10*). Thus, whereas the level of 7-OH-DPAT binding sites in the pituitary is about 10% that found in the olfactory tubercle by receptor autoradiography *(8)*, this signal may not represent D_3 receptor sites. A pharmacological D1-like receptor, not linked to adenylate cyclase, has been detected on lactotroph cells *(2)*. Although mRNA for neither of the D1-like receptors cloned to date, D_1 or D_5, is detected by Northern blot analysis of pituitary RNA *(11,12)*, D_5 receptor mRNA has been identified in the anterior pituitary by polymerase chain reaction *(13)*.

Pharmacological profiles have suggested the presence of more than one D2-like receptor in the anterior pituitary *(14,15)*. A recent report utilizing antisense oligonucleotide hybrid arrest in primary cultures of anterior pituitary suggests that a D2-like receptor distinct from the D_{2L} or D_{2S} can mediate dopaminergic inhibition of prolactin release in lactotrophs. Culturing cells in the presence of an oligonucleotide complementary to the initiation region of the D_2 cDNA markedly reduced [^{3}H]spiroperidol binding and eliminated dopaminergic inhibition of prolactin biosynthesis and adenylate cyclase. However, bromocriptine exposure was still able to inhibit prolactin release, and sulpiride antagonized this effect *(5)*. It is possible that these results reflect D_4 receptor-mediated inhibition of prolactin release.

The relative abundance in the pituitary of the two D_2 receptor isoforms has been investigated in several studies *(6,16–19)*. Using a subtype-specific

*As detailed in the Preface, the terms D1 or D1-like and D2 or D2-like are used to refer to the subfamilies of DA receptors, or used when the genetic subtype (D_1 or D_5; D_2, D_3, or D_4) is uncertain.

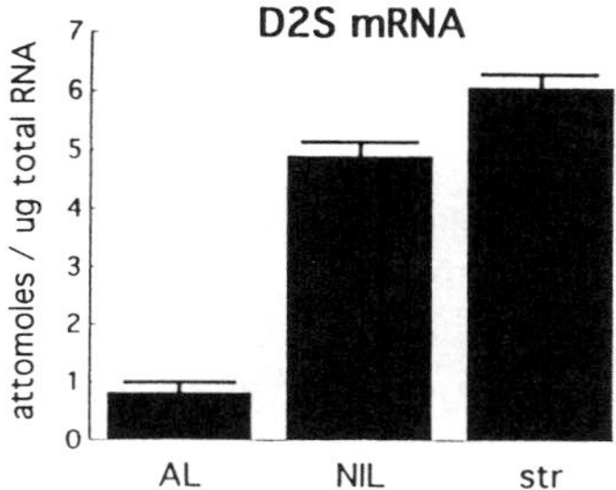

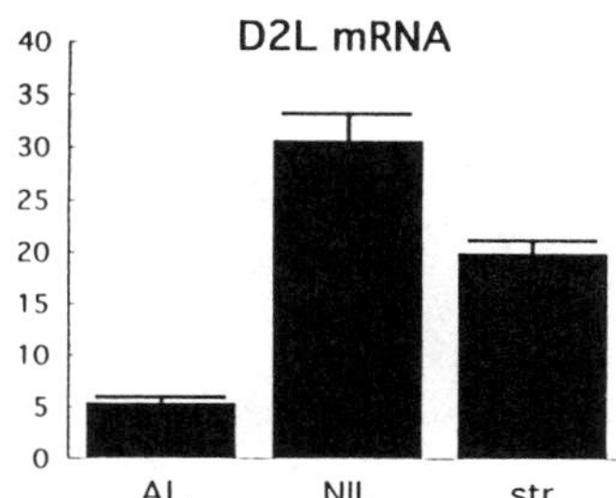

Fig. 1. Levels of D_{2S} and D_{2L} mRNAs in the rat anterior pituitary and neurointermediate lobe determined by solution hybridization/nuclease protection assay *(6)*.

solution hybridization/nuclease protection assay, we found that the level of expression of D_2 receptor mRNA is approx 5 atamol/µg total RNA in the rat anterior pituitary and 40 atamol/µg RNA in the neurointermediate lobe. In both tissues, the D_{2L} mRNA is more abundant than the D_{2S}, being sixfold higher in the anterior lobe and nearly eightfold higher in the neurointermediate lobe (Fig. 1; *6*). An *in situ* hybridization study of the intermediate lobe suggested that the relative amount of each D_2 isoform varies significantly in individual melanotropes *(20)*.

The D_{2L} isoform is more abundant in all brain regions as well as in the pituitaries of male and nonlactating female rats *(6)*. It is notable, therefore, that the only tissue in which the D_{2S} mRNA isoform has been found to be more abundant is in lactotrophs obtained from lactating rats *(21)*. Subpopulations of lactotrophs, which vary in their ratio of D_{2L} to D_{2S} mRNA *(21)* and in their physiological responses to dopamine *(22)*, have been characterized by sedimentation. In mixed anterior pituitary and in low density lactotrophs, dopamine causes a dramatic decrease of prolactin mRNA levels. However, a population of high-density lactotrophs has been identified in which dopamine induces an increase in prolactin mRNA expression *(22)*. The electrophysiological responses to dopamine also have been reported to vary in gradient-separated subpopulations of lactotrophs *(23)*. Similarly, these two subpopulations of lactotrophs have been found to vary in their D_{2S}/D_{2L} mRNA ratios. Whereas lactotrophs from less dense fractions have a high ratio of D_{2S}/D_{2L} mRNA, dense fraction lactotrophs have nearly even abundance of the two splice forms *(21)*.

3. Regulation of Pituitary Dopamine Receptors

The number of D2-like receptors in the rat anterior lobe is regulated by estrogen and during the estrous cycle. There is disagreement in the literature on the precise regulatory changes, with one study using equilibrium

[^{3}H]spiperone binding reporting an increase in D2 receptors on the afternoon of proestrus *(24)*, and others reporting that the lowest levels of D2-like binding occur during proestrus *(25,26)*. D_2 receptor mRNA levels also vary during the estrous cycle, being lowest during estrus and highest in diestrus II *(27)*. In contrast, no significant changes in D2-like binding sites were found by autoradiography in either the intermediate or neural lobes during the cycle *(26)*. In vivo administration of 17β-estradiol to ovariectomized rats induced a significant decrease in the number of D2-like receptors in the anterior pituitary at 2–4 d *(28)*, as did direct exposure of lactotrophs to estradiol in vitro *(29)*. Chronic 17β-estradiol increased the density of [^{3}H]spiperone binding sites *(30)* and the number of low-affinity domperidone sites *(31)*. In rats transplanted with dopamine receptor-expressing tumors, estradiol administration induced a marked decrease in the number of D2-like binding sites *(32)*. Chronic estradiol administration to ovariectomized rats results in an increase in the concentration of D_2 mRNA in the anterior pituitary *(33)*.

Removal of pups from lactating mothers decreases D2-like binding in the anterior lobe, although the changes were significant in only one study *(26,34)*. Short-term haloperidol administration in vivo was found to increase the D_2 receptor mRNA levels in the neurointermediate lobe of the rat without inducing any change in the anterior lobe *(20,35,36)*, whereas chronic administration was reported to induce increases of D_2 receptor mRNA and of the D_{2S} mRNA isoform *(35,37)*. Although bromocriptine has no effect on D_2 receptor mRNA levels in vivo *(36)*, dopamine exposure leads to an increase in D_2 receptor mRNA in pituitary cells in culture *(38)*. The D2-like receptor in the rat intermediate lobe has been found to be regulated by intraventricular administration of neurotensin, which induces a significant decrease in binding as determined by receptor autoradiography *(39)*.

The responsiveness of pituitary dopamine receptors to dopamine has also been found to show marked variation during the estrous cycle and with estrogen administration. In a perfusion study of the effect of dopamine on prolactin secretion, lactotrophs were insensitive to the inhibitory effect of dopamine during proestrus and diestrus I, and sensitive during estrus and diestrus II *(40)*. Direct exposure to estradiol also alters the sensitivity of the lactotroph to dopamine *(41,42)*. Chronic estrogen exposure induces a partial uncoupling of dopaminergic inhibition of adenylate cyclase in the anterior pituitary *(43,44)*.

An investigation of the regulation of the D_2 receptor mRNA splice forms in lactotroph populations from lactating rats, separated on a density gradient, found that the relative abundance of the splice forms was regulated by gonadal hormones *(21)*. When less dense fraction lactotrophs were treated with progesterone, the relative concentration of D_{2S} was significantly increased. Progesterone treatment induced an increase in the relative concentration of the D_{2L} splice form, similar to that seen in the male or nonlactating female anterior pituitary. Estradiol did not alter the splice form ratio but did lead to a reduction in D_2 receptor mRNA. This regulation by steroids has been proposed to account for the different relative abundance of the two forms observed in different reproductive states *(21)*.

4. Signal Transduction of Pituitary Dopamine Receptors

The lactotroph dopamine D2-like receptor has been found to couple to a plethora of signal transduction pathways. Activation of the receptor inhibits the release of prolactin and diminishes adenylate cyclase stimulation *(45–47)*. Both the inhibition of adenylate cyclase stimulation and of prolactin release are mediated by a G_i-type pertussis toxin-sensitive G protein *(45,48–51)*. Some researchers have suggested that activation of the pituitary lactotroph D2 receptor inhibits phosphoinositol hydrolysis *(48,51,52)*, whereas other workers have proposed that D2 receptors are not direct inhibitors of phosphoinositol metabolism *(53,54)*. Stimulation of the D2 receptor has also been reported to diminish arachidonate release *(55)*. In the growth hormone (GH) tumor line, dopaminergic agonists were found to promote phosphotyrosine phosphatase activity, although it is not known if this is a proximal signal transduction pathway of the receptor *(56)*.

The lactotroph dopamine receptor also modulates the activity of a variety of ion channels, including activation of potassium currents *(57–60)* and inhibition of distinct L- and T-type voltage-activated calcium currents *(61)*. Patch-clamp recording studies using antibodies to G protein subtypes suggest that the dopamine receptor-mediated inhibition of calcium channels and the coupling to potassium currents are mediated by distinct G proteins *(62)*. A recent study *(63)* found that D2 receptor activation had little effect on calcium influx in lactotrophs, and these authors proposed that dopaminergic inhibition of calcium currents is not a major contributor to the dopaminergic inhibition of prolactin secretion. Activation of dopamine D2-like receptors on the melanotrophs in the rat intermediate lobe has been shown to inhibit preferentially N- and T-type calcium currents, while not affecting L-type calcium currents *(64)*.

Since both D_2 and D_4 mRNA are present in the pituitary, some D2-agonist-induced signaling events may actually be mediated by the D_4 receptor. Functional studies suggest the presence of distinguishable D2-like receptors which may differ in their pharmacology and coupling to signal transduction *(5,13)*. Both the cloned D_2 receptor and the cloned D_4 receptor have been shown to inhibit L-type calcium channel activity when the receptors are expressed in the pituitary-derived GH_4C_1 cells *(65)*.

5. Regulation of Specific Gene Expression

As discussed, the two major cell types of the pituitary that express dopamine receptors are the lactotroph and the melanotroph; and as such, the data are most extensive on dopamine regulation of prolactin and POMC in these cell types. These studies were performed first in intact animals; to identify directly the site of action of dopamine agonists or antagonists, subsequent studies were done in primary culture. Interestingly, since cell lines derived from a lactotroph or melanotroph that also express the dopamine receptor were not available until very recently, the initial flurry of activities surrounding an understanding of dopamine regulation of pituitary gene expression was bypassed in reference to these two cell types and supplanted by analyses of direct regulation of prolactin and POMC gene expression by pharmacological second messenger mimicking agents. The recent cloning and expression of the dopamine receptors has allowed researchers to express them in lactotroph or melanotroph cell lines, and is now leading to a better understanding of how dopamine directly regulates gene expression.

Prolactin was one of the first hormonally regulated genes studied in the pituitary, and the drugs used for this analysis were the dopamine receptor agonist ergocryptine and antagonist haloperidol. Maurer *(66)* showed that ergocryptine is capable of inhibiting prolactin synthesis and showed that this is the result of a parallel decrease in prolactin mRNA. Subsequently, Maurer *(67)* was able to show that this dopamine agonist-induced loss of prolactin mRNA results from a decrease in transcription of the prolactin gene, and that this effect can be overcome by adding back cyclic adenosine 3',5' monophosphate (cAMP). This set of experiments led to the concept that dopamine regulates gene expression through its ability to modulate the level of cAMP being produced, and that it is the cAMP-activated processes that are actually involved in regulation of gene expression. Subsequent to this observation, research on regulation of prolactin gene expression shifted primarily to the prolactin-expressing rat GH cell lines, which do not naturally express the dopamine receptor. Several studies have been performed to identify the cAMP-regulated elements of the prolactin gene promoter *(68,69)* and the elements mediating tissue-specific gene expression *(70–72)*. Interestingly, sequence analysis of the promoter shows that it does not contain the typically characterized cAMP response element (CRE). Taken together, these studies suggest that cAMP is probably acting through short enhancer elements located in various regions of the proximal promoter region that bind the pituitary transcription factor Pit1.

With the cloning and expression of the dopamine receptor-encoding cDNA, it is now possible to study directly in the prolactin-expressing cell lines the regulation of prolactin expression by dopamine receptor-mediated events. McChesney et al. *(73)* used a transient transfection system to express both the D_{2L} and D_{2S} receptor in GH_3 cells and were able to show negative, dopamine-specific regulation of a prolactin promoter reporter construct expressing 2 kb of the rat prolactin gene. Elsholtz et al. *(74)* reported a more detailed study utilizing a GH_4 cell line stably transfected with the rat dopamine D_{2S} receptor, GH_4ZR_7 *(75)*. This study was able to show that, like positive regulation by cAMP, negative regulation by dopamine can be localized to the Pit1 elements of the proximal prolactin promoter region. A heterologous gene construct was used to show that the Pit1 binding domain from the rat GH gene is also sufficient to confer negative dopamine regulation in this cell line, strongly suggesting that the dopamine-mediated regulation can be attributed to this particular DNA element. Interestingly, the same study showed that the Pit1 gene promoter is negatively regulated by dopamine, and that this regulation is independent of the two CRE elements present in the Pit1 gene proximal promoter region. Thus, this observation suggests that the negative regulation by dopamine of the prolactin gene through the Pit1 elements may be in part owing to a negative regulation of Pit1 gene expression by dopamine.

Whereas cAMP may be a major mediator of dopamine receptor-regulated prolactin gene expression, some studies have suggested that other effects of the receptor, such as the modulation of ion channel activity, also may be involved in regulation of gene expression. Elsholtz et al. *(74)* were able to show that elevated concentrations of potassium chloride (KCl) that block dopamine-induced hyperpolarization of the GH_4ZR_7 cells and entry of calcium *(53)* can completely block dopamine inhibition of prolactin gene transcription and yet have no effect on production of cAMP. Thus, in the absence of the action of the ion channel activation/inactivation mediated by dopamine receptor modulation, the lowered level of cAMP is not capable of decreasing transcription of the prolactin gene. It appears, then, that proper regulation of prolactin gene expression via the dopamine receptors requires all of the signal transducing events mediated by the receptor, and not simply the cAMP modulation. A candidate for this type of regulation would be calcium ions, whose levels are suppressed by dopamine receptor activation. Elevations in intracellular calcium ion concentrations stimulate prolactin mRNA levels, due to enhanced transcription of the prolactin gene *(76,77)*. Promoter analysis shows that the calcium regulation of transcription is mediated by multiple elements in the distal and proximal promoters, and possibly through the Pit1 elements *(78,79)*. This concept is supported by the observations of Lew et al. *(80)* that dopamine treatment of a dopamine receptor expressing Ltk^- cell line,

LZR1, causes an elevation in intracellular calcium ions and also stimulates prolactin promoter activity, when the heterologous promoter is cotransfected with a Pit1 expression plasmid.

The intermediate lobe melanotroph is also subject to the regulation of gene expression by dopamine. It is unique from the pituitary perspective in that it is directly innervated by dopaminergic neuronal terminals from the tuberohypophysial dopaminergic neurons of the hypothalamus. Whereas some synapse-like structures in the intermediate lobe have been identified for the dopamine terminals, in general they release dopamine into the extracellular fluids surrounding the melanotrophs. As such, this has made the melanotroph an excellent cell to study possible unique aspects of regulation of gene expression by neurotransmitters, in a manner analogous to that which occurs in the brain. This work is also aided by the fact that the melanotroph appears to be a homogeneous population of POMC-expressing cells that respond in a concerted fashion to modulation by dopamine agonist and/or antagonist, similar to the lactotroph population. Although there is some indication that in the basal state the cells may exist in two different forms *(81,82)*, treatment with a dopamine receptor agonist or antagonist appears to put the cells in one state or the other. Thus, although the intermediate lobe is a small piece of tissue with only a few hundred thousand cells in the rodent intermediate lobe, the homogeneous nature, the ease of dissection, and the high level of gene expression of the POMC system have made it an ideal system for study of dopaminergic gene regulation.

As discussed, the dopamine receptor subtype present in melanotroph appears to be exclusively the D_2 type, and as such is negatively coupled to the production of cAMP. Since there is also adrenergic innervation of the intermediate lobe and expression of beta adrenergic receptors in the melanotrophs, it appears that once again the dopamine D_2 receptor is opposing the actions of this cAMP stimulatory system. Since the POMC gene system is known to be positively regulated by cAMP, one would expect negative regulation of POMC gene expression by dopamine agonists and a positive regulation of expression by dopamine antagonists. Dopamine also has been shown to be inhibitory to POMC gene expression in this tissue. Hollt et al. *(83)* found that treatment of male rats with haloperidol caused a reversible time-dependent increase in translatable POMC mRNA levels after 3 wk. Chen et al. *(84)* found that haloperidol elicited a time-dependent increase which required 6 h to be detected (also twofold above control), and that elevations in POMC mRNA continued for up to 7 d of hormone treatment to levels of 700% of the control value. Pritchett and Roberts *(85)* showed similar results by Northern blot analysis after 3 d of treatment. In several studies, dopamine agonists such as ergocriptine or CB 154 were shown to have two- to fourfold inhibitory effects on POMC mRNA levels in the melanotrophs *(84–87)*. Using the *in situ* hybridization technique, Chronwall et al. *(82)* reported that essentially all of the melanotrophs in the intermediate pituitary were stimulated by haloperidol treatment and inhibited by bromocriptine treatment, again in a time-dependent fashion. Thus, these studies show that stimulation of melanotroph POMC peptide secretion by blocking the inhibitory effect of the endogenous dopamine with haloperidol, or further suppressing release of POMC peptide by using a dopamine agonist, causes parallel changes in POMC peptide mRNA.

As was observed with prolactin, these dopaminergic-induced alterations in POMC mRNA levels are mediated to a large extent by changes in gene transcription. Nuclear run-on transcription assays showed a three- to fourfold stimulatory effect of haloperidol on melanotroph POMC gene transcription *(85)*. Again, it appears that the modulation of POMC gene transcription by dopamine, like the prolactin gene, is secondary to changes in cAMP and calcium ions. POMC mRNA levels in melanotrophs are positively regulated by cAMP *(88–91)*, and these changes are mediated by parallel changes in POMC gene transcription *(92–94)*. The POMC gene promoter also does not contain any of the well-characterized cAMP regulatory elements, and the regulation via this second messenger system appears to be mediated by multiple DNA elements *(94–96)*, including an AP1 element in the first exon *(97,98)*. POMC mRNA levels and gene transcription can also be stimulated by elevations in calcium ion levels *(89,92)*. Thus, cAMP may not be the only second messenger system mediating dopamine's effects on POMC transcription.

Recent advances in receptor pharmacology also provide a new perspective on earlier studies of regulation of pituitary gene expression by dopamine. We had always been puzzled by the observation that 30–60 min treatment with haloperidol has a stimulatory effect on POMC transcription in neurointermediate lobe primary cultures in the absence of any dopamine (Eberwine, Blum, and Roberts, unpublished observations). At the time of these experiments, an effect of antagonist in the absence of agonist was inexplicable. However, the recent discovery of the spontaneous activity of receptors and of inverse agonists, antagonists that turn off this spontaneous activation *(99)*, now provides a plausible explanation for this. The dopamine receptors could have some spontaneous inhibitory effect on POMC gene transcription that is reversed by haloperidol stabilizing the inactive state of the receptor.

6. Conclusion

Most of the investigations of the dopaminergic regulation of the prolactin and POMC genes were performed before the alternative signal transduction pathways of dopamine were elucidated. The robust suppression of POMC and prolactin gene expression by dopamine in vivo and in primary cultures seemed inconsistent with the modest effect on basal cAMP levels. Furthermore, the subsequent later promoter analysis of these genes failed to demonstrate the preponderance of CRE elements expected in a gene that are so dependent on cAMP for regulated gene expression. As suggested by the Elsholtz et al. *(74)* study, it seems likely that a major suppressive effect of dopamine on pituitary gene expression is probably mediated not only via altered cAMP levels, but also by the effect of dopamine on other signaling systems, in particular the modulation of channel activity. Thus the advances in signal transduction and molecular biology of the dopamine receptor system provide the framework for a more satisfying understanding of the processes underlying pituitary gene regulation by dopamine.

References

1. Mansour, A., Meador, W. J., Bunzow, J. R., Civelli, O., Akil, H., and Watson, S. J. (1990) Localization of dopamine D2 receptor mRNA and D1 and D2 receptor binding in the rat brain and pituitary: an in situ hybridization-receptor autoradiographic analysis. *J. Neurosci.* **10,** 2587–2600.
2. Schoors, D. F., Vauquelin, G. P., De Vos, H., Smets, G., Velkeniers, B., Vanhaelst, L., and Dupont, A. G. (1991) Identification of a D1 dopamine receptor, not linked to adenylate cyclase, on lactotroph cells. *Br. J. Pharmacol.* **103,** 1928–1934.
3. Goldsmith, P. C., Cronin, M. J., and Weiner, R. I. (1979) Dopamine receptor sites in the anterior pituitary. *J. Histochem. Cytochem.* **27,** 1205–1207.
4. Mudrick-Donnon, L. A., Williams, P. J., Pittman, Q. J., and Mac Vicar, B. A. (1993) Postsynaptic potentials mediated by GABA and dopamine evoked in stellate glial cells of the pituitary pars intermedia. *J. Neurosci.* **13,** 4660–4668.
5. Valerio, A., Alberici, A., Inti, C., Spano, P., and Memo, M. (1994) Antisense strategy unravels a dopamine receptor distinct from the D2 subtype, uncoupled with adenylyl cyclase, inhibiting prolactin release from rat pituitary cells. *J. Neurochem.* **62,** 1260–1266.
6. Snyder, L. A., Roberts, J. L., and Sealfon, S. C. (1991) Distribution of dopamine D2 receptor mRNA splice variants in the rat by solution hybridization/protection assay. *Neurosci. Lett.* **122,** 37–40.
7. Sokoloff, P., Giros, B., Martres, M. P., Bouthenet, M. L., and Schwartz, J. C. (1990) Molecular cloning and characterization of a novel dopamine receptor (D3) as a target for neuroleptics. *Nature* **347,** 146–151.
8. Levesque, D., Diaz, J., Pilon, C., Martres, M. P., Giros, B., Souil, E., Schott, D., Morgat, J. L., Schwartz, J. C., and Sokoloff, P. (1992) Identification, characterization, and localization of the dopamine D3 receptor in rat brain using 7-[^{3}H]hydroxy-N,N-di-n-propyl-2-aminotetralin. *Proc. Natl. Acad. Sci. USA* **89,** 8155–8159.
9. Freedman, J. E., Waszczak, B. L., Cox, R. F., Liu, J. C., and Greif, G. J. (1994) The dopamine D3 receptor and 7-OH-DPAT [letter]. *Trends Pharmacol. Sci.* **15,** 173,174.
10. Large, C. H. and Stubbs, C. M. (1994) The dopamine D3 receptor: Chinese hamsters or Chinese whispers? [letter]. *Trends Pharmacol. Sci.* **15,** 46,47.

11. Monsma, F. J., Mahan, L. C., McVittie, L. D., Gerfen, C. R., and Sibley, D. R. (1990) Molecular cloning and expression of a D1 dopamine receptor linked to adenylyl cyclase activation. *Proc. Natl. Acad. Sci. USA* **87,** 6723–6727.
12. Sunahara, R. K., Niznik, H. B., Weiner, D. M., Stormann, T. M., Brann, M. R., Kennedy, J. L., Gelernter, J. E., Rozmahel, R., Yang, Y., Israel, Y., Seeman, P., and O'Dowd, B. F. (1990) Human dopamine D1 receptor encoded by an intronless gene on chromosome 5. *Nature* **347,** 80–83.
13. Porter, T. E., Grandy, D., Bunzow, J., Wiles, C. D., Civelli, O., and Frawley, L. S. (1994) Evidence that stimulatory dopamine receptors may be involved in the regulation of prolactin secretion. *Endocrinology* **134,** 1263–1268.
14. Pizzi, M., Valerio, A., Benarese, M., Missale, C., Carruba, M., Memo, M., and Spano, P. F. (1990) Selective stimulation of a subtype of dopamine D-2 receptor by the azeprine derivative BHT 920 in rat pituitary. *Mol. Neuropharmacol.* **1,** 37–42.
15. Memo, M., Pizzi, M., Belloni, M., Benarese, M., and Spano, P. (1992) Activation of dopamine D2 receptors linked to voltage-sensitive potassium channels reduces forskolin-induced cyclic AMP formation in rat pituitary cells. *J. Neurochem.* **59,** 1829–1835.
16. DalToso, R., Sommer, B., Ewert, M., Herb, A., Pritchett, D. B., Bach, A., Shivers, B. D., and Seeburg, P. H. (1989) The dopamine D2 receptor: two molecular forms generated by alternative splicing. *EMBO J.* **8,** 4025–4034.
17. Giros, B., Sokoloff, P., Martres, M. P., Riou, J. F., Emorine, L. J., and Schwartz, J. C. (1989) Alternative splicing directs the expression of two D2 dopamine receptor isoforms. *Nature* **342,** 923–926.
18. Montmayeur, J. P., Bausero, P., Amlaiky, N., Maroteaux, L., Hen, R., and Borrelli, E. (1991) Differential expression of the mouse D2 dopamine receptor isoforms. *FEBS Lett.* **278,** 239–243.
19. O'Malley, K. L., Mack, K. J., Gandelman, K. Y., and Todd, R. D. (1990) Organization and expression of the rat D2A receptor gene: identification of alternative transcripts and a variant donor splice site. *Biochemistry* **29,** 1367–1371.
20. Chronwall, B. M., Dickerson, D. S., Huerter, B. S., Sibley, D. R., and Millington, W. R. (1994) Regulation of heterogeneity in D2 dopamine receptor gene expression among individual melanotropes in the rat pituitary intermediate lobe. *Mol. Cell. Neurosci.* **5,** 35–45.
21. Kukstas, L. A., Domec, C., Bascles, L., Bonnet, J., Verrier, D., Israel, J.-M., and Vincent, J.-D. (1991) Different expression of the two dopaminergic D2 receptors, D2415 and D2444, in two types of lactotroph each characterized by their response to dopamine, and modification of expression by sex steroids. *Endocrinology* **129,** 1101–1103.
22. Kazemzadeh, M., Velkeniers, B., Herregodts, P., Collumbien, R., Finne, E., Derde, M. P., Vanhaelst, L., and Hooghe-Peters, E. L. (1992) Differential dopamine-induced prolactin mRNA levels in various prolactin-secreting cell (sub)populations. *J. Endocrinol.* **132,** 401–409.
23. Lledo, P. M., Guerineau, N., Mollard, P., Vincent, J. D., and Israel, J. M. (1991) Physiological characterization of two functional states in subpopulations of prolactin cells from lactating rats. *J. Physiol.* **437,** 477–494.
24. Heiman, M. L. and Ben-Jonathan, N. (1982) Dopaminergic receptors in the rat anterior pituitary change during the estrous cycle. *Endocrinology* **111,** 37–41.

25. Pasqualini, C., Lenoir, V., El Abed, A., and Kerdelhue, B. (1984) Anterior pituitary dopamine receptors during the rat estrous cycle. A detailed analysis of proestrus changes. *Neuroendocrinology* **38,** 39–44.
26. Pazos, A., Stoeckel, M. E., Hindelang, C., and Palacios, J. M. (1985) Autoradiographic studies on dopamine D2 receptors in rat pituitary: influence of hormonal states. *Neurosci. Lett.* **59,** 1–7.
27. Zabavnik, J., Wu, W. X., Eidne, K. A., and McNeilly, A. S. (1993) Dopamine D2 receptor mRNA in the pituitary during the oestrous cycle, pregnancy and lactation in the rat. *Mol. Cell. Endocrinol.* **95,** 121–128.
28. Ali, S. F. and Peck, E. J. (1985) Modulation of anterior pituitary dopamine receptors by estradiol 17-β: dose–response relationship. *J. Neurosci. Res.* **13,** 497–507.
29. Pasqualini, C., Bojda, F., and Kerdelhue, B. (1986) Direct effect of estradiol on the number of dopamine receptors in the anterior pituitary of ovariectomized rats. *Endocrinology* **119,** 2484–2489.
30. Di Paolo, R. and Falardeau, P. (1985) Modulation of brain and pituitary dopamine receptors by estrogens and prolactin. *Prog. Neuro-Psychopharmacol.* **9,** 473–480.
31. Bression, D., Brandi, A. M., LeDafniet, M., Cesselin, F., Hamon, M., Martinet, M., Kerdelhue, B., and Peillon, F. (1983) Modifications of the high and low affinity pituitary domperidone-binding sites in chronic estrogenized rats. *Endocrinology* **113,** 1799–1805.
32. Andre, J., Marchisio, A. M., Morel, Y., and Collu, R. (1982) Dopamine receptors in the rat pituitary and the transplantable pituitary tumor MtTF4: effect of chronic treatment with oestradiol. *Biochem. Biophys. Res. Commun.* **106,** 229–235.
33. Levesque, D., Gagne, B., Barden, N., and Di Paolo, T. (1992) Chronic estradiol treatment increases anterior pituitary but not striatal D2 dopamine receptor mRNA levels in rats. *Neurosci. Lett.* **140,** 5–8.
34. Heiman, M. L. and Ben-Jonathan, N. (1982) Rat anterior pituitary dopaminergic receptors are regulated by estradiol and during lactation. *Endocrinology* **111,** 1057–1060.
35. Arnauld, E., Arsaut, J., and Demotes-Mainard, J. (1991) Differential plasticity of the dopaminergic D2 receptor mRNA isoforms under haloperidol treatment, as evidenced by in situ hybridization in rat anterior pituitary. *Neurosci. Lett.* **130,** 12–16.
36. Autelitano, D. J., Snyder, L., Sealfon, S. C., and Roberts, J. L. (1989) Dopamine D2-receptor messenger RNA is differentially regulated by dopaminergic agents in rat anterior and neurointermediate pituitary. *Mol. Cell. Endo.* **67,** 101–105.
37. Jaber, M., Tison, F., Fournier, M. C., and Bloch, B. (1994) Differential influence of haloperidol and sulpiride on dopamine receptors and peptide mRNA levels in the rat striatum and pituitary. *Brain Res. Mol. Brain Res.* **23,** 14–20.
38. Johnston, J. M., Wood, D. F., Read, S., and Johnston, D. G. (1993) Dopamine regulates D2 receptor gene expression in normal but not in tumorous rat pituitary cells. *Mol. Cell. Endocrinol.* **92,** 63–68.
39. von Euler, G., Meister, B., Hokfelt, T., Eneroth, P., and Fuxe, K. (1990) Intraventricular injection of neurotensin reduces dopamine D2 agonist binding in rat forebrain and intermediate lobe of the pituitary gland. Relationship to serum hormone levels and nerve terminal coexistence. *Brain Res.* **531,** 253–262.
40. Brandi, A. M., Joannidis, S., Peillon, F., and Joubert, D. (1990) Changes of prolactin response to dopamine during the rat estrous cycle. *Neuroendocrinology* **51,** 449–454.

41. Givuere, V., Meunier, H., Veilleux, R., and Labrie, F. (1992) Direct effects of sex steroids on prolactin release at the anterior pituitary level: interactions with dopamine, thyrotropin-releasing hormone, and isobutylmethylxanthine. *Endocrinology* **111,** 857–862.
42. Nansel, D. D., Gudelsky, G. A., Reymond, M. J., and Porter, J. C. (1981) Estrogen alters the responsiveness of the anterior pituitary gland to the actions of dopamine on lysosomal enzyme activity and prolactin release. *Endocrinology* **108,** 903–907.
43. Borgundvaag, B. and George, S. R. (1985) Dopamine inhibition of anterior pituitary adenylate cyclase is mediated through the high-affinity state of the D2 receptor. *Life Sci.* **37,** 379–386.
44. Munemura, M., Agui, T., and Sibley, D. R. (1989) Chronic estrogen treatment promotes a functional uncoupling of the D2 dopamine receptor in rat anterior pituitary gland. *Endocrinology* **124,** 346–355.
45. Cronin, M. J., Myers, G. A., Mac Leod, R. M., and Hewlett, E. L. (1983) Pertussis toxin uncouples dopamine agonist inhibition of prolactin release. *Am. J. Physiol.* **244,** 499–504.
46. De Camilli, P., Macconi, D., and Spada, A. (1979) Dopamine inhibits adenylate cyclase in human prolactin-secreting pituitary adenomas. *Nature* **278,** 252–254.
47. Enjalbert, A. and Bockaert, J. (1983) Pharmacological characterization of the D2 dopamine receptor negatively coupled with adenylate cyclase in rat anterior pituitary. *Mol. Pharmacol.* **23,** 576–584.
48. Enjalbert, A., Guilion, G., Mouiliac, B., Audinot, V., Rasolonjanahary, R., Kordon, C., and Bockaert, J. (1990) Dual mechanisms of inhibition by dopamine of basal and thyrotropin-releasing hormone-stimulated inositol phosphate production in anterior pituitary cells. *J. Biol. Chem.* **265,** 18,816–18,822.
49. Musset, F., Bertrand, P., Kordon, C., and Enjalbert, A. (1990) Differential coupling with pertussis toxin-sensitive G proteins of dopamine and somatostatin receptors involved in regulation of adenohypophyseal secretion. *Mol. Cell. Endocrinol.* **73,** 1–10.
50. Senogles, S. E., Amlaiky, N., Falardeau, P., and Caron, M. G. (1988) Purification and characterization of the D2-dopamine receptor from bovine anterior pituitary. *J. Biol. Chem.* **263,** 18,996–19,002.
51. Simmonds, S. H. and Strange, P. G. (1985) Inhibition of inositol phospholipid breakdown by D2 dopamine receptors in dissociated bovine anterior pituitary cells. *Neurosci. Lett.* **60,** 267–272.
52. Jarvis, W. D., Judd, A. M., and Mac Leod, R. M. (1988) Attenuation of anterior pituitary phosphoinositide phosphorylase activity by the D2 dopamine receptor. *Endocrinology* **123,** 2793–2799.
53. Vallar, L. and Meldolesi, J. (1989) Mechanisms of signal transduction at the dopamine D2 receptor. *Trends Pharmacol. Sci.* **10,** 74–77.
54. Vallar, L., Vicentini, L. M., and Meldolesi, J. (1988) Inhibition of inositol phosphate production is a late, Ca^{2+}-dependent effect of D2 dopaminergic receptor activation in rat lactotroph cells. *J. Biol. Chem.* **263,** 10,127–10,134.
55. Canonico, P. L. (1989) D-2 dopamine receptor activation reduces free [^{3}H] arachidonate release induced by hypophysiotropic peptides in anterior pituitary cells. *Endocrinology* **125,** 1180–1186.

56. Florio, T., Pan, M. G., Newman, B., Hershberger, R. E., Civelli, O., and Stork, P. J. (1992) Dopaminergic inhibition of DNA synthesis in pituitary tumor cells is associated with phosphotyrosine phosphatase activity. *J. Biol. Chem.* **267,** 24,169–24,172.
57. Einhorn, L. C., Gregerson, K. A., and Oxford, G. S. (1991) D2 dopamine receptor activation of potassium channels in identified rat lactotrophs: whole-cell and single-channel recording. *J. Neurosci.* **11,** 3727–3737.
58. Israel, J. M., Kirk, C., and Vincent, J. D. (1987) Electrophysiological responses to dopamine of rat hypophysial cells in lactotroph-enriched primary cultures. *J. Physiol.* **390,** 1–22.
59. Lledo, P. M., Legendre, P., Zhang, J., Israel, J. M., and Vincent, J. D. (1990) Effects of dopamine on voltage-dependent potassium currents in identified rat lactotroph cells. *Neuroendocrinology* **52,** 545–555.
60. Malgaroli, A., Vallar, L., Elahi, F. R., Pozzan, T., Spada, A., and Meldolesi, J. (1987) Dopamine inhibits cytosolic Ca^{2+} increases in rat lactotroph cells. Evidence of a dual mechanism of action. *J. Biol. Chem.* **262,** 13,920–13,927.
61. Lledo, P. M., Legendre, P., Israel, J. M., and Vincent, J. D. (1990) Dopamine inhibits two characterized voltage-dependent calcium currents in identified rat lactotroph cells. *Endocrinology* **127,** 990–1001.
62. Lledo, P. M., Homburger, V., Bockaert, J., and Vincent, J. D. (1992) Differential G protein-mediated coupling of D2 dopamine receptors to K^+ and Ca^{2+} currents in rat anterior pituitary cells. *Neuron* **8,** 455–463.
63. Rendt, J. and Oxford, G. S. (1994) Absence of coupling between D2 dopamine receptors and calcium channels in lactotrophs from cycling female rats. *Endocrinology* **135,** 501–508.
64. Williams, P. J., Mac Vicar, B. A., and Pittman, Q. J. (1990) Synaptic modulation by dopamine of calcium currents in rat pars intermedia. *J. Neurosci.* **10,** 757–763.
65. Seabrook, G. R., Knowles, M., Brown, N., Myers, J., Sinclair, H., Patel, S., Freedman, S. B., and McAllister, G. (1994) Pharmacology of high-threshold calcium currents in GH4C1 pituitary cells and their regulation by activation of human D2 and D4 dopamine receptors. *Br. J. Pharmacol.* **112,** 728–734.
66. Maurer, R. A. (1980) Dopaminergic inhibition of prolactin synthesis and prolactin messenger RNA accumulation in cultured pituitary cells. *J. Biol. Chem.* **255,** 8092.
67. Maurer, R. A. (1981) Transcriptional regulation of the prolactin gene by ergocryptine and cyclic AMP. *Nature* **294,** 94.
68. Keech, C. A. and Gutierrez-Hartmann, A. (1989) Analysis of rat prolactin promoter sequences that mediate pituitary-specific and 3',5'-cyclic adenosine monophosphate-regulated gene expression in vivo. *Mol. Endocrinol.* **3,** 832–839.
69. Day, R. N. and Maurer, R. A. (1989) The distal enhancer region of the prolactin gene contains elements conferring responses to multiple hormones. *Mol. Endocrinol.* **3,** 3–9.
70. Elzholtz, H. P., Mangalum, H. J., Potter, E., Albert, V. R., Supowit, S., Evans, R. M., and Rosenfeld, M. G. (1986) Two different *cis*-active elements transfer the transcriptional effects of both EGF and phorbol esters. *Science* **234,** 1552–1557.
71. Lufkin, T. and Bancroft, C. (1987) Identification by cell fusion of gene sequences that interact with positive *trans*-acting factors. *Science* **237,** 283–286.
72. Nelson, C., Crenshaw, E. B., III, Franco, R., Lira, S. A., Albert, V. R, Evans, R. M., and Rosenfeld, M. G. (1986) Discrete *cis*-active genomic sequences dictate the

pituitary cell type-specific expression of rat prolactin and growth hormone genes. *Nature* **322,** 557–562.

73. McChesney, R., Sealfon, S. C., Tsutsumi, M., Dong, K. W., Roberts, J. L., and Bancroft, C. (1991) Either isoform of the dopamine D2 receptor can mediate dopaminergic repression of the rat prolactin promoter. *Mol. Cell. Endocrinol.* **79,** R1–R7.
74. Elsholtz, H. P., Lew, A. M., Albert, P. R., and Sundmark, V. C. (1991) Inhibitory control of prolactin and Pit-1 gene promoters by dopamine. *J. Biol. Chem.* **266,** 22,919–22,925.
75. Albert, P. R., Neve, K. A., Bunzow, J. R., and Civelli, O. (1990) Coupling of a cloned rat dopamine-D2 receptor in inhibition of adenylyl cyclase and prolactin secretion. *J. Biol. Chem.* **265,** 2098–2104.
76. White, B. A., Bauerle, L. R., and Bancroft, F. C. (1981) Calcium specifically stimulates prolactin synthesis and mRNA sequences in GH3 cells. *J. Biol. Chem.* **256,** 5942–5945.
77. Gick, G. G. and Bancroft, C. (1985) Regulation by calcium of prolactin and growth hormone mRNA sequences in primary cultures of rat pituitary cells. *J. Biol. Chem.* **260,** 7614–7618.
78. Jackson, A. E. and Bancroft, C. (1988) Proximal upstream flanking sequences direct regulation of the rat prolactin gene. *Mol. Endocrinol.* **2,** 1139–1144.
79. Day, R. N. and Maurer, R. A. (1990) Pituitary calcium channel modulation and regulation of prolactin gene expression. *Mol. Endocrinol.* **4,** 736–742.
80. Lew, A. M., Yao, H., and Elsholtz, H. P. (1994) G(i) alpha 2- and G(o) alpha-mediated signaling in the Pit-1-dependent inhibition of the prolactin gene promoter. Control of transcription by dopamine D2 receptors. *J. Biol. Chem.* **269,** 12,007–12,013.
81. Beaulieu, M., Goldman, M. E., Miyazaki, K., Frey, E. A., Eskay, R. L., Kebabian, J. W., and Cote, T. E. (1984) Bromocriptine-induced changes in the biochemistry, physiology and histology of the intermediate lobe of the rat pituitary gland. *Endocrinology* **114,** 1871.
82. Chronwall, B. M., Millington, W. R., Griffin, W. S. T., Unnerstall, J. R., and O'Donohue, T. L. (1987) Histological evaluation of the dopaminergic regulation of proopiomelanocortin gene expression in the intermediate lobe of the rat pituitary, involving in situ hybridization and [^{3}H] thymidine uptake measurement. *Endocrinology* **120,** 1201.
83. Hollt, V., Haarmann, I., Seizinger, B. R., and Herz, A. (1982) Chronic haloperidol treatment increases the level of in vitro translatable messenger ribonucleic acid coding for the β-endorphin/adrenocorticotropin precursor proopiomelanocortin in the pars intermedia of the rat pituitary. *Endocrinology* **110,** 1885–1891.
84. Chen, C. L. C., Dionne, F. T., and Roberts, J. L. (1983) Regulation of the proopiomelanocortin mRNA levels in rat pituitary by dopaminergic compounds. *Proc. Natl. Acad. Sci. USA* **80,** 2211–2215.
85. Pritchett, D. B. and Roberts, J. L. (1987) Dopamine regulates expression of glandular-type kallikrein gene at transcriptional level in pituitary. *Proc. Natl. Acad. Sci. USA* **84,** 5545.
86. Cote, T. E., Felder, R., Kebabian, J. W., Sekura, R. D., Reisine, T., and Affolter, H. U. (1986) D-2 dopamine receptor-mediated inhibition of pro-opiomelanocortin synthesis in rat intermediate lobe. *J. Biol. Chem.* **261,** 4555–4561.

87. Loeffler, J.-P., Demeneix, B. A., Kley, N. A., and Hollt, V. (1988) Dopamine inhibition of proopiomelanocortin gene expression in the intermediate lobe of the pituitary. *Neuroendocrinology* **47,** 95–101.
88. Loeffler, J.-P., Kley, N., Pittius, C. W., and Hollt, V. (1985) Corticotropin-releasing factor and forskolin increase proopiomelanocortin messenger RNA levels in rat anterior and intermediate cell in vitro. *Neurosci. Lett.* **62,** 383–387.
89. Loeffler, J. P., Kley, N., Pittius, C. W., and Hollt, V. (1986) Calcium ion and cyclic adenosine 3',5'-monophosphate regulate proopiomelanocortin messenger ribonucleic acid levels in rat intermediate and anterior pituitary lobes. *Endocrinology* **119,** 2840–2847.
90. May, V., Stoffers, D. A., and Eipper, B. A. (1989) Proadrenocorticotropin/endorphin production and messenger ribonucleic acid levels in primary intermediate pituitary cultures: effects of serum, isoproterenol, and dibutyryl adenosine 3',5'-monophosphate. *Endocrinology* **124,** 157–166.
91. Autelitano, D. J., Blum, M., Lopingco, M., Allen, R. G., and Roberts, J. L. (1990) Corticotropin-releasing factor differentially regulates anterior and intermediate pituitary lobe proopiomelanocortin gene transcription, nuclear precursor RNA and mature mRNA in vivo. *Neuroendocrinology* **51,** 123–130.
92. Eberwine, J. H., Jonassen, J. A., Evinger, M. J. Q., and Roberts, J. L. (1987) Complex transcriptional regulation by glucocorticoids and corticotropin-releasing hormone of proopiomelanocortin gene expression in rat pituitary cultures. *DNA* **6,** 483–492.
93. Gagner, J. P. and Drouin, J. (1987) Tissue-specific regulation of pituitary proopiomelanocortin gene transcription by corticotropin-releasing hormone, 3',5'-cyclic adenosine monophosphate, and glucocorticoids. *Mol. Endocrinol.* **1,** 677–682.
94. Roberts, J. L., Lundblad, J. R., Eberwine, J. H., Fremeau, R. T., Salton, S. R. J., and Blum, M. (1987) Hormonal regulation of proopiomelanocortin gene expression in the pituitary. *Ann. NY Acad. Sci.* **512,** 275–285.
95. Therrien, M. and Drouin, J. (1991) Pituitary pro-opiomelanocortin gene expression requires synergistic interactions of several regulatory elements. *Mol. Cell. Biol.* **11,** 3492–3503.
96. Jin, W. D., Boutillier, A. L., Glucksman, M. J., Salton, S. R. J., Loeffler, J. P., and Roberts, J. L. (1994) Characterization of a corticotropin-releasing hormone-responsive element in the rat proopiomelanocortin gene promoter and molecular cloning of its binding protein. *Mol. Endocrinol.* **8,** 1377–1388.
97. Boutillier, A. L., Sassone-Corsi, P., and Loeffler, J. P. (1991) The protooncogene c-fos is induced by corticotropin-releasing factor and stimulates proopiomelanocortin gene transcription in pituitary cells. *Mol. Endocrinol.* **5,** 1301–1310.
98. Boutillier, A. L., Monnier, D., Lorang, D., Lundblad, J. R., Roberts, J. L., and Loeffler, J. P. (1995) CRH stimulates POMC transcription by cFos dependent and independent pathways: characterization of an AP1 site in exon 1. *Mol. Endocrinol.* **9,** 745–755.
99. Samama, P., Pei, G., Costa, T., Cotecchia, S., and Lefkantz, R. J. (1994) Negative antagonists promote an inactive conformation of the beta 2-adrenergic receptor. *Mol. Pharmacol.* **45,** 390–394.

CHAPTER 12

Mechanisms of Dopaminergic Regulation of Prolactin Secretion

Paul R. Albert, Mohammad H. Ghahremani, and Stephen J. Morris

1. Dopaminergic Regulation of Prolactin Secretion

Dopamine has been recognized as the primary regulator of prolactin (PRL) secretion in vivo for over 20 years *(1,2)*. Hypothalamic dopamine is secreted from tuberoinfundibular neurons at the median eminence of the hypothalamus into the hypophyseal portal blood flow, which directly perfuses the pituitary gland, delivering dopamine at high concentration to anterior pituitary cells. Several lines of evidence suggested that PRL, unlike other pituitary hormones, is under primarily inhibitory hypothalamic regulation, and that dopamine acts as the mediator. Interrupting the portal blood flow to the pituitary by pituitary stalk section results in elevated PRL secretion. The hypersecretion of PRL of the isolated pituitary was reversed by pituitary grafting to an intact portal blood flow *(3)*; by perfusion with hypothalamic extracts containing PIF (PRL inhibitory factor) activity, which was subsequently found to contain dopamine as the active principle *(3,4)*; and by perfusion with dopamine itself *(4,5)*. Dopamine antagonists (e.g., haloperidol) are known to induce hyperprolactinemia and block the PRL-inhibitory actions of dopamine *(6,7)*. Ultimately, dopamine levels measured in portal blood were shown to be sufficient to inhibit PRL secretion in vitro *(8)*, firmly establishing dopamine as the PIF, the only nonpeptide hypothalamic hormone. Thus, tonic inhibition of PRL secretion by dopamine controls the level of PRL in the organism: Decreasing dopamine release leads to enhanced PRL release, mediated in part by stimulatory hormones such as thyrotropin-releasing hormone (TRH) and vasoactive intestinal peptide (VIP). Enhancement of PRL release correlates with low hypophyseal portal dopamine concentration and is most pronounced

The Dopamine Receptors Eds.: K. A. Neve and R. L. Neve
Humana Press Inc., Totowa, NJ

during proestrous-estrous, pregnancy, and lactation in the female. The secreted PRL participates in breast and uterine development and milk generation in the breast. In males, PRL levels are low in part because of an absence of circulating estrogen, a powerful inducer of PRL gene transcription *(1,2)*. The present chapter summarizes the current understanding of the mechanisms by which dopamine regulates hormone secretion from the pituitary, focusing primarily on dopaminergic inhibition of PRL secretion.

2. Dopamine Receptors on Lactotrophs

2.1. D2* Receptor Family

Following the identification of dopamine as the major regulator of PRL secretion, studies of normal lactotrophs or adenomas in culture identified the presence of receptors for dopamine on these cells *(9,10)*. The pharmacology of the dopamine receptors corresponded to the D2 subtype, and dopamine was shown to inhibit adenylate cyclase activity and PRL secretion with a D2-like pharmacology. Further studies demonstrated that dopamine D2 receptors present on lactotrophs have both high- and low-affinity sites for agonist binding *(11)*, but a single affinity for antagonists. In the presence of guanosine triphosphate (GTP), high-affinity agonist binding at the dopamine receptor was converted to the low-affinity state, the earliest evidence of an inhibitory receptor that coupled to a GTP-sensitive process. The classic model of a ternary complex of receptor (high affinity)-G protein-guanosine diphosphate (GDP) was proposed for the dopamine D2 receptor: On binding of agonist to the receptor, a shift in the affinity of the G protein α subunit to bind GTP over GDP leads to dissociation of the ternary complex, releasing active GTP-bound α subunits and βγ subunits and the receptor in the low-affinity state *(12,13)*. The lack of receptor activation by antagonists renders antagonist binding of high affinity and insensitive to guanine nucleotides. Receptor-activated α and βγ subunits modulate the activity of a variety of effectors: for example, G_i/G_o proteins mediate inhibition of adenylate cyclase, inhibition of calcium channels, activation of potassium channels, and, in certain (non-pituitary) cells, phospholipase C activation. The coupling of dopamine D2 receptors to G_i/G_o proteins was further demonstrated using pertussis toxin (PTX) which selectively inactivates these "inhibitory" G proteins, preventing their activation by agonist-occupied receptor. Pretreatment with PTX abolishes most actions mediated by the dopamine D2 receptor, and shifts agonist

*As detailed in the Preface, the terms D1 or D1-like and D2 or D2-like are used to refer to the subfamilies of DA receptors, or used when the genetic subtype (D_1 or D_5; D_2, D_3, or D_4) is uncertain.

affinity to the low-affinity, uncoupled state *(14–16)*. Finally, G_i and G_o proteins copurify with the dopamine D2 receptor *(16,17)*, and addition of GTP to purified D2 receptor–G protein complex shifts the affinity of the receptor to the low-affinity state *(15)*. Thus, the dopamine D2 receptor expressed in lactotrophs appears to be a single receptor with two affinity states dictated by interaction with G proteins.

With the cloning of multiple dopamine receptor subtypes by low stringency screening *(18)* or reverse transcription-polymerase chain reaction (RT-PCR) with degenerate oligonucleotides directed to conserved regions *(19,20)*, the expression of the dopamine D2-related subtypes in the pituitary was examined. The dopamine D_2 receptor is alternatively spliced to generate two molecular forms (short [D_{2S}] and long [D_{2L}]) which differ by a 29-amino acid insert in the third cytoplasmic domain *(21–24)*, a region of potential importance for receptor function *(25)*. Both forms of the D_2 receptor are strongly expressed in the pituitary, although the ratio of the two forms varies depending on the study, probably owing to the nonquantitative nature of RT-PCR measurements. To date, no subtype-selective antibodies or ligands (but, *see* refs. *26,27*) have been developed to quantitate the expression of D_{2S} and D_{2L} receptors. These tools could be used to address whether both receptors (D_{2S} and D_{2L}) are in fact present on the same cell. The other D2-like receptors (i.e., D_3 and D_4) are weakly or not expressed in the pituitary *(19,20)*, and hence the inhibitory actions of dopamine in lactotrophs are predominantly mediated via dopamine D_2 receptors. Both forms of the D_2 receptor appear to be functionally similar, as discussed in Sections 3.2. and 3.3.

2.2. D1 Receptor Family

Recently, weak expression of dopamine D_5 receptors in pituitary cells has been detected by RT-PCR *(28)*. When transfected in pituitary GH_4C_1 cells, the dopamine D_1 and D_5 receptors are coupled to stimulation of adenylate cyclase and calcium channel opening *(29)*, which enhances hormone secretion *(28)*. The presence of D_5 receptors in pituitary has been suggested to mediate a paradoxical stimulatory action of low (10^{-12}–10^{-10}*M*) concentrations of dopamine on PRL secretion that has been observed in primary cultures of pituitary cells by certain groups *(30–32)*, but not universally *(33,34)*. However, the affinity of the dopamine D_5 receptor for dopamine is 10-fold lower (in the nanomolar range), and should not be activated by subnanomolar dopamine concentrations *(28)*. Thus, the mechanism of the stimulatory action of low dopamine concentrations on PRL secretion remains to be resolved. Furthermore, it is unclear whether sufficient levels of dopamine D_5 receptors are present to modulate pituitary hormone secretion *(28)*, and the functional role of the dopamine D_5 receptor in the pituitary remains to be demonstrated.

3. Dopamine D_2 Receptor Signaling in Lactotrophs

3.1. Lactotroph Systems

Two types of experimental systems have provided insight into the mechanisms of dopamine action in pituitary lactotrophs: primary cultures of enriched (e.g., by estrogen induction or by purification using centrifugal elutriation) or identified (e.g., by reverse hemolytic plaque assay *[35]*) normal rat lactotrophs; and the use of pituitary cell strains (e.g., GH_4C_1 cells) which represent a homogeneous population of lactotroph cells *(36,37)*. Growth hormone (GH) cells may be derived from a primitive pituitary cell, the mammosomatotroph, which secrete both GH and PRL and differentiate into somatotrophs or lactotrophs *(38,39)*. Both systems have advantages and disadvantages, but have generated results that are largely in agreement and provide independent validation for conclusions regarding dopamine action. The most extensively characterized cell lines, GH_3 and GH_4C_1 cells, can be grown in large quantities suitable for biochemical characterization. These cells have been studied for over 25 years *(37)*, and their biochemical characteristics have remained remarkably stable and reproducible over that period *(36,40)*. GH cells share many properties of normal lactotrophs: They synthesize and secrete PRL, processes that are regulated by several physiologically relevant receptors (e.g., TRH, somatostatin, glucocorticoid, thyroid, etc.). The membrane-bound receptors couple to effectors and ion channels in a manner characteristic of normal lactotrophs. Unlike normal lactotrophs, GH_4C_1 cells do not express dopamine D_2 receptors under normal culture conditions, i.e., in various serum-containing media *(41–43)*. However when exposed to medium containing epidermal growth factor (EGF) or nerve growth factor (NGF) for 2 d, GH_3 and GH_4C_1 cells express both D_{2S} and D_{2L} receptors *(43–45)*. The lack of D_2 receptors and dopamine responsiveness in the uninduced condition renders GH cells an excellent system in which to study the function of cloned dopamine receptors in isolation in a lactotroph cell environment *(40)*. However, since GH cells are transformed, it is important to replicate findings in normal lactotrophs where possible.

3.2. Signaling Pathways: D_{2S} vs D_{2L}

As indicated from in vivo studies, dopamine is a powerful inhibitor of PRL secretion in vitro, and research has focused mainly on elucidation of the pathways that lead to inhibition of PRL secretion *(10,14,46)*. Initially, dopamine was shown to inhibit cyclic [3',5']-adenosine monophosphate (cAMP) generation in adenomas, an action which was correlated with inhibition of PRL secretion *(47)*. However, with the realization that lactotrophs are excitable cells, which express voltage-gated ion channels and generate spontane-

Table 1
Shared Signaling Pathways of D_{2S} and D_{2L}[a]

Action	PTX	Cell line	Refs.
⇓ cAMP	+	GH_4, Ltk^-	*41,54–56*
⇓ Ca influx	+	GH_4	*57–59*
⇑ gK^+	+	GH_4, Ltk^-	*56,60,61*
⇑ YPase/⇓ DNA synth.	+	GH_4	*62,63*
⇓ PRL secretion	+	GH_4	*41,54,64*
⇓ PRL transcription	+	GH_4	*65,66*
⇑ PI/Ca mobilization	+	Ltk^-, CHO	*55,56,67*
⇑ Y-kin/⇑ DNA synth.	+	CHO	*68*
⇑ Na^+/H^+ exchange	–	C_6, Ltk^-	*69*
⇑ Ca-induced PLA2	+	CHO	*70,71*

[a]The shared signaling pathways and actions of transfected dopamine D_{2S} and D_{2L} receptors are tabulated in both pituitary (GH_4) and nonpituitary cells (Ltk^-, CHO, C_6, etc.). Sensitivity of receptor-mediated actions to PTX pretreatment is as indicated: +, sensitive; –, insensitive. Dopamine actions described include changes in potassium conductance (gK^+), tyrosine phosphatase (YPase), DNA synthesis (DNA synth., as measured by incorporation of ^3H-thymidine), PI turnover and calcium mobilization (PI/Ca mobilization), tyrosine kinase activation (Y-kin.), and enhancement of calcium-induced phospholipase A_2 activity (Ca-induced PLA2).

ous sodium- and calcium-dependent action potentials *(48–53),* a role for dopaminergic modulation of ionic currents in control of PRL secretion has been sought *(52)*. Thus, dopamine has multiple actions on lactotrophs, including inhibition of cAMP levels, opening of potassium channels, and inhibition of calcium channels (Table 1). These actions are all mediated via coupling of D_2 receptors to PTX-sensitive G proteins (e.g., G_i and G_o). Each of these actions of dopamine are discussed and related to dopaminergic inhibition of three states of PRL secretion in lactotrophs: basal, VIP-stimulated, and TRH-stimulated secretion (*see* Section 4.).

Stable transfection of dopamine D_{2S} and D_{2L} receptor subtypes separately into GH or fibroblast cells has revealed no clear differences between the two dopamine receptor subtypes in terms of pharmacology or function (Table 1). The two D_2 receptors share several properties in GH_4C_1 pituitary cells, including inhibition of cAMP, calcium influx, opening of potassium channels, inhibition of PRL secretion, and inhibition of cell proliferation. In nonpituitary cells, other signaling properties including enhancement of phosphatidyl inositol (PI) turnover, calcium mobilization, enhancement of calcium-induced phospholipase A2, increased sodium–proton exchange, and increased cell proliferation are also shared by D_{2S} and D_{2L} receptors. In both pituitary and

Table 2
Differences Between D_{2S} and D_{2L} Receptor Signaling[a]

Action	Cell	Efficacy	Refs.
⇓ cAMP	CHO	$D_{2S} > D_{2L}$	*67*
		$D_{2L} > D_{2S}$	*70*
	JEG	$D_{2S} > D_{2L}$	*72*
⇑ Ca-induced PLA_2	CHO	$D_{2L} > D_{2S}$	*70*

[a]Receptor-mediated actions which differ between dopamine D_{2S} and D_{2L} receptors are tabulated, and the relative efficacy of each receptor for the indicated action is given.

Table 3
Differential Modulation of D_{2S} and D_{2L}[a]

D2 Response	Kinase	Action	Efficacy	Refs.
⇓ cAMP (CHO)	C	Block	$D_{2S} = D_{2L}$	*70*
⇑ Ca-induced PLA_2	C	Enhance	$D_{2S} = D_{2L}$	*70*
⇓ cAMP (Ltk⁻)	C	N/E		*55*
⇑ PI (Ltk⁻)	C	Block	$D_{2S} >> D_{2L}$	*55*

[a]The acute actions of various protein kinases (C = PKC) on D2 receptor signaling is summarized, along with the relative selectivity of the kinases for D_{2S} or D_{2L} subtype.

nonpituitary cells, the D_2 receptor appears to couple almost exclusively to PTX-sensitive G proteins.

By comparing D_{2S} and D_{2L} receptor transfectants, small differences in potency or maximal effect have been reported, but no consistent pattern has emerged (Table 2). In Chinese hamster ovary (CHO) cells, one study found that D_{2S} receptors are better coupled to inhibition of cAMP than D_{2L} *(67)*, but another study showed the opposite *(70)*. Differences in the intrinsic responsiveness of the isolated clones have not always been examined and could explain some of the differences attributed to receptor subtype. For example, in cells with increased dopamine responses, a parallel increase in responsiveness of other receptors (e.g., somatostatin) would reflect a general increase in cellular responsiveness rather than an effect specific for the dopamine D_2 subtype tested. The most clear difference observed between the D_{2S} and D_{2L} receptors is not in function, but rather in modulation by protein kinase C (PKC, Table 3). The D_{2L} receptor expressed in Ltk⁻ fibroblast cells is resistant to acute uncoupling via PKC activation, whereas coupling of the D_{2S} receptor to calcium mobilization is potently blocked *(55)*. The difference between these two receptors may be caused by a predicted PKC pseudosubstrate domain that is present in the D_{2L} receptor but absent in the D_{2S}

receptor. The enhanced sensitivity of the D_{2S} receptor to PKC activation could account for decreased coupling of this receptor (vs the D_{2L} form) to calcium mobilization in certain transfected cell lines (e.g., in CHO cells *[67]*).

3.3. G Protein Coupling: D_{2S} vs D_{2L}

The sensitivity of dopamine actions in lactotrophs to pretreatment with PTX suggests that the D_2 receptor couples to the G_i/G_o class of "inhibitory" G proteins, composed of G_{i1}, G_{i2}, G_{i3}, G_{oA}, and G_{oB} *(12)*. Heterotrimeric G proteins are composed of an α subunit, which binds guanine nucleotides, and associated βγ subunits. On agonist binding to the high-affinity receptor–G protein complex, an exchange of GDP for GTP on the α subunit catalyzes the dissociation of the ternary complex, releasing receptor-activated α subunits and βγ dimers *(12,13,73)*. Both subunits modulate the activity of various effector enzymes and channels: In lactotrophs the G_i/G_o proteins inhibit adenylate cyclase (basal and G_s-stimulated activities), enhance opening of potassium channels to hyperpolarize the cell, and close calcium channels. The cycle is terminated by the hydrolysis of GTP on the α subunit, returning the complex to the inactive state.

Three general approaches have been used to examine the specificity of receptor–G protein effector interactions in the lactotroph: reconstitution with purified G proteins, antibodies to specific G proteins, and antisense approaches *(40,74)*. These approaches have demonstrated a remarkable specificity in the signaling network of the dopamine D_2 receptor. Perhaps the best characterized pathway is the coupling of multiple receptors, including D_{2S} and D_{2L}, to inhibition of calcium channels via G_o. Intracellular dialysis of normal lactotrophs under whole cell patch clamp condition with antibodies to α_o, but not α_{i1}, α_{i2}, or α_{i3}, largely attenuated dopamine-mediated inhibition of both T-type and L-type calcium channels *(75)*. Similarly, in GH_4C_1 cells stably transfected with antisense cDNAs, a specific and nearly complete knockout of α_o was achieved *(57)*. In these cells, both dopamine D_{2S} and D_{2L} receptors failed to inhibit dihydropyridine-induced opening of L-type calcium channels, although the effect of D_{2L} was not completely blocked. Depletion of G_o did not compromise receptor-mediated inhibition of basal or G_s-induced cAMP accumulation by D_{2S}, D_{2L}, somatostatin, or muscarinic-M4 receptors. On the other hand, reduction in G_{i1}, G_{i2}, or G_{i3} using antisense approaches did not inhibit the ability of D_2 receptors to inhibit calcium channel opening. Antisense knock-down experiments using the microinjection technique have demonstrated that somatostatin and muscarinic receptors in GH_3 cells recognize distinct G_o complexes, $\alpha_{oA}\beta_1\gamma_3$ and $\alpha_{oB}\beta_3\gamma_4$, respectively *(76–78)*. Thus, differences in efficacy among these receptors to inhibit calcium channels may be related to the different G_o subtypes involved *(57)*. The importance of G_o in coupling receptors to calcium channel inhibition is in agreement with findings in certain neuronal cells, such as the dorsal root ganglion cells where the $GABA_B$ receptor couples via G_o to calcium channels *(74)*.

The function of G_{i3} in lactotrophs has been another intensively studied pathway: Reconstitution experiments demonstrated that purified or recombinant α_{i3} could directly activate potassium channels in membrane patch preparations from pituitary and atrial preparations *(79,80)*. Intracellular dialysis of pituitary cells with antibody to α_{i3} but not α_o, α_{i1}, or α_{i2}, blocked dopamine-induced inhibition of both sustained (I_K) and transient (I_A) potassium currents *(75)*. Antisense oligonucleotide injection experiments in enriched lactotrophs have shown that the D_2 receptor is coupled via G_{i3}, but not G_o, to opening of potassium channels and consequent hyperpolarization of the membrane potential *(81)*. Whereas this coupling may represent "direct" coupling of the α subunit to the channel, recent reports have suggested that in some systems $\beta\gamma$ subunits mimic the action of G_{i3} to open potassium channels *(73)*. In addition, the activation of a serine phosphatase has been implicated as a possible intermediary step, based on pharmacological inhibition with okadaic acid, a serine phosphatase inhibitor *(82)*. Nevertheless, the evidence indicates that in lactotrophs G_{i3} is a major G protein involved in the coupling of multiple inhibitory receptors to open potassium channels.

Antisense experiments have shown that G_{i2} is important for the coupling of multiple receptors to inhibition of cAMP (both basal and Gs-stimulated) in lactotrophs, but is not involved in coupling to inhibition of calcium channel opening *(57)*. The effect of G_{i2} is receptor dependent: For example, D_{2L} receptor-mediated inhibition of cAMP is entirely blocked in G_{i2} knockout clones, however D_{2S} receptors retained efficacy by over 70%. Somatostatin actually induced a paradoxical increase (by 30%) in the VIP-stimulated level of cAMP in G_{i2}-knockout clones *(57,72)*, suggesting recruitment of $\beta\gamma$ subunits released from other G proteins (e.g., G_o, G_{i3}) to enhance adenylate cyclase activity (e.g., type II or IV proteins *[12,73]*). By contrast, PTX-insensitive G protein mutants were used to show that the D_{2S} receptor preferentially coupled via G_{i2} to inhibit forskolin-induced cAMP accumulation, but the D_{2L} receptor utilized G_{i1} and G_{i3} *(83)*. Perhaps the most intriguing finding is that receptors can choose one of several G_i subtypes to couple to inhibition of either basal, G_s-stimulated, or forskolin-stimulated adenylate cyclase activity.

4. Mechanisms of Inhibitory Regulation of PRL Secretion

4.1. Inhibition of cAMP Level

Elevation of cAMP by activation of adenylate cyclase (using forskolin), G_s (using fluoride or cholera toxin), or by G_s-coupled receptors (e.g., VIP or

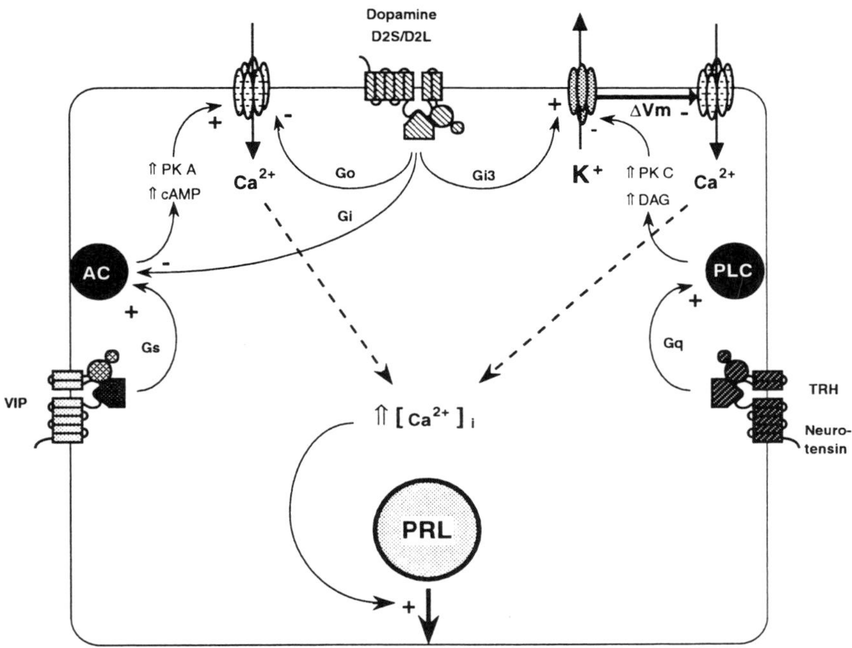

Fig. 1. Dopaminergic inhibition of PRL secretion. A figurative summary of the events in a lactotroph cell leading to enhanced PRL secretion, and their inhibition by D_2 receptor activation. At left, VIP receptor activation couples via G_s to enhance adenylate cyclase (AC) activity, leading to increased cAMP levels and activation of PKA, which in turn enhances calcium channel opening to increase $[Ca^{2+}]_i$ and PRL secretion. Acting via G_i, the D_2 receptors inhibit AC, the primary site of inhibition of VIP-induced changes; a secondary inhibition of calcium channel opening mediated via G_o is also possible. On the right, TRH or neurotensin receptors couple via G_q to enhance the activity of PLC to generate DAG which activates PKC. PKC may mediate an inhibition of potassium channel opening to depolarize the membrane (ΔVm) and enhance calcium channel opening, actions which are counteracted by dopamine-induced opening of potassium channels via G_{i3}. Additionally, opening of potassium channels appears to be the primary mechanism by which dopamine inhibits basal $[Ca^{2+}]_i$ and PRL secretion.

prostaglandin E_2 (PGE_2) or transfected D_1 receptors) results in enhanced PRL secretion *(29,40,41,84–87)*. The common pathway of these agents involves an obligatory elevation of cAMP, resulting in PKA-mediated enhancement of L-type calcium channel opening and calcium influx to increase $[Ca^{2+}]_i$ and stimulate hormone secretion (Fig. 1). Blockade of receptor-mediated coupling to G_s, increase in cAMP, or activation or blockade of PKA using pharmacological agents prevents receptor-mediated increases in $[Ca^{2+}]_i$ and PRL secretion, indicating the importance of cAMP-dependent mechanisms in the actions of G_s-coupled receptors in pituitary cells.

In contrast, there is abundant evidence that activation of D_2 receptors in adenomas, normal lactotrophs, or GH_4C_1 cells transfected with either D_{2S} or D_{2L} receptor inhibits adenylate cyclase leading to decreased cAMP levels *(10,41,46,57,83,88)*. Dopamine elicits a rapid (within minutes), concentration-dependent and complete inhibition of cAMP generation that correlates well with the time course and effect of dopamine on PRL secretion. Dopamine-mediated inhibition of cAMP generation in whole cells and in membranes is completely blocked by pretreatment with PTX, suggesting the involvement of G_i or G_o in coupling of the D_2 receptor to adenylate cyclase *(41,46,88,89)*. As described, antisense experiments indicate that G_o does not mediate inhibitory coupling to adenylate cyclase in GH cells, whereas the G_i family is implicated *(57,83)*. However, we have identified a receptor specificity within the G_i family; for example, knockout of G_{i2} completely blocks coupling of D_{2L}, but not D_{2S} receptor, to inhibition of VIP-stimulated cAMP levels *(57)*.

There is no evidence that dopamine induces a significant change in the resting cAMP level measured in the absence of phosphodiesterase inhibitors, or that a decrease in cAMP is required for inhibition of basal PRL secretion *(85,90,91)*. By contrast, inhibition of cAMP generation by hormones that couple to G_s correlates well with dopamine-induced inhibition of PRL secretion. This is consistent with data observed for other inhibitory hormones such as somatostatin *(84,85)*. For example, VIP-induced PRL secretion and cAMP accumulation are inhibited with similar time course and EC_{50} by somatostatin. Similarly, forskolin-induced increases in cAMP and secretion are also inhibited in parallel by somatostatin. However, PRL secretion induced by direct activation of PKA using cAMP analogs such as 8-BrcAMP is also sensitive to somatostatin. The evidence that a decrease in cAMP is the crucial step in inhibition of G_s-mediated PRL secretion rests mainly on the fact that the potencies for inhibition of cAMP and PRL secretion are greater than that for inhibition of 8-BrcAMP *(91–93)*.

As described earlier, treatment with EGF induces D_2 receptor expression in nontransfected GH_3 or GH_4C_1 cells *(43–45)*. In EGF-induced cells, the D_2 receptor retains the ability to enhance potassium channel opening *(43,44,94)*, and potently inhibits neurotensin-induced (cAMP-independent) PRL secretion *(44,95)*. However, D_2 receptor activation inhibits neither VIP- nor forskolin-induced increases in cAMP, and the reduction in PRL secretion is greatly attenuated. By contrast, in GH_4C_1 cells transfected with D_{2S} or D_{2L} receptors, but not treated with EGF, dopamine also inhibits VIP-stimulated cAMP accumulation and PRL secretion *(41,43)*. These results support the crucial role of dopamine-induced inhibition of adenylate cyclase in the attenuation of G_s-coupled PRL secretion.

The ability of inhibitory receptors to block secretion induced by 8-BrcAMP implies that an additional post-cAMP site is responsible for the observed inhibition *(85,93)*. Increased cAMP activates PKA, which in turn phosphorylates several target proteins leading to activation of voltage-dependent calcium channels in pituitary cells *(96)*. Since certain G proteins (especially G_o) mediate a direct (cAMP-independent) inhibitory action on calcium channel opening, this may be a post-cAMP site of inhibition. Alternatively, activation of potassium channels to hyperpolarize the membrane potential would produce a similar post-cAMP inhibition of calcium channel opening and PRL secretion.

4.2. Increase in Potassium Conductance

Early findings indicated that D_2 receptor agonists cause a profound hyperpolarization of lactotroph membrane potential associated with enhanced outward potassium current *(33,34,52,56,60,61)*. In a subset of lactotrophs that display spontaneous action potentials, D_2 receptor activation entirely blocks action potential firing. These actions have an immediate time course (within milliseconds of agonist addition) and are sustained minutes in the continued presence of agonist. Patch clamp analysis of identified lactotrophs indicates that in the presence of GTP, dopamine markedly enhances the frequency of opening of a potassium channel with unitary conductance of 42 ps, which accounts for the actions on potassium current and membrane hyperpolarization *(61)*. This action of dopamine is blocked by PTX or GDPβS, an inhibitor of G protein activation, and potentiated by the nonhydrolyzable GTP analog GTPγS, which activates G proteins. Receptor-mediated opening of potassium channels in GH cells can be reconstituted by addition of purified G_{i3} to membrane patches *(79)*, suggesting a direct action of G_{i3} on the channel(s). In support of this, dopamine-induced potassium currents and cellular hyperpolarization are independent of the concentration of cAMP or calcium in identified lactotrophs *(69,72)*. In EGF-treated GH_4C_1 cells, the induced D_2 receptor increases outward potassium current, but does not inhibit cAMP, again suggesting a cAMP-independent mechanism of action *(44,94)*. Uncoupling of the dopamine receptor from cAMP in this paradigm is associated with a small increase in the level of G_{i3} *(86)*, consistent with the importance of this G protein in coupling to potassium channels.

The rapid and sustained action of dopamine on potassium currents correlates well with the time course of inhibition of PRL secretion and suggests that dopamine-mediated opening of potassium channels may play an important role in blocking secretion *(33,34)*. Opening of potassium channels hyperpolarizes the cell, and decreases the spontaneous activity of voltage-dependent calcium channels, leading to reduced calcium influx and a decrease in $[Ca^{2+}]_i$ (Fig. 1). Reduction in $[Ca^{2+}]_i$ reduces the steady-state rate of hormone secretion *(97,98)*. In GH_4C_1 cells, somatostatin-induced inhibition of basal secretion is blocked by potassium channel blockers or conditions that neutralize the outward potassium gradient *(92,93)*. Although these manipulations affect the level of basal PRL secretion, they suggest that potassium efflux is necessary for receptor-mediated inhibition of basal secretion. In the presence of saturating concentrations of cAMP analogs, dopamine and somatostatin effectively open potassium channels and reduce basal PRL secretion, indicating that these events are cAMP independent *(33,34,61,90,92,93)*. Similarly, in EGF-treated GH_3 cells, which lack dopamine-mediated inhibition of VIP-stimulated cAMP accumulation and PRL secretion, dopamine-enhanced potassium channel opening and dopaminergic inhibition of basal PRL secretion or secretion induced by neurotensin or potassium-induced depolarization are not different from D_{2S}-transfected cells where inhibition of cAMP is intact. These arguments point to a cAMP-independent mechanism of receptor-mediated inhibition of basal and depolarization-induced PRL secretion associated with dopamine-induced opening of potassium channels.

4.3. Inhibition of PI Turnover

Hormones that stimulate PRL secretion via enhancement of PI turnover, such as TRH or neurotensin, are potently inhibited by dopamine *(14,41,43–45,54,95)*. The mechanism of TRH action has been the most intensively studied and is summarized *(40)*: TRH couples to PTX-insensitive G proteins G_q and G_{11} *(99)* to stimulate PI turnover leading to the generation of the dual second messengers, IP3 and diacylglycerol (DAG) *(100,101)*. An immediate spike increase in $[Ca^{2+}]_i$ is triggered by IP3, and DAG induces activation of PKC; the simultaneous increases in $[Ca^{2+}]_i$ and PKC synergistically activate the initial spike or burst phase of PRL secretion *(97,98,102–105)*. This initial phase (1–2 min) of TRH action is followed by a sustained plateau phase of increase in $[Ca^{2+}]_i$, which is the result of calcium influx (in part through voltage-activated calcium channels *[106,107]*). This sustained increase in $[Ca^{2+}]_i$ as well as sustained PKC activation are required for the sustained TRH-induced enhancement of PRL secretion rate *(103,105)*. It is this second sustained phase of increased $[Ca^{2+}]_i$ and secretion that represents the major contribution to TRH-induced PRL secretion. This phase is completely inhibited by dopamine, whereas the initial burst phase of TRH action is unaffected *(108–111)*. In normal lactotrophs and transfected GH_4C_1 cells, dopamine does not directly inhibit TRH-induced PI turnover in membranes or the acute increase in PI turnover in whole cells, nor the acute spike increase in $[Ca^{2+}]_i$ *(108–111)*. Whereas tonic PI turnover in the basal condition or during the second phase of TRH action is inhibited by dopamine, this is a consequence of the dopamine-induced reduction in $[Ca^{2+}]_i$, which decreases the activity of calcium-dependent forms of phospholipase C (PLC) *(111)*. Blockade of dopamine-induced calcium decrease (using ionomycin to induce calcium entry) prevents the dopamine-mediated decrease in PI turnover, whereas mimicking the effect (e.g., by blocking calcium influx) produces a decrease in PI turnover *(110)*. Thus, dopamine-induced inhibition of PI turnover is a secondary calcium-mediated effect. Similarly, inhibition of TRH-stimulated PRL secretion is related to inhibition of calcium entry by dopamine or other blockers (e.g., calcium channel blockers), suggesting that regulation of calcium influx plays a crucial role in the second phase of TRH-induced secretion *(97,98)*.

The mechanism by which dopamine blocks the sustained TRH-induced increase in $[Ca^{2+}]_i$ and PRL secretion is incompletely resolved (Fig. 1). The primary event appears to be caused by a decrease in calcium influx as argued previously, but the mechanism of this decrease is not clear *(106,107)*. Dopamine could act via activation of potassium efflux to hyperpolarize the cell and prevent opening of calcium channels, as described for the basal state *(106)*. Alternately, direct inhibition of calcium channels may be the primary mechanism of dopamine action *(107)*. Preliminary evidence using antibodies to G proteins introduced into streptolysin-permeabilized lactotrophs indicates that a G_{i3}-antibody, rather than G_o or G_{i1}/G_{i2} antibodies, selectively disrupts dopamine-induced inhibition of TRH-induced PRL secretion *(112)*. Preliminary antisense studies also implicate G_{i3} as the most important mediator of receptor-mediated inhibition of TRH-induced PRL secretion, although other G proteins participate depending on the inhibitory receptor studied *(113)*. Given the evidence that G_{i3} is the primary mediator of potassium channel activation in lactotrophs, these studies suggest that dopamine-mediated hyperpolarization may play the primary role in mediating inhibition of TRH-mediated PRL secretion. However, more direct studies must be done to confirm this hypothesis.

4.4. Decrease in Calcium Conductance

The expression of voltage-dependent ion channels in pituitary cells renders these secretory cells electrically excitable *(48,49,51–53,114)*. In particular, influx of calcium through voltage-dependent calcium channels is one of the crucial regulators of both basal and stimulated secretion (*50,97,98,115*; Fig. 1). Pharmacological blockade of these channels or prevention of calcium entry reduces the rate of basal secretion, and prevents stimulation of secretion by VIP and TRH (second phase). Both high threshold, slowly inactivating L-type channels and low threshold, rapidly inactivating T-type calcium channels are present in lactotrophs and GH cells *(58,59,96,114)*. Dopamine couples to inhibition of both L-type and T-type calcium channels in lactotroph and GH cell membrane patch preparations (but *see* ref. *115*), suggesting a direct action of receptor/G protein on the channel *(58,59)*. Both D_{2S} and D_{2L} receptors couple with equal efficacy to inhibit calcium influx induced by L-type calcium channel agonist BayK8644 *(115)*, however they have different sensitivities to antisense blockade of G_o: D_{2L} receptors inhibit the BAY-induced change by 30% in the absence of G_o, whereas D_{2S} action is abolished *(57)*. This difference in coupling could be mediated by activation of G_i, perhaps indirectly via membrane hyperpolarization owing to potassium channel opening. The finding that knockout of G_o does not affect dopamine-induced potassium channel opening or inhibition of cAMP levels indicates that the D_2 receptor may couple directly to calcium channels via G_o, and not by indirect mechanisms involving changes in cAMP or membrane potential.

The role of G_o-mediated inhibition of calcium channels in regulation of PRL secretion has yet to be clearly identified. At present, evidence favors a dominant role of G_{i3}-mediated potassium channel opening, rather than calcium channel inactivation, in coupling D_2 receptors to inhibition of secretion. However, experiments to clearly delineate a role for direct inhibition of calcium channels have yet to be reported. This dopamine-mediated pathway may play an important role in neurons, where action potential generation (sensitive to potassium channel opening) and transmitter release (sensitive to calcium channel closing) occur at distinct sites, cell body and nerve terminal, respectively.

4.5. PRL Rebound

One of the crucial aspects of dopaminergic regulation of PRL secretion in vivo is the pulsatile nature of PRL release, and the importance of dopamine rebound in initiating PRL release *(117)*. Normally, dopamine from the hypothalamus inhibits secretion of PRL from the pituitary. However, removal of dopamine results in a rebound effect wherein the level of PRL secretion rises quickly upwards and exceeds the predopamine state *(1,2)*. This effect has been reproduced in vitro using perifusion of pituitary cells, and a similar effect is observed: On dopamine removal, an exaggerated peak PRL secretion is induced, which is equivalent to a large stimulation above the predopamine baseline level of PRL *(33,34,117,118)*. This phenomenon is known as PRL rebound, and its mechanism is crucial to the physiological regulation of PRL secretion. For example, in infants, dopamine rebound is quite dramatic. Rebound begins within 20 min of dopamine removal and PRL levels increase 10-fold within 2 h to 1 d *(119)*.

One hypothesis is that PRL rebound is caused by low concentrations of dopamine which persist following dopamine washout and lead to stimulation of PRL secretion *(30,31)*. As discussed, the pharmacology and mechanisms of the response to low concentrations ($<10^{-10}M$) of dopamine, which appear too low to activate known receptors, have yet to be clearly described.

Experiments using primary cultures of identified lactotrophs have demonstrated that the rebound effect is correlated with increased calcium spiking and can be blocked by the calcium channel blocker verapamil *(34)*. Furthermore, PRL rebound is not directly related to either cAMP or inositol phosphate levels, although activation of cAMP levels by forskolin has been shown to augment the rebound *(118)*. Voltage-gated calcium channels are activated by a hyperpolarizing conditioning pulse, such as occurs during dopamine exposure. Dopamine stimulates the activity of single potassium channels *(60,61)* to hyperpolarize the cell membrane. It was shown that the extent of calcium channel activation and PRL rebound correlates with the extent of membrane hyperpolarization. More convincingly, valinomycin, which directly hyperpolarizes the cell, mimics dopamine by inducing a rebound in PRL secretion on removal; simply blocking the calcium channel (e.g., with verapamil) does not produce a rebound *(33)*. Therefore, it has been suggested that dopamine-mediated hyperpolarization acts to remove inhibition from verapamil-sensitive calcium channels and that these slowly inactivating channels continue to mediate a calcium influx on dopamine removal *(33,34)*. This calcium influx, in turn, may stimulate PRL release above basal levels. The concept that prolonged exposure to an inhibitory agent like dopamine can lead to alterations in the cell which augment its subsequent responsiveness can have important general implications, particularly in the nervous and endocrine systems where ion channel inactivation may play important roles.

5. Conclusion: Dopaminergic Regulation of Secretion

The actions of dopamine in pituitary cells can be summarized (*see* Fig. 1). Dopamine actions in the lactotroph are mediated via the two forms of D_2 receptor, D_{2S}, and D_{2L}. These receptors have similar signaling properties, however they differ in their regulation by phosphorylation, and in their G protein selectivities. The primary action of dopamine is to inhibit cAMP generation, which appears to mediate dopamine-induced inhibition of the actions of G_s-coupled receptors, particularly to enhance calcium influx and PRL secretion. Inhibition of cAMP does not, however, appear to influence the rate of basal PRL secretion, nor the sustained phase of enhanced secretion induced by receptors which couple to PI turnover pathways. Dopamine hyperpolarizes the membrane potential by opening potassium channels: This inhibits the firing of voltage-dependent calcium channels, decreases $[Ca^{2+}]_i$, and reduces the steady-state rate of PRL secretion, both basal and TRH-stimulated. Dopamine-induced hyperpolarization also leads to a loss of inactivation voltage-dependent calcium channels, which on dopamine removal leads to a hyperactivation, increased calcium influx, and PRL rebound of stimulated secretion. Finally, dopamine can also directly (via G_o) inhibit the opening of voltage-dependent calcium channels, but the importance of this pathway with regard to inhibition of secretion remains unclear. The possible roles of other reported actions of dopamine (e.g., activation of PKCε *[63]* and of tyrosine phosphatases *[62]*) with regard to regulation of pituitary hormone secretion are unclear at present. Similarly, the mechanisms by which $[Ca^{2+}]_i$ may regulate secretion *(120)* are complex and incompletely understood.

References

1. Ben-Jonathan, N. (1985) Dopamine, a prolactin-inhibiting hormone. *Endocrine Rev.* **6,** 564–589.
2. Leong, D. A., Frawley, L. S., and Neill, J. D. (1983) Neuroendocrine control of prolactin secretion. *Annu. Rev. Physiol.* **45,** 109–127.
3. Lu, K. H. and Meites, J. (1972) Effects of L-dopa on serum prolactin and PIF in intact and hypophysectomized, pituitary-grafted rats. *Endocrinology* **91,** 868–872.
4. Takahara, J., Arimura, A., and Schally, A. V. (1974) Suppression of prolactin release by a purified porcine PIF preparation and catecholamines infused in to a rat hypophysial portal vessel. *Endocrinology* **95,** 462–465.
5. Diefenbach, W. P., Carmel, P. W., Frantz, A. G., and Ferin, M. (1976) Suppression of prolactin secretion by L-dopa in the stalk sectioned rhesus monkey. *J. Clin. Endocrinol. Metab.* **43,** 638–642.
6. Leblanc, H., Lachelin, G. C. L., Abu-Fadil, S., and Yen, S. S. C. (1976) Effects of dopamine infusion on pituitary secretion in humans. *J. Clin. Endocrinol. Metab.* **43,** 668–674.
7. MacLeod, R. M. and Lehmeyer, J. E. (1974) Studies on the mechanism of the dopamine-mediated inhibition of prolactin secretion. *Endocrinology* **94,** 1077–1085.
8. Gibbs, D. M. and Neill, J. D. (1978) Dopamine levels in hypothalmic pituitary stalk blood in the rat are sufficient to inhibit prolactin secretion in vivo. *Endocrinology* **102,** 1895–1900.
9. Brown, G. M., Seeman, P., and Lee, T. (1976) Dopamine/neuroleptic receptors in basal hypothalamus and pituitary. *Endocrinology* **99,** 1407–1410.
10. Caron, M. G., Beaulieu, M., Raymond, J., Gagne, B., Drouin, J., Lefkowitz, R. J., and Labrie, F. (1978) Dopaminergic receptor in the anterior pituitary gland: correlation of [^{3}H]dihydroergocryptine binding with the dopaminergic control of prolactin release. *J. Biol. Chem.* **253,** 2244–2253.
11. DeLean, A., Kilpatrick, B. F., and Caron, M. G. (1982) Dopamine receptor of the porcine anterior pituitary gland: evidence for two affinity states discriminated by both agonists and antagonists. *Mol. Pharmacol.* **22,** 290–297.
12. Birnbaumer, L. (1992) Receptor-to-effector signaling through G proteins: roles for βγ dimers as well as α subunits. *Cell* **71,** 1069–1072.

13. Dolphin, A. C. (1987) Nucleotide binding proteins in signal transduction and disease. *Trends Neurosci.* **10,** 53–57.
14. Cronin, M. J., Myers, G. A., MacLeod, R. M., and Hewlett, E. L. (1983) Pertussis toxin uncouples dopamine agonist inhibition of prolactin release. *Am. J. Physiol.* **244,** E499–E504.
15. Ohara, K., Haga, K., Berstein, G., Haga, T., Ichiyama, A., and Ohara, K. (1988) The interaction between D-2 dopamine receptors and GTP-binding proteins. *Mol. Pharmacol.* **33,** 290–296.
16. Senogles, S. E., Benovic, J. L., Amlaiky, N., Unson, C., Milligan, G., Vinitsky, R., Spiegel, A., and Caron, M. G. (1987) The D2-dopamine receptor of anterior is functionally associated with a pertussis toxin-sensitive guanine nucleotide binding protein. *J. Biol. Chem.* **262,** 4860–4867.
17. Senogles, S. E., Spiegel, A. M., Padrell, E., Iyengar, R., and Caron, M. G. (1990) Specificity of receptor–G protein interactions. *J. Biol. Chem.* **265,** 4507–4514.
18. Bunzow, J. R., Van Tol, H. H. M., Grandy, D. K., Albert, P., Salon, J., Christie, M., Machida, C. A., Neve, K. A., and Civelli, O. (1988) Cloning and expression of a rat D2 dopamine receptor cDNA. *Nature* **336,** 783–787.
19. Civelli, O., Bunzow, J. R., Grandy, D. K., Zhou, Q.-Y., and Van Tol, H. H. M. (1991) Molecular biology of the dopamine receptors. *Eur. J. Pharmacol.* **207,** 277–286.
20. Civelli, O., Bunzow, J. R., and Grandy, D. K. (1993) Molecular diversity of the dopamine receptors. *Annu. Rev. Pharmacol. Toxicol.* **32,** 281–307.
21. Dal Toso, R., Sommer, B., Ewert, M., Herb, A., Pritchett, D. B., Bach, A., Shivers, B. D., and Seeburg, P. H. (1989) The dopamine D2 receptor: two molecular forms generated by alternative splicing. *EMBO J.* **8,** 4025–4034.
22. Giros, B., Sokoloff, P., Martres, M.-P., Riou, J.-F., Emorine, L. J., and Schwartz, J. C. (1989) Alternative splicing direct the two D2 dopamine receptor isoforms. *Nature* **342,** 923–926.
23. Grandy, D. K., Marchionni, M. A., Makam, H., Stofko, R. E., Alfano, M., Frothingham, L., Fischer, J. B., Burke-Howie, K. J., Bunzow, J. R., Server, A. C., and Civelli, O. (1989) Cloning of the cDNA and gene for a human D2 dopamine receptor. *Proc. Natl. Acad. Sci. USA* **86,** 9762–9766.
24. Monsma, F. J., McVittie, L. D., Gerfen, C. R., Mahan, S. C., and Sibley, D. R. (1989) Multiple D2 receptors produced by alternative RNA splicing. *Nature* **342,** 926–929.
25. Ostrowski, J., Kjelsberg, M. A., Caron, M. G., and Lefkowitz, R. J. (1992) Mutagenesis of the β_2-adrenergic receptor: how structure elucidates function. *Annu. Rev. Pharmacol. Toxicol.* **32,** 167–183.
26. Castro, S. W. and Strange, P. G. (1993) Differences in the ligand binding properties of the short and long versions of the D2 dopamine receptor. *J. Neurochem.* **60,** 372–375.
27. Malmberg, A., Jackson, D. M., Eriksson, A., and Mohell, N. (1993) Unique binding characteristics of antipsychotic agents interacting with human dopamine D2A, D2B, and D3 receptors. *Mol. Pharmacol.* **43,** 749–754.
28. Porter, T. E., Grandy, D., Bunzow, J., Wiles, C. D., Civelli, O., and Frawley, L. S. (1994) Evidence that stimulatory dopamine receptors may be involved in the regulation of prolactin secretion. *Endocrinology* **134,** 1263–1268.

29. Liu, Y. F., Civelli, O., Zhou, Q.-Y., and Albert, P. R. (1992) Cholera toxin-sensitive 3', 5'-cyclic adenosine monophosphate and calcium signals of the human dopamine-D1 receptor: selective potentiation by protein kinase A. *Mol. Endocrinol.* **6,** 1815–1824.
30. Burris, T. P., Nguyen, D. N., Smith, S. G., and Freeman, M. E. (1992) The stimulatory and inhibitory effects of dopamine on prolactin secretion involved different G-proteins. *Endocrinology* **130,** 926–932.
31. Burris, T. P. and Freeman, M. E. (1993) Low concentrations of dopamine increase cytosolic calcium in lactotrophs. *Endocrinology* **133,** 63–68.
32. Denef, C., Manet, D., and Dewals, R. (1980) Dopaminergic stimulation of prolactin release. *Nature* **285,** 243–246.
33. Gregerson, K. A., Golesorkhi, N., and Chuknyiska, R. (1994) Stimulation of prolactin release by dopamine withdrawal: role of membrane hyperpolarization. *Am. J. Physiol.* **267,** E781–E788.
34. Gregerson, K. A., Chuknyiska, R., and Golesorkhi, N. (1994) Stimulation of prolactin release by dopamine withdrawal: role of calcium influx. *Am. J. Physiol.* **267,** E789–E794.
35. Neill, J. D. and Frawley, L. S. (1983) Detection of hormone release from individual cells in mixed populations using a reverse hemolytic plaque assay. *Endocrinology* **112,** 1135–1137.
36. Tashjian, A. H., Jr. (1979) Clonal strains of hormone-producing pituitary cells. *Methods Enzymol.* **58,** 527–535.
37. Tashjian, A. H., Jr., Yasumura, Y., Levine, L., Sato, G. H., and Parker, M. L. (1968) Establishment of clonal strains of rat pituitary tumor cells that secrete growth hormone. *Endocrinology* **82,** 342–352.
38. Frawley, L. S. and Boockfor, F. R. (1991) Mammosomatotropes: presence and functions in normal and neoplastic pituitary tissue. *Endo. Rev.* **12,** 337–355.
39. Porter, T., Hill, J. B., Wiles, C. D., and Stephen, L. (1990) Is the mammosomatotroph a transitional cell for the functional interconversion of growth hormone- and prolactin-secreting cells? Suggestive evidence from virgin, gestating, and lactating rats. *Endocrinology* **127,** 2789–2794.
40. Albert, P. R. (1994) Heterologous expression of G protein-coupled receptors in pituitary and fibroblast cell lines. *Vitamins Hormones* **48,** 59–109.
41. Albert, P. R., Neve, K., Bunzow, J., and Civelli, O. (1990) Coupling of a rat dopamine D2 receptor to inhibition of adenylyl cyclase and prolactin secretion. *J. Biol. Chem.* **265,** 2098–2104.
42. Cronin, M. J., Faure, N., Martial, J. A., and Weiner, R. I. (1980) Absence of high affinity dopamine receptor in GH3 cells: a prolactin secreting clone resistant to the inhibitory action of dopamine. *Endocrinology* **106,** 718–723.
43. Missale, C., Boroni, F., Sigala, S., Castelletti, L., Falardeau, P., Dal Toso, R., Balsari, A., and Spano, P. (1994) Epidermal growth factor promotes uncoupling from adenylyl cyclase of the rat D2S receptor expressed in GH4C1 cells. *J. Neurochem.* **62,** 907–915.
44. Missale, C., Boroni, F., Castelletti, L., Dal Toso, R., Gabellini, N., Sigala, S., and Spano, P. (1991) Lack of coupling of D-2 receptors to adenylate cyclase in GH-3 cells exposed to epidermal growth factor. *J. Biol. Chem.* **266,** 23,392–23,398.
45. Missale, C., Boroni, F., Sigala, S., Zanellato, A., Dal Toso, R., Balsari, A., and Spano, P. (1994) Nerve growth factor directs differentiation of the bipotential cell line GH-3 into the mammotroph phenotype. *Endocrinology* **136,** 290–298.

46. Enjalbert, A. and Bockaert J. (1983) Pharmacological characterization of the D2 dopamine receptor negatively coupled with adenylate cyclase in rat anterior pituitary. *Mol. Pharmacol.* **23,** 576–584.
47. De Camilli, P., Macconi, D., and Spada, A. (1979) Dopamine inhibits adenylate cyclase in human prolactin-secreting pituitary tumors. *Nature* **278,** 252–254.
48. Biales, M., Dichter, M. A., and Tischler, A. (1977) Sodium and calcium action potential in pituitary cells. *Nature* **267,** 172–174.
49. Kidokoro, Y. (1975) Spontaneous calcium action potentials in a clonal pituitary cell line and their relationship to prolactin secretion. *Nature* **258,** 741,742.
50. Schlegel, W., Winiger, B. P., Mollard, P., Vacher, P., Wuarin, F., Zahnd, G. R., Wollheim, C. B., and Dufy, B. (1987) Oscillations of cytosolic Ca^{2+} in pituitary cells due to action potentials. *Nature* **329,** 719–721.
51. Taraskevich, P. S. and Douglas, W. W. (1977) Action potentials occur in cells of the normal anterior pituitary gland and are stimulated by the hypophysiotropic peptide thyrotropin-releasing hormone. *Proc. Natl. Acad. Sci. USA* **74,** 4064–4067.
52. Taraskevich, P. S. and Douglas, W. W. (1978) Catecholamines of supposed inhibitory hypophysiotropic function suppress action potentials in prolactin cells. *Nature* **276,** 832–834.
53. Taraskevich, P. S. and Douglas, W. W. (1980) Electrical behaviour in a line of anterior pituitary cells (GH cells) and the influence of the hypothalamic peptide, thyrotropin releasing factor. *Neuroscience* **5,** 421–431.
54. Burris, T. P. and Freeman, M. E. (1994) Comparison of the forms of the dopamine D2 receptor expressed in GH4C1 cells. *Proc. Soc. Exp. Med. Biol.* **205,** 226–235.
55. Liu, Y. F., Civelli, O., Grandy, D. K., and Albert, P. R. (1992) Differential sensitivity of the short and long human dopamine-D2 receptor subtypes to protein kinase C. *J. Neurochem.* **59,** 2311–2317.
56. Vallar, L., Claudia, M., Magni, M., Albert, P., Bunzow, J., Meldolesi, J., and Civelli, O. (1990) Differential coupling of dopaminergic D2 receptor expressed in different cell types. *J. Biol. Chem.* **265,** 10,320–10,326.
57. Liu, Y. F., Jakobs, K. H., Rasenick, M. M., and Albert, P. R. (1994) G protein specificity in receptor-effector coupling. Analysis of the roles of Go and Gi2 in GH4C1 pituitary cells. *J. Biol. Chem.* **269,** 13,880–13,886.
58. Lledo, P.-M., Legendre, P., Israel, J.-M., and Vincent, J.-D. (1990) Dopamine inhibits two characterized voltage-dependent calcium currents in identified rat lactotroph cells. *Endocrinology* **127,** 990–1001.
59. Seabrook, G. R., Knowles, M., Brown, N., Myers, J., Sinclair, H., Patel, S., Freedman, S. B., and McAllister (1994) Pharmacology of high-threshold calcium currents in GH4C1 pituitary cells and their regulation by activation of human D2 and D4 receptors. *Br. J. Pharmacol.* **112,** 728–734.
60. Einhorn, L. C., Gregerson, K. A., and Oxford, G. S. (1991) D2 dopamine receptor activation of potassium channels in identified rat lactotrophs: whole cell and single channel recording. *J. Neurosci.* **11,** 3727–3737.
61. Einhorn, L. C. and Oxford, G. S. (1993) Guanine nucleotide binding proteins mediate D2 dopamine receptor activation of a potassium channel in rat lactotrophs. *J. Physiol.* **462,** 563–578.

62. Florio, T., Pan, M., Newman, B., Hershberger, R. E., Civelli, O., and Stork, P. J. S. (1992) Dopaminergic inhibition of DNA synthesis in pituitary tumor cells is associated with phosphotyrosine phosphatase activity. *J. Biol. Chem.* **267,** 24,169–24,172.
63. Senogles, S. E. (1994) The D2 dopamine receptor mediates inhibition of growth in GH4ZR7 cells: involvement of protein kinase-Cε. *Endocrinology* **134,** 783–789.
64. Nilsson, C. and Eriksson, E. (1992) Partial dopamine D2 receptor agonists antagonize prolactin-regulating D2 receptors in a transfected clonal cell line (GH4ZR7). *Eur. J. Pharmacol.* **218,** 205–211.
65. Elsholtz, H. P., Lew, A. M., Albert, P. R., and Sundmark, V. C. (1991) Inhibitory control of prolactin and Pit-1 gene promoters by dopamine. Dual signaling pathways required for D2 receptor-regulated expression of the prolactin gene. *J. Biol. Chem.* **266,** 22,919–22,925.
66. Montmayeur, J.-P. and Borrelli, E. (1991) Transcription mediated by a cAMP-responsive promoter element is reduced upon activation of dopamine D2 receptors. *Proc. Natl. Acad. Sci. USA* **88,** 3135–3139.
67. Hayes, G., Biden, T. J., Selbie, L. A., and Shine, J. (1992) Structural subtypes of the dopamine D2 receptor are functionally distinct: expression of the cloned D2A and D2B subtypes in a heterologous cell line. *Mol. Endocrinol.* **6,** 920–926.
68. Lajiness, M. E., Chio, C. L., and Huff, R. M. (1993) D2 dopamine receptor stimulation of mitogenesis in transfected Chinese hamster ovary cells: relationship to dopamine stimulation of tyrosine phosphorylations. *J. Pharmacol. Exp. Ther.* **267,** 1573–1581.
69. Neve, K. A., Kozlowski, M. R., and Rosser, M. P. (1992) Dopamine D2 receptor stimulation of Na+/H+ exchange assessed by quantification of extracellular acidification. *J. Biol. Chem.* **267,** 25,748–25,753.
70. Di Marzo, V., Vial, D., Sokoloff, P., Schwartz, J.-C., and Piomelli, D. (1993) Selection of alternative Gi-mediated signaling pathways at the dopamine D2 receptor by protein kinase C. *J. Neurosci.* **13,** 4846–4853.
71. Kanterman, R. Y., Mahan, L. C., Briley, E. M., Monsma, F. J., Jr., Sibley, D. R., Axelrod, J., and Felder, C. C. (1990) Transfected D2 dopamine receptors mediate the potentiation of arachidonic acid release in Chinese hamster ovary cells. *Mol. Pharmacol.* **39,** 364–369.
72. Montmayeur, J.-P., Guiramand, J., and Borrelli, E. (1993) Preferential coupling between dopamine D2 receptors and G-proteins. *Mol. Endocrinol.* **7,** 161–170.
73. Clapham, D. E. and Neer, E. J. (1993) New roles for G-protein βγ-dimers in transmembrane signalling. *Nature* **365,** 403–406.
74. Albert, P. R. and Morris, S. J. (1994) Antisense knockouts: molecular scalpels for the dissection of signal transduction. *Trends Pharmacol. Sci.* **15,** 250–254.
75. Lledo, P.-M., Homburger, V., Bockeart, J., and Vincent, J.-D. (1992) Differential G protein-mediated coupling of D2 dopamine receptors to K^+ and Ca^{2+} currents in rat anterior pituitary cells. *Neuron* **8,** 455–463.
76. Kleuss, C., Hescheler, J., Ewel, C., Rosenthal, W., Schultz, G., and Wittig, B. (1991) Assignment of G-protein subtypes to specific receptors inducing inhibition of calcium currents. *Nature* **353,** 43–48.
77. Kleuss, C., Scherübl, H., Hescheler, J., Schultz, G., and Wittig, B. (1992) Different β-subunits determine G-protein interaction with transmembrane receptors. *Nature* **358,** 424–426.

78. Kleuss, C., Scherübl, H., Hescheler, J., Schultz, G., and Wittig, B. (1993) Selectivity in signal transduction determined by γ subunits of heterotrimeric G proteins. *Science* **259,** 832–834.
79. Yatani, A., Codina, J., Sekura, R. D., Birnbaumer, L., and Brown, A. M. (1987) Reconstitution of somatostatin and muscarinic receptor mediated stimulation of K^+ channels by isolated G_K protein in clonal rat anterior pituitary membranes. *Mol. Endocrinol.* **1,** 283–289.
80. Yatani, A., Mattera, R., Codina, J., Graf, R., Okabe, K., Padrell, E., Iyengar, R., Brown, A. M., and Birnbaumer, L. (1988) The G protein-gated atrial K^+ channel is stimulated by three distinct $G_i\alpha$ subunits. *Nature* **336,** 680–682.
81. Baertschi, A. J., Audigier, Y., Lledo, P.-M., Israel, J.-M., Bockaert, J., and Vincent, J.-D. (1992) Dialysis of lactotropes with antisense oligonucleotides assigns guanine nucleotide binding protein subtypes to their channel effectors. *Mol. Endocrinol.* **6,** 2257–2265.
82. White, R. E., Schonbrunn, A., and Armstrong, D. L. (1991) Somatostatin stimulates Ca^{2+}-activated K^+ channels through protein dephosphorylation. *Nature* **351,** 570–573.
83. Senogles, S. E. (1994) The D2 dopamine receptor isoforms signal through distinct Giα proteins to inhibit adenylyl cyclase. A study with site-directed mutant Giα proteins. *J. Biol. Chem.* **269,** 23,120–23,127.
84. Dorflinger, L. J. and Schonbrunn, A. (1983) Somatostatin inhibits vasoactive intestinal peptide-stimulated cyclic adenosine monophosphate accumulation in GH pituitary cells. *Endocrinology* **113,** 1541–1550.
85. Dorflinger, L. J. and Schonbrunn, A. (1983) Somatostatin inhibits basal and vasoactive intestinal peptide-stimulated hormone release by different mechanisms in GH pituitary cells. *Endocrinology* **113,** 1551–1560.
86. Gourdji, D., Bataille, D., Vauclin, N., Grouselle, D., Rosselin, G., and Tixier-Vidal, A. (1979) Vasoactive intestinal peptide (VIP) stimulates prolactin (PRL) release and cAMP production in a rat pituitary cell line (GH3/B6). Additive effects of VIP and TRH on PRL release. *FEBS Lett.* **104,** 165–168.
87. Guild, S. and Drummond, A. H. (1984) Vasoactive-intestinal-polypeptide-stimulated adenosine 3',5'-cyclic monophosphate accumulation in GH3 pituitary tumour cells. *Biochem. J.* **221,** 789–796.
88. Cooper, D. M. F., Bier-Laning, C. M., Halford, M. K., Ahlijannian, M. K., and Zahniser, N. R. (1986) Dopamine, acting through D2 receptors, inhibits rat striatal adenylate cyclase by a GTP-dependent process. *Mol. Pharmacol.* **29,** 113–119.
89. Bates, M. D., Senogles, S. E., Bunzow, J. R., Liggett, S. B., Civelli, O., and Caron, M. G. (1990) Regulation of responsiveness at D2 dopamine receptors by receptor desensitization and adenylyl cyclase sensitization. *Mol. Pharmacol.* **39,** 55–63.
90. Delbeke, D., Scammell, J. G., Martinez-Campos, A., and Dannies, P. S. (1986) Dopamine inhibits prolactin release when cyclic adenosine 3'5'-cyclic monophosphate levels are elevated. *Endocrinology* **118,** 1271–1277.
91. Koch, B. D., Dorflinger, L. J., and Schonbrunn, A. (1985) Pertussis toxin blocks both cyclic AMP-mediated and cyclic AMP-independent actions of somatostatin. Evidence for coupling of Ni to decreases in intracellular free calcium. *J. Biol. Chem.* **260,** 13,138–13,145.

92. Koch, B. D., Blalock, J. B., and Schonbrunn, A. (1988) Characterization of the cyclic AMP-independent actions of somatostatin in GH cells. I. An increase in potassium conductance is responsible for both the hyperpolarization and the decrease in intracellular free calcium produced by somatostatin. *J. Biol. Chem.* **263,** 216–225.
93. Koch, B. D. and Schonbrunn, A. (1988) Characterization of the cyclic AMP-independent actions of somatostatin in GH cells. II. An increase in potassium conductance initiates somatostatin-induced inhibition of prolactin secretion. *J. Biol. Chem.* **263,** 226–234.
94. Gardette, R., Rasolonjanahary, R., Kordon, C., and Enjalbert, A. (1994) Epidermal growth factor treatment induces D2 dopamine receptors functionally coupled to delayed outward potassium current (I_K) in GH4C1 clonal anterior pituitary cells. *Neuroendocrinology* **59,** 10–19.
95. Memo, M., Castelletti, L., Missale, C., Valerio, A., Carruba, M., and Spano, P. (1986) Dopaminergic inhibition of prolactin release and calcium influx induced by neurotensin in anterior pituitary is independent of cyclic AMP system. *J. Neurochem.* **47,** 1689–1695.
96. Cohen, C. J. and McCarthy, R. T. (1987) Nimodipine block of calcium channels in rat anterior pituitary cells. *J. Physiol.* **387,** 195–225.
97. Albert, P. R. and Tashjian, A. H., Jr. (1984) Thyrotropin-releasing hormone-induced spike and plateau in cytosolic free Ca^{++} concentrations in pituitary cells: relation to prolactin release. *J. Biol. Chem.* **259,** 5827–5832.
98. Albert, P. R. and Tashjian, A. H., Jr. (1984) Relationship of thyrotropin-releasing hormone-induced spike and plateau phases in cytosolic free Ca^{++} concentrations to hormone secretion: selective blockade using ionomycin and nifedipine. *J. Biol. Chem.* **259,** 15,350–15,363.
99. Aragay, A. M., Katz, A., and Simon, M. I. (1992) The $G\alpha_q$ and $G\alpha_{11}$ proteins couple the thyrotropin-releasing hormone receptor to phospholipase C in GH_3 rat pituitary cells. *J. Biol. Chem.* **267,** 24,983–24,988.
100. Berridge, M. J. (1993) Inositol trisphosphate and calcium signalling. *Nature* **361,** 315–325.
101. Nishizukà, Y. (1988) The molecular heterogeneity of protein kinase C and its implications for cellular regulation. *Nature* **334,** 661–665.
102. Aizawa, T. and Hinkle, P. M. (1985) Thyrotropin-releasing hormone rapidly stimulates a biphasic secretion of prolactin and growth hormone in GH4C1 rat pituitary tumor cells. *Endocrinology* **116,** 73–82.
103. Albert, P. R. and Tashjian, A. H., Jr. (1985) Dual actions of phorbol esters on cytosolic free Ca^{++} concentrations and reconstitution with ionomycin of acute thyrotropin-releasing hormone responses. *J. Biol. Chem.* **260,** 8746–8759.
104. Albert, P. R. and Tashjian, A. H., Jr. (1986) Ionomycin acts as an ionophore to release TRH-regulated Ca^{++} stores from GH4C1 cells. *Am. J. Physiol.* **251,** C887–C891.
105. Albert, P. R., Wolfson, G., and Tashjian, A. H., Jr. (1987) Diacylglycerol increases cytosolic free Ca^{++} concentration in rat pituitary cells. Relationship to thyrotropin-releasing hormone action. *J. Biol. Chem.* **262,** 6577–6581.
106. Dubinski, J. M. and Oxford, G. S. (1985) Dual modulation of K channels by thyrotropin-releasing hormone in clonal pituitary cells. *Proc. Natl. Acad. Sci. USA* **82,** 4282–4286.

107. Gollasch, M., Kleuss, C., Hescheler, J., Wittig, B., and Schulz, G. (1993) Gi2 and protein kinase C are required for thyrotropin-releasing hormone-induced stimulation of voltage-dependent Ca^{2+} channels in rat pituitary GH3 cells. *Proc. Natl. Acad. Sci. USA* **90,** 6265–6269.
108. Law, G. J., Pachter, J. A., and Dannies, P. S. (1988) Dopamine has no effect on thyrotropin-releasing hormone mobilization of calcium from intracellular stores in rat anterior pituitary cells. *Mol. Endocrinol.* **2,** 966–972.
109. Malgaroli, A., Vallar, L., Elahi, F. R., Pozzan, T., Spada, A., and Meldolesi, J. (1987) Dopamine inhibits cytosolic Ca^{2+} increases in rat lactotroph cells. *J. Biol. Chem.* **262,** 13,920–13,927.
110. Vallar, L., Vicentini, L. M., and Meldolesi, J. M. (1988) Inhibition of inositol phosphate production is a late, Ca^{2+}-dependent effect of D2 dopaminergic receptor activation in rat lactotroph cells. *J. Biol. Chem.* **263,** 10,127–10,134.
111. Vallar, L. and Meldelosi, J. (1989) Mechanisms of signal transduction at the dopamine D2 receptor. *Trends Pharmacol. Sci.* **10,** 74–77.
112. Kineman, R. D., Gettys, T. W., and Frawley, L. S. (1994) Paradoxical effects of dopamine (DA): Giα3 mediates DA inhibition of PRL release while masking its PRL-releasing activity. *Endocrinology* **135,** 790–793.
113. Albert, P. R. and Raquidan, D. (1995) Dopamine-D2 receptor routing through multiple G proteins for inhibition of TRH-stimulated prolactin secretion. *Proc. Endocrine Soc.* **77,** 113.
114. Armstrong, C. M. and Matteson, D. R. (1985) Two distinct populations of calcium channels in a clonal line of pituitary cells. *Science* **227,** 65–67.
115. Enjalbert, A., Musset, F., Chenard, C., Priam, M., Kordon, C., and Heisler, S. (1988) Dopamine inhibits prolactin secretion stimulated by the calcium channel agonist Bay-K-8644 through a pertussis toxin-sensitive G protein in anterior pituitary cells. *Endocrinology* **123,** 406–412.
116. Rendt, J. and Oxford, G. S. (1994) Absence of coupling between D2 dopamine receptors and calcium channels in lactotrophs from cycling female rats. *Endocrinology* **135,** 501–508.
117. Denef, C., Baes, M., and Schramme, C. (1984) Stimulation of prolactin secretion after short term or pulsatile exposure to dopamine in superfused anterior pituitary cell aggregates. *Endocrinology* **114,** 1371–1378.
118. Chen, C., Zheng, J., Israel, J. M., Clarke, I. J., and Vincent, J. D. (1993) Mechanism of the prolactin rebound after withdrawal in rat pituitary cells. *Am. J. Physiol.* **265,** E145–E152.
119. Van den Berghe, G., De Zegher, F., and Lauwers, P. (1994) Dopamine suppresses pituitary function in infants and children. *Crit. Care Med.* **22,** 1747–1753.
120. Lledo, P.-M., Vernier, P., Vincent, J.-D., Mason, W. T., and Zorec, R. (1993) Inhibition of Rab3B expression attenuates Ca^{2+}-dependent exocytosis in rat anterior pituitary cells. *Nature* **364,** 540–543.

CHAPTER 13

Regulation of Dopamine Receptor Function and Expression

David R. Sibley and Kim A. Neve

1. Introduction

Regulation of receptor responsiveness by neurotransmitters and hormones is a well-recognized phenomenon that has been demonstrated for most receptor systems *(1)*. Such regulation can involve desensitization, the tendency of receptor responsiveness to wane over time despite the presence of a stimulus of constant intensity; or amplification, in which the receptor system becomes supersensitive to agonist stimulation. Regulation of receptor function and expression may limit the efficacy of numerous pharmacological agents, and thus the investigation of such phenomena may have a major impact in refining and developing therapeutic agents.

One of the best studied receptor systems with respect to agonist-induced forms of regulation is the beta adrenergic receptor-coupled adenylate cyclase enzyme system. Desensitization of this receptor/enzyme system involves multiple events including covalent/functional modification of the receptor by cyclic adenosine monophosphate (cAMP)-dependent and receptor-specific protein kinases, sequestration or internalization of the receptors, as well as receptor degradation *(2)*. Long-term regulation of beta adrenergic receptor expression also can be controlled through transcriptional and translational mechanisms *(3)*. Although the beta adrenergic receptor has been a useful model system for the study of receptor regulation, recent findings indicate that the regulatory events controlling this system may not be applicable to all catecholamine receptor subtypes.

Altered dopamine receptor function has been implicated in numerous neurological and endocrine disorders including schizophrenia, Parkinson's disease, Tourette's syndrome, tardive dyskinesia, Huntington's chorea, and

The Dopamine Receptors Eds.: K. A. Neve and R. L. Neve
Humana Press Inc., Totowa, NJ

hyperprolactinemia *(4)*. In many instances, these disorders and/or their therapy have been suggested to involve aberrant regulatory mechanisms of the dopaminergic receptor systems *(5)*. A wide variety of in vivo investigations have, in fact, documented that dopamine receptors are subject to dynamic regulation in both a positive and negative fashion *(6)*. Although there is a large body of work describing regulation of dopamine receptors, the underlying biochemical or molecular mechanisms are just now beginning to be addressed. In large part, this has been the result of the inherent limitations of in vivo experimentation and the lack of suitable simple model systems for in vitro investigations of dopamine receptor function and regulation. Recently, this limitation has been largely overcome through the identification of cell lines that either express endogenous dopamine receptors or have been genetically engineered to express cloned dopamine receptor subtypes. The cloning and characterization of specific dopamine receptor subtypes has also provided useful protein and mRNA probes that facilitate in vitro and in vivo studies of dopamine receptor regulation. The aim of this chapter is to review these recent studies and to discuss what is known about the mechanisms of regulation for each of the dopamine receptor subtypes.

2. D_1* Receptors

2.1. In Vivo Regulation

Numerous in vivo studies have indicated that the D_1 receptor is subject to multiple forms of regulation. It is not the intent of this chapter to comprehensively review all of the literature in this field but to focus on recent studies that have examined the molecularly defined D_1 receptor subtype (also termed the D_{1A} receptor). The reader is referred to several recent reviews *(7–10)* as well as a previous review *(6)* that discuss some of the literature in this area. Given that the D_5 (also termed the D_{1B}) receptor is considerably less abundant and expressed in a more restricted pattern in rat brain when compared with the D_1 receptor *(11,12)*, it is likely that most of the in vivo experimentation in rodents has involved the latter receptor subtype. Thus, we use the D_1 designation for the genetic subtype, rather than using the terms D1 or D1-like, which refer to the pharmacological class.

2.1.1. Receptor Stimulation

The in vivo effects of receptor stimulation on subsequent D_1 receptor function have been explored through the administration of either direct acting

*As detailed in the Preface, the terms D1 or D1-like and D2 or D2-like are used to refer to the subfamilies of DA receptors, or used when the genetic subtype (D_1 or D_5; D_2, D_3, or D_4) is uncertain.

agonists or stimulant drugs, such as amphetamine or cocaine, that promote the release and/or block the reuptake of dopamine. For instance, the administration of amphetamine has been reported to result in desensitization of D_1 receptor-coupled adenylate cyclase activity *(13,14)*. However, since such stimulant drugs affect multiple receptor/neuronal systems in vivo, it cannot be assumed that their effects are entirely caused by activation of the D_1 receptor by dopamine. This discussion is thus limited to those few animal studies in which direct acting receptor agonists were administered.

Some early studies using L-DOPA administration to elevate brain dopamine levels in rats evaluated the effects on D_1 receptor-stimulated adenylate cyclase activity in brain slices or homogenates. These treatments were found to either decrease the potency of dopamine stimulation *(15)* or abolish it altogether *(16)*, suggesting that elevating synaptic levels of dopamine leads to desensitization of D_1 receptor coupling to adenylate cyclase. More recent studies have examined the effects of chronic treatment with the D1-selective agonist SKF 38393. McGonigle and colleagues have shown that daily treatment of rats with 5 mg/kg of SKF 38393 has no effect on D_1 receptor binding activity in normal animals, whereas this treatment decreases the density of D_1 receptors in dopamine-depleted reserpinized animals *(17)*. A higher dose of SKF 38393 (10 mg/kg) increases D_1 receptor binding in normal rats *(18)*. The up-regulation of D_1 receptors in response to chronic high dose SKF 38393 may be attributed to the partial agonist properties of SKF 38393 such that it functions to attenuate the action of endogenous dopamine. In contrast, in dopamine-depleted animals, the agonist properties of SKF 38393 act to down-regulate the receptor. Obviously, it will be important to repeat such studies using newly developed D1-selective agonists with full receptor efficacy (*see* Chapter 4).

2.1.2. Receptor Blockade

Subsequent to the introduction of SCH 23390 as the prototypic D1-selective antagonist in the mid 1980s, several studies examined the effects of chronic, selective blockade of D_1 receptors in rodents. Chronic SCH 23390 treatment up-regulates D_1 receptors in both brain *(19–23)* and retinal *(24)* tissue. In those studies that examined D_1 receptor-stimulated adenylate cyclase activity *(19,23,24)*, it was generally found that this enzyme activity is increased as well. Thus, it appears that chronic in vivo blockade of the D_1 receptor results in up-regulation of receptor expression which is accompanied by increased receptor function. Recently, mRNA analyses have shown that chronic SCH 23390 treatment is associated with increased D_1 receptor mRNA levels in striatum *(25)*. Similarly, Buckland et al. *(26,27)* showed that chronic antagonist treatment with high doses of loxapine and haloperidol, but not clozapine and sulpiride, leads to increases in whole brain D_1 receptor mRNA. Chronic blockade of D_1 receptors thus results in increased receptor synthesis through elevated D_1 receptor mRNA.

2.1.3. Dopamine Depletion

The most common method of producing a dopamine depletion is the use of the catecholamine neurotoxin, 6-hydroxydopamine (6-OHDA), although the dopamine neurotoxin, 1-methyl-4-phenyl-1,2,3,6-tetrahydropyridine (MPTP), and reserpine, which depletes vesicular dopamine but leaves dopaminergic fibers intact, also have been used. Such experimental paradigms are of great clinical relevance since, in humans, Parkinson's disease is associated with an idiopathic, progressive degeneration of brain dopaminergic neurons. It has been proposed that one compensatory mechanism in patients with Parkinson's disease that delays the appearance of behavioral symptoms until the loss of dopamine is severe, and that contributes to enhanced responsiveness to L-DOPA, is denervation supersensitivity *(28)*. Either D1- or D2-like receptors could contribute to this hypothetical compensatory denervation supersensitivity, and the density of both classes of dopamine receptors is elevated in the caudate-putamen in Parkinson's patients, as determined from postmortem tissue (*29,30*; *see also* ref. *31*) and by in vivo brain imaging studies *(32,33)*. The search for a model of dopaminergic supersensitivity to corroborate this hypothesis led to the demonstration in rats that depletion of forebrain dopamine through damage to the ascending dopaminergic fibers produces postsynaptic behavioral supersensitivity to dopamine receptor agonists *(34)*.

Among the first observations of denervation supersensitivity in the dopamine system were reports of enhanced dopamine-stimulated adenylate cyclase activity, a response mediated by D_1 receptors *(35,36)*. Following the development of the D1-selective radioligand [^{3}H]SCH 23390 *(37)* and probes for D_1 receptor mRNA, it was surprising that studies on the effect of dopamine depletion on the density of D_1 receptors and the abundance of D_1 receptor mRNA yielded apparently contradictory results. Some studies described denervation- or reserpine-induced increases in D_1 receptor density *(38,39)* and D_1 receptor mRNA *(40)*, whereas others have reported no change in the density of neostriatal D_1 receptors after chronic reserpine treatment *(41,42)* or inconsistent changes after 6-OHDA-induced dopamine depletion *(43)*. Jongen-Rêlo et al. *(44)* found increased D_1 binding but decreased D_1 mRNA in the rat nucleus accumbens and neostriatum after 6-OHDA-induced denervation. Joyce *(45)* determined that chronic reserpine treatment increased the density of neostriatal D_1 receptors, whereas extensive (>90%) 6-OHDA-induced depletions of neostriatal dopamine decreased D_1 receptor density by 15–18% *(46)*. In the study of Marshall et al. *(47)*, neostriatal D_1 receptor density was reduced by 10 and 18% at 2 wk and 11 mo, respectively, after 6-OHDA treatment. These results correspond closely with the magnitude of the decreased D_1 receptor binding and D_1 receptor mRNA observed in other studies *(48,49)*. Qin et al. *(49)* further determined that the decreased receptor density is associated with a reduced rate of synthesis of D_1 receptors, consistent with the reduced mRNA levels. In the study of Gerfen et al. *(48),* denervation-induced loss of D_1 receptor mRNA was reversed by intermittent, but not continuous, treatment with the agonist SKF 38393.

Thus, although there appears to be a functional (especially behavioral) supersensitivity of the D_1 receptor system following dopamine depletion, the results are somewhat mixed with respect to the effects on D_1 receptor density, with most of the data indicating a decrease in D_1 receptor synthesis.

2.2. *In Vitro Regulation*

In vitro studies of D_1 receptor regulation have involved a variety of model systems including brain tissue slices, primary cultures of neuronal cells, cell lines that endogenously express the D_1 subtype, and more recently, cells that have been transfected with the cloned D_1 receptor. As noted in Section 2.1., given the low abundance and limited distribution of the D_5 receptor subtype, it is probable that the following studies involving tissue slices and dispersed neuronal cells involved only the D_1 receptor subtype.

2.2.1. Regulation of Endogenous D_1 Receptors

One of the earliest reports of in vitro regulation of D_1 receptor function was that of Memo et al. *(50),* demonstrating that preincubation of striatal slices with dopamine or D1 receptor agonists reduces dopamine-stimulated adenylate cyclase activity by 50–60%. This desensitization response is maximal within 30 min of dopamine incubation and is caused by a decrease in both the potency and efficacy for dopamine stimulation of the enzyme. A more recent report using striatal and retinal slices showed that preincubating these tissues for 60 min with dopamine or D1 agonists results in desensitization of dopamine-stimulated adenylate cyclase activity, predominantly resulting from a decrease in the maximum response *(51)*. Preincubation of these tissues with forskolin, an agent used to elevate intracellular cAMP, mimics the action of dopamine, suggesting an involvement of cAMP in the desensitization response. Interestingly, under the preincubation conditions used in this study, there was no effect on the D_1 receptor density as detected using [^{3}H]SCH 23390 binding.

Dopamine-induced desensitization of the D_1 receptor also has been examined using striatal neurons in culture *(52)*. Incubating these cells with dopamine for 15 min results in a 50% decrease in the maximum stimulation of adenylate cyclase by dopamine, as well as a small decrease in potency. Further incubation with dopamine for 18 h almost totally abolishes the

D_1 response. Dopamine treatment also decreases stimulation of adenylate cyclase by beta adrenergic, serotonin, and vasoactive intestinal peptide receptors. Interestingly, the rate of resensitization of the dopamine response depends on the time of dopamine preincubation. Full recovery is observed within 15 min following a 15 min preincubation; however, after 18 h of dopamine treatment, the cells require 2 d of culture to recover fully. Recently, these investigators have shown that the rate of recovery from short term (15 min) desensitization can be accelerated by other transmitters *(53)*. For instance, glutamate accelerates the resensitization of the D_1 response in striatal slices whereas alpha-1 adrenergic agonists accelerate the recovery in cortical slices. The mechanisms underlying these effects are unknown.

Further progress in characterizing D_1 receptor desensitization has come from the use of cell lines that express the endogenous D_1 receptor. In the Sibley laboratory, we have utilized the mouse NS20Y cell which expresses moderate levels of D_1 receptors *(54)*. D1 agonist treatment of these cells results in decreased dopamine-stimulated adenylate cyclase activity that is owing to a decreased maximum response with no change in the potency of dopamine. The time course for the desensitization is rapid, occurring within minutes of dopamine exposure and is maximal by 1 h. The desensitization is accompanied by a loss in D_1 receptor binding activity, but this response occurs more slowly, not achieving maximal levels until after 3 h. The receptors remaining in the membrane appear to be uncoupled from their G protein as suggested by the absence of high affinity, guanine nucleotide-sensitive agonist binding. As described for the striatal neurons in culture, the rate of recovery of the NS20Y cells is dependent on the time of preincubation. After a 30 min treatment with dopamine, the NS20Y cells recover to control levels of activity within 2 h. After 3 h of pretreatment, however, the cells require up to 1 d for full recovery. Moreover, we found that the recovery from the longer (3 h) treatment is dependent on protein synthesis, whereas recovery from the shorter (30 min) treatment is not. Some of these basic findings have also been reported for D_1 receptors endogenously expressed in human D384 astrocytoma cells *(55)*, monkey COS-1 cells *(56)*, opossum kidney (OK) cells *(57)*, and human SK-N-MC neuroblastoma cells *(58)*. Interestingly, using the latter SK-N-MC cells, chronic antagonist treatment was shown to lead to increased D_1 receptor expression through a yet to be determined mechanism *(59)*.

These observations have led us to formulate the following scheme for agonist-induced desensitization of D_1 receptors:

$$R_N \Leftrightarrow R_U \Leftrightarrow R_L \Rightarrow R_D \tag{1}$$

In this scheme, the normal receptor (R_N) is rapidly converted on agonist exposure to a functionally uncoupled form (R_U) prior to a readily reversible loss in receptor–ligand binding activity (R_L). With further agonist exposure, the receptors undergo irreversible degradation (R_D) and require resynthesis for full recovery. We know from the time course of desensitization that the formation of R_U precedes that of R_L, and that both are readily reversible. The kinetics of the conversion of R_L to R_D, however, remain poorly defined. The actual state or location of the R_L receptor is similarly ill-defined and may represent the translocation or sequestration of the receptor into a membrane environment that is not readily accessible to hydrophilic ligands such as [^{3}H]SCH 23390. A similar mechanism has been suggested for the sequestration process that is involved in desensitization of the beta adrenergic receptor *(2)*.

A major issue concerning the agonist-induced desensitization of the D_1 receptor is the potential role of cAMP as a negative feedback modulator. In NS20Y cells, we have found that raising the intracellular cAMP levels by incubating the cells with membrane permeable analogs of cAMP, such as 8-(4-chlorophenylthio)-adenosine-3':5'-cyclic monophosphate (CPT-cAMP), can also produce desensitization and down-regulation of the D_1 receptor *(60)*. However, these responses differ somewhat from those induced by agonist preincubation. For instance, whereas preincubation with CPT-cAMP decreases both efficacy and potency for dopamine stimulation of adenylate cyclase, agonist pretreatment only decreases the maximum response, or efficacy. Similar results were found using cAMP analogs in OK cells *(57)*. Also, the time courses for the CPT-cAMP induced desensitization and down-regulation in the NS20Y cells are slower than those observed with agonists, and the recovery process is completely dependent on protein synthesis. Similar to the agonist-induced desensitization, however, is the demonstrated lack of high affinity, guanine nucleotide-sensitive agonist binding in the CPT-cAMP desensitized state. Based on these observations, we have suggested that cAMP is at least partially involved in the receptor uncoupling (R_U) and the down-regulation/degradation (R_D) events. In contrast, the agonist-induced reversible loss in receptor binding (R_L) appears not to be mimicked by cAMP.

Support for this hypothesis comes from studies performed in other cell lines. Using SK-N-MC cells, Zhou et al. *(58)* found that agonist treatment results in a decrease in both the potency and maximum response for dopamine stimulation of adenylate cyclase. Using permeabilized cells, these investigators showed that treatment with dopamine, ATP, and the cAMP-dependent protein kinase mimics the shift in potency observed with dopamine preincubation. Moreover, the dopamine-induced reduction in potency can be blocked by inhibitors of this kinase. These data argue for a role of cAMP-dependent phosphorylation in the D_1 receptor uncoupling/desensitization response. In another study, Bates et al. *(61)* found that overexpressing a cAMP phosphodiesterase in OK cells blunts dopamine-stimulated increases in cAMP

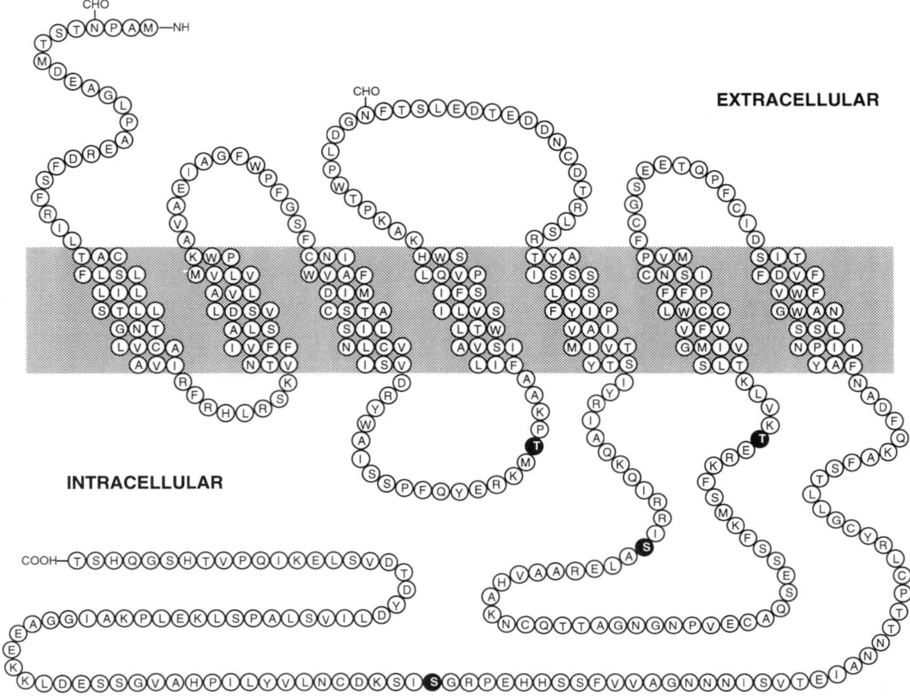

Fig. 1. Amino acid sequence of the rat D_1 dopamine receptor illustrating potential cAMP-dependent phosphorylation sites. Black residues indicate sites that may be phosphorylated by the cAMP-dependent protein kinase.

accumulation and activation of cAMP-dependent protein kinase. This was found to block dopamine-induced down-regulation of receptor binding activity, without preventing desensitization of the cAMP response. These results differ somewhat from ours and those of Zhou et al. *(58)* in that they would seem to preclude a major role for cAMP in the desensitization response, but are in good agreement with our data suggesting a role for cAMP in down-regulation of the D_1 receptor.

Although regulation of downstream effector proteins *(62)* as well as regulation occurring at the transcriptional and translation levels *(63)* must be considered, the most likely mechanism for regulating the functional coupling and down-regulation of the D_1 receptor is phosphorylation. With respect to the cAMP-dependent protein kinase, there are 3–4 potential phosphorylation sites on the receptor (Fig. 1) which approximate consensus recognition sequences for this enzyme *(64)*. Conceivably, phosphorylation of one or more of these sites could mediate the functional uncoupling and/or down-regulation events involved in the desensitization response of the D_1 receptor. Some of these sequences also correspond to potential protein kinase C phosphorylation sites *(64)*; at present, however, there is no evidence for a role of protein kinase C in desensitization of the D_1 receptor. We have found negligible effects of protein kinase C activation on D_1 receptor function and expression in NS20Y cells (unpublished observations) whereas a recent study reported small increases in D_1 receptor binding in vascular smooth muscle cells after activation of protein kinase C *(65)*. Recently, using a fusion protein approach, Zamanillo et al. *(66)* have provided evidence that purified cAMP-dependent protein kinase, but not protein kinase C, is capable of phosphorylating the COOH-terminus of the D_1 (Fig. 1) but not the D_5 receptor.

Although phosphorylation by the cAMP-dependent protein kinase may be involved in the desensitization of the D_1 receptor, it is probable that other protein kinases play an additional and perhaps even more important role. This is suggested by the observation of Bates et al. *(61)* that attenuation of cAMP accumulation does not alter agonist-induced desensitization of the D_1 receptor in OK cells. Similarly, we found that treatment of NS20Y cells with an antagonist Rp-adenosine-3',5'-monophosphorothioate (Rp-cAMPS) of the cAMP-dependent protein kinase does not block agonist-induced desensitization, although it attenuates the desensitization induced with CPT-cAMP *(60)*. Finally, Zhou et al. *(58)* showed that an inhibitor of a beta adrenergic receptor kinase (βARK) can partially attenuate the agonist-induced desensitization of the D_1 receptor in permeabilized SK-N-MC cells. These results suggest that βARK or a βARK-like enzyme is involved in agonist-induced desensitization of the D_1 receptor, particularly with respect to the initial uncoupling step. βARK belongs to a family of protein kinases (referred to as G protein-coupled receptor kinases or GRKs) that are involved in the phosphorylation and desensitization of a wide variety of G protein-coupled receptors *(2,67)*. There are several serine and threonine residues located in the COOH terminus of the D_1 receptor (Fig. 1) in close proximity to acidic residues (Asp/Glu), thereby making them favorable candidates for phosphorylation by βARK or other GRKs *(68)*. It is hypothesized that the D_1 receptor undergoes phosphorylation by one of the GRKs in an agonist-specific manner and that this plays a major role in the rapid uncoupling events which initiate the desensitization process.

It should be noted, however, that because of the overall low level of receptor expression in endogenous expressing cell lines, direct phosphorylation of the D_1 receptor by any protein kinase has yet to be demonstrated in these model systems. This difficulty can be overcome through the use of transfected cell systems which can be manipulated to express high levels of functional D_1 receptor proteins.

2.2.2. Regulation of Recombinant D_1 Receptors

In actuality, only a few studies have appeared investigating desensitization processes for the D_1 receptor in transfected cell lines. In the Sibley laboratory, we have begun to examine agonist-induced desensitization of the rat D_1 receptor in stably transfected Chinese hamster ovary (CHO) cells *(69)*. Preliminary experiments have revealed that agonist preincubation of these cells results in decreased dopamine-stimulated adenylate cyclase activity and a loss in D_1 receptor binding as detected using [^{3}H]SCH 23390. These responses are time-dependent with respect to dopamine preincubation, showing a $t_{1/2}$ of about 5 h and maximal effects at about 20 h. This time course for agonist-induced D_1 receptor desensitization is considerably slower than that observed using endogenous expressing cell lines (*see* Section 2.2.1.), particularly for the rapid uncoupling/desensitization response. However, the time course of agonist-induced desensitization in CHO cells correlates fairly well with the CPT-cAMP-induced desensitization/down-regulation of the D_1 receptor in NS20Y cells *(60)*. These data suggest that the CHO cells may predominantly exhibit the slower, potentially cAMP-mediated pathway for D_1 receptor desensitization/down-regulation. Agonist treatment (2–8 h) also down-regulates D_1 receptor binding in transfected C_6 glioma cells *(70)*. In these cells, dopamine treatment induces a more rapid decrease in dopamine-stimulated adenylate cyclase activity *(71)*. Thus, treatment with 1 μM dopamine for 2 h reduces the maximum response by ~90%.

D_1 receptor desensitization has also been evaluated in stably transfected Y1 adrenal cells *(72)*. These investigators reported that agonist treatment of the cells promotes a rapid (within minutes) loss in D_1 receptor stimulation of adenylate cyclase activity. Desensitization in these cells appears to be heterologous in nature as the adenylate cyclase responses to ACTH and NaF are also reduced. The effect of dopamine treatment on D_1 receptor binding was not evaluated. D_1 receptor desensitization was also examined in a Y1 mutant cell line which exhibits resistance to ACTH receptor desensitization. These mutant Y1 cells still support D_1 receptor desensitization, although the time course is considerably slower. The defect in the mutant Y1 cells has yet to be determined.

Desensitization of the D_1 receptor has additionally been examined in baculovirus-infected Sf9 insect cells. Ng et al. *(73)* reported that dopamine preincubation of D_1 receptor-expressing Sf9 cells decreases dopamine-sensitive adenylate cyclase activity. The agonist treatments were associated with an approximately twofold increase in the phosphorylation state of an epitope-tagged D_1 receptor. Evidence was also provided for agonist-induced internalization of the D_1 receptor in the Sf9 cells *(73,74)*. As noted, the protein kinase(s) associated with rapid agonist-induced uncoupling/desensitization of the D_1

receptor have yet to be identified. Recently, however, Tiberi et al. *(75)* used transiently transfected HEK-293 cells to show that GRK2 (βARK1), GRK3 (βARK2), and GRK5 all promote rapid agonist-induced phosphorylation of the D_1 receptor in association with a functional desensitization. In a separate study, Frail et al. *(76)* showed that long-term (24 h) treatment of D_1 receptor-transfected HEK-293 cells with dopamine produces a down-regulation of binding activity and a decline in adenylate cyclase stimulation. Interestingly, there were no differences in this regard between the human and goldfish D_1 receptors although the goldfish receptor has 80 fewer amino acids in the COOH terminus compared to the human. This suggests that this domain is not required for long-term down-regulation/desensitization of the D_1 receptor.

3. D_5 Receptors

In contrast to the D_1 receptor, few studies have appeared concerning the regulation of the recently cloned D_5 (or D_{1B}) receptor subtype. As noted, most, if not all, of the in vivo studies concerning regulation of D1-like receptors have probably involved the more predominant D_1 receptor subtype. Recently, however, two reports have appeared examining the level of D_5 receptor mRNA in rats that had been treated chronically with antipsychotic drugs *(27,77)*. These investigators found that administration of sulpiride (20 or 100 mg/kg/d), haloperidol (3 mg/kg/d), or loxapine (4 mg/kg/d) for up to 32 d has no effect on the D_5 receptor mRNA levels. In contrast, clozapine administration (30 mg/kg/d) results in a 50% increase in D_5 receptor mRNA after 4 d of treatment, but this returns to basal levels after 32 d. In these same animals, it was determined that after 32 d of treatment the D_1, D_2, and D_3 receptor mRNAs were all up-regulated by the haloperidol and loxapine treatments. These results suggest that the D_5 receptor may not be subject to identical in vivo regulatory mechanisms as are the D_1 or other dopamine receptor subtypes.

The D_5 subtype exhibits very high sequence homology with the D_1 receptor (*see* Chapter 2) and it might be expected that regulatory mechanisms occurring at the level of the receptor protein would be similar between these two receptor subtypes. For instance, both the D_1 (Fig. 1) and D_5 receptors contain several consensus sequence sites for the cAMP-dependent protein kinase *(64)* as well as multiple other serine and threonine residues within their cytoplasmic domains which may serve as additional sites of regulatory phosphorylation. A recent study using a fusion protein of the carboxy terminal tail of the mouse D_5 receptor suggests that this region of the receptor is not phosphorylated by either the cAMP-dependent protein kinase or protein kinase C *(66)*. This is not surprising given that this region does not contain consensus phosphorylation sites for the either of these protein kinases *(64)*.

Consistent with its overall low abundance, no cell lines have yet to be identified that express an endogenous D_5 receptor. Consequently, we have stably transfected the rat D_5 receptor into CHO cells to examine its regulatory properties *(69)*. We found that agonist pretreatment of the cells results in a desensitization of the ability of dopamine to stimulate intracellular cAMP accumulation. This is accompanied by a reduction in the D_5 receptor binding capacity in membranes prepared from these cells. These agonist-induced desensitization and down-regulation responses are similar to those demonstrated for the D_1 receptor in transfected cells (*see* Section 2.2.2.). However, when we compared D_1 and D_5 receptor-transfected CHO cells in parallel, we found that the D_5 receptor desensitization occurred more rapidly and to a greater extent than that for the D_1 receptor. These preliminary observations suggest that the D_5 receptor is more susceptible to agonist-induced desensitization than the D_1 receptor. Rapid desensitization of dopamine-stimulated adenylate cyclase activity is also observed for the human D_5 receptor expressed in Ltk^- cells *(78)*.

Our results demonstrating enhanced susceptibility to desensitization for the D_5 receptor in CHO cells are interesting in light of recent findings suggesting that this receptor exhibits properties of a constitutively active G protein-linked receptor *(79)*. One of the properties of such receptors, as demonstrated using a mutant beta-2 adrenergic receptor *(80)*, is constitutive engagement of cellular desensitization processes including phosphorylation, uncoupling, and down-regulation. Conceivably, if the D_5 receptor is indeed already partially desensitized in the basal state, then exposure to agonists could result in more rapid and extensive desensitization when compared with the D_1 receptor.

Recently, the promoter region of the D_5 receptor gene has been isolated and characterized *(81)*. Sequencing of this region has revealed multiple transcription factor binding motifs including sites for Sp1, AP1, estrogen, and glucocorticoid response element half-sites and an 11-base-long sequence identical to the binding site for the pituitary-specific transcription factor, GHF-1 (or Pit-1). Future experiments will be required to demonstrate that these elements are functionally active and involved in the transcriptional regulation of D_5 receptor expression.

4. D_2 Receptors

4.1. *In Vivo Regulation*

A vast number of investigations have been performed, using in vivo model systems, showing that the D2 receptor is subject to dynamic regulation. Much of this literature has been reviewed previously *(6–10)*. When D2 receptor changes have been quantified by radioligand binding, rather than by assessment of receptor mRNA levels, the results are generally a composite of effects on two or three of the D2-like receptor subtypes. The term D2 is used for this pharmacologically defined subtype, as opposed to the D_2 genetic subtype.

4.1.1. Receptor Stimulation

As for the D_1 receptor, the discussion here is limited to those studies that employed agonists that directly activate D2 receptors. Thus, the effects of stimulant drugs, such as amphetamine or cocaine, which promote the release and/or block the reuptake of dopamine, are not considered, because these agents are known to affect multiple receptor/neuronal systems and their effects on D2 receptor activity may be indirect.

Some of the earliest studies involving agonists utilized L-DOPA administration to increase the level of dopamine in the brain, as well as bromocriptine, an ergot alkaloid that is a D2 receptor agonist. Mishra et al. *(16)* reported that 3 wk of chronic L-DOPA or bromocriptine treatment reduced D2 receptor binding activity by 40% in rat striatal membranes. De Jesus et al. *(82)* also reported that chronic L-DOPA could decrease D2 receptor binding in mice and that this effect was dose dependent. List and Seeman *(83)* found that whereas 5 d of administration of either L-DOPA or bromocriptine to rats could partially reverse the increase in D2 receptor binding which was induced with chronic antagonist administration, these treatments had no effect in control animals. In contrast, Wilner et al. *(15)* reported that chronically treating rats with high doses of L-DOPA resulted in a doubling of D2 receptor binding in striatal homogenates. Finally, Rouillard et al. *(84)* found that in unilaterally 6-OHDA-lesioned rats, chronic bromocriptine reduced the density of D2 receptors in the intact striata, whereas on the lesioned side there was no change. In contrast, chronic L-DOPA induced an increase in D2 receptor binding in the lesioned striata in addition to that promoted by denervation, whereas on the intact side there was no effect. Thus, the effects of L-DOPA treatment are not clear, with reports of both D2 receptor down- and up-regulation, or no change in receptor binding at all. This situation is somewhat similar to that observed with agonist treatment of D2-like receptors in cultured cell lines.

In contrast to the L-DOPA data, the results obtained with bromocriptine and other ergot alkaloids appear more straightforward because D2 receptor down-regulation is usually observed on chronic treatment. In addition to those aforementioned studies, two other groups have found that chronic treatment of rodents with bromocriptine leads to decreased D2 receptor binding in striatal membranes *(85,86)*. Interestingly, treatment of rats with lisuride, another ergot alkaloid with D2 receptor agonist properties, reduces the ability of bromocriptine to inhibit adenylate cyclase activity in striatum, substantia nigra, and nucleus accumbens homogenates *(87)*. Thus far, this is the only study demonstrating agonist-induced functional desensitization of a second messenger response for D2 receptors in vivo. In contrast to these central nervous system (CNS) studies, somewhat different results have been found for D_2 receptors expressed in the pituitary gland. Two recent studies have demonstrated that chronic bromocriptine treatment has no effect on D_2 receptor binding activity or mRNA levels in the anterior *(88)* and neurointermediate *(86,89)* lobes of the pituitary. It is conceivable that different receptor regulatory mechanisms are operative in the CNS and pituitary.

More recently, two groups have examined the effects of chronically treating rodents with the D2-selective agonist quinpirole. Subramaniam et al. *(18)* found that daily injections of rats with quinpirole decreases D2 receptor binding in multiple brain regions as determined with quantitative autoradiography. Similarly, Chen et al. *(90)* continuously infused quinpirole into mice for 6 d and found that this results in a significant down-regulation of striatal D2 receptors as well as a decrease in D_2 receptor mRNA levels. Thus, although the data are somewhat mixed, particularly with regard to the effect of L-DOPA administration, the preponderance of data suggests that agonist administration in vivo results in a down-regulation of D2 receptor expression, and possibly a functional desensitization.

4.1.2. Receptor Blockade

Tardive dyskinesia refers to abnormal involuntary movements that often develop in humans after prolonged blockade of D2 receptors with antipsychotic drugs (*see* Chapter 15). Considerable interest in drug-induced regulation of dopamine D2 receptors has been stimulated by the hypothesis that neuroleptic-induced tardive dyskinesia results from supersensitivity of dopamine receptors *(91)*. As summarized by Muller and Seeman *(92)*, several lines of evidence support this dopamine receptor hypothesis. The symptoms of dyskinesia are aggravated by discontinuation of neuroleptic treatment and reduced by readministration of dopamine receptor antagonists, consistent with a model in which receptor blockade masks receptor supersensitivity that develops during chronic drug treatment. Furthermore, symptoms are exacerbated by treatment with the dopamine precursor L-DOPA. Finally, rodents treated with any of a variety of dopamine receptor antagonists, including haloperidol, thioridazine, chlorpromazine, trifluoperazine, and *cis*-flupentixol, display behavioral supersensitivity to a challenge with a dopamine agonist *(93–95)*. During prolonged (6 mo) treatment, rats display supersensitivity to apomorphine while still being administered the dopamine receptor antagonist *(93)*. Other data, however, are inconsistent with the dopamine receptor supersensitivity hypothesis of tardive dyskinesia *(96,97)*; for example, the lack of evidence for increased neostriatal dopamine receptor density *(98)* or dopaminergic supersensitivity in the neuroendocrine system *(99)* associated with tardive dyskinesia, and the differing time courses of the development of dopaminergic supersensitivity in rodents and tardive dyskinesia in humans. Nevertheless, many studies have been carried out with the goal of identifying neurochemical changes that result from chronic administration of antipsychotic drugs.

Burt et al. *(100)* first demonstrated in 1977 that treatment for 3 wk with haloperidol or fluphenazine induces an increase in the binding of the D2 receptor antagonist [^{3}H]haloperidol to membranes prepared from the rat neostriatum. Subsequent reports have demonstrated D2 receptor up-regulation (generally manifested as a change in the B_{max} of radioligand binding) in rodents in response to a variety of drugs and treatment regimens *(7,10),* not only in the neostriatum, but also in mesolimbic regions *(92),* nucleus accumbens *(101–104),* and in the frontal cortex *(105–107).*

Perhaps as a result of an increased density of D2-like receptors, chronic receptor blockade also enhances several neurochemical and cellular responses to stimulation of dopamine receptors, including D2 receptor-mediated inhibition of adenylate cyclase activity *(108,109),* stimulation of GTPase activity *(109),* inhibition of acetylcholine release *(110–112),* and inhibition of dopamine turnover *(113)*. Pituitary D2 receptors mediating inhibition of prolactin release also become supersensitive after chronic haloperidol administration *(114)*. In addition to those responses that involve postsynaptic receptors, presynaptic release-regulating D2 receptors in the neostriatum *(111,115,116),* and neuronal activity-regulating D2 receptors in the substantia nigra *(117)* also become supersensitive.

Because of the hypothesized relationship between dopamine receptor supersensitivity and tardive dyskinesia, it was of interest to determine the effects of chronic clozapine treatment on D2 receptor density. Clozapine is an atypical antipsychotic drug that is less likely to produce extrapyramidal side effects and is more efficacious than typical antipsychotics in the treatment of the negative symptoms of schizophrenia *(118,119)*. Numerous studies have documented that chronic treatment with clozapine does not increase the density of neostriatal D2-like receptors *(104,120–124)*. Interestingly, whereas D2 receptor density is not increased by clozapine in the basal forebrain, the density of D2 receptors in the medial prefrontal cortex is increased by chronic treatment with either haloperidol or clozapine *(105)*. The reduced or differing action of clozapine in the caudate-putamen, as compared to the action of typical antipsychotic drugs such as haloperidol, may be the basis for clozapine's relative lack of extrapyramidal side effects (*125)*. On the other hand, modulation of the frontal cortical dopamine system by clozapine may be the basis for clozapine's unique efficacy in the treatment of neuroleptic nonresponders and the negative symptoms of schizophrenia *(125,126)*.

One investigation into the mechanisms of antagonist-induced D2 receptor up-regulation determined that haloperidol treatment decreases the turnover of neostriatal D2 receptors, as assessed by recovery from irreversible blockade *(126a)*. Rates of synthesis and degradation are both reduced; the increased steady-state receptor density reflects an increase in the ratio of the rates of receptor synthesis and degradation. One possible mechanism for a drug-induced change in receptor synthesis is a change in availability of the cognate mRNA, a hypothesis that could be assessed after the cloning of a D_2 receptor cDNA *(127)*. Despite numerous studies, there is no consensus regarding the effect of chronic receptor blockade on the abundance of D_2 receptor mRNA *(10)*. Several techniques, including *in situ* hybridization histochemistry, Northern blot analysis, and solution hybridization, have produced results indicating that chronic blockade produces either no significant change *(128–133)* or increases in D_2 receptor mRNA ranging from 10 to >100% *(26,134–141)*. Haloperidol is the drug most commonly used, and although the route of administration, dose, and treatment duration varied substantially among studies, there does not seem to be a consistent relationship between the extent of the drug treatment (dose and duration) and the outcome. Furthermore, often no change in mRNA is observed after a treatment regimen that produces an increase in the density of D2-like receptors, assessed by radioligand binding. Two examples of thorough studies reporting no change in mRNA levels are those of Srivastava et al. *(131)* and Van Tol et al. *(132)*. The former group used all three of the methods listed and found no change in mRNA levels after treatment with haloperidol (3 mg/kg/d, ip) for 7–35 d, despite a 30–40% increase in the density of neostriatal D2-like receptors. The latter group detected no effect on the abundance of D_2 receptor mRNA, as assessed by both Northern blot and solution hybridization/RNase protection analyses, by haloperidol (0.5 mg/kg/d, ip) or raclopride (2.5 mg/kg/d) treatment for 21 d, although the density of neostriatal D2-like receptors was elevated by 20–30%. In contrast, Buckland et al. *(136)* found that message for both D_{2S} and D_{2L} receptors was increased approximately twofold after 32 d of treatment with haloperidol (1.5 mg/kg, ip, twice/d), and in the study of Rogue et al. *(141)*, a 25–50% increase in D_2 receptor mRNA was induced by 15 d treatment with haloperidol (2 or 4 mg/kg/d, ip).

Haloperidol treatment has also been shown to elevate D_2 receptor mRNA in the cingulate cortex *(128)* and the substantia nigra *(142)*. Both of these areas project to the neostriatum and an increase in D_2 receptor synthesis and transport to the striatum could conceivably account for the antagonist-induced increase in the density of striatal D_2 receptors in the absence of increased striatal D_2 receptor mRNA. An alternative hypothesis to explain these disparate results is that multiple mechanisms are involved in antagonist-induced up-regulation of D_2 receptors with the underlying variables being poorly understood.

There are also conflicting reports concerning modulation of D_2 receptor message in the pituitary. A haloperidol-induced increase in the anterior pituitary was found by Arnauld et al. *(143),* whereas Autelitano et al. *(88)* observed an increase in the neurointermediate lobe, but not in the anterior pituitary. The discrepancy could be resolved if the number of D_2 receptor-expressing cells is increased by haloperidol treatment without changing the expression per cell, since that would produce an increase assessed by *in situ* hybridization histochemistry *(143)* but not by solution hybridization *(88).*

4.1.3. D1/D2 Receptor Interactions

Several lines of evidence indicate that stimulation of D1 receptors can alter the turnover of D2 receptors, and vice versa. Thus, chronic treatment of rats with fluphenazine *(21)* or *cis*-flupenthixol *(20,144),* antagonists of both D1 and D2 receptors, leads to an increase in D2 but not D1 receptor binding. Similarly, treating rats concurrently with a D1-selective antagonist (SCH 23390) and a D2-selective antagonist (sulpiride or haloperidol) inhibits the D1 receptor up-regulation induced by D1 blockade alone *(21).* These data suggest that stimulation of D2 receptors by endogenous dopamine is necessary for drug-induced up-regulation of D1 receptors. On the other hand, stimulation of D1 receptors with the agonist SKF 38393 prevents haloperidol-induced up-regulation of D2 receptors in rats *(145)* and inhibits denervation-induced D2 receptor up-regulation in monkeys *(146),* suggesting a negative regulation of D2 receptor density by D1 receptors. This appears to conflict with data obtained by Cameron and Crocker *(147),* demonstrating that sparing of D1 receptors enhances the recovery of D2 receptors from irreversible blockade, presumably owing to stimulation of D1 receptors by endogenous dopamine. The mechanisms underlying these interactions are at present entirely unclear.

4.1.4. Dopamine Depletion

Following near-total destruction of the mesotelencephalic dopamine system in rats or mice, behavioral sensitivity to dopamine receptor agonists is enhanced up to 40-fold *(148,149).* Behavioral supersensitivity is accompanied by an equivalent supersensitivity of neostriatal cells to the inhibitory effects of apomorphine *(150),* and is also accompanied by enhanced neostriatal dopamine receptor inhibition of acetylcholine release *(151,152)* and stimulation of [^{14}C]2-deoxy-D-glucose incorporation *(153).* With the development of antagonist radioligand binding assays for D2 receptors, it was demonstrated that unilateral destruction of the ascending dopamine fibers increases the density of D2-like receptors by 40% in the denervated striatum, compared to the intact neostriatum *(154)*. Similarly, chronic (3 wk) daily treatment with reserpine increases D2 radioligand binding in membranes prepared from rat neostriatum by 25% *(100)*. Denervation also produces D2 receptor up-regulation in a nonhuman primate model of hemi-Parkinsonism, induced by unilateral intracarotid infusion of MPTP *(155–157)*.

Although there appeared to be a discrepancy between the relatively small magnitude of the lesion-induced increase in receptor density and the much greater enhancement of behavioral sensitivity to dopamine receptor agonists, the proliferation of D2 receptors was believed to contribute to the behavioral supersensitivity. Many subsequent studies were designed to assess this hypothesis. In contrast to the results of Creese et al. *(154),* Waddington et al. *(158)* found that the extent of D2 receptor up-regulation was significantly correlated with behavioral supersensitivity, as quantified by apomorphine-induced rotation. On the other hand, characterization of the time courses of the development of behavioral supersensitivity and D2 receptor up-regulation indicated that the onset and progression of the former are more rapid than the latter *(159,160)*. Indeed, behavioral supersensitivity to dopamine receptor agonists has been observed after dopamine-depleting lesions that do not induce D1 or D2 receptor up-regulation *(161,162)*. Perhaps most significantly, rats with dopamine-depleting lesions in one hemisphere that are treated chronically with D2 receptor antagonists have equivalent densities of D2 receptors in both striata, yet exhibit an imbalance in striatal sensitivity to D2 receptor agonists, with enhanced sensitivity in the denervated hemisphere manifested as agonist-induced rotation *(163,164)*.

It is now clear that many behavioral, electrophysiological, and cellular manifestations of denervation supersensitivity are due primarily to a breakdown in D1/D2 receptor synergism (*9*; *see* Chapter 7). The breakdown in synergism results from dopamine depletion, either by denervation or by reserpine-induced disruption of storage vesicles, and can be partially induced by very severe, continuous D2 receptor blockade *(165),* but not by once daily administration of antagonists *(166)*. An increase in receptor density of 30% would be expected to produce a twofold increase in sensitivity to an agonist, which is approximately the extent of behavioral supersensitivity observed after chronic receptor antagonism *(167)*. Thus, denervation-induced D2 receptor up-regulation, which is similar in magnitude to drug-induced receptor up-regulation, probably accounts for roughly a twofold increase in sensitivity.

In the search for mechanisms of drug- and denervation-induced receptor up-regulation, one controversy was stimulated by the work of Staunton et al. *(103)* which suggested that the effects of denervation and chronic receptor blockade on the density of D2 receptors in the rat nucleus accumbens were additive; that is, concurrent denervation and chronic treatment with haloperidol produced a greater D2 receptor up-regulation than did either manipulation alone. This finding, suggesting the existence of two distinct mechanisms for up-regulation, was replicated in the rat neostriatum by Reches et al. *(168)*. Marshall and colleagues, however, failed to find any evidence for additivity of denervation and chronic treatment with the D2 receptor antagonists haloperidol, eticlopride, or metoclopramide *(163,164,169)*, and speculated that apparent additivity in other studies resulted from incomplete denervation.

Receptor up-regulation could result from either an increase in the rate of synthesis or a decreased receptor degradation rate constant. Three studies were carried out to identify the mechanism of denervation- or reserpine-induced D2 receptor up-regulation *(49,170,171)*, with results indicating that dopamine depletion selectively increases the rate of synthesis of D2 receptors. These findings differ from the decreased rates of synthesis and degradation observed after chronic administration of haloperidol *(80)*, which would lend support to the hypothesis that there are two distinct mechanisms of receptor up-regulation. Unlike the conflicting data regarding the effect of chronic receptor blockade on the abundance of D_2 receptor mRNA, there is a fairly clear consensus that denervation- or reserpine-induced dopamine depletion elevates D_2 receptor mRNA levels in the basal forebrain. This was first shown by Gerfen et al. *(48)* and Brené et al. *(172)*, and confirmed by many other groups *(40,44,49,133,134,139,172a)*. Increases in both the density of D2-like receptors and D_2 receptor mRNA are most pronounced in the lateral portion of the neostriatum *(46,172)*. The few studies that have found message levels to be unchanged by denervation did not confirm by radioligand binding that receptor up-regulation had occurred *(173,174)*. In the rat neostriatum, dopamine depletion increases the abundance of both of the alternatively spliced forms of the D_2 receptor *(175)*. The denervation-induced increases in D_2 receptor density and mRNA levels can be reversed by chronic administration of agonists, suggesting that the loss of receptor stimulation by endogenous dopamine is the cause of the receptor proliferation *(48,176)*.

4.2. In Vitro Regulation

In contrast to in vitro studies of D_1 receptor regulation, such studies with D_2 receptors have produced quite variable results. For instance, agonist treatment paradigms have variously revealed functional D_2 receptor desensitization, down- or up-regulation of receptor expression, or no effects on expression or function at all. It is suspected that the D_2 receptor is subject to multiple positive and negative regulatory events (certainly more than the D_1 receptor), any of which can be operative depending on the tissue or cell type under study.

4.2.1. Regulation of Endogenous D_2 Receptors

One of the earliest in vitro studies on D_2 receptor regulation was performed using the intermediate lobe of the rat pituitary gland *(177)*. Preincubation of this tissue for 1 h with the agonist apomorphine reduces the ability of apomorphine to inhibit adenylate cyclase activity in subsequently homogenized tissue. The density of pituitary D_2 receptors is unchanged, although the ability of the receptor to form a high-affinity agonist binding state is decreased in the treated membranes. These initial data suggested that the D_2 receptor is susceptible to an agonist-induced desensitization response. In contrast, Drukarch et al. *(178)* reported that treatment for up to 2 h with high concentrations of either dopamine or the agonist quinpirole does not induce desensitization of the D2-like receptor-mediated inhibition of electrically evoked [^{3}H]acetylcholine release from rat striatal slices. A similar lack of effect of agonist treatment was observed on D_2 receptor mRNA levels in striatal neurons in primary culture *(179)*. In this study, striatal cells were cultured for up to 15 h with quinpirole with no change in D_2 receptor mRNA. Functional assessment of the D_2 receptor was not performed.

Because of the difficulties associated with the interpretation of data obtained using intact tissues (drug delivery, viability, etc.), Sibley and colleagues examined agonist-induced regulation of the D_2 receptor endogenously expressed in the human retinoblastoma Y-79 cell line *(180)*. Culturing the cells in the presence of agonists decreases the ability of the D_2 receptor to mediate inhibition of adenylate cyclase activity. This effect is time dependent ($t_{1/2}$ = 1 h) and involves both a reduction in the maximum response as well as a decrease in the potency of dopamine. Agonist preincubation also decreases radiolabeled antagonist binding to the D_2 receptor, as determined using both [^{3}H]methyspiperone and [^{125}I]iodosulpride. This response occurs more slowly ($t_{1/2}$ = 4 h) than the functional desensitization. Interestingly, the nature of the reduced binding activity differs for the two radioligands. For [^{125}I]iodosulpride, only the maximum binding capacity (B_{max}) is reduced, whereas for [^{3}H]methyspiperone, only the affinity (K_d) is decreased. The lack of change in the total receptor number assessed using [^{3}H]methylspiperone argues against a receptor degradation event. Instead, it is suggested that D_2 receptors undergo an agonist-induced translocation or sequestration which orients them into a membrane environment such that they are not readily exposed. Thus, [^{125}I]iodosulpride, which is a relatively hydrophilic ligand and probably membrane impermeable, can only detect those receptors remaining on the membrane surface following desensitization. Conversely, [^{3}H]methylspiperone is a very hydrophobic, membrane-permeable ligand and is able to detect the full complement of receptors associated with the membranes. The reduction in affinity of [^{3}H]methylspiperone following desensitization may be reflective of a *kinetic* barrier for penetration of this ligand into the sequestered membrane environment.

These observations have led us to formulate the following scheme for agonist-induced desensitization of D_2 receptor in Y-79 cells:

$$R_N \Leftrightarrow R_U \Leftrightarrow R_L \quad (2)$$

In this scheme, the normal receptor (R_N) is converted by agonist exposure to a functionally uncoupled form (R_U) prior to a loss in receptor binding activity (R_L), the latter of which is presumably owing to a sequestration process. These events appear to be readily reversible. The observation that the rate of desensitization is faster than the loss of antagonist binding suggests that the formation of R_U precedes that of R_L. Further evidence for the existence of R_U comes from the fact that a loss of high-affinity, guanine nucleotide-sensitive agonist binding is observed in dopamine competition for [^{3}H]methylspiperone and [^{125}I]iodosulpride binding after desensitization. Thus, the loss of receptor binding activity cannot completely explain the desensitization.

The biochemical mechanisms underlying these regulatory events are unknown but it is reasonable to speculate that phosphorylation of the D_2 receptor may be at least partially involved. The D_2 receptor protein contains multiple potential sites of regulatory phosphorylation (Fig. 2). There are several, albeit weak, consensus recognition sequences for the cAMP-dependent protein kinase *(64)* and most of these also represent reasonable substrates for protein kinase C. In addition, there are several serine and threonine residues within the third cytoplasmic loop in close proximity to acidic residues, thereby making them favorable candidates for phosphorylation by G protein receptor kinases *(68)*. Recent evidence has suggested that the D_2 receptor is indeed a phosphoprotein *(181)*. As described for the D_1 receptor (*see* Section 2.), it seems likely that agonist occupancy/activation of the D_2 receptor could result in its phosphorylation by a G protein receptor kinase, thereby leading to functional uncoupling and desensitization.

Although probably not related to agonist-induced desensitization, D2 receptors have been suggested to be regulated by both the cAMP-dependent protein kinase and protein kinase C. Elazar and Fuchs *(182)* showed that phosphorylation of striatal membranes with cAMP-dependent protein kinase results in a reduction of agonist affinity for D2 receptors and, in separate experiments, showed that this kinase can phosphorylate purified D2 receptors. Similarly, Rogue et al. *(183)* demonstrated that phosphorylation of striatal membranes with protein kinase C produces a reduction in high-affinity agonist binding to D2 receptors. This latter observation may be related to the finding that phorbol ester-induced protein kinase C activation in striatal slices can lead to a decreased ability of D2 agonists to inhibit dopamine release *(184)*.

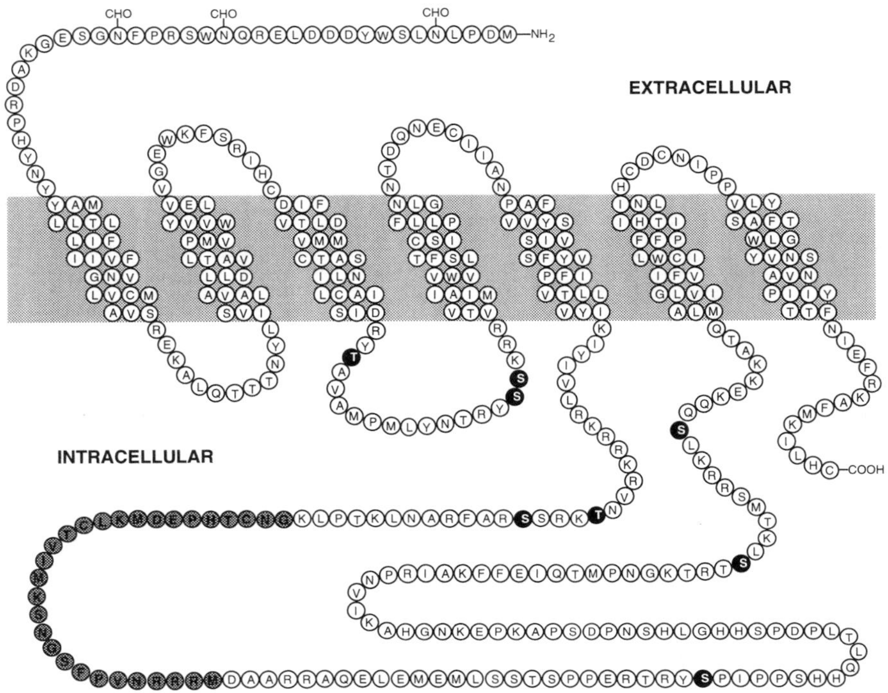

Fig. 2. Amino acid sequence of the rat D_{2L} dopamine receptor. Hatched residues represent the cassette exon sequence which is alternatively spliced in the two molecular subtypes, D_{2L} and D_{2S} (*see* Chapter 2). Black residues represent possible sites of phosphorylation by the cAMP-dependent protein kinase.

The only other study to appear on the regulation of D_2 receptors endogenously expressed in a cell line is that of Ivins et al. *(185)*. These authors used the rat SUP1 pituitary cell line, which is derived from the 7315a transplantable pituitary tumor. Culture of these cells for >6 h in the presence of dopamine increases the density of D_2 receptors. This up-regulation response can be prevented by cotreatment with an antagonist that by itself had no effect. The dopamine treatment also leads to increased basal intracellular cAMP levels in the cells; however, there is no effect on the ability of dopamine to diminish cAMP accumulation, suggesting an absence of receptor desensitization. In a similar study, Johnston et al. *(186)* found that dopamine treatment of rat pituitary cells in primary culture results in a 400% increase in D_2 receptor mRNA levels suggesting that, at least in some cell types, dopamine can induce increased D_2 receptor synthesis.

4.2.2. Regulation of Recombinant D_2 Receptors

Thus far, a variety of different cell lines have been utilized for D_2 receptor regulatory studies that have produced a variety of results. It is noteworthy that the vast majority of these studies have been performed using fibroblast-like cell lines, cell types that do not normally express the D_2 receptor. The earliest study is that of Bates et al. *(187)* using Ltk$^-$ cells transfected with the D_{2S} receptor. These authors found that treatment of the cells for 1 h with quinpirole reduces the potency for dopamine inhibition of cAMP accumulation but does not alter the maximal response. Longer treatment (>1 h) leads to increased basal levels of intracellular cAMP, as observed with the SUP1 cells. Interestingly, quinpirole treatment did not affect the level of D_2 receptor binding activity measured with [^{3}H]spiperone. A more recent study using Ltk$^-$ cells transfected with the D_{2L} receptor has reported that chronic elevation of intracellular cAMP levels with forskolin increases D_2 receptor binding with no change in the potency or efficacy for dopamine inhibition of cAMP accumulation *(188)*. The effect of activating protein kinase C on D_2 receptors in the Ltk$^-$ cells has also been studied *(189)*. Both the D_{2S} and D_{2L} receptors inhibit cAMP accumulation and stimulate Ca^{2+} mobilization in these cells. Activation of protein kinase C using phorbol esters has no effect on D_{2S}- or D_{2L}-mediated adenylate cyclase inhibition, but attenuates the mobilization of Ca^{2+}. Moreover, the D_{2S} receptor appears to be more sensitive to this effect than the D_{2L} receptor. Given the disparate results between signal transduction pathways, it seems likely that the locus of protein kinase C regulation is downstream of the receptor proteins.

In the Sibley laboratory, we have begun to investigate agonist-induced regulation of the D_{2S} and D_{2L} receptors in transfected CHO cells *(190)*. Agonist treatment of these cells was found to reduce both the potency and efficacy of dopamine inhibition of cAMP accumulation for both receptor forms. This desensitization response was time- and dose-dependent, pharmacologically specific and similar to that previously described for the Y-79 cells (*see* Section 4.2.1.). Somewhat different results were obtained, however, when receptor binding activities were measured with [^{3}H]methyspiperone. Whereas the D_{2S} receptor was down-regulated in response to agonist treatment, the D_{2L} receptor was up-regulated (by about twofold) as described for the SUP1 cells. The agonist-induced up-regulation of the D_{2L} receptor was blocked by prior treatment of the cells with pertussis toxin or cycloheximide, suggesting that G protein coupling and protein synthesis, respectively, are required for this effect. In contrast, cycloheximide did not block the agonist-induced desensitization. Interestingly, raising the CHO cell intracellular cAMP levels with membrane-permeable analogs of cAMP resulted in both desensitization and down-regulation of the D_{2L} receptor, in contrast to results obtained in the Ltk$^-$ cells *(188)*. RNA analyses further indicated that agonist treatment of the CHO cells resulted in a twofold increase in mRNA levels for the D_{2L} receptor, whereas minimal effects were observed on the D_{2S} receptor mRNA.

These data suggest that dopamine can promote several regulatory events simultaneously in CHO cells, including desensitization of both D_2 receptor forms, down-regulation of the D_{2S} receptor, and up-regulation of the D_{2L} receptor. The desensitization response appears similar to that observed for the Y-79 cells, albeit somewhat slower in onset ($t_{1/2}$ = 5 h in CHO cells). The D_{2S} receptor exhibits a decrease in the receptor binding activity as with the Y-79 cells, whereas, the D_{2L} receptor is paradoxically up-regulated. This appears to be caused by increased receptor synthesis resulting from increased levels of D_{2L} mRNA. Conceivably, the mRNA increase could come about from either increased transcription of the cDNA or agonist-induced stabilization of the mRNA. Because transcription of the cDNA in these transfected cells is under the control of a strong viral promoter, we favor the mRNA stabilization hypothesis. The signaling mechanism for this response is unclear, although the effect of pertussis toxin indicates that receptor/G protein coupling seems to be required. This might suggest that a reduction in intracellular cAMP levels is the stimulus for receptor up-regulation. However, the D_2 receptor is known to be linked to multiple second messenger systems in CHO cells *(191–193)*, suggesting that additional experimentation will be needed to address this point.

Agonist-induced up-regulation of D_2 receptors has also been demonstrated in transfected HEK-293 cells *(194,195)*. These authors found that saturating concentrations of agonists (i.e., >10 μM for dopamine) produced a time-dependent ($t_{1/2}$ = 6 h) increase in the level of D_{2L} receptor expression. D_{2S} receptor-expressing HEK-293 cells also exhibited agonist-induced receptor up-regulation. These treatments were associated with increased basal levels of intracellular cAMP as seen with the SUP1 and Ltk$^-$ cells, however, there was no effect on the ability of agonists to inhibit cAMP accumulation, as observed with the SUP1 cells but in contrast to the Ltk$^-$ and CHO cells. The up-regulation response was not blocked with pertussis toxin or actinomycin, indicating that neither G protein coupling nor transcription, respectively, are required. Treatment of the HEK-293 cells with antagonists also promoted an up-regulation of D_{2L} receptor binding activity after a time lag that was not observed with agonists. Raising intracellular cAMP levels was found to potentiated agonist-, but not antagonist-induced receptor up-regulation. In addition, blockade of mRNA translation with cycloheximide inhibited receptor up-regulation by both agonists and antagonists. In contrast to the CHO cells, however, agonist treatment did not elevate D_{2L} receptor mRNA levels in the HEK-293 cells. The authors concluded that the ligand-induced receptor up-regulation response is the result of an increased mRNA translation or increased insertion of receptors into the membrane.

Results similar to those obtained with both CHO and HEK-293 cells on D_{2L} receptor regulation have been obtained using C_6 glioma cells *(196)*. In these cells, saturating concentrations of agonists (>10 μ*M* for dopamine) produced a time-dependent ($t_{1/2}$ = 8 h) increase in the level of D_{2L} receptor expression. The time course for this effect was complex, however, and appeared to exhibit a lag or a slow initial phase. As with the CHO cells, but in contrast to the HEK-293 cells, these treatments were associated with a functional desensitization of the D_2 receptor response. Antagonist treatment also promoted an increase in D_2 receptor binding activity although this effect was much smaller than that observed with agonists. Similarly, agonists were observed to up-regulate D_{2S} receptors in transfected C_6 glioma cells and in Ltk^- cells, although this response was smaller than that of the D_{2L} receptor-expressing cells. Cycloheximide blocked the receptor up-regulation, whereas pertussis toxin treatment did not. In this report *(196),* several chimeric D_1/D_2 receptors were also examined for their ability to undergo agonist-induced up-regulation and it was found that this response correlated with the presence of the third cytoplasmic loop of the D_2 receptor.

Taking the CHO, HEK-293, and C_6 cell data together, it appears as if there are at least two and perhaps three different mechanisms by which D_2 receptors can undergo up-regulation. The first of these, best illustrated by the CHO cells, is operative at low concentrations of dopamine (EC_{50} = 100 n*M*), occurs relatively quickly ($t_{1/2}$ = 2 h) and appears to be selective for the D_{2L} receptor *(190)*. This requires receptor coupling with a pertussis toxin-sensitive G protein and is associated with increased levels of D_{2L} receptor mRNA leading to increased receptor synthesis. Increased levels of intracellular cAMP do not potentiate this response. The second mechanism, best illustrated by the HEK-293 cells, requires saturating concentrations of dopamine (EC_{50} > 10 μ*M*), is slower ($t_{1/2}$ = 6–8 h), and is similar for the D_{2S} and D_{2L} receptors. Receptor coupling to a pertussis toxin-sensitive G protein is not required. Although receptor mRNA levels are not altered, protein synthesis appears to be necessary. Antagonist treatment partially mimics the effects of agonists, although a time lag is observed, and cAMP analogs potentiate the action of agonists but not antagonists. The latter observations suggest that more than one mechanism may be operative in the HEK-293 cells. In general, the results obtained in the C_6 glioma cells are intermediate between those with CHO and HEK-293 cells in that the D_{2L} receptor was up-regulated to a greater extent than the D_{2S} receptor and that agonists were more effective than antagonists. As noted, the host cells used in these transfection studies do not represent cell types that normally express the D_2 receptor. It will thus be important to replicate these experiments using host cells with neuronal-like properties.

5. D_3 Receptors

The D_3 receptor is much less abundant and more restricted in its distribution in comparison to the D_1 and D_2 receptor subtypes (*see* Chapter 3). Despite this limitation, several groups have begun to examine the in vivo regulatory properties of this receptor subtype. Initial studies by Buckland and colleagues *(26,27)* found that chronic administration of haloperidol (3 mg/kg/d), loxapine (4 mg/kg/d), or sulpiride (100 mg/kg/d) to rats for up to 32 d increased D_3 receptor mRNA as detected using solution hybridization assays with whole brain mRNA. Clozapine treatment (30 mg/kg/d) produced biphasic results, showing significant increases in D_3 mRNA levels after 4 d of treatment, but this returned to baseline by 32 d. In contrast to these findings, Fishburn and colleagues *(138)* found that treating mice with haloperidol (4 mg/kg/d) for 16 d did not affect the level of D_3 receptor mRNA in the striatum, olfactory tubercle, or olfactory bulb, as determined using RNase protection assays. In these same animals, it was determined that the haloperidol treatment led to an increase in D_2 receptor binding and mRNA levels. The level of D_3 receptor mRNA in the striatum has also been determined using *in situ* hybridization after unilateral lesions of dopaminergic neurons using 6-OHDA *(197)*. No changes were noted at 15, 90, or 365 d after the lesion. In this same study, D_2 receptor mRNA and binding were increased, whereas D_1 receptor binding, but not mRNA, was elevated.

Perhaps the most comprehensive study published to date on in vivo regulation of D_3 receptor expression is that of Lévesque and colleagues *(139)*. In contrast to the studies of Buckland et al. *(26)*, but in agreement with those of Fishburn et al. *(138)*, these investigators found that treatment of rats with haloperidol (20 mg/kg/d) for 2 wk had no effect on D_3 receptor mRNA levels as determined using quantitative polymerase chain reaction and *in situ* hybridization, or on D_3 receptor binding using [^{3}H]7-OH-DPAT, a ligand with D_3-selective characteristics. This same treatment was found to enhance D_2 receptor density and mRNA levels by >50%. Further, in contrast to the Fornaretto et al. study *(197)*, Lévesque et al. found that interruption of dopaminergic transmission using 6-OHDA lesions resulted in a marked (up to 50%) decrease in D_3 receptor binding and mRNA in the nucleus accumbens. This effect was mimicked by transection of dopaminergic fibers in the medial forebrain bundle, but not by depletion of dopamine using treatment with reserpine (2 mg/kg/d). Interestingly, inhibition of axonal transport by the injection of colchicine into the substantia nigra also led to a decrease in D_3 receptor mRNA as well as binding activity. The data suggest that an anterograde factor that positively regulates the expression of the D_3 receptor may be released from dopaminergic neurons, and that dopamine itself is not the factor.

In vitro investigations of D_3 receptor regulation have been hampered because of the lack of known cell lines which endogenously express this receptor subtype as well as the lack of a well-characterized functional response to D_3 receptor activation. Although in some transfected cell systems the D_3 receptor produces modest inhibition of adenylate cyclase (*see* Chapter 6), this response is not believed to be the primary signaling pathway of the D_3 receptor. Recently, however, the D_3 receptor has been suggested to be coupled to the inhibition of Ca^{2+} currents as demonstrated in the transfected neuroblastoma NG108-15 cell line *(198)*. It is speculated that in vitro studies of the regulation of D_3 receptor function will most likely require the transfection of neural-type cells which express the appropriate transduction pathways linked to the D_3 receptor system.

Despite these limitations, the Neve laboratory has addressed the regulation of D_3 receptor expression in transfected cells *(199)*. We found that, in C_6 glioma cells, the D_3 receptor stimulates the rate of extracellular acidification by 9–16% through a pertussis toxin-insensitive G protein. Treatment of these cells with saturating concentrations of agonists causes a robust increase (up to sixfold) in the D_3 receptor binding activity, whereas antagonists have smaller and more variable effects. This agonist-induced up-regulation of D_3 receptor binding activity is not prevented by treatment with pertussis toxin, and does not involve increases in D_3 mRNA levels. However, inhibition of protein synthesis with cycloheximide completely prevents the increased receptor binding by agonists, suggesting that new receptor synthesis may be involved. These data further indicate that the D_3 receptor may be subject to at least some of the same regulatory control mechanisms as those being elucidated for the D_2 receptor (*see* Section 4.).

6. D_4 Receptors

As with the D_3 and D_5 receptors, little information is available concerning the regulation of the recently cloned D_4 receptor. A recent in vivo investigation has suggested that chronic treatment of rats with neuroleptics results in elevated numbers of D_4 receptors *(200)*. In this study, rats were treated with a single dose (25 mg/kg) of long-acting haloperidol-decanoate and after 1 mo sacrificed for analysis. D_4 receptor mRNA was subsequently quantitated in the striata and frontal cortices and D_4 receptor binding analyzed in striatal membranes. D_4 receptor mRNA and binding activity were both found to be elevated by about twofold. One problem with this study, however, relates to the fact that the D_4 receptor binding was defined as the difference between the binding of a radioligand that labels D_2, D_3, and D_4 receptors and another radioligand that only labels D_2 and D_3 receptors. These results must be considered tentative until confirmed with a D_4-selective radioligand.

Thus far, there have been no cell lines identified that express D_4 receptors in abundance; consequently, any in vitro investigations of D_4 receptor regulation will require the use of transfected cells. Both the rat and human D_4 receptors are coupled to inhibition of adenylate cyclase in transfected cell lines (*see* Chapter 6) and it will be interesting to determine if this (or perhaps other) signal transduction response(s) are attenuated on exposure to agonists. Another important question to be addressed will be whether or not the D_4 receptor undergoes an agonist-induced up-regulation response as do the other D2-like, D_2 and D_3 receptors. As noted in Chapter 2, the human D_4 receptor is highly polymorphic within the region of the third cytoplasmic loop. Thus far, there do not appear to be any major pharmacological *(201)* or transductional *(202)* differences between those polymorphic forms that have been evaluated. Consequently, it will be of interest to determine if any of the D_4 receptor variants differ in their regulatory properties as has been suggested for the D_{2S} and D_{2L} receptors *(190)*.

7. Summary

In this chapter we have described data that fit the classical model of receptor regulation in which overstimulation decreases and understimulation increases receptor responsiveness and/or density. Thus, D1-like and D2-like receptors are down-regulated in vivo by treatment with agonists and are up-regulated by chronic blockade with antagonists. Dopamine depletion through denervation also increases the density of D_2 receptors by increasing the abundance of cognate mRNA. In vitro, the responsiveness and density of D_1 and D_5 receptors are reduced by stimulation with agonists, with phosphorylation by both G protein-coupled receptor kinases and cAMP-dependent protein kinases being implicated in the desensitization. The responsiveness of D_2 receptors has also been shown to be decreased by agonist stimulation in vitro.

We have also described data that do not fit this classical model of receptor regulation. In vivo, for example, D_3 receptor density and mRNA are decreased by denervation, and there is evidence that the same is true of D_1 receptor density and mRNA levels. Also, in vitro both D_2 and D_3 receptors are up-regulated by treatment with high concentrations of agonists, and down-regulation of D_2 receptors by lower concentrations of agonists is not observed in all cell lines.

Our present understanding of the mechanisms of dopamine receptor regulation is still very limited. It is likely that the nature and sites of posttranslational modification of dopamine receptors will be identified in the near future. Furthermore, we are only in the preliminary stages of determining the factors that regulate dopamine receptor gene transcription and mRNA stability. Further characterization of the promoters and regulatory factors for the dopamine receptors is critical. Progress in these areas may help to resolve areas of controversy, such as the effect of D_2 receptor blockade on mRNA levels, and the effect of denervation on D_1 receptors, and may also help to resolve apparent differences between in vitro and in vivo responses of D2-like receptors to agonists.

References

1. Sibley, D. R. and Houslay, M. D., eds. (1994) *Molecular Pharmacology of Cell Regulation, Vol. 3: Regulation of Cellular Signal Transduction Pathways by Desensitization and Amplification.* Wiley, Chichester, UK.
2. Liggett, S. B. and Lefkowitz, R. J. (1994) Adrenergic receptor-coupled adenylyl cyclase systems: regulation of receptor function by phosphorylation, sequestration and down-regulation, in *Molecular Pharmacology of Cell Regulation, Vol. 3: Regulation of Cellular Signal Transduction Pathways by Desensitization and Amplification* (Sibley, D. R. and Houslay, M. D., eds.), Wiley, Chichester, UK, pp. 71–98.
3. Bahouth, S. and Malbon, C. C. (1994) Genetic transcriptional and post-transcriptional regulation of G-protein-linked receptor expression, in *Molecular Pharmacology of Cell Regulation, Vol. 3: Regulation of Cellular Signal Transduction Pathways by Desensitization and Amplification* (Sibley, D. R. and Houslay, M. D., eds.), Wiley, Chichester, UK, pp. 99–112.
4. Seeman, P. (1987) Dopamine receptors in human brain diseases, in *Receptor Biochemistry and Methodology, Vol. 8: Dopamine Receptors* (Creese, I. and Fraser, C. M., eds.), Liss, New York, pp. 233–245.
5. Mouradian, M. M., Juncos, J. L., Fabbrini, G., Schlegel, J., Bartko, J. J., and Chase, T. N. (1988) Motor fluctuations in Parkinson's disease: central pathophysiological mechanisms. *Ann. Neurol.* **24,** 372–378.
6. Creese, I. and Sibley, D. R. (1981) Receptor adaptations to centrally acting drugs. *Annu. Rev. Pharmacol. Toxicol.* **21,** 357–391.
7. Baker, G. B. and Greenshaw, A. J. (1989) Effects of long-term administration of antidepressants and neuroleptics on receptors in the central nervous system. *Cell. Mol. Neurobiol.* **9,** 1–44.
8. Kostrzewa, R. M. (1995) Dopamine receptor supersensitivity. *Neurosci. Biobehav. Rev.* **19,** 1–17.
9. LaHoste, G. J. and Marshall, J. F. (1993) New concepts in dopamine receptor plasticity. *Ann. NY Acad. Sci.* **702,** 183–196.
10. Srivastava, L. K. and Mishra, R. K. (1994) Dopamine receptor gene expression: effects of neuroleptics, denervation, and development, in *Dopamine Receptors and Transporters* (Niznik, H. B., ed.), Dekker, New York, pp. 401–415.
11. Meador-Woodruff, J. H., Mansour, A., Grandy, D. K., Damask, S. P., Civelli, O., and Watson, S. J., Jr. (1992) Distribution of D_5 dopamine receptor mRNA in rat brain. *Neurosci. Lett.* **145,** 209–212.
12. Tiberi, M., Jarvie, K. R., Silvia, C., Falardeau, P., Gingrich, J. A., Godinot, N., Bertrand, L., Yang-Feng, T. L., Fremeau, R. T., Jr., and Caron, M. G. (1991) Cloning, molecular characterization, and chromosomal assignment of a gene

encoding a second D_1 dopamine receptor subtype: differential expression pattern in rat brain compared with the D_{1A} receptor. *Proc. Natl. Acad. Sci. USA* **88,** 7491–7495.
13. Barnett, J. V. and Kuczenski, R. (1986) Desensitization of rat striatal dopamine-stimulated adenylate cyclase after acute amphetamine administration. *J. Pharmacol. Exp. Ther.* **237,** 820–825.
14. Roseboom, P. H. and Gnegy, M. E. (1989) Acute in vivo amphetamine produces a homologous desensitization of dopamine receptor-coupled adenylyl cyclase activities and decreases agonist binding to the D_1 site. *J. Pharmacol. Exp. Ther.* **34,** 148–156.
15. Wilner, K. D., Butler, I. J., Seifert, W. E., Jr., and Clement-Cormier, Y. C. (1980) Biochemical alterations of dopamine receptor responses following chronic L-DOPA therapy. *Biochem. Pharmacol.* **29,** 701–706.
16. Mishra, R. K., Wong, Y.-W., Varmuza, S. L., and Tuff, L. (1978) Chemical lesion and drug induced supersensitivity and subsensitivity of caudate dopamine receptors. *Life Sci.* **23,** 443–446.
17. Neisewander, J. L., Lucki, I., and McGonigle, P. (1991) Behavioral and neurochemical effects of chronic administration of reserpine and SKF-38393 in rats. *J. Pharmacol. Exp. Ther.* **257,** 850–860.
18. Subramanium, S., Lucki, I., and McGonigle, P. (1992) Effects of chronic treatment with selective agonists on the subtypes of dopamine receptors. *Brain Res.* **571,** 313–322.
19. Hess, E. J., Albers, L. J., Le, H., and Creese, I. (1986) Effects of chronic SCH23390 treatment on the biochemical and behavioral properties of D_1 and D_2 dopamine receptors: potentiated behavioral responses to a D_2 dopamine agonist after selective D_1 dopamine receptor up-regulation. *J. Pharmacol. Exp. Ther.* **238,** 846–854.
20. Hess, E. J., Norman, A. B., and Creese, I. (1988) Chronic treatment with dopamine receptor antagonists: behavioral and pharmacological effects on D_1 and D_2 dopamine receptors. *J. Neurosci.* **8,** 2361–2370.
21. McGonigle, P., Boyson, S. J., Reuter, S., and Molinoff, P. B. (1989) Effects of chronic treatment with selective and nonselective antagonists on the subtypes of dopamine receptors. *Synapse* **3,** 74–82.
22. Parashos, S. A., Barone, P., Tucci, I., and Chase, T. N. (1987) Attenuation of D-1 antagonist-induced D-1 receptor upregulation by concomitant D-2 receptor blockade. *Life Sci.* **41,** 2279–2284.
23. Schettini, G., Ventra, C., Florio, T., Grimaldi, M., Meucci, O., and Marino, A. (1992) Modulation by GTP of basal and agonist-stimulated striatal adenylate cyclase activity following chronic blockade of D_1 and D_2 dopamine receptors: involvement of G proteins in the development of receptor supersensitivity. *J. Neurochem.* **59,** 1667–1674.
24. Porceddu, M. L., DeMontis, G., Mele, S., Ongini, E., and Biggio, G. (1987) D_1 dopamine receptors in the rat retina: effect of dark adaptation and chronic blockade by SCH23390. *Brain Res.* **424,** 264–271.
25. Creese, I., Sibley, D. R., and Xu, S. X. (1992) Expression of rat striatal D_1 and D_2 dopamine receptor mRNAs: ontogenetic and pharmacological studies. *Neurochem. Int.* **20,** S45–S48.
26. Buckland, P. R., O'Donovan, M. C., and McGuffin, P. (1992) Changes in dopamine D_1, D_2, and D_3 receptor mRNA levels in rat brain following antipsychotic treatment. *Psychopharmacology* **106,** 479–488.

27. Buckland, P. R., O'Donovan, M. C., and McGuffin, P. (1993) Clopazine and sulpiride up-regulate dopamine D_3 receptor mRNA levels. *Neuropharmacology* **32,** 901–907.
28. Hornykiewicz, O. (1973) Parkinson's disease: from brain homogenate to treatment. *Fed. Proc.* **32,** 183–190.
29. Lee, T., Seeman, P., Rajput, A., Farley, I. J., and Hornykiewicz, O. (1978) Receptor basis for dopaminergic supersensitivity in Parkinson's disease. *Nature* **273,** 59–61.
30. Seeman, P., Bzowej, N. H., Guan, H. C., Bergeron, C., Reynolds, G. P., Bird, E. D., Riederer, P., Jellinger, K., and Tourtellotte, W. W. (1987) Human brain D_1 and D_2 dopamine receptors in schizophrenia, Alzheimer's, Parkinson's, and Huntington's diseases. *Neuropsychopharmacology* **1,** 5–15.
31. Rinne, J. O., Laihinen, A., Lönnberg, P., Marjamäki, P., and Rinne, U. K. (1991) A post-mortem study on striatal dopamine receptors in Parkinson's disease. *Brain Res.* **556,** 117–122.
32. Rinne, J. O., Laihinen, A., Någren, K., Bergman, J., Solin, O., Haaparanta, M., Ruotsalainen, U., and Rinne, U. K. (1990) PET demonstrates different behaviour of striatal dopamine D-1 and D-2 receptors in early parkinson's disease. *J. Neurosci. Res.* **27,** 494–499.
33. Rinne, U. K., Laihinen, A., Rinne, J. O., Någren, K., Bergman, J., and Ruotsalainen, U. (1990) Positron emission tomography demonstrates dopamine D_2 receptor supersensitivity in the striatum of patients with early Parkinson's disease. *Movement Disorders* **5,** 55–59.
34. Ungerstedt, U. (1971) Postsynaptic supersensitivity after 6-hydroxydopamine induced degeneration of the nigro-striatal system. *Acta Physiol. Scand.* **367(Suppl.),** 69–93.
35. Krueger, B. K., Forn, J., Walters, J. R., Roth, R. H., and Greengard, P. (1976) Stimulation by dopamine of adenosine cyclic 3',5'-monophosphate formation in rat caudate nucleus: effect of lesions of the nigroneostriatal pathway. *Mol. Pharmacol.* **12,** 639–648.
36. Mishra, R. K., Gardner, E. L., Katzman, R., and Makman, M. H. (1974) Enhancement of dopamine-stimulated adenylate cyclase activity in rat caudate after lesions in substantia nigra: evidence for denervation supersensitivity. *Proc. Natl. Acad. Sci. USA* **71,** 3883–3887.
37. Billard, W., Ruperto, V., Crosby, G., Iorio, L. C., and Barnett, A. (1984) Characterization of the binding of ^{3}H-SCH 23390, a selective D-1 receptor antagonist ligand, in rat striatum. *Life Sci.* **35,** 1885–1893.
38. Buonamici, M., Caccia, C., Carpentieri, L., Pegrassi, L., Rossi, A. C., and Di Chiara, G. (1986) D-1 receptor supersensitivity in the rat striatum after unilateral 6-hydroxydopamine lesions. *Eur. J. Pharmacol.* **126,** 347,348.
39. Porceddu, M. L., Giorgi, O., De Montis, G., Mele, S., Cocco, L., Ongini, E., and Biggio, G. (1987) 6-Hydroxydopamine-induced degeneration of nigral dopamine neurons: differential effect on nigral and striatal D-1 dopamine receptors. *Life Sci.* **41,** 697–706.
40. Butkerait, P. and Friedman, E. (1993) Repeated reserpine increases striatal dopamine receptor and guanine nucleotide binding protein RNA. *J. Neurochem.* **60,** 566–571.
41. Missale, C., Nisoli, E., Liberini, P., Rizzonelli, P., Memo, M., Buonamici, M., Rossi, A., and Spano, P. F. (1989) Repeated reserpine administration up-regulates the transduction mechanisms of D_1 receptors without changing the density of [^{3}H]SCH 23390 binding. *Brain Res.* **438,** 117–122.

42. Rubinstein, M., Muschietti, J. P., Gershanik, O., Flawia, M. M., and Stefano, F. J. E. (1990) Adaptive mechanisms of striatal D1 and D2 dopamine receptors in response to a prolonged reserpine treatment in mice. *J. Pharmacol. Exp. Ther.* **252,** 810–816.
43. Graham, W. C., Crossman, A. R., and Woodruff, G. N. (1990) Autoradiographic studies in animal models of hemi-parkinsonism reveal dopamine D_2 but not D_1 receptor supersensitivity. I. 6-OHDA lesions of ascending mesencephalic dopaminergic pathways in the rat. *Brain Res.* **514,** 93–102.
44. Jongen-Rêlo, A. L., Docter, G. J., Jonker, A. J., Vreugdenhil, E., Groenewegen, H. J., and Voorn, P. (1994) Differential effects of dopamine depletion on the binding and mRNA levels of dopamine receptors in the shell and core of the rat nucleus accumbens. *Mol. Brain Res.* **25,** 333–343.
45. Joyce, J. N. (1991) Differential response of striatal dopamine and muscarinic cholinergic receptor subtypes to the loss of dopamine. II. Effects of 6-hydroxydopamine or colchicine microinjections into the VTA or reserpine treatment. *Exp. Neurol.* **113,** 277–290.
46. Joyce, J. N. (1991) Differential response of striatal dopamine and muscarinic cholinergic receptor subtypes to the loss of dopamine. I. Effects of intranigral or intracerebroventricular 6-hydroxydopamine lesions of the mesostriatal dopamine system. *Exp. Neurol.* **113,** 261–276.
47. Marshall, J. F., Navarrete, R., and Joyce, J. N. (1989) Decreased striatal D_1 binding density following mesotelencephalic 6-hydroxydopamine injections: an autoradiographic analysis. *Brain Res.* **493,** 247–257.
48. Gerfen, C. R., Engber, T. M., Mahan, L. C., Susel, Z., Chase, T. N., Monsma, F. J., and Sibley, D. R. (1990) D_1 and D_2 dopamine receptor-regulated gene expression of striatonigral and striatopallidal neurons. *Science* **250,** 1429–1432.
49. Qin, Z.-H., Chen, J. F., and Weiss, B. (1994) Lesions of mouse striatum induced by 6-hydroxydopamine differentially alter the density, rate of synthesis, and level of gene expression of D_1 and D_2 dopamine receptors. *J. Neurochem.* **62,** 411–420.
50. Memo, M., Lovenberg, W., and Hanbauer, I. (1982) Agonist-induced subsensitivity of adenylate cyclase coupled with a dopamine receptor in slices from rat corpus striatum. *Proc. Natl. Acad. Sci. USA* **79,** 4456–4460.
51. Ofori, S., Bugnon, O., and Schorderet, M. (1993) Agonist-induced desensitization of dopamine D-1 receptors in bovine retina and rat striatum. *J. Pharmacol. Exp. Ther.* **266,** 350–357.
52. Chneiweiss, H., Glowinski, J., and Premont, J. (1990) Dopamine-induced homologous and heterologous desensitizations of adenylate cyclase-coupled receptors on striatal neurons. *Eur. J. Pharmacol.* **189,** 287–292.
53. Trovero, F., Marin, P., Tassin, J.-P., Premont, J., and Glowinski, J. (1994) Accelerated resensitization of the D_1 dopamine receptor-mediated response in cultured cortical and striatal neurons from the rat: respective role of α_1-adrenergic and N-methyl-D-aspartate receptors. *J. Neurosci.* **14,** 6280–6288.
54. Barton, A. and Sibley, D. R. (1990) Agonist-induced desensitization of D_1-dopamine receptors linked to adenylyl cyclase activity in cultured NS20Y neuroblastoma cells. *Mol. Pharmacol.* **38,** 531–541.
55. Balmforth, A. J., Warburton, P., and Ball, S. G. (1990) Homologous desensitization of the D_1 dopamine receptor. *J. Neurochem.* **55,** 2111–2116.
56. Steffey, M. E., Snyder, G. L., Barrett, R. W., Fink, J. S., Ackerman, M., Adams, P., Bhatt, R., Gomez, E., and MacKenzie, R. G. (1991) Dopamine D_1 receptor

stimulation of cyclic AMP accumulation in COS-1 cells. *Eur. J. Pharmacol.* **207,** 311–317.
57. Bates, M. D., Caron, M. G., and Raymond, J. R. (1991) Desensitization of DA_1 dopamine receptors coupled to adenylyl cyclase in opossum kidney cells. *Am. J. Physiol.* **260,** F937–F945.
58. Zhou, X.-M., Sidhu, A., and Fishman, P. H. (1991) Desensitization of the human D_1 dopamine receptor: evidence for involvement of both cyclic AMP-dependent and receptor-specific protein kinases. *Mol. Cell. Neurosci.* **2,** 464–472.
59. Gupta, S. K. and Mishra, R. K. (1994) Up-regulation of D_1 dopamine receptors in SK-N-MC cells after chronic treatment with SCH 23390. *Neurosci. Res. Comm.* **15,** 157–166.
60. Black, L. E., Smyk-Randall, E. M., and Sibley, D. R. (1994) Cyclic AMP-mediated desensitization of D_1 dopamine receptor-coupled adenylyl cyclase in NS20Y neuroblastoma cells. *Mol. Cell. Neurosci.* **5,** 567–575.
61. Bates, M. D., Olsen, C. L., Becker, B. N., Albers, F. J., Middleton, J. P., Mulheron, J. G., Jin, S.-L. C., Conti, M., and Raymond, J. R. (1993) Elevation of cAMP is required for down-regulation, but not agonist-induced desensitization, of endogenous dopamine D_1 receptors in opossum kidney cells. *J. Biol. Chem.* **268,** 14,757–14,763.
62. Gupta, S. K. and Mishra, R. K. (1993) Desensitization of D_1 dopamine receptors down-regulates the Gs alpha subunit of G protein in SK-N-MC neuroblastoma cells. *J. Mol. Neurosci.* **4,** 117–123.
63. Minowa, M. T., Minowa, T., Monsma, F. J., Jr., Sibley, D. R., and Mouradian, M. M. (1992) Characterization of the 5'-flanking region of the human D_{1A} dopamine receptor gene. *Proc. Natl. Acad. Sci. USA* **89,** 3045–3049.
64. Kennelly, P. J. and Krebs, E. G. (1991) Consensus sequences as substrate specificity determinants for protein kinases and protein phosphatases. *J. Biol. Chem.* **266,** 15,555–15,558.
65. Yasunari, K., Kohno, M., Murakawa, K.-I., Yokokawa, K., Horio, T., and Takeda, T. (1993) Interaction between a phorbol ester and dopamine DA_1 receptors on vascular smooth muscle. *Am. J. Physiol.* **264,** F24–F30.
66. Zamanillo, D., Casanova, E., Alonso-Llamazares, A., Ovalle, S., Chinchetru, M. A., and Calvo, P. (1995) Identification of a cyclic adenosine 3', 5'-monophosphate-dependent protein kinase phosphorylation site in the carboxy terminal tail of the human D_1 dopamine receptor. *Neurosci. Lett.* **188,** 183–186.
67. Inglese, J., Freedman, N. J., Koch, W. J., and Lefkowitz, R. J. (1993) Structure and mechanism of the G protein-coupled receptor kinases. *J. Biol. Chem.* **268,** 23,735–23,738.
68. Onorato, J. J., Palczewski, K., Regan, J. W., Caron, M. G., Lefkowitz, R. J., and Benovic, J. L. (1991) Role of acidic amino acids in peptide substrates of the β-adrenergic kinase and rhodopsin kinase. *Biochemistry* **30,** 5118–5125.
69. Atkinson, B. N., Burgess, L. H., and Sibley, D. R. (1994) Regulation of the D_{1A} and D_{1B} dopamine receptors in stably transfected chinese hamster ovary cells. *Soc. Neurosci. Abstr.* **20,** 521.
70. Machida, C. A., Searles, R. P., Nipper, V., Brown, J. A., Kozell, L. B., and Neve, K. A. (1992) Molecular cloning and expression of the rhesus macaque D_1 dopamine receptor gene. *Mol. Pharmacol.* **41,** 652–659.

71. Neve, K. A. and Kozell, L. B. (1993) Desensitization of recombinant dopamine D1 receptors. *Soc. Neurosci. Abstr.* **19,** 734.
72. Olson, M. F., and Schimmer, B. P. (1992) Heterologous desensitization of the human dopamine D_1 receptor in Y1 adrenal cells in a desensitization-resistant Y1 mutant. *Mol. Endocrinol.* **6,** 1095–1102.
73. Ng, G. Y. K., Mouillac, B., George, S. R., Caron, M., Dennis, M., Bouvier, M., and O'Dowd, B. F. (1994) Desensitization, phosphorylation and palmitoylation of the human dopamine D_1 receptor. *Eur. J. Pharmacol.* **267,** 7–19.
74. Trogadis, J. E., Ng, G. Y.-K., O'Dowd, B., George, S. G., and Stevens, J. K. (1995) Dopamine D_1 receptor distribution in Sf9 cells imaged by confocal microscopy: a quantitative evaluation. *J. Histochem. Cytochem.* **43,** 497–506.
75. Tiberi, M., Bertrand, L., Nash, S. R., and Caron, M. G. (1994) Cellular expression of G protein-coupled receptor kinase augments the agonist-dependent phosphorylation and desensitization of dopamine D_{1A} receptors. *Soc. Neurosci. Abstr.* **20,** 19.
76. Frail, D. E., Manelli, A. M., Witte, D. G., Lin, C. W., Steffey, M. E., and MacKenzie, R. G. (1993) Cloning and characterization of a truncated dopamine D_1 receptor from goldfish retina: stimulation of cyclic AMP production and calcium mobilization. *Mol. Pharmacol.* **44,** 1113–1118.
77. Buckland, P. R., O'Donovan, M. C., and McGuffin, P. (1992) Lack of effect of chronic antipsychotic treatment on dopamine D_5 receptor mRNA level. *Eur. Neuropsychopharm.* **2,** 405–409.
78. Jarvie, K. R., Tiveri, M., Silvia, C., Gingrich, J. A., and Caron, M. G. (1993) Molecular cloning, stable expression and desensitization of the human dopamine D1B/D5 receptor. *J. Recept. Res.* **13,** 573–590.
79. Tiberi, M. and Caron, M. (1994) High agonist-independent activity is a distinguishing feature of the dopamine D_{1B} receptor subtype. *J. Biol. Chem.* **269,** 27,925–27,931.
80. Pei, G., Samama, P., Lohse, M., Wang, M., Codina, J., and Lefkowitz, R. J. (1994) A constitutively active mutant β_2-adrenergic receptor is constitutively desensitized and phosphorylated. *Proc. Natl. Acad. Sci. USA.* **91,** 2699–2702.
81. Beischlag, T. V., Marchese, A., Meador-Woodruff, J., Damask, S. P., O'Dowd, B. F., Tyndale, R. F., Van Tol, H. H. M., Seeman, P., and Niznik, H. (1995) The human dopamine D_5 receptor gene: cloning and characterization of the 5'-flanking and promoter region. *Biochemistry* **34,** 5960–5970.
82. De Jesus, O. T., Van Moffart, G. J. C., Dinerstein, R. J., and Friedman, A. M. (1986) Exogenous L-DOPA alters spiroperidol binding, in vivo, in the mouse striatum. *Life Sci.* **39,** 341–349.
83. List, S. J. and Seeman, P. (1979) Dopamine agonists reverse the elevated ^{3}H-neuroleptic binding in neuroleptic-pretreated rats. *Life Sci.* **24,** 1447–1452.
84. Rouillard, C. Bedard, P. J., Falardeau, P., and DiPaulo, T. (1987) Behavioral and biochemical evidence for a different effect of repeated administration of L-DOPA and bromocriptine on denervated versus non-denervated dopamine receptors. *Neuropharmacology* **26,** 1601–1606.
85. Quick, M. and Iversen, L. L. (1978) Subsensitivity of the rat striatal dopaminergic system after treatment with bromocriptine: effects on [^{3}H]spiperone binding and dopamine-stimulated cyclic AMP formation. *Nauyn-Schmied. Arch. Pharm.* **304,** 141–145.

86. Wei-Dong, L., Xiao-Da, Z., and Guo-Zhang, J. (1988) Enhanced stereotypic behavior by chronic treatment with bromocriptine accompanies increase of D-1 receptor binding. *Life Sci.* **42,** 1841–1845.
87. Nisoli, E., Memo, M., Missale, C., Carruba, M. O. and Spana, P. F. (1990) Repeated administration of lisuride down-regulates dopamine D_2 receptor function in meso-striatal and in mesolimbocortical rat brain regions. *Eur. J. Pharmacol.* **176,** 85–90.
88. Autelitano, D. J., Snyder, L., Sealfon, S. C., and Roberts, J. L. (1989) Dopamine D_2-receptor messenger RNA is differentially regulated by dopaminergic agents in rat anterior and neurointermediate pituitary. *Mol. Cell. Endocrinol.* **67,** 101–105.
89. Chronwall, B. M., Dickerson, D. S., Huerter, B. S., Sibley, D. R., and Millington, W. R. (1994) Regulation of heterogeneity in D_2 dopamine receptor gene expression among individual melanotropes in the rat pituitary intermediate lobe. *Mol. Cell. Neurosci.* **5,** 35–45.
90. Chen, J. F., Aloyo, V. J., and Weiss, B. (1993) Continuous treatment with the D_2 dopamine receptor agonist quinpirole decreases D_2 dopamine receptors, D_2 dopamine receptor messenger RNA and proenkephalin messenger RNA, and increases mu opioid receptors in mouse striatum. *Neuroscience* **54,** 669–680.
91. Klawans, H. L. and McKendall, R. (1971) Observations on the effect of L-dopa on tardive lingual-facial-buccal dyskinesia. *J. Neurol. Sci.* **14,** 189–192.
92. Muller, P. and Seeman, P. (1977) Brain neurotransmitter receptors after long-term haloperidol: dopamine, acetylcholine, serotonin, α-noradrenergic and naloxone receptors. *Life Sci.* **21,** 1751–1758.
93. Clow, A., Jenner, P., Theodorou, A., and Marsden, C. D. (1979) Striatal dopamine receptors become supersensitive while rats are given trifluoperazine for six months. *Nature* **278,** 59–61.
94. Muller, P. and Seeman, P. (1978) Dopaminergic supersensitivity after neuroleptics: time-course and specificity. *Psychopharmacology* **60,** 1–11.
95. Tarsy, D. and Baldessarini, R. J. (1974) Behavioral supersensitivity to apomorphine following chronic treatment with drugs which interfere with the synaptic function of catecholamines. *Neuropharmacology* **13,** 927–940.
96. Baldessarini, R. J. and Tarsy, D. (1976) Mechanisms underlying tardive dyskinesia, in *The Basal Ganglia* (Yahr, M. D., ed.), Raven, New York, pp. 433–446.
97. Casey, D. E. (1995) Tardive dyskinesia: pathophysiology, in *Psychopharmacology: The Fourth Generation of Progress* (Bloom, F. E. and Kupfer, D. J., eds.), Raven, New York, pp. 1497–1502.
98. Crow, T. J., Cross, A. J., Johnstone, E. C., Owen, F., Owens, D. G. C., and Waddington, J. L. (1982) Abnormal involuntary movements in schizophrenia: are they related to the disease process or its treatment? Are they associated with changes in dopamine receptors? *J. Clin. Psychopharmacol.* **2,** 336–340.
99. Wolf, M. E., Bowie, L., Keener, S., and Mosnaim, A. D. (1982) Prolactin response in tardive dyskinesia. *Biol. Psychiat.* **17,** 485–490.
100. Burt, D. R., Creese, I., and Snyder, S. H. (1977) Antischizophrenic drugs: chronic treatment elevates dopamine receptor binding in brain. *Science* **196,** 326–328.
101. Laruelle, M., Jaskiw, G. E., Lipska, B. K., Kolachana, B., Casanova, M. F., Kleinman, J. E., and Weinberger, D. R. (1992) D_1 and D_2 receptor modulation in rat striatum and nucleus accumbens after subchronic and chronic haloperidol treatment. *Brain Res.* **575,** 47–56.

102. Memo, M., Pizzi, M., Missale, C., Carruba, M. O., and Spano, P. F. (1987) Modification of the function of D_1 and D_2 dopamine receptors in striatum and nucleus accumbens of rats chronically treated with haloperidol. *Neuropharmacology* **267,** 477–480.
103. Staunton, D. A., Magistretti, P. J., Koob, G. F., Shoemaker, W. J., and Bloom, F. E. (1982) Dopaminergic supersensitivity induced by denervation and chronic receptor blockade is additive. *Nature* **299,** 72–74.
104. Wilmot, C. A. and Szczepanik, A. M. (1989) Effects of acute and chronic treatments with clozapine and haloperidol on serotonin (5-HT_2) and dopamine (D_2) receptors in the rat brain. *Brain Res.* **487,** 288–298.
105. Janowsky, A., Neve, K. A., Kinzie, J. M., Taylor, B., de Paulis, T., and Belknap, J. K. (1992) Extrastriatal dopamine D2 receptors: distribution, pharmacological characterization, and region-specific regulation by clozapine. *J. Pharmacol. Exp. Ther.* **261,** 1282–1290.
106. Kazawa, T., Mikuni, M., Higuchi, T., Arai, I., Takahashi, K., and Yamauchi, T. (1990) Characterization of sulpiride-displaceable ^{3}H-YM-091512-2 binding sites in rat frontal cortex and the effects of subchronic treatment with haloperidol on cortical D-2 dopamine receptors. *Life Sci.* **47,** 531–537.
107. MacLennan, A. J., Atmadja, S., Lee, N., and Fibiger, H. C. (1988) Chronic haloperidol administration increases the density of D2 dopamine receptors in the medial prefrontal cortex of the rat. *Psychopharmacology* **95,** 255–257.
108. Memo, M., Pizzi, M., Nisoli, E., Missale, C., Carruba, M. O., and Spano, P. (1987) Repeated administration of (–)sulpiride and SCH 23390 differentially up-regulate D-1 and D-2 dopamine receptor function in rat mesostriatal areas but not in cortical-limbic brain regions. *Eur. J. Pharmacol.* **138,** 45–51.
109. Olianas, M. C. and Onali, P. (1987) Supersensitivity of striatal D_2 dopamine receptors mediating inhibition of adenylate cyclase and stimulation of guanosine triphosphatase following chronic administration of haloperidol in mice. *Neurosci. Lett.* **78,** 349–354.
110. Consolo, S., Ladinsky, H., Samanin, R., Bianchi, S., and Ghezzi, D. (1978) Supersensitivity of the cholinergic response to apomorphine in the striatum following denervation or disuse supersensitivity of dopaminergic receptors in the rat. *Brain Res.* **155,** 45–54.
111. Cubeddu, L. X., Hoffmann, I. S., James, M. K., and Niedzwiecki, D. M. (1983) Changes in the sensitivity to apomorphine of dopamine receptors modulating dopamine and acetylcholine release after chronic treatment with bromocriptine or haloperidol. *J. Pharmacol. Exp. Ther.* **226,** 680–685.
112. Miller, J. C. and Friedhoff, A. J. (1979) Dopamine receptor-coupled modulation of the K^+-depolarized overflow of ^{3}H-acetylcholine from rat striatal slices: alteration after chronic haloperidol and alpha-methyl-p-tyrosine pretreatment. *Life Sci.* **25,** 1249–1256.
113. Gianutsos, G., Hynes, M. D., and Lal, H. (1975) Enhancement of apomorphine-induced inhibition of striatal dopamine-turnover following chronic haloperidol. *Biochem. Pharmacol.* **24,** 581,582.
114. Lal, H., Brown, W., Drawbaugh, R., Hynes, M., and Brown, G. (1977) Enhanced prolactin inhibition following chronic treatment with haloperidol and morphine. *Life Sci.* **20,** 101–106.

115. Bannon, M. J., Bunney, E. B., Zigun, J. R., Skirboll, L. R., and Roth, R. H. (1980) Presynaptic dopamine receptors: insensitivity to kainic acid and the development of supersensitivity following chronic haloperidol. *Naunyn-Schmiedeberg's Arch. Pharmacol.* **312,** 161–165.
116. Stock, G., Steinbrenner, J., and Kummer, P. (1980) Supersensitivity of dopamine-autoreceptors: the effect of gammabutyrolactone in long-term haloperidol treated rats. *J. Neur. Transm.* **47,** 145–151.
117. Mereu, G., Lilliu, V., Vargiu, P., Muntoni, A. L., Diana, M., and Gessa, G. L. (1995) Depolarization inactivation of dopamine neurons: an artifact? *J. Neurosci.* **15,** 1144–1149.
118. Casey, D. E. (1989) Clozapine: neuroleptic-induced EPS and tardive dyskinesia. *Psychopharmacology* **99,** S47–S53.
119. Kane, J., Honigfeld, G., Singer, J., and Meltzer, H. (1988) Clozapine for the treatment-resistant schizophrenic. *Arch. Gen. Psychiat.* **45,** 789–796.
120. Boyson, S. J., McGonigle, P., Luthin, G. R., Wolfe, B. B., and Molinoff, P. B. (1988) Effects of chronic administration of neuroleptic and anticholinergic agents on densities of D_2 dopamine and muscarinic cholinergic receptors in rat striatum. *J. Pharmacol. Exp. Ther.* **244,** 987–993.
121. O'Dell, S. J., La Hoste, G. J., Widmark, C. B., Shapiro, R. M., Potkin, S. G., and Marshall, J. F. (1990) Chronic treatment with clozapine or haloperidol differentially regulates dopamine and serotonin receptors in rat brain. *Synapse* **6,** 146–153.
122. Rupniak, N. M. J., Hall, M. D., Mann, S., Fleminger, S., Kilpatrick, G., Jenner, P., and Marsden, C. D. (1985) Chronic treatment with clozapine, unlike haloperidol, does not induce changes in striatal D-2 receptor function in the rat. *Biochem. Pharmacol.* **34,** 2755–2763.
123. Seeger, T. F., Thal, L., and Gardner, E. L. (1982) Behavioral and biochemical aspects of neuroleptic-induced dopaminergic supersensitivity: studies with chronic clozapine and haloperidol. *Psychopharmacology* **76,** 182–187.
124. Severson, J. A., Robinson, H. E., and Simpson, G. M. (1984) Neuroleptic-induced striatal dopamine receptor supersensitivity in mice: relationship to dose and drug. *Psychopharmacology* **84,** 115–119.
125. Deutch, A. Y. (1995) Mechanisms of action of clozapine in the treatment of neuroleptic-resistant and neuroleptic-intolerant schizophrenia. *Eur. Psychiat.* **10(Suppl. 1),** 39S–46S.
126. Robertson, G. S. and Fibiger, H. C. (1991) Neuroleptics increase *c-fos* expression in the forebrain: contrasting effects of haloperidol and clozapine. *Neuroscience* **46,** 315–328.

126a. Pich, E. M., Benfenati, F., Farabegoli, C., Fuxe, K., Meller, E., Aronsson, M., Goldstein, M., and Agnati, L. F. (1987) Chronic haloperidol affects striatal D_2-dopamine receptor reappearance after irreversible blockade. *Brain Res.* **435,** 147–152.

127. Bunzow, J. R., Van Tol, H. H. M., Grandy, D. K., Albert, P., Salon, J., Christie, M., Machida, C. A., Neve, K. A., and Civelli, O. (1988) Cloning and expression of a rat D_2 dopamine receptor cDNA. *Nature* **336,** 783–787.
128. Fox, C. A., Mansour, A., and Watson, S. J., Jr. (1994) The effects of haloperidol on dopamine receptor gene expression. *Exp. Neurol.* **130,** 288–303.

129. Matsunaga, T., Ohara, K., Natsukari, N., and Fujita, M. (1991) Dopamine D_2-receptor mRNA level in rat striatum after chronic haloperidol treatment. *Neurosci. Res.* **12,** 440–445.
130. Qin, Z.-H., Zhou, L.-W., and Weiss, B. (1994) D_2 dopamine receptor messenger RNA is altered to a greater extent by blockade of glutamate receptors than by blockade of dopamine receptors. *Neuroscience* **60,** 97–114.
131. Srivastava, L. K., Morency, M. A., Bajwa, S. B., and Mishra, R. K. (1990) Effect of haloperidol on expression of dopamine D_2 receptor mRNAs in rat brain. *J. Mol. Neurosci.* **2,** 155–161.
132. Van Tol, H. H. M., Riva, M., Civelli, O., and Creese, I. (1990) Lack of effect of chronic dopamine receptor blockade on D_2 dopamine receptor mRNA level. *Neurosci. Lett.* **111,** 303–308.
133. Xu, S., Monsma, F. J., Sibley, D. R., and Creese, I. (1991) Regulation of D_{1A} and D_2 dopamine receptor mRNA during ontogenesis, lesion and chronic antagonist treatment. *Life Sci.* **50,** 383–396.
134. Angulo, J. A., Coirini, H., Ledoux, M., and Schumacher, M. (1991) Regulation by dopaminergic neurotransmission of dopamine D_2 mRNA and receptor levels in the striatum and nucleus accumbens of the rat. *Mol. Brain Res.* **11,** 161–166.
135. Bernard, V., Le Moine, C., and Bloch, B. (1991) Striatal neurons express increased level of dopamine D_2 receptor mRNA in response to haloperidol treatment: a quantitative *in situ* hybridization study. *Neuroscience* **45,** 117–126.
136. Buckland, P. R., O'Donovan, M. C., and McGuffin, P. (1993) Both splicing variants of the dopamine D_2 receptor mRNA are up-regulated by antipsychotic drugs. *Neurosci. Lett.* **150,** 25–28.
137. Chen, J. F., Aloyo, V. J., Qin, Z.-H., and Weiss, B. (1994) Irreversible blockade of D_2 dopamine receptors by fluphenazine-*N*-mustard increases D_2 dopamine receptor mRNA and proenkephalin mRNA and decreases D_1 dopamine receptor mRNA and mu and delta opioid receptors in rat striatum. *Neurochem. Int.* **25,** 355–366.
138. Fishburn, C. S., David, C., Carmon, S., and Fuchs, S. (1994) The effect of haloperidol on D2 dopamine receptor subtype mRNA levels in the brain. *FEBS Lett.* **339,** 63–66.
139. Lévesque, D., Martres, M.-P., Diaz, J., Griffon, N., Lammers, C. H., Sokoloff, P., and Schwartz, J.-C. (1995) A paradoxical regulation of the dopamine D_3 receptor expression suggests the involvement of an anterograde factor from dopamine neurons. *Proc. Natl. Acad. Sci. USA* **92,** 1719–1723.
140. Martres, M. P., Sokoloff, P., Giros, B., and Schwartz, J. C. (1992) Effects of dopaminergic transmission interruption on the D_2 receptor isoforms in various cerebral tissues. *J. Neurochem.* **58,** 673–679.
141. Rogue, P., Hanauer, A., Zwiller, J., Malviya, A. N., and Vincendon, G. (1991) Up-regulation of dopamine D_2 receptor mRNA in rat striatum by chronic neuroleptic treatment. *Eur. J. Pharmacol—Mol. Pharm.* **207,** 165–168.
142. Qin, Z.-H. and Weiss, B. (1994) Dopamine receptor blockade increases dopamine D_2 receptor and glutamic acid decarboxylase mRNAs in mouse substantia nigra. *Eur. J. Pharmacol.—Mol. Pharmacol.* **269,** 25–33.
143. Arnauld, E., Arsaut, J., and Demotes-Mainard, J. (1991) Differential plasticity of the dopaminergic D_2 receptor mRNA isoforms under haloperidol treatment, as evidenced by in situ hybridization in rat anterior pituitary. *Neurosci. Lett.* **130,** 12–16.

144. MacKenzie, R. G. and Zigmond, M. J. (1985) Chronic neuroleptic treatment increases D-2 but not D-1 receptors in rat striatum. *Eur. J. Pharmacol.* **113,** 159–165.
145. Marin, C. and Chase, T. N. (1993) Dopamine D_1 receptor stimulation but not dopamine D_2 receptor stimulation attenuates haloperidol-induced behavioral supersensitivity and receptor up-regulation. *Eur. J. Pharmacol.* **231,** 191–196.
146. Falardeau, P., Bouchard, S., Bédard, P. J., Boucher, R., and Di Paolo, T. (1988) Behavioral and biochemical effects of chronic treatment with D-1 and/or D-2 dopamine agonists in MPTP monkeys. *Eur. J. Pharmacol.* **150,** 59–66.
147. Cameron, D. L. and Crocker, A. D. (1988) Stimulation of D-1 dopamine receptors facilitates D-2 dopamine receptor recovery after irreversible receptor blockade. *Neuropharmacology* **27,** 447–450.
148. Mandel, R. J., Wilcox, R. E., and Randall, P. K. (1992) Behavioral quantification of striatal dopaminergic supersensitivity after bilateral 6-hydroxydopamine lesions in the mouse. *Pharmacol. Biochem. Behav.* **41,** 343–347.
149. Marshall, J. F. and Ungerstedt, U. (1977) Supersensitivity to apomorphine following destruction of the ascending dopamine neurons: quantification using the rotational model. *Eur. J. Pharmacol.* **41,** 361–367.
150. Schultz, W. and Ungerstedt, U. (1978) Striatal cells supersensitivity to apomorphine in dopamine-lesioned rats correlated to behavior. *Neuropharmacology* **17,** 349–353.
151. Enz, A., Goldstein, M., and Meller, E. (1990) Dopamine agonist-induced elevation of striatal acetylcholine: relationship between receptor occupancy and response in normal and denervated rat striatum. *Mol. Pharmacol.* **37,** 560–565.
152. Fibiger, H. C. and Grewaal, D. S. (1974) Neurochemical evidence for denervation supersensitivity: the effect of unilateral substantia nigra lesions on apomorphine-induced increases in neostriatal acetylcholine levels. *Life Sci.* **15,** 57–63.
153. Kozlowski, M. R. and Marshall, J. F. (1980) Plasticity of [^{14}C]2-deoxy-D-glucose incorporation into neostriatum and related structures in response to dopamine neuron damage and apomorphine replacement. *Brain Res.* **197,** 167–183.
154. Creese, I., Burt, D. R., and Snyder, S. H. (1977) Dopamine receptor binding enhancement accompanies lesion-induced behavioral supersensitivity. *Science* **197,** 596–598.
155. Chen, S. D., Zhou, X. D., Xu, D. L., Zhu, C. M., Zheng, J. X., Kuang, Q. F., Guo, Y. Z., and Li, B. (1993) "In-vivo" visualization by SPECT of the ipsilateral striatal dopamine D_2 receptor supersensitivity occurring in MPTP-induced hemiparkinsonism in primates. *Neurodegeneration* **2,** 147–151.
156. Graham, W. C., Clarke, C. E., Boyce, S., Sambrook, M. A., Crossman, A. R., and Woodruff, G. N. (1990) Autoradiographic studies in animal models of hemiparkinsonism reveal D2 but not D1 receptor supersensitivity. II. Unilateral intracarotid infusion of MPTP in the monkey *(Macaca fascicularis). Brain Res.* **514,** 103–110.
157. Joyce, J. N., Marshall, J. F., Bankiewicz, K. S., Kopin, I. J., and Jacobowicz, D. M. (1986) Hemiparkinsonism in a monkey after unilateral internal carotid artery infusion of 1-methyl-4-phenyl-1,2,3,6-tetrahydropyridine (MPTP) is associated with regional ipsilateral changes in striatal dopamine D_2 receptor density. *Brain Res.* **382,** 360–364.

158. Waddington, J. L., Cross, A. J., Longden, A., Owen, F., and Poulter, M. (1979) Apomorphine-induced rotation in the unilateral 6-OHDA-lesioned rat: relationship to changes in striatal adenylate cyclase activity and ^{3}H-spiperone binding. *Neuropharmacology* **18,** 643–645.
159. Neve, K. A., Kozlowski, M. R., and Marshall, J. F. (1982) Plasticity of neostriatal dopamine receptors after nigrostriatal injury: relationship to recovery of sensorimotor functions and behavioral supersensitivity. *Brain Res.* **242,** 33–44.
160. Staunton, D. A., Wolfe, B. B., Groves, P. M., and Molinoff, P. B. (1981) Dopamine receptor changes following destruction of the nigrostriatal pathway: lack of a relationship to rotational behavior. *Brain Res.* **211,** 315–327.
161. Breese, G. R., Duncan, G. E., Napier, T. C., Bondy, S. C., Iorio, L. C., and Mueller, R. A. (1987) 6-Hydroxydopamine treatments enhance behavioral responses to intracerebral microinjection of D_1- and D_2-dopamine agonists into nucleus accumbens and striatum without changing dopamine antagonist binding. *J. Pharmacol. Exp. Ther.* **240,** 167–176.
162. Mileson, B. E., Lewis, M. H., and Mailman, R. B. (1991) Dopamine receptor 'supersensitivity' occurring without receptor up-regulation. *Brain Res.* **561,** 1–10.
163. LaHoste, G. J. and Marshall, J. F. (1991) Chronic eticlopride and dopamine denervation induce equal nonadditive increases in striatal D_2 receptor density: autoradiographic evidence against the dual mechanism hypothesis. *Neuroscience* **41,** 473–481.
164. Neve, K. A. and Marshall, J. F. (1984) The effects of denervation and chronic haloperidol treatment on neostriatal dopamine receptor density are not additive. *Neurosci. Lett.* **46,** 77–83.
165. LaHoste, G. J. and Marshall, J. F. (1993) The role of dopamine in the maintenance and breakdown of D_1/D_2 synergism. *Brain Res.* **611,** 108–116.
166. LaHoste, G. J. and Marshall, J. F. (1992) Dopamine supersensitivity and D_1/D_2 synergism are unrelated to changes in striatal receptor density. *Synapse* **12,** 14–26.
167. Randall, P. K. (1985) Quantification of dopaminergic supersensitivity using apomorphine-induced behaviors in the mouse. *Life Sci.* **37,** 1419–1423.
168. Reches, A., Wagner, R. H., Jackson, V., Yablonskaya-Alter, E., and Fahn, S. (1983) Dopamine receptors in the denervated striatum: further supersensitivity by chronic haloperidol treatment. *Brain Res.* **275,** 183–185.
169. LaHoste, G. J. and Marshall, J. F. (1989) Non-additivity of D_2 receptor proliferation induced by dopamine denervation and chronic selective antagonist administration: evidence from quantitative autoradiography indicates a single mechanism of action. *Brain Res.* **502,** 223–232.
170. Neve, K. A., Loeschen, S., and Marshall, J. F. (1985) Denervation accelerates the reappearance of neostriatal D-2 receptors after irreversible receptor blockade. *Brain Res.* **329,** 225–231.
171. Norman, A. B., Battaglia, G., and Creese, I. (1987) Differential recovery rates of rat D_2 dopamine receptors as a function of aging and chronic reserpine treatment following irreversible modification: a key to receptor regulatory mechanisms. *J. Neurosci.* **7,** 1484–1491.
172. Brené, S., Lindefors, N., Herrera-Marschitz, M., and Persson, H. (1990) Expression of dopamine D2 receptor and choline acetyltransferase mRNA in the dopamine deafferented rat caudate-putamen. *Exp. Brain Res.* **83,** 96–104.

172a. Lisovoski, F., Haby, C., Borrelli, E., Schleef, C., Revel, M. O., Hindelang, C., and Zwiller, J. (1992) Induction of D_2 dopamine receptor mRNA synthesis in a 6-hydroxydopamine parkinsonian rat model. *Brain Res. Bull.* **28,** 697–701.

173. Chen, J. F., Qin, Z. H., Szele, F., Bai, G., and Weiss, B. (1991) Neuronal localization and modulation of the D_2 dopamine receptor mRNA in brain of normal mice and mice lesioned with 6-hydroxydopamine. *Neuropharmacology* **30,** 927–941.

174. Inoue, A., Ueda, H., Nakata, Y., and Misu, Y. (1994) Supersensitivity of quinpirole-evoked GTPase activation without changes in gene expression of D_2 and G_i protein in the striatum of hemi-dopaminergic lesioned rats. *Neurosci. Lett.* **175,** 107–110.

175. Neve, K. A., Neve, R. L., Fidel, S., Janowsky, A., and Higgins, G. A. (1991) Increased abundance of alternatively spliced forms of D-2 receptor mRNA after denervation. *Proc. Natl. Acad. Sci. USA* **88,** 2802–2806.

176. Reches, A., Wagner, H. R., Jackson-Lewis, V., Yablonskaya-Alter, E., and Fahn, S. (1984) Chronic levodopa or pergolide administration induces down-regulation of dopamine receptors in denervated striatum. *Neurology* **34,** 1208–1212.

177. Agui, T., Amlaiky, N., Caron, M. G., and Kebabian, J. W. (1988) Agonist-induced desensitization of the D-2 dopamine receptor in the intermediate lobe of the rat pituitary gland. *J. Biochem.* **103,** 436–441.

178. Drukarch, B., Schepens, E., and Stoof, J. C. (1991) Sustained activation does not desensitize the dopamine D_2 receptor-mediated control of evoked in vitro release of radiolabeled acetylcholine from rat striatum. *Eur. J. Pharmacol.* **196,** 209–212.

179. Maus, M., Vernier, P., Valdenaire, O., Homburger, V., Bockaert, J., Glowinski, J., and Mallet, J. (1993) D2-dopaminergic agonist quinpirole and 8-bromo-cAMP have opposite effects on $Go\alpha$ GTP-binding protein mRNA without changing D2 dopamine receptor mRNA levels in striatal neurones in primary culture. *J. Recept. Res.* **13,** 1–4.

180. Barton, A. C., Black, L. E., and Sibley, D. R. (1991) Agonist-induced desensitization of D_2 dopamine receptors in human Y-79 retinoblastoma cells. *Mol. Pharmacol.* **39,** 650–658.

181. Ng, G. Y. K., O'Dowd, B. F, Caron, M., Dennis, M., Brann, M. R., and George, S. R. (1994) Phosphorylation and palmitoylation of the human D_{2L} dopamine receptor in Sf9 cells. *J. Neurochem.* **63,** 1589–1595.

182. Elazar, Z. and Fuchs, S. (1991) Phosphorylation by cyclic AMP-dependent protein kinase modulates agonist binding to the D_2 dopamine receptor. *J. Neurochem.* **56,** 75–80.

183. Rogue, P., Zwiller, J., Malviya, A. N., and Vincendon, G. (1990) Phosphorylation by protein kinase C modulates agonist binding to striatal dopamine D_2 receptors. *Biochem. Int.* **22,** 575–582.

184. Cubeddu, L. X., Lovenberg, T. W., Hoffman, I. S., and Talmaciu, R. K. (1989) Phorbol esters and D_2 dopamine receptors. *J. Pharmacol. Exp. Ther.* **251,** 687–693.

185. Ivins, K. J., Luedtke, R. R., Artymyshyn, R. P., and Molinoff, P. B. (1991) Regulation of dopamine D_2 receptors in a novel cell line. *Mol. Pharmacol.* **39,** 531–539.

186. Johnston, J. M., Wood, D. F., Read, S., and Johnston, D. G. (1993) Dopamine regulates D_2 receptor gene expression in normal but not in tumorous rat pituitary cells. *Mol. Cell. Endocrinol.* **92,** 63–68.

187. Bates, M. D., Senogles, S. E., Bunzow, J. R., Liggett, S. B., Civelli, O., and Caron, M. G. (1988) Regulation of responsiveness at D_2 dopamine receptors by receptor desensitization and adenylyl cyclase sensitization. *Mol. Pharmacol.* **39,** 55–63.

188. Johansson, M. H. and Westlind-Danielson, A. (1994) Forskolin-induced up-regulation and functional supersensitivity of dopamine D_2 long receptors expressed by Ltk^- cells. *Eur. J. Pharmacol.—Mol. Pharmacol.* **269,** 149–155.
189. Liu, Y. F., Civelli, O., Grandy, D. K., and Albert, P. R. (1992) Differential sensitivity of the short and long human dopamine D_2 receptor subtypes to protein kinase C. *J. Neurochem.* **59,** 2311–2317.
190. Zhang, L.-J., Lachowicz, J. E., and Sibley, D. R. (1994) The D_{2S} and D_{2L} dopamine receptor isoforms are differentially regulated in chinese hamster ovary cells. *Mol. Pharmacol.* **45,** 878–889.
191. Di Marzo, V., Vial, D., Sokoloff, P., Schwartz, J. C. (1993) Selection of alternative Gi-mediated signaling pathways at the dopamine D_2 receptor by protein kinase C. *J. Neurosci.* **13,** 4846–4853.
192. Kanterman, R. Y., Mahan, L. C., Briley, E. M., Monsma, F. J., Jr., Sibley, D. R., Axelrod, J., and Felder, C. C. (1991) Transfected D_2 dopamine receptors mediate the potentiation of arachidonic acid release in chinese hamster ovary cells. *Mol. Pharmacol.* **39,** 364–369.
193. Piomelli, D. and Di Marzo, V. (1993) Dopamine D_2 receptor signaling via the arachidonic acid cascade: modulation by cAMP-dependent protein kinase A and prostaglandin E_2. *J. Lipid Mediat.* **6,** 433–443.
194. Filtz, T. M., Artymshyn, R. P., Guan, W., and Molinoff, P. B. (1993) Paradoxical regulation of dopamine receptors in transfected 293 cells. *Mol. Pharmacol.* **44,** 371–379.
195. Filtz, T. M., Guan, W., Artymyshyn, R. P., Pacheco, M., Ford, C. and Molinoff, P. B. (1994) Mechanisms of up-regulation of D_{2L} dopamine receptors by agonists and antagonists in transfected HEK-293 cells. *J. Pharmacol. Exp. Ther.* **271,** 1574–1582.
196. Starr, S., Kozell, L. B., and Neve, K. A. (1995) Drug-induced up-regulation of dopamine D2 receptors on cultured cells. *J. Neurochem.* **65,** 569–577.
197. Fornaretto, M. G., Caccia, C., Caron, M. G., and Fariello, R. G. (1993) Dopamine receptors status after unilateral nigral 6-OHDA lesion. *Mol. Chem. Neuropathol.* **19,** 147–162.
198. Seabrook, G. R., Kemp, J. A., Freedman, S. B., Patel, S., Sinclair, H. A., and McAllister, G. (1994) Functional expression of human D3 dopamine receptors in differentiated neuroblastoma X glioma NG108-15 cells. *Br. J. Pharmacol.* **111,** 391–393.
199. Cox, B. A., Rosser, M. P., Kozlowski, M. R., Duwe, K. M., Neve, R. L., and Neve, K. A. (1995) Regulation and functional characterization of a rat recombinant dopamine D3 receptor. *Synapse* **21,** 1–9.
200. Schoots, O., Seeman, P., Guan, H.-C., Paterson, A. D., and Van Tol, H. H. M. (1995) Long-term haloperidol elevates dopamine D_4 receptors by 2-fold in rats. *Eur. J. Pharmacol.—Mol. Pharmacol.* **289,** 67–72.
201. Asghari, V., Schoots, O., Van Kats, S., Ohara, K., Jovanovic, V., Guan, H.-C., Bunzow, J. R., Petronis, A., and Van Tol, H. H. M. (1994) Dopamine D_4 receptor repeat: analysis of different native and mutant forms of the human and rat genes. *Mol. Pharmacol.* **46,** 364–373.
202. Asghari, V., Sanyal, S., Buchwaldt, S., Paterson, S., Jovanovic, V., and Van Tol, H. H. M. (1995) Modulation of intracellular cyclic AMP levels by different human dopamine D_4 receptor variants. *J. Neurochem.* **65,** 1157–1165.

CHAPTER 14

Regulation of Motor Behavior by Dopamine Receptor Subtypes

An Antisense Knockout Approach

Ming Zhang, Abdel-Mouttalib Ouagazzal, Bao-Cun Sun, and Ian Creese

1. Introduction

The neurotransmitter dopamine mediates various behavioral functions in the central nervous system (CNS). The major dopamine pathway involved in motor function is the nigrostriatal pathway, which originates from the substantia nigra and is the primary source of dopaminergic innervation of the dorsal striatal neurons *(1–4)*. It plays an important role in regulating motor behavior and its deterioration is the major cause for the motor symptoms of Parkinson's disease *(5,6)*. The mesolimbic pathway, which originates from neurons of the ventral tegmental area (VTA), innervates the ventral striatum, nucleus accumbens, olfactory tubercle, and parts of the limbic system, and is most probably involved in emotional and motivational aspects of behavior *(7,8)*. It may well contribute to the etiology of schizophrenia and serve as the substrate for neuroleptic drug actions *(9,10)* along with the mesocortical pathway, which also originates in VTA and projects most densely to the prefrontal cortex *(11)*. This terminal area may be involved in certain aspects of learning and memory *(12,13)*. Dopamine agonists have been used to ameliorate the major symptoms of Parkinson's disease, whereas dopamine antagonists, the standard therapy for schizophrenia, often produce motor side effects; reminiscent of Parkinson's disease. Through its action in these different neural pathways in the CNS, dopamine activates a broad range of motor behaviors.

The Dopamine Receptors Eds.: K. A. Neve and R. L. Neve
Humana Press Inc., Totowa, NJ

Based on pharmacological studies using selective agonists and antagonists, two classes of dopamine receptors, D1 and D2,* were postulated in the late 1970s to explain the actions of dopaminergic drugs *(14–16)*. Studies to identify the role of each subtype in various behavioral functions led to results suggesting that the D2 receptor subtype is the site at which neuroleptic drugs bind to produce their antischizophrenic and also their motor side effects *(17,18)*. Additional studies showed that D2 receptors are involved in regulating motor activity, learning and/or performance of learned tasks, and motivated activities such as cocaine self-administration *(7,8,12,19)*. With the advent of selective D1 receptor antagonists, the primacy of D2 receptors in mediating dopamine's behavioral effects was questioned *(20)*. Indeed, the D1 antagonist SCH 23390 often produced identical behavioral effects in animals to those produced by D2 antagonists, although it appears to lack antipsychotic action in humans. On the other hand, specific behavioral effects associated with D1 receptor activation have been found with the use of selective D1 receptor agonists such as SKF 38393. SKF 38393 typically induces grooming behavior, which is distinct from the elements of behavior such as locomotor activation and sniffing induced by D2 receptor-selective agonists. More importantly, the coadministration of D1 and D2 receptor agonists seems to have a synergistic effect on the induction of stereotypy, suggesting that there is an interaction between the two classes of dopamine receptor subtypes. The general consensus now is that coactivation of both D1 and D2 receptors is often required for behavioral expression, and selective antagonism of either D1 or D2 receptors can disrupt behavioral output.

The recent cloning of more subtypes of dopamine receptors indicates a much more complicated picture and challenges the findings from the earlier pharmacological studies. Five distinct dopamine receptor subtypes (D_1, D_2, D_3, D_4, and D_5) have been cloned from the mammalian brain with some or all of these receptors coexpressed in the same brain regions *(21,22)*. The D_2, D_3, and D_4 receptor subtypes share a high degree of homology in their amino acid sequence and have a similar pharmacological profile with the previously defined D2 receptor: Thus they are considered to belong to the D2 receptor family *(23–25)*. The D_1 and D_5 receptor subtypes share a similar molecular and genetic structure and, hence, pharmacological profile, and are considered to be the members of the D1 receptor family *(26,27)*. The existence of the five dopamine receptor subtypes provides challenges for the study of the function

*D1 and D2 refer to pharmacologically defined receptor families. D_1, D_2, and so on refer to molecularly defined receptor subtypes.

of dopamine systems. The lack of selective drugs to distinguish the subtypes within each receptor family makes it impossible to attribute functional roles to the individual receptor subtypes. This is especially true because of the coexpression of dopamine receptors within the same neuron or brain region. Accordingly, the original conclusions about the functions of D1 and D2 receptors based on pharmacological studies have to be re-examined. An understanding of the functions of each receptor subtype requires the design of receptor subtype-selective drugs or the development of other methods with the capacity to selectively alter one receptor subtype without affecting the others.

Recently, our laboratory and others have used antisense oligodeoxynucleotides to study the functions of individual dopamine receptor subtypes *(28–31)*. In principle, the antisense "knockout" or "knockdown" strategy predicts that short synthetic oligodeoxynucleotides (DNA) designed to be uniquely complementary to a specific mRNA will bind by Watson-Crick base pairing to their target mRNA and thereby block the synthesis of a particular protein molecule. The efficiency and selectivity of this strategy both in vitro and in vivo have previously been examined extensively *(32,33)*, but only recently has it been applied to the CNS *(34,35)*. By taking advantage of the specificity of antisense knockout, it should be possible to selectively block the production, and thus the function, of only one dopamine receptor subtype with other neurotransmitter and dopamine receptors remaining unaltered. In this chapter we will discuss some recent dopamine receptor subtype antisense knockout studies by our laboratory and others, and compare the findings with related pharmacological studies using semiselective drugs.

2. The Antisense Strategy

Receptor proteins, like all other proteins, are translated by ribosomes from mRNA derived from gene transcription. The potential of small sequences of exogenous oligonucleotides to bind to mRNA and consequently inhibit gene expression and target protein synthesis was proposed many years ago *(36)*, but has only recently been brought to fruition *(37)*. In their unmodified form, oligodeoxynucleotides are small fragments of DNA that can be synthesized easily and in bulk. Theoretically, an oligodeoxynucleotide designed to have a base sequence complementary to a region of a target mRNA sequence is able to bind selectively to that specific sequence of bases of the target mRNA by the well-characterized hybridization process of base pairing. The melting temperature of such a duplex (with 15–25 bases) is sufficient for it to be stable at body temperature. The binding of the oligodeoxynucleotide to the target mRNA prevents translation by arresting protein synthesis. The major strength of this method is the specificity conferred by sequence complementarity: If an appropriate unique target sequence of bases in the mRNA can be identified, only the translation of that mRNA will be blocked by the oligodeoxynucleotide. Statistically, a specific nucleotide sequence of just 17 random nucleotides should be found only once in the mRNA complement of the entire human genome.

Antisense knockout has usually been studied in vitro *(32)*. Short synthetic oligodeoxynucleotides are thought to enter cells by receptor-mediated endocytosis and then to bind to mRNA *(38,39)*. However, despite the success of using oligodeoxynucleotides to inhibit gene expression in vitro, the precise molecular mechanisms involved are still not fully understood. The inhibition of translation of a target mRNA by bound antisense oligodeoxynucleotides might involve the ubiquitous enzyme RNase H, which hydrolyzes the RNA of the RNA–DNA duplex formed through the hybridization process. Usually, the antisense oligodeoxynucleotide is designed to be complementary to a sequence close to or overlapping the initiation codon. Cutting the mRNA at this point stops translation from occurring along the rest of the unhybridized sequence. Alternatively, the formation of the RNA–DNA duplex may serve to block the binding and translocation of ribosomes along the mRNA and thereby prevent the continued synthesis of the target protein.

Typically, an antisense oligodeoxynucleotide will comprise between 15 and 30 bases to optimize selectivity of hybridization. An oligonucleotide of this length should have only one unique target sequence in the genome and should hybridize well with the target mRNA at body temperature. Shorter sequences (<~14) will lack sequence uniqueness, whereas longer sequences (e.g., >50 bases) might bind to shorter alternative nontarget sequences where partial complementarity occurs and thus decrease the specificity of the knockout. The design of oligodeoxynucleotides as potential inhibitors of gene expression must avoid the selection of sequences that have a substantial amount of either external or internal complementarily, which can lead to either bimolecular self-association or intramolecular hairpin formation, inhibiting the ability of the DNA–RNA duplex to form. The bimolecular process is favored by high concentrations of oligomer, whereas the intramolecular process is concentration-independent and relatively fast, which makes it especially problematic. Both of these processes could also influence the transport and uptake of the oligomer by cells, as well as decrease the efficiency of binding to the target mRNA.

Initial attempts to inhibit gene expression utilized natural oligodeoxynucleotides with phosphodiester linkages. However, normal DNA is very sensitive to degradation by cellular nucleases. In order to meet the challenge of using this strategy in vivo, much effort has been devoted to modifying the oligodeoxynucleotides chemically to make them resistant to nuclease *(37)*. Most modifications have been introduced into the phosphodiester backbone. In our laboratory, we have chosen to use the popular backbone-modified phosphorothioate "S"-oligodeoxynucleotide which has increased nuclease resistance and has been shown to be taken up by cells. Such S-oligodeoxynucleotides exhibit extensive brain penetration and rapid neuronal uptake *(40,41)*. Continuous intraventricular infusion of the S-oligodeoxynucleotide at 1.5 nmol/h by implanted osmotic minipump can maintain the concentrations of intact phosphorothioate oligodeoxynucleotide at micromolar level in the cerebrospinal fluid (CSF) for at least 1 wk without obvious toxicity *(42)*. The success of the technique is dependent on the (receptor) protein of interest having a rapid turnover rate. For the dopamine receptors, inferential studies using irreversible antagonists suggest a receptor half-life of just a few days *(43)*, which is ideal for such studies.

3. Advantages and Disadvantages of the Antisense Techniques

As research tools, antisense oligonucleotides offer several potential advantages over classical antagonists. For example, antisense oligonucleotides are easy to design, based on the nucleotide sequence of the gene for the target protein. This eliminates the need to synthesize and screen innumerable compounds on many different assays to find one that binds selectively and with adequate affinity to the target receptor protein. With the discovery of receptor families for most neurotransmitters, this becomes an even more attractive and perhaps obligatory technique. Using the antisense approach, the critical receptor subtype for a given behavioral or physiological response can be quickly identified. Classical approaches along with more sophisticated molecular modeling approaches based on receptor tertiary structure then can be intensively applied for a single receptor subtype to design rationally a classical antagonist that may have therapeutic potential. Although antisense "drugs" appear unlikely in the near future for treating CNS disorders since they do not cross the brain-blood barrier, current synthetic approaches to modify the oligodeoxynucleotide backbone to made it more nuclease-resistant suggests that peripherally administrated oligodeoxynucleotides with central activity may ultimately be developed.

The antisense strategy contrasts with the classical technique of homologous recombination to produce transgenic knockouts in several aspects. The antisense approach can be applied at any stage of development. Transgenic animals typically lack the receptor of interest from conception, which may be lethal or induce compensatory mechanisms during development. Using a combination of different antisense oligodeoxynucleotides, a range of phenotypes can be created. Furthermore, the change in gene expression produced by antisense treatment is a reversible process, allowing for an animal to serve as its own control. The antisense strategy may also ultimately have some therapeutic uses if satisfactory delivery mechanisms can be developed—a very unlikely prospect for transgenic approaches.

Some potential problems have been associated with antisense oligodeoxynucleotides. Among them, the greatest drawback may be that oligodeoxynucleotide treatment often results in an incomplete knockout. To interpret receptor binding measurements of receptor loss after antisense treatment with respect to functional indices of receptor loss, we have to consider whether there is a linear relation between agonist binding at a receptor and the physiological effect of this binding. In the case where "spare" receptors exist for a particular response, when only a small fraction of receptors must bind with the agonist in order to saturate the effector mechanism, incomplete knockout of a certain receptor subtype may not result in an equivalent functional loss. Thus, we may find that oligodeoxynucleotide treatment, although reducing target receptors, may not affect the functional responses of the system because of a large receptor reserve. Obviously the lack of a physiological or behavioral effect following antisense knockout sometimes will mask the real relation between an individual receptor and these functions. Several studies have used N-ethoxycarbonyl-2-ethoxy-1,2 dihydroquinoline (EEDQ) to block irreversibly dopamine receptors to varying degrees in vivo to determine the receptor occupance-response curves *(18,44–47)*. These studies suggest that about 70–90% receptor occupation will produce a full behavioral response. Our dopamine receptor knockout results are in agreement with these findings.

4. Antisense Knockout of Dopamine D2 Receptor Subtypes

4.1. The Efficiency of Antisense Knockout

Successful antisense knockouts of the dopamine D2 receptor family in rodent brain have been achieved *(28–30)*. In our studies, an antisense S-oligodeoxynucleotide complementary to the mRNAs coding for the D_2 receptor was administered to the rat intraventricularly using subcutaneously implanted osmotic minipumps for various periods of time to determine whether receptor density was reduced by the administration. These studies show that antisense treatments for 72 h result in significant loss in receptor density without a change in the affinity of the remaining receptors. For D_2 receptors, both homogenate binding assay and autoradiographic studies indicate about a 50% decrease in the striatal and a 70% decrease in nucleus accumbens binding of [^{3}H]spiperone, a radioligand that binds to all D2 family receptor subtypes.

We also have achieved success in using another antisense oligodeoxynucleotide sequence to reduce D_3 receptors in rat brain. For D_3 receptors in the nucleus accumbens and islands of Calleja, autoradiography with the selective D_3 ligand [^{3}H]7-OH-DPAT indicated about a 40% decrease in both regions after 3 d administration.

Silvia et al. *(28)* delivered antisense phosphorothioate oligonucleotides against D_2 receptor mRNA unilaterally into the substantia nigra of rats via intracerebral cannula. This treatment caused a 40% decrease in the levels of nigral D_2 receptor relative to the control side, as measured by quantitative autoradiography. Administration of the antisense oligodeoxynucleotides for a period of several days resulted in marked contralateral rotational behavior consequent to a subcutaneous injection of cocaine.

4.2. The Specificity of Antisense Knockout

The major strength of the antisense strategy lies in the specificity it can provide in arresting protein synthesis. The importance of assuring the specificity of the antisense knockout can never be overemphasized in studying a complex system such as the brain. Four oligodeoxynucleotide controls can be used:

1. The same bases in the sense configuration: This has the problem of potentially hybridizing to another specific but unrelated sequence to which it is antisense;
2. Random oligodeoxynucleotides containing the same bases as in the antisense sequence but in a scrambled order which matches no known mRNA;
3. Mismatch oligodeoxynucleotides in which three or four bases are mismatched, a number sufficient to reduce the "melting" temperature of the duplex and prevent its formation: Since none of these control oligodeoxynucleotides has sequence complementary to the target mRNA, they should not affect the target protein synthesis; and
4. A different antisense sequence directed elsewhere along the mRNA of interest which should thus have the same specific antisense knockout of the target protein but different nonspecific target effects.

All control sequences must be checked in a DNA database to ensure that they are not antisense for another known sequence. However, only a small fraction of the entire genome has been sequenced to date and thus "specific" unwanted hybridization may occur. It is preferable at this early stage in the development of the antisense strategy to use all four possible controls until the technique is better understood. However, this is very expensive and time consuming. In our studies on D_2 or D_3 receptor antisense knockouts, we have used random oligodeoxynucleotides as controls and found that administration of the random oligodeoxynucleotides failed to reduce dopamine receptors or produce the behavioral effects of the antisense sequence. Neither set of oligodeoxynucleotides induced nonspecific toxicity. S-oligodeoxynucleotides have been reported to produce nonspecific toxicity. This may be the result of elemental sulfur nonspecifically bound to the oligodeoxynucleotide as a byproduct of a synthesis protocol. Toxic effects may also occur if several short oligodeoxynucleotide sequences are also present because of a poor synthesis protocol or lack of oligodeoxynucleotide purification. Some sequences of bases within an oligodeoxynucleotide (e.g., GGGG), also may be toxic and should be avoided *(48)*.

The specificity of D_2 antisense knockout was supported by several observations. To determine whether neuronal death might account for the D_2 receptor decrease, we measured striatal muscarinic receptors, a proportion of which are coexpressed on neurons with D_2 receptors *(49)*. In addition, 5-HT_2, D_1, and D_3 receptor densities were not reduced by the D_2 antisense treatment. These are all G protein-linked receptors and should have similar intracellular processing mechanisms.

In the D_3 antisense knockout study, we found that [^{3}H]spiperone binding, which labels D_2, D_3, and D_4 receptors, was not reduced in the dorsolateral striatum, an area devoid of the D_3 and D_4 mRNA. However, [^{3}H]spiperone binding in nucleus accumbens was reduced by about 20%. Considering the fact that D_3 receptor mRNA is expressed in the nucleus accumbens but not striatum *(24),* it is reasonable to assume that the 20% decrease in [^{3}H]spiperone binding may be caused by the loss of D_3 receptors while D_2 receptors in the striatum were unaffected, in keeping with the 40% loss of [^{3}H]7-OH-DPAT binding.

4.3. Effects of Dose and Time on Antisense Knockout

Antisense approaches involve a balance between two opposing factors: the extent of knockout and the specificity of knockout. Since the hybridization between the antisense oligodeoxynucleotide and the target mRNA is probably governed by the extent of secondary structure formation in the target mRNA, it is relatively inefficient. Thus, a high degree of hybridization requires an excess of antisense probe over its sense counterpart in the target mRNA. However, at high levels of antisense probe, cross-hybridization to related sequences might occur, limiting the specificity of the approach. Also, nonspecific neurotoxicity has been documented in several reports when excessive amount of oligodeoxynucleotide has been administrated. To determine the optimal range of dosage of antisense oligonucleotide, we administered various concentrations of oligodeoxynucleotide to knockout D_2 receptors. We found that the antisense knockout is somewhat dose dependent. At a concentration of 1 μg/μL of antisense oligonucleotide, D_2 receptors were reduced by 14% after 3 d continuous administration at 1 μL/h. At a concentration of 20 μg/μL, the reduction was 58%, slightly higher than at 10 μg/μL. At 20 μg/μL, we also did not observe any evidence of significant behavioral toxicity.

One key aspect of protein depletion by antisense approaches is the rate of target protein turnover. Antisense oligodeoxynucleotides block only the synthesis of new receptor protein. However, the existing receptor protein is not destroyed by the oligodeoxynucleotide. Even with 100% block of protein synthesis by the formation of antisense oligodeoxynucleotide and target sense mRNA duplex, depletion of the target protein will only occur as the remaining pool of the previously synthesized protein is degraded. Therefore, the half-life of degradation is an important consideration for transient antisense knockout, in which several days may be required for depletion of the target protein to produce optimal knockout for functional studies. Degradation is probably under the control of several different processes and therefore may vary across different experimental conditions.

Previous studies using the irreversible protein modifying reagent EEDQ, which blocks both D1 and D2 receptors, suggest that the half time for D2 receptors is 45–160 h, depending on the age of the rat *(43)*. To determine the effect of length of time of antisense treatment on the knockout, we administered 10 μg/μL D_2 antisense oligodeoxynucleotide at a delivery rate of 1 μL/h for 1, 3, 5, and 7 d, respectively. Not surprisingly, we found that the knockout is a time-dependent process. But even when the treatment was extended to 7 d, the apparent reduction in D2 receptor density was still not 100%. Since [^{3}H]spiperone also labels D_3 and D_4 receptors, it is likely that at least some of this residual binding is to these other dopamine receptor subtypes in the striatum.

5. Effects of D_2, D_3, and D_4 Antisense Knockouts on Motor Behaviors

Our earlier knowledge that suggested a direct role for CNS dopamine systems in motor behaviors came from several sources. Clinical observations indicated that some aspects of the movement disorder suffered by Parkinson's disease patients could be alleviated by the administration of dopamine agonists, whereas chronic administration of some dopamine antagonists as a treatment for schizophrenia resulted in side effects that mimicked the motor deficits of Parkinsonism. Laboratory studies showed that acute administration of dopamine antagonists induced hypoactivity and catalepsy *(50),* whereas the nonselective dopamine D1/D2 receptor agonist apomorphine and psychostimulant agent amphetamine, which releases dopamine, induced hyperactivity and stereotyped behaviors *(51–54)*. These observations suggest that increases in dopaminergic tone facilitate motor behavior. The search for the neurological substrates, more specifically the dopamine receptor subtypes involved in these motor behavioral effects and their CNS locations, has inspired many studies in which more selective approaches have been attempted. In the following section we discuss the effects of antisense knockout of dopamine receptor subtypes on some aspects of dopamine-related behavioral activity.

6. Dopamine Agonist-Induced Locomotor Activity

The mesolimbic dopaminergic neurons innervating the nucleus accumbens and olfactory tubercle are known to play a crucial role in the initiation and expression of the locomotor activity *(50,55)*. The microinjection of dopamine and dopamine agonists into the nucleus accumbens or the olfactory tubercle produces a robust locomotor stimulation in rodents *(56)*. Pharmacological blockade of dopamine receptors in the nucleus accumbens with haloperidol *(57,58)* or selective destruction of dopaminergic nerve terminals using local injection of the neurotoxin 6-hydroxydopamine (6-OHDA) in these structures *(52,57)*, markedly attenuates locomotor hyperactivity induced by systemic administration of psychostimulant drugs such as D-amphetamine and cocaine.

During the past few years, the development of specific dopamine receptor agonists and antagonists has made possible the investigation of the roles of the D1 and D2 receptor families in these dopaminergic behaviors. Both the selective D1 receptor antagonist, SCH 23390, and the selective D2 receptor antagonist, raclopride, counteract the locomotor hyperactivity induced by amphetamine in rats *(59,60)*. As shown in Fig. 1, pretreatment with small doses (0.005–0.02 mg/kg) of SCH 23390 completely suppressed the locomotor hyperactivity induced by amphetamine without reducing spontaneous locomotor activity, whereas pretreatment with raclopride (0.05–0.2 mg/kg) only reduced amphetamines' effects at a high dose (0.2 mg/kg), which was found to induce catalepsy, a response mediated by the striatum (Fig. 2). Similar effects of selective D1 and D2 antagonists were demonstrated previously in mice treated with cocaine *(62)*, and more recently in rats injected with amphetamine *(63)*. These results have been used to support the notion that since the D1 antagonist is more effective than the D2 antagonist in impairing mesolimbic dopaminergic activity *(2,64)*, D1 receptors may be more important in mediating locomotor activity. In accord with this hypothesis, recent studies on transgenic D_1 receptor knockout mice have shown that the congeni-

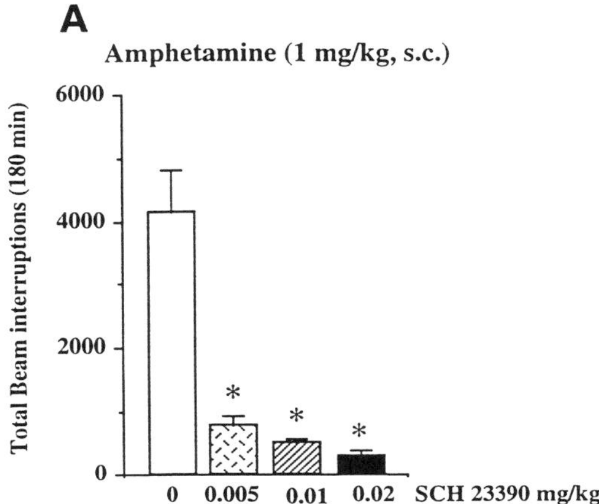

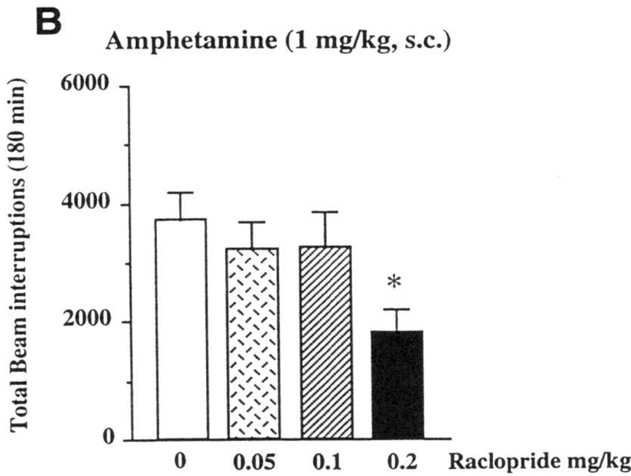

Fig. 1. The effects of different doses of the D1 antagonist, SCH 23390, and the D2 antagonist, raclopride, on the locomotor activity induced by subcutaneous (sc) 1.0 mg/kg of D-amphetamine. SCH 23390 (n = 8/dose) **(A)** or raclopride (n = 8/dose) **(B)** were injected sc 30 min prior to the amphetamine. The locomotor hyperactivity induced by amphetamine was then recorded for 180-min. *Significantly different from controls, $p < 0.05$, Duncan's test after significant analysis of variance (ANOVA). Redrawn from ref. *60* with permission.

tal lack of D_1 receptors completely prevented the locomotor hyperactivity induced either by cocaine or by the selective D1 agonist SKF 81297 *(65,66)*.

Although the D1 receptor subtype appears to be the major determinant for the expression of the locomotor activity induced by psychostimulants, a possible role of D2 receptors in this behavior cannot be ruled out. When

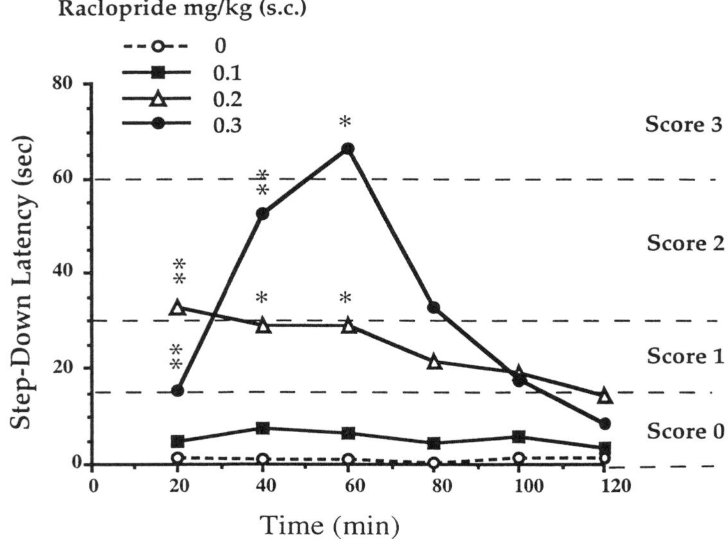

Fig. 2. Catalepsy (s) as measured in a bar test in rats treated with the D2 dopamine receptor antagonist raclopride (0, 0.1, 0.2, or 0.3 mg/kg sc, $n = 12$ rats at each dose). The animals were tested every 20 min after the raclopride injection. The results are expressed as medians. An arbitrary catalepsy score was assigned on the basis of the duration of the cataleptic posture according to Morelli and Di Chiara *(61)*. Significant differences with saline-treated controls were assessed using Mann-Whitney U-test for those 20-min periods for which Kruskal-Wallis test was statistically significant ($p < 0.05$). Reprinted from ref. *60* with permission.

injected at subthreshold doses, but in combination, SCH 23390 and raclopride attenuated the locomotor effect of amphetamine, indicating that D1 and D2 receptors function in a synergistic manner to regulate this behavior (Fig. 3). Similar functional linkage between D1 and D2 receptors was previously revealed after combined injection of selective D1 and D2 dopaminergic agonists either systemically or directly into the nucleus accumbens *(60,67,68)*.

In order to determine the relative roles of each of the D2 dopamine receptor subtypes in mediating the locomotor effect of stimulants, we have examined the effects of intraventricular administration of D_2, D_3, or D_4 antisense oligodeoxynucleotides on locomotor activation induced by amphetamine or by quinpirole, a nonselective agonist for the D2 receptor family. As shown in Fig. 4, whereas amphetamine-induced locomotor hyperactivity was not affected to any great extent by D_2 or D_3 antisense treatment, D_4 antisense treatment resulted in a dramatic inhibition of the effect of amphet-

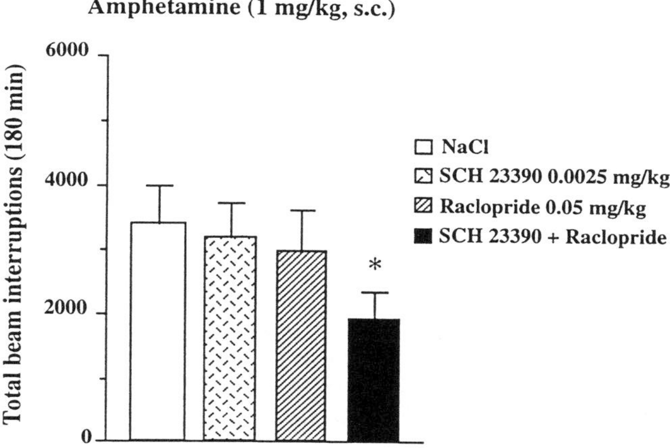

Fig. 3. Effects of simultaneous injection of subthreshold doses of SCH 23390 (0.0025 mg/kg) and raclopride (0.050 mg/kg) 30 min before amphetamine (1.0 mg/kg). The locomotor hyperactivity induced by amphetamine was then recorded for 180 min. *Significantly different from controls, $p < 0.05$, Duncan's test after significant ANOVA. Redrawn from ref. *59* with permission.

amine. These results suggest that D_4 receptors may play an essential role in mediating the activating action of amphetamine. Interestingly, all three antisense oligodeoxynucleotides significantly decreased the locomotor activity induced by quinpirole (Fig. 5), suggesting that the three members of the D2 receptor subfamily are all involved in this behavior, but perhaps mediated, in part, in different brain regions. Taking into account that the effect of amphetamine results from endogenous dopamine activating both D1 and D2 receptors, whereas the quinpirole effect results primarily from activation of the D2 receptor family, we can speculate that the differential effect of D_2, D_3, or D_4 antisense treatment on quinpirole- and amphetamine-induced locomotor hyperactivity may reflect a heterogeneous functional linkage between the D2 receptor subtypes and D1 receptors in the regulation of this behavior.

Weiss and colleagues *(69, 70)* applied the antisense knockout technique to study the function of dopamine D_2 receptors in mice. They found that multiple injections of a D_2 receptor antisense oligodeoxynucleotide into the cerebroventricle prevented some aspects of D2 agonist-mediated behavior. The model utilized mice with a unilateral 6-OHDA lesion of the striatum, in which acute administration of dopamine D2 receptor agonists such as quinpirole and N-0437 was followed by quantification of contralateral rotational behavior. This behavior is dependent on the denervation-induced super-

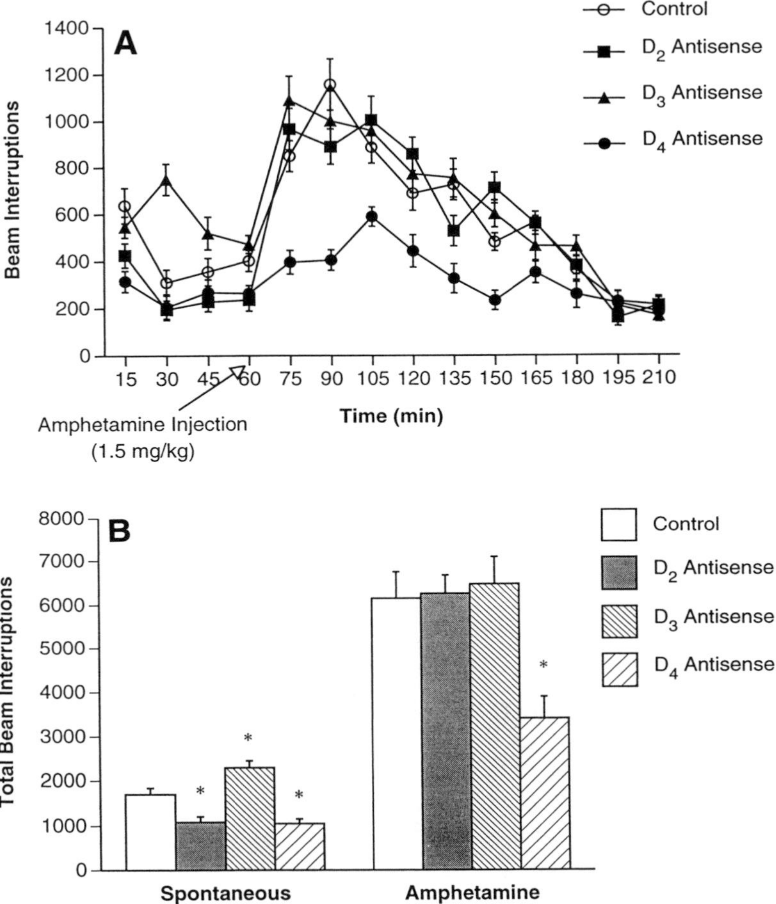

Fig. 4. **(A)** Effect of dopamine receptor antisense oligodeoxynucleotides on spontaneous and amphetamine-induced locomotor activation. Rats were intraventricularly infused with D_2, D_3, or D_4 antisense oligodeoxynucleotide (10 μg/μL/h) for 3 d. Amphetamine (1.5 mg/kg) was injected on d 4, 60 min after rats were placed in the cage. Locomotor activity was measured in a photobeam cage monitored by a computer. **(B)** Total beam interruptions during habituation period (60 min) and amphetamine-stimulation period (150 min). Data represent mean ± SEM (n = 6/group, *$p < 0.05$, ANOVA).

sensitivity of the D2 receptors in the lesioned striatum, which leads to a subsequent imbalance in agonist-induced stimulation of motor activity. D_2 antisense treatment prevented the occurrence of such D2-agonist-induced

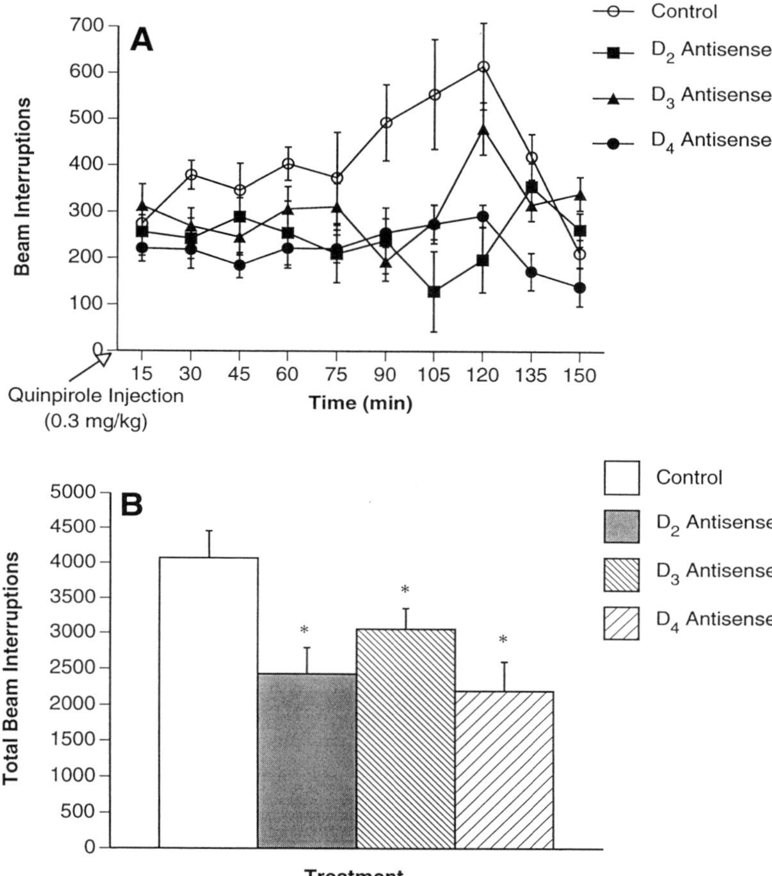

Fig. 5. **(A)** Effect of dopamine receptor antisense oligodeoxynucleotide on quinpirole-induced locomotor activation. Rats were intraventricularly infused with D_2, D_3, or D_4 antisense oligodeoxynucleotide (10 µg/µL/h) for 3 d. Quinpirole (0.3 mg/kg) was injected on d 4. Locomotor activity was measured in a photobeam cage monitored by a computer. **(B)** Total beam interruptions induced by quinpirole during 150-min observation period. Data represent mean ± SEM (n = 6 /group, $*p < 0.05$, ANOVA).

rotation in the 6-OHDA lesioned mice. However, the antisense treatment only produced a minor reduction in the D2 receptor binding in the lesioned striatum, although the decrease was statistically significant. The lower level of receptor losses they obtained may result from the discontinuous administration mode of the antisense oligodeoxynucleotide and the lower doses of the oligodeoxynucleotide or to species differences in receptor turnover. They have hypothesized that the marked behavioral effects indicate that the behavior may be mediated by a small pool of functional, rapidly turning over receptors. This same group has shown *(31)* that intracerebroventricular administration of D_1 dopamine receptor antisense oligodeoxynucleotides inhibits behaviors mediated by D_1 receptors in control and in 6-OHDA-lesioned mice.

7. Spontaneous Locomotor Activity

It is well established that the spontaneous locomotor activity shown by rats when they are placed in novel environments is dependent on the integrity of the dopaminergic mesolimbic system within the nucleus accumbens. 6-OHDA lesions of the nucleus accumbens produce clear decreases in exploratory activity in novel environments *(71–73)*. Hence, either systemic or direct infusion of selective D1 or D2 receptor antagonists into the nucleus accumbens reduces spontaneous activity, indicating that both dopamine receptors modulate this behavior *(74–76)*.

Recent studies of transgenic D_1 knockout mice showed that the congenital lack of the D_1 receptor not only failed to reduce spontaneous activity but even enhanced it *(65,66,77)*. One possible explanation of this data is that the congenital lack of D_1 receptors may result in compensatory mechanisms which lead to changes in basal activity. For instance, the loss of the D_1 gene was found to induce changes in the expression of some neuropeptides in the striatum *(66)*.

We have determined the effects of D_2, D_3, or D_4 receptor antisense knockout on the spontaneous locomotor activity in rats intraventricularly infused with the receptor-specific antisense oligodeoxynucleotides for 3 d. As shown in Fig. 4, whereas D_2 and D_4 antisense treatments inhibited spontaneous locomotor activity, D_3 antisense treatment increased spontaneous locomotor activity, especially in the nocturnal active period. These observations suggest that the three subtypes of dopamine receptors participate in the regulation of locomotor activity, but differ in their roles in the regulation of this behavior. Although the activation of D_4 and D_2 receptors appears to be critical for the expression of spontaneous locomotor activity, the normal stimulation of D_3 receptor seems to inhibit this behavior.

Previous observations showed that the D2 or mixed D1/D2 receptor agonists produce a biphasic effect on locomotion. In low doses they induce hypomotility, whereas at higher doses hyperlocomotion and stereotypy appears. The locomotor suppression seen after low doses has been suggested to be caused by a preferential stimulation of D2 autoreceptors, resulting in reduced release of dopamine. However, this hypothesis has recently been called into question by data showing that the suppression of the locomotion is not correlated temporally with the reduction of dopamine release as measured in vivo by microdialysis *(78)*. Therefore, it has been suggested that the locomotor suppression induced by dopaminergic agonists may be mediated primarily by a population of D2-like receptors localized at postsynaptic sites. More recently, it has been shown that low doses of the dopamine D_3 receptor-preferring agonist 7-OH-DPAT are more efficacious than apomorphine or quinpirole in reducing locomotion in rat *(79)*. Furthermore, unlike apomorphine and quinpirole, 7-OH-DPAT reduced locomotion at doses that did not affect brain dopamine synthesis rate (DOPA accumulation) or release (measured by in vivo dialysis experiments) *(79,80)*. On the basis of these results, it has been proposed that the dopamine D_3 receptor, as a postsynaptic receptor, exhibits an inhibitory influence on rat locomotor activity. However, in another study, 7-OH-DPAT was shown to display the profile of a classical dopamine D2 agonist *(81,82)*, raising the question whether the suppression of locomotion is related solely to its agonist actions at D_3 receptors. The possibility that the D_3 receptor has an inhibitory influence on locomotor activity has also been suggested by other studies using the new D_3 preferring antagonist, U99194A, which shows a 20-fold D_3 vs D_2 affinity in receptor binding assays *(83)*. U99194A was found to increase rat locomotor activity over a wide dose range *(83,84)*. Furthermore, the stimulation of locomotor activity was observed at doses that did not affect dopamine release, indicating that this behavioral effect may be mediated by a population of D_3 receptors located postsynaptically. Our data showing that selective knockout of dopamine D_3 receptors enhanced spontaneous activity is consistent with these findings.

It is worth noting, however, that D_3 antisense treatment did not potentiate the locomotor activity induced by amphetamine or quinpirole. Consequently, it appears that behavioral expression of D_3 receptor activation might actually be dependent on the level of mesolimbic dopaminergic activity or the test situation. Previous studies have suggested that the neural mechanisms underlying general motor activity (motility, locomotion, and rearing) induced by dopaminergic agonists differs subtly from that underlying motor activity of an exploratory nature *(75,85,86)*. This suggestion is also supported by recent reports on mutant mice showing that the loss of the D_1 receptor gene differentially affects spontaneous and drug-induced locomotor activity *(65,66)*.

8. Amphetamine-Induced Stereotyped Behavior

Amphetamine is known to induce different patterns of behavioral activation, depending on the doses used. Low doses of amphetamine produce a marked increase in locomotor activity with intensive sniffing and rearing, whereas high doses elicit repetition of invariant sequences of behavior (i.e., head-down sniffing, licking, biting, and gnawing) described as stereotyped behavior. This behavioral syndrome can be induced also by the nonselective direct dopamine D1/D2 receptor agonist, apomorphine, at high doses. The nigrostriatal dopaminergic system was reported to play a critical role in the expression of this behavior *(51)*. Recently, using local intracerebral injection of amphetamine, Kelley et al. *(53)* showed that subregions of the striatum are involved in the mediation of different aspects of stereotypy. Amphetamine injection into the ventrolateral striatum produced pure oral stereotypy of licking, biting, and self-gnawing, whereas the injection of this stimulant into the dorsal and lateral striatum did not induce oral stereotypy, but rather slightly increased locomotor activity and rearing.

In recent investigations, the abilities of SCH 23390 and raclopride to affect amphetamine-induced stereotypy have been examined *(59)*. Consistent with previous findings, pretreatment with SCH 23390 or raclopride reduced stereotypy induced by amphetamine indicating that both the D1 and D2 receptor families are involved in the mediation of this behavior, as previously reported in many studies (Fig. 6). Interestingly, raclopride blocked stereotypy induced by amphetamine at doses (i.e., 0.05 mg/kg) lower than that needed to block locomotor activity induced by this stimulant. In contrast, the dose of SCH 23390 needed to reduce stereotypy was higher than that needed to reverse the locomotor activity. This result is consistent with the idea that there is a more mesolimbic-like action of the D1 antagonist and more nigrostriatal-like action of the D2 antagonist *(64)*.

We have used antisense oligodeoxynucleotides to examine the relation between each D2 receptor subtype and the amphetamine-induced stereotypy using the stereotypy scale designed by our laboratory. As shown in Fig. 7, in the control rats treated with an oligodeoxynucleotide of random sequence, we observed that a high dose of amphetamine (5 mg/kg) reliably induced significant stereotypy for over 3 h. In contrast, both D_3 and D_4 antisense-treated groups show decreases in stereotypy, with the D_3-treated group expressing virtually no stationary stereotypy but constant activity for most of the time during the 3 h observation. D_2 antisense treatment failed to block the induction of stereotyped behavior by amphetamine. These observations imply that D_3 and D_4 receptors are critically involved in the processes underlying the induction of stereotyped behaviors.

The D2 antagonist raclopride was shown to possess higher affinity for the D_2 and D_3 receptor subtypes than for the D_4 receptor subtype. Therefore, we can assume that the effects of raclopride in animals mainly may reflect its action at D_2 and D_3 receptor subtypes rather than at the D_4 receptor subtype. In this case, the prediction would be that the effect of raclopride on amphet-

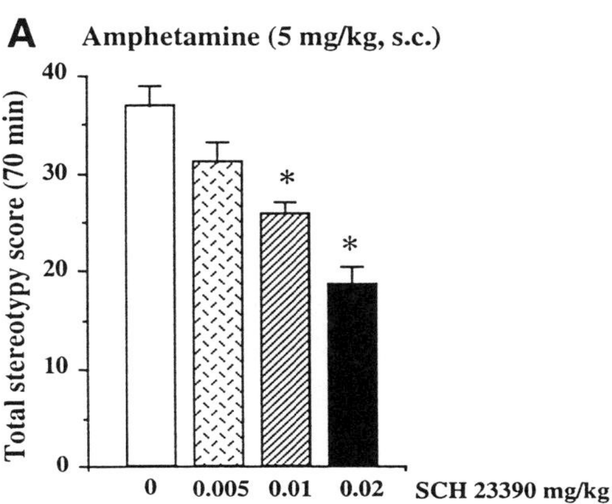

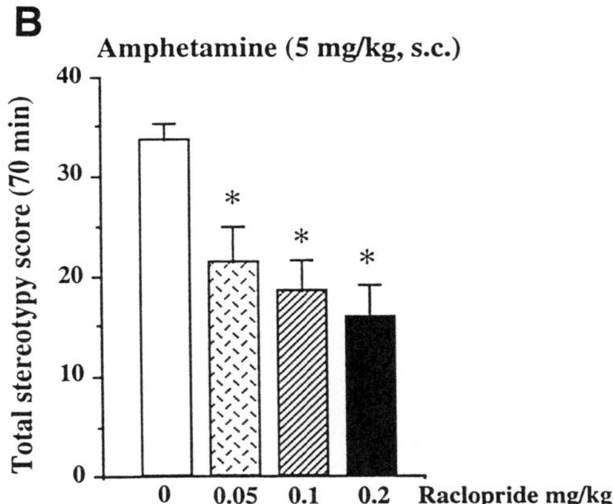

Fig. 6. The effects of different doses of D1 antagonist, SCH 23390, and the D2 antagonist, raclopride, on stereotypy induced by D-amphetamine (5.0 mg/kg, sc). SCH 23390 (n = 8/dose) **(A)** or raclopride (n = 8/dose) **(B)** were administered sc 30 min before amphetamine injection. Each rat was scored for degree of stereotypy every 10 min after amphetamine administration and each score was accumulated for 70 min. The rating scale used was: 1, no movement; 2, intermittent locomotor activity with sniffing; 3, continuous activity and sniffing; 4, intermittent stereotypy (i.e., pronounced sniffing and rearing); 5, continuous stereotypy over a wide area (head down and repetitive movements of the head and limbs); 6, pronounced continuous stereotypy over restricted area; 7, intermittent stereotyped licking directed at walls or floor; and 8, continuous licking and biting in restricted area. Significant differences with saline treated controls were assessed using the Mann-Whitney U-test after significant Kruskal-Wallis test ($p < 0.05$). Redrawn from ref. *59* with permission.

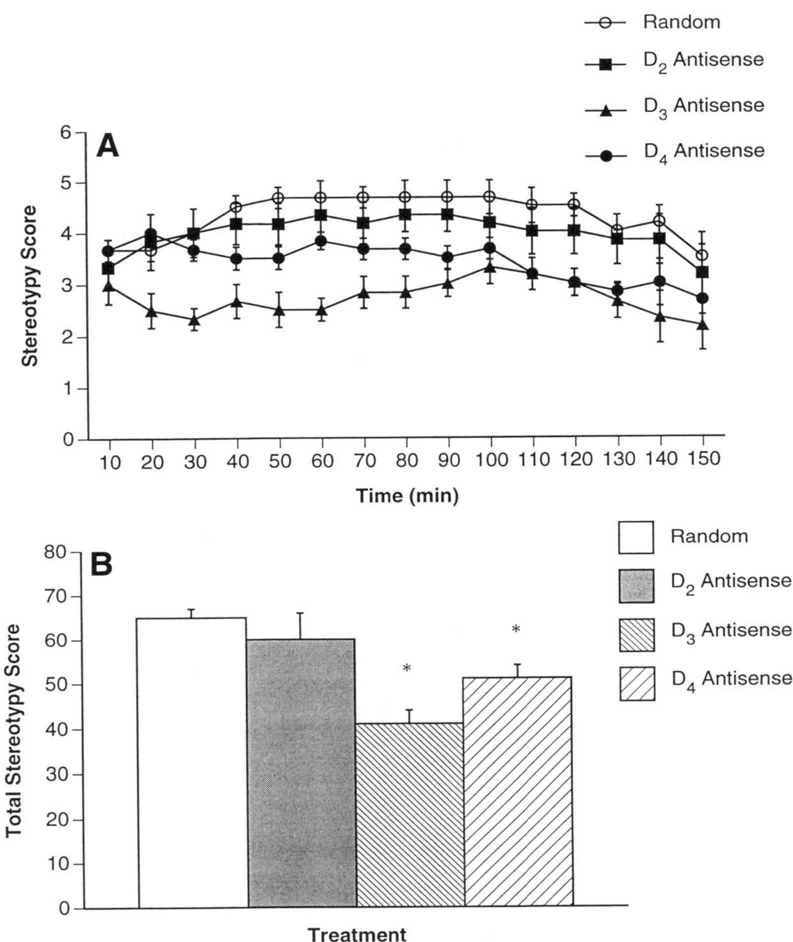

Fig. 7. **(A)** Effect of dopamine receptor antisense oligodeoxynucleotide on amphetamine-induced stereotyped behaviors. Rats were intraventricularly infused with D_2, D_3, or D_4 antisense oligodeoxynucleotide (10 μg/μL/h) for 3 d. Amphetamine (5 mg/kg) was injected on d 4. Stereotypy was evaluated using a scale created by Creese and Iversen *(51)*: 1, no movement; 2, intermittent locomotor activity with sniffing; 3, continuous activity and sniffing; 4, intermittent stereotypy (i.e., pronounced sniffing and rearing); 5, continuous stereotypy over a wide area (head down and repetitive movements of the head and limbs); and 6, pronounced continuous stereotypy over restricted area. **(B)** Total stereotypy score (sum) during 150-min observation following D-amphetamine injection. Data represent mean ± SEM ($n = 6$ /group, $*p < 0.05$, Mann-Whitney U-test).

amine-induced behavioral activation may be more comparable to the actions of D_2 and D_3 antisense treatments than to the action of D_4 antisense treatment. Consistent with this hypothesis, the D_4 antisense treatment appears to be more potent at blocking amphetamine-induced locomotor activation, whereas D_3 antisense treatment seems to affect amphetamine-induced stereotypy more as raclopride does. The D_2 antisense treatment failed to affect amphetamine-induced stereotypy.

In concluding that the dopamine D_2 receptor subtype has no role in the behavioral effects of amphetamine we must use some caution. Incomplete knockout of receptors may lead to a false-negative conclusion. Since in these studies only about a 50–60% receptor loss was obtained, one must consider the possibility that spare D_2 receptors exist for these responses and that the incomplete knockout of D_2 receptors is simply not sufficient to block the action of amphetamine. Since the degree of receptor reserve for different behavioral responses may be different, a positive finding is highly suggestive of a role for that receptor, but without complete knockout, a lack of effect is not informative.

9. Catalepsy

In the original description of the neuroleptic syndrome, it was understood that an akinetic state, often associated with rigidity, occurs as an integral part of the antipsychotic action of the drug. In experimental rodents, this behavior consists of the long-term maintenance of an unnatural position imposed on the animal by the experimenter, such as holding the forepaws over a bar. Traditionally, the cataleptogenic action of neuroleptics was attributed to the blockade of D2 receptors in the striatum. However, numerous studies have shown that this phenomenon also could be induced by specific antagonism of D1 receptors *(61,87–89)*. Recently, using intracerebral injection of dopamine receptor antagonists, Ossowska et al. *(89)* showed that both the striatum and the nucleus accumbens are involved in the mediation of cataleptogenic effects of dopaminergic antagonists. Interestingly, whereas within the striatum both D1 and D2 receptor antagonists can induce catalepsy, in the nucleus accumbens only D1 dopamine receptors seem to play an important role in this phenomenon *(87,89)*.

As newer neuroleptics were developed, however, it became evident that this motor effect was greatly variable and could even be absent in compounds, such as clozapine, that showed incontestable antipsychotic activity. It was suggested that this cataleptogenic action in animals was more closely linked to the induction of extrapyramidal side effects in humans, such as dystonia and pseudo-Parkinsonism, than of the antipsychotic effects.

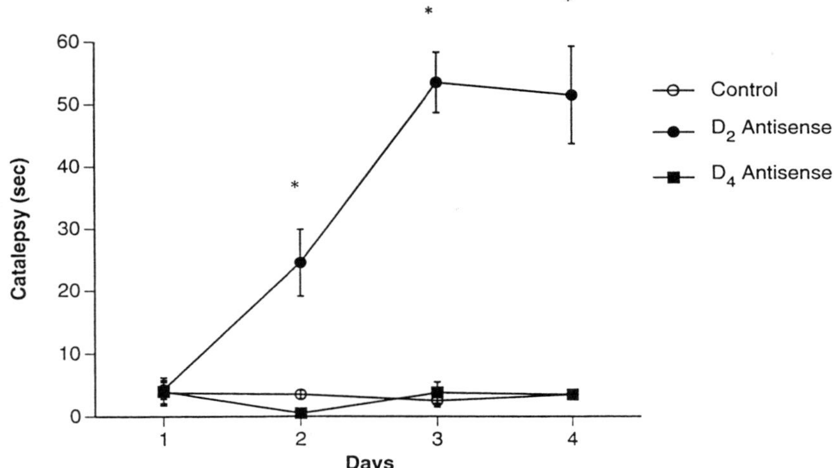

Fig. 8. Cataleptic behavior of rats treated with D_2 and D_4 antisense and random oligodeoxynucleotides. Dopamine D_2 antisense oligo-treated rats were significantly more cataleptic than D_4 or random oligo-treated rats from d 2 following the onset of infusion ($n = 6$/group, $^*p < 0.01$, ANOVA).

The cataleptogenic potency of various neuroleptics has been examined. Whereas most of them, including haloperidol, pimozide, chlorpromazine, and fluphenazine, have been shown to be very effective in inducing catalepsy, clozapine is devoid of this effect. Since clozapine possesses the ability to alleviate schizophrenic symptoms without these motoric side effects, it is termed atypical.

Anticholinergic drugs are typically used in Parkinson's disease and they can also antagonize catalepsy induced by typical neuroleptics. The lack of catalepsy in clozapine-treated rats has been suggested to result, in part, from the anticholinergic activity of clozapine *(90)*.

We have carried out recent antisense knockout studies to define the role of D2 receptor family subtypes in the induction of catalepsy. Catalepsy was measured by the amount of time that it took for the rat to remove both of its front paws from a 6-cm high bar after they were placed onto the horizontal bar by the investigator. We observed that, whereas the D_2 antisense oligodeoxynucleotide induced marked catalepsy during the 3 d of treatments, the D_3 or D_4 antisense oligodeoxynucleotide failed to induce strong catalepsy (Fig. 8). This is of great interest given that the atypical neuroleptic drug clozapine has a 5–10-fold higher affinity for D_4 than for D_2 or D_3 receptors. Clinical studies of clozapine show that it does not induce extrapyramidal motor side effects, suggesting that its therapeutic antipsychotic activity may be mediated by doses sufficient to block D_4 receptors, whereas motor side effects which are mediated by D_2 receptor blockade are not obtained at this level of D_2 receptor occupation. Our observations that the D_4 antisense oligodeoxynucleotide fails to induce catalepsy is consistent with this view.

10. Effect of Local Administration of Antisense Oligodeoxynucleotides on Motor Behavior

Within many brain areas, the same receptor subtype may have different neuronal localizations and mediate opposite effects on synaptic throughput. Electrophysiological and neurochemical data suggest that dopamine D2 receptors are located on the dopaminergic neuron soma and also on some of their terminals in the striatum, and may serve as presynaptic autoreceptors. Blockade of the postsynaptic striatal D2 receptors decreases dopaminergic input to the striatum, whereas blockade of the presynaptic autoreceptors increases dopaminergic neuron firing and dopamine release to the striatum *(91)*.

The autoreceptors are synthesized in the cell bodies of dopaminergic neurons (which contain a higher level of D_2 than D_3 mRNA, as demonstrated by *in situ* hybridization) in the substantia nigra, which is about 5–7 mm away from the striatum, whereas the postsynaptic D_2 receptors are synthesized by the intrinsic striatal neurons. Thus it should be possible to block selectively the production of postsynaptic receptors without reducing the presynaptic receptors if the D_2 receptor antisense oligodeoxynucleotide is delivered directly to the striatum. Conversely, local administration of the D_2 antisense oligodeoxynucleotide into substantia nigra should lead to a loss of autoreceptors in the nigral dopamine cell body and subsequently in striatal terminals, whereas striatal postsynaptic receptor synthesis should not be disrupted. Such a selective pre- or postsynaptic knockout of a given receptor subtype will make it possible to attribute functional roles for receptors of the same subtype in the same brain region, a result not possible to obtain with classical pharmacological techniques.

We used unilateral injection or infusion of D_2 or D_3 antisense oligodeoxynucleotides into striatum or substantia nigra to study the regional effects of D_2 or D_3 receptor loss on the modulation of motor behavior. Rats which received unilateral intrastriatal injection of D_2 antisense (1–20 µg/µL, 1 µL) twice daily for 3 d showed an antisense dose-dependent ipsilateral rotational response (toward the injected side) to the dopamine agonists apomorphine (2 mg/kg) or quinpirole (0.5 mg/kg) (Fig. 9), although they did not show significant spontaneous rotation. In contrast, rats which received unilateral

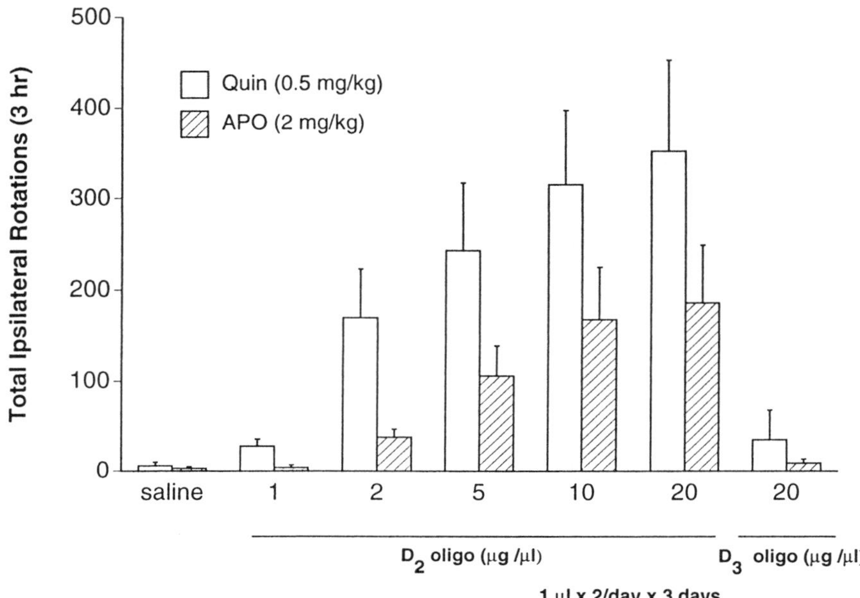

Fig. 9. Ipsilateral rotation induced by apomorphine (2 mg/kg) and quinpirole (0.5 mg/kg) in the rats which had received unilateral intrastriatal D_2 or D_3 antisense oligodeoxynucleotide injections (1 μL twice/d) for 3 d. The D_2 antisense oligodeoxynucleotide treatment induced a dose-dependent increase in ipsilateral rotations in rats when challenged with apomorphine or quinpirole, whereas the D_3 antisense oligodeoxynucleotide treatment failed to induce significant ipsilateral rotations.

intrastriatal injection of D_3 antisense oligodeoxynucleotide did not show a rotational response to the dopamine agonists. These results suggest that postsynaptic D_2, but not D_3, receptors in the striatum are essential for movement control.

The autoreceptors located in nigral dopamine neuron somatodendritic domains and in the striatal terminals were originally classified as D2 receptors. The somadendritic autoreceptors regulate the firing of the dopamine neurons, and activation of these receptors inhibits the firing of the dopamine neurons. The striatal terminal D2 receptors modulate the synthesis and release of dopamine, and activation of these receptors inhibits the release of dopamine. However, since mRNA coding the dopamine D_3 receptors is also present in nigral neurons, it is possible that some of the functions of the autoreceptors may be mediated by D_3 as well as D_2 receptors. To determine the role for the two receptor subtypes in modulating dopaminergic neurotransmission, we infused antisense oligodeoxynucleotides directly into the substantia nigra to block specifically the synthesis of the dopamine autoreceptors. Rats receiving unilateral intranigral infusion of either D_2 or D_3 antisense oligodeoxynucleotide (10 μg/μL × 0.1 μL/h) for at least 3 d showed spontaneous rotations away from the infusion side from the second day of infusion, suggesting increased dopamine release in the ipsilateral striatum as the result of D_2 or D_3 receptor knockout. Infusion of both D_2 and D_3 antisense oligodeoxynucleotides together seemed to produce an additive effect on the rotational behavior. These results suggest that both intranigral D_2 and D_3 antisense treatments block a tonic autoinhibition of nigrostriatal dopamine neurons by endogenous dopamine and that both D_2 and D_3 receptors in the nigral dopamine neurons are involved in the regulation of dopamine transmission. Indeed, the ability of apomorphine to inhibit dopamine neuron firing was blocked by either D_2 or D_3 antisense oligodeoxynucleotide treatment.

Receptor autoradiography showed that D_2 antisense oligodeoxynucleotide infusion selectively reduced D_2 receptor (specific [^{3}H]spiperone) binding sites, but not D3 ([^{3}H]7-OH-DPAT) binding in the ipsilateral substantial nigra. Similarly, D_3 antisense oligodeoxynucleotide infusion selectively decreased the [^{3}H]7-OH-DPAT binding. The effects appear to be highly localized, as evidenced by receptor autoradiography. Additionally, when challenged with the dopamine agonists apomorphine or quinpirole, the rats displayed no rotation behavior, indicating that the postsynaptic dopamine receptors in the striatum were not changed by intranigral antisense infusion. Compared with the results from the intrastriatal injection treatment in which ipsilateral rotations were induced, the contralateral rotational behavior following intranigral administration of antisense D_2 or D_3 antisense oligodeoxynucleotide implies that, depending on their specific brain locations, dopamine receptors of the same type can mediate opposite behavioral effects and thus, contribute differentially to the overall output of dopaminergically innervated systems.

11. Conclusion

The cloning of the five distinct dopamine receptor subtypes has produced a major challenge for the interpretation of dopamine-mediated biochemical, electrophysiological, as well as behavioral effects. The results of previous studies using drugs to define the roles for either D1 or D2 receptors in mediating behavioral functions need to be re-examined because of the more complicated receptor classification. The antisense knockout strategy is both efficient and selective in reducing dopamine receptor subtypes individually; and the receptor loss has observable and differential functional consequences. The efficacy of the strategy depends on several factors: the backbone modification of the oligodeoxynucleotide, the mode of administration of the oligodeoxynucleotide, the stability of the target receptor protein, and the degree of target receptor loss required for a functional change. Based on these preliminary studies, we can suggest that antisense strategies do indeed have the potential to block specifically the production of any given receptor subtype, thus uncovering its behavioral and/or biochemical function. These experiments demonstrate the utility of this approach for the D2 receptor family, and hence, should be applicable for determining the behavioral functions of the D1 dopamine receptor subtypes. We further suggest that the application of antisense strategies may be of general clinical use for diseases in which reducing CNS synaptic neurotransmission is therapeutic. For example, if the antipsychotic action of neuroleptic drugs is mediated by dopamine D_2 receptors, but their chronic use leads to a compensatory increase in D_2 receptor production and unwanted motor side effects such as tardive dyskinesia *(92),* D_2 antisense treatment could overcome this problem by arresting receptor synthesis.

References

1. Dahlstrom, A. and Fuxe, K. (1964) Evidence for the existence of monoamine-containing neurons in the central nervous system. I. Demonstration of monoamines in the cell bodies of brain stem neurons. *Acta Physiol. Scand.* **232(Suppl.),** 1–55.
2. Ferger, B., Kropf, W., and Kuschinsky, K. (1994) Studies on electroencephalogram (EEG) in rats suggest that moderate doses of cocaine or d-amphetamine activate D1 rather than D2 receptors. *Psychopharmacology* **114,** 297–308.
3. Graybiel, A. M. and Ragadale, C. W. (1983) Biochemical anatomy of the striatum, in *Chemical Neuroanatomy* (Emson, P. C., ed.), Raven, New York, pp. 427–504.
4. Lindvall, O. and Bjorklund, A. (1983) Dopamine and norephinephrine-containing neuron systems: their anatomy in the rat brain, in *Chemical Neuroanatomy* (Emson, P. C., ed.), Raven, New York, pp. 229–255.
5. Hornykiewicz, O. (1966) Dopamine and brain function. *Pharmacol. Res.* **18,** 925–964.
6. Marsden, C. D. (1992) Dopamine and basal ganglia in human. *Semin. Neurosci.* **4,** 171–178.
7. Koob, G. F. (1992) Dopamine, addition and reward. *Semin. Neurosci.* **4,** 139–148.
8. Koob, G. F. and Bloom, F. E. (1988) Cellular and molecular mechanisms of drug dependence. *Science* **242,** 715–723.
9. Carlsson, A. (1988) The current status of the dopamine hypothesis of schizophrenia. *Neuropsychopharmacology* **1,** 179–186.
10. Davis, K. L., Kahn, R. S., Ko, G., and Davidson, M. (1991) Dopamine in schizophrenia: a review and reconceptualization. *Am. J. Psychiatry* **148,** 1474–1486.
11. Bjorklund, A. and Lindvall, O. (1964) Dopamine-containing systems in the CNS, in *Classical Transmitters in the CNS. Handbook of Chemical Neuroanatomy* (Bjorklund, A. and Hokfelt, T., eds.), Elsevier, Amsterdam, pp. 55–122.
12. Beninger, R. J. (1993) Role of D1 and D2 receptors in learning, in *D1:D2 Dopamine Receptor Interaction* (Waddington, J., ed.), Academic, London, pp. 115–158.

13. Le Moal, M. and Simon, H. (1991) Mesocorticolimbic dopaminergic network: functional and regulatory roles. *Physiol. Rev.* **71,** 155–234.
14. Kebabian, J. W. and Calne, D. B. (1979) Multiple receptors for dopamine. *Nature* **277,** 93–96.
15. Kebabian, J. W., Petzold, G. L., and Greengard, P. (1972) Dopamine-sensitive adenylate cyclase in caudate nucleus of rat brain, and its similarity to the "dopamine receptor." *Proc. Natl. Acad. Sci. USA* **69,** 2145–2149.
16. Stoof, J. C. and Kebabian, J. (1984) Two dopamine receptors: biochemistry, physiology, and pharmacology. *Life Sci.* **35,** 2281–2296.
17. Creese, I., Burt, D. R., and Snyder, S. H. (1976) Dopamine receptor binding predicts clinical and pharmacological potencies of antischizophrenic drugs. *Science* **192,** 481–483.
18. Seeman, P., Lee, T., Chau-Wang, M., and Wang, K. (1976) Antipsychotic drug doses and neuroleptic/dopamine receptors. *Nature* **261,** 717–719.
19. Verma, A. and Kulkarni, S. K. (1993) On the D1 and D2 dopamine receptor participation in learning and memory in mice. *Methods Exp. Clin. Pharmacol.* **15,** 597–607.
20. Waddington, J. (ed.) (1993) *D1:D2 dopamine receptor interaction.* Academic, London.
21. Civelli, O., Bunzow, J. R., and Grandy, D. K. (1993) Molecular diversity of the dopamine receptors. *Annu. Rev. Pharmacol. Toxicol.* **32,** 281–307.
22. Sibley, D. R. and Monsma, F. J. (1992) Molecular biology of dopamine receptors. *Trends Pharmacol.* **131,** 61–69.
23. Bunzow, J. R., Van Tol, H. H. M., Grandy, D. K., Albert, P., Salon, J., Chisre, M., Machida, C. A., Neve, K. A., and Civelli, O. (1988) Cloning and expression of a rat D_2 dopamine receptor cDNA. *Nature* **336,** 783–787.
24. Sokoloff, P., Giros, B., Martres, M.-P., Bouthenet, M.-L., and Schwartz, J.-C. (1990) Molecular cloning and characterization of a novel dopamine receptor (D_3) as a target for neuroleptics. *Nature* **347,** 146–151.
25. Van Tol, H. H. M., Bunzow, J. R., Guan, H.-C., Sunahara, R. K., Seeman, P., Niznik, H. B., and Civelli, O. (1991) Cloning of a human dopamine D4 receptor gene with high affinity for the antipsychotic clozapine. *Nature* **350,** 614–619.
26. Grandy, D. K., Zhang, Y., Bouvier, C., Zhou, Q.-Y., Johnson, R. A., Allen, L., Buck, K., Bunzow, J. R., Salon, J., Civelli, O. (1991) Multiple human D_5 dopamine receptor genes: a functional receptor and two pseudogenes. *Proc. Natl. Acad. Sci. USA* **89,** 9175–9179.
27. Zhou, Q.-Y., Grandy, D. K., Thambi, L., Kushner, J. A., Van Tol, H. H. M., Cone, R., Pribnow, D., Salon, J., Bunzow, J. R., and Civelli, O. (1991) Cloning and expression of human and rat D_1 dopamine receptors. *Nature* **347,** 76–80.
28. Silvia, C. P., King, G. R., Lee, T. H., Xue, Z.-Y., Caron, M. G., and Ellinwood, E. H. (1994) Intranigral administration of D2 dopamine receptor antisense oligodeoxynucleotides establishes a role for nigrostriatal D2 autoreceptors in the motor actions of cocaine. *Mol. Pharmacol.* **46,** 51–57.
29. Zhang, M. and Creese, I. (1993) Antisense oligodeoxynucleotide reduces brain dopamine D2 receptors: behavioral correlates. *Neurosci. Lett.* **161,** 223–226.
30. Zhang, M., Tarazi, F., and Creese, I. (1994) Antisense knockout of rat-CNS dopamine D_3 receptors and its behavioral effects. *Soc. Neurosci. Abst.* **20,** 909.

31. Zhang, S.-P., Zhou, L.-W., and Weiss, B. (1994) Oligodeoxynucleotide antisense to the D_1 dopamine receptor mRNA inhibits D_1 dopamine receptor-mediated behaviors in normal mice and in mice lesioned with 6-hydroxydopamine. *J. Pharmacol. Exp. Ther.* **271,** 1462–1470.
32. Zamecnik, P. C. (1996) History of antisense oligonucleotides, in *Methods in Molecular Medicine: Antisense Therapeutics* (Agrawal, S., ed.), Humana Press, Totowa, NJ, pp. 1–11.
33. Chiasson, B. J., Hooper, M. L., Murphy, P. R., and Robertson, H. A. (1992) Antisense oligonucleotide eliminates in vivo expression of c-fos in mammalian brain. *Eur. J. Pharmacol.* **227,** 451–453.
34. Wahlestedt, C., Golanov, E., Yamamoto, S., Yee, F., Ericson, H., Yoo, H., Inturrisi, C. E., and Reis, D. J. (1993) Antisense oligodeoxynucleotides to NMDA-R1 receptor channel protect cortical neurons from excitotoxicity and reduce focal ischaemic infarctions. *Nature* **363,** 260–263.
35. Wahlestedt, C., Pich, M., Koob, G. F., Yee, F., and Heilig, M. (1993) Modulation of anxiety and neuropeptide Y–Y1 receptors by antisense oligodeoxynucleotides. *Science* **259,** 528–531.
36. Hoagland, M. B., Zamecnik, P. C., and Stephenson, M. L. (1959) A hypothesis concerning the roles of particulate and soluble ribonucleic acids in protein synthesis, in *A Symposium on Molecular Biology,* University of Chicago Press, Chicago, pp. 105–114.
37. Cohen, J. (1991) Oligonucleotides as therapeutic agents. *Pharmac. Ther.* **52,** 211–225.
38. Loke, S. L., Stein, C. A., Zhang, X. H., Mori, K., Nakanishi, M., Subasinghe, C., Cohen, J. S, and Neckers, L. M. (1989) Characterization of oligonucleotide transport into living cells. *Proc. Natl. Acad. Sci. USA* **86,** 3474–3478.
39. Yakubov, L. A., Deeva, E. A., Zarytova, V. F., Ivanova, E. M., Ryrte, S., Yurchenk, L. V., and Vlassov, V. V. (1989) Mechanism of oligonucleotide uptake by cells: involvement of specific receptors? *Proc. Natl. Acad. Sci. USA* **86,** 6454–6458.
40. Agrawal, S., Temsamani, J., and Tang, J. Y. (1991) Pharmacokinetics, biodistribution, and stability of oligodeoxynucleotide phosphorothioates in mice. *Proc. Natl. Acad. Sci. USA* **88,** 7595–7599.
41. Campbell, J. M., Bacon, T. A., and Wickstrom, E. (1990) Oligo-deoxynucleotide phosphorothioate stability in subcellular extracts, culture media, sera and cerbrospinal fluid. *J. Biochem. Biophys. Methods* **20,** 259–269.
42. Whitesell, L., Geselowitz, D., Chavany, C., Fahmy, B., Walbridge, S., Alger, J. R., and Neckers, L. M. (1993) Stability, clearance, and disposition of intraventricularly administered oligodeoxynucleotides: implications for therapeutic application within the central nervous system. *Proc. Natl. Acad. Sci. USA* **90,** 4665–4669.
43. Norman, A. B., Battaglia, G., and Creese, I. (1987) Differential recovery rates of rat D2 dopamine receptors as a function of aging and chronic reserpine treatment following irreversible modification: a key to receptor regulatory mechanisms. *J. Neurosci.* **7,** 1484–1491.
44. Hamblin, M. and Creese, I. (1983) Behavioral and radioligand binding evidence for irreversible dopamine receptor blockade by EEDQ. *Life Sci.* **32,** 2247–2255.
45. Meller, E., Bordi, F., and Bohmaker, K. (1989) Behavioral recovery after irreversible inactivation of D-1 and D-2 dopamine receptors. *Life Sci.* **44,** 1019–1026.

46. Neve, K. A., Loeschen, S., and Marshall, J. F. (1985) Denervation accelerates the reappearance of neostriatal D-2 receptors after irreversible receptor blockade. *Brain Res.* **329,** 225–231.
47. Saller, C. F., Kreamer, L. D., Adamovage, L. A., and Salama, A. I. (1989) Dopamine receptor occupancy in vivo: measurement using N-ethoxycarbonyl-2ethoxy-1,2-dihydroquinoline (EEDQ). *Life Sci.* **45,** 917–929.
48. Yaswen, P., Stampfer, M., and Ghosh, K. (1992) Effects of sequence of thioated oligonucleotides on cultured human mammary epithelial cells. *Antisense Res. Dev.* **3,** 67–77.
49. Weiner, D. M., Levey, A. I., and Brann, M. R. (1990) Expression of muscarinic acetylcholine and dopamine receptor mRNAs in rat basal ganglia. *Proc. Natl. Acad. Sci. USA* **87,** 7050–7054.
50. Worms, P., Broekkamp, C. L. E., and Lloyd, K. G. (1983) Behavioral effects of neuroleptics, in *Neuroleptics: Neurochemical, Behavioral, and Clinical Perspectives* (Coyle, J. T. and Enna, S. J., eds.), Raven, New York, 93–118.
51. Creese, I. and Iversen, S. D. (1975) The pharmacological and anatomical substrates of the amphetamine response in the rat. *Psychopharmacologium* **39,** 345–357.
52. Kelly, P. H., Seviour, P. W., and Iversen, S. D. (1975) Amphetamine and apomophine responses in the rat following 6-OHDA lesions of the nucleus accumbens septi and corpus striatum. *Brain Res.* **94,** 507–522.
53. Kelley, A. E., Lang, C. G., and Gauthier, A. M. (1988) Induction of oral stereotypy following amphetamine microinjection into a discrete subregion of the striatum. *Psychopharmacology* **95,** 556–559.
54. Seiden, L. S., Sabol, K. E., and Ricaurte, G. T. (1993) Amphetamine: effects on catecholamine systems and behavior. *Annu. Rev. Pharmacol. Toxicol.* **32,** 639–677.
55. Robbins, T. W. and Everitt, B. J. (1982) Functional studies of the central catecholamines. *Int. Rev. Neurobiol.* **23,** 303–365.
56. Amalric, M., Ouagazzal, A., Baunez, C., and Nieoullon, A. (1994) Functional interaction between glutamate and dopamine in the rat striatum. *Neurochem. Int.* **25,** 123–131.
57. Ouagazzal, A., Nieoullon, A., and Amalric, M. (1994) Locomotor activation induced by MK-801 in the rat: postsynaptic interaction with dopamine receptors in the ventral striatum. *Eur. J. Pharmacol.* **251,** 229–236.
58. Pijnenburg, A. J. J., Honig, W. M. M., and Van Rossum, J. M. (1975) Inhibition of D-amphetamine-induced locomotor activity by injection of haloperidol into the nucleus accumbens of the rat. *Psychopharmacology* **41,** 87–95.
59. Amalric, M., Koob, G., Ouagazzal, A., and Nieoullon, A. (1991) Low doses of SCH 23390 differentially modulate behaviors mediated by mesolimbic or nigrostriatal dopaminergic activation, in the rat. Eleventh European Winter Conference (EWBCR), Crans-Mantana, Switzerland.
60. Ouagazzal, A., Nieoullon, A., and Amalric, M. (1993) Effects of dopamine D1 and D2 receptor blockade on MK-801-induced hyperlocomotion in rats. *Psychopharmacology* **111,** 427–434.
61. Morelli, M. and Di Chiara, G. (1985) Catalepsy induced by SCH 23390 in rats. *Eur. J. Pharmacol.* **117,** 179–185.
62. Cabib, S., Castellano, C., Cestari, V., Filibeck, U., and Puglisi-Allegrs, S. (1991) D1 and D2 receptor antagonists differently affect cocaine-induced locomotor hyperactivity in the mouse. *Psychopharmacology* **105,** 335–339.

63. Jackson, D. M., Johansson, C., Lindgren, L.-M., and Bengtsson, A. (1994) Dopamine receptor antagonists block amphetamine and phencyclidine-induced motor stimulation in rats. *Pharmacol. Biochem. Behav.* **48,** 465–471.
64. Amalric, M. and Koob, G. (1993) Functionally selective neurochemical afferents and efferents of the mesocorticolimbic and nigrostriatal dopamine system. *Prog. Brain Res.* **99,** 209–226.
65. Xu, M., Hu, X.-T., Cooper, D. C., Moratalla, R., and Graybiel, A. M. (1994) Elimination of cocaine-induced hyperactivity and dopamine-mediated neurophysiological effects in dopamine D1 receptor mutant mice. *Cell* **79,** 945–955.
66. Xu, M., Moratalla, R., Gold, L. H., Hiroi, N., Koob, G. F., Graybiel, A. M., and Tonegawa, S. (1994) Dopamine D1 receptor mutant mice are deficient in striatal expression of dynorphin and in dopamine-mediated behavioral responses. *Cell* **79,** 729–742.
67. Clark, D. and White, J. F. (1987) Review: D1 dopamine receptor—the search for a function: a critical evaluation of the D1/D2 dopamine receptor classification and its functional implications. *Synapse* **1,** 347–388.
68. Phillips, G. D., Howes, S. R., Whitelaw, R. B., Robbins, T. W., and Everitt, B. J. (1995) Analysis of the effects of intra-accumbens SKF-38393 and LY-171555 upon the behavioural satiety sequence. *Psychopharmacology* **117,** 82–90.
69. Weiss, B., Zhou, L.-W., Zhang, S.-P., and Qin, Z.-H. (1993) Antisense oligodeoxynucleotide inhibits D_2 dopamine receptor-mediated behavior and D2 messenger RNA. *Neuroscience* **55,** 607–612.
70. Zhou, L.-W., Zhang, S.-P., Qin, Z.-H., and Weiss, B. (1994) *In vivo* administration of an oligodeoxynucleotide antisense to the D2 dopamine receptor messenger RNA inhibits D2 dopamine receptor-mediated behavior and the expression of D2 dopamine receptors in mouse striatum. *J. Pharmacol. Exp. Ther.* **268,** 1015–1023.
71. Iversen, S. D. (1977) Brain dopamine system and behavior, in *Handbook of Psychopharmacology* (Iversen, L. L., Iversen, S. D., and Snyder, S. H., eds.), Plenum, New York, pp. 334–384.
72. Joyce, E. M., Stinus, L., and Iversen, S. D. (1983) Effect of injections of 6-OHDA into either nucleus accumbens septi or frontal cortex on spontaneous and drug-induced activity. *Neuropharmacology* **22,** 1141–1145.
73. Taghzouti, K., Louilot, A., Herman, J. P., Le Moat, M., and Simon, H. (1985) Alternation behavior, spatial discrimination, and reversal disturbances following 6-hydroxydopamine lesions in the nucleus accumbens of the rat. *Behav. Neural Biol.* **44,** 354–363.
74. Hoffman, D. C. and Beninger, R. J. (1985) The D1 dopamine receptor antagonist, SCH 23390 reduces locomotor activity and rearing in rats. *Pharmacol. Biochem. Behav.* **22,** 341,342.
75. Boss, R., Cools, A. R., and Ogren, S. (1988) Differential effects of the selective D2-antagonist raclopride in the nucleus accumbens of the rat on spontaneous and d-amphetamine-induced activity. *Psychopharmacology* **95,** 447–451.
76. Plaznik, A., Stefanski, R., and Kostowski, W. (1989) Interaction between accumbens D1 and D2 receptors regulating rat locomotor activity. *Psychopharmacology* **99,** 558–562.
77. Drago, J., Gerfen, C. R., Lachowicz, J. E., Steiner, H., Hollon, T. R., Love, P. E., Ooi, G. T., Grinberg, A., Lee, E. J., Huang, S. P., Bartlett, P. F., Jose, P. A., Sibley,

D. R., and Westphal, H. (1994) Altered striatal function in a mutant mouse lacking D1A dopamine receptors. *Proc. Natl. Acad. Sci. USA* **91,** 12,564–12,568.

78. Stahle, L. (1992) Do autoreceptors mediate dopamine agonist-induced yawning and suppression of exploration? A critical review. *Psychopharmacology* **106,** 1–13.
79. Svensson, K., Carlsson, A., Huff, R. M., Kling-Petersen, T., and Waters, N. (1994) Behavioral and neurochemical data suggest functional differences between dopamine D2 and D3 receptors. *Eur. J. Pharmacol.* **263,** 235–243.
80. Svensson, K., Carlsson, A., and Waters, N. (1994) Locomotor inhibition by the D3 ligand R-(+)-7-OH-DPAT is independent of changes in dopamine release. *J. Neural Transm.* **95,** 71–74.
81. Ahlenius, S. and Salmi, P. (1994) Behavioral and biochemical effects of dopamine D3 receptor-selective ligand, 7-OH-DPAT, in normal and reserpined-treated rat. *Eur. J. Pharmacol.* **260,** 177–181.
82. Liu, J.-C., Cox, R. F., Greif, G. J., Freedman, J. E., and Waszczak, B. L. (1994) The putative dopamine D3 receptor agonist 7-OH-DPAT: lack of mesolimbic selectivity. *Eur. J. Pharmacol.* **264,** 269–278.
83. Waters, N., Lofberg, L., Haadsma-Svensson, S. R., Svensson, K., Sonesson, C, and Carlsson, A. (1994) Differential effects of dopamine D2 and D3 receptor antagonists in regard to dopamine release, in vivo receptor displacement and behaviour. *J. Neural. Transm.* **98,** 39–55.
84. Waters, N., Svensson, K., Haadsma-Svensson, S. R., Smith, M. W., and Carlsson, A. (1993) The dopamine D3-receptor: a postsynaptic receptor inhibitory on rat locomotor activity. *J. Neural Transm.* **94,** 11–19.
85. Hillegaart, V. and Ahlenius, S. (1987) Effect of raclopride on exploratory locomotor activity, treadmill locomotion, conditioned avoidance behaviour and catalepsy in rats: behavioural profile comparisons between raclopride, haloperidol and preclamol. *Pharmacol. Toxicol.* **60,** 350–354.
86. Mogenson, G. J. and Yang, C. R. (1991) The contribution of basal forebrain to limbic-motor integration and the mediation of motivation to action. *Adv. Exp. Med. Biol.* **295,** 267–290.
87. Fletcher, G. H. and Starr, M. S. (1988) Intracerebral SCH 23390 and catalepsy in the rat. *Eur. J. Pharmacol.* **149,** 175.
88. Meller, E., Kuga, S., Fiedhoff, A. J., and Golstein, M. (1985) Selective D2 dopamine receptor agonists prevent catalepsy induced by SCH 23390, a selective D1 antagonist. *Life Sci.* **36,** 1857–1864.
89. Ossowska, K., Karcz, M., Wardas, J., and Wolfarth, S. (1990) Striatal and nucleus accumbens D1/D2 dopamine receptors in neuroleptic catalepsy. *Eur. J. Pharmacol.* **182,** 327–334.
90. Miller, R. J. and Hiley, C. R. (1974) Anti-muscarinic properties of neuroleptics and drug-induced Parkinsonism. *Nature* **248,** 596,597.
91. Starke, K., Gothert, I., and Kilbinger, H. (1989) Modulation of neurotransmitter release by presynaptic autoreceptors. *Physiol. Rev.* **69,** 864–989.
92. Burt, D. R., Creese, I., and Snyder, S. H. (1977) Antischizophrenic drugs: chronic treatment elevates dopamine receptor binding in brain. *Science* **196,** 326–328.

Chapter 15

Dopamine Receptors and Clinical Medicine

Ross J. Baldessarini

1. Introduction

Other chapters in this volume document the explosive recent advances in the genetic, neuroanatomical, and molecular pharmacological understanding of both the traditional and most prevalent D_1* and D_{2L} dopamine (DA) receptors and their recently discovered subtypes, D_{2S}, D_3, D_4, and D_5 *(1–4)*. These advances promise to yield important contributions to clinical medicine. The localization of the genes for each of the identified DA receptors to specific regions of human chromosomes has stimulated many studies seeking evidence of genetic linkage in specific clinical conditions, particularly psychiatric and neurological disorders. There is also much progress in the development of DA receptor type-selective radioligands for application in postmortem neuropathological analyses and for positron emission tomography (PET) or single photon emission computed tomography (SPECT) of the brain for clinical applications. Finally, the discovery of the novel DA receptor types, and of apparently selective localization of types D_3 and D_4 to limbic or other nonextrapyramidal regions of forebrain, has stimulated a vigorous search for small-molecule ligands selective for these targets as potential psychotropic medicinal agents.

The present chapter considers the following major topics: associations of DA receptor genes with specific disorders of the human brain, applications of radioligand technologies to the study of postmortem brain tissue, applica-

*As detailed in the Preface, the terms D1 or D1-like and D2 or D2-like are used to refer to the subfamilies of DA receptors, or used when the genetic subtype (D_1 or D_5; D_2, D_3, or D_4) is uncertain.

The Dopamine Receptors Eds.: K. A. Neve and R. L. Neve
Humana Press Inc., Totowa, NJ

tions of neuroradiological methods to the clinical study of pathophysiology in neuropsychiatric disorders, extension of these methods to the clinical study of neuropharmacology, and promising leads to DA receptor-selective ligands and drug molecules. The material considered is limited largely to schizophrenia and other severe psychiatric and neuropsychiatric disorders that have been particularly intensively studied, and emphasizes the neuropharmacological aspects of available findings. This neuropsychiatric interest has been particularly strongly stimulated by the well-known association of antidopaminergic activity with the antipsychotic actions as well as the extrapyramidal neurological side effects of neuroleptic drugs *(5–11)*. Although these drugs represent the most effective treatment for schizophrenia and a variety of other psychotic disorders, their efficacy in mania and psychotic depression *(5,6)*, as well as other findings *(12)*, also implicate DA in the pathophysiology of severe mood disorders.

2. Clinical Genetics of Dopamine Receptors

Genes (DRD1–5) for the five currently recognized main subtypes of DA receptors (D_1 and D_5; D_2, D_3, and D_4) have been localized to specific segments of human chromosomes 5 and 4 for members of the D1-like subfamily (DRD1 and DRD5) or chromosomes 11 and 3 for members of the D2-like subfamily (DRD2, DRD4, and DRD3; Table 1). Both the more abundant long form of the D_2 receptor (D_{2L}) as well as its alternatively spliced shorter form lacking a 29-amino acid sequence from its third intracellular loop (D_{2S}) evidently are coded by the same gene localized to human chromosome 11 *(1)*. Most of the reported studies seeking associations between DA receptor genes and neuropsychiatric disorders pertain to schizophrenia, for which genes of all five DA receptors have been evaluated by several groups of investigators; much less information has come forth concerning other disorders.

Searches for a genetic basis of schizophrenia have been pursued for many decades. Most of the available evidence supports a familial contribution to risk; stronger support of a specific genetic contribution arises from twin and adoption studies *(22,23)*. Many problems complicate the search for specific genetic factors in schizophrenia. These include lingering suspicion that the syndrome does not represent a single disease, but instead may represent the expression of multiple etiologic factors as phenocopies *(7,8,24–29)*. Moreover, the descriptive and familial distinctions between the two major categories of idiopathic psychotic disorders (schizophrenia and manic-depressive illness) involves some crossing-over of phenotypes within pedigrees, particularly with respect to major depression as well as psychosis in pedigrees identified by probands diagnosed with schizophrenia *(22,28,30)*. Nevertheless,

Table 1
Location of Dopamine Receptor Genes on Human Chromosomes

Receptor	Gene	Locus[a]	Refs.
D1-like			
D_1	DRD1	5q–34-35	*13*
D_5	DRD5	4p–15.1-15.3	*14*
D2-like			
D_2	DRD2	11q–22-23	*15,16*
D_3	DRD3	3q–13.1	*17*
D_4	DRD4	11p–15.5	*18–20*

[a]p, long arm; q, short arm of the chromosome (*see also* ref. *21*).

the basic separation of major psychotic syndromes into schizophrenia (formerly dementia præcox) and manic-depressive disorders first proposed by Emil Kraepelin nearly a century ago *(31,32)* is generally well supported by contemporary epidemiological and clinical genetics research. Another limiting factor is the lack of a clearly identifiable neuropathological or molecular etiologic factor in most of the major psychiatric disorders, leaving open the entire human brain-selective subgenome of about 20,000 genes as potential candidates, possibly in complex polygenic patterns *(27)*.

As a means of limiting an incomprehensible array of potential molecular and genetic hypotheses, much of the activity in the biology of psychotic disorders in the past generation has focused on pharmacological leads. The most influential organizing concept has been that, since virtually all clinically effective antipsychotic agents share the property of antagonizing D_2 and other D2-like DA receptors, it seems plausible to consider such proteins as potential sites of genetic abnormality in schizophrenia *(6)*. However, the antipsychotic agents are clearly syndrome-nonspecific, exerting similar beneficial actions in mania, psychotic depression, and other disorders marked by severe agitation and psychotic features, as well as in schizophrenia *(5)*. Moreover, their development has been largely circular conceptually, since anti-D_2 activity has usually been a requirement for the consideration of candidate antipsychotic agents *(5,6)*. Given these considerations, it may not be surprising that efforts to find associations between schizophrenia and DRD genes, though extensive and technically impressive, have yielded little support for a relationship with any of the known DA receptor types (*33–57*; Table 2).

Evidence concerning an association of the gene (DRD1) for the D_1 DA receptor with schizophrenia is relatively limited but, so far, virtually entirely negative (Table 2). In addition, at least one study has found no evidence of linkage of the DRD1 gene with bipolar manic-depressive disorder *(37)*. Two

Table 2
Studies of Association of Dopamine Receptor Genes with Schizophrenia

Genes	Findings	Refs.
D1-like subfamily		
D1DR	No support for linkage	*33*
	No support for linkage	*34*
	No support for linkage	*35*
	No support for linkage	*36*
	No support for linkage	*37*
	No support for linkage	*38*
D5DR	No support for linkage	*39*
	No support for linkage	*40*
D2-like subfamily		
D2DR	No support for linkage	*33*
	No support for linkage	*34*
	No support for linkage	*41*
	No evidence of structural variants	*42*
	No support for linkage	*43*
	No support for linkage	*44*
	No association with chromosome 11q genes	*45*
	No support for linkage	*36*
	No evidence of structural variants	*46*
	No evidence for ser/gly polymorphism	*47*
	No support for linkage	*38*
	No evidence for ser/gly polymorphism	*48*
D3DR	No support for linkage	*49*
	No support for linkage	*50*
	No support for linkage	*51*
	No support for linkage	*38*
	Reviews mixed evidence for ser/gly polymorphism	*52*
D4DR	No support for linkage	*53*
	No support for linkage	*36*
	No support for linkage	*54*
	No support for linkage	*55*
	No association with response to clozapine	*56*
	Fewer 4-amino acid repeats in delusional patients	*57*
	No support for linkage	*48*
	No support for linkage	*38*

other studies have also found no association of schizophrenia with the gene for the other less prevalent D1-like DA receptor, DRD5 *(39,40)*.

Genetic studies seeking associations between the DRD2 gene and schizophrenia have also been entirely negative and have also not revealed unusual or specific structural variations of this gene in schizophrenia (Table 2). These studies have included searches for variations in the alleles that alternate in selection for serine or cysteine at amino acid 311 in the functionally critical third intracellular loop of the D_2 receptor peptide chain *(47,48)*. Moreover, no association was found between the DRD2 gene and manic-depressive disorders, even when patients with bipolar disorder were considered separately from those with unipolar major depression or the diagnostically more ambiguous schizoaffective disorders *(54,58)*. One of these studies also ruled out an association of manic-depressive illness with the gene for tyrosine hydroxylase, the rate-limiting enzyme in DA synthesis, which is located on human chromosome 11 with DRD2 *(54)*.

The DRD2 gene has also been particularly intensively investigated in other neuropsychiatric disorders. Notably, there were early indications of an association of alcoholism with a particular allele (A1), related to the marker enzyme *Taq-1*, found close to the DRD2 coding region of chromosome 11 *(59)*. In addition, evidence was developed that persons homozygous for the A1/A1 DRD2 genotype had the lowest density of D_2 receptors in their caudate nucleus tissue postmortem (with highest densities found with the A2/A2 genotype, and intermediate densities for the mixed genotypes), suggesting a relationship between low D_2 receptor levels and risk of alcoholism *(60)*. These initially promising associations with risk for alcoholism and the presence of the A1 allele have not held up well in other studies, nor have other particular structural features in the DRD2 gene been found in alcoholics *(46,61–63)*. A relationship between the intensity of alcohol addiction has also been suggested, and may be a valid observation *(63,64)*. To further complicate the relationship of DRD2 with alcoholism, a nonspecific association with other forms of addictive behaviors, including cocaine abuse, and particularly polysubstance abuse, has also been suggested *(65)*. Additional associations of DRD2 alleles with attention deficit disorder and conduct disorders of childhood, which sometimes are followed by substance abuse in adult life, have also been proposed *(65)*. These complex findings have been reviewed previously *(64–66)*. They seem to hint at a shared association of pathophysiological factors involving DA systems in a range of clinical disorders, perhaps related to arousal or reinforcement mechanisms associated with forebrain DA systems.

The neuropsychiatric syndrome of Gilles de la Tourette is marked by spasmodic tic-like body movements and vocalizations, and has evidence of a genetic component *(67)*. An association of the DRD2 gene with this disorder has been found by at least one group *(65)*, but has been disconfirmed by others *(21,68)*. Moreover, no relationship was found between this disorder and any other DA receptor gene, nor with the genes for the catecholamine synthesizing enzymes tyrosine hydroxylase and dopamine-β-hydroxylase *(21,69)*.

As a result of the selective localization of the D_3 DA receptor and its mRNA to the basal forebrain and other limbic structures, there is a great deal of interest in its potential as a lead to the pathophysiology of psychotic disorders as well as a target for improved antipsychotic drugs *(70,71)*. Several studies have sought evidence of linkage of the DRD3 gene with schizophrenia but, again, without supportive findings (Table 2). Additional work related to the D_3 receptor includes exploration of the alternative inclusion of serine or glycine near the extracellular N-terminal of the D_3 receptor peptide chain, a substitution sometimes known as the *Bal-1* polymorphism *(17)*. Initial findings suggested that homozygosity for either one of these amino acids (gly/gly or ser/ser) might be characteristic of schizophrenia *(72),* with some supportive findings added later *(73)*. Additional observations suggested that homozygosity for serine (ser/ser) at this site in the D_3 receptor might correspond to greater sensitivity to DA agonists, for example, as indicated by increases of circulating adrenocorticotrophic hormone (ACTH) and growth hormone (but not in reduction of prolactin) provoked with R(–)-apomorphine in humans, independent of psychiatric diagnosis *(52)*. Although these findings may have some physiological or pharmacological significance, the relationship of homozygosity of the gly/ser substitution in the D_3 peptide structure to schizophrenia has been disconfirmed by several groups *(74–79),* and the original findings weakened on accumulation of additional subjects *(52)*. However, at least one group of investigators suggested that the gly/gly genotype, specifically, may correspond to a higher risk of familial psychotic illness and an earlier onset of schizophrenia, presumably in association with reduced sensitivity to DA compared with the ser/ser genotype *(80)*. This topic has recently been reviewed in detail elsewhere *(52)*.

In general, the D_4 receptor, too, has failed to show a genetic association with schizophrenia (Table 2) or with manic-depressive disorders *(54)*. Other subtle aspects in its structure have also been investigated. In particular, this peptide is known to contain two to seven or more repeats of a 48-amino acid sequence located in its functionally important third intracellular loop *(81)*. The most common form of this polymorphism provides four of the repeated sequences in the D_4 receptor structure of normal human subjects (designated $D_{4.4}$), and seven or two repeats ($D_{4.7}$, $D_{4.2}$) account for most of the remaining variance; some variation in the affinity for DA and its antagonists is associated with the number of repeats *(49,56,82)*. So far, schizophrenic subjects do not appear to differ from the general population in the frequencies of the repeated amino acid sequence in the D_4 receptor or in the corresponding nucleic acid sequence of the DRD4 gene *(56,83)*. Interestingly, an additional mutation, involving only one instead of a more common double repeat of a four amino acid sequence in the N-terminus of the first extracytoplasmic loop of the D_4

receptor close to the first presumed transmembrane segment, has been found to occur more commonly in delusional disorder (paranoia) patients than in normal controls or even in schizophrenic subjects *(57)*.

A further basis of strong interest in the D_4 receptor is that the unusual antipsychotic agent clozapine is reported to have a preferential affinity for the D_4 over other DA receptors *(20,84,85)*, particularly in its most abundant human form, $D_{4.4}$ *(56)*. The lack of a full explanation for the unusually high efficacy and low risk of extrapyramidal side effects of this atypical antipsychotic agent *(86)* has greatly stimulated interest in its selective affinity for the D_4 receptor. Although the anatomical distribution and function of this D2-like receptor subtype remains uncertain, there is some hope that it may predominantly localize to the cerebral cortex or otherwise avoid involvement with extrapyramidal functions of the basal ganglia *(87–91)*. A specific hypothesis is that there may be a genetic association between this receptor and selective responsiveness to clozapine, for example, in comparison to either a placebo or a typical neuroleptic agent. However, this plausible and interesting idea lacks experimental support *(49,56,82)*.

Finally, even though it is not a DA receptor, the 12-fold membrane-crossing transporter protein believed to mediate the neuronal uptake of DA shares some similarities to the sevenfold transmembrane DA receptors *(92,93)*. The transport process is critical for the physiological inactivation of neuronally released DA and is a crucial site for the action of cocaine, amphetamines, and other stimulants. The human gene for the DA transporter has been localized to chromosome 5p-15.3. It does not appear to be associated with schizophrenia, and the relationship of this gene to stimulant abuse disorders is unknown *(94)*.

Overall, the results of genetics studies of the DA receptor types in specific neuropsychiatric disorders have been disappointing from the point of view of finding a simple genetic or pathophysiological factor in complex disorders such as schizophrenia. A general criticism is that there has been only limited consideration of the need for direct comparisons, not only of schizophrenic to normal subjects, but also to persons with other psychotic illnesses (e.g., catatonia, mania, paranoia, and schizoaffective disorders), from which schizophrenia is not always readily differentiated clinically. Despite the lack of evidence for linkage to specific DA receptor genes in human diseases, a great deal has been learned about the human genetics and molecular variation in this important gene family, with notable contributions to associating specific molecular variations in their peptide sequences and their physiology and pharmacology. The lack of robust or even tentative associations between DA receptor genes and particular clinical diagnoses may reflect the fundamental weakness of pharmacocentric hypotheses that attempt to associate the relatively nonspecific treatment of psychotic disorders with an anticipated pathophysiology *(6)*. Perhaps the genetics of DA receptors may provide other associations of interest, such as to variations in the physiology of the arousal or drive-modulating features of limbic forebrain dopaminergic systems and risks for a range of psychiatric disorders. These might include disorders of attention or motivation, substance dependence, mood, or reason, as well as vulnerability to neurological disorders of the extrapyramidal motor system.

3. Dopamine Receptors in Postmortem Human Brain Tissue

Since the late 1970s, many studies have used radiolabeled small molecules with high affinity and considerable selectivity for DA receptors to evaluate the relative abundance of these proteins in homogenates of postmortem brain tissue, particularly in patients with an antemortem history of schizophrenia. Some studies have employed saturation isotherms to estimate the apparent affinity (K_d) and maximum tissue binding site density (B_{max}) of the radioligand in vitro, and not merely binding at a single ligand concentration. Much less frequently, autoradiographic methods have been employed to analyze the distribution of DA receptors in sections of brain tissue *(95)*. Most of the available reports on this large and heterogeneous body of investigations spanning two decades *(96–117)* are summarized in Table 3 (pp. 466–467).

Most of this work has involved the application of antagonists which interact with D2-like receptors of all or most types, particularly labeled derivatives of neuroleptic agents such as spiperone and substituted benzamides; less often, DA agonist ligands, such as the DA-analog dihydroxyaminotetrahydronaphthalines (aminotetralins), have been employed. Several early reports concluded that, in the brains of patients considered to have had schizophrenia in life, the binding of radiolabeled butyrophenones or spiperones was elevated in the extrapyramidal basal ganglia (caudatoputamen) and in more limited samples of D2-rich limbic tissues such as the nucleus accumbens septi (accumbens) *(11,96–98,105,109,112,118–121)*. These studies found variable increases of D2-like receptors in association with schizophrenia, typically by >50%, but sometimes more than double that found in presumably normal age-matched brains. Subsequent studies have often, but not always, confirmed the initial findings, and none has found decreases of D2-like receptor binding (Table 3).

Although the general consistency of finding either increased or unchanged D2-like receptor binding in these studies is noteworthy, most studies have risked the probable confounding effect of up-regulation of D2-like receptors by antemortem drug treatment, a well-known adaptive response to prolonged exposure to most typical neuroleptic agents *(122)*. Such changes may contribute to the pathophysiology of tardive dyskinesia *(122)*; however, there is no indication that postmortem levels of DA receptor are unusually high in schizophrenia patients who have also had the abnormal movement disorder *(104,118)*. Findings with presumably drug-free patients have remained limited and inconsistent, but there is substantial evidence that prolonged periods without exposure to neuroleptics can diminish differences between the brain tissue of psychotic persons and normal controls *(99,100)*. An additional effect of prior drug exposure is suggested by occasional observations of decreases in apparent affinity of radioligand (higher apparent K_d), usually interpreted as suggesting the presence of unlabeled neuroleptic drug in the tissue samples carried over from antemortem exposure *(100,111)*. Presumably, the true effect of drug-associated increases in D2-like receptor binding would have been even higher if competing drug molecules had not been present. Interference with radioreceptor assays by residual neuroleptic drug in postmortem tissue would not, however, account for the sometimes large differences between studies of tissue from untreated control subjects. These include differences of more than twofold in estimates of the normal density of D2-like binding sites (B_{max}) with the same radioligand and otherwise apparently similar experimental methods and tissue specimens (*100,109*; Table 3). Finally, the presence of residual neuroleptic drug molecules may contribute artifactually to variance in the degree of increase in apparent D2-like binding depending on the nature of the radioligand employed. For example, although increases of more than twofold have been found in schizophrenia in brain binding site density assayed with [^{3}H]spiperone *(96)*, much smaller increases have been found in some studies employing a radioactive benzamide or a DA agonist *(97,98,104,112,115)*. The avidity of binding of these classes of radioligands differs markedly, with virtually irreversible binding found with spiperone-like compounds, and more readily displaceable binding with certain benzamides, such as raclopride, and most DA agonists *(123,124)*. Such pharmacological differences between radioligands may contribute to differences in observed densities of D2-like sites, particularly in the presence of interfering drugs acquired before death.

Attempts have also been made to assay concentrations of the more abundant but physiologically less well understood D1 receptors *(1–4)*. Initially, radioactive thioxanthenes were used in the presence of a highly selective D2 antagonist to occlude that component of their complex interactions with virtually all types of DA receptors *(102)*. Since the mid-1980s, D1-like sites have usually been labeled with radioactive preparation of the experimental phenylbenzazepine derivative SCH 23390 or other similar, D1-selective agents, ideally in the presence of a masking agent to limit interactions with serotonin S_2 sites to which these agents have some affinity, as do the spiperones

Table 3
Postmortem Studies of Brain Dopamine Receptor Levels in Schizophrenia[a]

Radioligand	Cases *(N)*	Findings	Refs.
$[^3H]$Spiperone	15	D2-like B_{max} increased 119% in striatum + accumbens, neuroleptic free ≥1 yr ($n = 5$)	*96*
$[^3H]$Haloperidol/apomorphine/ spiperone	30	D2-like increased 61% in striatum (44% without neuroleptic, n ≤ 11); single ligand concentration	*97,98*
$[^3H]$Spiperone	18	D2-like increased 83% in caudate, 74% accumbens, on neuroleptic; effect lost off neuroleptic ≥1 mo; other psychoses similar ($n = 9$ on, 3 off neuroleptics); control B_{max} = 5.8 pmol/g	*99,100*
$[^3H]$Spiperone	11	D2-like increased 45% in striatum; most neuroleptic-treated ($n = 8$)	*101*
$[^3H]$Flupenthixol (± D_2 mask)	18	D2-like increased 27% (45% if drug-free, $n = 5$); no change in D1-like component (D2 masked)	*102*
$[^3H]$Spiperone	7	D2-like not different from controls in striatum or cortex	*103*
$[^3H]$6,7-Dihydroxy-2-aminotetralin	17	DA agonist binding sites unchanged in striatum; most neuroleptic-treated ($n = 8$)	*104*
$[^3H]$Spiperone	53	D2-like increased 80% in striatum; most neuroleptic treated	*105*
$[^3H]$Spiperone/SCH 23390	8	D2-like increased 48%; D1-like unchanged (–14%); 1 ligand concentration; on neuroleptics	*106*
$[^3H]$Spiperone	11	D2-like increased 140% in caudate, 131% if neuroleptic free ≥3 mo ($n = 4$)	*107*
$[^3H]$Spiperone/SCH 23390	8	D2-like increased 56% in striatum; D1-like 43% decreased; most neuroleptic treated ($n = 7$)	*108*
$[^3H]$Spiperone/SCH 23390	92	D2-like increased 44% in striatum (wide variance), most neuroleptic treated; control D2 B_{max} = 12.9 pmol/g ($n = 122$); D1-like ($n = 25$) not different from controls	*109*
$[^3H]$Spiperone/SCH 23390	3	D2-like increased 93%, D1-like 15% in (striatum and accumbens in slices); 2 neuroleptic free; single ligand concentration	*95*
$[^3H]$SCH 23390	14	D1-like unchanged in caudate, probably neuroleptic treated	*110*
$[^3H]$Spiperone	27	D2-like B_{max} and K_d increased (B_{max} by 39%), but not if neuroleptic free ($n = 9$)	*111*

[^{3}H]Raclopride	14	D_{2-3} B_{max} (36%) and K_d (85%) increased, on neuroleptics	*112*
[^{3}H]SCH 23390	4	D1-like caudate high-affinity (K_d, not B_{max}) > 4 normals or 4 other diagnoses; treated cases	*113*
D_3 cDNA probe	18	D_3 mRNA decreased in cerebral cortex tissue	*114*
[^{3}H]Emonapride/raclopride/ spiperone	32	D_4 (as emonapride – raclopride) increased 5.6 times in striatum; raclopride not increased; emonapride (n = 32), spiperone (n = 90) 57–62% vs 208 controls; includes earlier cases	*91*
[^{3}H]Raclopride/SCH 23390	7	D2-like + D1-like both insignificant (+19%); 1 ligand concentration; no effect of neuroleptic or diagnosis	*115*
[^{3}H]Emonapride (± raclopride)	18	D_{2-3} increased in striatum only if on neuroleptic; no D_4 found	*90*
[^{123}I]Epidepride	7	D_{2-3} increased 40–100% in basal ganglia (neuroleptic treated) vs 8 normals; D_3 only 0–27%	*116*
[^{125}I]-(R)-*trans*-PIPAT	16	D_3 increased only if not treated with antipsychotics (n = 8)	*117*
[^{3}H]Emonopride/raclopride	7	D_4 75–100% > 7 normals = 8 neuroleptic-treated other diagnoses; D_{2-3} little altered.	*88*

[a]Most studies report binding as binding site density (B_{max}) from Scatchard analysis.

(106). In most reports, there have been either minimal changes in D1-like binding or even moderate decreases (Table 3). Although most antipsychotic agents appear to have limited ability to up-regulate D1 receptors, many of them (particularly thioxanthenes, phenothiazines, and clozapine) bind at D1 receptors and so might interfere with assays of the abundance of free D1 receptors by remaining in tissue to interfere with the binding of the experimental radioligand *(5,6)*. At least one study has found some evidence of increased D1-sensitive adenylate cyclase activity in postmortem forebrain tissue from schizophrenic subjects *(125)*, but this finding has not been verified and its possible significance remains unclear.

Several recent attempts have been made to use other radioligands to evaluate D_3 sites in brain tissue, including cDNA probes for the mRNA of D_3 receptors, as well as 7-hydroxyaminotetralins, which have some selectivity for D_3 receptors *(114,117)*. An interesting finding with respect to apparent D_3 receptor sites is that their density was increased in postmortem schizophrenia brain tissue only in the absence of prior neuroleptic treatment *(117)*. A possible interpretation of this finding is that neuroleptic treatment may fail to increase, or might even decrease, the abundance of limbic D_3 receptors and their mRNA, unlike their up-regulating and supersensitizing actions on D_2 receptors—both, presumably in response to a loss of access to DA or other presynaptic factors *(114,126)*. Availability of DA has also been found to sustain the abundance of D1 DA receptors in the immature mammalian brain, with reduced abundance when DA was removed *(127)*.

Very recently, combinations of agents with and without selectivity for certain D2-like receptors have been applied to the estimation of D_4 receptor abundance in human brain tissue. Most often, this conclusion has involved considering the difference between the total binding defined with a nonselective D2-like antagonist such as radiolabeled emonapride (the N-phenylbenzamide, YM-09151-2) and that remaining in the presence of a masking agent such as the N-alkylbenzamide raclopride, which has high affinity for D_2 and D_3 receptors but little interaction at D_4 sites *(88,90,91)*. These complex procedures have been required owing to the lack of highly D_4-selective radioligands. The findings have ranged from an initial claim of a sixfold increase of D_4 sites in basal ganglia in schizophrenia *(91)*, to a smaller increase averaging about 90% above levels in normal tissue or that from persons with other diagnoses but exposed to neuroleptics before death *(88)*, and a third study finding no detectable difference from controls *(90)*. Moreover, conclusions about these studies remain in some doubt because of uncertainties about the localization of D_4 receptor mRNA and proteins in brain tissue, including questions about their normal occurrence in human basal ganglia *(87,89,90)*.

Finally, the DA transporter protein can be labeled with various radiolabeled DA uptake antagonists, including the difluorophenylpiperazine derivative [^{3}H]GBR-12935 and several stable phenyltropane analogs of cocaine *(120,128–130)*. Such binding is diminished, as expected, in postmortem basal ganglia tissue of Parkinson's disease patients with degenerated DA neurons *(131)*. The DA transporter protein does not appear to be altered in such tissue from schizophrenia patients, however *(115,120,128,132)*. At least one report suggests that with advancing age, DA transporters may be more rapidly lost from the frontal cerebral cortex of schizophrenia patients than in normal controls *(128)*, although their natural abundance in the cortex is low, making such measurements difficult *(92,93)*.

In general, the attempt to study DA receptors in postmortem brain tissue has stimulated interesting applications of innovative technological developments, and has demonstrated many similarities between such proteins in normal postmortem human tissue and those in the brain tissues of laboratory animals. However, the initial hope that increases in DA receptors would represent a disease-specific marker for schizophrenia remains not so much disproved as in an only partially tested, and therefore inconclusive, state. Increases that have been found in D2-like receptors are suspected of representing adaptations to antemortem exposure to therapeutic antipsychotic agents to a large extent, and other artifacts owing to intercurrent illness or agonal events and postmortem changes commonly encountered in studies of human brain tissue can further complicate such analyses *(133–135)*. Despite these caveats, an additional and particularly noticeable deficiency in the studies reviewed is the rarity of critical comparisons across diagnostic categories in an effort to test for specificity of findings to schizophrenia, to psychotic disorders more generally, or perhaps even to antemortem neuroleptic drug treatment. Attempts to extend postmortem tissue analyses to newer DA receptors has only begun and awaits development of more selective indicator molecules, particularly for D_4 and D_5 receptors, as well as resolution of their normal anatomical distribution. Development of highly specific antibodies to DA receptors may have a special value in these endeavors *(136,137)*.

4. Clinical Neuropathophysiology of Dopamine Receptors

In principle, many of the potential problems associated with the chemical analysis of postmortem brain tissue can be avoided with clinical studies of brain function in living human subjects. It is possible to evaluate the abundance, ligand affinity, and some aspects of the pharmacology of DA receptors by brain imaging with PET as well as with the more accessible and less

Table 4
Dopamine Receptor Radioligands Used in Neuroradiology

Receptor radioligands	PET	SPECT
D1-like receptors		
Phenylbenzazepines	[^{11}C]SCH 23390	[^{123}I]SCH 23390 (TISCH)
	[^{76}Br]SKF 83566	[^{123}I]SCH 38840
	[^{18}F]N-Fluoroethyl-SCH 233390	
	[^{11}C]SCH 39166	
D2-like receptors		
Spiperones	[^{18}F]Spiperone	[^{123}I]Spiperone
	[^{11}C]N-Methylspiperone	
	[^{18}F]N-Methylspiperone	
	[^{18}F]N-Fluoroethylspiperone	
	[^{11}C]Methylbenperidol	
Benzamides	[^{11}C]Raclopride	[^{123}I]Iodobenzamide (IBZM)
	[^{11}C]Eticlopride	[^{123}I]Epidepride
	[^{18}F]FIDA-2	[^{123}I]FIDA-2
	[^{11}C]IMB	[^{125}I]IMB
		[^{123}I]IMBZ
		[^{123}I]IBF
Ergolines	[^{76}Br]Bromolisuride	[^{123}I]Iodolisuride

[a]The chemistry, pharmacology, and neuroradiological properties of these radioligands are reviewed elsewhere *(148,151)*.

expensive technique of SPECT *(120,123,138–146)*. Such studies have been made possible by the development of a growing series of radioligands that are relatively selective for the D1-like and D2-like DA receptor families *(144,147–152)*. Limitations of these methods include a lack of even more selective ligands with which to differentiate specific receptors within each family. There are also limits of spatial resolution intrinsic to the technology of the scanning methods and also the result of the relatively low density of DA receptors in many areas of human brain. Thus, for example, it is possible to visualize binding of radioligands such as the spiperones and benzamides to whatever D_2, D_3, and D_4 receptors are present in the human caudatoputamen, but not to specify reliably the D_3 subtype in limbic tissues of the basal forebrain. Moreover, most available radioligands are not entirely selective for DA receptors and commonly interact with serotonin S_2 sites or alpha adrenergic receptors *(123,151,153)*. A summary of representative radioligands that have been applied to clinical brain scanning to visualize binding sites believed to represent DA receptors is provided in Table 4.

A series of studies have examined binding to D1-like and D2-like DA receptors in the basal ganglia of patients with severe psychiatric disorders in comparison to normal controls, and many of these *(154–169)* are summarized in Table 5. In the late 1980s, several groups applied radioactively labeled spiperone derivatives to PET studies of psychotic patients and reported increases in apparent D2-like binding in brain in association with schizophrenia, ranging from an insignificant tendency to rise about 36% *(39),* to increases of over 50% *(156,162)* and more than twofold, even in small numbers of patients who had not been treated with neuroleptic agents *(163,165,170)*. However, later findings by some of the same investigators, who have compiled results with growing numbers of subjects, tended to yield smaller differences between schizophrenic and normal control subjects *(163)*. Inexplicably, elevations of only about 70% were found in recent cases, compared to nearly threefold increases among earlier subjects *(163)*.

Other investigators were unable to verify these PET findings with a radiolabeled benzamide ligand *(160,166,171,172)* and additional negative findings have arisen from the application of benzamide D2-like ligands in SPECT scanning *(140,168,169)*. Further disparities in the early observations of increased D2-like binding in the caudatoputamen in schizophrenia in PET studies have been reported, even with the application of radioligands similar to the spiperones originally used, and with presumably maximized opportunities to observe up-regulated D2 receptors soon after discontinuing neuroleptic treatment *(139,158)*. In addition, the application of other PET ligands, including the ergoline D2 partial agonist bromolisuride, found no evidence of significant increases of D2-like binding in schizophrenia patients who had been neuroleptic-free for more than 6 mo *(159)*. Some recent studies have, however, suggested more subtle differences that may distinguish psychotic subjects from one another, including an inverse correlation between D_2 binding and blunted affect and other so-called negative symptoms of schizophrenia *(167)*. There may also be differences in the rate of loss of D2-like binding sites in the basal ganglia with aging between persons with schizophrenia and normal control subjects *(158),* although these findings have not been consistent *(160)*. Again, such findings hint at a possible relationship between dopaminergic dysfunction and abnormal arousal in psychopathological states.

In efforts to resolve the disparities in these findings, consideration has been given to differences in the pharmacological characteristics of the radioligands used and to the assumptions and computations involved in the brain scanning methods employed *(123,139,173–175)*. For example, raclopride has high affinity for D_2 and D_3, but not D_4, receptors, whereas spiperones interact with all three of these D2-like receptors. Accordingly, spiperones may preferentially detect changes in D_4 receptors, which have been proposed to be selectively up-regulated in schizophrenia, as was discussed in Section 3. *(91,176)*. In addition, most benzamides are less avidly bound to D_2 receptors

Table 5

In Vivo Brain Scan Studies of Brain Dopamine Receptor Levels in Schizophrenia and Manic-Depressive Disorders[a]

Radioligand/method	Diagnosis	Cases	Findings	Refs.
[^{76}Br]Bromospiperone/PET	SZ	12	D2-like tended to be increased 36% (NS) as striatum/cerebellum ratio vs 13 normals	*154*
[^{11}C]N-Methylspiperone/PET	SZ	15	D2-like increased 2.5-fold, including 10 untreated cases	*155*
[^{11}C]N-Methylspiperone/PET	SZ	18	D2-like increased in 72%, but also in 53% of 15 BP and 100% of 3 with Tourette's; little gain of risk if exposed to neuroleptics or not	*156*
[^{76}Br]Bromospiperone/PET	SZ	8	D2-like not different in TD cases vs $n = 8$ neuroleptic controls	*157*
[^{11}C]Raclopride/PET [^{11}C]N-Methylspiperone	SZ	2	D2-like not different from $n = 4$ controls, all untreated	*139*
[^{76}Br]Bromospiperone/PET	SZ	12	D2-like not changed; recently off neuroleptics	*158*
[^{123}I]IMBZ/SPECT	SZ	27	D2-like unchanged if drug-free ($n = 8$) vs 13 normals; less D_2 with more neuroleptic; slight right > left density	*140*
[^{76}Br]Bromolisuride/PET	SZ	19	D2-like not changed, drug-free ≥6 mo; fall with age only in controls	*159*
[^{11}C]Raclopride/PET	SZ	18	D2-like not changed; untreated; binding fell with age as in controls	*160*
[^{11}C]SCH 23390/PET	BP	10	D1-like 32% lower in frontal cortex, not striatum; medications briefly stopped; no effect of current mood	*161*
[^{11}C]N-Methylspiperone	SZ	13	D2-like 55% increased in elderly schizophrenics vs 17 younger controls	*162*
[^{11}C]N-Methylspiperone/PET	SZ	25	D2-like increased 129%; all drug-free; inc. Wong et al. subjects (67% less in newer cases)	*163*
[^{123}I]IBZM/SPECT	MD	21	D2-like little changed (+11%, $p < 0.05$) vs 11 normals; left = right	*164*
[^{123}I]IBZM/SPECT	BP-II	10	D2-like not different from 5 normals; minor effect of sleep deprivation	*165*
[^{11}C]Raclopride/PET	SZ	13	D2-like unchanged in $n = 9$, slightly increased in $n = 4$; untreated; left = right	*166*
[^{76}Br]Bromospiperone/PET	SZ	10	D2-like correlated inversely with blunted affect ($r = -0.80$); neuroleptic free	*167*
[^{123}I]IBZM/SPECT	SZ	20	D2-like unchanged; left > right; fall with age greater in controls	*168*
[^{123}I]IBZM/SPECT	SZ	7	D2-like only 9% > 8 normals; no laterality; slightly lower with more psychotic severity	*169*

[a]BP, bipolar manic-depressive disorder (BP-II, with hypomania only); IBZM, iodobenzamides (a generic term); IMBZ, iodopyrrolidinemethylbenzamide; MD, major depression; NS, statistically nonsignificant; PET, positron emission tomography; SPECT, single photon emission computed tomography; SZ, schizophrenia.

than spiperones, and so may be more subject to competitive displacement by varying local concentrations of endogenous DA *(124,139)*. This concept of competition with endogenous DA, although plausible, has not yielded a satisfying explanation for the differences between scanning methods applied to the study of schizophrenia brain tissue. Nevertheless, it has encouraged recent consideration of employing PET or SPECT radioligands to the problem of estimating the dynamic availability of DA to cerebral D_2 receptors in varying physiological or pathophysiological states *(174,177,178)*. Changes in the ongoing activity of DA neurons, with varying output of DA, probably modulate the availability of D2-like receptors and might be detected with either benzamides or spiperones as radioligands. Such changes may be relevant to the pathophysiology of psychiatric or neurological disorders. For example, recent PET findings concerning the metabolic utilization of radiolabeled fluoro-DOPA suggest increased production of DA in the basal ganglia in schizophrenia as well as in temporal lobe epilepsy *(179)*, as had been suggested by earlier studies of postmortem brain tissue *(100,103)*.

An additional important consideration for brain scanning analyses of DA receptors in clinical subjects is that of specificity of the findings for diagnostic groups or selectivity for certain symptom patterns, treatment effects, or other conditions. Some studies have found only minor differences in D2-like binding in the basal ganglia of patients diagnosed with various types of manic-depressive disorders compared with normal controls, as well as small changes with sleep deprivation of mood-disordered subjects *(164,165)*. One study even found moderate decreases of D1-like binding in cerebral cortex which may or may not also include some interactions with cortical S_2 serotonin receptors labeled with the phenylbenzazepine [^{11}C]SCH 23390, but detected no changes in the basal ganglia *(151)*. In contrast, Wong and his colleagues *(156,180)* have reported preliminary observations of increased D2-like binding detected by [^{11}C]N-methylspiperone in a majority of manic-depressive patients. These elevations did not appear to covary with recent treatment but, intriguingly, were greater in subjects exhibiting psychotic features at the time of PET study.

Elevations of D2-like binding have been observed with PET imaging in other neuropsychiatric conditions as well, including a small number of subjects with Gilles de la Tourette's tic disorder *(156,180)*. The evidence concerning patients with tardive dyskinesia associated with long-term neuroleptic treatment is sparse and inconclusive, although a positive correlation of greater D2-like binding of radioactive bromospiperone with more severe dyskinetic movements has been suggested *(157)*. In supranuclear palsy, minor decreases in D2-like binding have been found in forebrain with a SPECT imaging technique *(181)*.

In Parkinson's disease, PET studies have indicated increased binding to D2-like receptors *(182),* perhaps owing to D2 receptor up-regulation *(122),* along with other findings of potential pathophysiological importance *(183).* In this condition of idiopathic degeneration of DA neurons projecting from mid- to forebrain, utilization of ^{18}F-DOPA, the immediate precursor of DA, was diminished, as expected. In addition, however, there was an inverse correlation between losses of the rate of utilization of radio-DOPA and increases of DA receptor binding in putamen (not in caudate) as detected by PET scanning with the benzamide D2-like antagonist [^{11}C]-raclopride; these relationships were particularly clear on the side opposite the more severe bradykinesia and tremor, consistent with the known anatomy and physiology of the basal ganglia *(183).* In other PET studies of D2-like receptors in Parkinson's disease, evidence of their up-regulation has been inconsistent *(120,142),* possibly the result of receptor-diminishing or normalizing actions of DA agonist treatments as is also suggested by radioreceptor assays of postmortem brain tissue in this disease *(120).*

Brain scanning methods have also employed radioligands, such as stable phenyltropane analogs of cocaine, which bind selectively to DA transporters *(130,148,184).* These methods have recently been used to demonstrate losses of DA neurons in Parkinson's disease by SPECT as well as PET scanning *(185,186).* These methods have an important potential for use in monitoring the progress of the disorder, particularly with long-term exposure to antioxidants or other treatments that may alter the rate of degeneration of DA neurons in the brains of Parkinson's disease patients.

Overall, the studies just reviewed indicate striking quantitative and qualitative inconsistencies in findings pertaining to the possibly increased abundance of D2-like binding sites in the basal ganglia of psychotic patients. These disparities do not appear to be readily ascribed to differences in the neuropharmacological properties of dissimilar radioligands since inconsistencies have been found with spiperone derivatives as well as between spiperones and benzamides. Though it is plausible to anticipate artifacts associated with the well known ability of neuroleptic treatment to up-regulate D2 receptors, regardless of the diagnosis involved, this potential source of artifacts also does not appear to account for the results reviewed (Table 5). It does seem clear that increased abundance of D2-like receptors is not specific to the brain tissue of persons diagnosed with schizophrenia, casting doubt on this approach to defining a diagnostic biological marker for that disorder. However, the preliminary impression concerning patients with major mood disorders that increased D2 binding may be limited to those with psychotic symptoms *(156,180)* is particularly intriguing and worthy of further study. If the effect does not depend on a greater risk of exposure to neuroleptic treatment, it may support long-standing pathophysiological speculations concerning a relationship of forebrain DA systems that extends beyond the actions of antipsychotic drugs *(6)*. Finally, the application of PET and SPECT brain scanning technologies to the evaluation of other types of DA receptors in psychotic and neuropsychiatric disorders has been very limited, even though appropriate methods for visualizing D1-like receptors exist and the search continues for appropriately selective radioligands for individual D2-like DA receptor subtypes, despite the daunting challenges of their evidently low relative abundance and relatively similar pharmacology.

5. Clinical Neuropharmacology of Dopamine Receptors

A powerful application of brain scanning methods has been to study the neuropharmacology of drugs that interact with central DA receptors. Most reported experiments have involved estimates of the levels of D2-like receptor occupation in basal ganglia in human subjects exposed to various doses of different antipsychotic agents, with some attempts also to correlate receptor occupation with the beneficial or side effects of these agents *(142)*. Both PET and SPECT technologies have been employed (*140,143,187–193*; Table 6). A consistent and reassuring finding has been that doses of most neuroleptic agents which have been found empirically to be clinically effective and adequately tolerated are in a range that provides a high proportion of occupation of D2-like receptors in the basal ganglia of human subjects, as indicated by a reduction of radiolabeling (*140,172,187–189,191,193–195*; *see* Table 6). A potentially powerful application of such methods is in predicting appropriate clinical dosing with similar new antipsychotic agents in a rational way early in the design and conduct of clinical trials. Indeed, the available studies of this type have provided important direct verification of generally unsatisfactory attempts to define dose–response relationships for antipsychotic agents over many years by clinical observation or the use of chemical assays of circulating drug concentrations *(5,6,196)*.

A particularly important insight arising from such studies is that the occupation of D2-like receptors by neuroleptic agents generally follows a hyperbolic dose-occupation function with rising dose or blood concentration of drug. Typical clinical doses and serum concentrations of neuroleptic agents correspond to the "shoulder" of such functions, in the region between about 50 and 90% of D_2 receptor occupation; higher levels of occupation are attained only with extraordinarily high doses that are not tolerated well clinically. Moreover, receptor occupation levels above 80–90% evidently are not necessary to provide an adequate balance between desirable antipsychotic benefits and an undesirably high risk of extrapyramidal side effects *(188,190–192,197)*.

Table 6

D2 Receptor Occupation in Human Caudatoputamen vs Dose or Plasma Concentration of Neuroleptics

Subjects	Ligand/technique	Neuroleptic	Dose or conconcentration	% D2 occupation	Refs.
6	[^{76}Br]Bromospiperone/PET	Various[a]	≤5[b]	50	*190*
			10[b]	60	
			50–100[b]	75–80	
1	[^{11}C]Raclopride/PET	Sulpiride	200[c]	38	*188*
			400[c]	55	
			800[c]	75	
			1600[c]	83	
26	[^{18}F]N-Methylspiperone/PET	Haloperidol	2.5[d]	50	*191*
			5[d]	70	
			10[d]	80	
			≥20[d]	85	
10	[^{123}I]Iodobenzamide/SPECT	Various[a]	1[e]	40	*192*
			2.5[e]	60	
			5[e]	70	
			10[e]	70	
13	[^{11}C]Racloride/PET	Raclopride			*193*

[a]Chlorpromazine-equivalent doses are provided. [b]μg/mL. [c]mg. [d]ng/mL. [e]mg/kg.

These scanning methods are sufficiently powerful to detect the D2-like receptor occupation that follows depot injections of long-acting ester prodrugs of neuroleptic agents, whose circulating concentrations approach the limits of contemporary chemical assays *(198)*. Their detection by brain scanning may reflect accumulation of drug in brain tissue, particularly at D_2 receptor sites. Several studies using either PET or SPECT scanning have also found that the atypical antipsychotic agent clozapine achieved only partial occupation of D2- and D1-like receptors, typically averaging about 50% at clinical doses *(187,192),* whereas occupation of S_2 serotonin receptors in cerebral cortex approached 80–90% *(195,199,200)*. Similarly, the mixed D_2/S_2 antagonist risperidone also occupied a high proportion of S_2 receptors *(153)*.

A remaining question is whether brain scanning techniques can provide an adequate basis for computation of elimination half-life ($t_{1/2\text{-}\beta}$) pertaining to the removal of neuroleptic agents from their site of action. Some contemporary studies of classical plasma pharmacokinetics have employed powerful analytical methods that can monitor blood drug concentrations for many days. Their results have indicated that some neuroleptics, particularly haloperidol, have complex patterns of elimination, with near-terminal apparent plasma elimination half-lives of more than a week; interestingly, these studies replicate estimates of apparent early elimination half-life, of about 1 d, as had been determined with less sensitive assays applied for shorter periods *(201,202)*. Such slow, multiphasic elimination accords very closely with both functional behavioral analyses and direct assays of brain concentrations of haloperidol in laboratory animals *(201)*. Some pilot data obtained by PET methods also suggest very slow elimination of haloperidol from human brain tissue, possibly even more slowly than from plasma, at least at early times after its administration *(188,194)*. However, several other groups have found indications that various types of neuroleptics appeared to be eliminated from human brain quite rapidly, so that residual D2-like receptor occupation became virtually undetectable within 10–14 d *(189,190,197)*. Moreover, they have not found different rates of elimination of butyrophenones compared to phenothiazines *(190,197),* whereas direct chemical assays of animal brain tissue clearly indicate marked dissimilarities, with more rapid and more classically biphasic distribution and elimination kinetics with phenothiazines *(201)*. Possibly, the technical limitations of detecting low levels of neuroleptic agents with PET or SPECT methods, particularly in presumably pharmacologically relevant D_2-receptor pools, may prevent adequate kinetic assessment of the late phase of elimination of the drugs from brain tissue. Since the matter of persistence of such agents in brain tissue is important to resolve so as to define apparently drug-free status for clinical and experimental purposes *(201),* it may be possible to design better experiments to test these questions more adequately.

6. Selective Agents for Specific Dopamine Receptors

A major challenge for contemporary neuropharmacology is to develop radioligands and drug molecules with a high degree of selectivity for the recently described subtypes within the D1- and D2-like receptor subfamilies. Thus far, there are few leads to agents with high levels of selectivity. The problem is complicated by the very limited state of knowledge of the precise histological and cellular localization of some of these DA receptor types, and even more limited understanding of their physiological roles. In addition, the high degree of molecular homology between members of each subfamily appears to accord with a challenging degree of pharmacological similarity as well. Thus, most neuroleptic agents interact with high affinity at D_3 and D_4 as well as D_2 receptors. Although an apparent bias toward high agonist affinity for D_3 and D_4 receptors may represent a lead to development of selective ligands, there is a risk of overlapping interactions with the agonist-preferring high-affinity state of D_2 receptors, as well as a potential for unfavorable competition between exogenous ligands and tightly bound endogenous DA *(71,203)*. A further challenge is to discover ligands with such high affinity and selectivity as to overcome the apparent disproportion in the natural abundance of D_2 over D_3 and D_4 receptors, and that of D_1 over D_5 receptors in many brain regions *(71)*. The desirability of selective ligands for specific DA receptor types is clear, both for use as experimental agents to advance the study of the physiology of the novel receptor subtypes, and as potentially clinically useful drugs, some of which might avoid the extrapyramidal neurological side effects which are common to almost all available neuroleptic agents *(5,6,204)*. Despite the potential difficulties in developing DA receptor-selective ligands and drugs, several promising leads have already been developed, particularly for the apparently limbic-selective D_3 and D_4 subtypes within the D2-like subfamily *(176,205)*. The structures of representative D_3- and D_4-selective agents are shown in Fig. 1.

For D_3 receptor proteins, whose functional role in the brain remains unclear, relatively selective agents of high affinity have been identified among analogs of DA which are known to have agonist actions at D_2 receptors. These include quinpirole, quinelorane, as well as R(+)-7-hydroxy-N,N-di-*n*-propyl-2-aminotetralin (7-OH-DPAT) and several structurally similar compounds such as R(+)-*trans*-7-hydroxy-2-[(N-*n*-propyl-N-3'-iodo-2'-propenyl)amino]-tetralin (7-OH-PIPAT), (+)-UH-232, and PD-128907 *(103,206–210)*. 7-OH-PIPAT can be readily prepared in a radioiodinated form as a radioligand *(211)*. Although such compounds can be applied usefully to the histological analysis of D_3 receptors in brain tissue in vitro, it is not clear that they are sufficiently

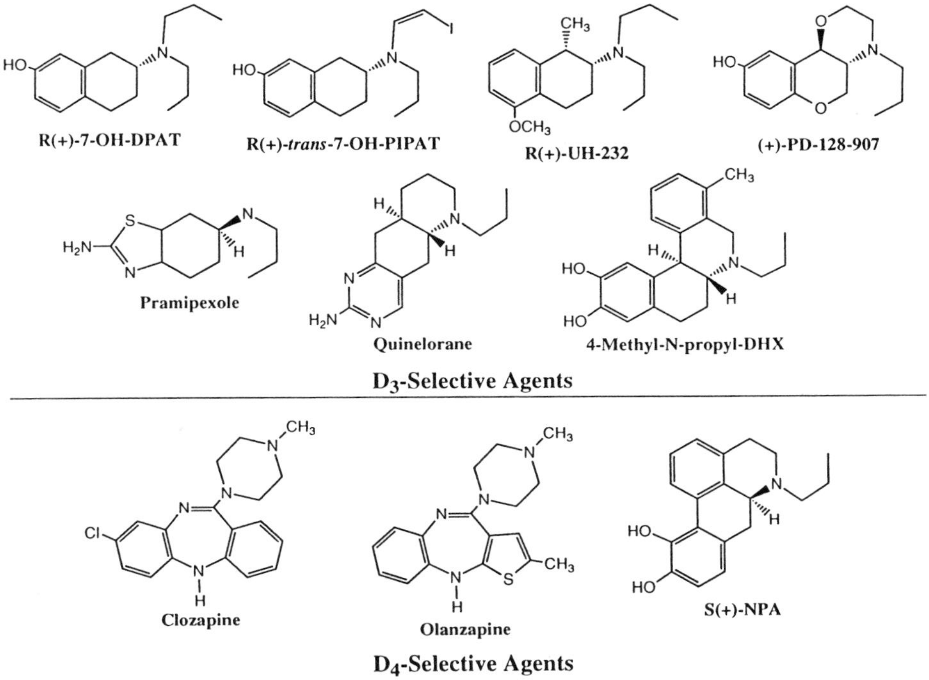

Fig. 1. Chemical structures of lead compounds with some selectivity for dopamine D_3 or D_4 receptors.

selective, particularly against the much more highly abundant D_2 receptors in their high-affinity state, and in competition with endogenous DA, for applications in PET or SPECT brain scanning. Moreover, initial experience with 7-OH-DPAT suggests that it shows little pharmacological selectivity for D_3- vs D_2-rich regions of mammalian forebrain when applied in vivo *(212)*. The 7-hydroxyaminotetralins may also interact with sigma sites in brain tissue *(176)*.

Several leads to selective D_4-receptor agents are also known. An especially provocative finding was a moderate degree of selectivity for clozapine at D_4 over D_{2L} and D_3 receptors *(20,84,85,176)*. This observation promised to provide a plausible basis for explaining the relatively high antipsychotic efficacy and lack of extrapyramidal side effects of this atypical antipsychotic agent. A similar tendency has been found in the experimental clozapine analog olanzapine, but not the new neuroleptic agent risperidone *(84,176)*. Since the precise anatomical distribution of D_4 receptors in human brain tissue is somewhat uncertain and their contributions to brain function remain unknown, it is not possible to suggest how the apparent D_4 selectivity of clozapine may contribute to the atypical clinical profile of activities of this pharmacologically extraordinarily complex agent *(86)*. Moreover, its reported D_4 over D_2 and D_3 selectivity may in part depend on the method of analysis employed. For example, comparison of the affinity of clozapine at D_4 receptors selectively expressed in the membranes of genetically transfected cells to the high-affinity component of binding to D_2 receptors, and particularly the less prominent D_{2S} (short transcript form) suggests rather similar D_4 and D_2 affinities *(213)*. Even this finding remains interesting, however, since the precise localization and function of D_{2S} receptors themselves remain to be defined and may be of neuropharmacological importance.

Finally, there is consistent evidence that the DA partial-agonist S(+) enantiomers of the 11-monohydroxy and 10,11-dihydroxy derivatives of N-propylnoraporphine may have some selectivity for D_4 receptors *(214,215)*. The S(+)-N-propylnoraporphines represent some of the most selective antagonists of DA placed locally in limbic vs extrapyramidal target sites of the forebrain of laboratory animals *(216)*. Since R(–) aporphines, which are potent DA agonists, also demonstrate D_4-over-D_2 selectivity *(214)*, the D_4 preference of both isomers may merely reflect the relatively high affinity of D_4 receptors for DA and its agonists, although such selectivity is not found with aporphines at D_3 receptors *(217)*. However, it is an interesting possibility that their D_4 preference may contribute to the apparent regional selectivity of anti-DA actions of the S(+)-aporphines for the limbic system *(215)*. These compounds also exhibit other desirable characteristics of potential atypical antipsychotic agents, including a lack of inducing signs of extrapyramidal dysfunction in animals, with neither tolerance to their anti-DA activity nor induction of DA supersensitivity, and a lack of ability to produce hyperprolactinemia *(218,219)*. Additional D_4-selective compounds (e.g., NGD-94-1 and others) are under development in several industrial pharmaceutical laboratories, and disclosure of their properties is anticipated in the near future.

In conclusion, the availability of several promising lead compounds pointing to selectivity for D_3 and D_4 receptors is very encouraging. Ironically, D_2-selective agents remain to be identified, since known D_2 agents also interact with D_3 or D_4 sites. Agents which are D_5-selective also remain elusive. Indeed, the functional and potential neuropsychopharmacological contributions of the D_1 and D_5 subfamily of DA receptors remains remarkably underdeveloped, despite the relative predominance of D1- over D2-like receptors in the mammalian forebrain *(3,4)*. Additional ligands selective for each of the DA receptor subtypes would contribute to further understanding of their anatomy and physiology and provide useful medicinal or radiopharmaceutical agents.

7. Conclusions

The recent discovery of several novel gene products in mammalian brain tissue which represent variations on the peptide sequences of the better known and more abundant D_1 and D_{2L} type DA receptors has greatly stimulated a renewed basic and clinical research interest in the study of this growing family of seven-transmembrane spanning proteins *(220)*. The molecular genetic analysis of these peptides led early to locating their genes on human chromosomes and to many attempts to find associations of their genes with human psychiatric and neurological disorders, particularly to schizophrenia, with less work on manic-depressive disorders. Most of these attempts have been disappointing in that they did not reveal a simple etiological factor in such complex disorders. Moreover, they have not yet contributed to the problem of differentiating phenotypically similar major psychiatric disorders for which objective means of diagnosis remain very limited. Indeed, the choice of DA receptors as a target for genetic analysis may be fundamentally somewhat presumptuous, since it arises largely from the clinical pharmacology of the notoriously nonspecific antipsychotic agents. Despite the lack of evidence for linkage to specific DA receptor genes in human neuropsychiatric diseases, much has been learned about the normal human genetics and molecular variation in this important gene family, with notable contributions to associating specific molecular variations in their peptide sequences, physiology, and pharmacology. The genetics of DA receptors may eventually help to clarify relationships between the limbic forebrain dopaminergic systems and risks for disorders of attention or motivation, substance dependence, or mood, as well as the psychotic disorders of reasoning; and the genetics of the extrapyramidal DA systems may help to account for vulnerability to neurological disorders of the basal ganglia, such as Parkinson's disease, tardive dyskinesia, Gilles de la Tourette's syndrome, and perhaps even obsessive-compulsive disorders.

Studies of DA receptors in postmortem brain tissue indicate many similarities between these proteins in normal postmortem human tissue and their homologs in brain tissues of laboratory animals. As with genetics studies, however, the hope that increases in DA receptor abundance would represent a marker for specific psychotic disorders remains unfulfilled, but only partially tested and therefore inconclusive. The evidently diagnostically nonspecific increases that have been found in D2-like receptors might reflect adaptations to antemortem exposure to antipsychotic agents, regardless of diagnosis. Additional studies aimed at critical evaluations of the relationship of DA receptor abundance to various types of psychotic disorders or to their treatment may be of value. Improved methods of specifically probing the newer DA receptor subtypes are urgently needed to help clarify their normal distribution and functions as well as their molecular pathology and pathophysiology.

The studies reviewed here indicate striking quantitative and qualitative inconsistencies in findings pertaining to the possibly altered abundance of DA receptors in the basal ganglia of psychotic patients as indicated in clinical brain scanning. These disparities are not satisfactorily ascribed to differences in the neuropharmacology of the radioligands employed nor to exposure to neuroleptic treatment. Very preliminary evidence also suggests an element of diagnostic nonspecificity in such studies which might be viewed as an additional basis for disappointment. Alternatively, the findings challenge consideration of the possible pathophysiological role of DA receptors in a range of disorders with certain features, such as psychotic symptoms and dysregulation of affect and arousal, all of which are plausibly referable to dopaminergic dysfunction. Also, technical improvements are required to permit visualization of the subtypes of DA receptors within the greater D1- and D2-like subfamilies by PET or SPECT techniques. These methods have already provided valuable information on the receptor occupation and dose–effect relationships of most antipsychotic agents and promise to further clarify their pharmacokinetics at their site of action in the brain.

The further development of radioligands which are effective in defining the location and properties of individual members of the D1 and D2 subfamilies is a highly challenging problem owing to the molecular and pharmacological similarities of DA receptors within the subfamilies and the relatively low natural abundance of D_{2S}, D_3, D_4, and D_5 receptors. Nevertheless, there are several encouraging leads to the development of selective ligands, particularly for D_3 and D_4 receptors. Additional work is required to develop agents that are actually D_2-selective, and agents that are D_5-selective remain elusive. As a final generalization, the cellular distribution, function, and pharmacology of the several recently discovered DA receptors remains much less well developed than their molecular genetics. Additional agonist, antagonist, and labeling agents with even greater selectivity for each of the DA receptor subtypes are urgently required, both to further understanding of their anatomy and physiology, and as potential medicinal agents.

In final summary, this review has considered some of the clinical implications of the recent discovery that the mammalian brain produces several variations on the peptide sequences of the more abundant D_1 and D_{2L} type DA receptors. Their discovery has greatly stimulated basic and clinical research interest in this important neuroreceptor family. Their genes have been localized on human chromosomes; attempts to associate them with specific disorders have been largely unsuccessful, though genetic studies are clarifying relationships between molecular variations in peptide sequences, physiology, and pharmacology. Predictions that DA receptor abundance in postmortem brain tissue would be increased in psychotic disorders remain only partially tested and results are inconclusive, with suggestions of diagnostic nonspecificity as well as contributions of antemortem drug treatment. Visualization of DA receptors in life by brain scanning has also yielded inconsistent results, not fully explained by the pharmacology of the radioligands employed or exposure to neuroleptic treatment. These several findings challenge reconsideration of possible pathophysiological contributions of DA receptors in a range of disorders characterized by dysregulation of arousal rather than as candidate diagnostic markers. Despite the formidable obstacles presented by the molecular similarities of receptors within the D1 and D2 subfamilies and the low natural abundance of D_{2S}, D_3, D_4 and D_5 receptors, additional agonist, antagonist, and labeling agents with even greater selectivity for specific DA receptors are urgently required, to further understanding of their anatomy and physiology, and as potential radiopharmaceutical and medicinal agents.

Acknowledgments

This work was supported in part by NIMH grants MH-31154, MH-34006, MH-47370, and awards from the Bruce J. Anderson Foundation and the Adam Corneel, Vivian Temte, and Richard Wahlin Research Funds. Rita Burke, Lyn Dietrich, and Debbie Wells provided expert bibliographical assistance; Nora Kula and John Neumeyer provided valuable advice.

References

1. Civelli, O., Bunzow, J. R., and Grandy, D. K. (1993) Molecular diversity of the dopamine receptors. *Annu. Rev. Pharmacol. Toxicol.* **32,** 281–307.
2. De Keyser, J. (1993) Subtypes and localization of dopamine receptors in human brain. *Neurochem. Int.* **22,** 83–93.
3. Gingrich, J. A. and Caron, M. G. (1993) Recent advances in the molecular biology of dopamine receptors. *Ann. Rev. Neurosci.* **16,** 299–321.
4. Hall, H., Sedvall, G., Manusson, O., Kopp, J., Halldin, C., and Farde, L. O. (1994) Distribution of D_1 and D_2 dopamine receptors, and dopamine and its metabolites in the human brain. *Neuropsychopharmacology* **11,** 245–256.
5. Baldessarini, R. J. (1996) Drugs and the treatment of psychiatric disorders, in *Goodman and Gilman's The Pharmacologic Basis of Therapeutics*, 9th ed. (Harden, W., Rudin, W., Molinoff, P. B., and Rall, T., eds.), McGraw-Hill, New York, pp. 399–459.
6. Baldessarini, R. J. (1996) *Chemotherapy in Psychiatry: Principles and Practice,* 3rd ed. Harvard University Press, Cambridge, MA, in press.
7. Davis, K. L., Kann, R. S., Ko, G., and Davidson, M. (1991) Dopamine in schizophrenia: a review and reconceptualization. *Am. J. Psychiatry* **148,** 1474–1486.

8. Lieberman, J. A. and Koreen, A. R. (1993) Neurochemistry and neuroendocrinology of schizophrenia: a selective review. *Schizophrenia Bull.* **19,** 371–429.
9. Meltzer, H. Y. and Stahl, S. M. (1976) The dopamine hypothesis of schizophrenia: a review. *Schizophrenia Bull.* **2,** 19–76.
10. Reynolds, G. P. (1989) Beyond the dopamine hypothesis: the neurochemical pathology of schizophrenia. *Br. J. Psychiatry* **155,** 305–316.
11. Seeman, P. (1987) Dopamine receptors and the dopamine hypothesis of schizophrenia. *Synapse* **1,** 133–152.
12. Gessa, G. L. and Serra, G (eds.) (1990) *Dopamine and Mental Depression.* Pergamon, Oxford, UK.
13. Sunahara, R. K., Niznik, H. B., Weiner, D. M., Stormann, T. M., Brann, M. R., Kennedy, J. L., Gelernter, J. E., Rozmahel, R., Yang, Y., Israel, Y., Seeman, P., and O'Dowd, B. F. (1990) Human dopamine D_1 receptor encoded by an intronless gene on chromosome 5. *Nature* **347,** 80–83.
14. Sherrington, R., Mankoo, B., Attwood, J., Kalsi, G., Curtis, D., Buetow, K., Povey, S., and Gurling, H. (1993) Cloning of the human dopamine D_5 receptor gene and identification of a highly polymorphic microsatellite for the DRD5 locus that shows tight linkage to the chromosome 4p reference marker RAF-1P1. *Genomics* **18,** 423–425.
15. Grandy, D. K., Litt, M., Allen, L., Bunzow, J. R., Marchionni, M., Makam, H., Reed, L., Megenis, R. E., and Civelli, O. (1989) The human dopamine D_2 receptor gene is located on chromosome 11 at q22–q23 and identifies a *Taq*-1 RFLP. *Am. J. Hum. Genetics* **45,** 778–785.
16. Huage, X. Y., Grandy, D. R., Eubanks, J. H., Evans, G. A., Civelli, O., and Litt, M. (1991) Detection and characterization of additional DNA polymorphisms in the dopamine D_2 receptor gene. *Genomics* **10,** 527–530.
17. Lannfelt, L., Sokoloff, P., Martres, M.-P., Pilon, C., Giros, B., Jönsson, E., Sedvall, G., and Schwartz, J.-C. (1992) Amino acid substitution in the dopamine D_3 receptor as a useful polymorphism for investigating psychiatric disorders. *Psychiatr. Genetics* **2,** 249–256.
18. Kennedy, J. L., Sidenberg, D. G., Van Tol, H. H., and Kidd, K. K. (1991) A Hinc-II RFLP in the human D_4 dopamine receptor locus (DRD4). *Nucleic Acids Res.* **19,** 5801,5802.
19. Gelernter, J., Kennedy, J. L., Van Tol, H. H., Civelli, O., and Kidd, K. K. (1992) The D_4 dopamine receptor (DRD4) maps to distal 11p close to HRAS. *Genomics* **13,** 208–210.
20. Van Tol, H. H., Bunzow, J. R., Guan, H.-C., Sunahara, R. K., Seeman, P., Niznik, H. B., and Civelli, O. (1991). Cloning of the gene for a human dopamine D_4 receptor with high affinity for the antipsychotic clozapine. *Nature* **350,** 610–614.
21. Brett, P. M., Curtis, D., Robertson, M. M., and Gurling, H. M. (1995) The genetic susceptibility to Gilles de la Tourette syndrome in a large multiplex affected British kindred: linkage analysis excludes a role for the genes coding for dopamine D_1, D_2, D_3, D_4, D_5 receptors, dopamine beta-hydroxylase, tyrosinase, and tyrosine hydroxylase. *Biol. Psychiatry* **37,** 533–540.
22. Kendler, K. S. and Diehl, S. R. (1993) The genetics of schizophrenia: a current genetic-epidemiologic perspective. *Schizophrenia Bull.* **19,** 261–285.
23. Kety, S. S., Wender, P. H., Jacobsen, B., Ingraham, L. J., Jansson, L., Faber, B., and Kinney, D. K. (1994) Mental illness in the biological and adoptive relatives of schizophrenic adoptees. *Arch. Gen. Psychiatry* **51,** 442–455.

24. Carpenter, W. T. and Buchanan, R. W. (1994) Schizophrenia. *N. Engl. J. Med.* **330,** 681–690.
25. Crow, T. J. (1980) Molecular pathology of schizophrenia: more than one disease process. *Br. Med. J.* **280,** 60–68.
26. Crow, T. J. (1986) The continuum of psychosis and its implications for the structure of the gene. *Br. J. Psychiatry* **149,** 419–429.
27. Crowe, R. R. (1993) Candidate genes in psychiatry: an epidemiological perspective. *Am J. Med. Genetics (Neuropsychiatric)* **48,** 74–77.
28. Taylor, M. A., Berenbaum, S. A., Jampala, V. C., and Cloninger, C. R. (1993) Are schizophrenia and affective disorder related? Preliminary data from a family study. *Am. J. Psychiatry* **150,** 278–285.
29. Wright, I. and Woodruff, P. (1995) Aetiology of schizophrenia: a review of theories and their clinical and therapeutic implications. *CNS Drugs* **3,** 126–144.
30. Loranger, A. W. (1981) Genetic independence of manic-depression and schizophrenia. *Acta Psychiatr. Scand.* **63,** 444–452.
31. Kraepelin, E. (1919) *Dementia Præcox and Paraphrenia* (Barclay, R. M. and Robertson, G. M., trans.), Livingstone, Edinburgh, UK.
32. Kraepelin, E. (1921) *Manic-Depressive Insanity and Paranoia* (Barclay, R. M. and Robertson, G. M., trans.), Livingstone, Edinburgh, UK.
33. Byerley, W., Plaetke, R., Jensen, S., Holik, J., Hoff, M., Myles-Worsley, M., Wender, P., Reimherr, F., Leppert, M., O'Connell, P., Lalouel, J., and White, R. (1991) Linkage analysis of six schizophrenia pedigrees with 150 DNA markers. *Schizophrenia Res.* **4,** 274,275.
34. Wildenauer, D. B., Schwab, S., Würl, D., Ertl, M., Ackenheil, M., Schmidt, S., Drews, B., Schmidt, F., Hallmayer, J., and Maier, W. (1991) Linkage analysis in schizophrenia: exclusion of 5q11–q13, 5q34-qtr, 11q22, 23, Xpter and chromosome 19 in 15 systematically ascertained European families. *Am. J. Hum. Genetics* **49(Suppl. 4),** 363.
35. Jensen, S., Plaetke, R., Holik, J., Hoff, M., Myles-Worsley, M., Leppert, M., Coon, H., Vest, K., Freedman, R., Waldo, M., Zhou, Q.-Y., Litt, M., Civelli, O., and Byerley, W. (1993) Linkage analysis of schizophrenia: the D_1 dopamine receptor gene and several flanking DNA markers. *Hum. Heredity* **43,** 58–62.
36. Campion, D., d'Amato, T., Bastard, C., Laurent, C., Guedj, F., Jay, M., Dollfus, S., Thibault, F., Petit, M., Gorwood, P., Babron, M. C., Waksman, G., Martinez, M., and Mallet, J. (1994) Genetic study of dopamine D_1, D_2, and D_4 receptors in schizophrenia. *Psychiatry Res.* **51,** 215–230.
37. Cichon, S., Nöthen, M. M., Rietschel, M., Körner, J., and Propping, P. (1994) Single-strand conformation analysis (SSCA) of the dopamine D_1 receptor gene (DRD1) reveals no significant mutation in patients with schizophrenia and manic-depression. *Biol. Psychiatry* **36,** 850–853.
38. Dollfus, S., Campion, D., Vasse, T., Preterre, P., Laurent, C., d'Amato, T., Thibault, F., Mallet, J., and Petit, M. (1995) Association between dopamine D_1, D_2, D_3, and D_4 receptor genes and schizophrenia defined by 13 diagnostic systems. *Schizophrenia Res.* **15(1–2),** 37,38.
39. Coon, H., Byerley, W., Holik, J., Hoff, M., Myles-Worsley, M., Lannfelt, L., Sokoloff, P., Schwartz, J.-C., Waldo, M., Freedman, R., and Plaetke, R. (1993) Linkage analysis of schizophrenia with five dopamine receptor genes in nine pedigrees. *Am. J. Hum. Genetics* **52,** 327–334.

40. Ravindranathan, A., Coon, H., DeLisi, L., Holik, J., Hoff, M., Brown, A., Shields, G., Crow, T., and Byerley, W. (1994) Linkage analysis between schizophrenia and a microsatellite polymorphism for the D_5 dopamine receptor gene. *Psychiatr. Genetics* **4,** 77–80.
41. Moises, H. W., Gelernter, J., Guiffa, L. A., Zarcone, V., Wetterberg, L., Civelli, O., Kidd, K. K., Cavalli-Szforza, L. L., Grandy, D. K., Kennedy, J. L., Vinogradov, S., Mauer, J., Litt, M., and Sjögren, B. (1991) No linkage between D_2 dopamine receptor gene region and schizophrenia. *Arch. Gen. Psychiatry* **48,** 643–647.
42. Sarkar, G., Kapelner, S., Grandy, D. K., Marchionni, M., Civelli, O., Sobell, J., Heston, L., and Sommer, S. S. (1991) Direct sequencing of the dopamine D_2 receptor (DRD2) in schizophrenics reveals three polymorphisms but no structural changes in the receptor. *Genomics* **11,** 8–14.
43. Gill, M., McGuffin, P., Parfitt, E., Mant, R., Asherson, D., Collier, D., Vallada, H., Powell, J., Shaikh, S., Taylor, C., Sargeant, M., Clements, A., Nanko, S., Takazawa, N., Llewellyn, D., Williams, J., Whatley, S., Murray, R., and Owen, M. (1993) A linkage study of schizophrenia with DNA markers from the long arm of chromosome 11. *Psychol. Med.* **23,** 27–44.
44. Su, Y., Burke, J., O'Neill, F. A., Murphy, B., Nie, L., Kipps, B., Bray, J., Skinkwin, R., Nuallain, M. N., MacLean, C., Walsh, D., Diehl, S. R., and Kendler, J. S. (1993) Exclusion of linkage between schizophrenia and the D_2 dopamine receptor gene region of chromosome 11q in 112 Irish multiplex families. *Arch. Gen. Psychiatry* **50,** 205–211.
45. Wang, Z., Black, D., Andreasen, N. C., and Crowe, R. R. (1993) A linkage study of chromosome 11q in schizophrenia. *Arch. Gen. Psychiatry* **50,** 212–216.
46. Gejman, P. V., Ram, A., Gelernter, J., Friedman, E., Cao, Q., Pickar, D., Blum, K., Noble, E. P., Kranzler, H. R., O'Malley, S., Hamer, D. H., Whitsitt, F., Rao, P., DeLisi, L. E., Virkkunen, M., Linnoila, M., Goldman, D., and Gershon, E. S. (1994) No structural mutation in the dopamine D_2 receptor gene in alcoholism or schizophrenia. *JAMA* **271,** 204–208.
47. Hattori, M., Nanko, S., Dai, X. Y., Fukuda, R., and Kazamatsuri, H. (1994) Mismatch PCR RFLP detection of DRD2 Ser311Cys polymorphism and schizophrenia. *Biochem. Biophys. Res. Commun.* **202,** 757–763.
48. Shaikh, S., Collier, D. A., Arranz, M., Crocq, M.-A., Gill, M., and Kerwin, R. (1995) Examination of the DRD2 (Ser-Cys) variant in schizophrenia. *Schizophrenia Res.* **15(1–2),** 48.
49. Nanko, S., Hattori, M., Ikeda, K., Sasaki, T., Kazamatzuri, H., and Kuwata, S. (1993) Dopamine D_4 receptor polymorphism and schizophrenia. *Lancet* **341,** 689,690.
50. Wiese, C., Lannfelt, L., Kristbjarnarson, H., Yang, L., Zoega, T., Sokoloff, P., Ivarsson, O., Schwartz, J.-C., Moises, H. W., and Helgason, T. (1993) No evidence of linkage between schizophrenia and D_3 receptor gene locus in Icelandic pedigrees. *Psychiatry Res.* **46,** 69–78.
51. Sabaté, O., Campion, D., d'Amato, T., Martres, M. P., Sokoloff, P., Giros, B., Leboyer, M., Jay, M., Guedj, F., Thbault, F., Dollfus, S., Preterre, P., Petit, M., Bobron, M.-C., Waksman, G., Mallet, J., and Schwartz, J.-C. (1994) Failure to find evidence for linkage or association between the dopamine D_3 receptor gene and schizophrenia. *Am. J. Psychiatry* **151,** 107–111.
52. Crocq, M. A., Duval, F., Mayerova, A., Sokoloff, P., Mokrani, M. C., and Macher, J. P. (1995) Clinical and functional correlates of a dopamine D_3 receptor polymorphism. *Hum. Psychopharmacol.* **10,** 19–24.

53. Barr, C. L., Kennedy, J. L., Lichter, J. B., Van Tol, H. H., Wetterberg, L., Livak, K. J., and Kidd, K. K. (1993) Alleles at the dopamine D_4 receptor locus do not contribute to the genetic susceptibility to schizophrenia in a large Swedish kindred. *Am. J. Med. Genetics (Neuropsychiatric)* **48,** 218–222.
54. Debryn, A., Mandelbaum, K., Sandkuijl, L. A., Delvenne, V., Hirsch, D., Staner, L., Mendlewicz, J., and Van Broeckhoven, C. (1994) Nonlinkage of bipolar illness to tyrosine hydroxylase, tyrosinase, and D_2 and D_4 dopamine receptor genes on chromosome 11. *Am. J. Psychiatry* **151,** 1202–1206.
55. Macciardi, F., Petronis, A., Van Tol, H. H., Marino, C., Cavallini, C., Smeraldi, E., and Kennedy, J. L. (1994) Analysis of the D_4 dopamine receptor gene variant in an Italian schizophrenia kindred. *Arch. Gen. Psychiatry* **51,** 288–293.
56. Rao, P. A., Pickar, D., Gejman, P. V., Ram, A., Gershon, E. S., and Gelernter, J. (1994) Allelic variation in the D_4 dopamine receptor (DRD4) gene does not predict response to clozapine. *Arch. Gen. Psychiatry* **51,** 912–917.
57. Catalano, M., Nobile, M., Novelli, E., Nöthen, M. M., and Smeraldi, E. (1993) Distribution of a novel mutation in the first exon of the human dopamine D_4 receptor gene in psychotic patients. *Biol. Psychiatry* **34,** 459–464.
58. Holmes, D., Brynjolfsson, J., Brett, P., Curtis, D., Pertursson, H., Sherrington, R., and Gurling, H. (1991) No evidence of a susceptibility locus predisposing to manic-depression in the region of the dopamine (D_2) receptor gene. *Br. J. Psychiatry* **158,** 635–641.
59. Blum, K., Noble, E. P., Sheridan, P. J., Montgomery, A., Ritchie, T., Jagadeeswaran, P., Nogami, H., Briggs, A. H., and Cohn, J. B. (1990) Allelic association of human dopamine D_2 receptor gene in alcoholism. *JAMA* **263,** 2055–2060.
60. Noble, E. P., Blum, K., Ritchie, T., Montgomery, A., and Sheridan, P. J. (1991) Allelic association of the D_2 dopamine receptor gene with receptor-binding characteristics in alcoholism. *Arch. Gen. Psychiatry* **48,** 648–654.
61. Bolos, A. M., Dean, M., Lucas-Derse, S., Ramsburg, M., Brown, G. L., and Goldman, D. (1990) Population and pedigree studies reveal a lack of association between the dopamine D_2 receptor gene and alcoholism. *JAMA* **264,** 3156–3160.
62. Gelernter, J., O'Malley, S., Risch, N., Kranzler, H. R., Krystal, J., Merikangas, K., Kennedy, J. L., and Kidd, K. K. (1991) No association between an allele at the D_2 dopamine receptor gene (DRD2) and alcoholism. *JAMA* **266,** 1801–1807.
63. Parsian, A., Todd, R. D., Devor, E. J., O'Malley, K. L., Suarez, B. K., Reich, T., and Cloninger, C. R. (1991) Alcoholism and alleles of the human D_2 dopamine receptor locus. *Arch. Gen. Psychiatry* **48,** 655–663.
64. Noble, E. P. (1993) The D_2 dopamine receptor gene: a review of association studies in alcoholism. *Behav. Genetics.* **23,** 119–129.
65. Comings, D. E. (1994) The dopamine D_2 receptor gene (DRD2) and neuropsychiatric disorders. *CNS Drugs* **1,** 1–5.
66. Conneally, P. M. (1991) Association between the D_2 dopamine receptor gene and alcoholism. *Arch. Gen. Psychiatry* **48,** 664–666.
67. Devor, E. J. (1984) Complex segregation analysis of Gilles de la Tourette syndrome: further evidence for a major locus mode of transmission. *Am. J. Hum. Genetics* **36,** 704–709.
68. Gelernter, J., Pakstis, A., Pauls, D. L., Kurlan, R., Grancher, S. T., Civelli, O., Grandy, D., and Kidd, K. K. (1990) Tourette syndrome is not linked to D_2 dopamine receptor. *Arch. Gen. Psychiatry* **47,** 1073–1077.

69. Gelernter, J., Kennedy, J. L., Grandy, D. K., Zhou, Q.-Y., Civelli, O., Pauls, D. L., Pakstis, A., Kurlan, R., Sunahara, R. K., Niznik, H. B., O'Dowd, B., Seeman, P., and Kidd, K. K. (1993) Exclusion of close linkage of Tourette's syndrome to D_1 dopamine receptor. *Am. J. Psychiatry* **150,** 449–453.
70. Landwehrmeyer, B., Mengod, G., and Palacios, J. M. (1993) Differential visualization of dopamine D_2 and D_3 receptor sites in rat brain: a comparative study using in situ hybridization histochemistry and ligand binding autoradiography. *Eur. J. Neurosci.* **5,** 145–153.
71. Lévesque, D., Diaz, J., Pilon, C., Martres, M.-P., Giros, B., Souil, E., Schott, D., Morgat, J.-L., Schwartz, J.-C., and Sokoloff, P. (1992) Identification, characterization, and localization of the dopamine D_3 receptor in rat brain using 7[^{3}H]hydroxy-*N,N*-di-*n*-propyl-2-aminotetralin. *Proc. Natl. Acad. Sci. USA* **89,** 8155–8159.
72. Crocq, M. A., Mant, R., Asherson, P., Williams, J., Hode, Y., Mayerova, A., Collier, D., Lannfelt, L., Sokoloff, P., Gill, M., Macher, J. P., McGuffin, P., and Owen, M. J. (1992) Association between schizophrenia and homozygosity at the dopamine D_3 receptor gene. *J. Med. Genetics* **29,** 858–860.
73. Mant, R., Williams, J., Asherson, P., Parfitt, E., McGuffin, P., and Owen, M. J. (1994) Relationship between homozygosity at the dopamine D_3 receptor gene and schizophrenia. *Am. J. Med. Genetics (Neuropsychiatric)* **54,** 21–26.
74. Jönsson, E., Lannfelt, L., Sokoloff, P., Schwartz, J.-C., and Sedvall, G. (1993) Lack of association in a Bal-I polymorphism in the dopamine D_3 receptor gene in schizophrenia. *Acta Psychiatr. Scand.* **87,** 345–347.
75. Laurent, C., Savoye, C., Samolyk, D., Meloni, R., Mallet, J., Campion, D., Martinez, M., d'Amato, T., Bastard, C., and Dollfus, S. (1994) Homozygosity at the dopamine D_3 receptor locus is not associated with schizophrenia. *J. Med. Genetics* **31,** 260–264.
76. Nanko, S., Sasaki, T., Fukuda, R., Hattori, M., Dai, X. Y., Kazamatzuri, H., Kuwata, S., Juri, T., and Gill, M. (1993) A study of the association between schizophrenia and the dopamine D_3 receptor gene. *Hum. Genetics* **92,** 336–338.
77. Nimgaonkar, V. L., Zhang, X. R., Caldwell, J. G., Ganguli, R., and Chakravarti, A. (1993) Association study of schizophrenia with dopamine D_3 receptor gene polymorphisms: probable effects of family history of schizophrenia? *Am. J. Med. Genetics (Neuropsychiatric)* **48,** 214–217.
78. Nöthen, M. M., Cichon, S., Propping, P., Fimmers, R., Schwab, S. G., and Wildenauer, D. P. (1993) Excess of homozygosity at the dopamine D_3 receptor gene not confirmed. *J. Med. Genetics* **30,** 708–712.
79. Yang, L., Li, T., Wiese, C., Lannfelt, L., Sokoloff, P., Xu, C. T., Zeng, Z., Schwartz, J. C., and Moises, H. W. (1993) No association between schizophrenia and homozygosity at the D_3 dopamine receptor gene. *Am. J. Med. Genetics (Neuropsychiatric)* **48,** 83–86.
80. Nimgaonkar, V. L., Zhang, X. R., Brar, J., DeLeo, M., Hogge, W., Ganguli, R., and Chakravarti, A. (1995) Association between the dopamine D_3 receptor gene locus and liability to schizophrenia, as well as its age of onset. *Schizophrenia Res.* **15(1–2),** 45.
81. Van Tol, H. H., Wu, C. M., Guan, H.-C., Ohara, K., Bunzow, J. R., Civelli, O., Kennedy, J., Seeman, P., Niznik, H. B., and Janovic, V. (1992) Multiple dopamine D_4 receptor variants in the human population. *Nature* **358,** 149–152.
82. Shaikh, S., Collier, D. A., Kerwin, R. W., Pilowsky, L. S., Gill, M., Xu, W.-M., and Thornton, A. (1993) Dopamine D_4 receptor subtypes and response to clozapine. *Lancet* **341,** 116.

83. Shaikh, S., Gill, M., Owen, M., Asherson, P., McGuffin, P., Nanko, S., Murray, R. M., and Collier, D. A. (1994) Failure to find linkage between a functional polymorphism in the dopamine D_4 receptor gene and schizophrenia. *Am. J. Med. Genetics* **54,** 8–11.
84. Baldessarini, R. J., Gardner, D. M., and Garver, D. L. (1995) Conversions from clozapine to other antipsychotics. *Arch. Gen. Psychiatry* **52,** 1071,1072.
85. Seeman, P. (1992) Dopamine receptor sequences: therapeutic levels of neuroleptics occupy D_2 receptors, clozapine occupies D_4. *Neuropsychopharmacology* **7,** 261–284.
86. Baldessarini, R. J. and Frankenburg, F. R. (1991) Clozapine—a novel antipsychotic agent. *N. Engl. J. Med.* **324,** 746–754.
87. Lahti, R. A., Roberts, R. C., and Tamminga, C. A. (1995) D_2-Family receptor distributions in human postmortem tissue: an autoradiographic study. *NeuroReport* **6,** 2505–2512.
88. Murray, A. M., Hyde, T. M., Knable, M. B., Herman, M. M., Bigelow, L. B., Carter, J. M., Weinberger, D. R., and Kleinman, J. E. (1995) Distribution of putative D_4 dopamine receptors in postmortem striatum from patients with schizophrenia. *J. Neurosci.* **15,** 2186–2191.
89. O'Malley, K. L., Harmon, S., Tang, L., and Todd, R. D. (1992) The rat dopamine D_4 receptor: sequence, gene structure, and demonstration of expression in the cardiovascular system *New Biologist* **4,** 137–146.
90. Reynolds, G. P. and Mason, S. L. (1994) Are striatal dopamine D_4 receptors increased in schizophrenia? *J. Neurochem.* **63,** 1576–1577.
91. Seeman, P., Guan, H.-C., and Van Tol, H. H. (1993) Dopamine D_4 receptors elevated in schizophrenia. *Nature* **365,** 441–445.
92. Amara, S. G. and Kuhar, M. J. (1993) Neurotransmitter transporters: recent progress. *Annu. Rev. Neurosci.* **16,** 73–93.
93. Schloss, P., Wayser, W., and Betz, H. (1992) Neurotransmitter transporters: a novel family of integral plasma membrane proteins. *FEBS Lett.* **307,** 76–80.
94. Persico, A. M., Wang, Z. W., Black, D. W., Andreasen, N. C., Uhl, G. R., and Crowe, R. R. (1995) The dopamine transporter gene and schizophrenia spectrum disorders: exclusion of close linkage. *Am. J. Psychiatry* **152,** 134–136.
95. Joyce, J. N., Lexow, N., Bird, E. D., and Winokur, A. (1988) Organization of dopamine D_1 and D_2 receptors in human brain: receptor autoradiographic studies in Huntington's disease and schizophrenia. *Synapse* **2,** 546–557.
96. Owen, F., Crow, T. J., Poulter, M., Cross, A. J., Longden, A., and Riley, G. J. (1978) Increased dopamine receptor sensitivity in schizophrenia. *Lancet* **2,** 223–225.
97. Lee, T. and Seeman, P. (1980) Elevation of brain neuroleptic/dopamine receptors in schizophrenia. *Am. J. Psychiatry* **137,** 191–197.
98. Lee, T., Seeman, P., Tourtelotte, W. W., Farley, U. J., and Hornykiewicz, O. (1978) Binding of ^{3}H-neuroleptics and ^{3}H-apomorphine in schizophrenic brains. *Nature* **274,** 897–900.
99. Mackay, A. V., Bird, E. D., Spokes, E. G., Rossor, M., Iversen, L. L., Creese, I., and Snyder, S. H. (1980) Dopamine receptors and schizophrenia: drug effect or illness? *Lancet* **2,** 915,916.
100. Mackay, A. V., Iversen, L. L., Rossor, M., Spokes, E., Bird, E. D., Arregui, A., Creese, I., and Snyder, S. H. (1982) Increased brain dopamine and dopamine receptors in schizophrenia. *Arch. Gen. Psychiatry* **39,** 991–997.

101. Reisine, T. D., Rossor, M., and Spokes, E. G. (1980) Opiate and neuroleptic receptor alterations in human schizophrenic brain tissue, in *Receptors for Neurotransmitters and Peptide Hormones* (Pepeu, G., Kuhar, M. J., and Enna, S. J., eds.), Raven, New York, pp. 443–450.
102. Cross, A. J., Crow, T. J., and Owen, F. (1981) ^{3}H-Flupenthixol binding to postmortem brains of schizophrenics: evidence for a selective increase in dopamine D_2 receptors. *Psychopharmacology* **74,** 122–124.
103. Toru, M., Nishikawa, T., Semba, J., Mataga, N., Takashima, M., Noda, K., and Shibuya, H. (1982) Increased dopamine metabolism in the putamen and caudate in schizophrenic patients, in *Psychobiology of Schizophrenia* (Namba, M. and Kaiya, H., eds.), Pergamon, Oxford, UK, pp. 235–240.
104. Cross, A. J., Crow, T. J., Ferrier, I. N., Johnstone, E. C., McCreadie, R. M., Owen, F., Owens, D. G., and Pouter, M. (1983) Dopamine receptor changes in schizophrenia in relation to the disease process and movement disorder. *J. Neural Transm.* **18(Suppl.),** 265–272.
105. Owen, F., Cross, A. J., and Crow, T. J. (1983) Ligand-binding studies in brains of schizophrenics, in *Cell Surface Receptors* (Strange, P. G., ed.), Wiley, New York, pp. 163–183.
106. Pimoule, C., Schoemaker, H., Reynolds, G. P., and Langer, S. Z. (1985) [^{3}H]SCH-23390 labeled D_1 dopamine receptors are unchanged in schizophrenia and Parkinson's disease. *Eur. J. Pharmacol.* **114,** 235–237.
107. Mita, T., Hanada, S., Nishino, N., Kuno, T., Nakai, H., Yamadori, T., Mizoi, Y., and Tanaka, C. (1986) Decreased serotonin S_2 and increased dopamine D_2 receptors in chronic schizophrenia. *Biol. Psychiatry* **21,** 1407–1414.
108. Hess, E. J., Bracha, H. S., Kleinman, J. E., and Creese, I. (1987) Dopamine receptor subtype imbalance in schizophrenia. *Life Sci.* **40,** 1487–1497.
109. Seeman, P., Bzowej, N. H., Guan, H. C., Bergeron, C., Reynolds, G. P., Bird, E. P., Riederer, P., Jellinger, K., and Tourtellote, W. W. (1987) Human brain D_1 and D_2 dopamine receptors in schizophrenia, Alzheimer's, Parkinson's, and Huntington's diseases. *Neuropsychopharmacology* **1,** 5–15.
110. Reynolds, G. P. and Gzudek, C. (1988) Status of the dopaminergic system in postmortem brain in schizophrenia. *Psychopharmacol. Bull.* **24,** 345–347.
111. Kornhuber, J., Riederer, P., Reynolds, G. P., Beckmann, H., Jellinger, K., and Gabriel, E. (1989) ^{3}H-Spiperone binding sites in postmortem brains from schizophrenic patients: relationship to drug treatment, abnormal movements, and positive symptoms. *J. Neural Transm.* **75,** 1–10.
112. Seeman, P., Niznik, H. B., Guan, H.-C., Booth, G., and Ulpian, C. (1989) Link between D_1 and D_2 dopamine receptors is reduced in schizophrenia and Huntington's diseased brain. *Proc. Natl. Acad. Sci. USA* **86,** 10,156–10,160.
113. Mamelak, M., Chiu, S., and Mishra, R. K. (1993) High and low affinity states of dopamine D_1 receptors in schizophrenia. *Eur. J. Pharmacol.* **233,** 175,176.
114. Schmauss, C., Hartounian, V., Davis, K. L., and Davidson, M. (1993) Selective loss of dopamine D_3-type receptor mRNA expression in parietal and motor cortices of patients with chronic schizophrenia. *Proc. Natl. Acad. Sci. USA* **90,** 8942–8946.
115. Knable, M. B., Hyde, T. M., Herman, M. M., Carter, J. M., Bigelow, L., and Kleinman, J. E. (1994) Quantitative autoradiography of dopamine D_1 receptors, D_2 receptors, and dopamine uptake sites in postmortem striatal specimens from schizophrenic patients. *Biol. Psychiatry* **36,** 827–835.

116. Kessler, R. M., Whetsell, W. O., Sib Ansari, M. S., Tamminga, C. A., and Mason, N. S. (1995) [^{125}I]Epidepride binding in postmortem brain of schizophrenic subjects. *Schizophrenia Res.* **15(1–2),** 62.
117. Joyce, J. N., Gurevich, E. V., Kung, H. F., Kung, M.-P., Bordelon, Y., Shapiro, P., Arnold, S. E., and Gur, R. E. (1995) Dopamine D_3 receptors are elevated in schizophrenic brain and decreased by neuroleptic treatment. *Schizophrenia Res.* **15(1–2),** 61.
118. Crow, T. J., Cross, A. J., Johnstone, E. C., Owen, F., Owens, D. G., and Waddington, J. L. (1982) Abnormal involuntary movements in schizophrenia: are they associated with changes in dopamine receptors? *J. Clin. Psychopharmacol.* **2,** 336–340.
119. Crow, T. J., Owen, F., Cross, A. J., Ferrier, N., Johnstone, E. C., McCreadie, R. M., Owens, D. G., and Poulter, M. (1981) Neurotransmitter enzymes and receptors in postmortem brain in schizophrenia: evidence that an increase in D_2 dopamine receptors in associated with the type I syndrome, in *Transmitter Biochemistry of Human Brain Tissue* (Riederer, P. and Usdin, E., eds.), MacMillan, London, pp. 85–96.
120. Seeman, P. and Niznik, H. B. (1990) Dopamine receptors and transporters in Parkinson's disease and schizophrenia. *FASEB J.* **4,** 2737–2744.
121. Seeman, P., Ulpian, C., Bergeron, C., Riederer, P., Jellinger, K., Gabriel, E., Reynolds, G. P., and Tourtelotte, W. W. (1984) Bimodal distribution of dopamine receptor densities in brains of schizophrenics. *Science* **225,** 728–731.
122. Baldessarini, R. J., and Tarsy, D. (1979) Relationship of the actions of neuroleptic drugs to the pathophysiology of tardive dyskinesia. *Intl. Rev. Neurobiol.* **21,** 1–45.
123. Andreasen, N. C., Carson, R., Diksic, M., Evans, A., Farde, L. O., Gjedde, A., Hakin, A., Lal, S., Nair, N., Sedvall, G., Tune, L., and Wong, D. F. (1988) Workshop on schizophrenia, PET and dopamine D_2 receptors in the human neostriatum. *Schizophrenia Bull.* **14,** 471–484.
124. Baldessarini, R. J., Kula, N. S., Campbell, A., Bakthavachalam, V., Yuan, J., and Neumeyer, J. L. (1992) Prolonged D_2 antidopaminergic activity of alkylating and nonalkylating derivatives of spiperone in rat brain. *Mol. Pharmacol.* **42,** 856–863.
125. Memo, M., Kleinman, J. E., and Hanbauer, I. (1983) Coupling of dopamine D_1 recognition sites with adenylate cyclase in nuclei accumbens and caudatus of schizophrenics. *Science* **221,** 1304–1307.
126. Lévesque, D., Martres, M.-P., Diaz, J., Griffon, N., Lammers, C. H., Sokoloff, P., and Schwartz, J.-C. (1995) A paradoxical regulation of the dopamine D_3 receptor expression suggests the involvement of an anterograde factor from dopamine neurons. *Proc. Natl. Acad. Sci. USA* **92,** 1719–1723.
127. Gelbard, H., Teicher, M. H., Gallitano, A., Marsh, E. R., Zorc, J., Faedda, G., and Baldessarini, R. J. (1990) Dopamine D_1 receptor development depends on endogenous dopamine. *Dev. Brain Res.* **56,** 137–140.
128. Hitri, A., Casanova, M. F., Kleinman, J. E., Weinberger, D. R., and Wyatt, R. J. (1995) Age-related changes in [^{3}H]GBR-12935 binding site density in the prefrontal cortex of controls and schizophrenics. *Biol. Psychiatry* **37,** 175–182.
129. Madras, B. K., Spealman, R. D., Fahey, M. A., Neumeyer, J. L., Saha, J. K., and Milius, R. A. (1989) Cocaine receptors labeled by [^{3}H]2β-carbomethoxy-3β-(4-fluorophenyl)tropane. *Mol. Pharmacol.* **36,** 518–524.
130. Neumeyer, J. L., Wang, S., Milius, R. A., Baldwin, R. M., Zea-Ponce, Y., Hoffer, P. B., Sybirska, E., Al-Tairiti, M., Charney, D. S., Malison, R. T., Larnelle, M., and Innis, R. B. (1991) [^{123}I]2β-Carbomethoxy-3β-(4-iodophenyl)-tropane: high affinity SPECT radiotracer for monoamine reuptake sites in brain. *J. Med. Chem.* **34,** 3144–3146.

131. Janowsky, A., Vocci, F., Berger, P., Angel, I., Zelnck, N., Kleinman, J. E., Skolnick, P., and Paul, S. M. (1987) [^{3}H]GBR-12935 binding to the dopamine transporter is decreased in the caudate nucleus in Parkinson's disease. *J. Neurochem.* **49,** 617–621.
132. Czudek, C. and Reynolds, G. P. (1989) [^{3}H]GBR-12935 binding to the dopamine uptake sites in postmortem brain tissue in schizophrenia. *J. Neural Transm.* **77,** 227–230.
133. Bogerts, B. (1991) The neuropathology of schizophrenia: pathophysiologic and neurodevelopmental implications, in *Fetal Neural Development and Adult Schizophrenia* (Mednick, S. A., Cannon, T. D., and Barr, C. E., eds.), Cambridge University Press, London, pp. 153–173.
134. Shapiro, R. M. (1993) Regional neuropathology in schizophrenia: where are we; where are we going? *Schizophrenia Res.* **10,** 187–239.
135. Trojanowski, J. Q. and Arnold, S. E. (1995) In pursuit of the molecular neuropathology of schizophrenia. *Arch. Gen. Psychiatry* **52,** 274–276.
136. Ariano, M. A. and Sibley, D. R. (1994) dopamine receptor distribution in the rat CNS: elucidation using antipeptide antisera directed against D_{1A} and D_3 subtypes. *Brain Res.* **649,** 95–110.
137. Boundy, V. A., Luedtke, R. R., Gallitano, A. L., Smith, J. E., Filtz, T. M., Kallen, R. G., and Molinoff, P. B. (1993) Expression and characterization of the rat D_3 dopamine receptor: pharmacologic properties and development of antibodies. *J. Pharmacol. Exp. Ther.* **264,** 1002–1011.
138. Baron, J. C., Comar, D., Farde, L. O., Martinot, J. L., and Mazoyer, B. (eds.) (1991) *Brain Dopaminergic Systems: Imaging with Positron Tomography.* Kluwer, Dordrecht, Germany.
139. Farde, L. O., Nordström, A.-L., Eriksson, L., Halldin, C., and Sedvall, G. (1990) Comparison of methods used with [^{11}C]raclopride and [^{11}C]N-methylspiperone for the PET-determination of central D_2 dopamine receptors. *Clin. Neuropharmacol.* **13(Suppl. 2),** 87,88.
140. Konig, P., Benzer, M. K., and Fritsche, H. (1991) SPECT technique for visualization of cerebral dopamine D_2 receptors. *Am. J. Psychiatry* **148,** 1607,1608.
141. Reba, R. C. (1993) PET and SPECT: opportunities and challenges for psychiatry. *J. Clin. Psychiatry* **54(Suppl. 11),** 26–32.
142. Ring, H. A. (1995) the value of positron emission tomography in psychopharmacology (1995) *Hum. Psychopharmacology* **10,** 79–87.
143. Sedvall, G. (1992) The current status of PET scanning with respect to schizophrenia. *Neuropsychopharmacology* **7,** 41–54.
144. Verhoeff, N. P. (1991) Pharmacological implications for neuroreceptor imaging. *Eur. J. Nucl. Med.* **18,** 482–502.
145. Weinberger, D. R. (1993) SPECT imaging in psychiatry: introduction and overview. *J. Clin. Psychiatry* **54(Suppl. 11),** 3–5.
146. Wong, D. F. (1992) PET studies of neuroreceptors in schizophrenia: commentary on the current status of PET scanning with respect to schizophrenia. *Neuropsychopharmacology* **7,** 69–72.
147. Billings, J. J., Kung, M.-P., Chumpradit, S., Mozley, D., Alavi, A., and Kung, H. F. (1992) Characterization of radioiodinated TISCH: a high-affinity and selective ligand for mapping CNS D_1 dopamine receptor. *J. Neurochem.* **58,** 277–286.
148. Kung, H. F. (1993) SPECT and PET ligands for CNS imaging. *RBI Neurotransmissions* **9(4),** 1–6.

149. Kung, M.-P., Kung, H. F., Billings, J., Yang, Y., Murphy, R. A., and Alavi, A. (1990) Characterization of IBF as a new selective dopamine D_2 receptor imaging agent. *J. Nucl. Med.* **31,** 648–654.
150. Mazière, B., Coenen, H. H., Halldin, C., Någren, K., and Pike, V. W. (1992) PET radioligands for dopamine receptors and reuptake sites: chemistry and biochemistry. *Nucl. Med. Biol.* **19,** 497–512.
151. Stöcklin, G. (1992) Tracers for metabolic imaging of brain and heart. *Eur. J. Nucl. Med.* **19,** 527–551.
152. Wong, D. F., Wilson, A. A., Chen, C., Minkin, E., Dannals, R. F., Ravert, H. T., Sanchez-Roa, P., Villemagne, V., and Wagner, J. N., Jr. (1992) In vivo studies of [^{125}I]iodobenzamide and [^{11}C]iodobenzamide: a ligand suitable for positron emission tomography and single photon emission tomography imaging of cerebral D_2 dopamine receptors. *Synapse* **12,** 236–241.
153. Farde, L. O., Nyberg, S., Oxenstierna, G., Nakashima, Y., Halldin, C., and Ericsson, B. (1995) Positron emission tomography studies on D_2 and 5-HT_2 receptor binding in risperidone-treated schizophrenic patients. *J. Clin. Psychopharmacol.* **15(Suppl. 1),** 19–23.
154. Crawley, C. W., Crow, T. J., Johnstone, E. C., Oldland, S. R., Owen, F., Owens, D. G., Smith, T., Veall, N., and Zanelli, G. D. (1986) Uptake of ^{76}Br-spiperone in the striata of schizophrenic patients and controls. *Nucl. Med. Commun.* **7,** 599–607.
155. Wong, D. F., Wagner, H. N., Jr., Tune, L. E., Dannals, R. F., Pearlson, G. D., Links, J. M., Tamminga, C. A., Broussolle, E. P., Ravert, H. T., Wilson, A. A., Toung, J. K., Malat, J., Williams, J. A., O'Tauma, L. A., Snyder, S. H., Kuhar, M. J., and Gjedde, A. (1986) Positron emission tomography reveals elevated D_2 dopamine receptors in drug-naïve schizophrenics. *Science* **234,** 1558–1563.
156. Wong, D. F., Singer, H., Pearlson, G., Tune, L., Ross, C., Villemagne, V., Dannals, R. F., Links, J. M., Wilson, A., Ravert, H., Wagner, H. N., Jr., and Gjedde, A. (1988) D_2 dopamine receptors in Tourette's syndrome and manic-depressive illness. *J. Nucl. Med.* **29,** 820,821.
157. Blin, J., Baron, J. C., Cambon, H., Bonnet, A. M., DuBois, D., Loc'h, C., Mazière, B., and Agid, Y. (1989) Striatal dopamine D_2 receptors in tardive dyskinesia: PET study. *J. Neurol. Neurosurg. Psychiatry* **52,** 1248–1252.
158. Martinot, J.-L., Peron-Magna, P., Huret, J.-D., Mazoyer, B., Baron, J.-C., Boulenger, J.-P., Loc'h, C., Mazière, B., Caillard, V., Loo, H., and Syrota, A. (1990) Striatal D_2 dopaminergic receptors assessed with positron emission tomography and [^{76}Br]bromospiperone in untreated schizophrenic patients. *Am. J. Psychiatry* **147,** 44–50.
159. Martinot, J.-L., Paillère-Martinot, M. L., Loc'h, C., Hardy, P., Poirier, M. F., Mazoyer, B., Beaufils, B., Mazière, B., Allilaire, J. F., and Syrota, A. (1991) The estimated density of D_2 striatal receptors in schizophrenia: a study with positron emission tomography and [^{76}Br]bromolisuride. *Br. J. Psychiatry* **158,** 346–350.
160. Nordström, A.-L., Farde, L. O., Pauli, S., Litton, J.-E., and Halldin, C. (1992) PET analysis of central [^{11}C]raclopride binding in healthy young adults and schizophrenic patients: reliability and age effects. *Hum. Psychopharmacology* **7,** 157–165.
161. Suhara, T., Nakayama, K., Inoue, O., Fukuda, H., Shimizu, M., Mori, A., and Tateno, Y. (1992) D_1 dopamine receptor binding in mood disorders measured by positron emission tomography. *Psychopharmacology* **106,** 14–18.

162. Pearlson, G. D., Tune, L. E., Wong, D. F., Aylward, E. H., Barta, P. E., Powers, R. E., Tien, A. Y., Chase, G. A., Harris, G. J., and Rabins, P. V. (1993) Quantitative D_2 dopamine receptor PET and structural MRI changes in late-onset schizophrenia. *Schizophrenia Bull.* **19,** 783–795.
163. Tune, L. E., Wong, D. F., Pearlson, G., Strauss, M., Young, T., Shaya, E. K., Dannals, R. F., Wilson, A. A., Ravert, H. T., Sapp, J., Cooper, T., Chase, G. A., and Wagner, H. N., Jr. (1993) Dopamine D_2 receptor density estimates in schizophrenia: a positron emission tomography study with ^{11}C-N-methylspiperone. *Psychiatry Res.* **49,** 219–237.
164. D'haenen, H. A. and Bossuyt, A. (1994) Dopamine D_2 receptors in depression measured with single photon emission computed tomography. *Biol. Psychiatry* **35,** 128–132.
165. Ebert, D., Feistel, H., Kaschka, W., Barocka, A., and Pirner, A. (1994) Single photon emission computerized tomography assessment of cerebral dopamine D_2 receptor blockade in depression before and after sleep deprivation—preliminary results. *Biol. Psychiatry* **35,** 880–885.
166. Hietala, J., Syvälahti, E., Vuorio, K., Någren, K., Lehikoinen, P., Ruotsalainen, U., Räkkäpläinen, V., Lehtinen, V., and Wegelius, U. (1994) Striatal D_2 dopamine receptor characteristics in neuroleptic-naïve schizophrenic patients studied with positron emission tomography. *Arch. Gen. Psychiatry* **51,** 116–123.
167. Martinot, J.-L., Paillère-Martinot, M. L., Loc'h, C., Lecrubier, Y., Dao-Castellana, M. H., Aubin, F., Allilaire, J. F., Mazoyer, B., Mazière, B., and Syrota, A. (1994) Central D_2 receptors and negative symptoms of schizophrenia. *Br. J. Psychiatry* **164,** 27–34.
168. Pilowsky, L. S., Costa, D. C., Ell, P. J., Murray, R. M., Verhoeff, N. P., and Kerwin, R. W. (1994) D_2 dopamine receptor binding in the basal ganglia of antipsychotic-free schizophrenic patients: an ^{123}I-IBZM single photon emission computerized tomographic study. *Br. J. Psychiatry* **164,** 16–26.
169. Knable, M. B., Gonzalez, J., Coppola, R., Jones, D. W., Nawroz, S., Gorey, J., and Weinberger, D. R. (1995) ^{123}I-IBZM SPECT in neuroleptic-free schizophrenic patients. *Schizophrenia Res.* **15(1–2),** 88.
170. Tune, L. E., Wong, D. F., and Pearlson, G. D. (1992) Elevated dopamine D_2 receptor density in 23 schizophrenic patients: a positron emission tomography study with [^{11}C]N-methylspiperone. *Schizophrenia Res.* **6,** 147.
171. Farde, L. O., Wiesel, F.-A., Hall, H., Halldin, C., Stone-Elander, S., and Sedvall, G. (1987) No D_2 receptor increase in PET study of schizophrenia. *Arch. Gen. Psychiatry* **44,** 671–675.
172. Farde, L. O., Wiesel, F.-A., Stone-Elander, S., Halldin, C., Nordström, A.-L., Hall, H., and Sedvall, G. (1990) D_2 dopamine receptors in neuroleptic-naïve schizophrenic patients. *Arch. Gen. Psychiatry* **47,** 213–219.
173. Hall, H., Wedel, I., Halldin, C., Kopp, J., and Farde, L. (1990) Comparison of the in vitro receptor binding properties of [^{3}H]-N-methylspiperone and [^{3}H]raclopride to rat and human brain membranes. *J. Neurochem.* **55,** 2048–2057.
174. Logan, J., Dewey, S. L., Wolf, A. P., Fowler, J. S., Brodie, J. D., Angrist, B., Wolkow, N. D., and Gatley, S. J. (1991) Effects of endogenous dopamine on measures of [^{18}F]N-methylspiroperidol binding in the basal ganglia: comparison of simulations and experimental results from PET studies in baboons. *Synapse* **9,** 195–207.

175. Young, L. T., Wong, D. F., Goldman, S., Minkin, E., Chen, C., Matsumura, K., Scheffel, U., and Wagner, H. N., Jr. (1991) Effects of endogenous dopamine on kinetics of [^{3}H]N-methylspiperone and [^{3}H]raclopride binding in the rat brain. *Synapse* **9,** 188–194.
176. TenBrink, R. E. and Huff, R. M. (1994) Recent advances in dopamine D_3 and D_4 receptor ligands and pharmacology. *Ann. Rep. Med. Chem.* **2,** 43–52.
177. Dewey, S. L., Logan, J., Wolf, A. P., Brodie, J. D., Angrist, B., Fowler, J. S., and Volkow, N. D. (1991) Amphetamine induced decreases in [^{18}F]N-methylspiroperidol binding in the baboon brain using positron emission tomography (PET). *Synapse* **7,** 324–327.
178. Dewey, S. L., Smith, G. S., Logan, J., Brodie, J. D., Simkowitz, P., MacGregor, R. R., Fowler, J. S., Volkow, N. D., and Wolf, A. P. (1993) Effects of central cholinergic blockade on striatal dopamine release measured with positron emission tomography in normal human subjects. *Proc. Natl. Acad. Sci. USA.* **90,** 11,816–11,820.
179. Reith, J., Benkelfat, C., Sherwin, A., Yasuhara, Y., Kuwabara, H., Andermann, F., Bachneff, S., Cumming, P., Diksic, M., Dyve, S. E., Etienne, P., Evans, A. C., Lal, S., Shevell, M., Savard, G., Wong, D. F., Chouinard, G., and Gjedda, A. (1994) Elevated DOPA decarboxylase activity in living brain of patients with psychosis. *Proc. Natl. Acad. Sci. USA* **91,** 11,651–11,654.
180. Wong, D. F., Pearlson, G. D., Young, L. T., Singer, H., Villemagne, V., Tune, L., Ross, C., Dannals, R. F., Links, J. M., Chan, B., Wilson, A. A., Ravert, H. T., Wagner, H. N., Jr., and Gjedde, A. (1989) D_2 dopamine receptors are elevated in neuropsychiatric disorders other than schizophrenia. *J. Cereb. Blood Flow Metabol.* **9(Suppl. 1),** 593S.
181. Chabriat, H., Levasseur, M., Vidailhet, M., Loc'h, C., Mazière, B., Bourguignon, M. H., Bonnet, A. M., Zilbovicius, M., Raynaud, C., Agid, Y., Syrota, A., and Samson, Y. (1992) In vivo SPECT imaging of D_2 receptor with ^{123}I-iodolisuride: results in supranuclear palsy. *J. Nucl. Med.* **33,** 1481–1485.
182. Rutgers, A. W., Lakke, J. P., Paans, A. M., Vaalburg, W., and Korf, J. (1987) Tracing of dopamine receptors in hemiparkinsonism with positron emission tomography (PET). *J. Neurol. Sci.* **80,** 237–248.
183. Sawle, G. V., Playford, E. D., Brooks, D. J., Quin, N., and Fackowiack, R. S. (1993). Asymmetrical presynaptic and postsynaptic changes in the striatal dopamine projection in DOPA-naïve parkinsonism. *Brain* **116,** 853–867.
184. Carroll, F. I., Lewin, A. H., Boja, J. W., and Kuhar, M. J. (1992) Cocaine receptor: biochemical characterization and structure–activity relationships of cocaine analogs at the dopamine transporter. *J. Med. Chem.* **35,** 969–981.
185. Frost, J. J., Rosier, A. J., Reich, S. G., Smith, J. S., Ehlers, M. D., Snyder, S. H., Ravert, H. T., and Dannals, R. F. (1993) Positron emission tomographic imaging of the dopamine transporter with [^{11}C]WIN-35428 reveals marked declines in mild Parkinson's disease. *Ann. Neurology* **34,** 423–431.
186. Innis, R. B., Seibyl, J. P., Scanley, B. E., Laruelle, M., Abi-Dargham, A., Wallace, E., Baldwin, R. M., Zea-Ponce, Y., Zoghbi, S., Wang, S., Gao, Y., Neumeyer, J. L., Charney, D. S., Hoffer, P. B., and Marek, K. L. (1993) Single photon emission computed tomographic imaging demonstrates loss of striatal monoamine transporters in Parkinson's disease. *Proc. Natl. Acad. Sci. USA* **90,** 11,965–11,969.

187. Farde, L. O., Nordström, A.-L., Wiesel, F.-A., Pauli, S., Halldin, C., and Sedvall, G. (1992) Positron emission tomographic analysis of central D_1 and D_2 receptor occupancy in patients treated with classical neuroleptics and clozapine. *Arch. Gen. Psychiatry* **49,** 538–544.
188. Farde, L. O., Wiesel, F.-A., Halldin, C., and Sedvall, G. (1988) Central D_2-dopamine receptor occupancy in schizophrenic patients treated with antipsychotic drugs. *Arch. Gen. Psychiatry* **45,** 71–76.
189. Smith, M., Wolf, A. P., Brodie, J. D., Arnett, C. D., Barouche, F., Shiue, C.-Y., Fowler, J. S., Russell, J. A., MacGregor, R. R., Wolkin, A., Angrist, B., Rotrosen, J., and Peselow, E. (1988) Serial [^{18}F]N-methylspiroperidol PET studies to measure changes in antipsychotic drug D_2 receptor occupancy in schizophrenic patients. *Biol. Psychiatry* **23,** 653–663.
190. Cambon, H., Baron, J. C., Boulenger, J. P., Loc'h, C., Zarifian, E., and Mazière, B. (1987) In vivo assay for neuroleptic receptor binding in the striatum. *Br. J. Psychiatry* **151,** 824–830.
191. Wolkin, A., Brodie, J. D., Barouche, F., Rotrosen, J., Wolf, A. P., Smith, M., Fowler, J. S., and Cooper, T. B. (1989) Dopamine receptor occupancy and plasma haloperidol levels. *Arch. Gen. Psychiatry* **46,** 482–484.
192. Brücke, T., Roth, J., Podreka, I., Strobl, R., Wegner, S., and Asenbaum, S. (1992) Striatal dopamine D_2-receptor blockade by typical and atypical neuroleptics. *Lancet* **339,** 497.
193. Nordström, A.-L., Farde, L. O., Wiesel, F.-A., Forslund, K., Pauli, S., Halldin, C., and Uppfeldt, G. (1993) Central D_2-dopamine receptor occupancy in relation to antipsychotic drug effects: a double-blind PET study of schizophrenia patients. *Biol. Psychiatry* **33,** 227–235.
194. Nordström, A.-L., Farde, L. O., and Halldin, C. (1992) Time course of D_2-dopamine receptor occupancy examined by PET after single oral doses of haloperidol. *Psychopharmacology* **106,** 433–438.
195. Nordström, A.-L., Farde, L. O., and Halldin, C. (1993) High 5-HT_2 receptor occupancy in clozapine treated patients demonstrated by PET. *Psychopharmacology* **110,** 365–367.
196. Baldessarini, R. J., Cohen, B. M., and Teicher, M. H. (1988) Significance of neuroleptic dose and plasma level in the pharmacological treatment of psychoses. *Arch. Gen. Psychiatry* **45,** 79–91.
197. Baron, J. C., Martinot, J. L., Cambon, H., Boulenger, J. P., Poirir, M. F., Caillard, V., Blin, J., Huret, J. D., Loc'h, C., and Mazière, B. (1989) Striatal dopamine receptor occupancy during and following withdrawal from neuroleptic treatment: correlative evaluation by positron emission tomography and plasma prolactin levels. *Psychopharmacology* **99,** 463–472.
198. Nyberg, S., Farde, L. O., Halldin, C., Dahl, M.-L., and Bertilson, L. (1995) D_2 dopamine receptor occupancy during low-dose treatment with haloperidol decanoate. *Am. J. Psychiatry* **152,** 173–178.
199. Pickar, D., Su, T.-P., Coppola, R., Lee, C. S., Hsiao, J. K., Breier, A., and Weinberger, D. R. (1995) D_2 occupancy and dopamine release determined by ^{123}I-IBZM SPECT following clozapine dose reduction. *Schizophrenia Res.* **15(1–2),** 96.
200. Pilowsky, L. S., Costa, D. C., Ell, P. J., Murray, R. M., Verhoeff, N. P., and Kerwin, R. W. (1992) Clozapine, single photon emission tomography, and the D_2 dopamine receptor blockade hypothesis of schizophrenia. *Lancet* **340,** 199–202.

201. Cohen, B. M., Tsuneizumi, T., Baldessarini, R. J., Campbell, A., and Babb, S. (1992) Differences between antipsychotic drugs in persistence of brain levels and behavioral effects. *Psychopharmacology* **108,** 338–344.
202. Hubbard, J. W., Ganes, D., and Midha, K. K. (1987) Prolonged pharmacologic activity of neuroleptic drugs. *Arch. Gen. Psychiatry* **44,** 99,100.
203. Schotte, A., Janssen, P. F., Gommeren, W., Luyten, W. H., and Leysen, J. E. (1992) Autoradiographic evidence for the occlusion of rat brain dopamine D_3 receptors in vivo. *Eur. J. Pharmacol.* **218,** 373–375.
204. Meltzer, H. Y. (ed.) (1992) *Novel Antipsychotic Drugs.* Raven, New York.
205. Sokoloff, P., Martres, M. P., Giros, B., Bouthenet, M. L., and Schwartz, J.-C. (1992) The third dopamine receptor (D_3) as a novel target for antipsychotics. *Biochem. Pharmacol.* **43,** 659–666.
206. Baldessarini, R. J., Kula, N. S., McGrath, C., Kebabian, J. W., and Neumeyer, J. L. (1993) Isomeric selectivity of D_3 dopamine receptors. *Eur. J. Pharmacol.* **239,** 269,270.
207. Chumpradit, S., Kung, M.-P., and Kung, H. F. (1993) Synthesis and optical resolution of (R) and (S)-*trans*-7-hydroxy-2-[N-propyl-N-(3'-iodo'-2'-propenyl) amino]tetralin: a new D_3 dopamine receptor ligand. *J. Med. Chem.* **36,** 4308–4312.
208. DeWald, H. A., Heffner, T. G., Jaen, J. C., Lustgarten, D. M., McPhail, A. T., Meltzer, L. T., Pugsley, T. A., and Wise, L. D. (1990) Synthesis and dopamine agonist properties of (±)-*trans*-3,4,4a,10b-tetrahyro-4-propyl-2*H*,5*H*-[1]benzopyrano[4,3-*b*]-1,4-oxazin-9-ol and its enantiomers. *J. Med. Chem.* **33,** 445–450.
209. Sokoloff, P., Giros, B., Martres, M. P., Bouthenet, M. L., and Schwartz, J.-C. (1990) Molecular cloning and characterization of a novel dopamine receptor (D_3) as a target for neuroleptics. *Nature* **347,** 146–151.
210. Waters, N., Lagerkvist, S., Löfberg, L., Piercey, M., and Carlsson, A. (1993) The dopamine D_3 receptor and autoreceptor preferring antagonists (+)-AJ76 and (+)-UH232: a microdialysis study. *Eur. J. Pharmacol.* **242,** 151–163.
211. Burris, K. D., Filtz, T. M., Chumpradit, S., Kung, M.-P., Foulon, C., Hensler, J. G., Kung, H., and Molinoff, P. B. (1994) Characterization of [^{125}I]-*trans*-7-hydroxy-2-[(N-propyl-N-3'-iodo-2'-propenyl)amino]tetralin binding to dopamine D_3 receptors in rat olfactory tubercle. *J. Pharmacol. Exp. Ther.* **268,** 935–942.
212. Booth, R. G., Baldessarini, R. J., and Marsh, E. R. (1994) Effects of a D_3-selective aminotetralin on dopamine metabolism in rat brain regions in vitro and in vivo. *Brain Res.* **662,** 283–288.
213. Malmberg, A., Jackson, D. M., Eriksson, A., and Mohell, N. (1993) Unique binding characteristics of antipsychotic agents interacting with human dopamine D_{2A}, D_{2B}, and D_3 receptors. *Mol. Pharmacol.* **43,** 749–754.
214. Lahti, R. A., Evans, D. L., Stratman, N. C., and Figur, L. M. (1993) Dopamine D_4 vs. D_2 receptor selectivity of dopamine receptor antagonists: possible therapeutic implications. *Eur. J. Pharmacol.* **236,** 483–486.
215. Seeman, P. and Van Tol, H. H. (1993) Dopamine D_4 receptors bind inactive (+)-aporphines, suggesting a neuroleptic role: sulpiride not stereoselective. *Eur. J. Pharmacol.* **233,** 173,174.
216. Campbell, A., Yeghiayan, S., Baldessarini, R. J., and Neumeyer, J. L. (1991) Selective antidopaminergic effects of S(+) N-*n*-propylnoraporphines in limbic vs. extrapyramidal sites in rat brain: comparisons with typical and atypical antipsychotic agents. *Psychopharmacology* **103,** 323–329.

217. Kula, N. S., Baldessarini, R. J., Kebabian, J. W., and Neumeyer, J. L. (1994) S(+)aporphines are not selective for human D_3 dopamine receptors. *Cell. Mol. Neurobiology* **14,** 185–191.
218. Baldessarini, R. J., Campbell, A., Ellingboe, J., Zong, R., and Neumeyer, J. L. (1994) Effects of aporphine isomers on rat prolactin. *Neurosci. Lett.* **176,** 269–271.
219. Campbell, A., Baldessarini, R. J., and Neumeyer, J. L. (1993) Altered spontaneous behavior and sensitivity to apomorphine in rats following pretreatment with S(+)-aporphines or fluphenazine. *Psychopharmacology* **111,** 351–358.
220. Baldessarini, R. J. and Tarazi, F. I. (1996) Brain dopamine receptors: a primer on their current status, basic and clinical. *Harvard Rev. Psychiatry* **3,** 301–325.

CHAPTER 16

Dopamine Receptors and Cognitive Function in Nonhuman Primates

Patricia S. Goldman-Rakic, Clare Bergson, Ladislav Mrzljak, and Graham V. Williams

1. Introduction

The long-range goal of relating the neurotransmitter dopamine (DA) to specific receptors and specified cognitive processes could barely have been considered even a few years ago. The cloning of five distinct DA receptors, the development of receptor-specific ligands, the anatomical precision of immunohistochemistry and *in situ* hybridization, and not least, the development of sophisticated behavioral paradigms, are among the major advances that have made understanding DA's role in cognition a reasonable goal. All of these approaches have been applied to the analysis of the anatomical and functional architecture of the DA innervation of the prefrontal cortex in macaque monkeys. The prefrontal cortex as a model system has been a focus of study because of its evolutionary expansion and differentiation in humans and its ample DA innervation in primates in general. Experimental studies in animal models, along with a plethora of neuropsychological studies in clinical populations, suggest that specifiable cognitive processes are impaired when DA transmission is altered in the prefrontal cortex. This chapter primarily reviews findings on the anatomical and receptor mechanisms that could underlie this association in nonhuman primates. New findings indicate that a major action of DA in the prefrontal association cortex is to modulate directly excitatory neurotransmission in pyramidal neurons and regulate their ability to integrate their high level sensory input. As is described in this chapter, these effects appear to be mediated in part via

The Dopamine Receptors Eds.: K. A. Neve and R. L. Neve
Humana Press Inc., Totowa, NJ

D1* receptors in axospinous synapses on pyramidal neurons that are engaged in maintaining information in short-term or working memory. To the extent that the neuronal architecture of prefrontal cortex is repeated in other regions of the cerebral cortex, the mechanisms elucidated may reveal general principles of receptor function in cortical circuits.

2. Overview of the Organization of DA Innervation of Cerebral Cortex

2.1. Light Microscopic Analysis of DA–Fiber Systems in Cerebral Cortex

Views of DA's functions in the cerebral cortex have been influenced in part by our understanding of its anatomical distribution, in part by the actions of drugs on behaviors considered to be mediated by cortical circuitry, and importantly by drug–receptor interactions in diseases like Parkinsonism and schizophrenia. These various avenues of investigation have led to different conceptions and misconceptions about the key neural structures and pharmacological mechanisms that could explain DA's prominent role in a multiplicity of behaviors (for review, *see* ref. *1*). Through the many decades since the discovery of a mesolimbic DA system *(2),* its function has often been conceptualized in global terms—motivation, reward, affect, and/or for translating motivation into action *(1)*. Much of the work directed toward explaining the neural basis of these effects has focused on subcortical structures such as the caudate nucleus and/or nucleus accumbens. Studies of the cortical DA innervation have also focused on elucidating the cortical regulation of subcortical mechanisms *(3–5)*. This is reasonable in view of the greater density of DA fibers in the nigrostriatal pathway compared to that in the cerebral cortex, the clear and powerful role of DA on involuntary motor responses, and the response of these symptoms to L-DOPA therapy. In consequence, DA's influence on specific cortical functions has been much less studied and much less understood.

Nevertheless, the independent role of DA as a neurotransmitter in prefrontal areas has been well appreciated for decades (for review, *see* ref. *6*), and this recognition has spawned a variety of seminal pharmacological and anatomical observations on the cortical DA innervation. Studies of the DA innervation in primate cortex have established that a meshwork of DA fibers is concentrated in the upper cortical layers I, II, and IIIc and in deep layers V and VI of the prefrontal, premotor, motor, and cingulate cortex in macaque monkeys (Fig. 1) *(9)*. Similar findings have been observed in biopsy samples from

*As detailed in the Preface, the terms D1 or D1-like and D2 or D2-like are used to refer to the subfamilies of DA receptors, or used when the genetic subtype (D_1 or D_5; D_2, D_3, or D_4) is uncertain.

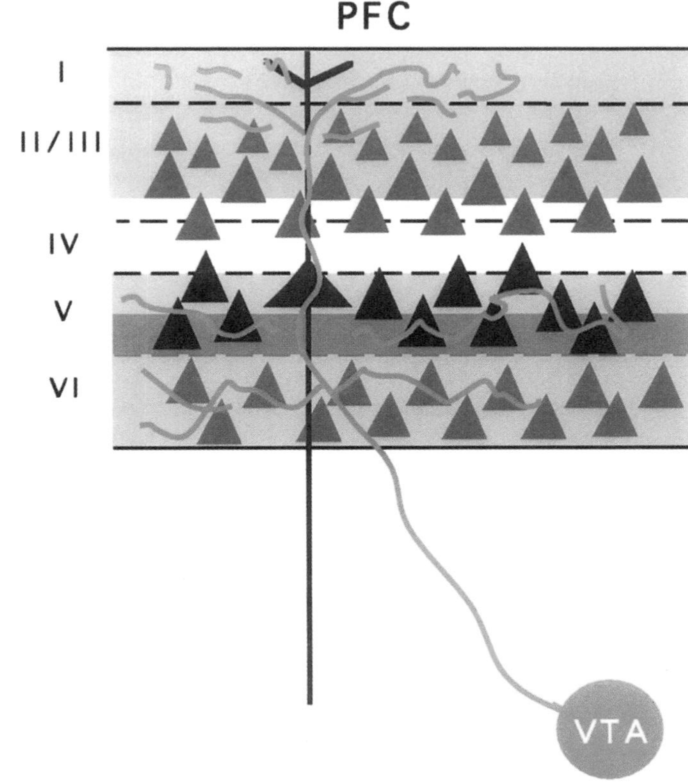

Fig. 1. Diagram of a patch of prefrontal cortex, illustrating the cell layers I–VI and by color code, the predominant pattern of target specificity of pyramidal neurons residing in individual layers. Also indicated (gold wavy lines) is the general distribution of the dopamine meshwork in superficial and deep layers. Light blue cells in layers II/III are cortico–cortical neurons. Dark blue cells in layer V project to striatum, tectum, and spinal cord. Purple cells in layer VI project to the thalamus. The light shade of gray in the supragranular layers I, II, and IIIa as well as infragranular layers V and VI represents the location of highest density of the D1 receptor family as revealed by autoradiographic analysis of SCH 23390 binding in the presence of ketanserin. The darker band of gray through layer V represents D2 binding sites established by labeling with radiolabeled epidepride. Layer V contains a high number of both D1 and D2 receptors, though even in this layer, D2 receptors represent only a small fraction of the combined D1 and D2 complement of proteins. An important difference between the two receptor sites should be noted: The D1 receptor sites were approx 10-fold more abundant than D2 sites. Data on receptors compiled from refs. *7* and *8*. Reprinted with permission from ref. *9a*.

the frontal and/or temporal lobes of Parkinson patients and epilepsy patients *(10)*. Further, receptor autoradiographic studies have localized the D1 and D2 receptor families of DA receptors to precisely the same laminae and regions where the innervation is most prominent *(7,8)*. The matching bilaminar distributions of DA terminals and DA receptors imply some degree of target specificity and indicate that DA is not indiscriminate in its functional or pharmacological consequences.

2.2. Electron Microscopic Evidence of DA Synaptic Triads and D_1 Receptor Localization in Spines

Although the preferential DA innervation of the frontal lobe is generally acknowledged, until quite recently, the integration of DA axons in cortical circuitry was largely obscure. Electronmicroscopic studies have now shown that DA-immunopositive terminals selectively target the pyramidal neurons of the cortex *(11–14)*. In the primate, DA axon terminals in prefrontal areas and elsewhere have been found to form symmetrical synapses indistinguishable from traditional Gray Type II synaptic specializations *(10,11,15,16)*. The DA terminals in the prefrontal cortex are largely found on the distal dendrites and spines of pyramidal cells in both monkeys *(11,15)* and humans *(10)*. In rodents, where DA synapses are also symmetrical, there seem to be fewer of them on spines relative to dendrites and soma *(12,13,14)*, whereas in humans, we have estimated that about 60% of DA synapses are axospinous. As illustrated in Fig. 2, a percentage of spines on pyramidal neurons are invariably the targets of both DA afferents and another unspecified asymmetrical (presumed excitatory) synapse *(10,11,15)*, an arrangement referred to as a synaptic triad because it involves three circuit elements—two afferents and a target neuron. Similar synaptic triads are found in prefrontal, premotor, and motor cortex, suggesting that this anatomical arrangement may be widespread and common to many cortical areas. They have been observed in rats *(14)*, monkeys *(11,15)*, and humans *(10)*. Synaptic triads in which the dendritic spines of medium spiny neurons receive both a DA- or tyrosine hydroxylase-positive bouton and an asymmetrical synaptic specialization are not unique to the cortex, however. A similar arrangement has previously been described for the medium spiny cell in the rodent neostriatum *(17)*. In both the cortex and striatum, the DA terminal endings are strategically placed within the local circuitry to alter local spine responses to excitatory inputs, and through these, to make a contribution to the overall excitability of the projection neurons in both structures. Most recently, as is described in Section 3.2., the spines and distal dendrites of cortical pyramidal cells have been shown to harbor D_1 receptors *(16,18)*. Many of these D_1 positive spines are more often observed in contact with asymmetric excitatory terminals and less often in contact with symmetric

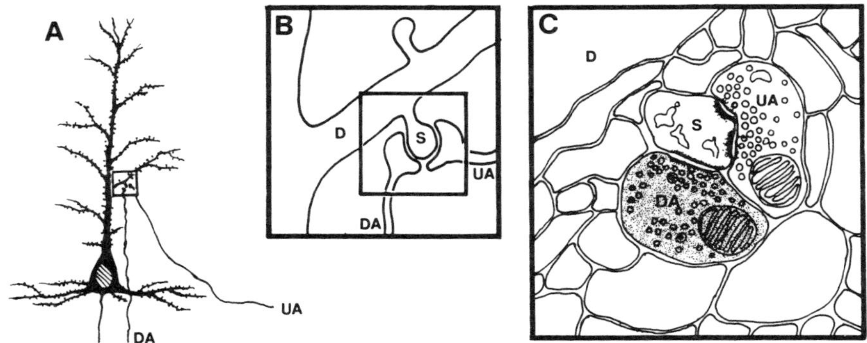

Fig. 2. Diagram of synaptic arrangements involving the dopamine input to the cortex. **(A)** Afferents labeled with a DA-specific antibody terminate on the spine(s) of a pyramidal cell in the prefrontal cortex, together with an unidentified axon (UA); **(B)** enlargement of axospinous synapses illustrated in (A) showing opposition of the DA input and an unidentified excitatory afferent (UA) that makes an asymmetrical synapse on the same dendritic (D) spine; **(C)** diagram of ultrastructural features of the axospinous synapses illustrated in (B); the DA terminal (darkened profile representing DA immunoreactivity) forms a symmetrical synapse; the unidentified profile forms an asymmetrical synapse with the postsynaptic membrane. Reprinted with permission from ref. *9a*.

dopaminergic afferents. These findings suggest that there may be both synaptic and nonsynaptic modes of D_1 modulation of excitatory transmission.

3. Localization of DA Receptor Subtypes

3.1. Subtype-Specific Antibody Probes for DA Receptors

In recent years, the dopaminergic D1-specific ligand, SCH 23390, and the D2-specific antagonist, raclopride, have been used to label D1 and D2 receptor families in the prefrontal cortex of both monkeys and humans. Quantitative autoradiographic studies have established that the combined distribution of D1 and D2 receptors generally match the concentration of afferents on a layer by layer basis. Importantly, however, the number of D1 DA binding sites in the primate prefrontal cortex far outnumber the number of D2 binding sites *(7,8,19)*.

Expression studies of the five cloned mammalian DA receptor subtypes (D_1–D_5) and their variants (arising from alternative splicing and genomic polymorphisms) (reviewed in ref. *20*) have made it clear that the existing repertoire of dopaminergic ligands cannot distinguish D_1 from D_5 receptor sites nor D_2 from D_3 and D_4 receptor sites. Knowledge of the receptors' nucle-

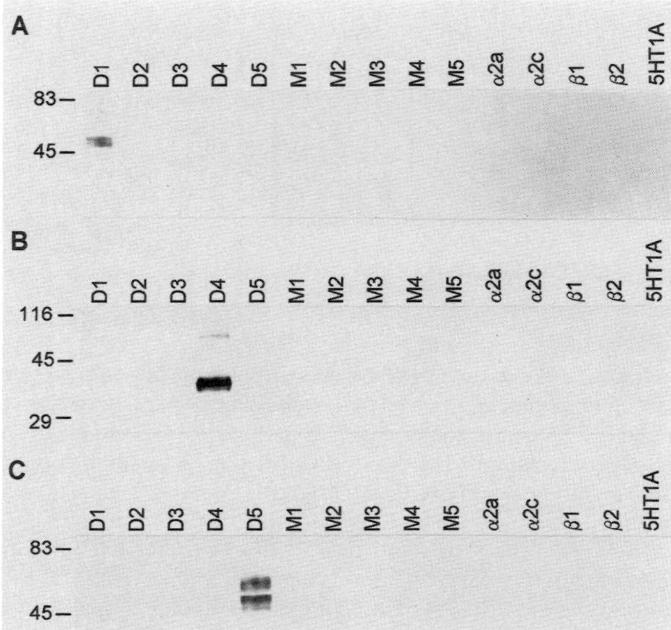

Fig. 3. Subtype specificity of D_1, D_4, and D_5 antibodies. D_1 **(A),** D_4 **(B),** or D_5 **(C)** antibodies reacted with immunoblots of crude membrane fractions (5 μg/lane) prepared from recombinant Sf9 cells expressing h (human) D_1, hD_2, rat D_3, $hD_{4.2}$, and hD_5 dopaminergic; hM1, hM2, hM3, hM4, and M5 muscarinic, hα2a, hα2c, hβ1, and hβ2 adrenergic; and $h5HT_{1A}$ serotonergic receptors. Proteins were fractionated on SDS-containing 10% (A,C) or 12.5% (B) polyacrylamide gels, and transferred to polyvinylidene difluoride filters. The positions of molecular mass markers in kilodaltons are indicated on the left.

otide and amino acid sequences, however, makes it possible to design subtype specific antisense RNA probes and fusion proteins for antibody production. For example, although the D_1 and D_5 receptor subtypes exhibit >80% amino acid sequence identity in transmembrane segments, we *(18,21)* and others *(22,23)* have taken advantage of the sequence divergence present in the C-terminal intracellular segment of the respective polypeptides to design recombinant fusion proteins that have led to the production of D_1 and D_5 subtype-specific antibodies (Fig. 3A,C). We have taken a similar approach with sequences in the N-terminal segment of the D_4 receptor (Fig. 3B) *(24),* and in the third cytoplasmic loop of the D_3 receptor variant, D_{3nf} *(25).*

With these antibody probes in hand, it has been possible for us to localize the D_1, D_{3nf}, D_4, and D_5 DA receptor proteins at the light and electron microscopic level using standard immunohistochemical techniques. We review our findings on DA receptor localization in the primate prefrontal cortex in the remaining portion of this section. In Section 4., we discuss how these anatomical studies bear on recent cognitive findings.

3.2. Localization of the D1 Family of DA Receptors in Prefrontal Cortex

Although less abundant than in striatum, D_1 receptor expression is readily detectable in the primate prefrontal cortex *(16,18)*. D_1-immunoreactive neurons are visible in all cortical layers, but particularly in layers II, III, and V. Labeling with D_1 antibodies is predominantly associated with pyramidal neurons. In addition to pyramidal cell body and apical dendrite labeling, heavy labeling of neuropil in layers Ib–II and V–VI is also observed. The bilaminar appearance of the neuropil labeling is similar to the distribution of D1-like binding sites as determined by autoradiography *(7,8)*.

Electronmicroscopic analysis reveals both presynaptic and postsynaptic localization of D_1 receptors in the prefrontal cortex (area 46), although postsynaptic localization is more frequently observed. The dominant ultrastructural feature of D_1 antibody staining is labeling of spines (Fig. 4). Approximately 20% of the total number of spines counted (147 of 735) in micrographs of layer III, area 46, are labeled by D_1 antibodies. Within the labeled spines, D_1 receptor immunoreactivity is concentrated in the neck and head of dendritic spines and typically is slightly displaced from the asymmetric synapse or completely fills the spine head. Tyrosine hydroxylase (TH) and D_1 receptor double-labeling studies reveal that D_1-immunoreactive spines are often in close proximity to TH-positive axons *(16)*. When localized presynaptically, D_1 receptor immunoreactivity appears as small patches of reaction product in axon terminals. These terminals often form asymmetric synaptic specializations and the immunoreaction product is most often found at a distance from the synaptic specialization.

3.3. Localization of the D_5 Receptor

In contrast to the D_1 receptor, the D_5 receptor appears to be more abundant in cortical areas than in subcortical areas including striatum. Light microscopic analysis of D_5 antibody staining of cortex reveals expression of D_5 receptors in pyramidal neuronal populations in the prefrontal cortex as well as in all other neo-, meso-, and archicortical areas of monkey brain examined. D_5 antibodies predominantly label the soma and apical dendrites of pyramidal neurons (Fig. 5). The only apparent difference between the D_5 and D_1 antibody staining pattern is a more prominent staining of dendrites with D_5 receptor

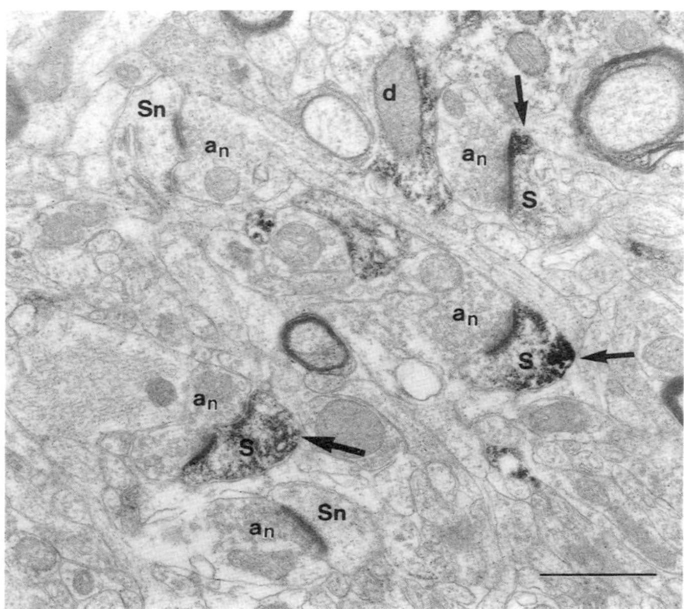

Fig. 4. Postsynaptic localization of the D_1 receptor in macaque prefrontal cortex. Note labeling of multiple dendritic spines (s) receiving asymmetric input from unlabeled axons (a_n). Arrows point to immunoreaction product near or slightly displaced from the synapses. d, labeled dendrite; s_n, unlabeled spine. Bar = 0.5 mm. Reprinted with permission from ref. *18*.

antibodies. In the prefrontal (Walker areas 9–12 and 46) cortex, D_5-immunoreactive neurons are visible in all cortical layers. However, labeled cells are most prominent in layers II, III, and V. The similarity of the staining patterns in cortex obtained with the D_1 and D_5 receptor antibodies raised the possibility that the two receptors may be coexpressed within pyramidal neurons. This was confirmed by double-label immunofluorescence microscopy (Fig. 4). The vast majority of D_5-labeled neurons also contain D_1 receptors, but not all D_1-labeled cells appear to contain D_5 receptors.

At the electronmicroscopic level, D_5 immunoreactivity in the prefrontal cortex is predominantly localized in dendritic shafts, and to a lesser extent, in dendritic spines. By comparison with antibodies to D_1, D_5 receptor antibodies label fewer spines (37 of 796, 4.6%) in layer III of area 46 of the prefrontal cortex. The different proportions of D_1- vs D_5-labeled spines in the cortex is highly significant ($\chi^2 = 85.17$, $p = .0001$). Like the D_1-labeled spines, D_5 receptor-containing spines are nearly always targets of asymmet-

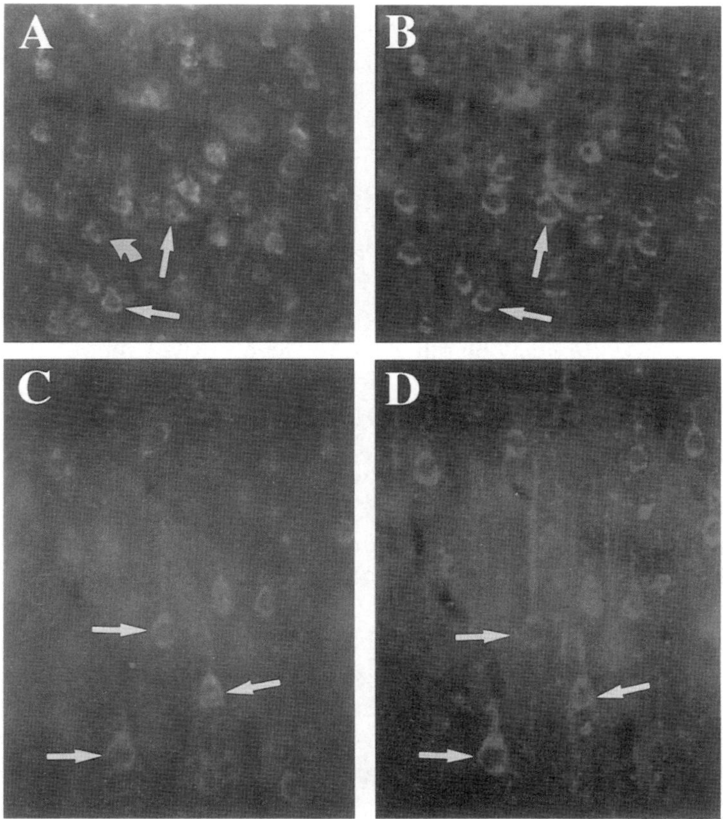

Fig. 5. Double labeling immunofluorescence of D_1 **(A,C)** and D_5 **(B,D)**. (A,B) D_1 and D_5 receptor in layer III of area 46. The two receptors are colocalized in numerous pyramidal cells (thin arrows); D_1 receptor-positive and D_5-negative neurons are also visible (thick arrow in A); (C,D) D_1 and D_5 receptor in deep third of layer III and in layer V in area 6. Two of the double-labeled pyramidal neurons are labeled with thin arrows.

ric synaptic input. The immunoreaction product in D_5-positive spines is either distant from the synaptic specialization or diffusely distributed in the spine. D_5 antibody staining of dendritic shafts is either distributed throughout the cytoplasm, or found in patches. In contrast to the D_1 receptor labeling, presynaptic D_5 antibody labeling is observed not only in terminals, but also in the initial axonal segments of pyramidal neurons. D_5 receptors are also present in axon terminals forming both symmetric and asymmetric synapses.

As an evolutionary aside, we *(16,18)* observe significantly greater D_1 receptor immunoreactivity in the primate prefrontal cortex compared to the rodent cortex (*23*; Bergson et al., unpublished observations). We also observe an abundance of D_5 receptor immunoreactivity in the macaque cerebral cortex compared to rodent cortex, as has been reported for D_5 mRNA expression *(27)*. Interestingly, phylogenic differences in DA innervation of the cerebral cortex have also been observed (reviewed in ref. *28*), with primates and humans receiving much more extensive dopaminergic input than rodents. The higher levels of the D_1 and D_5 receptor expression and DA innervation in monkey compared to rodent cortex raise the possibility that levels of DA receptor expression in cortex may be related to the levels of DA input.

3.4. Localization of the D_{3nf} and D_4 Members of the D2 Family of DA Receptors in Prefrontal Cortex

The D_{3nf} DA receptor appears to be expressed by primates, but not rodents. In humans, D_{3nf} presumably arises from a splicing event that deletes 98 nucleotides of the D_3 receptor coding sequence and also shifts the D_3 receptor's open reading frame *(29)*. Consequently, the primary sequence of the D_{3nf} polypeptide is expected to be identical to that of the full length D_3 receptor from its N terminus through the first five transmembrane segments and 76 amino acids of the predicted third cytoplasmic loop. Beyond this point, however, the D_3 and D_{3nf} proteins share no sequence homology. Furthermore, protein secondary structure algorithms predict that the novel C-terminal 55 amino acids of the D_{3nf} protein are unlikely to form a sixth and seventh transmembrane segment. Studies with other members of the seven-transmembrane-segment receptor superfamily suggest that residues in these membrane-bound segments, as well as in the corresponding intracellular domains, play a role in ligand binding and/or G protein coupling *(30–32)*. Therefore, it is unclear what role the D_{3nf} polypeptide may play in vivo. Nevertheless, the D_{3nf} protein looms as a potentially intriguing link in the proposed role of DA in schizophrenia, as levels of D_{3nf} transcripts are maintained in postmortem neocortical tissues obtained from schizophrenic patients, whereas those for D_3 are dramatically reduced relative to the levels observed in tissue from unaffected individuals *(29)*.

We immunized rabbits with a TrpE fusion protein containing the 55 C-terminal amino acids unique to D_{3nf} *(25)*. D_{3nf} rabbit antiserum prominently labels apical dendrites in all layers of the macaque prefrontal cortex, whereas preimmune serum does not label these elements. In contrast to D_1 and D_5 antibody labeling of pyramidal neurons in the prefrontal cortex, the triangular-shaped perikarya from which the D_{3nf} immunoreactive dendrites

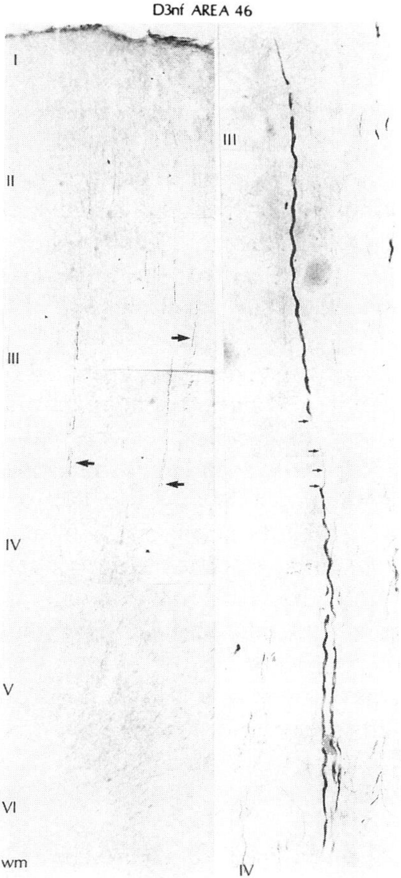

Fig. 6. Immunohistochemical staining of D_{3nf} polypeptide in area 46 of the macaque neocortex. Neurons and apical dendrites in layers II, III, and V are the most prominently stained. Note D_{3nf}-labeled dendrites are often detected in bundles (arrows).

presumably originate are only faintly labeled (Fig. 6). The densest packing of labeled dendrites is observed in layers I–III and V and VI. However, dendrites in layer IV exhibit the most intense staining and are often present in irregularly spaced bundles. Some of these boldly stained, relatively straight dendrites project across several cortical layers. In contrast, the labeled dendrites in the upper layers are shorter and more frequently branched. The subcellular localization of the D_{3nf} polypeptide has not yet been determined.

3.5. Localization of the D_4 Receptor

The D_4 receptor is thought to be one of the major targets of the atypical neuroleptic, clozapine, and is possibly the receptor most important for its beneficial effects on negative symptoms *(33)*. We have analyzed the DA D_4 receptor distribution in the macaque monkey cerebral cortex, hippocampus, and subcortical structures at both light and electron microscopic levels with D_4 receptor-specific antibodies *(24)*. The D_4 antibody labeling pattern is similar to the laminar distribution of epidepride binding sites in the primate cortex *(7,8)*, in that D_4 antibodies label neurons in infragranular layers of the prefrontal cortex more frequently than they label neurons in supragranular layers. Like D_1 and D_5 receptors, D_4 receptors localize to pyramidal neurons in cerebral cortex. However, in contrast to the D_1 and D_5 receptors, prominent D_4 staining is observed in a considerable population of nonpyramidal neurons (presumably GABAergic local circuit neurons) in all layers of the cerebral cortex. The localization of D_4 receptors to interneurons is even more striking in the hippocampus. Numerous nonpyramidal neurons in the stratum oriens and hilar area of the hippocampus exhibit the strongest D_4 immunoreactivity, whereas labeling of pyramidal cells in the CA1 field is at background levels (Fig. 7). In certain subcortical structures such as the globus pallidus and the pars reticulata of the substantia nigra, neurons with morphologies characteristic of GABAergic neurons are also intensely labeled by D_4 antibodies. At the electron microscopic level, the D_4 receptor is also frequently observed on postsynaptic structures apposed to asymmetric synapses. Like the D1 family of receptors, the D_4 receptor appears to modulate excitatory transmission via volume transmission of DA *(34)*.

4. The Role of D1 Receptors in Mnemonic Functions

4.1. Parkinsonism, Schizophrenia, and Working Memory

The DA innervation of the prefrontal cortex in rodents and monkeys has been known for decades but, as mentioned, its precise role in this and other cortical areas has remained elusive. Cognitive deficits are well-documented in many Parkinsonian patients *(35–37)* and in schizophrenic patients *(38–40)*, though these are usually overshadowed by problems of motor control in the former, and positive (hallucinations, delusions) or negative symptoms (e.g., anhedonia) in the latter group. The cognitive deficits presented by these patients on a wide range of neuropsychological tests appear to involve impairments in working memory and consequently in their ability to use internalized instructions or cues to guide their behavioral responses, much as is the case with patients that have sustained prefrontal cortical damage *(41,42)*, and monkeys with direct surgical excisions of the prefrontal cortex *(43,44)* or

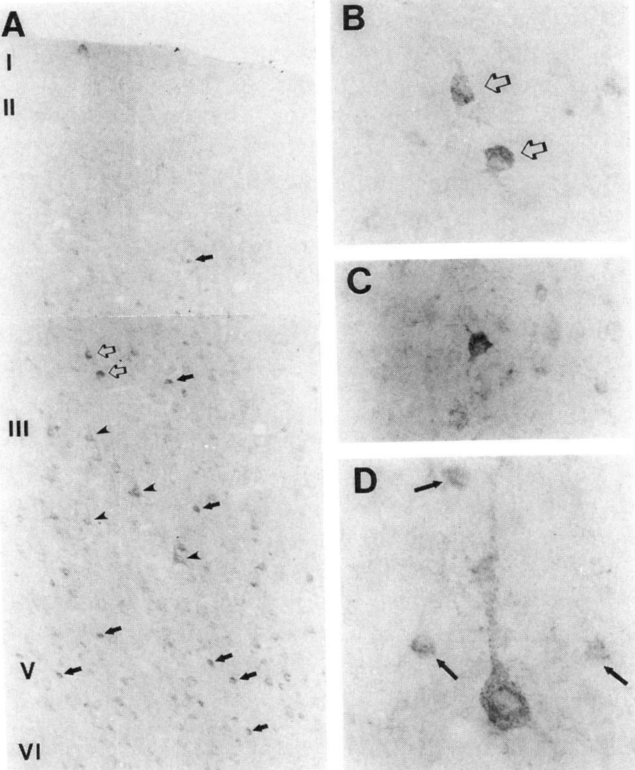

Fig. 7. **(A–D)** D_4 receptor immunoreactivity in the macaque monkey frontal lobe (Walker's area 6). Both pyramidal (arrowheads) and nonpyramidal neurons (arrows) exhibit D_4 immunoreactivity. Open arrows point to intensely labeled nonpyramidal neurons shown at higher magnification in (B). (C) Strongly labeled nonpyramidal neuron in layer V of area 9. (D) D_4 receptor-labeled pyramidal neuron. Note granular appearance of immunoreaction product in the cell body and dendrites. Arrows point to neighboring D_4 positive nonpyramidal neurons.

experimental localized depletion of neurotransmitter *(45)*. The working memory deficits in schizophrenics and Parkinson patient groups add support to the view that a highly specific cognitive process is vulnerable to the influence of DA dysfunction and that a common process may underlie the cognitive loss in these diseases and frontal lobe dysfunction. This process has been characterized and widely studied in humans in the context of linguistic processing *(46,47)*, and its core features have been studied in the nonhuman primate by the use of a variety of delay-response tasks *(44,48)*.

Treatment of Parkinson's disease and schizophrenia, in particular, has emphasized the significance of the D2 receptors in drug medication *(49–51)* as well as in the pathophysiology of the disease process (e.g., *51*). The effectiveness of the atypical neuroleptic, clozapine, with its low affinity for D_2 receptors, as well as several findings in experimental animals, including our own, which have implicated the D1 family of receptors in the working memory process (*see* Section 4.2.), have opened the door to a more inclusive view of receptor mechanisms in mental illnesses and cognitive dysfunction.

4.2. In Vivo Examination of the D1 and D2 Receptors on the Memory Fields of Prefrontal Neurons

Studies combining single cell recording with iontophoretic application of drugs in trained monkeys have begun to identify and isolate the effects of selective DA receptor antagonists on the dissociable mnemonic and sensorimotor processes displayed by prefrontal neurons that increase their firing when specific stimuli are removed from view and sustain that firing until a response is initiated *(52–54)*. These "memory" neurons are considered the putative cellular basis of working memory *(43,48,54)*. The question addressed by the iontophoretic application of drug was whether and how their neuronal firing is altered by blocking specific receptors.

Recordings from a prefrontal neuron from a monkey performing a delayed-response task are shown in Fig. 8. To obtain such a record, monkeys have to be trained to fixate a central spot on a TV monitor, while stimuli (targets) are briefly presented in different locations of the visual field. On any given trial, the target is extinguished after 0.5 s and a delay lasting 3 s follows. At the end of the delay, the extinction of the fixation spot signals the animal to direct its gaze to the location where the target had been presented seconds before. As the position of the target varies randomly from trial to trial, the monkey has to continually update its memory of target position. The neuron shown in Fig. 8 is consistently activated only during the delay period of all trials in which the monkey is required to hold "on line" the memory of a particular location, for example, the target at position #2. This selective pattern of firing has been termed the "memory field" of the neuron, consistent with the role of prefrontal cortex in memory guided behavior.

The specific influence of D1 and D2 receptors on memory fields has now been examined with specific antagonists; and the results have provided direct evidence that DA in part regulates the cellular basis of working memory via D1 receptors. In a recent study, both D1 and D2 antagonists were at different times iontophoresed onto neurons with spatially tuned "memory fields"; that is, on neurons that responded maximally during the delay period for targets

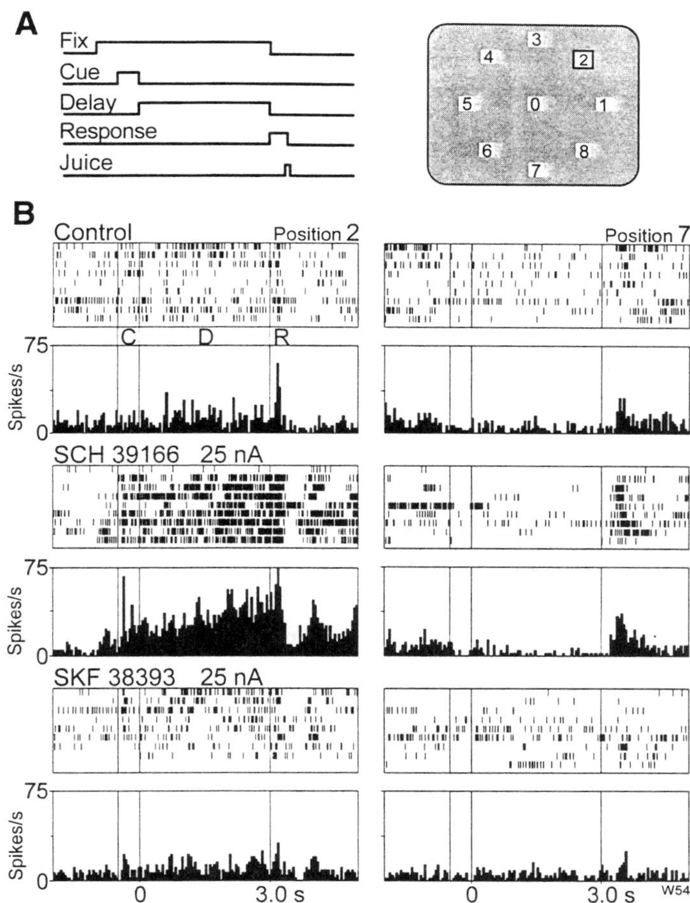

Fig. 8. Single-cell recording from a prefrontal neuron during performance of a working memory task under various control and drug conditions. **(A)** Diagram of the task events and their temporal relationships; **(B)** rasters and histograms showing the neuron's response during the cue (C), delay (D), and response periods of a delayed response trial. Panels on the left represent activity recorded whenever the animal had to remember a target presented at position 2 (see upper right-hand corner). Panels on the right represent the neuron's activity whenever the monkey had to recall the target at position 7. The top row are recordings obtained under normal conditions without drug. The middle row shows recordings from the same neuron obtained with the D1 receptor antagonist SCH 39166. The bottom row depicts neuronal activity in the same neuron when a DA agonist, SKF 38393 was applied. See text for further details; physiology from ref. *55*. Reprinted with permission from ref. *55*.

in one or a few adjacent target locations and not above baseline for any other location of a delayed response task *(55)*. Iontophoresis of the D1-selective antagonist, SCH 39166 *(56,57)*, consistently enhanced the memory fields of prefrontal neurons when applied at low ejection currents (20–25 nA). For example, as shown in Fig. 8, neuron W54 displayed an increase in activity for target position #2 (top left) but not for other target positions, including position #7 (top right). Delay activity (and the phasic cue and response period activity occurring in conjunction with it) was enhanced by SCH 39166 for target #2 (Fig. 8B, right) and not for any target locations outside the memory field (Fig. 8B, left). Indeed, SCH 39166 accentuated the reduction in activity during the delay period which often occurs on trials with targets in the location opposite to the memory field (position #7, Fig. 8B, middle panel, right). Finally, the enhanced activation for target #2 during the delay could be reversed by iontophoresis of the partial D1 agonist SKF 38393 onto the cell; and the reversal was also selective to the neuron's memory field (Fig. 8B). The diminution of cell firing consequent to D1 stimulation is in keeping with DA's reported inhibitory actions in rodent cortex *(58–60)* and indicates that the D1 receptor plays a role in DA-stimulated inhibition.

The effects described were specific to prefrontal neurons with memory-related activation and memory fields *(55)*. Doses of SCH 39166, which enhanced delay activity in neurons with sustained delay-period activity, produced no increase in neurons with phasic cue- or response-related neuronal activity, as might be expected from the well-established sparing of sensory-guided performance in various tasks after surgical and reversible lesions of prefrontal cortex (reviewed in ref. *44*).

The modulation of delay-period activation by the D1 antagonist appears to be pharmacologically specific. Iontophoretic application of raclopride, a D_2/D_3 antagonist, produced only a generalized inhibition of neuronal responsivity in all phases of the task, including the delay period. On the other hand, until D_1- and D_5-specific ligands are developed, the relative contributions of the D_1 vs D_5 receptor to prefrontal cellular function cannot be separated. As discussed, both receptors are present in pyramidal cells and in spines, though the incidence of D_1 receptors in spine processes is higher. Both receptors are therefore associated with terminals of specific afferents associated with glutamatergic transmission via *N*-methyl-D-aspartate (NMDA) and non-NMDA receptors. Recently, it has become evident that spines are important sites of calcium influx during electrical stimulations *(61)*; theoretical considerations suggest that spines provide a "microenvironment for calcium and other second messengers" *(62)*. These characteristics support our proposal that the synaptic triads provide a unique microcircuit for mediating spatially selective memory processes and the uniquely precise dopaminergic modulation that regulates a specific component of the neuron's afferent excitatory input without affecting its general excitability.

The finding that D1 receptor blockade could enhance the neuronal signal in the delay period provides evidence of a memory-enhancing action at select doses. Yet previous studies from this laboratory have indicated that D1 blockade is detrimental to working memory performance *(63,64)*. The explanation for these discrepancies may be related to dosage *(55)*. Presumably, high doses of drug more completely block DA's synaptic actions and produce effects equivalent to DA depletion, as in the experimental and clinical studies described in Section 4.1. The impact of dose may explain the deficits observed when D1 antagonists are injected locally or systemically at higher doses. Current evidence in both humans *(65)* and monkeys *(66)* has documented the detrimental effects of excess DA on cognitive process, including working memory.

4.3. *Mechanisms of DA Modulation*

Extracellular recordings in rat have consistently shown that DA inhibits spontaneous *(58)* and evoked *(59,67)* activity in the prefrontal cortex. This inhibition has generally been interpreted in terms of a DA action on cortical interneurons *(68,69)*. However, the D1 effects on memory fields described in Section 4.2., coupled with anatomical localization of the D1 receptor in pyramidal neurons, suggest that DA may modulate pyramidal cell firing directly, via D1 receptors. This direct mechanism thus provides an alternate and complementary mode of control of cell firing in prefrontal cortex and perhaps one of particular importance in cognitive processing. Similar results on D1 modulation of excitatory transmission have recently been obtained in rat cortical slices. Law-Tho et al. *(60)* showed that bath application of DA attenuates both NMDA- and non-NMDA-mediated excitatory postsynaptic potential (EPSPs) and that this response can be blocked by D1 but not D2 antagonists. Pralong and Jones *(70)* obtained similar results in the rat entorhinal cortex. These results and the in vivo data in rhesus monkey prefrontal cortex are mutually supportive. Both show that pyramidal cell firing can be directly inhibited by modulation of excitatory synapses and both show that this effect can be reversed by D1 receptor blockade. In the behaving monkey, the removal of DA-mediated inhibition by D1 blockade results in an accentuation of the neuron's response to its excitatory input, presumably at spine triads *(55)*. We have hypothesized that this effect is mediated in prefrontal neurons via reversal of the attenuating effects of D1 stimulation on the slow inward sodium currents in DA receptive neurons as described by Calabresi et al. *(71)*.

The view that DA produces inhibition of cortical neurons does not explain, however, the fact that too little DA depresses neuronal activity *(55)* and leads to behavioral impairments. Experimental depletion of DA in prefrontal cortex by intracerebral injection of 6-OHDA results in deficits similar to that produced by direct ablation of this area, and high doses of D1 antagonists injected either locally or systemically clearly impair delayed-response performance. Thus, a certain level of DA appears to be essential for the cellular mechanism of working memory and somehow DA must facilitate synaptic input or neuronal output in prefrontal cortex. Several lines of evidence support this enabling function of DA. First, it has been shown that under certain conditions, DA can dramatically facilitate the NMDA component of excitatory synaptic input to pyramidal neurons in slices of human cortex *(72)*. Second, DA can powerfully enhance the frequency and duration of cell firing, in response to continued suprathreshold depolarization, by attenuating the calcium-activated potassium conductance of the neuron which is normally responsible for limiting action potential generation *(73)*. Both these facilitatory actions of DA are probably D1 receptor-dependent. Finally, we have explicated the importance of DA as an enabling neurotransmitter by showing that iontophoresis of D1 antagonists at levels that probably produce dramatic reduction in receptor occupancy leads to almost complete cessation in the firing of prefrontal neurons engaged in working memory function *(55)*. The anatomical localization of D1 receptors to the spines of pyramidal neurons, which also receive asymmetric inputs, may provide the essential substrate for this neuromodulatory action. What is clear, however, is that DA's actions in the cortex vary depending on its prevailing concentration and its specific site of action; for working memory, there may be an optimal level of D1 receptor occupancy.

5. Future Directions

The findings described in this chapter raise many questions for future study, regarding not only the interaction of DA and glutamate systems, but also their interaction with GABAergic neurons in the prefrontal cortex. As shown, the D_4 receptor has been localized to an as yet undefined subpopulation of GABA cells in the primate prefrontal cortex and it is tempting to speculate that D2 effects on cognitive behavior may be mediated via stimulation of interneurons. Other neurotransmitter systems such as the cholinergic system have also been explored *(74)* and must be fit into the developing picture of cortical circuitry and cortical function.

Even though the cortical DA innervation seems negligible compared to the levels in the striatum, anatomical and receptor-binding data in the prefrontal cortex have provided a picture of the intricate manner in which cortical circuits can be influenced by DA. If 60% of DA synapses are onto spines, as studies in our laboratory have estimated *(15)*, and a given neuron receives hundreds of DA synapses, the effect may be powerful. It has recently been shown by computational and mathematical modeling that small changes in the responsivity of a network of individual elements can improve signal detection in the network as a whole, although the responsivity of any individual element may be unaltered *(75)*.

Several considerations have led investigators to expect an effect of DA on GABA interneurons in cortex, because iontophoretic application of DA can produce inhibition of spontaneous firing *(68,69)*. So far, however, it appears that DA synapses on GABAergic neurons in prefrontal cortex are less frequent than those on pyramidal, excitatory neurons. The triadic arrangement of DA synapses opposed to excitatory synapses on the same spine provides a subtle way of influencing the degree and functional polarity of cell firing. Inhibitory effects could be produced indirectly by dampening a cell's response to its excitatory inputs rather than directly by feedforward inhibition. Probably both direct and indirect mechanisms obtain, however. Cortical neurons in vitro show multiphasic responses *(60)*, as do prefrontal neurons recorded in monkeys during memory tasks *(53,54)*; and for which all phases of the neuronal response need to be accounted. Our finding of higher density of D1 receptors in prefrontal cortex draws attention to the potential functional significance of these receptors for cognitive processes in normal individuals and for deficits such as the thought fragmentation characteristic of diseases like schizophrenia.

An important quest for the future will be to learn which specific cells and which specific receptors are the targets of most of the DA terminals, and whether these target cells have special physiological characteristics and influence specific intrinsic or extrinsic circuits. For example, neurons that discharge tonically during mnemonic processing and thus hold on-line information that would otherwise be lost to conscious consideration may bear one receptor subtype, whereas those that discharge phasically after a response is made may bear another subtype. If these neurons preferentially express specific receptor subtypes, then understanding the cellular basis of short-term memory would become a more realistic goal and may lead to therapeutic advances in the treatment of neuropathological conditions.

Acknowledgments

This work was supported by grants MH38546, MH00298, MH44866 from the National Institute of Mental Health to P. S. Goldman-Rakic; National Association for Research in Schizophrenia and Depression (NARSAD) Young Investigator Award to C. Bergson and L. Mrzljak.

References

1. Willner, P. and Scheel-Kruger, J. (1991) *The Mesolimbic Dopamine System: From Motivation to Action.* Wiley, New York.
2. Anden, N.-E., Dahlstrom, A., Fuxe, K., Larsson, K., Olson, L., and Ungerstedt, U. (1966) Ascending monoamine neurons to the telencephalon and diencephalon. *Acta Physiol. Scand.* **67,** 313–326.
3. Grace, A. A. (1991) Phasic vs. tonic dopamine release and modulation of dopamine system responsivity: a hypothesis for the etiology of schizophrenia. *Neuroscience* **41,** 1–24.
4. Robbins, T. W. (1990) The frontostriatal dysfunction in schizophrenia. *Schizophrenia Bull.* 6, 391–402.
5. Weinberger, D. R., Berman, K. F., and Illowsky, B. P. (1988) Physiological dysfunction of dorsolateral prefrontal cortex in schizophrenia iii. A new cohort and evidence for a monoaminergic mechanism. *Arch. Gen. Psychiatry* **49,** 609–615.
6. Thierry, A. M., Tassin, J. P., Blanc, G., Stinus, L., Scatton, B., and Glowinski, J. (1977) Discovery of the mesocortical dopaminergic system: some pharmacological and functional characteristics, in *Advances in Biochemical Psychopharmacology, Vol. 16* (Costa, E. and Gessa, G. L., eds.), Raven, New York, pp. 5–12.
7. Lidow, M. S., Goldman-Rakic, P. S., Gallager, D. W., and Rakic, P. (1991) Distribution of dopaminergic receptors in the primate cerebral cortex: quantitative autoradiographic analysis using [^{3}H]raclopride, [^{3}H]spiperone and [^{3}H]SCH23390. *Neuroscience* **40,** 657–671.
8. Goldman-Rakic, P. S., Lidow, M. S., and Gallager, D. W. (1990) Overlap of dopaminergic, adrenergic, and serotoninergic receptors and complementarity of their subtypes in primate prefrontal cortex. *J. Neurosci.* **10,** 2125–2138.
9. Williams, S. M. and Goldman-Rakic, P. S. (1993) Characterization of the dopaminergic innervation of the primate frontal cortex using a dopamine-specific antibody. *Cereb. Cortex* **3,** 199–222.

9a. Goldman-Rakic, P. S. (1992) Dopamine-mediated mechanisms of the prefrontal cortex. *Sem. Neurosci.* **4,** 149–159.

10. Smiley, J. F., Williams, S. M., Szigeti, K., and Goldman-Rakic, P. S. (1992) Light and electron microscopic characterization of dopamine-immunoreactive processes in human cerebral cortex. *J. Comp. Neurol.* **321,** 325–335.
11. Goldman-Rakic, P. S., Leranth, C., Williams, M. S., Mons, N., and Geffard, M. (1989) Dopamine synaptic complex with pyramidal neurons in primate cerebral cortex. *Proc. Natl. Acad. Sci. USA* **86,** 9015–9019.
12. Seguela, P., Watkins, K. C., and Descarries, L. (1988) Ultrastructural features of dopamine axon terminals in the anteromedial and suprarhinal cortex of adult rat. *Brain Res.* **442,** 11–22.
13. Van Eden, C. G., Hoorneman, E. M. D., Buijs, R. M., Matthissen, M. A. H., Geffard, M., and Uylings, H. B. M. (1987) Immunocytochemical localization of dopamine in the prefrontal cortex of the rat at the light and electron microscopical level. *Neuroscience* **22,** 849–862.
14. Verney, C., Alvarez, C., Geffard, M., and Berger, B. (1990) Ultrastructural double-labelling study of dopamine terminals and GABA-containing neurons in rat anteromedial cerebral cortex. *Eur. J. Neurosci.* **2,** 960–972.

15. Smiley, J. F. and Goldman-Rakic, P. S. (1993) Heterogeneous targets of dopamine synapses in monkey prefrontal cortex demonstrated by serial section electron microscopy: a laminar analysis using the silver-enhanced diaminobenzidine sulfide (SEDS) immunolabeling technique. *Cereb. Cortex* **3,** 223–238.
16. Smiley, J. F., Levey, A. I., Ciliax, B. J., and Goldman-Rakic, P. S. (1994) D1 dopamine receptor immunoreactivity in human and monkey cerebral cortex: predominant and extrasynaptic localization in dendritic spines. *Proc. Natl. Acad. Sci. USA* **91,** 5720–5724.
17. Freund, T. F., Powell, J. F., and Smith, A. D. (1984) Tyrosine hydroxylase-immunoreactive boutons in synaptic contact with identified striatonigral neurons, with particular reference to dendritic spines. *Neuroscience* **13,** 1189–1215.
18. Bergson, C., Mrzljak, L., Smiley, J. F., Pappy, M., Levenson, R., and Goldman-Rakic, P. S. Regional, cellular, and subcellular variations in the distribution of D1 and D_5 dopamine receptors in primate brain. *J. Neurosci.* **15,** 7821–7836.
19. Lidow, M. S., Goldman-Rakic, P. S., Rakic, P., and Innis, R. (1989) Dopamine D_2 receptors in the cerebral cortex: distribution and pharmacological characterization with [^{3}H]raclopride. *Proc. Natl. Acad. Sci. USA* **86,** 6412–6416.
20. Gingrich, J. A. and Caron, M. G. (1993) Recent advances in the molecular biology of dopamine receptors. *Annu. Rev. Neurosci.* **16,** 299–321.
21. Bergson, C., Mrzljak, L., Lidow, M. S., Goldman-Rakic, P. S., and Levenson, R. (1995) Characterization of subtype-specific antibodies to the human D_5 dopamine receptor: studies in primate brain and transfected mammalian cells. *Proc. Natl. Acad. Sci. USA* **92,** 3468–3472.
22. Huang, Q., Zhou, D., Chase, K., Gusella, J. F., Aronin, N., and DiFiglia, M. (1992) Immunohistochemical localization of the D1 dopamine receptor in rat brain reveals its axonal transport, pre-and postsynaptic localization, and prevalence in the basal ganglia, limbic system, and thalamic reticular nucleus. *Proc. Natl. Acad. Sci. USA* **89,** 11,988–11,992.
23. Levey, A. I., Hersch, S. M., Rye, D. B., Sunahara, R., Niznik, H. B., Kitt, C. A., Price, D. L., Maggio, R., Brann, M. R., and Ciliax, B. J. (1993) Localization of D1 and D2 dopamine receptors in brain with subtype-specific antibodies. *Proc. Natl. Acad. Sci. USA* **90,** 8861–8865.
24. Mrzljak, L., Bergson, C., Pappy, M., Levenson, R., and Goldman-Rakic, P. S. (1996) Characterization of the human D_4 dopamine receptor and its localization in the primate brain. *Nature,* **381,** 245–248.
25. Liu, K., Bergson, C., Levenson, R., and Schmauss, C. (1994) On the origin of mRNA encoding the truncated D_3-type receptor D_{3nf} and detection of D_{3nf}-like immunoreactivity in human brain. *J. Biol. Chem.* **269,** 29,220–29,226.
26. Sesack, S. R., Aoki, C., and Pickel, V. M. (1994) Ultrastructural localization of D_2 receptor-like immunoreactivity in midbrain dopamine neurons and their striatal targets. *J. Neurosci.* **14,** 88–106.
27. Laurier, L., O'Dowd, B. F., and George, S. R. (1994) Heterogeneous tissue-specific transcription of dopamine receptor subtype messenger RNA in rat brain. *Mol. Brain Res.* **25,** 344–350.
28. Berger, B., Gaspar, P., and Verney, C. (1991) Dopaminergic innervation of the cerebral cortex: unexpected differences between rodents and primates. *TINS* **14,** 21–27.

29. Schmauss, C., Haroutunian, V., Davis, K. L., and Davidson, M. (1993) Selective loss of dopamine D_3-type receptor mRNA expression in parietal and motor cortices of patients with chronic schizophrenia *Proc. Natl. Acad. Sci. USA* **90,** 8942–8946.
30. Chen, J., Makino, C. L., Peachey, N. S., Baylor, D. A., and Simon, M. I. (1995) Mechanisms of rhodopsin inactivation *in vivo* as revealed by a COOH-terminal truncation mutant. *Science* **267,** 374–377.
31. Dahl, S., Edvardsen, O., and Sylte, I. (1991) Molecular dynamics of dopamine at the D_2 receptor. *Proc. Natl. Acad. Sci. USA* **88,** 8111–8115.
32. Wess, J. (1993) Molecular basis of muscarinic acetylcholine receptor function. *Trends Pharmacol. Sci.* **14,** 308–313.
33. Lee, M. A., Thompson, P. A., and Meltzer, H. Y. (1994) Effects of clozapine on cognitive function in schizophrenia. *J. Clin. Psychiatry* **55(Suppl. B),** 82–87.
34. Fuxe, K. and Agnati, L. F. (1991) *Volume Transmission in the Brain: Novel Mechanisms for Neuronal Transmission.* Raven, New York.
35. Bradley, V. A., Welch, J. L., and Dick, D. J. (1989) Visuospatial working memory in Parkinson's disease. *J. Neurol. Neurosurg. Psychiatry* **52,** 1228–1235.
36. Brown, R. G. and Marsden, C. D. (1988) Internal versus external cues and the control of attention in Parkinson's Disease. *Brain* **111,** 323–345.
37. Lange, K. W., Robbins, T. W., Marsden, C. D., James, M., Owen, A. M., and Paul, G. M. (1992) L-dopa withdrawal in Parkinson's disease selectively impairs cognitive performance in tests sensitive to frontal lobe dysfunction. *Psychopharmacology* **107,** 394–404.
38. Andreasen, N. C., Rezai, K., Alliger, R., Swayze, V. W., II, Flaum, M., Kirchner, P., Cohen, G., and O'Leary, D. S. (1992) Hypofrontality in neuroleptic-naive patients and in patients with chronic schizophrenia: assessment with xenon 133 single-photon emission computed tomography and the Tower of London. *Arch. Gen. Psychiatry* **49,** 943–958.
39. Park, S. and Holzman, P. S. (1992) Schizophrenics show spatial working memory deficits. *Arch. Gen. Psychiatry* **49,** 975–982.
40. Schooler, C., Neumann, E., Caplan, L., and Roberts, B. R. (1996) A time course analysis of Stroop interference and facilitation: comparing normal and schizophrenic individuals. *J. Exp. Psychol.* in press.
41. Milner, B. (1964) Some effects of frontal lobectomy in man, in *The Frontal Granular Cortex and Behavior* (Warren, J. M. and Akert, K., eds.), McGraw-Hill, New York, pp. 313–334.
42. Stuss, D. T. and Benson, D. F. (1986) *The Frontal Lobes.* New York, Raven.
43. Fuster, J. M. (1989) *The Prefrontal Cortex,* 2nd ed., Raven, New York.
44. Goldman-Rakic, P. S. (1987) Circuitry of primate prefrontal cortex and regulation of behavior by representational memory, in *Handbook of Physiology, The Nervous System, Higher Functions of the Brain* (Plum, F., ed.), American Physiological Society, Bethesda, MD, pp. 373–417.
45. Brozoski, T., Brown, R. M., Rosvold, H. E., and Goldman, P. S. (1979) Cognitive deficit caused by regional depletion of dopamine in prefrontal cortex of rhesus monkey. *Science* **205,** 929–932.
46. Baddeley, A. D. (1986) *Working Memory.* Oxford University Press, London.
47. Carpenter, P. A. and Just, M. A. (1989) The role of working memory in language comprehension, in *Complex Information Processing: The Impact of Herbert A. Simon* (Klahr, D. and Kotovsky, K., eds.), Erlbaum, Hillsdale, NJ, pp. 31–68.

48. Goldman-Rakic, P. S. (1995) Cellular basis of working memory. *Neuron* **14,** 477–485.
49. Creese, I. and Fraser, C. M. (1987) *Dopamine Receptors*. Liss, New York.
50. Ogren, S. O. and Hogberg, T. (1988) Novel dopamine D_2 antagonists for the treatment of schizophrenia. *ISI Atlas of Science Pharmacology* **2,** 141–147.
51. Seeman, P. and Van Tol, H. H. M. (1994) Dopamine receptor pharmacology. *Trends Pharmacol. Sci.* **15,** 264–270.
52. Funahashi, S., Bruce, C. J., and Goldman-Rakic, P. S. (1989) Mnemonic coding of visual space in the monkey's dorsolateral prefrontal cortex. *J. Neurophysiol.* **61,** 331–349.
53. Fuster, J. M. (1973) Unit activity in prefrontal cortex during delayed-response performance: neuronal correlates of transient memory. *J. Neurophysiol.* **36,** 61–78.
54. Goldman-Rakic, P. S., Funahashi, S., and Bruce, C. J. (1990) Neocortical memory circuits. *Q. J. Quant. Biol.* **55,** 1025–1038.
55. Williams, G. V. and Goldman-Rakic, P. S. (1995) Blockade of dopamine D1 receptors enhances memory fields of prefrontal neurons in primate cerebral cortex. *Nature* **376,** 572–575.
56. Chipkin, R. E., Iorio, L. C., Coffin, V. L., McQuade, R. D., Berger, J. G., and Barnett, A. (1988) Pharmacological profile of SCH39166: a dopamine D1 selective benzonaphthazepine with potential antipsychotic activity. *J. Pharmacol. Exp. Ther.* **247,** 1093–1102.
57. Taylor, J. R., Sladek, J. R., Jr., Roth, R. H., and Redmond, D. E. (1990) Cognitive and motor deficits in the performance of an object retrieval task with a barrier detour in monkeys (cercopithecus aethiops sabeus) treated with MPTP: long-term performance and effect of transparency of the barrier. *Behavioral Neuorsci.* **104,** 564–576.
58. Bunney, B. S. and Agajanian, G. K. (1976) Dopamine and norepinephrine innervated cells in the rat prefrontal cortex: pharmacological differentiation using microiontophoretic techniques. *Life Sci.* **19,** 1783–1792.
59. Ferron, A., Thierry, A. M., Le Douarin, C., and Glowinski, J. (1984) Inhibitory influence of the mesocortical dopaminergic system on spontaneous activity or excitatory response induced from the thalamic mediodorsal nucleus in the rat medial prefrontal cortex. *Brain Res.* **302,** 257–265.
60. Law-Tho, D., Hirsch, J. C., and Crepel, F. (1994) Dopamine modulation of synaptic transmission in rat prefrontal cortex: an *in vitro* electrophysiological study. *Neurosci. Res.* **21,** 151–160.
61. Muller, W. and Connor, J. A. (1991) Dendritic spines as individual neuronal compartments for synaptic Ca^{2+} responses. *Nature* **354,** 73–76.
62. Koch, C. and Zador, A. (1993) The function of dendritic spines: devices subserving biochemical rather than electrical compartmentalization. *J. Neurosci.* **13,** 413–422.
63. Arnsten, A. F. T., Cai, J. X., Murphy, B. L., and Goldman-Rakic, P. S. (1994) Dopamine D1 receptor mechanisms in the cognitive performance of young adult and aged monkeys. *Psychopharmacology* **116,** 143–151.
64. Sawaguchi, T. and Goldman-Rakic, P. S. (1991) D1 dopamine receptors in prefrontal cortex: involvement in working memory. *Science* **251,** 947–950.
65. Ruzicka, E., Roth, J., Spackova, N., Mecir, P., and Jech, R. (1994) Apomorphine induced cognitive changes in Parkinson's disease. *J. Neurol. Neurosurg. Psychiatry* **57,** 998–1001.

66. Murphy, B. L., Arnsten, A. F. T., Goldman-Rakic, P. S., and Roth, R. H. (1996) Increased dopamine turnover in the prefrontal cortex impairs spatial working memory performance. *Proc. Natl. Acad. Sci. USA* **93,** 1325–1329.
67. Mantz, J., Milla, C., Glowinski, J., and Thierry, A. M. (1988) Differential effects of ascending neurons containing dopamine and noradrenaline in the control of spontaneous and evoked responses in the rat prefrontal cortex. *Neuroscience* **27,** 517–526.
68. Penit-Soria, J., Audinat, E., and Crepel, F. (1987) Excitation of rat prefrontal cortical neurons by dopamine: an *in vitro* electrophysiological study. *Brain Res.* **425,** 263–274.
69. Pirot, S., Godbout, R. Mantz, J., Tassin, J. P., Glowinski, J., and Thierry, A. M. (1992) Inhibitory effect of ventral tegmental area stimulation on the activity of prefrontal cortical neurons: evidence for the involvement of both dopaminergic and GABAergic components. *Neuroscience* **49,** 857–865.
70. Pralong, E. and Jones, R. S. G. (1993) Interactions of dopamine with glutamate- and GABA-mediated synaptic transmission in the rat entorhinal cortex *in vitro*. *Eur. J. Neurosci.* **5,** 760–767.
71. Calabresi, P., Mercuri, N., Stanzione, P., Stefani, A., and Bernardi, G. (1987) Intracellular studies on the dopamine-induced firing inhibition of neostriatal neurons in vitro: evidence for D1 receptor involvement. *Neuroscience* **20,** 757–771.
72. Cepeda, C., Radisavljevic, Z., Peacock, W., Levine, M. S., and Buchwald, N. A. (1992) Differential modulation by dopamine of responses evoked by excitatory amino acids in human cortex. *Synapse* **11,** 330–341.
73. Rutherford, A., Garcia-Munoz, M., and Arbuthnott, G. W. (1988) An afterhyperpolarization recorded in striatal cells "*in vitro*": effect of dopamine administration. *Expl. Brain Res.* **71,** 399–405.
74. Mrzljak, L., Pappy, M., Leranth, C., and Goldman-Rakic, P. S. (1995) Cholinergic synaptic circuitry in the macaque prefrontal cortex. *J. Comp. Neurol.* **357,** 603–617.
75. Servan-Schreiber, D., Printz, H., and Cohen, J. D. (1990) A network model of catecholamine effects: gain, signal-to-noise ratio, and behavior. *Science* **249,** 892–895.

Index

D

H

I

J

K

L

M

O

P

Q